T0222959

Lecture Notes in Computer Science   10762

Commenced Publication in 1973
Founding and Former Series Editors:
Gerhard Goos, Juris Hartmanis, and Jan van Leeuwen

More information about this series at http://www.springer.com/series/7407

Alexander Gelbukh (Ed.)

Computational Linguistics and Intelligent Text Processing

18th International Conference, CICLing 2017
Budapest, Hungary, April 17–23, 2017
Revised Selected Papers, Part II

 Springer

Editor
Alexander Gelbukh
CIC, Instituto Politécnico Nacional
Mexico City, Mexico

ISSN 0302-9743   ISSN 1611-3349   (electronic)
Lecture Notes in Computer Science
ISBN 978-3-319-77115-1   ISBN 978-3-319-77116-8   (eBook)
https://doi.org/10.1007/978-3-319-77116-8

Library of Congress Control Number: 2018934347

LNCS Sublibrary: SL1 – Theoretical Computer Science and General Issues

This Springer imprint is published by the registered company Springer Nature Switzerland AG
The registered company address is: Gewerbestrasse 11, 6330 Cham, Switzerland

Preface

CICLing 2017 was the 18th International Conference on Computational Linguistics and Intelligent Text Processing. The CICLing conferences provide a wide-scope forum for discussion of the art and craft of natural language processing research, as well as the best practices in its applications.

This set of two books contains four invited papers and a selection of regular papers accepted for presentation at the conference. Since 2001, the proceedings of the CICLing conferences have been published in Springer's *Lecture Notes in Computer Science* series as volumes 2004, 2276, 2588, 2945, 3406, 3878, 4394, 4919, 5449, 6008, 6608, 6609, 7181, 7182, 7816, 7817, 8403, 8404, 9041, 9042, 9623, and 9624.

The set has been structured into 18 sections representative of the current trends in research and applications of natural language processing:

General
Morphology and Text Segmentation
Syntax and Parsing
Word Sense Disambiguation
Reference and Coreference Resolution
Named Entity Recognition
Semantics and Text Similarity
Information Extraction
Speech Recognition
Applications to Linguistics and the Humanities
Sentiment Analysis
Opinion Mining
Author Profiling and Authorship Attribution
Social Network Analysis
Machine Translation
Text Summarization
Information Retrieval and Text Classification
Practical Applications

This year our invited speakers were Marco Baroni (Facebook Artificial Intellgence Research), Iryna Gurevych (Ubiquitous Knowledge Processing Lab, TU Darmstadt), Björn W. Schuller (University of Passau, Imperial College London, Harbin Institute of Technology, University of Geneva, Joanneum Research, and EERING GmbH), and Hinrich Schuetze (Center for Information and Language Processing, University of Munich). They delivered excellent extended lectures and organized lively discussions. Full contributions of these invited talks are included in this book set.

After careful reviewing, the Program Committee selected 86 papers for presentation, out of 356 submissions from 60 countries.

To encourage providing algorithms and data along with the published papers, we selected three winners of our Verifiability, Reproducibility, and Working Description Award. The main factors in choosing the awarded submission were technical correctness and completeness, readability of the code and documentation, simplicity of installation and use, and exact correspondence to the claims of the paper. Unnecessary sophistication of the user interface was discouraged; novelty and usefulness of the results were not evaluated, instead, they were evaluated for the paper itself and not for the data.

The following papers received the Best Paper Awards, the Best Student Paper Award, as well as the Verifiability, Reproducibility, and Working Description Awards, respectively:

Best Verifiability Award, First Place:
"Label-Dependencies Aware Recurrent Neural Networks"
by Yoann Dupont, Marco Dinarelle, and Isabelle Tellier

Best Paper Award, Second Place, and Best Presentation Award:
"Idioms: Humans or Machines, It's All About Context"
by Manali Pradhan, Jing Peng, Anna Feldman, and Bianca Wright

Best Student Paper Award:
"Dialogue Act Taxonomy Interoperability Using a Meta-Model"
by Soufian Salim, Nicolas Hernandez, and Emmanuel Morin

Best Paper Award, First Place:
"Gold Standard Online Debates Summaries and First Experiments Towards Automatic Summarization of Online Debate Data"
by Nattapong Sanchan, Ahmet Aker, and Kalina Bontcheva

Best Paper Award, Third Place:
"Efficient Semantic Search over Structured Web Data: A GPU Approach" by Ha-Hguyen Tran, Erik Cambria, and Hoang Giang Do.

A conference is the result of the work of many people. First of all I would like to thank the members of the Program Committee for the time and effort they devoted to the reviewing of the submitted articles and to the selection process. Obviously I thank the authors for their patience in the preparation of the papers, not to mention the very development of their scientific results that form this book. I also express my most cordial thanks to the members of the local Organizing Committee for their considerable contribution to making this conference become a reality.

January 2018   Alexander Gelbukh

Organization

CICLing 2017 was hosted by the Pázmány Péter Catholic University, Faculty of Information Technology and Bionics, Budapest, Hungary, and organized by the CICLing 2017 Organizing Committee in conjunction with the Pázmány Péter Catholic University, Faculty of Information Technology and Bionics, the Natural Language and Text Processing Laboratory of the CIC, IPN, and the Mexican Society of Artificial Intelligence (SMIA).

Organizing Committee

Attila Novák (Chair)	MTA-PPKE Language Technology Research Group, Pázmány Péter Catholic University
Gábor Prószéky	MTA-PPKE Language Technology Research Group, Pázmány Péter Catholic University
Borbála Siklósi	MTA-PPKE Language Technology Research Group, Pázmány Péter Catholic University

Program Committee

Bayan Abushawar	Gregory Grefenstette	Inderjeet Mani
Galia Angelova	Tunga Gungor	Alexander Mehler
Alexandra Balahur	Eva Hajicova	Farid Meziane
Sivaji Bandyopadhyay	Yasunari Harada	Rada Mihalcea
Leslie Barrett	Karin Harbusch	Evangelos Milios
Roberto Basili	Koiti Hasida	Ruslan Mitkov
Pushpak Bhattacharyya	Ales Horak	Dunja Mladenic
Christian Boitet	Veronique Hoste	Marie-Francine Moens
Nicoletta Calzolari	Diana Inkpen	Hermann Moisl
Nick Campbell	Hitoshi Isahara	Masaki Murata
Michael Carl	Aminul Islam	Preslav Nakov
Violetta Cavalli-Sforza	Guillaume Jacquet	Costanza Navarretta
Niladri Chatterjee	Milos Jakubicek	Joakim Nivre
Dan Cristea	Sylvain Kahane	Kjetil Norvag
Walter Daelemans	Alma Kharrat	Attila Novák
Mike Dillinger	Philipp Koehn	Nir Ofek
Samhaa El-Beltagy	Valia Kordoni	Kemal Oflazer
Michael Elhadad	Mathieu Lafourcade	Constantin Orasan
Anna Feldman	Elena Lloret	Ivandre Paraboni
Robert Gaizauskas	Bente Maegaard	Saint-Dizier Patrick
Alexander Gelbukh	Cerstin Mahlow	Maria Teresa Pazienza
Dafydd Gibbon	Suresh Manandhar	Ted Pedersen

Software Reviewing Committee

Best Paper Award Committee

Contents – Part II

Sentiment Analysis

A Comparison Among Significance Tests and Other Feature Building
Methods for Sentiment Analysis: A First Study.   3
   Raksha Sharma, Dibyendu Mondal, and Pushpak Bhattacharyya

BATframe: An Unsupervised Approach for Domain-Sensitive
Affect Detection   20
   Kokil Jaidka, Niyati Chhaya, Rahul Wadbude, Sanket Kedia,
   and Manikanta Nallagatla

Leveraging Target-Oriented Information for Stance Classification   35
   Jiachen Du, Ruifeng Xu, Lin Gui, and Xuan Wang

Sentiment Polarity Classification of Figurative Language: Exploring
the Role of Irony-Aware and Multifaceted Affect Features   46
   Delia Irazú Hernández Farías, Cristina Bosco, Viviana Patti,
   and Paolo Rosso

Sarcasm Annotation and Detection in Tweets   58
   Johan G. Cyrus M. Ræder and Björn Gambäck

Modeling the Impact of Modifiers on Emotional Statements   71
   Valentina Sintsova, Margarita Bolívar Jiménez, and Pearl Pu

CSenticNet: A Concept-Level Resource for Sentiment Analysis
in Chinese Language   90
   Haiyun Peng and Erik Cambria

Emotional Tone Detection in Arabic Tweets. 105
   Amr Al-Khatib and Samhaa R. El-Beltagy

Morphology Based Arabic Sentiment Analysis of Book Reviews 115
   Omar El Ariss and Loai M. Alnemer

Adaptation of Sentiment Analysis Techniques to Persian Language 129
   Kia Dashtipour, Amir Hussain, and Alexander Gelbukh

Verb-Mediated Composition of Attitude Relations Comprising Reader
and Writer Perspective. 141
   Manfred Klenner, Simon Clematide, and Don Tuggener

Customer Churn Prediction Using Sentiment Analysis and Text
Classification of VOC   156
   Yiou Wang, Koji Satake, Takeshi Onishi, and Hiroshi Masuichi

Benchmarking Multimodal Sentiment Analysis   166
   Erik Cambria, Devamanyu Hazarika, Soujanya Poria, Amir Hussain,
   and R. B. V. Subramanyam

Machine Learning Approaches for Speech Emotion Recognition: Classic
and Novel Advances   180
   Panikos Heracleous, Akio Ishikawa, Keiji Yasuda, Hiroyuki Kawashima,
   Fumiaki Sugaya, and Masayuki Hashimoto

Opinion Mining

Mining Aspect-Specific Opinions from Online Reviews Using a Latent
Embedding Structured Topic Model   195
   Mingyang Xu, Ruixin Yang, Paul Jones, and Nagiza F. Samatova

A Comparative Study of Target-Based and Entity-Based
Opinion Extraction   211
   Joseph Lark, Emmanuel Morin, and Sebastián Peña Saldarriaga

Supervised Domain Adaptation via Label Alignment
for Opinion Expression Extraction.   224
   Yifan Zhang, Arjun Mukherjee, and Fan Yang

Comment Relevance Classification in Facebook   241
   Chaya Liebeskind, Shmuel Liebeskind, and Yaakov HaCohen-Kerner

Detecting Sockpuppets in Deceptive Opinion Spam.   255
   Marjan Hosseinia and Arjun Mukherjee

Author Profiling and Authorship Attribution

Invited Paper:

Reading the Author and Speaker: Towards a Holistic and Deep Approach
on Automatic Assessment of What is in One's Words   275
   Björn W. Schuller

Improving Cross-Topic Authorship Attribution: The Role
of Pre-Processing   289
   Ilia Markov, Efstathios Stamatatos, and Grigori Sidorov

Author Identification Using Latent Dirichlet Allocation   303
   Hiram Calvo, Ángel Hernández-Castañeda, and Jorge García-Flores

Personality Recognition Using Convolutional Neural Networks.   313
    Maite Giménez, Roberto Paredes, and Paolo Rosso

Character-Level Dialect Identification in Arabic Using Long
Short-Term Memory   324
    *Karim Sayadi, Mansour Hamidi, Marc Bui, Marcus Liwicki,
    and Andreas Fischer*

A Text Semantic Similarity Approach for Arabic Paraphrase Detection   338
    Adnen Mahmoud, Ahmed Zrigui, and Mounir Zrigui

Social Network Analysis

Curator: Enhancing Micro-Blogs Ranking by Exploiting User's Context   353
    Hicham G. Elmongui and Riham Mansour

A Multi-view Clustering Model for Event Detection in Twitter.   366
    Di Shang, Xin-Yu Dai, Weiyi Ge, Shujiang Huang, and Jiajun Chen

Monitoring Geographical Entities with Temporal Awareness in Tweets   379
    Koji Matsuda, Mizuki Sango, Naoaki Okazaki, and Kentaro Inui

Just the Facts: Winnowing Microblogs for Newsworthy Statements
using Non-Lexical Features   391
    Nigel Dewdney and Rachel Cotterill

Impact of Content Features for Automatic Online Abuse Detection   404
    *Etienne Papegnies, Vincent Labatut, Richard Dufour,
    and Georges Linarès*

Detecting Aggressive Behavior in Discussion Threads Using Text Mining . . .   420
    Filippos Karolos Ventirozos, Iraklis Varlamis, and George Tsatsaronis

Machine Translation

Combining Machine Translation Systems with Quality Estimation.   435
    László János Laki and Zijian Győző Yang

Evaluation of Neural Machine Translation for Highly Inflected
and Small Languages.   445
    Mārcis Pinnis, Rihards Krišlauks, Daiga Deksne, and Toms Miks

Towards Translating Mixed-Code Comments from Social Media.   457
    Thoudam Doren Singh and Thamar Solorio

Multiple System Combination for PersoArabic-Latin Transliteration   469
    Nima Hemmati, Heshaam Faili, and Jalal Maleki

Building a Location Dependent Dictionary for Speech
Translation Systems.   482
 Keiji Yasuda, Panikos Heracleous, Akio Ishikawa, Masayuki Hashimoto,
 Kazunori Matsumoto, and Fumiaki Sugaya

Text Summarization

Best Paper Award, First Place:

Gold Standard Online Debates Summaries and First Experiments Towards
Automatic Summarization of Online Debate Data   495
 Nattapong Sanchan, Ahmet Aker, and Kalina Bontcheva

Optimization in Extractive Summarization Processes Through
Automatic Classification   506
 Angel Luis Garrido, Carlos Bobed, Oscar Cardiel, Andrea Aleyxendri,
 and Ruben Quilez

Summarizing Weibo with Topics Compression   522
 Marina Litvak, Natalia Vanetik, and Lei Li

Timeline Generation Based on a Two-Stage Event-Time
Anchoring Model   535
 Tomohiro Sakaguchi and Sadao Kurohashi

Information Retrieval and Text Classification

Best Paper Award, Third Place:

Efficient Semantic Search Over Structured Web Data: A GPU Approach   549
 Ha-Nguyen Tran, Erik Cambria, and Hoang Giang Do

Efficient Association Rules Selecting for Automatic Query Expansion   563
 Ahlem Bouziri, Chiraz Latiri, and Eric Gaussier

Text-to-Concept: A Semantic Indexing Framework for Arabic
News Videos   575
 Sadek Mansouri, Chahira Lhioui, Mbarek Charhad, and Mounir Zrigui

Approximating Multi-class Text Classification Via Automatic Generation
of Training Examples   585
 Filippo Geraci and Tiziano Papini

Practical Applications

Generating Appealing Brand Names   605
 Gaurush Hiranandani, Pranav Maneriker, and Harsh Jhamtani

Radiological Text Simplification Using a General Knowledge Base 617
  Lionel Ramadier and Mathieu Lafourcade

Mining Supervisor Evaluation and Peer Feedback
in Performance Appraisals 628
  Girish Keshav Palshikar, Sachin Pawar, Saheb Chourasia,
  and Nitin Ramrakhiyani

Automatic Detection of Uncertain Statements in the Financial Domain 642
  Christoph Kilian Theil, Sanja Štajner, Heiner Stuckenschmidt,
  and Simone Paolo Ponzetto

Automatic Question Generation From Passages 655
  Karen Mazidi

Author Index 667

Contents – Part I

General

Invited Paper:

Overview of Character-Based Models for Natural Language Processing   3
Heike Adel, Ehsaneddin Asgari, and Hinrich Schütze

Pooling Word Vector Representations Across Models   17
*Rajendra Banjade, Nabin Maharjan, Dipesh Gautam, Frank Adrasik,
Arthur C. Graesser, and Vasile Rus*

Strategies to Select Examples for Active Learning with Conditional
Random Fields   30
Vincent Claveau and Ewa Kijak

Best Verifiability Award, First Place:

Label-Dependencies Aware Recurrent Neural Networks   44
Yoann Dupont, Marco Dinarelli, and Isabelle Tellier

Universal Computational Formalisms and Developer Environment
for Rule-Based NLP   67
Svetlana Sheremetyeva

Morphology and Text Segmentation

Several Ways to Use the Lingwarium.org Online MT Collaborative
Platform to Develop Rich Morphological Analyzers   81
*Vincent Berment, Christian Boitet, Jean-Philippe Guilbaud,
and Jurgita Kapočiūtė-Dzikienė*

A Trie-structured Bayesian Model for Unsupervised Morphological
Segmentation   87
Murathan Kurfalı, Ahmet Üstün, and Burcu Can

Building Morphological Chains for Agglutinative Languages   99
Serkan Ozen and Burcu Can

Joint PoS Tagging and Stemming for Agglutinative Languages   110
Necva Bölücü and Burcu Can

Hungarian Particle Verbs in a Corpus-Driven Approach   123
Ágnes Kalivoda

HANS: A Service-Oriented Framework for Chinese Language Processing . . .   134
  Lung-Hao Lee, Kuei-Ching Lee, and Yuen-Hsien Tseng

Syntax and Parsing

Learning to Rank for Coordination Detection   145
  *Xun Wang, Rumeng Li, Hiroyuki Shindo, Katsuhito Sudoh,
  and Masaaki Nagata*

Classifier Ensemble Approach to Dependency Parsing   158
  Silpa Kanneganti, Vandan Mujadia, and Dipti M. Sharma

Evaluation and Enrichment of Stanford Parser Using an Arabic Property
Grammar   170
  *Raja Bensalem Bahloul, Nesrine Kadri, Kais Haddar,
  and Philippe Blache*

Word Sense Disambiguation

SenseDependency-Rank: A Word Sense Disambiguation Method
Based on Random Walks and Dependency Trees   185
  *Marco Antonio Sobrevilla-Cabezudo, Arturo Oncevay-Marcos,
  and Andrés Melgar*

Domain Adaptation for Word Sense Disambiguation Using Word
Embeddings   195
  *Kanako Komiya, Shota Suzuki, Minoru Sasaki, Hiroyuki Shinnou,
  and Manabu Okumura*

Reference and Coreference Resolution

Invited Paper:

"Show Me the Cup": Reference with Continuous Representations   209
  Marco Baroni, Gemma Boleda, and Sebastian Padó

Improved Best-First Clustering for Coreference Resolution in Indian
Classical Music Forums   225
  Joe Cheri Ross and Pushpak Bhattacharyya

A Robust Coreference Chain Builder for Tamil   233
  R. Vijay Sundar Ram and Sobha Lalitha Devi

Named Entity Recognition

Structured Named Entity Recognition by Cascading CRFs   249
  Yoann Dupont, Marco Dinarelli, Isabelle Tellier, and Christian Lautier

Arabic Named Entity Recognition: A Bidirectional GRU-CRF Approach   264
  Mourad Gridach and Hatem Haddad

Named Entity Recognition for Amharic Using Stack-Based
Deep Learning   276
  Utpal Kumar Sikdar and Björn Gambäck

Semantics and Text Similarity

Best Paper Award, Second Place, and Best Presentation Award:

Idioms: Humans or Machines, It's All About Context   291
  Manali Pradhan, Jing Peng, Anna Feldman, and Bianca Wright

Best Student Paper Award:

Dialogue Act Taxonomy Interoperability Using a Meta-model   305
  Soufian Salim, Nicolas Hernandez, and Emmanuel Morin

Textual Entailment Using Machine Translation Evaluation Metrics   317
  *Tanik Saikh, Sudip Kumar Naskar, Asif Ekbal,
  and Sivaji Bandyopadhyay*

Supervised Learning of Entity Disambiguation Models by Negative
Sample Selection.   329
  *Hani Daher, Romaric Besançon, Olivier Ferret, Hervé Le Borgne,
  Anne-Laure Daquo, and Youssef Tamaazousti*

The Enrichment of Arabic WordNet Antonym Relations   342
  Mohamed Ali Batita and Mounir Zrigui

Designing an Ontology for Physical Exercise Actions   354
  *Sandeep Kumar Dash, Partha Pakray, Robert Porzel,
  Jan Smeddinck, Rainer Malaka, and Alexander Gelbukh*

Visualizing Textbook Concepts: Beyond Word Co-occurrences.   363
  *Chandramouli Shama Sastry, Darshan Siddesh Jagaluru,
  and Kavi Mahesh*

Matching, Re-Ranking and Scoring: Learning Textual Similarity by
Incorporating Dependency Graph Alignment and Coverage Features   377
  Sarah Kohail and Chris Biemann

Text Similarity Function Based on Word Embeddings for Short
Text Analysis.   391
  Adrián Jiménez Pascual and Sumio Fujita

Information Extraction

Domain Specific Features Driven Information Extraction from Web Pages
of Scientific Conferences   405
    Piotr Andruszkiewicz and Rafał Hazan

Classifier-Based Pattern Selection Approach for Relation
Instance Extraction   418
    Angrosh Mandya, Danushka Bollegala, Frans Coenen,
    and Katie Atkinson

An Ensemble Architecture for Linked Data Lexicalization   435
    Rivindu Perera and Parma Nand

A Hybrid Approach for Biomedical Relation Extraction Using Finite State
Automata and Random Forest-Weighted Fusion   450
    Thanassis Mavropoulos, Dimitris Liparas, Spyridon Symeonidis,
    Stefanos Vrochidis, and Ioannis Kompatsiaris

Exploring Linguistic and Graph Based Features for the Automatic
Classification and Extraction of Adverse Drug Effects   463
    Tirthankar Dasgupta, Abir Naskar, and Lipika Dey

Extraction of Semantic Relation Between Arabic Named Entities Using
Different Kinds of Transducer Cascades.   475
    Fatma Ben Mesmia, Kaouther Bouabidi, Kais Haddar,
    Nathalie Friburger, and Denis Maurel

Semi-supervised Relation Extraction from Monolingual Dictionary
for Russian WordNet.   488
    Daniil Alexeyevsky

Speech Recognition

ASR Hypothesis Reranking Using Prior-Informed Restricted
Boltzmann Machine.   503
    Yukun Ma, Erik Cambria, and Benjamin Bigot

A Comparative Analysis of Speech Recognition Systems
for the Tatar Language   515
    Aidar Khusainov

Applications to Linguistics and the Humanities

Invited Papers:

Interactive Data Analytics for the Humanities   527
Iryna Gurevych, Christian M. Meyer, Carsten Binnig,
Johannes Fürnkranz, Kristian Kersting, Stefan Roth,
and Edwin Simpson

Language Technology for Digital Linguistics: Turning the Linguistic
Survey of India into a Rich Source of Linguistic Information   550
Lars Borin, Shafqat Mumtaz Virk, and Anju Saxena

Classifying World Englishes from a Lexical Perspective:
A Corpus-Based Approach   564
Frank Z. Xing, Danyuan Ho, Diyana Hamzah, and Erik Cambria

Towards a Map of the Syntactic Similarity of Languages   576
Alina Maria Ciobanu, Liviu P. Dinu, and Andrea Sgarro

Romanian Word Production: An Orthographic Approach
Based on Sequence Labeling   591
Liviu P. Dinu and Alina Maria Ciobanu

Author Index   605

Sentiment Analysis

A Comparison Among Significance Tests and Other Feature Building Methods for Sentiment Analysis: A First Study

Raksha Sharma[✉], Dibyendu Mondal, and Pushpak Bhattacharyya

Department of Computer Science and Engineering, Indian Institute of Technology Bombay, Mumbai, India
{raksha,dibyendu,pb}@cse.iitb.ac.in

Abstract. Words that participate in the sentiment (positive or negative) classification decision are known as *significant words* for sentiment classification. Identification of such significant words as features from the corpus reduces the amount of irrelevant information in the feature set under supervised sentiment classification settings. In this paper, we conceptually study and compare various types of feature building methods, *viz., unigrams, TFIDF, Relief, Delta-TFIDF,* χ^2 test and *Welch's t-test* for sentiment analysis task. Unigrams and TFIDF are the classic ways of feature building from the corpus. Relief, Delta-TFIDF and χ^2 test have recently attracted much attention for their potential use as feature building methods in sentiment analysis. On the contrary, *t*-test is the least explored for the identification of significant words from the corpus as features.

We show the effectiveness of significance tests over other feature building methods for three types of sentiment analysis tasks, *viz.*, in-domain, cross-domain and cross-lingual. Delta-TFIDF, χ^2 test and Welch's *t*-test compute the significance of the word for classification in the corpus, whereas unigrams, TFIDF and Relief do not observe the significance of the word for classification. Furthermore, significance tests can be divided into two categories, bag-of-words-based test and distribution-based test. Bag-of-words-based test observes the total count of the word in different classes to find significance of the word, while distribution-based test observes the distribution of the word. In this paper, we substantiate that the distribution-based Welch's *t*-test is more accurate than bag-of-words-based χ^2 test and Delta-TFIDF in identification of significant words from the corpus.

1 Introduction

A wide variety of feature sets have been used in sentiment analysis, for example, unigrams, bigrams, Term Frequency Inverse Document Frequency (TFIDF), *etc.* However, none of these feature sets computes the significance of a feature (word) for classification before considering it as a part of the feature set. However, all the

© Springer Nature Switzerland AG 2018
A. Gelbukh (Ed.): CICLing 2017, LNCS 10762, pp. 3–19, 2018.
https://doi.org/10.1007/978-3-319-77116-8_1

words available in the corpus do not equally participate in the classification decision. For example, words like *high-quality, unreliable, cheapest, faulty, defective, broken, flexible, heavy, hard, etc.*, are prominent features for sentiment analysis in the *electronics* domain. It is possible to compute association of a word with a particular class in the sentiment annotated corpus. A word which shows statistical association with a class in the corpus is essentially a significant word for classification. In this paper, we propose that a feature set which consists of only those words that are significant for classification is more promising for sentiment analysis than any other feature set. We provide a comparison between various feature building methods, *viz.*, unigrams, TFIDF, Relief, Delta-TFIDF, χ^2 and Welch's *t*-test for sentiment analysis task. χ^2 test, Delta-TFIDF and Welch's *t*-test determine the significance of words in the corpus unlike unigrams, TFIDF and Relief.

χ^2 test has been fairly used in the literature for the identification of significant words from the corpus [1–3]. This test takes decisions on the basis of the overall count of the word in the corpus. It does not observe the distribution of the word in the corpus, which in turn may lead to spurious results [4,5]. Similarly, Delta-TFIDF takes significance decision by observing the overall count of the word in the positive and negative corpora. The test which takes total count of the word from the corpora as input is known as the bag-of-words-based test [6], hence χ^2 and Delta-TFIDF are bag-of-words-based tests. However, it is possible to represent the data differently and employ other significance tests. *t*-test is a distribution-based significance test, which takes into consideration the distribution of the word in the corpus. Observation of the distribution of the word in the corpus helps to identify the biased words. The distribution-based tests have not been explored well in Natural Language Processing (NLP) applications. We show that a distribution-based test, *i.e.*, Welch's *t*-test is more effective than χ^2 test and Delta-TFIDF in the identification of words which are significant for sentiment classification in a domain. The major contributions of this research are as follows:

– Feature building methods which are able to identify association of a word with a particular class give a better solution for sentiment classification than existing feature-engineering techniques. We show that the results possible with significance tests, *viz.*, Delta-TFIDF, χ^2 test or *t-test* give a less computationally expensive and more accurate sentiment analysis system in comparison to unigrams, TFIDF or Relief.
– Welch's *t*-test is able to capture poor dispersion of words, unlike χ^2 test and Delta-TFIDF, as it considers frequency distribution of words in the positive and negative corpora. We substantiate that distribution-based *t*-test is better than bag-of-words-based Delta-TFIDF and χ^2 test.

In this paper, we have shown the effectiveness of significance tests over other feature building methods for three types of Sentiment Analysis (SA) tasks, *viz.*, in-domain, cross-domain and cross-lingual SA. Essentially, in this paper, we have emphasized the need for a correct significance test with an example in sentiment analysis. The road-map for rest of the paper is as follows. Section 2 describes the related work. Section 3 conceptually compares and formulates the considered feature building methods. Section 4 elaborates the dataset used in the paper. Section 5 presents the experimental setup. Section 6 depicts the results and provides discussion on the results. Section 7 concludes the paper.

2 Related Work

Though deep learning based approaches perform reasonably well for the overall sentiment analysis task [7,8], they do not perform explicit identification and visualization of prominent features in the corpus. On the other hand, feature engineering is proved to be effective for sentiment analysis [9–12]. Pang et al. [9] showed variation in accuracy with varying feature sets. They showed that unigrams with presence perform better than unigrams with frequency, bigrams, combination of unigrams and bigrams, unigrams with parts of speech, adjectives and top-n unigrams for sentiment analysis. On the other hand, TFIDF is popularly used for information retrieval task [13].

χ^2 test has been widely used to identify significant words in the corpus. Oakes and Ferrow [14] showed the vocabulary differences using χ^2 test, which reveals the linguistic preferences in various countries in which English is spoken. Al-Harbi et al. [15] used χ^2 test to find out significant words for the purpose of document classification. They presented results with seven different Arabic corpora. Rayson and Garside [16] showed the differences between the corpora using χ^2 test. There are a few instances of the use of χ^2 test in the sentiment classification. Sharma et al. [17] showed that χ^2 test can be used to create a compact size sentiment lexicon. Cheng et al. [12] compared significant words by χ^2 test with popular feature sets like unigrams and bigrams. They proved that χ^2 test produces better results than unigrams and bigrams for sentiment analysis. Relief is a classic feature building method proposed by Kira and Rendell [18] which assigns weights to the words based on their distance from the randomly selected instances of different classes. However, it does not discriminate between redundant features, and a smaller number of training instances may fool the algorithm. Recently, Delta-TFIDF has come out as an emergent feature building method for sentiment analysis [19,20]. Delta-TFIDF also computes the belongingness of a feature to a particular class in the sentiment annotated corpus. It discards the features which do not belong to any class. χ^2 test and Delta-TFIDF are the bag-of-words-based significance tests, while Welch's t-test is a distribution-based test. Distribution-based tests are very less explored for feature building from the corpus.

Though all the feature building methods have been used in various NLP applications independently, they are not relatively studied with respect to the sentiment analysis task to the best of our knowledge. In this work, we show that the use of significant words given by significance tests provide a good feature-engineering option for sentiment analysis applications. In addition, we have conceptually compared bag-of-words-based tests, viz., χ^2 test and Delta-TFIDF with distribution-based t-test and have shown that the use of t-test is more effective for sentiment analysis than χ^2 test and Delta-TFIDF.

3   Conceptual Comparison and Formulation of Feature Building Methods

This section elaborates the preparation of a feature vector according to different feature building methods for supervised classification.

Unigrams: In this case, feature set is made up of all the unique words in the corpus. The feature value corresponding to a feature in a feature vector is set to 1, if the feature is present in the document, else it is set to 0.[1]

Term Frequency Inverse Document Frequency (TFIDF): This is a numerical statistic that is intended to reflect how important a word is to a document in a collection or corpus. In case of TFIDF, feature value in the feature vector increases proportionally to the number of times a word appears in the document, but is offset by the frequency of the word in the corpus, which helps to adjust for the fact that some words appear more frequently in general [21]. The value of the feature in the feature vector of a document is given by the following TFIDF formula.

$$TFIDF(w,d,D) = tf(w,d) * \log \frac{N}{|\{d \in D : w \in d\}|} \qquad (1)$$

where $tf(w,d)$ gives the count of the word w in the document d, N is the total number of documents in the corpus $N = |D|$ and $|\{d \in D : w \in d\}|$ gives the count of documents where the word w appears (*i.e.*, $tf(w,d)! = 0$).

Relief: It is a feature building algorithm proposed by Kira and Rendell [18] for binary classification. Cehovin and Bosnic [22] showed that the features selected by Relief enable the classifiers to achieve the highest increase in classification accuracy while reducing the number of unnecessary attributes. We have used java-based machine learning library (java-ml)[2] to implement Relief. Relief decreases weight of any given feature if it differs from that feature in nearby instances of the same class more than nearby instances of the other class, or increases in the reverse case.[3] In other words, the quality estimate of a feature depends on the context of other features. Hence, Relief does not treat words independently like Delta-TFIDF, χ^2 test and t-test. Due to inter dependence among words, Relief is susceptible to the data sparsity problem. It produces erroneous results when the dataset is small.

---

[1] We also observed the performance of unigrams with the frequency in the document as feature value, but we did not find any improvement in SA accuracy over the unigram's presence.

[2] Available at: http://java-ml.sourceforge.net/.

[3] More detail about the implementation of Relief can be obtained from Liu and Hiroshi [23].

Delta-TFIDF: The problem with the TFIDF-based feature vector is that it fails to differentiate between terms from the perspective of the conveyed sentiments, as it doesn't utilize the annotation information available with the corpus. Delta-TFIDF assigns feature value to a word w for a document d by computing the difference of that word's TFIDF scores in the positive and the negative training corpora D [19]. The value of the feature in the feature vector of a document is given by the following Delta-TFIDF formula:

$$Delta - TFIDF(w, d, D) = tf(w, d) * \log \frac{N_w}{P_w} \qquad (2)$$

where N_w and P_w are the number of documents in the negatively labeled and positively labeled corpus with the word w. Features that are more prominent in the negative training corpus than the positive training corpus will get a positive score by Delta-TFIDF, and features that are more prominent in the positive training corpus will get a negative score. Features which have equal occurrences in positive and negative corpora will get a zero value in the feature vector. Hence, Delta-TFIDF makes a linear division between the positive sentiment features and the negative sentiment features. Since Delta-TFIDF observes the association of a word with a particular class, it also considers only those words as features which are significant for classification.

χ^2 **test:** It is a statistical significance test, which is based on computing the probability (P-value) of a test statistic given that the data follows the null hypothesis. In the case of comparing the frequencies of a given word in different classes of a corpus, the test statistic is the difference between these frequencies and the null hypothesis is that the frequencies are equal. If the P-value is below a certain threshold, then we reject the null hypothesis. χ^2 test and Delta-TFIDF are bag-of-words-based tests as they consider the total frequency count of the word in the positive and negative corpora. To employ χ^2 test, data is represented in a $2 * 2$ table, as illustrated in Table 2. This representation does not include any information on the distribution of the word w in the corpus. Table 1 lists the notations used in Tables 2 and 3. χ^2 test takes into consideration the labels (classes) associated with the words and is formulated as follows.

$$\chi^2(w) = ((C_p^w - \mu^w)^2 + (C_n^w - \mu^w)^2)/\mu^w \qquad (3)$$

Here, μ^w represents an average of the word's count in the positive and negative corpora. If a word w gives χ^2 value above a certain threshold value, we hypothesize that the word w belongs to a particular class, hence it is significant for classification.[4] In this way, χ^2 test gives a compact set of significant words from the corpus as features for sentiment classification.

---

[4] χ^2 value and P-value have inverse correlation, hence a high χ^2 value corresponds to a low P-value. The correlation table is available at: http://sites.stat.psu.edu/~mga/401/tables/Chi-square-table.pdf.

Table 1. Notations used in Tables 2 and 3

Symbol	Description
C_p^w	Count of w in the positive corpus
C_n^w	Count of w in the negative corpus
C_p	Total number of words in the positive corpus
C_n	Total number of words in the negative corpus
C_{pi}^w	Count of w in i^{th} positive document
C_{ni}^w	Count of w in i^{th} negative document

Welch's t-test: It is evident from the formulation that Delta-TFIDF and χ^2 test do not account for the uneven distribution of the word, as it relies only on the total number of occurrences in the corpus. Therefore, it underestimates the uncertainty. On the contrary, Welch's t-test assumes independence at the level of texts rather than an individual word and represents data differently. It considers the number of occurrences of a word per text, and then compares a list of counts from one class against a list of counts from another class. The representation of the input data for Welch's t-test is illustrated in Table 3. Welch's t-test generates a P-value corresponding to a t value for the null hypothesis that the mean of the two distributions are equal. Let x_p^w be the mean of the frequency of w over texts in positive documents and let s_p^w be the standard deviation. Likewise, let x_n^w be the mean of the frequency of w over texts in negative documents, and let s_n^w be the standard deviation. The symbols $|p|$ and $|n|$ represent the total number of positive and negative documents in the corpus. t-test is formulated as follows:

$$t(w) = \frac{x_p^w - x_n^w}{\sqrt{\frac{(s_p^w)^2}{|p|} + \frac{(s_n^w)^2}{|n|}}} \tag{4}$$

If a word w gives t value above a certain threshold value, we hypothesize that the word w belongs to a particular class, hence it is significant for classification.[5] In this way, t-test gives a compact set of significant words from the corpus as features.

Table 2. The data representation to employ χ^2 test

Word	Corpus-pos	Corpus-neg
Word w	C_p^w	C_n^w
Not Word w	$C_p - C_p^w$	$C_n - C_n^w$

---

[5] t value and P-value have inverse correlation, hence a high t value corresponds to a low P-value. The correlation table is available at: http://www.sjsu.edu/faculty/gerstman/StatPrimer/t-table.pdf.

Table 3. The data representation to employ t-test

Corpus-Pos	text$_1$	text$_2$	....	text$_M$
Frequency of word w	C_{p1}^w	C_{p2}^w	....	C_{pM}^w
Corpus-Neg	**text$_1$**	**text$_2$**	....	**text$_M$**
Frequency of word w	C_{n1}^w	C_{n2}^w	....	C_{nM}^w

An Example from Literature Comparing χ^2 and Welch's t-test: Lijffijt et al. [6] assessed the difference between χ^2 test and Welch's t-test to answer the question 'Is the word Matilda more frequent in male conversation than in female conversation?'. Here, null hypothesis was that the name *Matilda* is used at an equal frequency by male and female authors in the pros fiction sub-corpus of the British National Corpus. χ^2 test gave P-value less than 0.0001 for the word *Matilda*, while Welch's t-test gave P-value of 0.4393. The P-value given by t-test is greater than the threshold P-value 0.05 unlike χ^2 test, which indicates that the probability of the null hypothesis being true is greater than 5%. Hence, the word *Matilda* is used at an equal frequency by male and female authors as per Welch's t-test. Welch's t-test proved that the observed frequency difference between male and female conversation is not significant. On the other hand, χ^2 test substantiated that the word *Matilda* is used more frequently by male authors than female authors. The reason behind the disagreement between tests is that the word *Matilda* is used in only 5 of 409 total texts with an uneven frequency distribution: one text written by male author contains 408 instances and the other 4 texts written by female authors contain 155 instances, 11 instances, 2 instances, and 1 instance, respectively. χ^2 test does not account for this uneven distribution, as it makes use of the total frequency count of the word in the corpus. Therefore, χ^2 test *erroneously* substantiates that male authors use the name *Matilda* significantly more often than female authors. Therefore, bag-of-words-based tests like Delta-TFIDF and χ^2 test are not an appropriate choice when comparing corpora.

The accuracy in results of significance tests matters more when it has to be used as input for some other application. χ^2 test, Delta-TFIDF and Welch's t-test, all three can be used to identify significant words available in the corpus for sentiment analysis. However, Delta-TFIDF differs from χ^2 test and Welch's t-test statistically. Delta-TFIDF makes a linear division between positive features and negative features by assigning a value of opposite sign in the feature vector. On the other hand, χ^2 test and Welch's t-test are hypothesis testing tools as they have a distribution for P-value corresponding to the score given by the test. If a word depicts a P-value less than a threshold of 0.05[6], we reject the null hypothesis, *i.e.*, we reject the uniform use of the word in positive and negative class. Consequently, we consider that the word is used significantly more often in one class (positive or negative), hence it is significant for classification.

---

[6] The threshold 0.05 on P-value is a standard value in statistics as it gives 95% confidence in the decision.

A few examples of words which are found significant by χ^2 test, but not by t-test in the electronics domain are shown in Table 4. The symbols C_{pos} and C_{neg} represent the total count of the word in the positive and negative review corpora respectively. The P-values given by χ^2 test are less than the threshold 0.05, hence words are significant for sentiment classification in the electronics domain by χ^2 test. However, Welch's t-test gives P-value greater than the threshold 0.05 for all the examples. Words which have very few total occurrences in the corpus are found significant by χ^2 test, like *flaky* is wrongly declared significant by χ^2 test. On the other hand, words which have sufficient occurrences in the corpus, but don't have sufficient difference in the distribution of the word in two classes (*eg.*, *experience, wrong and heavy*), are also erroneously found significant by χ^2 test. However, Welch's t-test observes the difference in the distribution of the word in the two classes, which makes it statistically more accurate. Hence, a distribution-based test like Welch's t-test is a better choice than bag-of-words-based tests like χ^2 test and Delta-TFIDF. Table 7 shows that Welch's t-test gives an accuracy of 87% in the electronics domain, which is 2.75% higher than the accuracy obtained with χ^2 test and 5% higher than the accuracy obtained with Delta-TFIDF.

Table 4. P-value for χ^2 and t tests with χ^2 value and t value in the electronics domain

Word	C_{pos}	C_{neg}	χ^2 value	P-value	t value	P-value
Flaky	0	4	4	0.04	−1.38	0.16
Experience	27	49	6.37	0.01	−0.81	0.41
Wrong	28	56	9.3	0.00	0.79	0.43
Heavy	29	15	4.45	0.03	0.79	0.43

4   Dataset

We validate our hypothesis that significance tests give a more promising and robust solution in comparison to existing feature engineering techniques for three types of SA tasks, *viz.*, in-domain, cross-domain and cross-lingual SA.

For in-domain and cross-domain SA, we have shown the results with four different domains, *viz.*, Movie (M), Electronics (E), Housewares (H) and Books (B). The movie review dataset is taken form the IMDB archive [24].[7] Data for the other three domains is taken from the amazon archive [25].[8] Each domain has 1000 positive and 1000 negative reviews.

---

[7] Available at: http://www.cs.cornell.edu/people/pabo/movie-review-data/.
[8] Available at: http://www.cs.jhu.edu/~mdredze/datasets/sentiment/index2.html. This dataset has one more domain, that is, DVD domain. The contents of reviews in the DVD domain are very similar to the reviews in the movie domain; hence, to avoid redundancy, we have not reported results with the DVD domain.

Table 5. Dataset statistics

Domain	No. of Reviews	Avg. Length
Movie (M)	2000	745 words
Electronic (E)	2000	110 words
Housewares (H)	2000	93 words
Books (B)	2000	173 words
Language	No. of Reviews	Avg. Length
English (en)	5000	201 words
French (fr)	5000	91 words
German (de)	5000	77 words
Russian (ru)	1400	40 words

Balamurali et al. [26] showed that a small set of manually annotated corpus in the language gives a better sentiment analysis system in the language than a machine-translation-based cross-lingual system. We have used the same dataset used by Balamurali et al. [26] to show the impact of significant words in cross-lingual sentiment analysis. The dataset contains movie review corpus in the four different languages, *viz.*, English (en), French (fr), German (de) and Russian (ru). Table 5 shows the statistics of all the dataset used for this work.

5 Experimental Setup

Unigrams, TFIDF and Delta-TFIDF are coded as per their definitions to obtain the feature vector of a document. In case of unigrams, TFIDF and Delta-TFIDF, we have selected those words as features whose count is greater than 3 in the corpus to avoid the misspelled or very low impact words. Though the feature set size is the same, the feature value in the feature vector is as per the definition of unigrams, TFIDF and Delta-TFIDF (Sect. 3). To implement Relief, we have used the publicly available java-based machine learning library (java-ml). Relief assigns a score to features based on how well they separate the instances in the problem space. We set a threshold on score assigned by Relief to filter out the low score features.[9] In the case of Relief, feature value in the feature vector is the presence (1) or absence (0) of the feature (word) in the document.

To implement statistical significance tests, *viz.*, Welch's t-test and χ^2 test, we have used a java-based statistical package, that is, Common Math 3.6.[10] We opted for Welch's t-test over Student's t-test, because the former test is more general than Student's t-test. Student's t-test assumes equal variance in the two

---

[9] A threshold on score is set empirically to filter out the words about which tests are not very confident, where the low confidence is visible from the low score assigned by Relief.

[10] Available at: https://commons.apache.org/proper/commons-math/download_math.cgi.

populations which have to be compared, which is not true in the case of Welch's t-test. χ^2 test and Welch's t-test result into a P-value (Probability-value), which is probability of the data given null hypothesis is true. Threshold on P-value gives confidence in the significance decision. The value 0.05 is a standard threshold value, which gives 95% confidence in the significance decision. In the case of t-test and χ^2 test, features are the words which satisfy the test at threshold of 0.05. The feature value in the feature vector is 1, if the significant word given by the test is present in the document, else 0. Table 6 depicts the variation in feature set size obtained from the training data in Movie (M), Electronics (E), Books (B) and Housewares (H) domains under various features building methods. Application of statistical significance tests, specifically t-test reduces the feature vector size substantially. SVM algorithm [27] is used to train a classifier with all the mentioned feature building methods in the paper.[11]

Table 6. Feature vector size

	Unigrams	TFIDF	Relief	Delta-TFIDF	χ^2 test	t-test
M	19152	19152	17232	19152	4877	2157
E	4235	4235	3125	4235	1039	522
B	7835	7835	6810	7835	1727	583
H	3649	3649	2650	3649	912	493

6   Results and Discussion

We validate the effectiveness of significant words as features for three types of sentiment analysis tasks, *viz.*, in-domain, cross-domain and cross-lingual. The data in all three cases is divided into two parts, *viz.*, train data (80%) and test data (20%). Accuracy is the popularly used measure for evaluation in sentiment analysis [9,11,12,24,28]. We report the accuracy for all the below mentioned systems on the test data. The reported accuracy is the ratio of the correctly predicted documents to that of the total number of documents.

6.1   In-Domain Sentiment Classification

In case of in-domain SA, the domain of the test and training dataset remains the same. Table 7 shows the in-domain SA accuracies obtained with SVM algorithm in the four domains, *viz.*, Electronics (E), Movie (M), Books (B) and Housewares (H). Significant words as features obtained by Delta-TFIDF, χ^2 test and Welch's t-test outperform unigrams, TFIDF and Relief in all the four domains. The performance of Delta-TFIDF and χ^2 test is approximately equal as they are

---

[11] We use SVM package libsvm, which is available in java-based WEKA toolkit for machine learning. Available at: http://www.cs.waikato.ac.nz/ml/weka/downloading.html.

bag-of-words-based significance tests. On the other hand, Welch's t-test which is a distribution-based test performs consistently better than χ^2 test and Delta-TFIDF.[12] Table 8 shows the confusion matrices obtained with unigrams, TFIDF and Relief in the movie domain. Table 9 shows the confusion matrices obtained with Delta-TFIDF, χ^2 test and t-test in the movie domain.[13]

Table 7. In-domain sentiment classification accuracy in % using SVM

	Unigrams	TFIDF	Relief	Delta-TFIDF	χ^2 test	t-test
M	84.5	84	85.5	87	88.75	89
E	81	76	82.5	82	84.25	87
B	76	75	82	83	82.5	87.5
H	84	84	86	86.5	87	88.5

Table 8. Confusion matrices for Unigrams, TFIDF and Relief using SVM in the Movie domain

	neg	pos			neg	pos			neg	pos
neg	171	29		**neg**	171	29		**neg**	171	29
pos	33	167		**pos**	35	165		**pos**	29	171
(a) Unigrams				(b) TFIDF				(c) Relief		

Table 9. Confusion matrices for Delta-TFIDF, χ^2 test and t-test using SVM in the Movie domain

	neg	pos			neg	pos			neg	pos
neg	172	28		**neg**	181	19		**neg**	180	20
pos	24	176		**pos**	26	174		**pos**	24	176
(a) DTFIDF				(b) χ^2 test				(c) t-test		

6.2 Cross-Domain Sentiment Classification

Training a classifier in a labeled source domain and testing it on an unlabeled target domain is known as cross-domain sentiment analysis [25,29]. Identification of significant words in the source domain restricts the transfer of irrelevant

---

[12] Application of significance test (Delta-TFIDF or χ^2 test or t-test) reduces the feature set size substantially, which stimulates a less computationally expensive SA system in comparison to unigrams, TFIDF and Relief.

[13] Since movie domain has the highest average length of the document (review), we have selected movie domain to show the variation among confusion matrices obtained with different feature building methods.

information to the target domain, which in turn leads to an improvement in the cross-domain classification accuracy. Figure 1 shows the sentiment classification accuracy obtained in the target domain for 12 pairs of source and target domains. TFIDF performed the worst for all domain pairs and significant words consistently performed better than unigrams, TFIDF and Relief. In addition, on an average, t-test performs better than significant words obtained using χ^2 test and Delta-TFIDF.

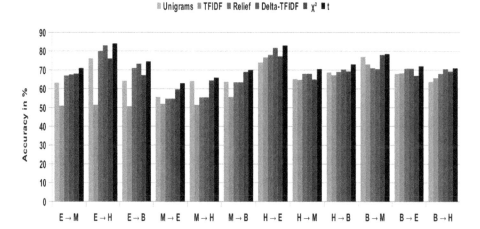

Fig. 1. Cross-domain sentiment classification accuracy in % for 12 pairs of Source (s) → Target (t) domains

6.3   Cross-Lingual Sentiment Classification

In Cross-Lingual Sentiment Analysis (CLSA), the task is to build a classifier for a resource deprived language [30,31]. By resource deprived, we mean that a language in which labeled review corpus is not available. Though Balamurali et al. [26] claimed that obtaining a small set of manually annotated data is a better option than using machine translation systems for CLSA, collecting an annotated corpus will always remain a challenging task.[14] We translate labeled data in the source language into the target language to obtain labeled data in the target language.[15] Language translation is done using Google translator API[16] available on the Web.

---

[14] CLSA results are reported using the four different languages, *viz.*, English (en), French (fr), German (de) and Russian (ru). The more detail about the dataset is given in Table 5.

[15] In all CLSA experiments, training data is obtained by translating source language data, while test data is taken from the available manually tagged non-translated data.

[16] Available at: http://crunchbang.org/forums/viewtopic.php?id=17034.

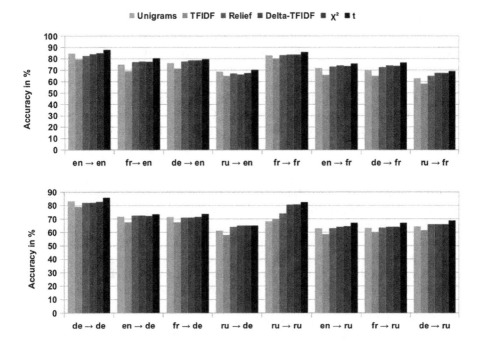

Fig. 2. Results for cross-lingual SA using common unigrams, TFIDF, Relief, Delta-TFIDF, significant words by χ^2 test and t-test as features

Though the translation process does not alter labels (positive or negative) of review documents, it introduces errors in the content of the data due to translation challenges. Exclusion of irrelevant words from the feature set by significance tests decreases the ratio of wrongly translated words in the feature set. Essentially, the use of significant words overcomes the deficiency introduced by the use of machine translation system in CLSA. Figure 2 presents the cross-lingual accuracy obtained for 12 pairs of source and target languages.[17] It depicts that TFIDF performs the worst for all language pairs. On the other hand, significant words consistently perform better than unigrams, TFIDF and Relief. In addition, on an average, t-test performs better than significant words obtained using χ^2 test and Delta-TFIDF. To observe the impact of machine translation in CLSA, we computed Pearson product moment correlation between BLEU score of translation and the CLSA accuracy obtained with t-test for all 16 pairs.[18] The BLEU score of translation for each pair is taken from Koehn [32]. We observed a strong *positive correlation* of 0.89 between the BLEU score and the CLSA accu-

---

[17] For pairs $en{\to}en$, $fr{\to}fr$, $de{\to}de$ and $ru{\to}ru$, source and target languages are the same and training data is not the translated data, it is the original manually tagged dataset in the language.

[18] In case of in-language pairs, for example, en→en we assumed a BLEU score of 100 considering that this pair has 100% correct translation as there is no translation process involved.

racy obtained with t-test, which indicates that the reduction in noise caused by translation leads to a high accuracy cross-lingual sentiment analysis system.

Discussion: In literature, unigrams (bag-of-words) are considered to be the best visible features in the corpus for sentiment analysis [9,33]. Unigrams-based model does not differentiate between relevant and irrelevant words, but the presence of irrelevant features affects the classifier negatively. The product of term frequency and inverse document frequency (TF * IDF) of a word gives a measure of how frequent this word is in the document with respect to the entire corpus of documents. A word in the document with a high TFIDF score occurs frequently in the document and provides the most information about that specific document. Finding the feature value using TFIDF has been proven to be very helpful for Information Retrieval (IR) [21,34]. However, a high frequency of a word in the document relative to the corpus does not give any information about the polarity of the document. Hence, TFIDF is not a good measure for sentiment analysis. On the other hand, Relief assigns weight to a word based on the weights of other context words in the corpus. It restricts the information gain to a fixed number of context words of the input word, which makes Relief a less informative method. In addition, dependence on the context words to assign score makes it susceptible to the data sparsity problem.

Delta-TFIDF is mainly associated with sentiment classification or polarity detection of text [35–37]. Delta-TFIDF filters out the words which are evenly distributed in positive and negative classes of the corpus. In this way, Delta-TFIDF score better represents the word's true importance in the document for sentiment classification. Similarly, χ^2 test and t-test extract words from the corpus which are important for sentiment classification, but these significance tests have a probability distribution associated with the test's score. This probability distribution allows us to select the significant words efficiently as per the desired confidence level. It is noticeable that Welch's t-test appears more promising in comparison to Delta-TFIDF and χ^2 test. t-test compares the distribution of the word in positive and negative corpora instead of the total frequency, which makes it more foolproof for significant words detection from the sentiment annotated corpus. Therefore, the set of significant words given by the t-test is less erroneous, which encourages a less erroneous sentiment analysis system.

6.4   Statistical Comparison of Different Feature Building Methods with t-test

To observe the difference among reported feature building methods statistically, we applied t-test on the accuracy distribution produced by various methods for in-domain SA (Table 7). Table 10 reports only those combinations where method-X is found to be statistically different from method-Y.[19] It depicts the t value, P-value with respect to t value and the confidence interval for t value. Table 10 shows that the results produced by t-test are significantly better than unigrams,

---

[19] Here, the P-value for the t value is less than 0.05. Significance of difference in accuracy is observed at $P < 0.05$, which gives 95% confidence in decision.

TFIDF, Relief and Delta-TFIDF. Negative sign before the t value indicates that method-2 is better than method-1. No other combination of methods showed a significant difference in accuracy as per t-tests. However, the consistent improvement in 4 domains (Table 7) asserts that Relief is better than unigrams, while Delta-TFIDF and χ^2 are better than relief. It is difficult to compare Delta-TFIDF and χ^2 test in terms of superiority. On the other hand, t-test is consistently better than any other feature building method for all the considered cases, which asserts our hypothesis that the feature set produced by t-test is more accurate than any other feature building method.

Table 10. In all rows, method-2 is significantly better than method-1 as P-value for the observed t value is less than 0.05

Method-1 *vs.* Method-2	t value	P-value	Confidence Interval
Unigrams *vs.* t-test	-3.30	0.01	$(-11.52, -1.72)$
TFIDF *vs.* t-test	-3.29	0.01	$(-14.37, -2.12)$
Relief *vs.* t-test	-3.50	0.01	$(-6.73, -1.26)$
Delta-TFIDF *vs.* t-test	-2.54	0.04	$(-6.62, -0.12)$

7   Conclusion

In this paper, we have shown that the methods which analyze class (positive or negative) or significance of a feature before considering the feature into feature set are more promising for sentiment analysis. We have conceptually studied and compared various types of feature building methods, *viz.*, unigrams, TFIDF, Relief, Delta-TFIDF, χ^2 and *t-test*. We have shown the impact of significance tests over other feature building methods for three types of sentiment analysis tasks, *viz.*, in-domain, cross-domain and cross-lingual sentiment analysis. Results show that the significance tests, *viz.*, Delta-TFIDF, χ^2 and t-test give a better feature set than the existing standard feature building methods, *viz.*, unigrams, TFIDF and Relief for sentiment analysis task. In addition, we showed that the distribution-based significance test, *i.e.*, Welch's t-test is better than the bag-of-words-based χ^2 test and Delta-TFIDF. Essentially, in this paper, we have emphasized the need for a correct significance test with an example in sentiment analysis. The future work consists of extending the observations to other NLP tasks.

References

1. Oakes, M., Gaaizauskas, R., Fowkes, H., Jonsson, A., Wan, V., Beaulieu, M.: A method based on the chi-square test for document classification. In: Proceedings of the 24th Annual International ACM SIGIR Conference on Research and Development in Information Retrieval, pp. 440–441. ACM (2001)

2. Jin, X., Xu, A., Bie, R., Guo, P.: Machine learning techniques and chi-square feature selection for cancer classification using SAGE gene expression profiles. In: Li, J., Yang, Q., Tan, A.-H. (eds.) BioDM 2006. LNCS, vol. 3916, pp. 106–115. Springer, Heidelberg (2006). https://doi.org/10.1007/11691730_11
3. Moh'd, A., Mesleh, A.: Chi square feature extraction based SVMS arabic language text categorization system. J. Comput. Sci. **3**, 430–435 (2007)
4. Kilgarriff, A.: Comparing corpora. Int. J. Corpus Linguist. **6**, 97–133 (2001)
5. Paquot, M., Bestgen, Y.: Distinctive words in academic writing: a comparison of three statistical tests for keyword extraction. Lang. Comput. **68**, 247–269 (2009)
6. Lijffijt, J., Nevalainen, T., Säily, T., Papapetrou, P., Puolamäki, K., Mannila, H.: Significance testing of word frequencies in corpora. Digital Scholarsh. Humanit. (2014) (fqu064)
7. Glorot, X., Bordes, A., Bengio, Y.: Domain adaptation for large-scale sentiment classification: a deep learning approach. In: Proceedings of the 28th International Conference on Machine Learning (ICML-11), pp. 513–520 (2011)
8. Zhou, J.T., Pan, S.J., Tsang, I.W., Yan, Y.: Hybrid heterogeneous transfer learning through deep learning. AAAI, 2213–2220 (2014)
9. Pang, B., Lee, L., Vaithyanathan, S.: Thumbs up?: sentiment classification using machine learning techniques. In: Proceedings of Conference on Empirical Methods in Natural Language Processing, pp. 79–86 (2002)
10. Meyer, T.A., Whateley, B.: Spambayes: effective open-source, bayesian based, email classification system. In: CEAS. Citeseer (2004)
11. Kanayama, H., Nasukawa, T.: Fully automatic lexicon expansion for domain-oriented sentiment analysis. In: Proceedings of Conference on Empirical Methods in Natural Language Processing, pp. 355–363 (2006)
12. Cheng, A., Zhulyn, O.: A system for multilingual sentiment learning on large data sets. In: Proceedings of International Conference on Computational Linguistics, pp. 577–592 (2012)
13. Leskovec, J., Rajaraman, A., Ullman, J.D.: Mining of massive datasets. Cambridge University Press, Cambridge (2014)
14. Oakes, M.P., Farrow, M.: Use of the chi-squared test to examine vocabulary differences in english language corpora representing seven different countries. Lit. Linguist. Comput. **22**, 85–99 (2007)
15. Al-Harbi, S., Almuhareb, A., Al-Thubaity, A., Khorsheed, M., Al-Rajeh, A.: Automatic Arabic text classification (2008)
16. Rayson, P., Garside, R.: Comparing corpora using frequency profiling. In: Proceedings of the workshop on Comparing Corpora, Association for Computational Linguistics, pp. 1–6 (2000)
17. Sharma, R., Bhattacharyya, P.: Detecting domain dedicated polar words. In: Proceedings of the International Joint Conference on Natural Language Processing, pp. 661–666 (2013)
18. Kira, K., Rendell, L.A.: The feature selection problem: traditional methods and a new algorithm. AAAI **2**, 129–134 (1992)
19. Martineau, J., Finin, T.: Delta TFIDF: an improved feature space for sentiment analysis. ICWSM **9**, 106 (2009)
20. Martineau, J., Finin, T., Joshi, A., Patel, S.: Improving binary classification on text problems using differential word features. In: Proceedings of the 18th ACM conference on Information and knowledge management, pp. 2019–2024. ACM (2009)
21. Wu, H.C., Luk, R.W.P., Wong, K.F., Kwok, K.L.: Interpreting TF-IDF term weights as making relevance decisions. ACM Trans. Inf. Syst. (TOIS) **26**, 13 (2008)

22. Čehovin, L., Bosnić, Z.: Empirical evaluation of feature selection methods in classification. Intell. Data Anal. **14**, 265–281 (2010)
23. Liu, H., Motoda, H.: Computational methods of feature selection. CRC Press, Boca Raton (2007)
24. Pang, B., Lee, L.: A sentimental education: Sentiment analysis using subjectivity summarization based on minimum cuts. In: Proceedings of Association for Computational Linguistics, pp. 271–279 (2004)
25. Blitzer, J., Dredze, M., Pereira, F., et al.: Biographies, bollywood, boom-boxes and blenders: domain adaptation for sentiment classification. In: Proceedings of Association for Computational Linguistics, pp. 440–447 (2007)
26. Balamurali, A.R., Khapra, M.M., Bhattacharyya, P.: *Lost in translation*: viability of machine translation for cross language sentiment analysis. In: Gelbukh, A. (ed.) CICLing 2013. LNCS, vol. 7817, pp. 38–49. Springer, Heidelberg (2013). https://doi.org/10.1007/978-3-642-37256-8_4
27. Tong, S., Koller, D.: Support vector machine active learning with applications to text classification. J. Mach. Learn. Res. **2**, 45–66 (2001)
28. Sharma, R., Bhattacharyya, P.: Domain sentiment matters: a two stage sentiment analyzer. In: Proceedings of the International Conference on Natural Language Processing (2015)
29. Pan, S.J., Ni, X., Sun, J.T., Yang, Q., Chen, Z.: Cross-domain sentiment classification via spectral feature alignment. In: Proceedings of the 19th International Conference on World Wide Web, pp. 751–760. ACM (2010)
30. Wan, X.: Co-training for cross-lingual sentiment classification. In: Proceedings of the Joint Conference of the 47th Annual Meeting of the ACL and the 4th International Joint Conference on Natural Language Processing of the AFNLP, vol. 1, pp. 235–243. Association for Computational Linguistics (2009)
31. Wei, B., Pal, C.: Cross lingual adaptation: an experiment on sentiment classifications. In: Proceedings of the ACL 2010 Conference Short Papers, Association for Computational Linguistics, pp. 258–262 (2010)
32. Koehn, P.: Europarl: a parallel corpus for statistical machine translation. MT Summit. **5**, 79–86 (2005)
33. Ng, V., Dasgupta, S., Arifin, S.: Examining the role of linguistic knowledge sources in the automatic identification and classification of reviews. In: Proceedings of the COLING/ACL on Main Conference Poster Sessions, pp. 611–618. Association for Computational Linguistics (2006)
34. Salton, G., Buckley, C.: Term-weighting approaches in automatic text retrieval. Inf. Process. Manag. **24**, 513–523 (1988)
35. Lin, Y., Zhang, J., Wang, X., Zhou, A.: An information theoretic approach to sentiment polarity classification. In: Proceedings of the 2nd Joint WICOW/AIRWeb Workshop on Web Quality, pp. 35–40. ACM (2012)
36. Demiroz, G., Yanikoglu, B., Tapucu, D., Saygin, Y.: Learning domain-specific polarity lexicons. In: 2012 IEEE 12th International Conference on Data Mining Workshops, pp. 674–679. IEEE (2012)
37. Habernal, I., Ptáček, T., Steinberger, J.: Sentiment analysis in czech social media using supervised machine learning. In: Proceedings of the 4th Workshop on Computational Approaches to Subjectivity, Sentiment and Social Media Analysis, pp. 65–74 (2013)

BATframe: An Unsupervised Approach for Domain-Sensitive Affect Detection

Kokil Jaidka[1], Niyati Chhaya[2(✉)], Rahul Wadbude[3], Sanket Kedia[4], and Manikanta Nallagatla[5]

[1] University of Pennsylvania, Philadelphia, PA, USA
jaidka@sas.upenn.edu
[2] Adobe Research Big Data Experience Lab, Bangalore, India
nchhaya@adobe.com
[3] Indian Institute of Technology, Kanpur, India
rahulwadbude2@gmail.com
[4] Indian Institute of Technology, Kharagpur, India
kediasanket11121993@gmail.com
[5] Indian Institute of Technology, Roorkee, India
manikanta001nallagatla@gmail.com

Abstract. Generic sentiment and emotion lexica are widely used for the fine–grained analysis of human affect from text. In order to accurately detect affect, there is a need for domain intelligence, that enables understanding of the perceived interpretation of the same words in varied contexts. Recent work has focused on automatically inducing the polarity of given terms in changing contexts. We propose an unsupervised approach for the construction of domain–specific affect lexica along these lines. The algorithm is seeded with existing standard lexica and expanded based on context–relevant word associations. Experiments show that our lexicon provides better coverage than standard lexica on both short as well as long texts, and corresponds well with human–annotated affect values. Our framework outperforms the state–of–the–art generic and domain–specific approaches with a precision of over 70% for the emotion detection task on the SemEval 2007 Affect Corpus.

Keywords: Sentiment mining · Affect lexicon · ANEW
Domain adaptation · Convex optimization

1 Introduction

The growth of social media has created large archives of digital opinion data comprising reviews, forums, discussions, blogs, micro-blogs, and social networks, which are a valuable resource for analyzing and understanding people's **affects**:

K. Jaidka—This work was done when all authors were at Adobe Research. All authors have equal contribution in this paper.

© Springer Nature Switzerland AG 2018
A. Gelbukh (Ed.): CICLing 2017, LNCS 10762, pp. 20–34, 2018.
https://doi.org/10.1007/978-3-319-77116-8_2

their emotional reactions or feelings towards different topics. For accurate affect mining, there is a need to develop tools which require little or no training data, and may be bootstrapped onto supervised methods or general–purpose lexica, in order to improve their effectiveness for domain–specific applications.

Most general–purpose affect lexica are insensitive to the changing affects of words, for varying domains. For instance, in a crime news story, 'slay' would imply a negative affect. On the other hand, in a social event news story, 'slay' would imply a positive affect, which is impossible to infer without considering its context within the textual description. The approach is based on the intuition that 'the affective value of a word changes across domains, and it can be measured through its co–occurring words'.

In this paper, we propose an unsupervised method to construct domain–specific affect lexica. We propose the **Biblion of Affective Topics framework (BATframe)**, an optimization framework that is applicable to unlabeled opinionated corpus in any domain, and does not make assumptions about the availability of human–judged labels (which are usually expensive to obtain in a new domain). Our paper makes the following contributions:

1. It provides a framework to construct domain–specific affect lexica using general–purpose affective lexica, with no prior human labeling.
2. It identifies the dependency relations that can cause relatively neutral words to gain affective value, such as 'child' in a crime news story.
3. It demonstrates its efficacy in emotion detection on the SemEval 2007 Affect detection task, where it outperforms the state of the art generic and domain–specific lexica in detecting emotions from text.

A qualitative evaluation against a Crime News dataset annotated with affective labels by human annotators is conducted to demonstrate its qualitative merits. Further, on two datasets varying in their topics, document lengths, and intended purpose: Crime News and Beauty Industry Email subject lines; we demonstrate the consistently better coverage of the BAT framework than standard, general purpose lexica. Finally, a comparison against the state–of–the–art general and domain–specific emotion classifiers on the SemEval 2007 Affect detection task demonstrates its superiority in detecting emotions in the held–out test set.

1.1   Glossary of Terms

Affect: Affect is the experience of an emotion or a feeling.

Topic Word: Representative words from the text corpus are termed as topic words. These words will be entries in the output lexicon (BAT) with corresponding affect scores in terms of a PAD (Pleasure, Arousal, Dominance) score.

Affect Word: The ANEW lexicon [1] is used for this work. The words present in the ANEW lexicon (along with the corresponding affects <word,P,A,D>) are referred to as 'affect words'. Note that the affect words are only used for affect score calculation of topic words and are not present in the output lexicon.

Topic–Affect Tuples: These tuples are defined as <topic word, affect word>, where topic and affect words are as already defined. These tuples are mined from text and are required for our framework. The tuple represents the existence of a relationship between the topic word and affect word. These tuples will not be present in the output lexicon.

BAT Lexicon: It is our output lexicon, which has domain–specific affect values for a given domain. It will contain entries of the form <word, P,A,D>.

2 Related Work

A recent body of work has explored approaches to adapt general–purpose lexica for specific contexts. Studies have recognized the limited applicability of general purpose lexica such as ANEW to identify affect in verbs and adverbs, as they focus heavily on adjectives (see Fig. 2). Recognizing that general–purpose lexica often detect sentiment which is incongruous with context, Ribeiro et al. [2] proposed a sentiment damping method which utilizes the average sentiment strength over a document to damp any abnormality in derived sentiment strength. Similarly, Blitzer et al. [3] argued that words like 'predictable' induced a negative connotation in book reviews, while 'must–read' implied a highly positive sentiment. Muhammad et al. [4] illustrated the need to adapt sentiment methods trained for longer documents, by creating the TEC–lexicon for detecting eight emotions from one–liner social media texts. Neilson (2011) [5] also created a new ANEW specifically geared towards detecting sentiment in microblog posts, but it only comprises 2477 words, scored manually on a scale of +2 (positive) to −2 (negative). It ignores the Arousal and Dominance dimensions of the ANEW lexicon.

These studies identified a few challenges, which the present study addresses. Firstly, the studies dependent on crowdsourcing, such as the work by Blitzer et al. [3] and the NRC EmoLex tool developed by Mohammad et al. [6] are usually a product of experiments on a very specific domain. Secondly, due to the high relative cost of human labor, the outcomes are limited in size, and are built based on a small hand–annotated corpus; hence, they cannot be easily expanded or generalized to other domains, or even to non–document forms of text [4] [such as social media posts]. Thirdly, studies have recommended future work to focus on making affect lexica more *syntactically representative* of parts–of–speech other than adjectives, such as adverbs and verbs, which has not seen much progress so far [7].

Several approaches have been used in order to adapt general–purpose lexica for research problems in specific domains. Some studies have used bootstrapping to start with an initial seed lexicon [8], such as the work by Miller et al. [9] which used WordNet to assign positive or negative polarity to words based on synonyms and antonyms for a small set of seed words; however, this method is heavily biased by the number and set of seed words chosen. Our approach is similar to the previous studies which have explored the use of syntactic structure – such as parsing rules, linguistic patterns [3,8,10], and Latent Semantic Analysis [11–13] and collocations [14] – to identify topical affect words. Some studies

have also used semantic structure, such as the presence of synsets [15], to identify the polarity of unlabeled words. These approaches have shown some success;
however, they are dependent on the availability of human labels, furthermore
show low agreement against humans labels. Their experiments aim at polarity detection on a un–dimensional sentiment scale, while ignoring the Arousal
and Dominance aspects of affect. Also, a lot of these studies aim at mining
opinion words, and are evaluated on opinion mining tasks for product reviews,
which are not relevant to our purpose of **improving context–sensitive affect
detection**. While we were able to adapt existing techniques for our research
problem, there were only a few systems which could be compared head–to–head
against our implementation, which have been covered in our Evaluation section
(see Sect. 5). On the SemEval Affect Task, we show that our three–dimensional
affect lexicon can also be applied - and outperforms the state–of–the–art - in
detecting eight dimensions of emotional categories from text.

3   The BATframe

Figure 1 outlines our Biblion of Affective Topics Framework (BATframe) for
obtaining a domain–specific affect lexicon. First, the Topic words Extractor
identifies the topic words, by removing stopwords from the frequently occurring
words. Next, the Topic–affect Tuple Extractor mines the affect words around
each topic word, by using n-grams and syntactic rules, and then represents them
as a topic–affect tuple <topic word, affect word> word pair. Finally, the Optimization Framework tunes the domain–specific Pleasure, Arousal, and Dominance (P,A,D) scores for the topic word on the basis of a set of constraints.
These steps are explained here:

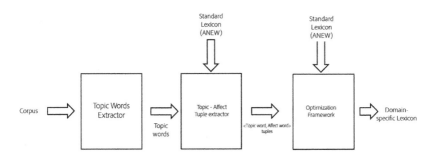

Fig. 1. The BAT framework

3.1   Topic Words Extractor

Topic Words are selected by identifying the high–frequency part–of–speech tags
which were relevant for the affect mining task. The corpus is tagged with

the POS using the NLTK [16] PoS tagger. All nouns **(NN)**, adjectives **(JJ)**, adverbs **(RB)**, gerund verbs **VBG**, non–modal verbs, and non–auxiliary action verbs **(VB)** are retained. This forms our initial topic words set. In the next step, the elbow point of the frequency distribution of these words is used to prune this set, retaining only the high–frequency words. Examples provided in Sect. 5.

3.2 Topic–Affect Tuple Extractor

This step identifies the affective words (words with an affective value and belonging to the ANEW lexicon) occurring in context of the topic words, and constructs pairwise tuples of topic and affective words. We use the Affective Norms for English Words (ANEW) lexicon [1], which comprises words rated on the dimensions of Pleasure (P), Arousal (A), and Dominance (D) on a scale of 1 to 9. Two methods are used to identify the relevant affective words: Context window and Dependency parsing. While chosing the context window; experiments for varied window size are conducted. Further, to implement the dependency parser, the Stanford NLP Parser [17] is used. Finally 4 dependency relations, which are provided in Table 1 are selected. Example tuples from our datasets are provided in Table 3. In Table 1, **A** denotes the affective words, which fulfilled one or more of the following dependency relations with a topic word **T**. The first two rules do not constrain the part–of–speech for the affect word. After obtaining the topic–affect tuples, we score each tuple for its importance using the following approaches:

Table 1. Data statistics: dependency rules

Dependency relation	Syntactic representation
Affect word is in a clause where the Subject: Topic word	**(A, nsubj, T)**
Topic word is the Subject of a passive clause with: Affect word	**(A, nsubjpass, T)**
Affect word is an adjective where Noun: Topic word	**(T, amod, A)**
Affect word is the adjective in a clause, which modifies: Topic word	**(A, nmod, T)**

1. **Correlation between Topic and Affect word:** For a given topic word, once the affect words have been identified, they are scored to indicate the correlation with the topic word. We use Pointwise Mutual Information (PMI) to capture this correlation. The PMI for each <u,v> tuple is given by:

$$PMI = log\frac{O_{11}}{E_{11}} \tag{1}$$

where, O_{11} denotes the frequency of co-occurrence of the topic–affect tuple and E_{11} denotes the expected frequency in the case that the topic and affect words always co–occur.

2. **Separation of High and Low Collocation Affect words:** For a given topic word, the affect words that are related to it are split into two classes: affect words with high probability of occurring with the topic word and affect words with low co–occurrence probability with the topic word. These two categories of affect words are formulated as two terms in the proposed optimization equation.

3.3 Optimization Approach

Consider a collection of m text documents $D = d_1, d_2, ..., d_m$ for a given domain. Let n be the number of topic words obtained. Our goal is to compute S, a $n \times 4$ matrix, where each row S_j is a tuple <topic word, P,A,D> indicating the Pleasure, Arousal, and Dominance scores of the topic word. Let S_{jp} denote the Pleasure score of jth topic word. Similarly S_{ja} denote the arousal and S_{jd} the dominance score.

Constraints for Affect Prior: Provided with a general–purpose affect lexicon L, we define two $n \times 3$ vectors G and I^G. $I^G_{w_j}$ is an indicator function to determine whether the word w_j has prior affect score or not. While G_{w_j} is the score when the prior is present. For each topic word w_j, we set $G_{w_j} = L(w_j)$ and $I^G_{w_j} = (1\ 1\ 1)$ if w_j exists in L; otherwise, $G_{w_j} = (0\ 0\ 0)$ and $I^G_{w_j} = (0\ 0\ 0)$. Thus, the first part of our objective function is as follows:

$$\min \sum_{j=1}^{n} I^G_{w_j} ||S_{w_j} - G_{w_j}||_2 \tag{2}$$

These equations apply for each of Pleasure, Arousal, and Dominance. This component in the objective function favors a domain–specific affect score of S that is closest to the general-purpose affect lexicon, i.e. G.

Constraints for High Collocation Affect Words: Let HF_j denote the set of high frequency affect words for the word w_j. For $a_k \in HF_j$, and let G_{a_k} denote its affect score. The second part of our objective function is:

$$\min \sum_{j=1}^{n} \sum_{a_k \in HF_j} \alpha_{jk} ||S_{w_j} - G_{a_k}||_2 \tag{3}$$

where α_{jk} denotes the normalized correlation between w_j and a_k. This equation applies for Pleasure, Arousal and Dominance.

Constraints for Low Collocation Affect Words: Let LF_j denote the set of low frequency affect words for the word w_j. For $b_k \in LF_j$, and let G_{b_k} denote its affect score. The third part of our objective function becomes:

$$\min \sum_{j=1}^{n} \sum_{b_k \in LF_j} \alpha_{jk} ||S_{w_j} - G_{b_k}||_2 \tag{4}$$

where α_{jk} denotes the normalized correlation between w_j and a_k. This equation applies for Pleasure, Arousal and Dominance.

Full Objective Function: Combining all the constraints defined above, we have the final objective function for ω_{PAD}, where PAD is suitably substituted to calculate each of the Pleasure(p), Arousal(a) and Dominance(d) separately:

$$\omega_{PAD} = \lambda_1 \sum_{j=1}^{n} I_{w_j}^{G} ||S_{w_j} - G_{w_j}||_2 + \lambda_2 \sum_{j=1}^{n} \sum_{a_k \in HF_j} \alpha_{jk} ||S_{w_j} - G_{a_k}||_2$$
$$+ \lambda_3 \sum_{j=1}^{n} \sum_{b_k \in LF_j} \alpha_{jk} ||S_{w_j} - G_{b_k}||_2 \tag{5}$$

Now the optimization problem is given by

$$S_p = \min \omega_{PAD} \tag{6}$$

subject to:

$$1 \leq S_{jp} \leq 9 \; ; \; 1 \leq S_{ja} \leq 9 \; ; \; 1 \leq S_{jd} \leq 9 \tag{7}$$

where λ_1, λ_2 are weighting parameters which should be set to the degree that we trust each source of information, and λ_3 can be set to a small non–zero value such as 0.002.

4   Datasets and Experimental Setup

Results on three varied datasets are presented here. The first dataset comprises of full–length newspaper articles reporting crime news in a prominent British newspaper. The second dataset comprises one-liners about personal care products, which were the subject lines of commercial marketing emails by prominent beauty brands. The third dataset comprises the SemEval 2007 Corpus for Affect Detection [18]. The datasets are described here in Table 2. Figure 2 shows the parts–of–speech distribution of the three corpora as compared to ANEW. We see that the nouns comprise the largest proportion of all three corpora. Furthermore, since the Beauty Subject Lines corpus has a notably larger proportion of nouns and a smaller proportion of verbs than the Crime News corpus, we expect that nouns would play a more important role in short texts. Other parts-of-speech, such as verbs, comprise a major proportion of the Crime and the SEMEval corpora, but are notably less represented in the ANEW lexicon. We therefore expect that a domain–specific affect lexicon would better capture the nuances of affect from adverbs and verbs, especially in short text, than its generic counterpart. Key parameters used for the experiments are described here:

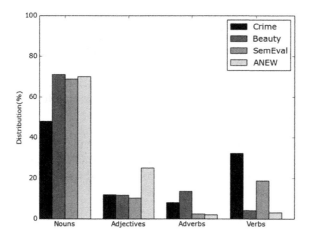

Fig. 2. POS distribution of the Corpora

Table 2. Dataset description

Domain	No. of documents	No. of words	Avg. sent length (words)	Remarks
Crime news	875	160485	14.3	Article selection based on affective analysis from Brett et al. 2013 [19]
Beauty subject lines	10000	83348	8.33	Short text corpus, similar to microblog corpus
SemEval affect corpus	1000	6900	7	Tagged for Ekman's emotions [20]

- Dataset sampling: For lexicon creation, we conducted our experiments on an 80% sample of the total sentences from either corpus. The remaining 20% was held out as the test set, and was used in subsequent evaluations.
- Topic Words Extractor: All high frequency nouns, verbs, adjectives, and adverbs above the elbow point cutoff are used as candidate topic words.
- Topic-Affect Tuple Extractor: The context window size is empirically set to sentence-level for this computation.
- Topic-Affect Correlation computation: Tenfold random sampling of 30% of the sentences containing the topic word is done in order to divide affect words into high collocation (affective words which occurred in at least 7 out of 10 samples) and low collocation words (affective words which occurred less than 7 out of 10 times with the topic word). For instance, for the topic word 'people', the high collocation words identified were 'vulnerable', 'abuse', and 'normal'. Some low frequency words identified were 'local', 'white', and 'bad'.

– Convex Optimization: The optimization function is solved as a linear programming problem using Python's Pulp module [21] and scipy.optimize [22].
The parameters used in the proposed optimization framework(OPT) are
empirically set to $\lambda_1 = 0.5$, $\lambda_2 = 0.5$ and $\lambda_3 = 0.002$. The parameters λ_1
and λ_2 are chosen to give equal weight to the corresponding constraints in
the optimization equation. λ_3 is chosen small to give significantly lower weight
to low co–occurring words.

Table 3 illustrates some resulting domain–specific topic words from the three
corpora, that were not present in ANEW in any lemma form. The third column
illustrates the topic–affect tuples that were extracted from the corpora.

5 Evaluation

This section presents the evaluation of our framework on various aspects of
quality and application. Following the approach suggested by Lu et al. 2011 [23]
and Mohammed et al. 2012 [24], we conducted experiments to find answers to
the following questions:

– Evaluation of Lexicon Quality: What is the accuracy of our domain–specific
affective lexicon? In particular, when examining the extracted lexica, how
many entries are correct and what is the fraction of false entries?
– Domain–specific Coverage: What is the improvement in domain coverage,
provided by the newly constructed lexica as compared to standard lexica?
How representative is a lexicon of a document's vocabulary?
– Extrinsic Evaluation: When applying a BAT lexicon, how well does it perform
in the task of predicting the affect in a given text? In particular, what is the
utility of augmenting a standard sentiment lexicon with entries extracted by
our approach?

5.1 Evaluation of Lexicon Quality

The purpose of this experiment is to evaluate the accuracy of the affect values
obtained from the BAT lexicon. This is done by comparing the domain–specific
scores obtained from the BAT framework against the scores assigned by human
annotators. We conducted an Amazon Mechanical Turk Task to construct our
own gold-standard annotated corpus for BAT-Crime. Fifty high–frequency nouns
were chosen for the human annotation, and 5 representative documents were
chosen for each of the nouns based on their inverse–document frequency [25].
Annotators were provided a similar set of instructions as in the original ANEW
annotation task [1].

  We filtered out those annotations which had no inter-coder agreement, or
where the annotators took less than a minimum amount of time (25s) to rate
a topic word after reading a passage. We were finally left with 453, 440, and
362 annotations respectively for Pleasure, Arousal, and Dominance. The BAT

Table 3. The <topic,affect> tuples created during the experiments with the Crime and the Beauty corpora, comprising domain-specific topic words which are not present in ANEW or in the extended ANEW.

Corpus	Topic words	Topic-Affect tuples
Crime news	Detectives, evidence, stabbing, policeman	<workers,vulnerable>;
		<children,scared>;
		<safety,doubt>;
		<children,frightened>
Beauty subject lines	Skin, skincare, makeup	<gift,free>;
		<service,bliss>;
		<skin,beautiful>
SemEval news headlines	Assault, flood, judge, hacker	<flood,storm>;
		<victory,delight>;
		<judge,scandal>

lexicon was then evaluated as an inter-coder agreement score between automatically scored topic words, and the labels provided by humans. Table 4 shows the mean absolute error (MAE) and the root mean squared error (RMSE) between the Pleasure, Arousal, and Dominance scores assigned by the BAT lexicon and human annotations, on a 5-point scale. The results indicate that BAT lexica corresponds well with human annotations and perform well above the random baseline.

Table 4. Lexicon quality metric computed against the Gold–standard Dataset

Affect	Pleasure		Arousal		Dominance	
Lexica	MAE	RMSE	MAE	RMSE	MAE	RMSE
BAT-Crime	1.2	1.63	0.74	1.01	0.84	1.11
Random	1.18	1.49	1.09	1.35	1.12	1.39

5.2 Domain–Specific Coverage

The aim of this experiment is to show the importance of the non–ANEW words in representing a given corpus. Coverage is defined as the number of distinct words that are common to both the lexicon and corpus normalized by the total number of words in the corpus. We compare the coverage of lexica developed using the BAT framework, from other affective and topical lexica such as basic ANEW [1], the General Inquirer (GI) [26], and Linguistic Inquiry and Word

Count Lexica [27]. Let L be the lexicon whose coverage is to be computed. Let C be the vocabulary of the corpus. The coverage ψ is defined as:

$$\psi = \frac{|L \cap C|}{|C|} \tag{8}$$

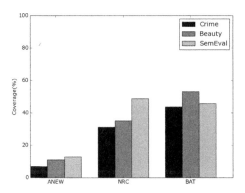

Fig. 3. Coverage Statistics on the Crime News, Beauty Subject Lines, and the SemEval corpora. This figure shows that domain-specific lexica generated from BAT can represent up to 55% of the total vocabulary of a corpus, as compared to ANEW (10%) and NRC (50%).

Figure 3 shows that our domain–specific BAT lexica offer more consistent coverage in all three corpora, after excluding stop words. Coverage can be interpreted as how representative any lexicon is of the overall vocabulary of the test set. The results highlight the importance of a domain–dependent lexicon for the accurate affect analysis of short texts. This supports our argument that a domain-specific lexicon would be more representative of text than a general-purpose lexicon.

5.3   Extrinsic Evaluation

The purpose of this evaluation is to test the performance of the BAT framework in detecting emotion in unseen text. We used the SemEval Affect Corpus of 1000 news headlines, posing a classification problem, where six binary classifiers had to determine whether or not a text expresses a particular emotion each. Unlike our previous gold standard dataset, here, the hand annotations comprise a headline-emotion label pair, along with a score on a scale of 0 to 100 to reflect the presence of each of Ekman's six discrete categories of emotion in the head-line [20]. In this experiment, we posited that our BAT lexica can improve on the state–of–the–art, for the classification of discrete emotions.

Features: We posit that by producing new features along the three dimensions of Pleasure, Arousal and Dominance, our lexicon will improve on the performance of standard lexica to classify discrete emotions. Following the experimental setup of [24] for evaluating the TEC discrete emotion lexicon, we created the BAT-SemEval lexicon over the entire corpus. We chose Support Vector Machines (SVM) as our machine learning approach. The following features were used for the classification:

1. **High-Low-Medium Pleasure, Arousal and Dominance words:** The mean and standard deviation were calculated for all 3 dimension of the words in subject line lexicon and low, high cutoff were decided by taking 0.33 standard deviation on both side of mean. This divided words in 3 categories on each of pleasure, arousal and dominance scale which produced 3^2 i.e.27 categories to which a word could belong.
2. **Statistical means:** Eighteen($3 * 6$) features depicting the mean, median, mode, kurtosis, skew, and standard deviation of the Pleasure, Arousal, and Dominance values for a single text are also considered.

We provide the results of the BAT-SemEval emotion classifier constructed on the features of the training set and evaluated against the test set. For comparison, we also provide the results reported for this task, by generic emotion classifiers. Table 5 is adapted from [24] and provides the overall performance of the following classifiers:

- The TEC Lexicon: Classifier trained on all the n-grams occurring in the training set and the strength of their association with emotion labels [24]
- The WordNet Affect Lexicon: 1536 words with associations to the six Ekman emotions, developed by Strapparava and Valitutti [28].
- The NRC Emotion Lexicon: Words with associations to the eight Plutchik emotions [29] and positive and negative sentiment, developed by [6].
- Random classifier: Classifier that randomly assigns a binary value as an outcome.

Table 5 also provides the number of features used in training the classifier as well as the Precision(P), Recall(R) and F–Score(F) results. The results show a small improvement in the F–score, which makes it the best–performing classifier among the other contenders. This is facilitated by a large improvement in the precision score of the classifier trained on BAT features, while it performs at par with others in terms of recall. These results show the utility of our lexicon for real–world applications, such as to detect discrete emotions in text.

Table 5. Comparison against state-of-the-art classifiers on the SemEval Affect Corpus (source: [24])

System	No. of features	P	R	F
BAT-SemEval	2396 + 45	**71.6**	32.2	**42.3**
TEC Lexicon	1181 + 6	44.4	35.3	39.3
WordNet Affect	1181 + 6	39.7	30.5	34.5
NRC Emotion	1181 + 10	46.7	38.6	42.2
Random classifier	–	27.8	50.0	35.7

6   Contribution and Future Work

In this paper we provide an approach for generating domain–specific affect lexica, from an unlabeled text collection. The proposed BAT framework is capable of identifying new affect words unique to a domain, finding other affect words associated with them, and incorporating all this information in an unsupervised approach for constructing domain–specific affective lexica. It reduces the dependency on human effort, and suggests the potential of using contextual information in affective analysis. Our results have several implications, which are summarized below:

- A domain–sensitive BAT lexicon can better score the affect of words in a domain, and detect affects in unseen text, than traditional supervised approaches.
- A domain–sensitive BAT lexicon identifies contextual affect words that generic lexica may miss. It has better coverage than the standard ANEW lexicon in different domains, in both short and long text.
- A domain–sensitive BAT lexicon, though generated by an unsupervised method, is scalable and precise, as it produces affect scores that agree with the scores assigned by human annotators.
- The BAT framework outperforms the state–of–the–art generic and domain–specific lexica in detecting emotions from text.

As future work, we plan to test this on more domains, and explore the inter-relationships between Pleasure, Arousal, and Dominance scores. Further, we plan to test the affective interdependence between topic words, and the influence of the overall affect of a domain on the perceived affect of its topic words.

References

1. Bradley, M.M., Lang, P.J.: Affective norms for english words (anew): instruction manual and affective ratings. Technical report C-1, The Center for Research in Psychophysiology, University of Florida (1999)
2. Ribeiro, F.N., Araújo, M., Gonçalves, P., Gonçalves, M.A., Benevenuto, F.: Sentibench-a benchmark comparison of state-of-the-practice sentiment analysis methods. EPJ Data Sci. **5**, 1–29 (2016)

3. Blitzer, J., Dredze, M., Pereira, F.: Biographies, bollywood, boom-boxes and blenders: Domain adaptation for sentiment classification. ACL **7**, 440–447 (2007)
4. Muhammad, A., Wiratunga, N., Lothian, R., Glassey, R.: Domain-based lexicon enhancement for sentiment analysis. In: SMA@ BCS-SGA I, pp 7–18 (2013)
5. Nielsen, F.Å.: A new anew: evaluation of a word list for sentiment analysis in microblogs. arXiv preprint arXiv:1103.2903 (2011)
6. Mohammad, S.M., Turney, P.D.: NRC emotion lexicon. NRC Technical Report (2013)
7. Hatzivassiloglou, V., McKeown, K.R.: Predicting the semantic orientation of adjectives. In: Proceedings of the Eighth Conference on European Chapter of the Association for Computational Linguistics, pp. 174–181. Association for Computational Linguistics (1997)
8. Qiu, G., Liu, B., Bu, J., Chen, C.: Opinion word expansion and target extraction through double propagation. Comput. Linguist. **37**, 9–27 (2011)
9. Miller, G.A.: Wordnet: a lexical database for english. Commun. ACM **38**, 39–41 (1995)
10. Chikersal, P., Poria, S., Cambria, E., Gelbukh, A., Siong, C.E.: Modelling public sentiment in twitter: using linguistic patterns to enhance supervised learning. In: International Conference on Intelligent Text Processing and Computational Linguistics, pp. 49–65. Springer (2015)
11. Turney, P.D., Littman, M.L.: Measuring praise and criticism: inference of semantic orientation from association. ACM Trans. Inform. Syst. (TOIS) **21**, 315–346 (2003)
12. Bestgen, Y.: Building affective lexicons from specific corpora for automatic sentiment analysis. In: LREC (2008)
13. Bestgen, Y., Vincze, N.: Checking and bootstrapping lexical norms by means of word similarity indexes. Behav. Res. Methods **44**, 998–1006 (2012)
14. Wiebe, J., Wilson, T., Bell, M.: Identifying collocations for recognizing opinions. In: Proceedings of the ACL-01 Workshop on Collocation: Computational Extraction, Analysis, and Exploitation, pp. 24–31 (2001)
15. Dragut, E.C., Yu, C., Sistla, P., Meng, W.: Construction of a sentimental word dictionary. In: Proceedings of the 19th ACM International Conference on Information and Knowledge Management, CIKM '10, pp. 1761–1764. ACM, New York (2010)
16. Bird, S.: Nltk: the natural language toolkit. In: Proceedings of the COLING/ACL on Interactive presentation sessions, pp. 69–72. Association for Computational Linguistics (2006)
17. Manning, C.D., Surdeanu, M., Bauer, J., Finkel, J.R., Bethard, S., McClosky, D.: The stanford corenlp natural language processing toolkit. In: ACL (System Demonstrations), pp. 55–60 (2014)
18. Strapparava, C., Mihalcea, R.: Semeval-2007 task 14: Affective text. In: Proceedings of the 4th International Workshop on Semantic Evaluations, pp. 70–74. Association for Computational Linguistics (2007)
19. Brett, D., Pinna, A.: The distribution of affective words in a corpus of newspaper articles. Procedia - Soc. Behav. Sci. **95**, 621–629 (2013)
20. Ekman, P.: An argument for basic emotions. Cogn. Emot. **6**, 169–200 (1992)
21. Mitchell, S., OSullivan, M., Dunning, I.: Pulp: a linear programming toolkit for python. The University of Auckland, Auckland, New Zealand. http://www.optimization-online.org/DB_FILE/2011/09/3178.pdf (2011)
22. Jones, E., Oliphant, T., Peterson, P., et al.: SciPy: open source scientific tools for Python (2001). Accessed 2017 Jan 17

23. Lu, Y., Castellanos, M., Dayal, U., Zhai, C.: Automatic construction of a context-aware sentiment lexicon: An optimization approach. In: Proceedings of the 20th International Conference on World Wide Web, pp. 347–356. ACM (2011)
24. Mohammad, S.M.: # emotional tweets. In: Proceedings of the First Joint Conference on Lexical and Computational Semantics-Volume 1: Proceedings of the main conference and the shared task, and Volume 2: Proceedings of the Sixth International Workshop on Semantic Evaluation, pp. 246–255. Association for Computational Linguistics (2012)
25. Robertson, S.: Understanding inverse document frequency: on theoretical arguments for idf. J. Doc. **60**, 503–520 (2004)
26. Stone, P.J., Dunphy, D.C., Smith, M.S.: The general inquirer: a computer approach to content analysis (1966)
27. Pennebaker, J.W., Francis, M.E., Booth, R.J.: Linguistic inquiry and word count: Liwc 2001, vol. 71. Lawrence Erlbaum Associates, Mahway (2001)
28. Strapparava, C., Valitutti, A.: Wordnet affect: an affective extension of wordnet. LREC **4**, 1083–1086 (2004)
29. Plutchik, R.: A general psychoevolutionary theory of emotion. Theor. Emot. **1**, 3–31 (1980)

Leveraging Target-Oriented Information for Stance Classification

Jiachen Du[1], Ruifeng Xu[1,2(✉)], Lin Gui[1], and Xuan Wang[1]

[1] Harbin Institute of Technology Shenzhen Graduate School, Shenzhen, China
dujiachen@stmail.hitsz.edu.cn, xuruifeng@hit.edu.cn, guilin.nlp@gmail.com,
wangxuan@cs.hitsz.edu.cn
[2] Guangdong Provincial Engineering Technology Research Center for Data Science,
Guangzhou, China

Abstract. Classifying the stance expressed in text towards specific target, namely stance detection, is a challenging task. The biggest distinction between stance detection and ordinary sentiment classification is that the determination of the stance is dependent on target while the target might not be explicitly mentioned in text. This indicates that the stance detection is not only dependent on the text content but also highly determined by the concerned target. To this end, we propose a neural network based model for stance detection, which leverages target-oriented information by utilizing target-augmented embedding and attention mechanism. The attention mechanism here is expected to locate the important parts of a text. The evaluation on SemEval 2016 Task 6 Twitter Stance Detection dataset shows that our proposed model achieves the state-of-the-art results.

Keywords: Stance detection · Neural attention · Sentiment analysis

1 Introduction

With the rapidly development of social network like Twitter[1], the mass User Generated Content (UGC) become available. To retrieve the valuable subjective content from these text, sentiment analysis and opinion mining [8,14,19] have become a hot research topic in natural language processing. Various techniques were investigated to identify the sentiment and opinion from the text and to determine the polarity of the text. However, for many practical applications, people are interested to learn the attitude of the author to a specific target topic rather than the sentiment of the text [1]. For example, in the topic of US Election, people want to know the attitude of the author to Trump, namely support or not. We call this attitude as stance to a target. Oftentimes, people focus on not only author's stance towards a specific target, but also the argumentations which the author used to express this stance. Considering the above example of

---

[1] www.twitter.com.

© Springer Nature Switzerland AG 2018
A. Gelbukh (Ed.): CICLing 2017, LNCS 10762, pp. 35–45, 2018.
https://doi.org/10.1007/978-3-319-77116-8_3

US election, besides knowing the stance of a sentence towards Trump, we are keen on the reason why the author express this attitude.

Stance detection is formalized as the task of assigning stance label to a piece of texts with respect to a specific target [18], i.e. whether a text is in favor of or against the given target, or neither of them. A major difference between stance detection and traditional aspect-level sentiment classification [11] is that stance detection is dependent on both the sentiment expression and target topic while the target of the stance might not be explicitly mentioned in text. It puzzles the methods in traditional sentiment classification to correctly predict the stance labels.

Most Previous studies on stance detection and argumentation mining are always on corpora from online debates [9, 18] or news [7]. Spurred by the growth in the use of microblogging platforms such as Twitter and Microblog, companies and media organizations are increasingly seeking ways to mine Twitter for information about what people think and feel about their products and services. Studying how stance is expressed on microblogging platforms can be beneficial for a lot of areas.

In the stance detection research, several models were proposed. Some of them used feature-engineering to extract manual features [9], and some used classical neural-network based models like RNN [2] and CNN [17]. However, as known that the stance detection is determined by both the sentiment expressed in the text content and the concerned target. Most of the above models concentrate on the features extracted from the text to be predicted rather than the given targets. This makes these models hard to focus on the target-related parts in text, especially when the text expresses stance towards other targets instead of the given one.

To address this problem, we propose a neural-network based model to make full use of the target information in stance detection. Our model utilizes a novel target-oriented attention extractor to focus the important parts in text which are highly related to the targets. Firstly, we concatenate the embedding vectors of text and target to represent target-specific embedding for modelling both text and target. We then use a fully-connected network to learn attention signal for driving the classifier to focus the salient parts in text and finally to determine the stance. Experimental results on Semeval Stance Detection dataset show that the proposed model achieved the highest performance, based on our knowledge. The main contributions of our work can be summarized as follows:

- We propose to use a novel embedding to represent text with target-oriented information.
- We propose a neural attention model to extract target-related information for stance detection. This model is able to extract core parts of given text when different targets are concerned.
- Experimental results on dataset of Semeval-2016 Stance Detection Task show that our model outperforms several strong existing models including the first-place system in Semeval-2016. Furthermore, the visualization of selected instances illustrates why the proposed model works well.

The rest of this paper is organized as follows: Sect. 2 briefly reviews the related works and Sect. 3 presents our model. Section 4 gives the evaluation and discussions. Finally, Sect. 5 concludes.

2 Related Works

In this section, we will review related works on stance detection and deep learning for sentiment analysis briefly.

Stance Detection: Previous work mostly considered stance detection in debates [9,18] or student essays [6]. There is a growing interest in performing stance classification on microblogging media. SemEval-2016 Task 6 [13] involved two stance detection subtasks in tweets in supervised and weakly supervised settings. The majority of current approaches attempt to detecting the stance label of the entire sentence, regardless of the target information. Augenstein et al. [2] uses two bidirectional RNN to model both target and text for stance detection. However this model concentrates on weakly-supervised tasks.

Deep Learning for Sentiment Analysis: In the general domain of sentiment analysis, there has been an increasing amount of attention on deep learning approaches. Tang et al. [16] used gated recurrent neural network (RNN) to model documents for sentiment classification. Tai et al. [15] explored the structure of a sentence and used a tree-structured recurrent neural network with long-short term memory (LSTM) for sentiment classification. The advantage of RNN is its ability to better capture the contextual information, especially the semantics of long texts. However, RNN model cannot pay attention to the salient parts of text. This limitation influences the performance of RNN when it is applied to text classification. To address this problem, a new direction of neural attention has emerged. Neural networks with attention mechanism showed promising results on sequence-to-sequence (seq2seq) processing tasks in NLP, including machine translation [3], caption generation [20] and opinion expression extraction [4]. In the area of text classification, [21] applied the attention model used in seq2seq document-level classification. However, there is no neural attentional models for stance detection task up to now.

3 Model

As discussed, the performance of stance detection may be improved by considering both text content features and target related features. Motivated by this, we propose an RNN-based model which concentrates the salient parts in text corresponding to given target. The overall architecture of our model is shown in Fig. 1. It consists of two main components: a recurrent neural network (RNN) as the feature extractor for text and a fully-connected network as the target-specific attention selector. These two components are combined by an element-wise multiplication operation in the classification layer. We describe the details of these two components in the following subsections.

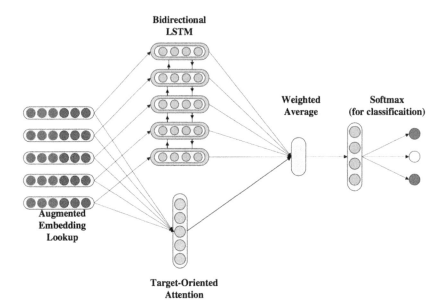

Fig. 1. Overall architecture of our model

3.1 Target-Augmented Embedding

The target information is vital for determining the stance polarity. To combine the information of target and text to be predicted, we propose to learn a target-augmented embedding for each target. In neural-network based models, a text sequence of length T (padded where necessary) is normally represented as $[x_1, x_2, \ldots, x_T]$, where $x_t \in \mathbb{R}^d$ ($t = \{0, 1, \ldots, T - 1\}$) corresponds to the d-dimensional vector representation of the t-th word in the text sequence. A target sequence of length S is represented as $[z_1, z_2, \ldots, z_S]$ where $z_t \in \mathbb{R}^{d'}$ is the d'-dimensional vector of the i-th word in the target sequence. Since the common word embedding representations exhibit linear structure that make it possible to meaningfully combine words by an element-wise addition of their vector representations, we use the average vector $\bar{z}$ to obtain a more compact target representation:

$$\bar{z} = \frac{1}{S} \sum_{t=1}^{S} z_t \tag{1}$$

In order to better take the advantage of target information, we append the target representation to the embedding of each word in original text. The target-augmented embedding of a word i for specific target z is $e_t^z = x_t \oplus \bar{z}$ where $\oplus$ is the vector concatenation operation. Notice that the dimension of e_t^z is $(d + d')$.

3.2   Recurrent Neural Network with Long-Short Time Memory

An Recurrent Neural Network (RNN) [5] is a kind of neural network that processes sequences of arbitrary length by recursively applying a function to its hidden state vector $h_t \in \mathbb{R}^d$ of each element in the input sequences. The hidden state vector at time-step t depends on the input symbol x_t and the hidden state vector at last time-step h_{t-1} is:

$$h_t = \begin{cases} 0 & t = 0 \\ g(g_{t-1}, x_t) & \text{otherwise} \end{cases} \tag{2}$$

A fundamental problem in traditional RNN is that gradients propagated over many steps tend to either vanish or explode. It affects RNN to learn long-dependency correlations in a sequence. Long short-term memory network (LSTM) was proposed by [10] to alleviate this problem. LSTM has three types of gate: an input gate i_t, a forget gate f_t, an output gate o_t as well as a memory cell c_t. They are all vectors in $\mathbb{R}^d$. The LSTM transition equations are:

$$\begin{aligned} i_t &= \sigma(W_i e_t^z + U_i h_{t-1} + V_i c_{t-1}), \\ f_t &= \sigma(W_f e_t^z + U_f h_{t-1} + V_f c_{t-1}), \\ o_t &= \sigma(W_o e_t^z + U_o h_{t-1} + V_o c_{t-1}), \\ \tilde{c}_t &= \tanh(W_c e_t^z + U_c h_{t-1}), \\ c_t &= f_t \odot c_{t-1} + i_t \odot \tilde{c}_t, \\ h_t &= o_t \odot \tanh(c_t) \end{aligned} \tag{3}$$

where e_t^z is the target-augmented embedding at the current time step, σ is the sigmoid function and $\odot$ is the elementwise multiplication operation, $W_{\{i,f,o,c\}}, U_{\{i,f,o,c\}}, V_{\{i,f,o\}}$ are all sets of learned weight parameters. In our model, we use the hidden-state vector of each time step as the representation of corresponding word in the sentence.

In this study, we employ bi-directional LSTM model to capture the information in the text. The bi-directional LSTM has a forward and a backward LSTM. The annotation for each word are obtained by concatenating the forward hidden state and the backward state.

3.3   Target-Specific Attention Extraction

Traditional RNN model cannot capture the important parts in sentences. To address this problem, we design an attention mechanism which drives the model to concentrate the salient parts in text with respect to specific target. To make full use of target information, this model uses a bypass network which take the target-augmented embedding discussed in 2.2 as input to extract target-specific attention signal. Here, we use a linear transformation to map the $(d+d')$-dimensional target-augmented embedding of each word to a scalar value:

$$c_t' = W_a e_t^z + b_a \tag{4}$$

where W_a and b_a are learned set of weights and bias terms for attention extraction.

To obtain more stable attention signal, we then feed the attention vector $[c'_1, c'_2, \ldots, c'_T]$ into a softmax transformation to get the final attention signal for each word:

$$c_t = \textbf{softmax}(c_t) = \frac{e^{c'_t}}{\sum_{i=1}^{T} e^{c'_i}} \tag{5}$$

3.4 Stance Classification

We use the product of attention signal c_t and the corresponding hidden state vector of RNN h_t to represent the word t in a sequence with attention signal. The representation of the whole sequence can be obtained by averaging the word representations:

$$s = \frac{1}{T} \sum_{t=0}^{T-1} c_t h_t \tag{6}$$

where $s \in \mathbb{R}^d$ is the vector representation of the text sequence and it can be used as features for text classification:

$$p = \textbf{softmax}(W_c s + b_c) \tag{7}$$

where $p \in \mathbb{R}^C$ is the vector of predicted probability for stance polarity, Where C is the number of classes of stance polarity, And W_c and b_c are parameters of the classification layer.

3.5 Model Training

We use cross-entropy loss to train our model in end-to-end manner by giving a set of training data $\{x^i, z^i, y^i\}$, where x^i is the i-th text to be predicted, z^i is the corresponding target and y^i is one-hot representation of the ground-truth stance polarity for target z^i and text x^i. We represent this model as a black-box function $f(x, z)$ whose output is a vector representing the probability of stance polarity. The goal of training is to minimize the loss function:

$$\mathcal{L} = -\sum_i \sum_j y_j^i \log f_j(x_i, z_i) + \lambda \|\theta\|^2 \tag{8}$$

where i is the index of data and j is the index of class. $\lambda \|\theta\|^2$ is the L_2-regularization term and θ is the parameter set.

Except the parameter sets of standard LSTM $\{W_{\{i,f,o,c\}}, U_{\{i,f,o,c\}}, V_{\{i,f,o\}}\}$ and softmax classification $\{W_c, b_c\}$, our model only has additional parameters $\{W_a, b_a\}$ for attention extractor.

4   Evaluation and Discussions

In this section, we evaluate our proposed model and several strong baselines on stance detection task. We will firstly describe the experiment settings. The experimental results will be compared to baseline methods. Finally, some attention signals selected from text will be visualized to show the validity of the proposed attention extractor.

4.1   Experiment Settings

In this section, we will describe the settings of experiments including datasets, evaluation metrics and baseline methods used in the evaluation. The details of training process of the proposed model will also be discussed.

Dataset. Our experiment is performed on the dataset of Semeval-2016 Task 6 [13]. In this dataset, more than 4000 tweets are annotated for whether one can deduce favorable or unfavorable stance towards one of five targets "Atheism", "Climate Change is a Real Concern", "Feminist Movement", "Hillary Clinton", and "Legalization of Abortion". Task 6 has two subtasks including subtask-A supervised learning and subtask-B unsupervised learning. In this evaluation, we only use the dataset of subtask-A in which the targets provided in test set are all seen in the training set. Table 1 gives the statistics of this dataset. We also illustrate the corresponding distribution of instances in Table 2.

Table 1. Statistics of dataset

Target	#All	#Train	#Test
Atheesim	733	513	220
Climate Change is Concern	564	395	169
Feminist Movement	949	664	285
Hillary Clinton	984	689	295
Legalization of Abortion	933	653	280
Total	4163	2914	1249

Metrics. The micro average of $F1$-score across targets which is utilized in Semeval evaluation is adopted as the metrics. Firstly, the $F1$-score for *Positive* and *Negative* categories for all instances in the dataset is calculated as:

$$F_{positive} = \frac{2P_{positive}R_{positive}}{P_{positive} + R_{positive}}$$
$$F_{negative} = \frac{2P_{negative}R_{negative}}{P_{negative} + R_{negative}} \qquad (9)$$

Table 2. Distribution of instances in dataset

Target	% of stances in Train			% of stances in Test		
	Pos	Neg	None	Pos	Neg	None
Atheesim	60.43	35.09	4.48	59.09	35.45	5.45
Climate Change is Concern	31.65	49.62	18.73	29.59	51.48	18.93
Feminist Movement	17.92	77.26	4.82	19.30	76.14	4.56
Hillary Clinton	32.08	64.01	3.92	25.76	70.17	4.07
Legalization of Abortion	28.79	66.16	5.05	20.36	72.14	7.5
Total	33.05	60.47	6.49	29.46	63.33	7.20

where P and R are precision and recall. Then the average of $F_{positive}$ and $F_{negative}$ is calculated as the final metrics:

$$F_{average} = \frac{F_{postive} + F_{negative}}{2} \tag{10}$$

Note that the final metrics does not disregard the *None* class. By taking the average F-score for only the *Positive* and *Negative* classes, we treat *None* as a class that is not of interest.

Baselines. We compare the following baseline methods:

– Neural Bag-Of-Words (NBOW): The NBOW sums the word vectors within the sentence and applies a softmax classifier.
– LSTM without target-specific attention(LSTM): LSTM with target augmented embedding but without attention.
– MITRE: The top-performance model in Semeval-2016 stance detection shared task. This model uses two recurrent neural network: the first one is trained to predict task-relevant hashtags on a very large unlabeled Twitter corpus. This network is used to initialize the second RNN classifier, which was trained on the provided subtask-A data.

Training Details. We use *ad-hoc* strategy to train one model for each target. The final result is obtained by concatenating all the predicted results of these models. Although different models are used for different targets, they all share the same sets of hyper-parameters. All hyper-parameters are tuned to obtain the best performance through 5-fold cross validation on training set. In our experiment, all word vectors are initialized by word2vec [12]. The word embedding vectors are pre-trained on unlabeled corpora which is crawled from Twitter. The other parameters are initialized using a uniform distribution $U(-0.01, 0.01)$. The dimension of word and target embeddings are 300 and the size of units in LSTM is 100. Adam is used for our optimization method, and its learning rate is $5e-4$, β_1 is 0.9, β_2 is 0.999, ϵ is $1e-8$. All models are trained by mini-batch of 50 instances.

4.2   Results

The overall performance of all baselines and our proposed model are listed in Table 3. Firstly, it is observed that *NBOW* performs the worst among all baselines, since *NBOW* only use the average of embedding vectors of text as the discriminative feature which is not enough to obtain satisfactory performance. It is also observed that our method performs much better than traditional LSTM without target-specific attention. It shows that our method can capture the target information to improve the performance of stance detection. Our method also outperforms *MITRE* which is the first-place model on this shared task. Especially, it is shown that the performance of our method is higher than *MITRE*'s by 5.7% on target *Hillary Clinton*. For this target, most tweets always compare other candidates of president with Hillary Clinton. This obviously affects the performance of the models which cannot find the important words corresponding to given target. Our method applies the novel attention mechanism to extract key words corresponding to targets, and applies the information obtained from the stance polarity using back-propagation to determine the factional relation between them and targets. Overall, our methods outperforms all of the baseline methods obviously. This empirically results show that target-specific attention could benefit stance detection task.

Table 3. Performance of competing methods

Target	NBOW	LSTM	MITRE	Our method
Athesim	55.12	56.22	*61.47*	58.33
Climate	39.93	40.30	41.63	*47.59*
Feminist	50.21	49.06	*62.09*	52.77
Hillary	55.98	61.84	57.67	*63.38*
Abortion	55.07	50.21	57.28	*59.72*
Overall	60.19	62.38	67.82	*68.29*

4.3   Visualization of Attention

In order to validate that our model is able to select target-specific parts in a text sequence, we visualize the attention layers for several sentences whose labels are correctly predicted by our model in Fig. 2. It is shown that our model selects the words that have strong relation with the given targets. For example, in the first sentence, our model highlights *Hillary, Warren* and *Karl Marx* which are all politicians. In the second sentence, *Life* and *Human right* are selected by our model, they are all highly related to the topic of abortion and pre-life.

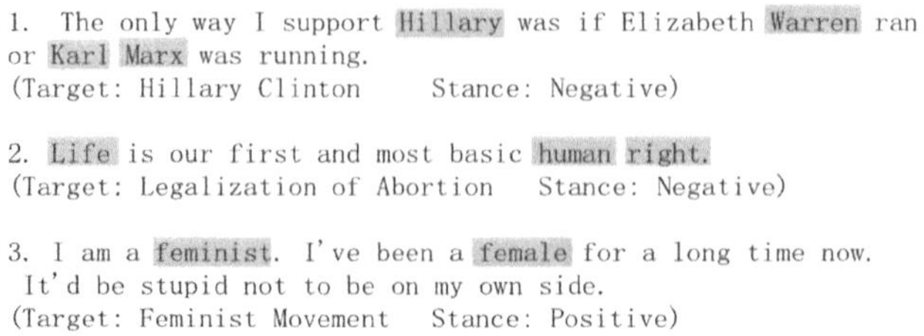

1. The only way I support Hillary was if Elizabeth Warren ran
or Karl Marx was running.
(Target: Hillary Clinton   Stance: Negative)

2. Life is our first and most basic human right.
(Target: Legalization of Abortion   Stance: Negative)

3. I am a feminist. I've been a female for a long time now.
It'd be stupid not to be on my own side.
(Target: Feminist Movement   Stance: Positive)

Fig. 2. Visualization of learned attention. Red patches highlight the words strongly related to given targets.

5   Conclusion

In this paper, we proposed an attentional-based neural network for stance detection. The main contribution of this model is to learn target-augmented embedding for text and use attention mechanism to extract target-specific parts in text to improve classification performance. Experimental results show that our model outperforms several strong baselines. Meanwhile, the visualization of some attentions extracted by our model shows the impressive capability of our model to extract important parts are helpful to improve stance detection. In the future works, we will mainly focus on combining the proposed attention mechanism with other state-of-art models in stance detection. Moreover, we have interest on exploring the potential of our model on attitude identification task.

Acknowledgments. This work was supported by the National Natural Science Foundation of China 61370165, U1636103, 61632011, National 863 Program of China 2015AA015405, Shenzhen Foundational Research Funding JCYJ20150625142543470 and Guangdong Provincial Engineering Technology Research Center for Data Science 2016KF09.

References

1. Anand, P., Walker, M., Abbott, R., Tree, J.E.F., Bowmani, R., Minor, M.: Cats rule and dogs drool!: classifying stance in online debate. In: Proceedings of the 2nd Workshop on Computational Approaches to Subjectivity and Sentiment Analysis, pp. 1–9. Association for Computational Linguistics (2011)
2. Augenstein, I., Rocktäschel, T., Vlachos, A., Bontcheva, K.: Stance detection with bidirectional conditional encoding. In: Proceedings of the 2016 Conference on Empirical Methods in Natural Language Processing, pp. 876–885 (2016)
3. Bahdanau, D., Cho, K., Bengio, Y.: Neural machine translation by jointly learning to align and translate (2014). arXiv:1409.0473
4. Du, J., Gui, L., Xu, R.: Extracting opinion expression with neural attention. In: Li, Y., Xiang, G., Lin, H., Wang, M. (eds.) SMP 2016. CCIS, vol. 669, pp. 151–161. Springer, Singapore (2016). https://doi.org/10.1007/978-981-10-2993-6_13
5. Elman, J.L.: Finding structure in time. Cogn. Sci. **14**(2), 179–211 (1990)

6. Faulkner, A.: Automated classification of stance in student essays: an approach using stance target information and the wikipedia link-based measure. Science **376**(12), 86 (2014)
7. Ferreira, W., Vlachos, A.: Emergent: a novel data-set for stance classification. In: Proceedings of the 2016 Conference of the North American Chapter of the Association for Computational Linguistics: Human Language Technologies. ACL (2016)
8. Gui, L., Xu, R., He, Y., Lu, Q., Wei, Z.: Intersubjectivity and sentiment: from language to knowledge. In: Proceedings of International Joint Conference on Artificial Intelligence, pp. 2789–2795 (2016)
9. Hasan, K.S., Ng, V.: Stance classification of ideological debates: data, models, features, and constraints. In: International Joint Conference on Natural Language Processing, pp. 1348–1356 (2013)
10. Hochreiter, S., Schmidhuber, J.: Long short-term memory. Neural Comput. **9**(8), 1735–1780 (1997)
11. Liu, B.: Sentiment analysis and opinion mining. In: Synthesis Lectures on Human Language Technologies, vol. 5, no. 1, pp. 1–167 (2012)
12. Mikolov, T., Chen, K., Corrado, G., Dean, J.: Efficient estimation of word representations in vector space (2013). arXiv:1301.3781
13. Mohammad, S.M., Kiritchenko, S., Sobhani, P., Zhu, X., Cherry, C.: Semeval-2016 task 6: detecting stance in tweets. In: Proceedings of SemEval, vol. 16 (2016)
14. Pang, B., Lee, L., Vaithyanathan, S.: Thumbs up?: sentiment classification using machine learning techniques. In: Proceedings of the ACL-02 Conference on Empirical Methods in Natural Language Processing, vol. 10, pp. 79–86. Association for Computational Linguistics (2002)
15. Tai, K.S., Socher, R., Manning, C.D.: Improved semantic representations from tree-structured long short-term memory networks. In: Proceedings of the 53rd Annual Meeting of the Association for Computational Linguistics and the 7th International Joint Conference on Natural Language Processing. pp. 1556–1566. Association of Computational Linguistics (2015)
16. Tang, D., Qin, B., Liu, T.: Document modeling with gated recurrent neural network for sentiment classification. In: Proceedings of the 2015 Conference on Empirical Methods in Natural Language Processing, pp. 1422–1432 (2015)
17. Vijayaraghavan, P., Sysoev, I., Vosoughi, S., Roy, D.: Deepstance at semeval-2016 task 6: detecting stance in tweets using character and word-level CNNs (2016). arXiv:1606.05694
18. Walker, M.A., Anand, P., Abbott, R., Grant, R.: Stance classification using dialogic properties of persuasion. In: Proceedings of the 2012 Conference of the North American Chapter of the Association for Computational Linguistics: Human Language Technologies, pp. 592–596. Association for Computational Linguistics (2012)
19. Wang, Y., Huang, M., Zhao*, L., Zhu, X.: Attention-based lstm for aspect-level sentiment classification. In: Proceedings of the 2016 Conference on Empirical Methods in Natural Language Processing, pp. 606–615. Association of Computational Linguistics (2016)
20. Xu, K., et al.: Show, attend and tell: neural image caption generation with visual attention. In: Proceedings of the 32nd International Conference on Machine, pp. 77–81 (2015)
21. Yang, Z., Yang, D., Dyer, C., He, X., Smola, A., Hovy, E.: Hierarchical attention networks for document classification. In: Proceedings of the 2016 Conference of the North American Chapter of the Association for Computational Linguistics: Human Language Technologies (2016)

Sentiment Polarity Classification of Figurative Language: Exploring the Role of Irony-Aware and Multifaceted Affect Features

Delia Irazú Hernández Farías[1,2]([✉]), Cristina Bosco[1],
Viviana Patti[1], and Paolo Rosso[2]

[1] Dipartimento di Informatica, University of Turin, Turin, Italy
{bosco,patti}@di.unito.it
[2] PRHLT Research Center, Universitat Politècnica de València, Valencia, Spain
{dhernandez1,prosso}@dsic.upv.es

Abstract. The presence of figurative language represents a big challenge for sentiment analysis. In this work, we address the task of assigning sentiment polarity to Twitter texts when figurative language is employed, with a special focus on the presence of ironic devices. We introduce a pipeline model which aims to assign a polarity value exploiting, on the one hand, irony-aware features, which rely on the outcome of a state-of-the-art irony detection model, on the other hand a wide range of affective features that cover different facets of affect exploiting information from various sentiment and emotion lexical resources for English available to the community, possibly referring to different psychological models of affect. The proposed method has been evaluated on a set of tweets especially rich in figurative language devices proposed as a benchmark in the shared task on "Sentiment Analysis of Figurative Language" at SemEval-2015. Experiments and results of feature ablation show the usefulness of irony-aware features and the impact of using different affective lexicons for the task.

1 Introduction

Twitter has provided a huge volume of data containing judgments, attitudes, and beliefs of people. Opinions and their related concepts such as sentiments and emotions are the subjects addressed by Sentiment Analysis (SA) [1]. Figurative language devices, such as, for instance, irony and sarcasm, represent one of the main challenges for SA [2]. The presence of these kinds of expressions could indeed undermine the accuracy of SA systems [3]. Therefore, identifying irony and sarcasm become crucial for a SA system.

Among the definitions proposed in theoretical pragmatics for irony there is that of Grice [4], which refers to the speaker's intention to express the opposite meaning of what it is literally said. When irony becomes offensive with a specific target to attack it is considered as a form of sarcasm [5,6]. Irony detection in

© Springer Nature Switzerland AG 2018
A. Gelbukh (Ed.): CICLing 2017, LNCS 10762, pp. 46–57, 2018.
https://doi.org/10.1007/978-3-319-77116-8_4

social media has become a hot research topic and many research works have been carried out recently on this topic, with a special focus on Twitter data [7–12]. Most of the current approaches consider irony as an umbrella term that covers also sarcasm.

A SA system may fail when applied to inferring the polarity in sentences like:

(1) *Thanks for this birthday card. I'm really glad you didn't put any money in it.*[1]

(2) *My level of annoyance is at an all time high right now. Thanks to this wonderful @Starbucks experience. #wah*[2].

(3) *RT GregCooper: These annoying home buyers want to purchase my listings before the sign actually goes up. How inconvenient. #sarcasm #grate.*[3]

In (1) the overall polarity of the tweet is negative but the presence of three positive terms ("Thanks", "birthday" and "glad") could be misinterpreted by sentiment analysis systems, which often rely on information included in sentiment lexicons. Also in sarcastic posts like the one reported in (2) [8], where there is a contrast between a positive sentiment expressed and a situation which is typically negative, the presence of two positive terms ("thanks" and "wonderful") and of a negative one ("annoyance") could cause a problem to SA system in assigning the correct polarity. In such cases, indeed, tweets could be identified as positive under a basic approach for SA which simply considers the presence and frequency of positive and negative terms to assign polarity. However, both tweets convey a meaning far from being positive: the authors use irony/sarcasm to express their evaluation towards a target, by using a literally positive sentence to point out their real negative opinion on the specific target.

While the use of sarcasm to convey a negative sentiment is the most common, it is very rare to use it the other way around. Theoretical accounts state that expressing positive attitudes in a negative mode is rare and harder to process for humans [14]. This seems to be confirmed by an analysis of the SemEval-2015 Task 11 corpus, where tweets marked with the hashtag #sarcasm and tagged with a positive polarity score were very few (only 18 out of 2,260 posts). Among them, only three tweets expressing a literally negative statement, that finally reverted to an intended positive one, were identified by manual analysis [15], see for instance post (3) reported above.

Currently, even if SA systems are able to understand the most salient polarity of words, they do not have a well-established methodology to deal with the presence of figurative language expressions [16]. In this sense, in order to develop

---

[1] This tweet is part of the dataset used in the SemEval-2015 Task 11: Sentiment Analysis of Figurative Language in Twitter [13]. It was labeled as having negative polarity (−1.8).

[2] This tweet is part of the sarcastic tweets in the dateset of Riloff et al. [8].

[3] This tweet is part of the dateset used in the SemEval-2015 Task 11: Sentiment Analysis of Figurative Language in Twitter [13]. It was labeled as having positive polarity (+0.63).

an irony-aware system which correctly identifies the sentiment behind a text, it is needed to recognize whether the sentence contains some figurative device, such as irony, before deciding on sentiment polarity. In general, irony detection and SA have been addressed individually. However, there are some efforts devoted to integrate both tasks in the framework of evaluation campaigns, where the main objective is to perform Twitter SA considering the presence of irony [17–19].

In order to investigate whether the performance of a SA system improves or not when it takes into account the presence of ironic content, we propose an approach based on a pipeline that incorporates two modules: irony detection and SA for polarity assignment. To the best of our knowledge, exploiting an irony detection module in a sentiment analysis pipeline has not been investigated in depth before. In our approach, the irony detection module was trained by using a set of tweets labeled as ironic. Whereas the sentiment analysis one was trained by using tweets with figurative language manually annotated with their polarity degree.

The paper is organized as follows. Section 2 introduces the SA task on figurative language. Section 3 describes our method to perform irony-aware SA. Section 4 describes the evaluation and results. Finally, Sect. 5 draws some conclusions.

2   Sentiment Analysis and Figurative Language

The *SemEval-2015 Task 11: Sentiment Analysis of Figurative Language in Twitter*[4] was the first SA task attempting to identify the sentiment score in texts featured by the occurrence of figurative language devices. The goal of the task was to determine the degree in which a sentiment was communicated in a fine-grained scale ranging from -5 (very negative) to $+5$ (very positive) over a set of tweets rich in metaphorical, sarcastic and ironical content. Overall, the dataset included more than 13,000 tweets (SE15-Task11 dataset, henceforth).

Fifteen teams participated in the task on SA of figurative language [17]. Their systems were evaluated using the cosine similarity (cosSim) measure. The best ranked system, called ClaC [20], exploited n-grams, some SA resources as well as linguistic features such as negations and modality. ClaC achieved a cosSim measure of 0.758. The UPF-taln [21] system considered a set of features to detect the style and the unexpectedness in tweets combined with textual features such as bigrams, skipgrams and other word patterns. It achieved the second place in the ranking with a 0.711 in cosSim terms.

LLT_PolyU [22] and EliRF, [23], ranked as third (0.687) and fourth (0.658), respectively, considering features such as n-grams, negations as well as some SA resources. LT3 [24] ranked as fifth (0.687) and ValenTo [25], ranked as sixth (0.634), systems included in their sets of features the presence of punctuation marks, emoticons and hashtags. LT3 took advantage of features such as contrasting, contradictory and polysemic words. ValenTo system exploited SA resources

---

[4] http://alt.qcri.org/semeval2015/task11/.

as well as some emotional and psycholinguistic information. Besides, it considered the presence of sarcastic content in a tweet by exploiting specific hashtags.

3   Our Proposal

The aim of our approach is to perform irony-aware SA by exploiting different facets of affective information. Our irony-aware SA attempts to incorporate two strongly related tasks: irony detection and sentiment analysis. The importance of considering the presence of ironic content before performing sentiment analysis has been recognized by several authors [17,26]. Our main objective is not only identifying the presence of ironic content but rather assigning a polarity value consistent with its detected presence. The overall process in our irony-aware SA system can be briefly summarized as follows:

Given a tweet, we first identify the presence of ironic content. Then, both the tweet and its irony-aware features are processed by a sentiment analysis model in order to calculate a polarity value for the post.

Unlike the best ranked systems at SemEval-2015 Task 11, our approach does not exploit n-grams as features. Instead our irony-aware SA system mainly relies on affective information for both identifying irony and calculating a polarity degree.

We propose a pipeline involving two main phases:

I. *Irony-aware features.* As output of this step we have two irony-aware features: the first one depends on the possible presence of explicit irony-related hashtags, whereas the second one is obtained by using the irony detection model described in Sect. 3.1. In Twitter messages hashtags such as "#irony", "#sarcasm", and "#not" can indeed be recognized as labels used to point out user's ironical intention [15,27]. However, this is not enough to identify irony. When a SA system is dealing with a tweet as the one mentioned in Sect. 1, in which no explicit hashtag indicating the user ironic intention is present, it is needed to apply a model able to identify irony without considering potentially ironic hashtags. Thus, we exploit an irony detection model (see Sect. 3.1) by using a set of 10,000 ironic tweets retrieved by Reyes et al. [7], 10,000 sarcastic and 20,000 non-ironic tweets retrieved by Ptáčeket al. [10]. For this purpose, we trained the Weka [28] implementation of a set of classifiers (Naïve Bayes, Decision Tree and Support Vector Machine) with default parameters. Then, the "ironic" or "non-ironic" label was determined by a majority vote between these classifiers.

II. *Polarity assignment.* The polarity degree of a tweet is the output of this step. It is determined by a SA model that exploits a set of features that covers not only textual markers but also affective information as well as the irony-aware features obtained from the previous step. Since the irony detection model we exploited here does not distinguish between different types of figurative language, such as irony and sarcasm, we decided to use the presence of ironic content only as a feature for assigning polarity rather than for reverting the polarity of a given tweet. This is motivated by the results of the analysis in Sulis et al. [15], which highlights that tweets tagged with #irony and #sarcasm behave differently with respect to the polarity reversal phenomenon. In fact, with respect to the twist of the polarity in tweets tagged with #irony and #sarcasm, it has been observed that, when the #sarcasm hashtag is used, it is common to have a full polarity reversal (from a polarity to its opposite, mostly from positive to negative polarity), while, when #irony is used, there is often just an attenuation of the polarity (mostly from negative to neutral). See also [3] for a similar study about this issue on the Italian corpus Senti-TUT.

3.1   Irony Detection Model

Irony has been recognized as a linguistic device strongly connected with the expression of feelings, emotions, attitudes and evaluations [4,29,30]. We relied on a state-of-the-art irony detection model: *emotIDM* described in [31]. emotIDM detects irony in Twitter taking advantage of different facets of affective content by exploiting a wide range of resources available for English. Such facets include sentiment and emotional aspects in a finer-grained way by capturing information from both categorical and dimensional models of emotions. Besides, it considers textual markers (such as punctuation marks, part-of-speech labels, emoticons, and specific Twitter's markers) that have been recognized as reliable clues for identifying ironic intent in social media. *emotIDM* considers irony as an umbrella term that covers also sarcasm. It outperforms the state-of-the-art results validating the importance of affect-related information for detecting ironic content in tweets.

3.2   Sentiment Analysis Model

The SA model takes ValenTo system [25] as starting point and improve its performance by adding lexical resources with the aim to capture affective information[5]. We chose to use ValenTo system for two reasons: (1) It does not include bag-of-words (BOW) as features to perform SA. Such features can be highly topic and domain dependent. We are instead interested in proposing a model exploiting mainly affective information, and therefore it considers features able to capture this kind of information disregarding domain. (2) It includes a feature to identify ironic content by exploiting the presence of hashtags.

The SA module in our pipeline is then composed by seven groups of features:

(i) *Structural*: punctuation marks, POS labels, uppercase chars, URL, and emoticons.
(ii) *Twitter markers*: hashtags, mentions and retweets.
(iii) *Sentiment modifiers*: elongated words, interjections and negations.
(iv) *Sentiment Analysis lexica*: AFINN [32], Hu&Liu (HL) [33], SentiWordNet (SWN) [34], SenticNet polarity (SNpol) [35], Emolex polarity (EmoLexPol) [36], General Inquirer (GI) [37], Sentiment140 (S140) and NRC Hashtag

---

[5] https://github.com/ironyAware-SA/sentimentAnalysisFeatures.

Sentiment Lexicon (NRC-Hash) [38], MPQA [39] and Sentiment-Pattern[6] (sPat).

(v) *Categorical models of emotions*: Emolex emotions (EmoLexEmot), EmoSenticNet (EmoSN) [40], Linguistic Inquiry and Word Count (LIWC) [41], and DepecheMood (DM) [42].

(vi) *Dimensional models of emotions*: ANEW [43], Dictionary of Affect in Language (DAL) [44], and SenticNet (SNemot).

(vii) *Irony-aware*: two binary features are also considered in order to take into account the presence of ironic intent (*ironyIDM*) as well the presence of an ironic hashtag (*ironyHashtag*). These features are obtained in the first phase of our pipeline.

The polarity assignment is carried out by building a regression model. We used the Weka implementation of M5P, a decision tree regressor. We experimented with other algorithms, and found that the results were worst than those obtained using M5P.

4   Evaluation

We experimented with the SE15-Task11 dataset; it is distributed in *training* (8,000 tweets), *trial* (1,000 tweets), and *test* (4,000 tweets). The organizers of the task retrieved tweets rich on figurative language by considering either the presence of specific hashtags (such as #irony, #sarcasm, and #not) or words commonly associated with the use of metaphor (such as "literally" and "virtually"). We present experimental results for the *test* set used in SE15-Task11. For the training phase, we used the remaining tweets. As evaluation measures we used the cosine similarity (cosSim) and the Mean-Squared-Error (MSE) as were defined in [17]. cosSim is calculated as the cosine between the vector containing the golden labels in the *test* set and the vector with the results obtained by our pipeline. A score of 1 is achieved when a given system provides all the same scores than in the *test* set. For what concerns to MSE, lower measures of it indicates better performance.

In order to evaluate the effectiveness of our method, we trained the SA module in the pipeline by using each group of features described in Sect. 3.2 individually as well as different combinations among of them. It is important to highlight that we applied the same irony detection model in all the experiments. To further investigate the importance of the different lexica considered in our model, we evaluated the sentiment analysis, categorical models of emotions, and dimensional models of emotions groups of features by removing an affective resource each time.

Finally, we also are interested in to find how well different groups of features performed when bag-of-words are also exploited (unigrams with binary representation were used as BOW features). Our experimental setting was two-fold: (i)

---

[6] http://www.clips.ua.ac.be/pages/pattern-en#sentiment.

To demonstrate the robustness of our method in assigning polarity, by exploiting high-level features comprising mainly affective information from different aspects; and (ii) To compare the performance of our model when n-grams are combined with the set of features described in Sect. 3.2.

4.1 Results

Table 1 shows the results of our system in cosSim and MSE terms. *All features* label in the first row of the table refers to all the features described in Sect. 3.2 (composed by a total of 140 features). The second row shows the performance of our sentiment analysis module when the irony-aware features are removed from *All features*. As can be noticed, there is a drop, although small, in the performance of our system. This result could provide an insight useful to validate our hypothesis about the usefulness of recognizing irony before performing SA. Therefore, in the rest of the experiments the irony-aware features were always considered.

Table 1. Comparison of the performance of our approach when it is evaluated with and without irony-aware features. Both results are statistically significant.

Features	cosSim	MSE
All features	0.689	2.640
All features without irony-aware features	0.673	2.836

Table 2 shows the performance of the pipeline when the SA module is trained with different sets of features. From Table 2 it can be appreciated that, in general, our model outperforms the official result before achieved by ValenTo (0.634 in cosSim). As can be noticed, the result obtained in the experiment involving saLex group of features together with irony-aware features is the best one with respect to all the groups of features in the sentiment analysis module. Our best result in conSim terms slightly outperforms the one obtained by the second place in the official ranking in the SemEval-2015 Task 11 (0.710 in cosSim). It uses irony-aware features in addition to some widely known features for sentiment analysis related tasks.

In order to investigate the performance of the resources exploited in our approach, we performed feature ablation by removing one resource in the sentiment analysis, categorical models of emotions and dimensional models of emotions groups of features. Figure 1 shows the results of this experiment. As can be noticed the performance of our irony-aware system when it exploits each group of features individually is still competitive. It seems that removing the resource *S140* from the saLex group provokes the biggest drop in the performance of our pipeline. *LIWC* could be considered as one of the most important resource in the eCat group. Furthermore, when *SenticNet* is not considered, the performance of our pipeline decreases with respect to using all the features in the eDim group.

Table 2. Performance of the proposed pipeline in cosine similarity and MSE terms by using different features in the sentiment analysis module. All the experiments use also the features belonging to the irony-aware group.

Features	cosSim	MSE
Structural (Str)	0.588	3.381
Sentiment modifiers (SentiM)	0.558	3.498
Twitter markers (TwM)	0.589	3.458
Sentiment analysis lexica (saLex)	**0.67**	**2.836**
Categorical models of emotions (eCat)	0.63	3.070
Dimensional models of emotions (eDim)	0.60	3.296
Str + TwM + SentiM + saLex (group1)	**0.711**	**2.504**
Str + TwM + saLex + eCat (group2)	0.687	2.663
TwM + saLex (group3)	0.669	2.797
TwM + saLex + eCat (group4)	0.68	2.705
TwM + saLex + eDim (group5)	0.678	2.737
TwM + saLex + eCat + eDim (group6)	0.674	2.771
TwM + eCat (group7)	0.653	2.879
TwM + eDim (group8)	0.635	3.070
saLex + eCat + eDim (group9)	0.665	2.856

On the other hand, there are some resources that when are removed allow us for a slight improvement of the performance in terms of cosSim.

Additionally, we carried out experiments by adding bag-of-words (more than 10,000 features composed the set of those coming from n-grams) together with our set of features. Figure 2 shows the obtained results. When we experimented by using BOW combined only with irony-aware features, the cosSim achieved was 0.61. Our evaluation shows that the proposed pipeline achieves comparative performance at assigning polarity degree even without exploiting BOW. Besides, the dimensionality of the feature space in our model is noticeably lower when compared with BOW. This means that by using our set of features it is possible to obtain a lower computational cost with a set of relevant features for assigning polarity in tweets with figurative language. Our two best results 0.71 (by using group1) and 0.74 (by using group3 + BOW) are not higher than the one of the best ranked system in the task (0.758 in cosSim), but they are still competitive and reach the second position in the official ranking, showing that affective information helps.

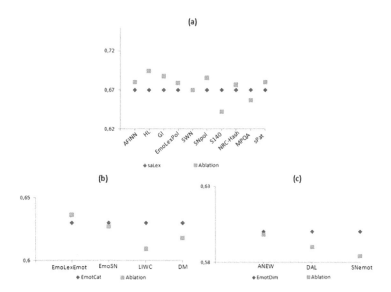

Fig. 1. Ablation experiment results in cosine similarity terms for the (a) saLex, (b) eCat, and (c) eDim groups of features.

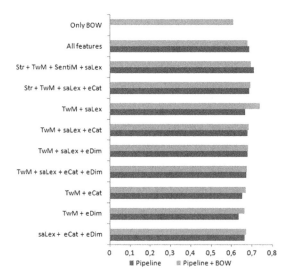

Fig. 2. Cosine similarity results of applying different groups of features together with bag-of-words.

5   Conclusions

In this paper we have shown that including irony detection is a relevant step for sentiment analysis. The experiments described were conducted on a Twitter dataset including a rich variety of figurative language devices, labeled with sentiment at a fine-grained level. We obtained comparable results to the best systems in the SE15-Task11 and show that features related to affective content play an important role. As future work, it would be interesting to distinguish between irony and sarcasm [15,45] in order to reason on the possibility to apply different polarity reversal criteria. Moreover, we consider also to employ other kinds of resources for improving the coverage of our approach such as the one described in [46] that is based on ANEW. Besides, we are planning to improve the irony detection module by exploiting not only affective information but also aspects related to pragmatic context [47]. Furthermore, we are interested in evaluating our system using datasets coming from different social media as well as in other languages.

Acknowledgments. The National Council for Science and Technology (CONACyT Mexico) has funded the research work of the first author (Grant No. 218109/313683 CVU-369616). The work of Paolo Rosso has been partially funded by the SomEMBED TIN2015-71147-C2-1-P MINECO research project and by the Generalitat Valenciana under the grant ALMAMATER (PrometeoII/2014/030).

References

1. Liu, B.: Sentiment analysis and opinion mining. In: Synthesis Lectures on Human Language Technologies, vol. 5, pp. 1–167 (2012)
2. Mohammad, S.M.: Sentiment analysis: detecting valence, emotions, and other affectual states from text. In: Meiselman, H. (ed.) Emotion Measurement. Elsevier (2016)
3. Bosco, C., Patti, V., Bolioli, A.: Developing corpora for sentiment analysis: the case of irony and Senti-TUT. IEEE Intell. Syst. **28**, 55–63 (2013)
4. Grice, H.P.: Logic and conversation. In: Cole, P., Morgan, J.L. (eds.) Syntax and Semantics: Vol. 3: Speech Acts, pp. 41–58. Academic Press, San Diego, CA (1975)
5. Bowes, A., Katz, A.: When sarcasm stings. Discourse Process. Multidiscip. J. **48**, 215–236 (2011)
6. Lee, C., Katz, A.: The differential role of ridicule in sarcasm and irony. Metaphor Symb. **13**, 1–15 (1998)
7. Reyes, A., Rosso, P., Veale, T.: A multidimensional approach for detecting irony in Twitter. Lang. Resour. Eval. **47**, 239–268 (2013)
8. Riloff, E., Qadir, A., Surve, P., Silva, L.D., Gilbert, N., Huang, R.: Sarcasm as contrast between a positive sentiment and negative situation. In: Proceedings of the 2013 Conference on Empirical Methods in Natural Language Processing, (EMNLP 2013), Seattle, Washington, USA, pp. 704–714. ACL (2013)
9. Barbieri, F., Saggion, H., Ronzano, F.: Modelling sarcasm in Twitter, a novel approach. In: Proceedings of the 5th Workshop on Computational Approaches to Subjectivity, Sentiment and Social Media Analysis, pp. 50–58. ACL (2014)
10. Ptáček, T., Habernal, I., Hong, J.: Sarcasm detection on Czech and English Twitter. In: Proceedings of COLING 2014, the 25th International Conference on Computational Linguistics, pp. 213–223. Dublin City University and ACL (2014)
11. Karoui, J., Benamara, F., Moriceau, V., Aussenac-Gilles, N., Hadrich-Belguith, L.: Towards a contextual pragmatic model to detect irony in tweets. In: Proceedings of the 53rd ACL-IJCNLP 2015 (vol. 2: Short Papers), Beijing, China, pp. 644–650. ACL (2015)

12. Poria, S., Cambria, E., Hazarika, D., Vij, P.: A deeper look into sarcastic tweets using deep convolutional neural networks. CoRR abs/1610.08815 (2016)
13. Nakov, P., Zesch, T., Cer, D., Jurgens, D. (eds.): Proceedings of the 9th International Workshop on Semantic Evaluation (SemEval 2015). ACL (2015)
14. Attardo, S.: Irony as relevant inappropriateness. In: Colston, H., Gibbs, R. (eds.) Irony in Language and Thought: A Cognitive Science Reader, pp. 135–172. Lawrence Erlbaum (2007)
15. Sulis, E., Hernández Farías, D.I., Rosso, P., Patti, V., Ruffo, G.: Figurative messages and affect in Twitter: differences between #irony, #sarcasm and #not. Knowl. Based Syst. **108**, 132–143 (2016)
16. Maynard, D., Greenwood, M.: Who cares about sarcastic tweets? investigating the impact of sarcasm on sentiment analysis. In: Proceedings of the 9th International Conference on Language Resources and Evaluation (LREC 2014), pp. 4238–4243. ELRA (2014)
17. Ghosh, A., et al.: Semeval-2015 Task 11: sentiment analysis of figurative language in Twitter. In: Navok et al. (2015), pp. 470–478 (2015)
18. Basile, V., Bolioli, A., Nissim, M., Patti, V., Rosso, P.: Overview of the Evalita 2014 SENTIment POLarity classification task. In: Proceedings of the 4th Evaluation Campaign of Natural Language Processing and Speech tools for Italian (EVALITA 2014), Pisa, Italy, pp. 50–57. Pisa University Press (2014)
19. Barbieri, F., Basile, V., Croce, D., Nissim, M., Novielli, N., Patti, V.: Overview of the Evalita 2016 SENTIment POLarity classification task. In: Proceedings of 3rd Italian Conference on Computational Linguistics (CLiC-it 2016) & Fifth Evaluation Campaign of Natural Language Processing and Speech Tools for Italian. Final Workshop (EVALITA 2016), vol. 1749 (2016) (CEUR-WS.org)
20. Özdemir, C., Bergler, S.: CLaC-SentiPipe: SemEval2015 Subtasks 10 B, E, and Task 11. In: Navok et al. (2015), pp. 479–485 (2015)
21. Barbieri, F., Ronzano, F., Saggion, H.: UPF-taln: SemEval 2015 tasks 10 and 11. Sentiment analysis of literal and figurative language in Twitter. In: Navok et al. (2015), pp. 704–708 (2015)
22. Xu, H., Santus, E., Laszlo, A., Huang, C.R.: LLT-PolyU: identifying sentiment intensity in ironic tweets. In: Navok et al. (2015), pp. 673–678 (2015)
23. Giménez, M., Pla, F., Hurtado, L.F.: ELiRF: A SVM approach for SA tasks in Twitter at SemEval-2015. In: Navok et al. (2015), pp. 574–581 (2015)
24. Van Hee, C., Lefever, E., Hoste, V.: LT3: Sentiment analysis of figurative tweets: piece of cake #notreally. In: Navok et al. (2015), pp. 684–688 (2015)
25. Hernández Farías, D.I., Sulis, E., Patti, V., Ruffo, G., Bosco, C.: ValenTo: sentiment analysis of figurative language tweets with irony and sarcasm. In: Navok et al. (2015), pp. 694–698 (2015)
26. Hernández Farías, D.I., Rosso, P.: Irony, sarcasm, and sentiment analysis. Chapter 7. In: Pozzi, F.A., Fersini, E., Messina, E., Liu, B. (eds.) Sentiment Analysis in Social Networks, pp. 113–127. Morgan Kaufmann (2016)
27. Wang, A.P.: #irony or #sarcasm—a quantitative and qualitative study based on Twitter. In: Proceedings of the PACLIC: the 27th Pacific Asia Conference on Language, Information, and Computation, pp. 349–356 (2013)
28. Hall, M., Frank, E., Holmes, G., Pfahringer, B., Reutemann, P., Witten, I.H.: The WEKA data mining software: an update. SIGKDD Explor. Newsl. **11**, 10–18 (2009)
29. Wilson, D., Sperber, D.: On verbal irony. Lingua **87**, 53–76 (1992)

30. Alba-Juez, L., Attardo, S.: The evaluative palette of verbal irony. In: Thompson, G., Alba-Juez, L. (eds.) Evaluation in Context, pp. 93–116. John Benjamins Publishing Company, Amsterdam/Philadelphia (2014)

31. Hernández Farías, D.I., Patti, V., Rosso, P.: Irony detection in Twitter: the role of affective content. ACM Trans. Internet Technol. **16**, 19:1–19:24 (2016)

32. Nielsen, F.Å.: A new ANEW: evaluation of a word list for sentiment analysis in microblogs. In: Proceedings of the ESWC2011 Workshop on 'Making Sense of Microposts': Big things come in small packages. Volume 718 of CEUR Workshop Proceedings., pp. 93–98 (2011) (CEUR-WS.org)

33. Hu, M., Liu, B.: Mining and summarizing customer reviews. In: Proceedings of the 10th ACM SIGKDD International Conference on Knowledge Discovery and Data Mining. KDD 2004, pp. 168–177. ACM (2004)

34. Baccianella, S., Esuli, A., Sebastiani, F.: SentiWordNet 3.0: an enhanced lexical resource for sentiment analysis and opinion mining. In: Proceedings of the LREC 2010, pp. 2200–2204. ELRA (2010)

35. Cambria, E., Olsher, D., Rajagopal, D.: SenticNet 3: a common and common-sense knowledge base for cognition-driven sentiment analysis. In: Proceedings of AAAI Conference on Artificial Intelligence, vol. I, pp. 1515–1521. AAA (2014)

36. Mohammad, S.M., Turney, P.D.: Crowdsourcing a word-emotion association lexicon. Comput. Intell. **29**, 436–465 (2013)

37. Stone, P.J., Hunt, E.B.: A computer approach to content analysis: studies using the general inquirer system. In: Proceedings of the May 21–23, 1963, Spring Joint Computer Conference. AFIPS 1963 (Spring), pp. 241–256. ACM (1963)

38. Mohammad, S., Kiritchenko, S., Zhu, X.: NRC-Canada: building the state-of-the-art in sentiment analysis of tweets. In: Proceedings of the 7th International Workshop on Semantic Evaluation Exercises (SemEval-2013), USA (2013)

39. Wilson, T., Wiebe, J., Hoffmann, P.: Recognizing contextual polarity in phrase-level sentiment analysis. In: Proceedings of the Conference on HLT and Empirical Methods in Natural Language Processing, pp. 347–354. ACL (2005)

40. Poria, S., Gelbukh, A., Hussain, A., Howard, N., Das, D., Bandyopadhyay, S.: Enhanced SenticNet with affective labels for concept-based opinion mining. IEEE Intell. Syst. **28**, 31–38 (2013)

41. Pennebaker, J.W., Francis, M.E., Booth, R.J.: Linguistic Inquiry and Word Count: LIWC 2001, vol. 71, pp. 2–23. Lawrence Erlbaum Associates, Mahway (2001)

42. Staiano, J., Guerini, M.: DepecheMood: A lexicon for emotion analysis from crowd-annotated news. CoRR abs/1405.1605 (2014)

43. Bradley, M.M., Lang, P.J.: Affective norms for English words (ANEW): instruction manual and affective ratings. Technical report, Center for Research in Psychophysiology, University of Florida, Gainesville, Florida (1999)

44. Whissell, C.: Using the revised dictionary of affect in language to quantify the emotional undertones of samples of natural languages. Psychol. Rep. **2**, 509–521 (2009)

45. Khokhlova, M., Patti, V., Rosso, P.: Distinguishing between irony and sarcasm in social media texts: linguistic observations. In: Proceedings of ISMW FRUCT, pp. 1–6. IEEE Xplore (2016)

46. Warriner, A.B., Kuperman, V., Brysbaert, M.: Norms of valence, arousal, and dominance for 13,915 English lemmas. Behav. Res. Methods **45**, 1191–1207 (2013)

47. Karoui, J., Benamara, F., Moriceau, V., Patti, V., Bosco, C., Aussenac-Gilles, N.: Exploring the impact of pragmatic phenomena on irony detection in tweets: a multilingual corpus study. In: Proceedings of EACL 2017 In Press. (2017)

Sarcasm Annotation and Detection in Tweets

Johan G. Cyrus M. Ræder and Björn Gambäck[(✉)] [iD]

Department of Computer Science,
Norwegian University of Science and Technology, NO–7491 Trondheim, Norway
`gamback@ntnu.no`

Abstract. Identifying sarcasm in text is a challenging task which can be difficult also for humans, in particular in very short texts with little explicit context, such as tweets (Twitter messages). The paper presents a comparison of three sets of tweets marked for sarcasm, two annotated manually and one annotated using the common strategy of relying on the authors correctly using hashtags to mark sarcasm. To evaluate the difficulty of the datasets, a state-of-the-art system for automatic sarcasm detection in tweets was implemented. Experiments on the two manually annotated datasets show comparable results, while deviating considerably from results on automatically annotated data, indicating that using hashtags is not a reliable approach to creating Twitter sarcasm corpora.

1 Introduction

In 2010 a man was arrested for writing a sarcastic tweet [5]. To avoid such incidents, the United States Department of Homeland Security expressed interest in a sarcasm detector for Twitter [7]. Unfortunately, sarcasm is not always easy to detect, in particular in text, where cues such as facial expressions and change of vocal pitch are lost. At the same time, the amount of text generated is growing rapidly, especially text on social media such as Twitter, where the messages (*tweets*) are subject to a 140 character limit, forcing the author to get to the point, but also encouraging the use of abbreviations and non-standard language, as well as URLs, mentions, and hashtags, i.e., words prefixed with the hashmark symbol "#", which can be used for grouping tweets by topic. There are currently 310 million user monthly active on Twitter and 1.65 billion on Facebook.[1] The texts they produce can contain information valuable for various stake holders, e.g., companies wondering what people think of newly launched products or police trying to interpret potential threats in the posts of gang members. In such cases it is vital to find ways of dealing with figurative language to understand the users' actual intentions and opinions.

Here we introduce a state-of-the-art system for automatic sarcasm detection in tweets, and discuss its application to three different datasets, two manually and one automatically annotated for sarcasm. The rest of the paper is laid out as

---

[1] https://about.twitter.com/company, http://newsroom.fb.com/company-info/.

© Springer Nature Switzerland AG 2018
A. Gelbukh (Ed.): CICLing 2017, LNCS 10762, pp. 58–70, 2018.
https://doi.org/10.1007/978-3-319-77116-8_5

follows: The next section gives an introduction to the state-of-the-art in Twitter sarcasm detection. Then the training and testing datasets are detailed in Sect. 3, including information about the creation of a dataset manually annotated for sarcasm, as well as the dataset created automatically based on hashtag information. In Sect. 4 the sarcasm detection system setup and the experiments are presented, while Sect. 5 discusses the results of the experiments with two datasets and a third set, previously annotated manually by Riloff et al. [27]. Section 6 concludes and gives suggestions for future work.

2   Related Work

SemEval 2015 (the annual international workshop on semantic evaluation) included a task on "Sentiment Analysis of Figurative Language in Twitter" [10], with the goal to determine the sentiment of tweets containing one of four types of figurative language: irony, sarcasm, metaphor, and 'other'. Systems were penalized if they did not assign a score to a tweet. All the 15 participating systems achieved a lot better results for irony and sarcasm than for metaphor and 'other'. The setups used by the SemEval participants will be taken as starting point for defining the state-of-the-art, together with some related approaches.

In addition to standard features such as word and character n-grams and skipgrams, several researcher have reported good sarcasm detection results using features derived from part-of-speech tagging, e.g., [2,16], with Carvalho et al. [3] suggesting to look specifically at the number of interjections. The number and type of punctuations are also possible features, since heavy use of punctuation can be a good indicator of sarcasm [3], as can certain punctuation combinations, such as several exclamation and question marks together, or ellipsis marked by three consecutive dots [6]. Onomatopoeic expressions for laughter and emoticons (such as the smiley ':-)') are also helpful both for automatic sarcasm detection [6] and for humans trying to identify sarcasm in tweets [12]. However, Wang and Castanon [28] showed that only some emoticons are used very consistently (i.e., always convey the same emotion).

A common strategy is to include the use of sentiment dictionaries, e.g., [2,24]. These dictionaries usually contain sentiment values for individual words, but can also contain sentiment values for other things, e.g., hashtags. Xu et al. [29] used several dictionaries to look at the overall sentiment of tweets and for sentiment polarity shifts, e.g., when a positive verb is used with reference to a negative clause. Similarly, the system by Riloff et al. [27] learns a set of positive verb phrases (e.g., "love", "excited", "can't wait"), and a set of negative situation phrases, (e.g., "being ignored", "not getting", "doing homework"), and then looks for these in tweets, assuming that a polarity shift is an indicator of sarcasm. Joshi et al. [15] used this idea to look for incongruity in tweets, i.e., when a positive word is followed by a negative word and vice versa.

Some attempts have been made to use author data and tweet history to detect sarcasm. If the sentiment of a tweet does not match the sentiment of an author's previous tweets on a topic, the chance of sarcasm increases [17]. Bamman and

Table 1. Words that often are used sarcastically, from Ghosh et al. [11]

love	like	great	good	really	best	better	glad
yeah	nice	happy	cool	amazing	favorite	perfect	super
fantastic	joy	cute	beautiful	shocked	interested	brilliant	genius
mature	right	fun	attractive	lovely	proud	awesome	excited
always	sweet	hot	wonderful	wonder			

Smith [1] looked at other user information and found that individual authors tend to use certain terms when being sarcastic, and not use those terms when not being sarcastic. Ghosh et al. [11] in contrast identified a set of 37 words that are often used sarcastically by everyone, shown in Table 1.

Several other features have been mentioned, such as excessive use of upper case letters and use of ambiguous words [2], and temporal imbalance [14]. Karanasou et al. [16], motivated by the work of Hao and Veale [13], looked for patterns in the language such as *"as * as a/an *"*. However, the contribution of these features seems only marginal.

3   Data and Annotation

A common way of getting a Twitter dataset is to use the Twitter API to download a large amount of tweets and filter out those relevant to the task at hand. For the case of sarcastic messages, tweets with the hashtags #sarcasm and #sarcastic have often be assumed to be relevant, see e.g. [8,11,19]. However, even if the authors would label their tweets correctly, some of these tweets might be about sarcasm instead of containing sarcasm [6]. Hence González-Ibáñez et al. [12] suggest only using tweets where the hashtag of interest appears at the very end, although this does not entirely eliminate the problem. In addition, this data might be biased towards the hardest forms of sarcasm, where the reader needs an explicit marker to understand that it is sarcasm [6]. However, Bamman and Smith [1] found that users are more likely to explicitly state sarcasm when they are not very familiar with their audience. The bias, if there is any, might therefore lie more towards sarcasm used between people that do not know each other very well, rather than towards difficult forms of sarcasm.

Cliche [4] collected such an automatically annotated dataset consisting of about 150,000 tweets with the hashtags #sarcasm or #sarcastic, and 330,000 without. This dataset will also be used here, but for this work tweets containing URLs were removed as were tweets where the sarcasm hashtag is placed somewhere other than at the end. If a tweet ends with a sequence of hashtags, the entire sequence is considered the end of the tweet. For instance, the following tweet would *not* be discarded: *"i love waking up at 5 am #sarcasm #notreally #fml"*. 100,000 of the remaining tweets were chosen, of which 20,000 have a sarcasm hashtag. This dataset will below be referred to as 'Automatic' (automatically annotated dataset). As a rough quality check of the dataset, 100 randomly

Table 2. Rough quality check of the automatically annotated dataset

	Tweets with #sarcasm	Tweets without #sarcasm
Number of tweets	100	200
Not Sarcastic	9	199
Sarcastic	64	1
Difficult to judge without hashtag	27	N/A

selected tweets with the sarcasm hashtag and 200 without were manually investigated. Only 1 of the 200 tweets without sarcasm hashtags was sarcastic, while 9 of the 100 with sarcasm hashtags clearly were not sarcastic, 64 were sarcastic, and 27 were difficult to identify without the hashtag, as shown in Table 2. If this division would be valid for the whole automatically annotated dataset, about 87% of the tweets in it would be annotated as not sarcastic, and about 13% as sarcastic. Given the small sample size and single annotator, this obviously is a very rough assumption. However, it seems reasonable to assume that tweets without a sarcasm hashtag are not sarcastic, but the converse is not quite as good. This is in line with Kunneman et al. [18], who manually checked 500 French tweets annotated with sarcasm hashtags, finding 63% of those tweets to actually be sarcastic, but they [19] also report that of 250 sarcasm-tagged Dutch tweets, 90% were judged to be sarcastic by manual annotation.

An alternative to using hashtags to automatically select sarcastic tweets is manual annotation. Due to time and human resource constraints, this will obviously lead to a smaller dataset, but tentatively one with higher quality. There is no guarantee, however, that the dataset will be flawless. Humans can make mistakes about and disagree on the meaning of what someone has said or written, and as shown by González-Ibáñez et al. [12], detecting sarcasm in tweets is a difficult task for humans. McGillion et al. [21] noted that the dataset used in SemEval 2015 contains some repeated tweets that had not always been consistently annotated. Normally, duplicate tweets and retweets should be removed before annotation. Forslid and Wikén [8] also suggest eliminating any tweets containing URLs or pictures, as their contents might be necessary to identify any figurative language present in the tweets.

Two manually annotated datasets will be used for the experiments reported in the next section. The first dataset was created by Riloff et al. [27] and originally consisted of 3,200 tweets, half of which having the sarcasm hashtag. The tweets were individually annotated by three annotators, who judged only 742 of the tweets to be sarcastic. Unfortunately, some of the tweets are no longer available for download from Twitter, so the dataset as used in this work (called 'Riloff' below) consists of 2,116 tweets, out of which 1,047 have the sarcasm hashtag, and 459 were judged to be sarcastic.

The second manually annotated set was collected as part of this work in a similar fashion to the first: 2,205 tweets, including 1,115 with the sarcasm

Table 3. Statistics for the different data sets

	Automatic	Riloff	Manual
Number of tweets	100,000	2,116	2,205
Tweets with #sarcasm	20%	49%	51%
Tweets without #sarcasm	80%	51%	49%
Annotated as sarcastic	N/A	22%	25%
Annotated as not sarcastic	N/A	78%	75%

hashtag, were taken from a larger set of about 103 million tweets collected using the Twitter API during the period December 2015–March 2016 [9]. Tweets containing URLs were not included in the set. The tweets were converted to lower case and individually annotated by three annotators, with 554 being judged to be sarcastic. The annotation guidelines included that for a tweet to be judged as sarcastic must be sarcastic in and of itself: trying to guess the context of a tweet should be avoided. Tweets that can be interpreted as either sarcastic or not sarcastic should be judged as not sarcastic. The reasoning being that if a tweet needs the right context or interpretation to be sarcastic, the language (words, phrasing, emoticons, etc.) is not clear enough to be considered sarcastic.

To prevent the annotators from being influenced by knowing whether or not a tweet is meant sarcastically, the sarcasm hashtags were removed prior to annotation. The tweets were given in random order, but the annotators were aware that roughly half the tweets used to have #sarcasm in them. In total, 35 tweets were brought up for discussion to clarify their content, and were then annotated. The final judgement for whether or not a tweet is sarcastic was done by a majority rules scheme. The pairwise inter-annotator agreement scores for this manually annotated set (called 'Manual' below) calculated using Cohen's kappa are 0.59, 0.60, and 0.71. For comparison, they are 0.80, 0.81, and 0.82 for the Riloff corpus (measured on 200 of the original 3200 tweets, 100 with and 100 without the sarcasm hashtag).

Hence three datasets were used in the experiments reported in this paper, one automatically created and two manually annotated. The statistics for the different datasets are summarized in Table 3. The two manually annotated datasets have similar distributions to each other, with about 49% of the tweets in the first and 51% in the second containing sarcasm hashtags, while 22% resp. 25% were actually annotated as being sarcastic. The automatically annotated data differs from these sets: if it had contained equal amounts of each kind of tweet, it would tentatively rather had 68% not sarcastic and 32% sarcastic.

4   Sarcasm Detection

To properly compare the automatically and manually annotated sarcasm corpora, a state-of-the-art sarcasm detection system was built, as shown in Fig. 1.

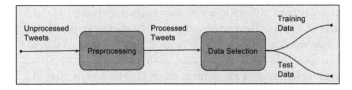

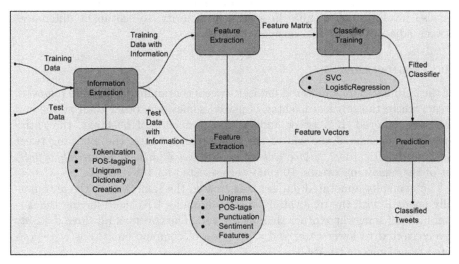

Fig. 1. An overview of the system architecture.

The system implements a range of features, and a grid search was performed to optimize its performance. The following four types of features were used:

Unigrams: Certain words might be used more or less frequently in sarcastic tweets, so a dictionary containing all the unigrams in the dataset was created, and each unigram considered a feature. To reduce the dictionary size, unigrams appearing fewer than 200 times were discarded, and only unigrams whose distributions are skewed more than a ratio of 1.5 were included. The threshold and skewed ratio values were chosen empirically by testing on the automatically annotated data.

Part-of-Speech: Sarcastic sentences might have characteristic structures such as excessive use of interjections, so each POS-tag is used as a feature.

Punctuation: Certain usages of punctuation such as heavy use of full stops might be characteristic of sarcasm. For each of a set of punctuation marks, a ratio of occurence of that character to the length of the tweet is considered a feature. The punctuation characters used were exclamation marks, question marks, colons, semicolons, commas, full stops, quotation marks, and ellipsis.

Sentiment: Various measures using sentiment values are used as features. Since sarcastic tweets are often negative, the sentiment of the whole tweet is considered as a feature. The sentiment values are calculated using the AFINN [23] and

NRC [22] lexica. Several sentiment differences are also considered, e.g., the difference between the words of a tweet and the emoticons. In sarcastic sentences, there are sometimes one key word or phrase which is used sarcastically. All verbs and nouns are regarded as possible such words, and their sentiment value compared to the tweet's overall sentiment. The same is done for the words that [11] identified as good candidates for being used sarcastically, see Table 1. Sarcasm can also involve sudden shifts in sentiment polarity, so sentiment differences between adjacent words are considered a feature.

4.1 Preprocessing

For the automatically annotated dataset, tweets containing URLs were removed. Tweets where the sarcasm hashtag is placed somewhere other than at the end were also removed. If a tweet ends with a sequence of hashtags, the entire sequence is considered the end of the tweet. For instance, the following tweet would *not* be discarded: *"i love waking up at 5 am #sarcasm #notreally #fml"*. Out of the remaining tweets, 100,000 were selected for the dataset.

The manually annotated dataset was treated the same way as the automatically created, with the removal of tweets containing URLs and having the sarcasm hashtag somewhere other than at the end. The tweets of all three datasets were converted to lower case, and a selection of common emoticons were converted from Unicode to ASCII.

4.2 Workflow

After the tweets are preprocessed, they are fed to the system. The initial step is tokenization and POS-tagging. The tokenization is done by a special tokenizer from NLTK [20] called `TweetTokenizer`, which is calibrated for more casual language, like the kind of language often found in tweets. The POS-tagging is done by the standard NLTK POS-tagger, which uses the "University of Pennsylvania Treebank Tag-set",[2] with a few additions for special characters and symbols. The following set of tags were not used as features due to either being deemed unimportant, or being part of the punctuation features: {'$', '"', '(', ')', ',', '–', '.', ':', 'CD', 'FW', 'LS', 'POS', '"', '#'}. The system then makes a pass through all the tokens to make the dictionary of unigrams. Since the hashtag #sarcasm is not removed from the tweets in the preprocessing stage, it is explicitly removed from the dictionary after it has been created.

Based on the chosen feature combination, the system generates a matrix, which is used as input to a classifier. One of two classifiers from `Scikit-learn` [25] are used, either the `LogisticRegression` (LReg) or the `SVC` Support Vector Machine (SVM) classifier. Decision Trees and Naïve Bayes classifiers were also tested, but LReg and SVM gave the best performances. Features are calculated for the test data, and the prediction function yielded by the classifier is used to classify the data. The result is then compared to the gold standard (automatic/manual annotation) for the test data to evaluate classifier performance.

---

[2] www.ling.upenn.edu/courses/Fall_2003/ling001/penn_treebank_pos.html.

Table 4. SVC and LogisticRegression classifier performance, using all features

		Automatic	Riloff	Manual	R-auto	M-auto
	Correct sarcastic	52%	32%	31%	26%	25%
SVM	Correct non-sarcastic	98%	88%	90%	93%	96%
	Incorrect sarcastic	2%	12%	10%	7%	4%
	Incorrect non-sarcastic	48%	68%	69%	74%	75%
	Correct sarcastic	40%	54%	55%	45%	46%
LReg	Correct non-sarcastic	96%	77%	79%	86%	87%
	Incorrect sarcastic	4%	23%	21%	14%	13%
	Incorrect non-sarcastic	60%	46%	45%	55%	54%

4.3 Grid Search

To improve the performance of the classifiers, a grid search was executed using the automatically annotated data with 10-fold cross validation, on randomly selected, stratified folds, i.e., all folds had the same distribution of sarcastic (20%) and non-sarcastic (80%) tweets as in the 'Automatic' dataset. The grid search process was performed twice, first on a coarse level, and then on a finer level, based on the best coarse search result. The parameters used in the grid search for LReg were 'C', which controls the regularization strength, and 'penalty', which dictates the penalization norm. For SVM the parameters were 'C' and 'kernel', which dictates the kernel, and 'γ', a parameter used by certain kernels.

To gauge the effect of the different features, three 10-fold cross validations were run using the best parameters found in the grid search. The first used all the features, the second only the sentiment features, and the third only the lexical features (unigrams, POS-tags, punctuation). The classifier was then trained on the automatically annotated data and tested using all the features on the two manually annotated datasets. Last, instead of using the manual annotation, those datasets were automatically annotated based on #sarcasm and the system was tested with all features on these datasets too.

4.4 Results

Table 4 summarizes the results for the experiments using all features, with 'Automatic' being the automatically annotated dataset, 'Riloff' the dataset manually annotated by Riloff et al. [27], 'R-auto' the result of annotating that dataset automatically using hashtags, 'Manual' the manually annotated dataset introduced in this work, and 'M-auto' its automatically annotated cousin.

A comparison between the results for SVC and LogisticRegression (LReg) with all features and the most-frequent baseline is given in Table 5.

Table 5. Performance (F_1-score) on the datasets: Most frequent baseline vs. classifiers

	Automatic	Riloff	Manual	R-auto	M-auto
Most Frequent	0.44	0.44	0.34	0.43	0.34
SVC	**0.79**	0.61	0.55	0.62	0.55
LReg	0.71	**0.64**	**0.64**	**0.66**	**0.65**

5 Discussion

The average inter-annotator agreement score for the manually annotated dataset, as given by Cohen's kappa, was 0.63. This value falls between two values reported by others; Riloff et al. [27] report 0.81 and Ptáček et al. [26] report 0.54. The number of tweets annotated, tweet language, and the instructions given to the annotators vary for these three results, but sarcasm generally appears difficult to agree upon. As long as humans do not agree, how well a system performs is somewhat subjective. It is interesting to note that the 'Manual' dataset presented here and that of Riloff et al. [27] have very similar sarcastic vs. non-sarcastic distributions, with slightly more than half the tweets containing the hashtag #sarcasm either not being sarcastic or needing context to be identified by the annotators. Creating datasets by automatically annotating tweets based on the presence of #sarcasm seems inaccurate.

When using just the sentiment features, only a small amount of tweets are classified as sarcastic; the behaviour is thus very close to the most-frequent classifier. This shows that the sentiment features are quite bad at separating sarcastic and non-sarcastic tweets from each other. Using just lexical features performs much better, being only slightly worse than using all features together. Both the SVC and the LogisticRegression classifiers perform best when using all the features, with the SVC classifier being better. Both classifiers are conservative, classifying a large portion of the smaller class as the larger class.

When tested with automatically annotated versions of the manual datasets, the performance for the SVC classifier drops compared to the results on the automatically annotated data. Logistic regression assigns weights to the features, and it seems like it has emphasized features that are not so specific to the 'Automatic' dataset. For the 'Manual' data, the major difference is that non-sarcastic tweets are classified as sarcastic at a much higher rate than for 'Automatic'. Both classifiers identify a higher percentage of sarcastic tweets for 'Manual' than for 'M-auto'. This might be because the manually annotated dataset contains more stereotypical sarcasm than 'M-auto', which makes them easier to group together. 'Riloff' and 'R-auto' behave in the same way, since the respective performance on those sets is almost identical to that on 'Manual' and 'M-auto', as can be seen in Table 4. This is an interesting result in and of itself, as it implies that both 'Riloff' and 'Manual' are well annotated and reasonably representative.

Comparing the results for manual datasets to the baseline, Table 5 shows that both SVC and LogisticRegression beats the most-frequent classifier in F-scores. They also score better in both precision and recall, but tie or get beaten in accuracy. While the accuracy for manual data fails to beat the most-frequent baseline, this is not the case for automatically tagged versions of those datasets, where it is beaten by 15% for `LogisticRegression` and 9% for `SVC`.

Forslid and Wikén [8] and Cliche [4] use similar data for training, and report F-scores of 60% and 71%, respectively, but the test data is different so it is difficult to compare directly. Riloff et al. [27], Joshi et al. [15], and Khattri et al. [17] used Riloff's corpus for testing (but with more of the tweets available for download, see Sect. 3), and report F-scores of 51%, 61%, and 88%, respectively. The best F-score obtained on that corpus here is 64%, by the `LogisticRegression` classifier. Hence it seems fair to claim that the system outlined in Sect. 4 is state-of-the-art, and that the results obtained thus are highly relevant as basis for the discussion.

6   Conclusion and Future Work

The paper has described how a dataset of tweets was manually annotated with respect to the presence of sarcasm. The result was very similar to that of a previously made set [27], and both show considerable deviation from automatic annotation. This implies that using automatically annotated data for the task of sarcasm detection in tweets is a mediocre approximation.

Experiments with both manually annotated datasets yield comparable results, even though the sets are completely independent. This shows that the sets are well annotated and reasonably representative for sarcasm detection in tweets. When experimenting with different algorithms and features, it became evident that a naïve implementation of sentiment features is not sufficient, although it did slightly improve overall performance. The experiments also showed that with the chosen features, both the SVM and Logistic Regression algorithms beat most-frequent baseline F-scores by 13–23% when tested with previously unseen manually annotated data. They also beat F-scores from some of the other work using the same data by 3–13%. However, the best score from this work for Riloff et al.'s manually annotated data is outperformed by 24% when compared to that of Khattri et al. [17], who make use of user data.

The annotators mentioned that context would have helped a lot with some of the tweets. This is not surprising, as sarcasm is often used in response to some occurrence or utterance. Taking into account the context of a tweet is a natural next step in the quest for an automatic sarcasm detection system. Another issue brought up by the annotators was domain knowledge. Some tweets would refer to people or incidents that they had no knowledge about, making them difficult to judge. Classifying tweets by topic and then using domain knowledge is another area to explore. This could be of particular interest when dealing with tweets about current topics and news stories.

Liebrecht et al. [19] discussed that the amount of sarcasm on Twitter is far less than the amount typically used in training data. The sarcastic tweets used for our manual annotation were taken from a set of about 103 million tweets, out of which roughly 2,000 had the sarcasm hashtag. Not all sarcastic tweets have this hashtag, but it gives some impression of the difference between real data and training data. Even a rather conservative system like the one presented here would almost certainly have a poor precision score when dealing with realistically distributed data.

References

1. Bamman, D., Smith, N.A.: Contextualized sarcasm detection on Twitter. In: Proceedings of the 9th International Conference on Web and Social Media, pp. 574–577. Oxford, United Kingdom (2015)
2. Barbieri, F., Ronzano, F., Saggion, H.: UPF-taln: SemEval 2015 Task 10 and 11 sentiment analysis of literal and figurative language in Twitter. In: Proceedings of the 9th International Workshop on Semantic Evaluation (SemEval 2015), pp. 704–708. Denver, Colorado (2015)
3. Carvalho, P., Sarmento, L., Silva, M.J., de Oliveira, E.: Clues for detecting irony in user-generated contents: Oh...!! it's "so easy"; -). In: Proceedings of the 1st International CIKM Workshop on Topic-Sentiment Analysis for Mass Opinion, pp. 53–56. Hong Kong, China (2009)
4. Cliche, M.: The Sarcasm Detector. http://www.thesarcasmdetector.com/about/ (2014)
5. Daily Mail: Frustrated air passenger arrested under Terrorism Act after Twitter joke about bombing airport. http://www.dailymail.co.uk/news/article-1244091/Man-arrested-Twitter-joke-bombing-airport-Terrorism-Act.html (2010)
6. Davidov, D., Tsur, O., Rappoport, A.: Semi-supervised recognition of sarcastic sentences in Twitter and Amazon. In: Proceedings of the 14th Conference on Computational Natural Language Learning, pp. 107–116. Uppsala, Sweden (2010)
7. Department of Homeland Security: computer based annual social media analytics subscription. https://www.fbo.gov/?s=opportunity&mode=form&id=8aaf9a50dd4558899b0df22abc31d30e&tab=core&_cview=0 (2014)
8. Forslid, E., Wikén, N.: Automatic irony- and sarcasm detection in Social media. Master's thesis, Uppsala University, Uppsala, Sweden (2015)
9. Fredriksen, V., Jahren, B.E.: Twitter sentiment analysis: exploring automatic creation of sentiment lexica. Master's thesis, Norwegian University of Science and Technology, Trondheim, Norway (2016)
10. Ghosh, A., Li, G., Veale, T., Rosso, P., Shutova, E., Reyes, A., Barnden, J.: SemEval-2015 Task 11: sentiment analysis of figurative language in Twitter. In: Proceedings of the 9th International Workshop on Semantic Evaluation (SemEval 2015), pp. 470–478. Denver, Colorado (2015a)
11. Ghosh, D., Guo, W., Muresan, S.: Sarcastic or not: word embeddings to predict the literal or sarcastic meaning of words. In: Proceedings of the 2015 Conference on Empirical Methods in Natural Language Processing, pp. 1003–1012. Lisbon, Portugal (2015b)
12. González-Ibáñez, R., Muresan, S., Wacholder, N.: Identifying sarcasm in Twitter: a closer look. In: Proceedings of the 49th Annual Meeting of the Association for Computational Linguistics:shortpapers, pp. 581–586. Portland, Oregon (2011)
13. Hao, Y., Veale, T.: An ironic fist in a velvet glove: creative mis-representation in the construction of ironic similes. Minds Mach. **20**(4), 635–650 (2010)

14. Hee, C.V., Lefever, E., Hoste, V.: LT3: sentiment analysis of figurative tweets: piece of cake #NotReally. In: Proceedings of the 9th International Workshop on Semantic Evaluation (SemEval 2015), pp. 684–688. Denver, Colorado (2015)
15. Joshi, A., Sharma, V., Bhattacharyya, P.: Harnessing context incongruity for sarcasm detection. In: Proceedings of the 53rd Annual Meeting of the Association for Computational Linguistics and the 7th International Joint Conference on Natural Language Processing, pp. 757–762. Beijing, China (2015)
16. Karanasou, M., Doulkeridis, C., Halkidi, M.: DsUniPi: an SVM-based approach for sentiment analysis of figurative language on Twitter. In: Proceedings of the 9th International Workshop on Semantic Evaluation (SemEval 2015), pp. 709–713. Denver, Colorado (2015)
17. Khattri, A., Joshi, A., Bhattacharyya, P., Carman, M.J.: *Your Sentiment Precedes You*: using an author's historical tweets to predict sarcasm. In: Proceedings of the 6th Workshop on Computational Approaches to Subjectivity, Sentiment and Social Media Analysis (WASSA 2015), pp. 25–30. Lisbon, Portugal (2015)
18. Kunneman, F., Liebrecht, C., van Mulken, M., van den Bosch, A.: Signaling sarcasm: from hyperbole to hashtag. Inf. Process. Manag. **51**(4), 500–509 (2015)
19. Liebrecht, C., Kunneman, F., van den Bosch, A.: The perfect solution for detecting sarcasm in tweets #not. In: Proceedings of the 4th Workshop on Computational Approaches to Subjectivity, Sentiment and Social Media Analysis (WASSA 2013), pp. 29–37. Atlanta, Georgia (USA) (2013)
20. Loper, E., Bird, S.: NLTK: the natural language toolkit. In: Proceedings of the 40th Annual Meeting of the Association for Computational Linguistics. Workshop on Effective Tools and Methodologies for Teaching Natural Language Processing and Computational Linguistics, pp. 63–70. ACL, Philadelphia, Pennsylvania (2002)
21. McGillion, S., Alonso, H.M., Plank, B.: CPH: sentiment analysis of figurative language on Twitter #easypeasy #not. In: Proceedings of the 9th International Workshop on Semantic Evaluation (SemEval 2015), pp. 699–703. Denver, Colorado (2015)
22. Mohammad, S.M., Kiritchenko, S., Zhu, X.: NRC-Canada: building the state-of-the-art in sentiment analysis of tweets. In: Proceedings of the 7th International Workshop on Semantic Evaluation (SemEval-2013), pp. 321–327. Atlanta, Georgia (USA) (2013)
23. Årup Nielsen, F.: A new ANEW: evaluation of a word list for sentiment analysis in microblogs. In: Proceedings of the ESWC2011 Workshop on 'Making Sense of Microposts': Big Things come in Small Packages, pp. 93–98. Heraklion, Crete (2011)
24. Özdemir, C., Bergler, S.: ClaC-SentiPipe: SemEval2015 Subtasks 10 B, E, and Task 11. In: Proceedings of the 9th International Workshop on Semantic Evaluation (SemEval 2015), pp. 479–485. Denver, Colorado (2015)
25. Pedregosa, F., Varoquaux, G., Gramfort, A., Michel, V., Thirion, B., Grisel, O., Blondel, M., Prettenhofer, P., Weiss, R., Dubourg, V., Vanderplas, J., Passos, A., Cournapeau, D., Brucher, M., Perrot, M., Duchesnay, E.: Scikit-learn: machine learning in Python. J. Mach. Learn. Res. **12**, 2825–2830 (2011)
26. Ptáček, T., Habernal, I., Hong, J.: Sarcasm detection on Czech and English Twitter. In: Proceedings of COLING 2014, the 25th International Conference on Computational Linguistics: Technical Papers, pp. 213–223. Dublin, Ireland (2014)
27. Riloff, E., Qadir, A., Surve, P., Silva, L.D., Gilbert, N., Huang, R.: Sarcasm as contrast between a positive sentiment and negative situation. In: Proceedings of the 2013 Conference on Empirical Methods in Natural Language Processing (EMNLP 2013). Seattle, Washington (2013)

28. Wang, H., Castanon, J.A.: Sentiment expression via emoticons on social media. In: IEEE International Conference on Big Data (Big Data 2015), pp. 2404–2408. Santa Clara, California (2015)
29. Xu, H., Santus, E., Laszlo, A., Huang, C.R.: LLT-PolyU: identifying sentiment intensity in ironic tweets. In: Proceedings of the 9th International Workshop on Semantic Evaluation (SemEval 2015), pp. 673–678. Denver, Colorado (2015)

Modeling the Impact of Modifiers on Emotional Statements

Valentina Sintsova[1](✉), Margarita Bolívar Jiménez[2], and Pearl Pu[1]

[1] Ecole Polytechnique Fédérale de Lausanne, Lausanne, Switzerland
valentina.sintsova@alumni.epfl.ch, pearl.pu@epfl.ch
[2] Universidad Politécnica de Madrid, Madrid, Spain
m.bolivar@alumnos.upm.es

Abstract. Humans use a variety of modifiers to enrich communications with one another. While this is a deliberate subtlety in our language, the presence of modifiers can cause problems for emotion analysis by machines. Our research objective is to understand and compare the influence of different modifiers on a wide range of emotion categories. We propose a novel data analysis method that not only quantifies how much emotional statements change under each modifier, but also models how emotions shift and how their confidence changes. This method is based on comparing the distributions of emotion labels for modified and non-modified occurrences of emotional terms within labeled data. We apply this analysis to study six types of modifiers (negation, intensification, conditionality, tense, interrogation, and modality) within a large corpus of tweets with emotional hashtags. Our study sheds light on how to model negation relations between given emotions, reveals the impact of previously under-studied modifiers, and suggests how to detect more precise emotional statements.

Keywords: Emotion analysis · Text mining · Modifiers · Twitter

1 Introduction

Emotions are present in our everyday actions and influence our decisions, behaviors, and relationships. For that reason, emotion identification is becoming increasingly important for developing marketing strategies [4], inferring user interests [15], and understanding personal well-being [27]. A common strategy for text-based emotion recognition is to learn the associations of lexical terms to the given emotions and to classify text based on their occurrences [2,19,32]. However, the true feeling expressed in the text can change under a variety of modifiers. Even the most explicit emotional terms, such as the word 'happy', can relate to another emotion when they occur in the scope of a modifier, such as in the phrase 'not happy'. Examples from Table 1 illustrate how different modifiers can lead to different effects on emotional statements. In order to detect emotions in text more correctly, we should be able to properly model the effects of such modifiers on emotions. Addressing this challenge is the subject of this paper.

© Springer Nature Switzerland AG 2018
A. Gelbukh (Ed.): CICLing 2017, LNCS 10762, pp. 71–89, 2018.
https://doi.org/10.1007/978-3-319-77116-8_6

Table 1. Illustrative examples of the effects of modifiers on emotional statements.

Example	Modifier	Effect
I'm **not** *ashamed* to say it	Negation	Shifts to another emotion
I feel **so** *relieved* now	Intensifier	Increases emotion intensity
I feel **a little** *sad* tonight	Diminisher	Decreases emotion intensity
I know I **should** be *happy*	Modality	Eliminates the presence of emotion
I'll be *sad* **if** you leave	Conditionality	Refers to a non-experienced emotion
Do you *love* her**?**	Interrogation	Refers to a non-confirmed emotion
I **was** *happy* then	Past Tense	Refers to a non-present emotion

Previous studies do not fully address how these common modifiers affect specific emotions. Most related works tend to treat the strongest modifiers, e.g. negation and intensification, only in terms of the change of polarity and intensity [8,9,13]. The effects of other modifiers are disregarded or blocked by removing the modified statements [29]. When models specify per-emotion effects, they are hand-coded [1], which makes their adaptation to other emotion categories or data from another domain more difficult.

This paper proposes a unified data-driven analytical framework for modeling the effects of different modifier types on fine-grained emotion categories that are defined as sets of associated emotional expressions. We quantify the impact of each modifier on each specific emotion using a novel data analysis method, which is based on investigating the distributions of emotion labels for modified and non-modified occurrences of emotional terms in social media. The source of our data is Twitter, from which we collect tweets having an emotional hashtag, viewed as the author's self-revealed emotion label.

This data-driven method derives the model of the modifiers' effects from the patterns of their usage in the large corpus of linguistic data that are automatically labeled with emotions. This makes our method *easily adaptable* to different emotion categories and modifiers, hence giving us a significant advantage over the hand-coding method.

Another contribution of our work lies in *detailed modeling* of the modifiers' effects. Our model not only quantifies the difference between emotions associated with modified and non-modified emotional statements, but also describes how each modifier shifts emotions and changes our confidence in the emotions' presence. In this way, our method produces a fine-grained emotion-based model of modifiers' effects, describing how each emotion changes under each modifier.

We applied this method to analyze the effects of six types of linguistic modifiers. We discovered that the effects of all these modifiers are emotion-specific, confirming that we need a detailed, per-emotion model when treating modifiers. Furthermore, our analysis demonstrated how emotions shift under negations and revealed that some largely ignored modifiers, such as modality and interrogation, can also shift emotions. Finally, we showed the potential of the proposed modeling to find more precise emotional statements. All these findings lead to important implications for developing a modifier-aware emotion classification system.

In the remainder of this paper, we describe the modifiers that we study and the related works that model their effects. Then we present our quantitative framework for analyzing modifiers and introduce the used emotion model and lexicon, modifiers' detection methods, and Twitter data. The two result sections present the extracted effects of modifiers and analyze their modeling within emotion classification. Finally, we summarize our work and discuss future directions.

2   Studied Types of Modifiers and Related Work

This paper studies six linguistic modifiers that were previously discussed in the context of sentiment analysis and that were shown to affect the meaning of emotional statements at least for some of them.

Negation. Negation is the most studied modifier of sentiment polarity. In the simplest approach, researchers consider negation as a polarity reversal [24]. Several other studies have concluded that negation affects both polarity and intensity in the words that are within its scope [3,8,10,13]. Other researchers found that negation's effects depend on the prior polarity of words and the used negation expressions [33]. In automatically learned systems for polarity and emotion classification, negation is treated by considering each negated term as a separate feature [12,23]. Our model follows an under-studied emotion-based approach, where the effect of negation is modeled separately for each emotion category of modified terms. The antonym-based reversal of emotions under negation was shown to increase the accuracy of polarity classification [1]. However, the reversal of emotion is not as simple because of complex relations between emotional concepts and between their linguistic expressions. For example, the phrase "I don't love you anymore" implies rather *Sadness* than either original emotion of *Love* or its antonym emotion of *Hate*. We assume that negation of an emotional term may express the absence of any specific emotion, may refer to another emotion category, or may just change the intensity (or confidence) of the given emotion.

Intensification. Intensification terms change the intensity of emotional words. They form two classes: *intensifiers* that increase the intensity, such as 'very' or 'really' (also called *amplifiers*), and *diminishers* that decrease the intensity, such as 'less' or 'little' (also called *downtoners*). To treat intensification, some methods add (subtract) points from the valence score of sentiment terms if they are preceded by an intensifier (a diminisher) [11,24]. Other methods associate each intensification term with a hand-coded multiplication coefficient representing its strength [28]. We hypothesize that neither intensifiers nor diminishers can change the original emotion category. However, we assume that the confidence in the emotion presence would change according to the direction of an intensity shift (an increase or decrease). For instance, in the sentence "I love you so much", *Love* emotion is intensified, and we can be more confident that it is present.

Modality. Modality is a linguistic construction used to distinguish non-factual situations (*irrealis* events) from situations that happened or are happening (*realis* events). Modal operators can express a degree of uncertainty or possibility, and can also be used to express desires and needs [1,24]. Consider, for instance, the phrases "I will regret it" and "I should be angry with you". In these examples the presence of modal verbs conceals whether the writer actually experienced the referenced emotion or not. Some modal expressions can directly imply the absence of the referenced emotion, as in "I would have loved to see you". Most researchers that consider modality in sentiment analysis treat it as a polarity blocker, ignoring the occurrence of sentiment terms in its scope [24,28]. Benamara et al. [3] show that modality affects the strength and the degree of certainty of the opinion words that are within its scope. Others suggest hand-coding coefficients for the change in certainty or confidence [21]. Liu et al. [16] further argue for including detected modality classes as separate classification features. In a manually crafted model of modality's effects on emotional expressions, researchers found that some modal verbs can even reverse emotions [1]. This suggests we need to further investigate the effects of modality on specific emotions.

Conditionality. Conditional sentences can also describe *irrealis* events, i.e. potential or hypothetical situations that are not yet known to happen. For instance, in the sentence "I'll be sad if you leave", the emotion *Sadness* is not yet experienced. Conditionality is rarely treated in sentiment analysis. One example work suggests that classifying sentiment in conditional sentences is challenging and argues for training a tailored classifier to deal with them [20]. To shed light on how to treat conditionality in emotion recognition, we will study its exact effects on fine-grained emotions.

Interrogation. Interrogation represents sentences where a question is asked. Similarly to conditionality, we cannot be certain whether the states or events mentioned in questions actually happened. Thus, interrogative sentences can change our confidence in detected emotion, as shown by the example question "Do you love me?" However, interrogation can also shift an original emotion to another one: e.g. the sentence "Are you mad at me?" implies rather *Worry* than *Anger*. As interrogative emotional statements are not common in the review texts, the effects of interrogation are traditionally neglected in sentiment analysis. One of the few exceptions is the work of Tabaoda et al. [28], who consider interrogation as a polarity blocker, along with modality and conditionality. Nevertheless, interrogation is frequent in personal communications and deserves further investigation.

Past Tense. Past tense describes situations that happened in the past. Therefore, we may be more certain that the stated emotion was experienced (a potential confidence increase). Yet, expressing emotion in the past may also mean that currently it is not experienced anymore (a potential for confidence decrease). These two phrases illustrate these effects: "I was happy with you" and "I loved you so much before". Thus, we investigate the effects of past tense to identify which case is more frequent and to conclude whether this under-studied modifier is relevant for consideration in emotion recognition.

Summary. Our analysis is different from the previous ones in the following aspects. First, we study six different linguistic modifiers using the same analytical framework, thus giving an advantage to compare their relative impact. Second, we consider their effects using a fine-grained emotion model of up to 20 categories. This choice helps us to model more precisely the emotional shifts of expressions that employ modifiers. Third, our modeling technique is data-driven and automatic, hence overcoming the costly nature of manual approaches. It also enables researchers to easily adapt, validate, and extend our analysis.

3   Quantitative Analysis of Modifiers

This section describes the method we have developed to quantify and analyze the effects of modifiers on emotional expressions.[1] We introduce below the model of emotion categories and the corresponding lexicon of explicit emotional terms that we use in our study. We describe the collected data that are automatically labeled with those emotions and present our approach to detect modifiers for the input emotional terms. Finally, we explain how these components are employed altogether to quantify the impact of different modifiers' types on the original emotion expressed by an emotional term.

3.1   Input Data

Emotion Model. In order to analyze a wide range of emotions, we chose the fine-grained emotion model of the Geneva Emotion Wheel (GEW, version 2.0 [25,26]), which has twenty categories of emotions. The high number of categories enables a detailed modeling of the modifiers' effects, where we are likely to detect which emotions correspond to statements with modifiers. GEW was developed by the psychologists in order to categorize self-reported emotional states. It contains 10 positive and 10 negative emotion categories, each one represented by two close category names (e.g. Amusement/Laughter). We will use only the first names throughout the paper.

Lexicon of Explicit Emotional Terms. A list of explicit emotional terms is associated with this model—the Geneva Affect Label Coder (GALC) [25]. It is an affective lexicon that enumerates for each emotion category the stemmed words expressing it, i.e. each stemmed word from the GALC lexicon is associated with an emotion category. Overall, there are 212 stems for 20 GEW emotions, 10.9 in average per each emotion category. However, we discovered that using

---

[1] The input data, used linguistic resources, and code for analysis method are available at www.cicling.org/2017/data/263/.

stems with a wildcard token * at the end is undesirable, as sometimes non-related terms would be also matched. For instance, one of the most frequently mapped instances of the GALC stemmed term *happ** (*Happiness* emotion) is *happy*, which is the correct association, but the instance *happen* is also frequent while it does not correspond to this emotion category. This is why we instantiate the stemmed words into actual linguistic tokens by matching those stems in the dataset of around 15 million random tweets collected with the Twitter Sample API in Nov. and Dec. 2014. Then, we manually discovered correctly matched emotional terms among the most frequent instances. The new revised lexicon GALC-R consists of 1026 terms, 52.9 in average per emotion category.

Twitter Data Labeled by Emotional Hashtags. To perform our data-driven quantitative analysis, we require a large dataset. Furthermore, this dataset must be labeled with emotions. As manual annotation is not achievable at the desired scale, we resort to using the pseudo-annotated dataset of tweets. To obtain such dataset labeled with GEW emotion categories, we follow the distant supervision idea of using the emotional hashtags appearing at the end of the text as a self-reported emotion label for the tweet [7,18,31]. Concerning the quality, we rely on the previous evaluations of similar hashtag-based labeling, which showed that the emotion of the hashtag correctly corresponded to the tweet content in 83% of tweets for a large set of emotional and mood-descriptive hashtags [6].

We specify the list of 167 emotional hashtags assigned to the GEW categories based on previously introduced GALC lexicon [25]. 17.6 millions of English tweets with those hashtags were collected via Twitter Streaming API in Mar.–May of 2014. After cleaning, we extracted $1,729,980$ tweets that had those hashtags at the end of the text, were not repeated, were not retweets, did not contain URLs, and were assigned to only one emotion category. All these tweets were converted to lower-case and preprocessed to correctly separate emoticons, user-names, and punctuation marks from other tokens. We randomly sampled 1.5 million of such tweets to be used for studying the effects of the modifiers on emotional terms (analysis dataset D_A). The remaining $229,980$ tweets will be used in the emotion classification experiments (test dataset D_T).

3.2   Data Preparation

Our data preparation process consists of three main steps as shown in Fig. 1. We overview each of them and present our terminology.

First, we take as input the analysis dataset D_A, which consists of the ***tweets with emotional hashtags***. For each hashtagged tweet, we consider the emotion category associated with the emotional hashtag to be a true emotion label for the full tweet (one per tweet), and refer to it as a *hashtag emotion*.

We define the *emotion distribution* of a subset of the hashtagged tweets as the distribution of the tweets' *hashtag emotion* labels over the GEW's 20 categories. For each category, we compute the proportional amount of tweets

1. Collect tweets with emotional hashtags

2. Detect lexicon emotional terms and their modifiers

	TERM EMOTION	DETECTED MODIFIER	HASHTAG EMOTION
(a) I am <u>happy</u> you are here #joy	Happiness	No modifier	Happiness
(b) *Not* <u>ashamed</u> to admit it #proud	Shame	Negation	Pride
(c) I <u>love</u> you *so much* #love	Love	Intensifier	Love

3. Aggregate distributions of hashtag emotions
for each term emotion and modifier class

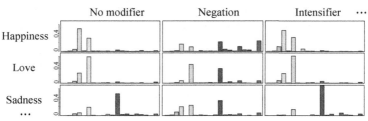

Fig. 1. The process of data collection, preparation, and extraction of emotion distributions.

whose emotional hashtags belong to that category. The emotion distribution of all tweets in the analysis dataset D_A is called the *baseline distribution* P_{BASE}.

Next, we **detect lexical emotional terms and their modifiers** in the text of the input tweets. We identify a subset of the above dataset with tweets containing exactly one emotional term from the emotion lexicon GALC-R. We look for such terms in the content of the tweets while disregarding their emotional hashtags. There are 245,591 such tweets, some of them containing modified emotional terms and some non-modified ones.

For each tweet in this set, we construct a triplet data representation with the following elements: (1) the GEW category of the detected emotional term (henceforth called *term emotion*), (2) whether that term is modified and with which modifier (*non-overlapping modifier class*), and (3) the true emotion as revealed by the tweet's hashtag (*hashtag emotion*). For example, from the tweet "I don't love it #sad" we will identify *Love* as its *term emotion* (based on the term *love*), Negation as the *modifier class* (based on the negation term *don't*), and *Sadness* as the *hashtag emotion* label (based on hashtag #sad).

Finally, we **compute the emotion distributions for each term emotion and each modifier class**. For every *term emotion* such as *Happiness*, we construct emotion distributions of 20 *hashtag emotions* for every *non-overlapping modifier class*, starting with a distribution for tweets without any modifiers, for tweets with negations, for tweets with intensifiers, etc. Note that for each *term emotion* we aggregate all the tweets with the lexicon terms corresponding to that emotion.

We further provide details about detecting modifiers, separating modifier classes, and extracting emotion distributions.

Detection of Modifiers that Affect Lexicon Terms. We apply the modifier detection module to discover how emotional terms are modified by the respective modifier types. We detect each of six modifiers' types based on the presence of specific words and multi-word expressions from a modifier's list. Depending on the modifier's scope, these terms can appear either some words before or after an emotional term in question. We additionally ensure that no punctuation marks or emoticons appear in between the modifier and emotion terms to avoid splitting sentences. Also, to avoid detection errors, we compile lists of frequent false positive expressions and ignore modifier terms that appear within them.

Negation. The list of negation expressions contains common negation words, such as *wasn't, not,* or *no,* and their misspelled variants, such as *weren't* or *didn't* (taken from [1]). It also includes 38 verbs, such as *pretend* or *fail,* implying that a modified statement is not experienced or does not happen (taken from [17], where they are marked as having a negative signature). To be detected, a negation word should appear up to three words before the emotional term. We additionally extract 202 false positive expressions for negation, such as *nothing but* and *can't help,* among which 83 are marked as intensifiers, e.g. *couldn't be more.* We also deal with double negations, which are marked as not negated.

Intensification. We compile the lists of 93 intensifiers (e.g. *much*) and 38 diminishers (e.g. *a bit*) from the related literature [1,9], and extend them with manually validated frequent n-grams containing those words. We further classify each term according to its position: whether it can appear before (e.g. *lots of*) or after (e.g. *very much*) the emotional term, or both (e.g. *less*). The scope of an intensification modifier is then defined as one word directly before or after it, depending on its classification. 9 false positive phrases, such as *that kind of* and *at least,* were also added.

Modality. Our list of modal expressions has 143 terms. The significant part of it consists of modal verbs, such as *should, might,* or *can. Will, 'll, wont* are also in this list, i.e. the emotional terms in the future tense will be detected as modified with modality [22]. Additionally, our list contains the expressions of desire (e.g. *wish, want*) and of uncertainty (e.g. *maybe, seems*). To be detected, a modal expression should appear up to 4 words before an emotional term. Note that we avoided including modal expressions of high certainty or 'trueness', such as *sure* or *indeed,* because they are assumed to have a different effect on emotions than other considered modal expressions [21].

Interrogation, Conditionality, and Past Tense. We detect interrogation by inspecting whether there is the interrogation sign '*?*' after the emotional term or the question-specific patterns, such as *am i* and *why does*, before the term. Conditionality is detected by finding the word '*if*' before the emotional term. The sentence boundaries are checked in both cases. To detect past tense, we search for the part-of-speech tag specific for verbs in the past tense, using Stanford POS Tagger [30]. The emotional term is considered to be in past tense if this tag appears up to four tokens before it.

Separation of Modifier Classes. Several modifiers can modify the same term in the text, e.g. both negation and intensification are present in the phrase "not very interested". To exclude confounding effects between modifiers from our analysis, we split the entries of modified terms into *non-overlapping modifier classes* using the following rules. We recognize Past tense modifier only if it does not overlap with any other modifier, otherwise we assign the overlapping modifier alone (i.e. Past Tense plus Negation will be assigned to the Negation class). The case of Modality and Conditionality is assigned to Conditionality only. The same is for Modality and Interrogation (assigned to Interrogation only). We also separate a class of Mixed Negation containing all the cases where other modifiers (except for Past Tense) overlap with Negation. All other overlapping cases of found modifiers are placed into the Mixed class and are not considered in the analysis of the modifiers' effects.

We note that 34% of the emotional terms from GALC-R are modified by at least one modifier, with Intensifiers being the most common modifier (14.9% of the entries), followed by Past Tense (5.2%). Negations in total modify 3.6% of the terms, while 32% of them are Mixed Negation cases. Mixed class covers only 2.2% of the terms.

Extracted Emotion Distributions. Our data preparation step produces three types of *emotion distributions* (where each of the emotion distributions represents the proportions of hashtag emotions in the corresponding subsets of tweets):

(1) The *baseline distribution* P_{BASE}, which is the emotion distribution of the entire dataset D_A.
(2) The *modified emotion distribution* $P_M(E)$, which is the emotion distribution of the tweets with the term emotion E and with the non-overlapping modifier class M. We count only those tweets where an emotional term for emotion E is within the scope of the modifier M.
(3) The *non-modified emotion distribution* $P(E)$, which is the emotion distribution of the tweets with the term emotion E and having no modifiers. In order to neutralize potential mistakes of the modifiers' scope detection process, we exclude from this distribution the tweets that contain any modifiers' terms, ignoring whether emotional terms are within their scope or not.

Our analysis of the effects of each modifier class will be based on comparing these emotion distributions.

3.3   Quantification of Modifiers' Impact

For each modifier class, we study how it affects each *term emotion* by estimating the change in the emotion distributions of the tweets with modified and non-modified terms. We quantify the influence of the modifiers by comparing the corresponding distributions of hashtag emotion labels using the Kullback-Leibler (KL) divergence [5,14].

The KL divergence is an asymmetrical measure of the difference between two probability distributions S and Q. In our discrete case, it is computed as follows:

$$D(S||Q) = \sum_i s_i \, \log \frac{s_i}{q_i}$$

where s_i and q_i are the corresponding percentage of hashtag emotion E_i in the emotion distributions S and Q. The KL divergence measures how well the distribution S could be approximated by the distribution Q. The closer it is to zero, the better is the approximation. As our goal is to analyze the modified distributions, we consider more restrictive modified distributions $P_M(E)$ as S in the formula, and take more general non-modified or baseline distributions as Q.

To obtain representative *modified emotion distributions*, we include in the analysis of a modifier only those emotions for which at least 50 tweets contain their modified terms. Also, to avoid division by zero in the KL computation when emotion distributions are sparse, we add a smoothing constant of 0.05 to each emotion label count before normalizing the distributions to percentage values.

Our analysis of modifiers' effects aims to answer the following three questions regarding the effects of each modifier class M on a specific term emotion E, which we will refer to as original term emotion.

Question 1. To what extent does the modified emotion differ from the original non-modified emotion? (**modifier divergence**)

We answer this question by comparing the emotion distributions of the modified cases with the ones without any modifier (i.e. modified distribution vs. non-modified distribution for the original term emotion E). This means we compute the KL divergence $D(P_M(E)||P(E))$. We refer to this metric as a *modifier divergence*. It can help quantify how much impact each modifier has. In this comparison, we assume that people who express their emotions with and without using modifiers assign emotion labels to their statements in a similar manner.

Question 2. Does the original emotion change under the modifier into another outcome emotion, or does it stay the same? (**shift** or **no shift**)

To detect which non-modified emotion approximates the best the extracted modified emotion, we compare the distribution of the modified emotion $P_M(E)$ with distributions of each non-modified emotion $P(E_i)$. The emotion E_i that provides the minimal KL divergence will be referred to as the *outcome emotion* E_{out} under that modifier, i.e.

$$E_{out} = \arg \min_{E_i} D\big(P_M\left(E\right)||P\left(E_i\right)\big).$$

We say that the modifier *shifts* the emotion E if the outcome emotion is different from the original emotion, i.e. if $E_{out} \neq E$. Otherwise, we say the emotion remains the same or *no shift* has been detected under the modifier ($E_{out} = E$). This knowledge is necessary for properly modeling the modifier's effect within emotion classification.

Question 3. How confident are we that the discovered outcome emotion is actually expressed in the modified text? (**confidence coefficient**)

Regardless of whether there was a shift of emotion or not, it is likely that the modified distribution $P_M(E)$ differs from the closest non-modified distribution of the outcome emotion $P(E_{out})$. It can differ in two ways: the modified emotion distribution can be more pronounced than the non-modified one, e.g. due to a higher peak for the outcome emotion; or it can have a more random distribution, corresponding to a more mixed state of emotions or an absence of them. The first case intuitively increases our confidence that the outcome emotion is present, while the second one decreases it.

Following this intuition, we compute a *confidence coefficient* (CC) that measures a change of confidence in the presence of the outcome emotion in modified distribution relative to such confidence in the non-modified case. To compute it, we additionally compare both modified and non-modified distributions with the baseline emotion distribution of all analysis tweets P_{BASE}. We define the *confidence coefficient* (CC) as a ratio of two KL divergences: one between the modified and baseline distributions and one between non-modified distribution of the outcome emotion E_{out} and the baseline distribution, i.e.

$$CC = \frac{D\big(P_M(E)\|P_{BASE}\big)}{D\big(P(E_{out})\|P_{BASE}\big)}$$

The confidence decrease ($CC < 1$) implies that the modified emotion distribution $P_M(E)$ is more random than the non-modified $P(E_{out})$. And the confidence increase ($CC > 1$) implies that the modified emotion distribution $P_M(E)$ is more pronounced than the non-modified one.

To illustrate the suggested analysis method, we visualize the corresponding emotion distributions in the case of analyzing how Negation modifier affects original emotion of *Pride* (Fig. 2). It can be observed that the distribution for negated Pride (B) is considerably different from the non-modified Pride distribution (A). More particularly, it has the peak on *Shame* instead of *Pride*, which leads to a high modifier divergence value of 1.96. Furthermore, this makes the non-modified Shame distribution (C) to have the smallest KL divergence to the negated Pride distribution. We thus infer that *Shame* is the outcome emotion of *Pride* under Negation. However, negated Pride distribution (B) has higher percentage of *Pride* and *Love* than non-modified Shame distribution (C), showing that it does not follow it exactly. In result, (B) is closer to the baseline distribution P_{BASE} than (C) (1.02 vs. 1.11), which in its turn results in decreased confidence ($CC = 0.92$).

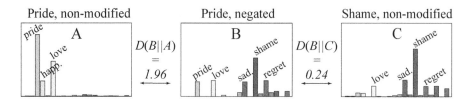

Fig. 2. Examples of non-modified (A) and modified (B) emotion distributions for analyzing the effects of Negation on Pride. (C) visualizes the non-modified distribution of the outcome emotion Shame, which has the smallest KL divergence to (B).

3.4   General Remarks

The presented method of analysis does not aim at building a general theory of modifiers' effects on emotions. Instead, it provides a data-driven linguistic approach where emotion categories are detected based on corresponding sets of emotional terms or expressions. Because of that, the extracted model of modifiers' effects is purely linguistic and the exact computed effects depend on the used set of emotional expressions, applied methods of modifiers' detection, and input linguistic data. The goal of this analysis is to better understand and model how modifiers affect emotional statements and how such effects could be properly treated within emotion classification.

4   Computed Effects of Modifiers

Our method models the effects of each modifier class M on each emotion E in terms of four characteristics: modifier divergence, the outcome emotion, whether there is an emotion shift, and the confidence coefficient of the outcome emotion. In this section, we summarize the detected effects of modifiers.

To show how each modifier affects the explicit emotional statements in general, we present the aggregated effects of modifiers in Table 2. We report for each modifier the average of *modifier divergences* across all emotion categories along with the names of the original and outcome emotions corresponding to the highest modifier divergence. We also summarize the behavior of shifts and confidence changes: what proportion of the original emotions shifts into other outcome emotions with either increase or decrease of confidence, and what proportion of emotions remains the same under the modifier. Note that we use in our analysis (and thus in this aggregation) only the emotion categories for which enough modified entries are detected (≥ 50).

These results confirm the expected differences in the impact of different modifiers, with Intensifiers being the least influential modifiers and Negation—the most. At the same time, they show that the effects of each modifier differ depending on the emotion category it modifies. This is reflected in the facts that every modifier shifts at least one emotion and that no modifiers have the same effect on all emotions. According to the overall shifting pattern, we can separate three groups of effects: no shift, mixed, and shift.

Table 2. Comparison of the different non-overlapping modifier classes using metrics aggregated across emotion categories. The modifier divergence (MD) for an emotion category is the KL divergence between modified and non-modified distributions. We count the percentage of shifted and non-shifted emotions under each modifier, aggregated by the confidence coefficient (CC) behavior.

Modifier class	MD mean	$E \rightarrow E_{out}$ for max MD	% of shifted		% of no-shift		Summary of effects
			$CC > 1$	$CC < 1$	$CC > 1$	$CC < 1$	
Intensifiers	0.12	Nostal. $\rightarrow$ Regr.	0%	11%	**78%**	11%	no shift, $CC > 1$
Past tense	0.17	Guilt $\rightarrow$ Guilt	0%	6%	19%	**75%**	no shift, $CC < 1$
Modality	0.19	Involv. $\rightarrow$ Worry	6%	13%	6%	**75%**	no shift, $CC < 1$
Conditionality	0.26	Involv. $\rightarrow$ Sadn.	0%	27%	18%	**55%**	no shift, $CC < 1$
Diminishers	0.28	Nostal. $\rightarrow$ Regr.	11%	11%	**56%**	22%	no shift, $CC > 1$
Interrogation	0.40	Awe $\rightarrow$ Involv.	29%	24%	**35%**	12%	mixed
Mix.negation	0.52	Pleas. $\rightarrow$ Regr.	8%	**50%**	0%	42%	mixed
Negation	0.80	Pride $\rightarrow$ Shame	0%	**75%**	0%	25%	shift, $CC < 1$

Modifiers with No-Shift Effects. The smallest value of average modifier divergence belongs to Intensifiers. As expected, for most emotions they do not shift the original emotion, but increase its confidence ($CC > 1$).

Past Tense, Modality, and Conditionality modifiers have another behavior. They mostly decrease confidence ($CC < 1$) without shifting the original emotion. This means that these modifiers can introduce uncertainty on whether a specific emotion is expressed.

Our model further makes an interesting but counter-intuitive observation about Diminishers: they increase confidence for most of emotions, while preserving the original emotion. This is explained by the fact that when a person states he or she is "kinda/a little/only/a bit" "sad/in love/worry/disappointed" we can be more confident that the stated emotion is actually being experienced.

Yet, there are exceptions from these main patterns of behavior. For example, some negative emotions expressed in the past tense are linked more confidently to the associated emotions, i.e. with $CC > 1$ (an example is "I was disappointed"). Also, each of these modifiers with general no-shift effects shifts some emotions. For example, Modality shifts 19% of emotions (3 out of 16 analyzed ones) and Conditionality—27% (3 out of 11). We note that some of these shifts reflect the specifics of using the studied emotional expressions in English. For instance, Modality shifts *Involvement* into *Worry/Fear* with an increased confidence due to the widespread of the phrase "this will/shall/gonna/should be interesting" that expresses the author's worry about what is going to happen.

Modifiers with Mixed Effects. Not every modifier has a clear overall shifting behavior. More particularly, we discover that Interrogation has its own pattern. It shifts many positive emotions into *Involvement*. This may be explained by the fact that asking other people questions about their positive emotions is an expression of *Involvement/Interest* by itself. Meanwhile, negative emotions mostly do not shift in interrogative sentences.

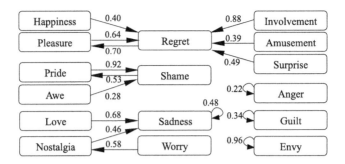

Fig. 3. The extracted model of emotion shifts under negation. The arrows point to the outcome emotion of negating the original emotion. They are labeled with the confidence coefficients (CCs) of the outcome emotions.

Mixed Negation has a mixed effect as well. For example, it shifts several positive emotions, including *Happiness* and *Pleasure*, into *Regret*, because of the dominance of Negation mixed with Modality (an example is "I can't be happy"). At the same time, many other emotions stay the same with the lower confidence.

Modifiers with Shifting Effects. The highest modifier divergence value corresponds, as expected, to Negation. In line with the previous findings in sentiment analysis, we observe that negation tends to shift emotions (it happens for 12 out of 16 analyzed emotions, i.e. for 75%) and decreases the confidence in the outcome for all emotions, even non-shifted (average CC is $0.56 < 1$). However, our analytical method allows establishing which emotion the original emotion shifts to (i.e. the outcome emotion), and discovering emotions that do not shift even after being modified. Figure 3 summarizes such per-emotion shifting effects of negation. We can observe several clusters of these effects.

(1) Five of the positive emotions shift towards *Regret*, while *Regret* itself shifts back towards *Pleasure*. This cluster represents a standard notion of negation influence, where "not happy" and "not amused" are considered to have negative sentiment. It is noteworthy that *Happiness* does not shift to its direct antonym *Sadness*. Also, we do not have direct antonyms of *Amusement* and *Involvement* in the emotion model, thus under negation they shift into the most appropriate emotion category among the given ones (i.e. *Regret* in this case).

(2) We discover a reciprocal negation relationship along the antonym pair *Pride-Shame*. *Awe* also shifts towards *Shame*, which can be attributed to the frequently negated expression "no wonder".

(3) Negation of *Love* and *Nostalgia* becomes *Sadness*, as in the tweet "Nobody loves me enough to hang out with me". At the same time, *Worry* shifts into *Nostalgia*. However, the KL divergence between modified and baseline emotion distributions is small and thus negated *Worry* might rather represent a mixture of emotions than *Nostalgia*, even with a lower confidence.

(4) There are four negative emotions, namely *Sadness, Anger, Envy*, and *Guilt*, for which there is no emotion shift under negation (i.e. they remain the same). This can be illustrated by the examples "I'm not normally an angry ranty person" and "I'm trying not to get sad". The confidence coefficients are small for all of these emotions, except *Envy*, for which it is close to one, meaning that "not envious" has almost the same meaning as "envious".

Overall, all positive emotions shift towards negative emotions, and several negative emotions shift towards positive ones. This confirms the expected power of negation to reverse polarity of emotions. Yet, we find no shift under negation for several negative emotions. This once again shows the importance of treating the effects of modifiers individually for each emotion.

5   Classification Quality of Modified Statements

We evaluate in this section how the classification quality of emotional statements depends on the presence of modifiers and the type of their modeled effects.

The extracted quantitative model of the modifiers' effects specifies, for each emotion category and a modifier, what is the outcome emotion after modification and what is its confidence coefficient. For example, it specifies that a negated term of *Sadness* remains assigned to the category *Sadness* with a confidence coefficient of 0.48.

As the basis of classification, we use the GALC−R lexicon of explicit emotional terms. For each occurrence of a lexicon term, we detect which of the studied modifiers are present. We again use the non-overlapping classes of detected modifiers (as described in Sect. 3.2). Based on the presence of modifiers and their extracted effects, we separate three cases of emotional terms' occurrences:

(1) **Not Modified**—No modifiers are detected. We return the original emotion associated with the emotional term.

(2) **No Shift**—Exactly one modifier is detected for the term, and it produces no shift of the emotion associated with the term. The term's emotion is returned.

(3) **Shift**—Exactly one modifier is detected for the term, and it shifts the original term emotion into another outcome emotion. We separate two scenarios for treating this shift: whether to return the outcome emotion or the original emotion of the emotional term.

We exclude the mixed cases, where several non-overlapping modifiers are detected, and the cases where the modifier's effect is not modeled because not enough of such modified statements appeared in the analysis dataset D_A.

To compute the classification quality of each case of modified emotional statements, we use a test dataset of hashtagged tweets D_T, containing $229,980$ tweets with one of the emotional hashtags for 20 GEW emotion categories. These tweets did not participate in the extraction of the modifiers' effects. We again consider the emotion category of a hashtag to be a ground-truth label, and remove the

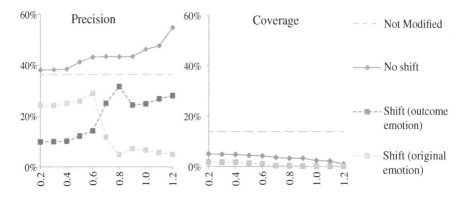

Fig. 4. The precision and coverage of different modified cases of emotional terms depending on the confidence threshold τ. Only emotional terms with $CC \geq \tau$ are included in each case.

hashtags themselves from the tweet text. We use precision and coverage as performance metrics. Precision is the ratio of correctly found hashtag labels among all labels returned based on the considered statements. Coverage is the percentage of tweets in which the considered statements are found.

We investigate the quality of classification depending on the level of filtering: we ignore the modified statements with the confidence coefficient CC lower than a confidence threshold τ. Figure 4 shows the dependency of precision and coverage on the confidence threshold τ for the three considered modified cases. When $\tau = 0.2$ all corresponding modified statements are used without filtering. Notice that the non-modified case is independent of τ values.

The results show that no-shift modified cases have a higher precision than non-modified emotional entries for any values of τ. This means that when an emotional term appears in the scope of a no-shift modifier the precision of its association with the corresponding emotion is higher. Also, we can observe that the higher the confidence threshold τ is, the higher the precision of the no-shift modified cases is, but the lower their coverage is. We can thus identify more precise emotional statements by increasing the τ value.

Considering the shifted modified cases, we observe that their precision is lower than of the non-modified case, regardless of what emotion (original or outcome) is returned. This means that we can exclude such shifted cases altogether in order to obtain more precise classification results. The plot also shows a shift in dominance between two options to return emotion at $\tau = 0.7$. This suggests how to potentially increase the overall precision without excluding shifted modified cases: we can return the original emotion for lower CC values, and the shifted, outcome emotion for higher CC values.

In essence, knowledge about the shifting and non-shifting behavior of modified cases helps us find more precise emotional statements. Therefore, we can construct higher-precision classifiers, which can be then used to initialize distant supervision algorithms or to identify more reliable classification examples within an application.

Limitations. In the current analysis, we considered only lexicon-based classification approach and only one lexicon (GALC-R) for which the modifiers' model was computed. Further research is required to understand how to incorporate such modifiers' model within machine learning-based classification methods, such as Support Vector Machines or Multinomial Naïve Bayes, and how this approach relates to other automatic techniques of treating modifiers, such as coding them as separate features.

6   Conclusion and Future Work

This paper proposes a data analysis method to model the effects of different linguistic modifiers on fine-grained emotional statements based on their usage in social media. It analyses how the modifiers change respective emotion distributions, how they shift the term emotions, and how they affect the confidence in the outcome emotions. With this method, we study the effects of six different linguistic modifiers, such as negation, intensification, and modality, on the explicit emotional terms that are within their scope. As labeled data, we use a large number of tweets with author's self-revealed emotions, identified via emotional hashtags. This work, to our best knowledge, is the first systematic study of the effects of different modifiers at a fine-grained level of emotion categories.

Our analysis reveals multiple interesting patterns of modifiers' impact. First of all, the effects of modifiers are non-uniform across emotion categories, suggesting that to more effectively treat modifiers and their effects we need to model the fine-grained per-emotion effects. For example, we show that some under-studied modifiers can even shift emotion categories: conditionality and modality shift *Involvement* to *Sadness* and *Worry* correspondingly. Second, our data confirms that negations are the most notorious modifiers, shifting 75% of the emotion categories. More interestingly, our model shows how the original emotions shift in the presence of negation and other modifiers. Third, we show the potential of incorporating the computed modifier model along with its confidence coefficients to identify more precise emotional statements. Such profound, detailed understanding of the modifiers' effects is essential for building emotion classifiers of superior quality.

The proposed method aims at helping researchers to treat modifiers for classification purposes, not at universal modeling of emotion-modifiers relations. Nevertheless, our data-driven modeling method can extract the different modifiers' effects within any data where bootstrapping a large quantity of high-quality emotional data is feasible, e.g. using hashtags or emoticons. It also allows updating the modifier model for new modifier types or emotion categories, which would help to test another hypothesized impact. Because of these properties, our analytical framework could facilitate future research on automatic discovery of new modifier expressions, investigation of other linguistic or contextual modifiers, and construction of modifier-aware emotion classification systems.

References

1. Carrillo-de Albornoz, J., Plaza, L.: An emotion-based model of negation, intensifiers, and modality for polarity and intensity classification. J. Am. Soc. Inf. Sci. Technol. **64**(8), 1618–1633 (2013)
2. Aman, S., Szpakowicz, S.: Identifying expressions of emotion in text. In: Matoušek, V., Mautner, P. (eds.) TSD 2007. LNCS (LNAI), vol. 4629, pp. 196–205. Springer, Heidelberg (2007). https://doi.org/10.1007/978-3-540-74628-7_27
3. Benamara, F., Chardon, B., Mathieu, Y., Popescu, V., Asher, N.: How do negation and modality impact on opinions? In: Proceedings of the Workshop on Extra-Propositional Aspects of Meaning in Computational Linguistics, pp. 10–18. ACL (2012)
4. Consoli, D.: A new concept of marketing: The emotional marketing. BRAND. Broad Res. Account. Negot. Distrib. **1**(1), 52–59 (2010)
5. Cover, T.M., Thomas, J.A.: Elements of Information Theory. Wiley (2006)
6. De Choudhury, M., Counts, S., Gamon, M.: Not all moods are created equal! Exploring human emotional states in social media. In: Proceedings of the International AAAI Conference on Weblogs and Social Media (ICWSM), pp. 66–73 (2012)
7. De Choudhury, M., Gamon, M., Counts, S.: Happy, nervous or surprised? Classification of human affective states in social media. In: Proceedings of the International AAAI Conference on Weblogs and Social Media (ICWSM), pp. 435–438 (2012)
8. Hogenboom, A., Van Iterson, P., Heerschop, B., Frasincar, F., Kaymak, U.: Determining negation scope and strength in sentiment analysis. In: Proceedings of Systems, Man, and Cybernetics (SMC), pp. 2589–2594. IEEE (2011)
9. Hutto, C.J., Gilbert, E.: VADER: A parsimonious rule-based model for sentiment analysis of social media text. In: Proceedings of the Eighth International AAAI Conference on Weblogs and Social Media (ICWSM), pp. 216–225 (2014)
10. Jia, L., Yu, C., Meng, W.: The effect of negation on sentiment analysis and retrieval effectiveness. In: Proceedings of the 18th ACM Conference on Information and Knowledge Management (CIKM), pp. 1827–1830. ACM (2009)
11. Kennedy, A., Inkpen, D.: Sentiment classification of movie reviews using contextual valence shifters. Comput. Intell. **22**(2), 110–125 (2006)
12. Kiritchenko, S., Zhu, X., Mohammad, S.M.: Sentiment analysis of short informal texts. J. Artif. Intell. Res. **50**, 723–762 (2014)
13. Kou, X.: The effect of modifiers for sentiment analysis. In: Su, X., He, T. (eds.) CLSW 2014. LNCS (LNAI), vol. 8922, pp. 240–250. Springer, Cham (2014). https://doi.org/10.1007/978-3-319-14331-6_24
14. Kullback, S., Leibler, R.A.: On information and sufficiency. Ann. Math. Stat. **22**(1), 79–86 (1951)
15. Lewenberg, Y., Bachrach, Y., Volkova, S.: Using emotions to predict user interest areas in online social networks. In: Proceedings of International Conference on Data Science and Advanced Analytics (DSAA), pp. 1–10. IEEE (2015)
16. Liu, Y., Yu, X., Chen, Z., Liu, B.: Sentiment analysis of sentences with modalities. In: Proceedings of UnstructureNLP, pp. 39–44. ACM (2013)
17. Lotan, A., Stern, A., Dagan, I.: TruthTeller: Annotating predicate truth. In: Proceedings of the 2013 Conference of the North American Chapter of the Association for Computational Linguistics: Human Language Technologies (NAACL-HLT), pp. 752–757. ACL (2013)

18. Mohammad, S.M.: # Emotional tweets. In: Proceedings of the First Joint Conference on Lexical and Computational Semantics (*SEM), pp. 246–255. ACL (2012)
19. Mohammad, S.M., Turney, P.D.: Crowdsourcing a word-emotion association lexicon. Comput. Intell. **29**(3), 436–465 (2013)
20. Narayanan, R., Liu, B., Choudhary, A.: Sentiment analysis of conditional sentences. In: Proceedings of the Conference on Empirical Methods in Natural Language Processing, vol. 1, pp. 180–189. ACL (2009)
21. Neviarouskaya, A., Prendinger, H., Ishizuka, M.: SentiFul: A lexicon for sentiment analysis. IEEE Trans. Affect. Comput. **2**(1), 22–36 (2011)
22. Palmer, F.R.: Modality and the English Modals. Routledge, London (2014)
23. Pang, B., Lee, L., Vaithyanathan, S.: Thumbs up? Sentiment classification using machine learning techniques. In: Proceedings of the Conference on Empirical Methods in Natural Language Processing, vol. 10, pp. 79–86. ACL (2002)
24. Polanyi, L., Zaenen, A.: Contextual valence shifters. In: Computing Attitude and Affect in Text: Theory and Applications, pp. 1–10. Springer, Dordrecht (2006)
25. Scherer, K.R.: What are emotions? And how can they be measured? Soc. Sci. Inf. **44**(4), 695–729 (2005)
26. Scherer, K.R., Shuman, V., Fontaine, J.R.J., Soriano, C.: The GRID meets the wheel: Assessing emotional feeling via self-report. Components of emotional meaning: a sourcebook, pp. 281–298 (2013)
27. Schwartz, H.A., et al.: Characterizing geographic variation in well-being using tweets. In: Proceedings of the International AAAI Conference on Weblogs and Social Media (ICWSM), pp. 583–591 (2013)
28. Taboada, M., Brooke, J., Tofiloski, M., Voll, K., Stede, M.: Lexicon-based methods for sentiment analysis. Comput. Linguist. **37**(2), 267–307 (2011)
29. Thelwall, M., Buckley, K., Paltoglou, G., Cai, D., Kappas, A.: Sentiment strength detection in short informal text. J. Am. Soc. Inf. Sci. Technol. **61**(12), 2544–2558 (2010)
30. Toutanova, K., Klein, D., Manning, C.D., Singer, Y.: Feature-rich part-of-speech tagging with a cyclic dependency network. In: Proceedings of the Conference of the North American Chapter of the Association for Computational Linguistics on Human Language Technology-Volume 1 (NAACL-HLT), pp. 173–180. ACL (2003)
31. Wang, W., Chen, L., Thirunarayan, K., Sheth, A.P.: Harnessing Twitter "Big data" for automatic emotion identification. In: Proceedings of the International Conference on Social Computing (SocialCom), pp. 587–592. IEEE (2012)
32. Yang, C., Lin, K.H.Y., Chen, H.H.: Building emotion lexicon from weblog corpora. In: Proceedings of the 45th Annual Meeting of the ACL on Interactive Poster and Demonstration Sessions, pp. 133–136. ACL (2007)
33. Zhu, X., Guo, H., Mohammad, S., Kiritchenko, S.: An empirical study on the effect of negation words on sentiment. In: Proceedings of the 52nd Annual Meeting of the Association for Computational Linguistics (ACL), pp. 304–313. ACL (2014)

CSenticNet: A Concept-Level Resource for Sentiment Analysis in Chinese Language

Haiyun Peng[(✉)] and Erik Cambria

School of Computer Science and Engineering, Nanyang Technological University,
Singapore, Singapore
{PENG0065,cambria}@ntu.edu.sg

Abstract. In recent years, sentiment analysis has become a hot topic in natural language processing. Although sentiment analysis research in English is rather mature, Chinese sentiment analysis has just set sail, as the limited amount of sentiment resources in Chinese severely limits its development. In this paper, we present a method for the construction of a Chinese sentiment resource. We utilize both English sentiment resources and the Chinese knowledge base NTU Multi-lingual Corpus. In particular, we first propose a resource based on SentiWordNet and a second version based on SenticNet.

1 Introduction

The development of artificial intelligence (AI) has been rather rapid in recent years. As a major branch of AI, natural language processing (NLP) attracts much attention in both research and industrial fields [6]. One of the hottest topic in NLP is sentiment analysis, a 'suitcase' research problem [4] that requires tackling many NLP sub-tasks, including aspect extraction [24], subjectivity detection [8], concept extraction [27], named entity recognition [18], and sarcasm detection [25], but also complementary tasks such as personality recognition [19] and user profiling [22]. However, research in the area of sentiment analysis can hardly progress much without a good pool of sentiment resources.

There are currently numerous English-language sentiment knowledge bases already in existence, such as SenticNet [5] and SentiWordNet [1]. When it comes to Chinese language, however, the numbers of similar resources are insufficient. Two major sentiment lexicons are currently available in Chinese: HowNet [10] and NTUSD [14]. However, both have their own drawbacks: HowNet only provides a positive or negative label for words. The labeling polarity does not give users information as to what extent a word expresses a sentiment. The entries in HowNet are basically simple words or idioms. As the fundamental elements (word level) in Chinese sentences and passages, their contribution to the overall sentiment is trivial compared with multi-word phrases. Furthermore,

A. Gelbukh (Ed.): CICLing 2017, LNCS 10762, pp. 90–104, 2018.
https://doi.org/10.1007/978-3-319-77116-8_7

HowNet lacks semantic connections between its words. Their words are simply listed in pronunciation order, which makes it impossible to infer sentiment from semantics.

Although bigger than HowNet in size, NTUSD contains all the above drawbacks. To conclude, they are all word-level polarity lexicons. Because of these problems in the existing lexicons, this paper proposes a method to construct a concept-level sentiment resource in simplified Chinese to tackle the above issues, taking advantage of existing English sentiment resources and multi-lingual corpus.

The rest of the paper is organized as follows: Sect. 2 proposes the literature review of Chinese sentiment resources; Sect. 3 presents our framework for the construction of CSenticNet; Sects. 4 and 5 explain in detail the first and second version of the Chinese-language sentiment resource, respectively; Sect. 6 presents evaluations of our methods; finally, Sect. 7 concludes the paper.

2 Literature Review

Two forms of sentiment resources are corpus and lexicon. A corpus is a collection of texts, especially if complete and self-contained: the corpus of Anglo-Saxon verse [20]. Due to the lack of large, expressively labeled Chinese language corpus, Chinese sentiment classification is very much hindered in its development. As such, some researchers decided to expand on or modify existing Chinese corpora. A relative fine-grained scheme was proposed by annotating emotion in text on three levels: document, paragraph and sentence [26]. Eight emotion classes (can be mapped to sentiment classes) were used to annotate the corpus and explore different emotion expressions in Chinese. Later, a Chinese Sentiment Treebank over social data was introduced [17]. 13550 sentences of movie reviews from social websites were crawled and manually labeled. Zhao et al. [31] created a fine-grained corpus with complex and manual annotation procedure.

Two issues exist in the above and current sentiment corpora. Firstly, they were manually built which is time and human-resource consuming. Secondly, they were annotated at sentence level. Sentiment corpus annotated at sentence-level is not enough. Because a corpus is usually utilized in a machine learning way. Words and phrases within a sentence play more important role in machine learning methodology compared with sentence itself. For instance in the negative sentiment sentence "I would prefer to read the novel after watching the movie" no negative words or phrases appeared. However, the words and phrases will wrongly be given a negative label due to sentence level annotation.

Another form of sentiment resource is sentiment lexicon. There are basically three types of sentiment lexicons in all [30]: (1) The ones only containing sentiment words, such as The Never-Ending Language Learner (NELL) [7]; (2) The ones containing both sentiment words and sentiment polarities (sentiment orientation), such as National Taiwan University Sentiment Dictionary (NTUSD) [14] and HowNet [10]; (3) The ones containing words and relevant sentiment polarity values (sentiment orientation and degree), such as SentiWordNet [1] and SenticNet [5]. In the first type, the lexicon only contains words for certain sentiments.

It can help distinguish texts with sentiments from those that without. However, it is not able to tell whether the texts have positive or negative sentiments.

Furthermore, it is an English language corpus and not Chinese sentiment-related. The second, HowNet [10], is an on-line common-sense knowledge base which represents concepts in a connected graph. In terms of its sentiment resources, it has two lists which sentiment words are classified under: positive and negative. The problem this poses is a three-fold one. Firstly, it lacks semantic relationship among the words, as words are listed in alphabetical order. Secondly, it lacks multi-word phrases. Thirdly, it cannot distinguish the extent of the sentiment expressed by the words. For example, *uneasy* and *indignant* are both negative-connotation words but to different extents. HowNet classified these two words as equals in the 'negative' list with no discrepancy between them. NTUSD also has the above disadvantages.

With regards to the third type, both SentiWordNet and SenticNet provide polarity values for each entry in the lexicon. They are currently the most state-of-the-art sentiment resources available. However, their drawback is that they are only available in the English language, and hence do not support Chinese language sentiment analysis. Thus, some researchers seek to build sentiment resources via multi-lingual approach. Mihalcea et al. [21] tried projections between languages, but they have the problem of sense ambiguity during translation and time consuming annotation.

Hence, we propose a method that utilizes multi-lingual resources to construct a Chinese sentiment resource (third type above) which does not need manual labeling and solves the sense ambiguity issue. Its concepts are in connected graph and have both sentiment polarity and sentiment extent. Unlike existing cross-lingual approach [9,12,13,15], there is no machine translation or mapping function learning step in the method. It discovers latent connection between two resources to map the English entity to Chinese in a dedicated way.

3   Framework

In this section, we introduce our proposal in general by listing the resources we are using and discussing the main steps we are taking. Our goal is to construct a Chinese sentiment resource, termed CSenticNet.

The CSenticNet should contain firstly sentiment words or phrases in simplified Chinese. The words and phrases should be organized in the form of synsets: a set of one or more synonyms. Under each synset node, we have words or phrases contributing to a similar meaning and a sentiment polarity value (between -1 and $+1$) they share. Figure 1a illustrates the data structure of CSenticNet.

3.1   Resources

By constructing the sentiment resource, we take advantage of existing resources available on the Internet within copyright/ethical guidelines. We present the different resources utilized in our resource below: *SenticNet* [5], *Princeton Word-Net* [11], *NTU multi-lingual corpus* [28] and *SentiWordNet* [1].

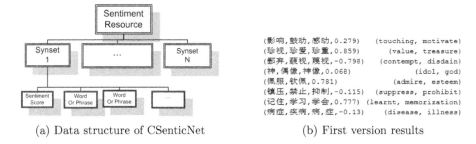

| (a) Data structure of CSenticNet | (b) First version results |

Fig. 1. Data structure and examples of CSenticNet

SenticNet [5] is an English resource for concept-level sentiment analysis. It consists of 17 k concept entries. Five affiliated semantic nodes are listed following each concept. These nodes are connected by semantic relations as illustrated in Fig. 2. There are also four sentics and a sentiment polarity value. The four sentics are a detailed emotional description of the concept they belong to (Fig. 4). The sentiment polarity value is an integrated evaluation of the concept sentiment based on the four parameters. Figure 3b gives an illustration of one such concept. *Princeton WordNet* [11] is a large lexical database of English. It contains four part-of-speech (POS) categories: Nouns, Verbs, Adjectives and Adverbs. Each category is a set of synsets. It totals 117 k synsets, which are connected with each other by *conceptual relations*. It is the most popular English resource for its comprehensiveness and friendly access.

NTU MC (*NTU multi-lingual corpus*[1]) [28] translates Princeton *WordNet* into as many different languages as possible. NTU MC is a multilingual corpus that was built by Nanyang Technology University, and it contains 375,000 words (15,000 sentences) in 6 languages (English, Chinese, Japanese, Korean, Indonesian and Vietnamese) [28]. It has 42 k Chinese concepts in the corpus and are linked by corresponding English translations in WordNet. Most importantly, concepts that are similar in English and Chinese were manually aligned, and such an approach makes NTU MC as the ideal referent for the mutual mapping of the concepts. Moreover, it is no longer merely a lexicon resource, because the translations comprise human semantic translation, like multi-word expressions and phrases. *SentiWordNet* is a lexical resource. It has one-to-one relations with *WordNet*, because it assigns each synset in *WordNet* with a positive score, a negative score and an objective score. The positive score represents the extent to which the word expresses a positive emotion, and vice versa for the negative score. With the above resources, we illustrate basic steps to show how to construct the Chinese sentiment resource.

---

[1] http://compling.hss.ntu.edu.sg/ntumc/.

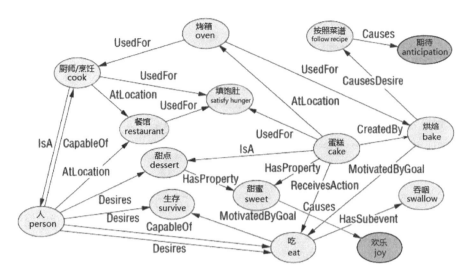

Fig. 2. CSenticNet graph

3.2   Two Versions

Among all the resources we are using, only NTU MC is in Chinese language. Therefore, it serves as the kernel of our resource. However, it does not have any information on sentiment, so our idea is to add affective information to this corpus to make it a sentiment resource.

As for sentiment resources, we have SentiWordNet and SenticNet. Since these are independent of each other, we can use either of them to construct the sentiment resource. As such, we used SentiWordNet in the first version and then SenticNet in the second version.

In the first version, we map the sentiment information from SentiWordNet to NTU MC. Because SentiWordNet has corresponding sentiment polarity to each sense of WordNet, and NTU MC is manually translated from WordNet, we extract sentiment polarity from SentiWordNet and give to their Chinese translations in NTU MC via WordNet.

In the second version, we map the sentiment information from SenticNet to NTU MC. We first try to match all the single and multi-word concepts from SenticNet to WordNet. This is called direct mapping. We also proposed an enhanced version, which combines POS analysis and extended Lesk algorithm to deal with concepts and semantics that were not matched in the direct mapping. The increased number of matches is added to those derived through direct mapping. Finally, we find the overlap between the matched items and NTU MC.

In the following sections, we introduce these two versions in detail and present our evaluations.

```
<LexicalEntry id ='w240003'>
   <Lemma writtenForm='售完' partOfSpeech='v'/>
   <Sense id='w240003_02208409-v' synset='cmn-10-02208409-v'/>
</LexicalEntry>
<LexicalEntry id ='w223142'>
   <Lemma writtenForm='高玲' partOfSpeech='a'/>
   <Sense id='w223142_00444519-a' synset='cmn-10-00444519-a'/>
   <Sense id='w223142_00444984-a' synset='cmn-10-00444984-a'/>
   <Sense id='w223142_00447472-a' synset='cmn-10-00447472-a'/>
</LexicalEntry>
<LexicalEntry id ='w229294'>
   <Lemma writtenForm='神经过敏+地' partOfSpeech='r'/>
   <Sense id='w229294_00409327-r' synset='cmn-10-00409327-r'/>
</LexicalEntry>
```

```
<text>delicious meal</text>
<semantics casserole />
<semantics meatloaf />
<semantics  hot_dog_bun />
<semantics hamburger />
<semantics  hot_dog />
<pleasantness>0.028</pleasantness>
<attention>-0.073</attention>
<sensitivity>0</sensitivity>
<aptitude>0</aptitude>
<polarity>0.034</polarity>
```

(a) NTU MC data   (b) SenticNet data

Fig. 3. Example of used sentiment resources

4   First Version: SentiWordNet + NTU MC

As we explained previously, the role of the first version is to map the sentiment information from SentiWordNet to NTU MC. Because WordNet serves as the bridge that links SentiWordNet to NTU MC, we start by mapping both NTU MC and SentiWordNet to WordNet individually.

We begin by studying the structure of NTU MC. The knowledge base was organized in a lexical structure. The root hierarchy is 'LexicalResource'. Under the root node, there are two children branches: 'Lexicon' and 'SenseAxes'. 'Lexicon' is the mother of 61,536 'LexicalEntry(ies)'. Each 'LexicalEntry' has a Chinese word, its POS, its Sense ID and synset. Because some Chinese words can have different meanings in English, these 'LexicalEntries' sometimes have more than one pair of Sense ID and synset. Figure 3a below gives an example. The key clue that links NTU MC to WordNet is the synset ID. $synset = cmn\text{-}10\text{-}02208409\text{-}v$ is a synset. The combination of $\text{-}02208409$ and $\text{-}v$ uniquely distinguish each synset(sense) in the NTU MC and in WordNet. Naturally, we re-organize the structure of this knowledge base by grouping all the words by synsets with unique synset ID. After processing, we have obtained 42,312 synsets and each synset has at least one Chinese word. The data was stored in a python dictionary.

Then we move to SentiWordNet. We firstly combine *POS* and *ID* of each synset and write them into the same format like NTU MC. Then we compute the sentiment polarity value of each synset. As each synset has a positive score and a negative score, we subtract the absolute value of negative score from positive score and treat the result as the sentiment polarity score. The range of final score is between -1 and $+1$, where polarity stands for sentiment orientation and absolute value means sentiment degree.

In some cases, the calculation results can be 0. This is due to either the synset having neither positive nor negative sentiment or the synset having equal positive and negative scores. We eliminate these synsets since they express no sentiment. Even though this reduces the size the resulting resource, the elimination of these synsets prevents introducing false information. However, the second reason may be a future topic to study. The final version is in a text file format. Each line of the file has a synset (omitted in Figure) with its sentiment polarity score and the relevant Chinese words. Figure 1b shows some examples of the results.

5   Second Version: SenticNet + NTU MC

In the second version, we map the sentiment information from SenticNet to NTU MC. Because NTU MC is directly correlated with WordNet and WordNet is much bigger than SenticNet, it is better to map SenticNet to NTU MC rather than doing it the other way around. Thus, the complete mapping contains these three steps: map NTU MC to WordNet, map SenticNet to WordNet, then find and extract the overlap between SenticNet's and NTU MC's mappings in WordNet. As the first step of mapping NTU MC to WordNet was already finished in the first version, we directly inherit from there. The last step of finding and extracting the overlap is relatively straightforward and does not need much emphasis. Thus, in this second version, we mainly focus on the second step of how to map SenticNet to WordNet. Before that, we present an analysis of SenticNet below.

5.1   SenticNet and Preprocessing

As we can see from the Fig. 3b, the sentiment value of the multi-word concept is *0.034*, which is a positive sentiment. The five semantics *casserole*, *meatloaf*, *hot_dog_bun*, *hamburger* and *hot_dog* all contribute to the concept of a delicious meal. We consider each of the semantics alone as sharing a similar sentiment value with the concept it describes, but we give each concept a higher priority than its semantics. From SenticNet, we have extracted about 17,000 concepts. Before mapping, we need to preprocess SenticNet. We extract every concept, its five semantics and its sentiment score and then put them in a python dictionary. The key of the dictionary is the concept, and the value is the corresponding semantics and sentiment score.

5.2   Mapping SenticNet to WordNet

After the preprocessing is done, we start step 2: mapping SenticNet to WordNet. Due to the diversity of SenticNet (single word, multi-word phrase, semantics), we have proposed two solutions to the problem: direct mapping and enhanced mapping. Direct mapping tries to map SenticNet to WordNet by word-to-word matching. Enhanced mapping integrate direct mapping with keyword extraction based on POS and extended Lesk algorithm.

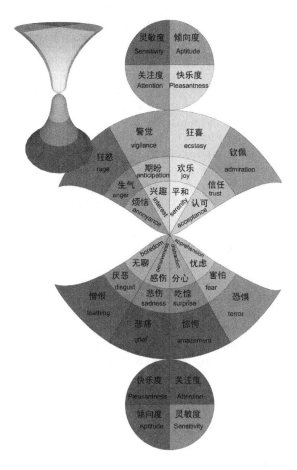

Fig. 4. Hourglass model for Chinese language

Direct Mapping. Since we have covered both SenticNet and WordNet in the python dictionary, we can conduct mapping directly. With WordNet, we have obtained a python dictionary which key is the word or phrase in WordNet and value is a list of synset ID.

For WordNet, a key-value pair may look like this (concept followed by synset IDs): {abandoned : [cmn-10-01313004-a, cmn-10-01317231-a], ...}. For SenticNet, the key is concept and the value is its semantics, like {bank : [coffer, bank_vault, finance, government_agreement, money], ...}. We match each key in SenticNet dictionary to each key from the WordNet dictionary. If a key was matched, the hypernyms of each synset ID in the value from WordNet dictionary would be retrieved. Hypernyms are retrieved from WordNet itself. Synsets (hyponyms) are subordinates of their hypernyms. Then hypernyms of each synset ID will be matched with the words (both concept and semantics) in key-value pair from SenticNet.

If hypernyms from only one synset ID were matched, then this matched synset from WordNet shares the same meaning with the concept-semantics pair from SenticNet. Thus, the sentiment score of this concept from SenticNet will be given to this synset ID. If hypernyms from more than one synset ID were matched, we compute how many words are matched with hypernyms for each synset ID and choose the synset that has most matched words as final matched synset, which will be given the sentiment score from SenticNet. The hypernym of synset ID is considered as layer 1. Hypernyms of the previous hypernyms are considered as layer 2 so on and so forth. If nothing was matched through the whole concept-semantics list in layer 1, we proceed to layer 2. If nothing was matched after layer 3, a concept is scraped. In the end, we accomplish mapping and obtain a dictionary whose key is the synset ID and value is the sentiment score.

The dictionary has 12,042 key-value pairs, which means we have mapped 12,042 synsets from SenticNet to WordNet, a size about one fourth of that of NTU MC. However, one issue that direct mapping failed to solve is the accuracy of matches. For example, referring to Fig. 3b, we have a concept *delicious meal* and a sentiment score of 0.034. We can see that the sentiment score strongly represents the word *delicious* rather than *meal*. However, due to its non-exact match to WordNet, we lose the sentiment score of *delicious meal*, as well as the word *delicious*. In order to figure out the above-mentioned issue, we have developed an enhanced mapping method on top of direct mapping.

Enhanced Mapping with POS Analysis and Extended Lesk Algorithm. As direct mapping has above problems, we develop POS analysis to tackle the exact match problem when concept was not matched, and combine extended Lesk algorithm to settle the problem of sense disambiguation when matching hypernyms failed. In this section, we first provide a review of the techniques we use and then introduce our methods. Before POS analysis, we tokenize the phrases first. This means breaking a string of short phrase into a string of tokens. Each token is a word from the phrase and this token can be read and analyzed by computer algorithms. Because we use python programming in our experiments, we apply the most popular third party tool *Natural Language Toolkit* to do the tokenization. After that is done, we annotate the tokens with POS tag. It helps to extract the key meaning in terms of sentiment and to distinguish the usage of a word in its different senses. We again take the example from Fig. 3b. The concept *delicious meal* has a word *delicious* that is a POS adjective and a word *meal* which is a POS noun. The sentiment of this concept is expressed more by the adjective than the noun. By annotating the POS of each token, we have a better understanding of the sentiment of concept.

The Lesk algorithm is a word sense disambiguation algorithm developed by Michael Lesk in 1986 [16]. The algorithm is based on the idea that the sense of a word is in accordance with the common topic of its neighborhood. A practical example used in word sense disambiguation may look like this. Given an ambiguous word, each of its sense definition in the dictionary is fetched and compared with its neighborhood text. The number of common words that appear in both the sense definition and neighborhood text is recorded. At the end, the sense that has the biggest number of common words is the sense of this ambiguous word.

However, the ambiguous word may sometimes not have enough neighborhood text, so, people have developed ways to extend this algorithm. Timothy [2] explores different tokenization schemes and methods of definition extension. Inspired by their paper, we also developed a way of extension in our experiments. The extended algorithm can solve the ambiguous mapping problem in our direct mapping method.

In our experiments, all single words from SenticNet were easily matched to WordNet. The difficulty mainly falls in mapping multi-word phrases. We put a higher priority on the concepts in SenticNet and lower priority on its semantics. The reason is that sentiment scores in SenticNet are specifically computed for the concepts. Its semantics carry close-related meaning of the concept, so they share the same sentiment score. In a strict sense, this is not ideal.

Therefore, like direct mapping, we decide to match each concept in SenticNet to WordNet first. If it was not matched, we annotate the concept (if it is multi-word phrase) tokens with POS tags before sorting them by POS tag priority. The POS tag priority, from top to bottom, is: Verbs, Adjective, Adverb and Noun. This order of priority is based on the heuristics that top POS tags are more emotionally informative [23,29]. The next step is to extend the contexts. We tokenize all five semantics of a concept and concatenate them with the concept token string to form a large token string. This string is considered as our extended context. At this point, we have prepared the necessary inputs for the Lesk algorithm.

The prioritized tokens with POS tags are considered as the ambiguous words while the large token string is the neighborhood text. We then treat the concept tokens one by one as ambiguous words, based on their POS priority, and apply these to the Lesk algorithm to compute the sense. Once the sense was matched to a sense in WordNet, the processing of this concept is finished and this sense and sentiment score is stored. If it was not matched after iterating through the concept tokens, then one of its semantics is POS tagged and the earlier listed procedures repeated. This process will not stop until a match is found in WordNet or all five semantics have been iterated. Figure 5 summarizes the framework of our two-version method.

In the end, we obtained a dictionary with 18,781 key-value pairs of synsets mapped from SenticNet to WordNet. This gave us 6,739 more pairs than the direct mapping method.

5.3   Find and Extract the Overlap

From the previous section, we obtained a python dictionary whose key-value pair is synset ID-sentiment score by mapping SenticNet to WordNet. In this section, we combine the dictionary with the NTU MC python dictionary we got in the first version and find their overlap. Altogether, 5,677 synsets were overlapping,

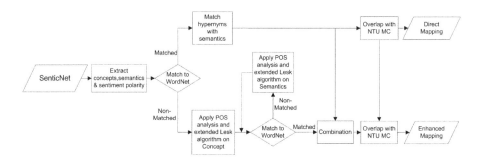

Fig. 5. Mapping framework of SenticNet version

Table 1. Accuracy of SentiWordNet and SenticNet version (column 2–7) and accuracy of small value sentiment synsets (last 3 columns)

Annotator	SentiWordNet version			SenticNet version			[−0.25, 0)	[0, 0.25]	Overall
	Positive	Negative	Overall	Positive	Negative	Overall			
1	48%	64%	56%	82%	80%	**81%**	75%	81%	78%
2	50%	58%	54%	78%	76%	77%	75%	83%	79%
Kappa measure	0.96	0.79	–	0.88	0.88	–	0.73	0.70	–

which meant they had corresponding Chinese translations in NTU MC. Over 15,000 overlapped synsets with their sentiment score and Chinese translations were eventually written into a text file.

6   Evaluation

In this section, we conduct three evaluations of our mapping. For manual validation, we asked two native Chinese speakers to each evaluate 200 entries in our final text files for the two versions of Chinese sentiment resource. Particularly for each of the two versions, 50 positive and 50 negative entries were randomly selected. Both experts were asked to label 200 entries from two versions as either positive or negative independently. We treat their manual labels as ground truth and compute the accuracies of our mapped sentiment resources. The results and inter-annotator agreement measures are in columns 2–7 of Table 1.

The results shown in the tables suggest that the SenticNet version outperforms the SentiWordNet version by almost 50%. This also validates our assumption that SenticNet is more reliable than SentiWordNet in terms of sentiment accuracy. As can be seen, the highest accuracy rate is over 80%. Moreover, there is still space to make improvements to this in the future.

In our mapping procedure, we assume synonyms and hypernyms share similar sentiment orientation with their root word. We believe this is true for the majority of words in the corpora. However, some words or expressions could have opposite sentiment orientation with their synonyms and hypernyms. As illustrated by the Hourglass model in [3], we know that words or expressions that have ambiguous sentiment orientation tend to have small absolute sentiment values. In order to validate our assumptions, we firstly inspect the sentiment value distribution of our SenticNet version sentiment resource and conduct manual validations.

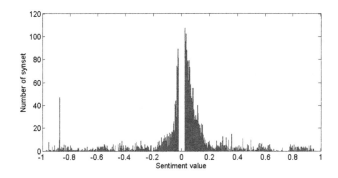

Fig. 6. Distribution of sentiment values

Table 2. Comparisons between CSenticNet and state-of-art sentiment lexicons

Sentiment resource	Chn2000			It168			Weibo		
	P	R	F1	P	R	F1	P	R	F1
NTUSD	50.08%	**99.18%**	66.55%	54.51%	**97.66%**	69.97%	51.17%	**99.39%**	67.56%
HowNet	53.29%	98.68%	69.21%	**61.07%**	96.79%	**74.89%**	50.76%	98.66%	67.03%
CSenticNet (SenticNet version)	**54.85%**	96.18%	**69.86%**	59.04%	94.19%	72.58%	**55.90%**	87.11%	**68.10%**

Figure 6 presents the distribution of all synsets based on their sentiment values. An empty interval exists in the sentiment axis around zero value. This suggests no synsets have very small absolute sentiment values. It partially proves our initial assumptions. However we also notice the high intensity of synsets with small values just beyond the empty interval. The sentiment of these synsets could be wrongly mapped due to our synonym and hypernym assumptions. Thus, we randomly picked up five subsets of synsets from sentiment value ranges (−0.25, 0] and (0, 0.25], respectively. Each subset contains 20 synsets. Then we asked the two native Chinese speakers to label sentiment orientation of the 200 chosen synsets and treat their labels as ground truth. Results are shown in last 3 columns of Table 1. Accuracies within the chosen intervals keep abreast with that of the whole axis. According to the second expert, the intervals even outperform the whole axis in sentiment orientation prediction. Furthermore, we also find that kappa measures of these intervals are less confident than that of the whole axis(columns 3–7 in Table 1). These results further support our initial assumptions and guaranteed the accuracy of our proposed sentiment resources.

Last but not least, shown in Table 2, we conduct sentiment analysis experiments to compare our CSenticNet (SenticNet version) with state-of-art baselines, HowNet and NTUSD. Three datasets we used are: Chn sentiment corpus 2000 (Chn2000[2]), It168[3] and Weibo dataset from NLP&CC[4]. The first dataset contains 1000 positive and 1000 negative reviews from hotel customers. We preprocess this dataset by manually selecting only one sentence which has clear sentiment orientation from each review. The second dataset contains 886 reviews of digital product downloaded and mannually labeled from a Chinese digital product website. The third dataset was micro-blogs originally used for opinion mining. We manually selected and labeled 1900 positive and negative sentences, respectively. We use a simple rule-based keyword matching classifier. Specifically for a test sentence, we match each of its words in sentiment lexicon and sum up the sentiment polarity of matched words in the sentence. For the baselines, positive words have +1 polarities and negative words have −1 polarities. If the final sum is above zero, then the sentence is positive and vice versa.

We see that CSenticNet outperforms the other two baselines in Chn2000 and Weibo datasets, at it has both higher precision and F1 score. However, it narrowly falls behind HowNet in the It168 dataset. We believe this is because of the highly domain biased dataset. It168 reviews are mostly in digital fields, but CSenticNet is not tuned for that domain. Thus, it was not supposed to defeat the other two baselines, but even thought it still performs better than NTUSD. We also find that the recall of CSenticNet is not high, and this gives us a chance to further enlarge the resource by using new versions of SenticNet in the future.

7   Conclusion

In this paper, we presented a method to construct the first concept-level Chinese sentiment resource. Instead of using machine translation, we mapped English sentiment resources to the Chinese corpus using a multilingual corpus. Special techniques were designed to solve issues such as ambiguity. We provide two versions of Chinese sentiment resource: one based on SentiWordNet, the other based on SenticNet. The SenticNet version outperforms state-of-the-art Chinese sentiment lexicons in our evaluations. Moreover, the proposed method can also be applied to other languages in NTU MC.

In the near future, we will focus on the unmatched cases and utilize other sources to enlarge the size of the proposed sentiment resource.

---

[2] http://searchforum.org.cn/tansongbo/corpus/ChnSentiCorp_htl_ba_2000.rar.

[3] http://product.it168.com.

[4] NLP&CC is an annual conference of Chinese information technology professional committee organized by Chinese computer Federation (CCF). More details are available at http://tcci.ccf.org.cn/conference/2013/index.html.

References

1. Baccianella, S., Esuli, A., Sebastiani, F.: Sentiwordnet 3.0: an enhanced lexical resource for sentiment analysis and opinion mining. In: LREC, vol. 10, pp. 2200–2204 (2010)
2. Baldwin, T., Kim, S., Bond, F., Fujita, S., Martinez, D., Tanaka, T.: A reexamination of MRD-based word sense disambiguation. ACM Trans. Asian Lang. Inf. Process. (TALIP) **9**(1), 4 (2010)
3. Cambria, E., Hussain, A.: Sentic Computing: A Common-Sense-Based Framework for Concept-Level Sentiment Analysis. Springer, Cham (2015)
4. Cambria, E., Poria, S., Gelbukh, A., Thelwall, M.: Sentiment analysis is a big suitcase. IEEE Intell. Syst. **32**(6), 74–80 (2017)
5. Cambria, E., Poria, S., Hazarika, D., Kwok, K.: SenticNet 5: discovering conceptual primitives for sentiment analysis by means of context embeddings. In: AAAI, pp. 1795–1802 (2018)
6. Cambria, E., Hussain, A., Havasi, C., Eckl, C.: Sentic computing: exploitation of common sense for the development of emotion-sensitive systems. In: Esposito, A., Campbell, N., Vogel, C., Hussain, A., Nijholt, A. (eds.) COST 2102 Int. Training School 2009. LNCS, vol. 5967, pp. 148–156. Springer, Heidelberg (2010). https://doi.org/10.1007/978-3-642-12397-9_12
7. Carlson, A., Betteridge, J., Kisiel, B., Settles, B., Hruschka Jr, E.R., Mitchell, T.M.: Toward an architecture for never-ending language learning. In: AAAI, vol. 5, p. 3 (2010)
8. Chaturvedi, I., Cambria, E., Vilares, D.: Lyapunov filtering of objectivity for Spanish sentiment model. In: IJCNN, pp. 4474–4481. Vancouver (2016)
9. Chen, Q., Li, W., Lei, Y., Liu, X., He, Y.: Learning to adapt credible knowledge in cross-lingual sentiment analysis. In: ACL (2015)
10. Dong, Z., Dong, Q.: HowNet and the Computation of Meaning. World Scientific (2006)
11. Fellbaum, C.: WordNet: An Electronic Lexical Database. Bradford Books (1998)
12. Gui, L., et al.: Cross-lingual opinion analysis via negative transfer detection. In: ACL, vol. 2, pp. 860–865 (2014)
13. Jain, S., Batra, S.: Cross-lingual sentiment analysis using modified brae. In: EMNLP. Association for Computational Linguistics, pp. 159–168 (2015)
14. Ku, L.W., Liang, Y.T., Chen, H.H.: Opinion extraction, summarization and tracking in news and blog corpora. In: AAAI Spring Symposium: Computational Approaches to Analyzing Weblogs, vol. 100107 (2006)
15. Lambert, P.: Aspect-level cross-lingual sentiment classification with constrained SMT. In: Proceedings of the 53rd Annual Meeting of the Association for Computational Linguistics and the 7th International Joint Conference on Natural Language Processing (Short Papers). Association for Computational Linguistics, pp. 781–787 (2015)
16. Lesk, M.: Automatic sense disambiguation using machine readable dictionaries: how to tell a pine cone from an ice cream cone. In: Proceedings of the 5th Annual International Conference on Systems Documentation, pp. 24–26. ACM (1986)
17. Li, C., et al.: Recursive deep learning for sentiment analysis over social data. In: Proceedings of the 2014 IEEE/WIC/ACM International Joint Conferences on Web Intelligence (WI) and Intelligent Agent Technologies (IAT)-Volume 02. IEEE Computer Society, pp. 180–185 (2014)

18. Ma, Y., Cambria, E., Gao, S.: Label embedding for zero-shot fine-grained named entity typing. In: COLING, pp. 171–180. Osaka (2016)
19. Majumder, N., Poria, S., Gelbukh, A., Cambria, E.: Deep learning based document modeling for personality detection from text. IEEE Intell. Syst. **32**(2), 74–79 (2017)
20. McArthur, T., McArthur, F.: The Oxford Companion to the English Language. Oxford Companions Series. Oxford University Press, Oxford (1992)
21. Mihalcea, R., Banea, C., Wiebe, J.M.: Learning multilingual subjective language via cross-lingual projections (2007)
22. Mihalcea, R., Garimella, A.: What men say, what women hear: finding gender-specific meaning shades. IEEE Intell. Syst. **31**(4), 62–67 (2016)
23. Pavlenko, A.: Emotions and the body in Russian and English. Pragmat. Cogn. **10**(1), 207–241 (2002)
24. Poria, S., Cambria, E., Gelbukh, A.: Aspect extraction for opinion mining with a deep convolutional neural network. Knowl.-Based Syst. **108**, 42–49 (2016)
25. Poria, S., Cambria, E., Hazarika, D., Vij, P.: A deeper look into sarcastic tweets using deep convolutional neural networks. In: COLING, pp. 1601–1612 (2016)
26. Quan, C., Ren, F.: Construction of a blog emotion corpus for Chinese emotional expression analysis. In: EMNLP. Association for Computational Linguistics, pp. 1446–1454 (2009)
27. Rajagopal, D., Cambria, E., Olsher, D., Kwok, K.: A graph-based approach to commonsense concept extraction and semantic similarity detection. In: WWW, Rio De Janeiro, pp. 565–570 (2013)
28. Tan, L., Bond, F.: Building and annotating the linguistically diverse NTU-MC (NTU-multilingual corpus). Int. J. Asian Lang. Proc. **22**(4), 161–174 (2012)
29. Wierzbicka, A.: Preface: bilingual lives, bilingual experience. J. Multiling. Multicult. Develop. **25**(2–3), 94–104 (2004)
30. Wu, H.H., Tsai, A.C.R., Tsai, R.T.H., Hsu, J.Y.: Building a graded Chinese senti-ment dictionary based on commonsense knowledge for sentiment analysis of song lyrics. J. Inf. Sci. Eng. **29**(4), 647–662 (2013)
31. Zhao, Y., Qin, B., Liu, T.: Creating a fine-grained corpus for Chinese sentiment analysis. IEEE Intell. Syst. **30**(1), 36–43 (2015)

Emotional Tone Detection in Arabic Tweets

Amr Al-Khatib and Samhaa R. El-Beltagy$^{(\boxtimes)}$

Center of Informatics Sciences, Nile University, Giza, Egypt
a.mehasseb@nu.edu.eg, samhaa@computer.org

Abstract. Emotion detection in Arabic text is an emerging research area, but the efforts in this new field have been hindered by the very limited availability of Arabic datasets annotated with emotions. In this paper, we review work that has been carried out in the area of emotion analysis in Arabic text. We then present an Arabic tweet dataset that we have built to serve this task. The efforts and methodologies followed to collect, clean, and annotate our dataset are described and preliminary experiments carried out on this dataset for emotion detection are presented. The results of these experiments are provided as a benchmark for future studies and comparisons with other emotion detection models. The best results over a set of eight emotions were obtained using a complement Naïve Bayes algorithm with an overall accuracy of 68.12%.

Keywords: Emotion detection · Arabic · Twitter · NLP

1 Introduction

Emotion tone detection in Arabic text is a research area which has only recently started to attract attention. In the context of social media, emotion tone detection can be an important tool in detecting and gauging public or customers' feelings towards a product, a marketing campaign, an event, a political figure, a government decision, etc. This can provide invaluable information to decision makers in a wide spectrum of sectors and assist them in understanding their audience. This can further help in identifying critical issues before they erupt into full blown problems. Physiologists and sociologists can also analyze the changing moods of people and factors that impact those moods.

The continuous rise of Arabic usage on social media, presents an opportunity for tackling this research topic for the Arabic language [1]. Work that has been carried out to address this topic in other languages has made use of supervised approaches which employ datasets annotated with emotion and a machine learning classifier to associate input texts with a set of predefined emotions. Since emotion detection in Arabic text is a relatively new research area, a very limited number of datasets is available for carrying out this task as detailed in the related work section. This paper presents an Arabic tweet dataset that we have built to serve this task as well as preliminary experiments carried out on this dataset for emotion detection. To our knowledge, this dataset is the biggest Arabic dataset annotated with emotion both in terms of size (10,000+ tweets) and diversity of emotions (eight emotions). The dataset is meant to be used by anyone doing work in the area of emotion detection from Arabic social media.

© Springer Nature Switzerland AG 2018
A. Gelbukh (Ed.): CICLing 2017, LNCS 10762, pp. 105–114, 2018.
https://doi.org/10.1007/978-3-319-77116-8_8

The rest of this paper is organized as follows: Sect. 2 presents background related to the targeted research area, Sect. 3 briefly overviews related work, Sect. 4 describes the data collection process as well as the preprocessing and annotation steps carried out on the data, Sect. 5 presents the carried out experiments and their results, while Sect. 6 concludes this paper and offers future research directions.

2   Background

A survey prepared by Canales and Martínez-Barco [2] about recent efforts in the area of emotion detection in text, summed up the psychological models that have been used to represent human emotions. There are two main approaches that have been vastly used: emotional categories and emotional dimensions. In the emotional categories model, it is assumed that there are discrete basic emotion classes. One of the most used emotional category models is the Ekman's model [3]. According to Ekman's Atlas of emotions, there are six basic emotions: happiness, sadness, fear, anger, surprise, and disgust. Another popular model is the Plutchik bi-polar emotional model, which consists of Ekman's six emotions in addition to trust and anticipation [4]. Plutchik emotions are divided into 4 bipolar groups: happiness/joy vs. sadness, trust vs. disgust, anger vs. fear, and surprise vs. anticipation.

In "emotional dimension" approaches, emotions take places in an affect space. Russell's Circumplex model is one of the affect space models, where emotions are located in a circular space of two dimensions (valence and arousal) [5]. The valence dimension defines the level of pleasantness or unpleasantness associated with an emotion, whereas the arousal dimension detects the activation or deactivation status of an emotion.

According to Canales and Martínez [2], machine learning based approaches in emotion detection are divided into supervised and unsupervised. Supervised learning approaches rely heavily on annotated datasets for emotions, which require a long tedious process of collection, annotation, for building clean structured datasets. We believe this to be one of the main reasons for the current poverty in emotion annotated Arabic corpora.

3   Related Work

As stated before, supervised learning approaches for text classification, require annotated datasets for training a model. In the area of emotion detection from text, very few of those can be found in literature. Furthermore, the datasets described, are very limited in size.

For example, El Gohary et al. [6] annotated a dataset taken from children's stories. The data was composed of 100 documents that were further broken down to 2,514 non-overlapping sequential sentences. Six emotion classes were used when annotating the data: surprise, disgust, anger, fear, sadness, and happiness (سعادة, حزن, خوف, غضب, اشمئزاز, مفاجأة). In addition, a neutral class was introduced for cases where a sentence did not convey any emotion. Another 35 documents were used for testing. A lexicon mapping individual terms to the six target categories was used to classify each sentence. The emotional classes and sentences were represented as vectors, where each vector was composed of the co-occurrence frequencies of emotional words. The cosine similarity metric was used to compute the similarity between sentences and classes to detect the appropriate class for each sentence. A sentence was labeled as "neutral" if its cosine similarity with all emotional class vectors did not exceed a certain threshold. This method achieved an average f-measure of 65%.

In the work presented in [7], a corpus of 1776 tweets was collected by Omneya Rabie and Christian Sturm. The corpus was sampled over the period from 25 Jan 2011 to 11 Feb 2011. The Egyptian revolution in 2011 was the topic of the collected corpus, while #jan25 was the identification hashtag. The data was prepared for annotation by eliminating non-Arabic tweets, retweets, videos, and photo tweets. The data annotation process was accomplished in 3 rounds; in the first round 1012 tweets were annotated out of 1130 tweets. In the second round, 609 tweets were annotated out of 646 tweets, while the third round was confirmatory. On average, each tweet was labeled by around 15 individuals. Any tweet with 50% or less agreement among annotators, was excluded. The final dataset was composed of **1605** tweets divided amongst six emotions (anger, disgust, fear, happiness, sadness, surprise). Data preprocessing included stop-word removal and stemming by the Lucene light Arabic stemmer [8], the Khoja stemmer [9], and a modified Khoja Arabic stemmer. The preprocessing phase also included non-Arabic letter removal, and multiple space and punctuation removal. A subset of 1012 tweets was used to compare the performance of Sequential Minimal Optimization (SMO) and Naïve Bayes classifiers. This subset was distributed among the emotion classes as follows: 271 happiness, 259 anger, 149 fear, 127 disgust, 110 sadness, and 96 surprise. The SMO algorithm outperformed Naïve Bayes in terms of the f-measure, as it scored 44.1%, while Naïve Bayes scored 39.4%. The whole dataset test was then evaluated using 10-fold cross validation. The whole dataset's distribution among classes was as follows: 409 anger, 340 happiness, 285 fear, 204 disgust, 201 sadness, and 166 surprise. The result of this test using the SMO classifier was: 53.1% weighted average balanced f-measure, 53.5% weighted average precision, and 53.5% weighted average recall. A simple word-emotion lexicon was formed by applying a feature selection algorithm to the first 1012 labeled tweets. The selected features formed the base for the word-emotion lexicon, which was extended with manually selected emotion related words. This word-emotion lexicon was used in emotion detection by counting emotion related words in each tweet and selecting the emotion's category based on the highest encountered count for each tweet. Another test was conducted to compare the performance of SMO, and the sample word-emotion lexicon search and frequency (SF). Towards this end, the data was split into two sets: 1012 tweets of the first run for training, and 609 tweets of the second run for testing. The results revealed that SF outperformed SMO in terms of precision, recall, and f-measure for all emotion categories.

Abdul-Mageed et al. [10] collected a dataset of 3000 tweets and annotated them with emotions. The annotated dataset was divided into six main categories (happiness, sadness, anger, disgust, surprise, and fear). The tweets were collected by querying Twitter using a specified a set of seeds for each class. Each set of seeds was composed of less than 10 phrases of multiple Arabic dialects formed with a personal pronoun followed by an emotion expressing word (e.g. "سعيد انا" = "I am + happy"). Duplicate tweets were removed in two distinct steps. In the first step, duplicate IDs were detected and deleted. In the second step, tweets were compared to each other after being pre-processed, and duplicates were removed. Pre-processing for duplicate checking included removal of white spaces, the "rt" string, non-alphabetical characters, and usernames. 500 tweets per category were selected from the tweet pool to form a corpus of 3000 tweets. Besides being annotated with the tweets' emotional class label, each tweet was also assigned another label to describe the emotion intensity as low, medium, or high. Tweets with mixed emotions or no-emotions at all were given an intensity level of zero, so the overall intensity levels were (zero, low, medium, and high). Two native Arabic speaking annotators with high proficiency in modern standard Arabic were selected for the annotation process. In case of encountering unfamiliar dialects, annotators were advised to consult each other, search online, or to ask friends. At the end of the annotation process, the inter-annotator agreement with respect to intensity labels was 92.48%, which reflected the effectiveness of the seeding approach in the querying phase. For emotion labels, the annotators' agreement was higher in certain classes than in others; the agreement level was 0.71 in happiness, and 0.23 in fear, but on average it was 0.51, which reflects a difficulty inherent of emotion annotation.

The target of the work presented by Hussien et al. [11] was to automatically annotate data with emotions by using emojis. 134,194 Arabic tweets were fetched by using trending hashtags during the period between Aug 2015 and Feb 2016. The authors used only four emotion classes (joy, sadness, anger, and disgust), and focused only on the most used emojis in each class. Tweets were labeled based on emojis in each tweet. To assign weights to the emojis, the AFINN lexicon [12] was used. The AFFIN lexicon contains a weighted list of English words, as well as some emojis with their weights. For emojis that were not found in the lexicon, a search for the emoji's corresponding name in the lexicon was conducted and the emoji was assigned the same weight as that of the matching term. From the collected set of tweets, 22,752 tweets that contain emojis were selected and automatically labeled based on the emojis found in them into the 4 target categories. The automatic annotation process resulted in 10,467 joy tweets, 7878 sadness tweets, 2874 anger tweets, and 1533 disgust tweets. Another dataset consisting of 2025 emoji free tweets was annotated manually. The dataset was split into a training set consisting of 1620 tweets and a testing set consisting of 405 tweets. Two different models were built: one trained using the automatically annotated dataset and another trained using the manually annotated training dataset. Both models were tested using the manually annotated test set. The models were built using both Support vector machines and Multinomial Naïve Bayes algorithms. For both classification algorithms, better results were obtained using the models built using automatically labeled data. The SVM results were (0.6976 precision, 0.6904 recall, and 0.6852 f-measure) using the model based on manually annotated data (MMD), and (0.748 precision, 0.7224 recall, and 0.7226 f-measure) using the model based on automatically annotated data (MAD). Similarly, the MNB results were (0.666 precision, 0.6608 recall, and 0.6606 f-measure) using MMD, and (0.757 precision, 0.7526 recall, and 0.7534 f-measure) using MAD.

In conclusion, all surveyed manually created datasets for Arabic emotion detection are of a modest size. The only reasonably sized dataset is the one that has been automatically created by Hussien et al. [11], but it covers only four emotion classes. All reviewed Arabic emotion detection datasets including that automatically created one, are imbalanced in nature and are often skewed towards a certain category. In addition, it is not clear whether those datasets are readily available to researchers or not.

4 The Arabic Emotions Twitter Dataset

In this paper, we present a manually created, balanced Arabic emotions Twitter dataset. The dataset has more than 10,000 tweets and was created with the aim of covering the most frequently used emotion categories in Arabic tweets. In the following subsections, we describe how the dataset was collected and annotated. We also present a set of experiments that were carried out on the data. We believe that these experiments can serve as a benchmark for researchers wishing to experiment with this dataset.

4.1 Data Collection

Data in the presented Arabic emotions dataset was aggregated from multiple resources. The first resource was a corpus composed of 1167 tweets previously collected and labeled for polarity (i.e. positive, negative, and neutral) by the text mining research group at Nile University (NU) [13, 14]. This corpus was then re-annotated for emotions by a group of graduate students at NU. To annotate this dataset, the students used Ekman's emotional model with its six emotional classes (happiness/joy, sadness, anger, disgust, fear, and surprise). The second resource consisted of 2807 tweets harvested using Twitter's search API by the same students and annotated with the same set of emotions. Collected tweets were filtered by Egypt's geo-location between the period from 31/Jul/2016 to 20/Aug/2016. The tweets were downloaded using the word "اوليمبياد" (Olympics) as a search term as during that timeframe people were expected to tweet about the Olympics using a diverse set of emotions. The fetching process resulted in the collection of 16482 tweets, but after elimination of retweets, advertisements, and duplicates, only 2807 tweets remained. When annotating the tweets, the term "none" was used to label a tweet that conveyed no emotion.

The third resource consisted of a dataset that was collected after searching the Twitter API with terms from the NileULex sentiment lexicon [15]. The search resulted in the collection of more than 500,000 tweets. From those, a random subset was selected for the labeling process. The labeling process took into consideration observations made during the previous annotation tasks. These observations can be summarized as follows:

- Tweets often contain more than one emotion.
- A lot of Arabic tweets convey emotions not covered by any category of Ekman's model (e.g. sympathy and love).
- The disgust category is quite rare to come across and is often confused with anger.

So before starting the annotation process for the new corpus, we expanded our emotions categories to cover the most prevalent emotions in Arabic tweets. The modified emotion categories were (joy/happiness, anger, sympathy, sadness, fear, surprise, love, none). The annotation process was conducted through the design and implementation of a web based front end that facilitated the process of emotion annotation. Multiple labels were allowed to cater for tweets that contain more than one emotion. Annotations were carried out by a paid annotator, and revised by a graduate student. Tweets on which the paid annotator and the student disagreed, were flagged and a judge decided the final emotion. A corpus of 1905 tweets was created using this methodology.

Combing this dataset with the previous two datasets resulted in a corpus of 5879 tweets. The breakdown of emotions covered by this unified dataset was as follows: 806 sadness tweets, 1444 anger tweets, 1281 happiness/joy tweets, 452 surprise tweets, 214 love tweets, 40 sympathy tweets, 92 fear tweets, and 1550 tweets with no emotions (none label). Since one of our goals was to create a non-skewed dataset, the next step consisted of balancing the dataset categories to avoid the consequent bias caused by skewed data. A category oriented search was conducted for the following under-represented categories (sadness, surprise, love, sympathy, and fear). Towards this end, the Twitter API was queried using hashtags that were expected to return tweets with the desired emotions (e.g. عشق# شوق# غزل# حب# for love, and معكم_قلوبنا# مؤثر# الرحمه# for sympathy). Once the data was collected, those hashtags were removed from the tweets. Retrieved tweets were revised manually by an annotator to ensure that they all fall within the target categories. This approach was not very effective for the fear category, as it was difficult to find tweets for people expressing their fears by hashtags. To overcome this limitation, multiple queries with groups of terms about fear were invoked (e.g. "خايف OR خايفه OR مرعب OR مخاوف OR مخيف OR قلق OR فوبيا OR رعب") which can be translated to ("afraid OR terrifying OR fears OR scary OR worry OR phobia OR horror"). In addition, other searches were conducted using certain events for which people were expected to express fear (e.g. horror movies, or the civil war in Syria).

This dataset was then combined with the previous one. The previous aggregated dataset was thus balanced by adding 450 sadness tweets, 593 surprise tweets, 1006 love tweets, 1022 sympathy tweets, and 1115 fear tweets resulting in a total of 10,065 tweets annotated with 8 categories as shown in Table 1.

Table 1. Breakdown of emotions in the final dataset.

Emotion	Count
Sadness	1256
Anger	1444
Joy	1281
Surprise	1045
Love	1220
Sympathy	1062
Fear	1207
None	1550
Total	**10,065**

5 Experiments and Results

The aim of the presented experiments is to provide benchmark results for future comparisons between our benchmarked dataset and any introduced methodology or model. The experiments were conducted using the WEKA workbench [16]. Before carrying out any experiments, input data had to be preprocessed. The pre-processing steps as well as the subsequent experiments are presented in the following sub-sections.

5.1 Data Preprocessing

The following preprocessing steps were carried out on all tweets:

– **Normalization:** The first phase in data preprocessing was the normalization of Arabic characters, where "ٱ", "ﺇ", and "ﺃ" characters were replaced with "ﺍ", and "ﻯ" was replaced by "ﻱ", while "ﺓ" was replaced by "ﻩ". Hindi numerals were also replaced by Arabic.
– **Diacritics Removal:** In this step, Arabic diacritics (e.g. "ﺅ" "ﺅ" "ﺅ" "ﻕ" or "ﻕ") were removed.
– **Links, Mentions, and Retweet Indicators Removal:** In this step, hyperlinks, mentions and the retweet indicator ("RT") were removed.

5.2 Experiments

In the carried-out series of the experiments for detecting emotion in Arabic social media text, a simple Naïve Bayes Classifier, a Complement Naïve Bayes Classifier [17] and a Sequential Minimal Optimization classifier were used. The input to any of the classifiers consisted of a bag of words (BOW) representation of the stemmed input tweets. The tweets were stemmed using an Arabic Light Stemmer [18], and the BOW model consisted of n-grams with a minimum of 1 and a maximum of 3 g. For a term to appear in the feature vector, it had to have appeared at least three times in the input corpus. In all shown experiments, evaluation was carried out using 10 fold cross validation.

Results using the Naïve Bayes Classifier

In this experiment, we used a Naïve Bayes classifier to distinguish between the different emotion classes. This model scored **55.67%** overall **accuracy**, as the correctly classified instances were 5603, and the incorrectly classified ones were 4462. Table 2, displays the precision, recall, and f-measure results for this model.

Table 2. Naïve Bayes algorithm results by class using 10 FCV

Class	Precision	Recall	F-measure
Sadness	0.356	0.208	0.262
Anger	0.479	0.522	0.500
Joy	0.370	0.306	0.335
Fear	0.879	0.775	0.824
Love	0.571	0.702	0.630
Surprise	0.432	0.344	0.383
Sympathy	0.712	0.783	2.746
None	0.569	0.781	0.658
Weighted average	**0.542**	**0.557**	**0.542**

Results using the Complement Naïve Bayes Classifier

The same features that were used as input for the simple Naïve Bayes classifier, were input to a Complement Naïve Bayes one. The Complement Naïve Bayes classifier outperformed the conventional Naïve Bayes model with an overall **accuracy** of **68.12%.** The correctly classified instances were 6856, while the incorrectly classified instances were 3209. Details of other measures are shown in Table 3.

Table 3. Complement Naïve Bayes algorithm results by class class using 10 FCV

Class	Precision	Recall	F-measure
Sadness	0.661	0.304	0.417
Anger	0.682	0.727	0.704
Joy	0.691	0.375	0.487
Fear	0.820	0.911	0.863
Love	0.650	0.832	0.730
Surprise	0.679	0.397	0.501
Sympathy	0.787	0.917	0.847
None	0.575	0.928	0.710
Weighted average	**0.688**	**0.681**	**0.658**

5.3   Sequential Minimal Optimization

Again, the same features were input to the Sequential Minimal Optimization algorithm with a linear kernel. This classifier scored an overall **accuracy** of **63.43%.** The correctly classified instances were 6384, while incorrectly classified ones were 3681. Details of other measures are shown in Table 4. While this result is better than that achieved by the basic Naïve Bayes classifier, it is still not as high as that achieved by the Complement Naïve Bayes model.

Table 4. Sequential minimal optimization algorithm results by class class using 10 FCV

Class	Precision	Recall	F-measure
Sadness	0.415	0.452	0.433
Anger	0.599	0.578	0.588
Joy	0.471	0.530	0.499
Fear	0.944	0.871	0.906
Love	0.713	0.658	0.684
Surprise	0.509	0.449	0.477
Sympathy	0.871	0.815	0.842
None	0.653	0.718	0.684
Weighted average	**0.643**	**0.634**	**0.637**

6 Conclusion

In this paper, we introduced a Twitter based dataset for Arabic emotion detection. Various phases for building this dataset were also detailed. Initial and very basic experiments with this dataset have shown that the Complement Naïve Bayes classifier yields the best results with an overall accuracy of 68.1%. While the presented benchmark experiments show promising results, more efforts are needed to achieve better results. As part of our future work, we intend to experiment with deep learning approaches, which have been very successful in English sentiment analysis. We also plan to expand our dataset to include more diverse data from multiple dialects as the current dataset consists mostly of Egyptian tweets.

Acknowledgements. The authors would like to thank Seif El-Din Sameh, Ahmed Mosharafa, Muhammed Abdul-Aziz, and Mustafa Alaa for their efforts in dataset annotation. The authors also wish to thank Abu Bakr Soliman for building the web based annotation tool. This work was partially supported by ITAC [PRP]2015.R19.9.

References

1. Semiocast: Semiocast Web Site (2012). http://semiocast.com/en/publications/2012_07_30_Twitter_reaches_half_a_billion_accounts_140m_in_the_US. Accessed 14 Dec 2016
2. Canales, L., Martínez-Barco, P.: Emotion detection from text: a survey. In: 11th International Workshop on Natural Language Processing and Cognitive Science—NAACL 2014, Venice (2014)
3. Ekman, p.: An argument for basic emotions. Cognit. Emot. **6**(3–4), 169–200 (1992)
4. Plutchik, R.: The nature of emotions. Am. Sci. **89**(4), 344–350 (2001)
5. Russell, J.A.: A circumplex model of affect. J. Personal. Soc. Psychol., 1161–1178 (1980)
6. El Gohary, A.F., Sultan, T.I. Hana, El Dosoky, M.: A computational approach for analyzing and detecting emotions in Arabic text. Int. J. Eng. Res. Appl. (IJERA) **3**(3) (2013)
7. Rabie, O., Sturm, C.: Feel the heat: emotion detection in Arabic social media content. In: Proceedings of the International Conference on Data Mining, Internet Computing, and Big Data, Kuala Lumpur (2014)
8. Apache Software Foundation: Apache Lucene Web Site. http://lucene.apache.org/core/3_0_3/api/contrib-analyzers/org/apache/lucene/analysis/ar/ArabicAnalyzer.html
9. Khoja: Pacific University, Oregon. http://zeus.cs.pacificu.edu/shereen/research.htm
10. Abdul-Mageed, M., Alhuzali, H., Abu-Elhij'a, D., Diab, M.: DINA: a multi-dialect dataset for arabic emotion analysis. In: Proceedings of the LREC 2016 2nd Workshop on Arabic Corpora and Processing Tools, Portorož, Slovenia, pp. 29–37 (2016)
11. Hussien, WA, Tashtoush, Y.M., Al-Ayyoub, M., Al-Kabi, M.N.: Are emoticons good enough to train emotion classifiers of Arabic tweets? In: 7th International Conference on Computer Science and Information Technology (CSIT), Amman, Jordan (2016)
12. Nielsen, FÅ.: A new ANEW: evaluation of a word list for sentiment analysis in microblogs. In: Proceedings of the ESWC 2011 Workshop on 'Making Sense of Microposts': Big Things Come in Small Packages, Heraklion, Crete, pp. 93–98 (2011)
13. Khalil, T., Halaby, A., Hammad, M.H., El-Beltagy, S.R.: Which configuration works best? An experimental study on supervised Arabic twitter sentiment analysis. In: Proceedings of

the First Conference on Arabic Computational Liguistics (ACLing 2015), Co-located with CICLing 2015, Cairo, Egypt, pp. 86–93 (2015)

14. El-Beltagy, Samhaa R., Khalil, T., Halaby, A., Hammad, M.: Combining lexical features and a supervised learning approach for arabic sentiment analysis. In: Gelbukh, A. (ed.) CICLing 2016. LNCS, vol. 9624, pp. 307–319. Springer, Cham (2018). https://doi.org/10.1007/978-3-319-75487-1_24

15. El-Beltagy, S.R.: NileULex: A phrase and word level sentiment Lexicon for Egyptian and modern standard Arabic. In: Proceedings of the Tenth International Conference on Language Resources and Evaluation (LREC 2016), Portorož, Slovenia, pp. 2900–2905 (2016)

16. Witten, I.H., Frank, E., Trigg, L., Hall, M., Holmes, G., Cunningham, S.J.: Weka: practical machine learning tools and techniques with java implementations. Seminar **99**, 192–196 (1999)

17. Rennie, J.D.M., Shih, L., Teevan, J., Karger, D.R.: Tackling the poor assumptions of Naïve Baye text classifiers. In: Proceedings of the Twenty-First International Conference on Machine Learning, vol. 20, pp. 616–623 (2003)

18. Saad, M.K., Ashour, W: Arabic morphological tools for text mining. In: 6th ArchEng International Symposiums, EEECS 2010 the 6th International Symposium on Electrical and Electronics Engineering and Computer Science, European University of Lefke, Cyprus, pp. 112–117 (2010)

Morphology Based Arabic Sentiment Analysis of Book Reviews

Omar El Ariss[1] and Loai M. Alnemer[2(✉)]

[1] Computer & Mathematical Sciences, The Pennsylvania State University,
Harrisburg, PA, USA
oelariss@psu.edu
[2] Computer Information Systems, The University of Jordan, Amman, Jordan
l.nemer@ju.edu.jo

Abstract. Sentiment analysis is a fundamental natural language processing task that automatically analyzes raw textual data and infer from it semantic meaning. The inferred information focuses on the author's attitude or opinion towards a written text. Although there is extensive research done on sentiment analysis on English language, there has been little work done that targets the morphologically rich structure of the Arabic language. In addition, most of the research done on Arabic either focus on introducing new datasets or new sentiment lexicons. We propose a supervised sentiment analysis approach for two tasks: positive/negative classification and positive/negative/neutral classification. We focus on the morphological structure of the Arabic language by introducing filtering, segmentation and morphological processing specifically for this language. We also manually create an emoticon sentiment lexicon in order to stress the expressed emotions and improve on the sentiment analyzer.

Keywords: Sentiment analysis · Arabic language · Morhological processing
Classification

1 Introduction

Until recently the main source of opinions were friends, relatives, books, newspapers, or websites. Now with the advent of social media such as Twitter and Facebook, electronic news websites, and specialized websites such as tripadvisor.com, hotels.com and amazon.com provide we can find useful sources of people's opinions almost on everything. Furthermore, our reliance on these resources is growing evermore where it is affecting our daily decisions. This can be seen from various studies done one the effects of online reviews [1, 2]. As a result, mining these unstructured information is an important task for many researchers and e-commerce companies.

Sentiment analysis, which falls under the general concept of opinion mining [3], is a large problem domain that detects people's opinions and attitudes about entities and services such as hotels, products, books, movies and many others [4]. In the last decade, much of the work has been done to analyze people's opinion where current sentiment analysis techniques are moving towards a better understanding of raw text. For example, Poria et al. [5] introduced sentic computing, which is a concept-level

© Springer Nature Switzerland AG 2018
A. Gelbukh (Ed.): CICLing 2017, LNCS 10762, pp. 115–128, 2018.
https://doi.org/10.1007/978-3-319-77116-8_9

sentiment analysis technique that exploits social sciences and common-sense knowledge. Through the use of dependency relation of linguistic patterns, the authors were able to improve the performance of sentiment analysis and get good results even when applied at the sentence level. On the other hand, Poria et al. [6] presented the first deep learning approach for the extraction and polarity analysis of aspects in reviews. The authors through the use of a convolutional neural network algorithm focused on the polarity of the target opinion towards a particular aspect of a service or product. They also showed that the introduction of linguistic patterns and part of speech tags further improved their results and outperformed previous work. However, sentiment analysis for the Arabic language is still not sufficiently studied because of the limited resources. In addition, the nature of the Arabic language and its morphological structure needs more manipulation and processing in order to get acceptable results.

Arabic as a language although has a standardized form varies considerably according to its intended use and its geographical location. Arabic that is used in daily communication differs from Arabic that is used in books, magazines, newspapers, or TV. While most of the Arabic text in magazines and books are the same in all the Arab world, there is no standardization for text on the internet. There is at least four different styles of Arabic found on the web, where any of these styles can be used interchangeably by the user. For example, if you look at the posts of a particular forum then you can find at least two different styles of writing. Therefore, it is important that our choice of corpus supports different styles. The most popular forms of Arabic writing are as follows:

- Classical Arabic which is the old form of the language that was used in early recorded Arabic literature such as Jahili poetry.
- Modern Standard Arabic (MSA): is a modernized version of classical Arabic with additional vocabulary, and additional symbols such as hamzah and diacritics, or tashkil, to improve the readability of the Arabic text. It is the form of language that that is taught in all Arabic speaking countries.
- Colloquial or dialectical Arabic: Although colloquial Arabic dialects are spoken languages and are not intended to be written in publications and formal transcription, more written dialectical material are appearing now thanks to the internet as a growing medium of communication. There are numerous Arabic dialects that differ considerably from each other and from the Modern Standard Arabic. Dialectical differences are not the sole product of the differences in country where many dialects exist even in one country. Other factors such as rural or urban regions play important roles in the way the dialect is formed. For example in Lebanon, dialects are different in the south, north, Beirut, and the mountains, further dialectical subdivisions can also be made.
- Arabic chat alphabet (also known as Arabizi) where the Arabic text, which can either be standard or colloquial, is written using Latin alphabet. This is very popular format, mostly with young people, to express ideas on the internet with ease without the need to have an Arabic keyboard. One main problem with this form of textual writing is the lack of standardization since each person decides how an Arabic word is written. For example three popular Latin forms of the name إدريس in Arabic is: "Idris", "Edrees" or "Idrees". In this paper we will not focus on this form of writing

since it requires further research and the need for a large corpus that supports this style of writing.

One of the early works on sentiment analysis focused on English reviews from the Internet Movie Database (IMDB) [7]. The authors treated the sentiment analysis problem as a classification one where the goal is to map a movie review to a class label, such as positive, negative or neutral. Recently, Ali and Atiya [8] introduced the largest Arabic dataset for book reviews (LABR). In their paper, the authors evaluate their dataset using different machine learning algorithms with different feature selection methods. What is interesting about this dataset, in addition to its size, is that it is not specific to a certain form or dialect. Although the majority of the reviews are written in MSA there are still some written text in colloquial dialect such as the Egyptian dialect. In addition, LABR has some reviews with diacritics and some without them. This causes a high vocabulary growth rate and as a result makes opinion mining harder to achieve. Finally, the presence of emoticons in the dataset provide useful emotional hints. All of these reasons make this dataset a perfect choice to train and evaluate our proposed system.

Synthetic languages or morphologically rich languages such as Arabic introduce some challenges to the sentiment analysis problem since a large number of word forms can match a single root form [9]. This problem is usually addressed by using a stemmer. Although stemming will definitely reduce the number of distinct words in a text, yet the difficulty lies in the effectiveness of Arabic stemmers, which still have lots of challenges to overcome, on the sentiment problem. In this paper, we introduce filtering, segmentation and morphological processing specifically for the Arabic language. We evaluate the effectiveness of our proposed technique using three different classifiers; Naïve Bayes, Bernoulli Naïve Bayes and support vector machines. The experimental results show an enhancement in the accuracy of classification of processed dataset compared to the unprocessed dataset and to approaches that use a stemmer preprocessing step. The focus here is to show that a small subset of filtering, segmentation and morphological processing on Arabic text has a determinantal effect on the sentiment analysis process. Our contribution can be summarized as follows:

- We investigate the benefits of Arabic preprocessing such as filtering diacritics and tashkil, formatting of Arabic dates, numbers and punctuations, and simple lemmatization on sentiment analysis.
- We investigate the performance of three supervised classifiers on balanced and unbalanced two class dataset and three class dataset.
- We introduce an emoticon sentiment lexicon of 63 emoticons belonging to nine emotional categories.

The rest of this paper is organized as follows. In Sect. 2 we give a survey of related work in Arabic sentiment analysis then review some of the popular Arabic corpora. We next describe the preprocessing of reviews such as emoticon handling and extraction, text filtering, word and sentence segmentation and morphological processing. In Sect. 4 we study the effectiveness of three supervised machine learning techniques on the Arabic sentiment classification problem where we present the results of the experiments. In addition, we conduct two studies to investigate the effectiveness of the preprocessing steps on sentiment analysis. First we compare our results with unprocessed data. Second we compare our result with another work that makes use of an Arabic stemmer and morphological analyzer. Finally in Sect. 5 we conclude the proposed work and suggest future directions.

2   Related Work

Although there is, and still continues, extensive research and advancements in sentiment analysis and opinion mining on the English language, there has been little research done on the Arabic language [10]. Previous work can be divided into two main groups, where the first group focused on introducing new datasets and new sentiment lexicons while the other group focuses on sentiment analysis techniques.

Al-Ayyoub et al. [11] introduce a new Arabic tweet corpus and a new sentiment lexicon for the Arabic language. The lexicon is automatically built, and contains 120 k terms. The corpus is composed of a balanced dataset of 900 tweets divided into three categories: positive, negative and neutral. The authors then use an unsupervised approach for sentiment analysis. On the other hand, Nabil et al. [12] present a corpus of Arabic tweets that can be used either for sentiment analysis of three class labels or for subjectivity analysis.

Al-Smadi et al. [13] present an annotated dataset for book reviews. Analysis on this dataset can be used for aspect extraction and polarity detection. ElSahar and El-Beltagy [14] on the other hand, propose a corpus of 33 k reviews of movies, hotels, restaurants and miscellaneous products. They also introduce a new sentiment lexicon for the proposed corpus. The reviews are of three categories: positive, negative, or neutral and the total lexicon size is around 1,900 words. The only downside about the proposed corpus is that some of the reviews are not written in Arabic and that the text has no diacritics.

SANA [15] is a subjectivity and sentiment analysis lexicon of 225 K entries for two colloquial regions Egypt and the Levantine. The entries are labelled without preprocessing, so multiple entries might refer to the same lemma, and include useful information such as part of speech, gender, and singular, or plural. The most recent lexicon for the Arabic language is proposed by Eskander and Rambow [16]. It is composed of 35 k lemmas with their part of speech. It links AraMorph with SentiWordNet, where the lexicon construction is based on semi-supervised learning and heuristic-based approaches.

Most of the work on Arabic sentiment analysis are applied to different datasets of various sizes ranging from 625 reviews [17] to 63,000 reviews [8], while targeting different Arabic dialects. The majority of the work focuses on the classification of reviews whether they are positive or negative. This can be seen from the comparison of thirty two techniques from 2009 to 2015 [10]. The two major approaches are usually supervised or corpus based approaches and lexicon or dictionary based approaches. Next, we briefly outline relevant sentiment analysis research on Arabic reviews.

One of the interesting work on Arabic sentiment [18] is the translation of reviews from Arabic into English and then the use of English sentiment analyzer to figure out the opinion of the text. The authors use two data sets, Levantine and Syrian dialects, and a support vector machine classifier as a sentiment analyzer then compare their results with Arabic sentiment analyser. Abdul-Mageed et al. [19] proposed a sentiment analysis approach at the sentence level based on a Modern Standard Arabic corpus of news articles. Since news articles can either be written subjectively/objectively, or with a positive/negative attitude, the authors proposed a two stage classification process. The first stage focuses on subjectivity classification while the second stage classifies subjective sentences into either positive or negative sentiment.

Aly and Atiya [8] proposed a corpus of 63,000 book reviews where each review is labelled with a rating value that ranges from 1 to 5. The authors then applied three classifiers: Naïve Bayes, Bernoulli Naïve Bayes and support vector machines to the problems of two class classification and five class classification. Al Shboul et al. [20] worked on the same book reviews dataset that the authors in [8] proposed. They focused on the five class classification problem where decision tree, decision table, support vector machine, k-nearest neighbour, Naïve Bayes, and an ensemble of three classifiers were used.

3   Preprocessing

Comments and reviews posted on the internet are written by people from different backgrounds, geographical locations and age range. These reviews are unstructured information and may contain various formatting and content that might drastically affect the performance of our classifiers. Therefore, it is important to preprocess and normalize the reviews and remove any unnecessary data that needs to be filtered out. Before we proceed with normalization, we need to extract all the emoticons so they are not modified by processing punctuations and as a result cannot be extracted anymore.

Our preprocessing of reviews include the following activities:

(1) Emoticon Extraction
(2) Word Normalization
     (a) Word tokenization
     (b) Word format normalization and Filtering of text
(3) Morphological Processing

3.1   Emoticon Extraction

Emoticons are textual representations of facial expressions, and are very useful emotional cues in a text. The corpus that we are using has been filtered by the original authors where Latin and any non-Arabic alphabet has been removed. The problem with having a text where all the Latin characters have been deleted is that some of the emoticons, such as X D, are altered and therefore cannot be used anymore. As a result, we were only able to construct an emoticon dictionary of 63 emoticons. We then categorize these emoticons into nine emotional categories. We then assign each emoticon that we find in a review the category label. Table 1 shows a subset of the emoticon dictionary:

Table 1. Emoticon dictionary

Emoticon	Emotional category
;-) ;) *)	Wink
:'-(:'(	Cry
:-)) :-) :)	Smiley face
=−3 =3	Laugh
:-‖ :@ >:(	Angry
: [:[:-<	Sad
>:\ >:/ :-/	Annoyed
:* (‘}{‘)	Kisses
:$	Embarrassed

3.2 Word Normalization

The first step in normalization is to tokenize the words. Here we concentrate on the tokenization of dates and numbers. Both Arabic and Indian numerals are used in the reviews so we normalize them. Next, we substitute all the Arabic punctuations, such as comma, semicolon and question mark, to its English equivalent. We also treat multiple occurrences of exclamation mark as its own feature that is separate from the occurrence of a single exclamation mark.

Next, we need to normalize the same words that are written differently, for example if we are dealing with a corpus that contains both English and American text then we need to normalize color and colour since they are referring to the same property. In Arabic this situation is very frequent due to tashkil: diacritics, hamzah and maddah. So we need to normalize all the variants of the word into one. For example, here are three common variations of the word "felt" in Arabic:

أحسسسْت   أحسسُت   أحسسِت   احسسست

We also remove multiple occurrences of the same character, which is frequently used in Arabic to stress a concept. For example, if we want to stress that the weather is very cold, then one way we can do that is by repeating one of the characters in the word *very* multiple times: verrrrrrrrrrrrry (جدددددددددددددددا). We need to be careful during this process not to remove valid repetitive characters. For example, in English we cannot remove *c* and *r* in *occurrence*.

The final step in preprocessing is to filter the text and get rid of extra symbols that do not add any impact on the sentiment process. We remove all punctuations in the text except from the ones that are useful during sentiment analysis, such as exclamation and question marks.

3.3 Morphological Processing

Arabic is a morphologically rich language and it is important to simplify the structure of the word if possible. In this study we perform a simple morphological processing, which is the removal of the definite article (ال), *the* in English. Articles in Arabic are

Table 2. Removal of the definite article

Article+Term	Term	English term
الأزرق	ازرق	Blue
الأرمله	أرمله	Widow
الأسئلة	أسئلة	Questions

treated as prefixes and are part of the word. Table 2 shows three examples of Arabic words before the removal of the article, after its removal and their English translation.

4   Experimental Setup and Evaluation

In this work, we use the LABR book reviews dataset [8], which contains over 63,000 book reviews with a rating that ranges from 1 to 5, where 1 is the lowest rating. We focus in this Section on two tasks:

(1) Sentiment classification of two classes (Positive: 1 and Negative: 0) where ratings 1 and 2 are considered as Negative class while ratings 4 and 5 are considered as Positive class. For this task we remove all the rating that have a value of 3.

(2) Sentiment classification of three classes (Positive: 1, Negative: -1, and Neutral: 0): ratings 1 and 2 are considered as a Negative class, rating 3 is considered as Neutral class and ratings 4 and 5 are considered as a Positive class.

In literature, several feature extraction methods are used for sentiment classification of text. Some of the popular features are word existence, word count, term frequency–inverse document frequency (tf-idf), and part of speech. In this work, our classifiers make use of the word count and tf-idf methods. For each classification method, we try word counts at three different levels: Uni-gram, Bi-gram and Tri-gram. We also conduct the experiments using both balanced and unbalanced datasets.

For classification, three well known and heavily used classifiers are used from the scikit-learn library in python [21]; Bernoulli Naïve Bayes (BNB), Naïve Bayes (NB) and Support Vector Machines (SVM). Our choice of these classifiers are mainly to allow us to compare our work with previous work done on this dataset. After all, we need to determine the effectiveness of the preprocessing step on Arabic sentiment analysis using previous work. We report the accuracy and the f-measure for each classifiers and feature extraction methods. The comparison with the unprocessed datasets is also reported in terms of its accuracy for both balanced and unbalanced datasets.

4.1   Sentiment Classification of Two Classes

Table 3 shows the experimental results of our proposed method for all classifiers and feature extraction methods when two classes data set is used. It can be seen from the table that the SVM classifier outperforms both Naïve Bayes and Bernoulli Naïve Bayes classifiers for all feature extraction methods. The highest accuracy achieved, highlighted in bold, for both balanced and unbalanced datasets are 88.75% and 90.98% respectively.

Table 3. The experimental results of the proposed method when applied to two class task.

	Feature	Classifier	Balanced		Unbalanced	
			Accuracy	F-measure	Accuracy	F-measure
2 classes	Count unigram	BNB	0.816709	0.897424	0.837351	0.909547
		NB	0.848252	0.916471	0.880854	0.933076
		SVM	0.86104	0.917385	0.885166	0.932011
	Count bigram	BNB	0.83035	0.907226	0.834607	0.909665
		NB	0.841858	0.913419	0.858417	0.921997
		SVM	0.865303	0.920723	0.895552	0.938651
	Count trigram	BNB	0.833333	0.909091	0.836371	0.910857
		NB	0.840153	0.912526	0.856849	0.92121
		SVM	0.864024	0.920588	0.896042	0.939124
	Tfidf unigram	BNB	0.835038	0.910105	0.837449	0.911534
		NB	0.835038	0.910105	0.838036	0.911826
		SVM	0.887042	0.934969	0.902018	0.943175
	Tfidf bigram	BNB	0.835038	0.910105	0.837449	0.911534
		NB	0.835038	0.910105	0.837547	0.911583
		SVM	**0.887468**	0.935798	**0.909759**	0.94791
	Tfidf trigram	BNB	0.835038	0.910105	0.837449	0.911534
		NB	0.835038	0.910105	0.837449	0.911534
		SVM	0.886616	0.935311	0.908975	0.947374

It is interesting to notice that bi-gram models gave the best results and that the use of tri-grams over bi-grams did not add any significant improvement. One reason is due to the morphological structure of the Arabic language that introduces larger number of new words thus making tri-gram ineffective without the use of a larger dataset.

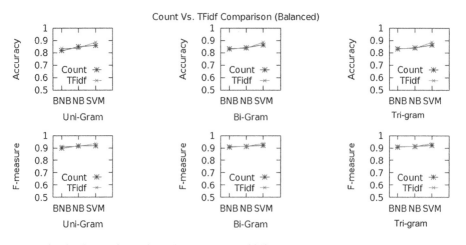

Fig. 1. Comparison of word count versus tf-idf on balanced datasets (2 classes)

Another observation we can see from Table 3, Figs. 1 and 2 is that the addition of tf-idf method showed good improvement on the results of the balanced datasets while it did not have a significant effect on the unbalanced datasets, where even in some cases the accuracy decreased with the use of tf-idf. This observation is different than that was reported in [8] which we believe it happened because of the preprocessing of the dataset.

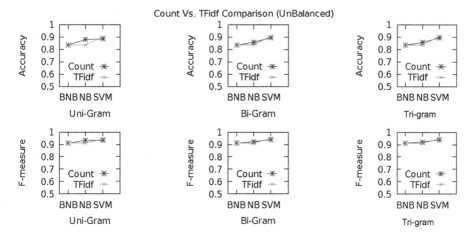

Fig. 2. Comparison of word count versus tf-idf on unbalanced datasets (2 classes)

Figure 2 shows the corresponding results of our proposed method when applied to unbalanced dataset. The observations can be seen on this figure in term of the feature extraction methods. The overall classification accuracy is better than the accuracy of the balanced dataset. This is expected because the positive examples are higher than the negative examples.

4.2   Sentiment Classification of Three Classes

Table 4 shows the corresponding results of our proposed method when three class dataset is used. It can be seen from the table below that the SVM classifier gives better results than NB and BNB classifiers. In addition, the tfidf accuracy outperforms the word count one. In both tasks the balanced dataset accuracy is lower than the unbalanced dataset which is expected due to the nature of the data. The highest accuracy achieved, highlighted in bold, for both balanced and unbalanced datasets are 80.26% and 83.57% respectively. The tri-gram model outperformed the other models although by a small margin when compared to bi-gram. Although the accuracy percentages are promising for a 3 class sentiment classification problem, the f-measure values need further improvement.

Table 4. The experimental results of the proposed approach when applied to three class task

	Feature	Classifier	Balanced		Unbalanced	
			Accuracy	F-measure	Accuracy	F-measure
3 classes	Count unigram	BNB	0.75354912	0.438044	0.774774063	0.454478
		NB	0.77149347	0.460961	0.803862672	0.520739
		SVM	0.763997729	0.515591	0.798013332	0.559702
	Count bigram	BNB	0.75786485	0.380681	0.780201829	0.345854
		NB	0.769903464	0.478196	0.79885648	0.522051
		SVM	0.772856332	0.514451	0.817721919	0.590726
	Count trigram	BNB	0.758319137	0.259537	0.781571944	0.356083
		NB	0.768313458	0.473991	0.799172661	0.527658
		SVM	0.779897785	0.524648	0.818670461	0.586726
	Tfidf unigram	BNB	0.759000568	0.259791	0.781993518	0.268179
		NB	0.759227712	0.41456	0.782204305	0.397673
		SVM	0.791027825	0.534313	0.819513609	0.581466
	Tfidf bigram	BNB	0.759000568	0.259791	0.781993518	0.268179
		NB	0.759454855	0.415121	0.782046215	0.417417
		SVM	0.799659284	0.547749	0.834268701	0.611285
	Tfidf trigram	BNB	0.759000568	0.259791	0.781993518	0.268179
		NB	0.759227712	0.41456	0.782046215	0.417417
		SVM	**0.802612152**	0.557792	**0.835691513**	0.61668

4.3   Effectiveness of Preprocessing

A comparison of the accuracy of the results after applying our preprocessing technique with the accuracy of the data without preprocessing is presented in Figs. 3 and 4 for two classes and three classes respectively. Both figures show that the results of the preprocessed dataset outperform the unprocessed dataset for both two and three class problems. In addition, we can observe that the f-measure of the 2-class task out performs the f-measure of the 3-class task. This is because if a neutral review contains more positive words than negative and neutral words, the false positive rate of the positive class will increase and the false negative rate for the neutral class will also increase which will decrease both precision and the recall and therefore the f-measure value.

Next we investigate the benefits of morphological preprocessing and compare it with Aly and Atiya's work [8], which we label as LABR. The authors of LABR used Qalsadi stemmer and morphological analyzer. The first thing that came to our mind is that stemming should give better results since Arabic has a high vocabulary growth rate due to its templatic and concatenative morphology. To our surprise, the results in Table 5 showed the opposite of that. Our proposed approach outperformed LABR in all the balanced experiments. On the other hand, we were not able to achieve the same accuracies on the unbalanced sentiment classification problem comparable to those reported in LABR. Out of the 18 unbalanced experiments, the use of our preprocessing

Average Accuracy of Processed Vs. Un-Processed (2 Classes)

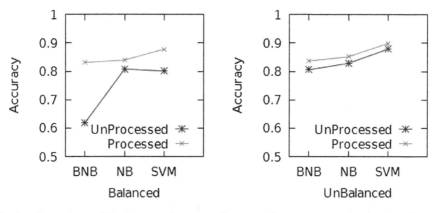

Fig. 3. Comparison of the Average Accuracy of processed vs. unprocessed method (2 classes)

Average Accuracy of Processed Vs. Un-Processed (3 Classes)

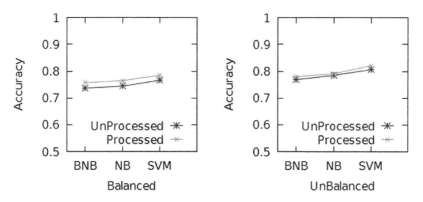

Fig. 4. Comparison of the Average Accuracy of processed vs. unprocessed method (3 classes)

techniques outperformed tokenization only six times. Although the F-measure value is not included here for simplicity of the presentation, all our experiments achieved higher values than LABR's.

Table 5. The experimental results of morphological preprocessing compared with LABR [8]

	Feature	Classifier	Balanced		Unbalanced	
			LABR	Our approach	LABR	Our approach
2 classes	Count uniGram	BNB	0.807	**0.816709**	0.889	0.837351
		NB	0.801	**0.848252**	0.887	0.880854
		SVM	0.766	**0.86104**	0.880	**0.885166**
	Count biGram	BNB	0.821	**0.83035**	0.891	0.834607
		NB	0.821	**0.841858**	0.893	0.858417
		SVM	0.789	**0.865303**	0.892	**0.895552**
	Count triGram	BNB	0.823	**0.833333**	0.886	0.836371
		NB	0.821	**0.840153**	0.889	0.856849
		SVM	0.786	**0.864024**	0.893	**0.896042**
	Tfidf uniGram	BNB	0.529	**0.835038**	0.838	0.837449
		NB	0.809	**0.835038**	0.838	**0.838036**
		SVM	0.801	**0.887042**	0.903	0.902018
	Tfidf biGram	BNB	0.513	**0.835038**	0.837	**0.837449**
		NB	0.822	**0.835038**	0.838	0.837547
		SVM	0.818	**0.887468**	**0.910**	0.909759
	Tfidf triGram	BNB	0.511	**0.835038**	0.837	**0.837449**
		NB	0.827	**0.835038**	0.838	0.837449
		SVM	0.821	**0.886616**	0.910	0.908975

5 Conclusion

In this paper, we presented a supervised sentiment analysis approach on a large Arabic book reviews dataset. The proposed approach focuses on morphological structure of the Arabic language, also an emoticon sentiment lexicon was created to stress the expressed emotions in order to improve the sentiment analyser. Three classification algorithms were used to evaluate the proposed method; Naïve Bayes, Bernoulli Naïve Bayes and Support vector machines. The experimental results showed that filtering, segmentation and morphological processing for the Arabic language enhanced the classification results when applied to both positive/negative classification task and positive/negative/neutral classification task. What is interesting is that our morphological modifications, although simple in nature, outperformed the use of Arabic stemmer. We also evaluated the effectiveness of the proposed approach on both balanced and unbalanced dataset and showed that data balancing decreases the accuracy of the classification results.

Future work will focus on two main directions. The first will concentrate on improving the performance of the three class sentiment analyzer. Although our results were acceptable since we only relied on the preprocessed words, additional features such as handling negation will definitely help. The second direction will build on our preprocessing techniques introduced here and develop a lemmatizer and stemmer for the Arabic language. We will then evaluate their effectiveness on other NLP techniques.

References

1. Dellarocas, C., Zhang, X. (Michael), Awad, N.F.: Exploring the value of online product reviews in forecasting sales: the case of motion pictures. J. Interact. Mark. **21**, 23–45 (2007)
2. Duan, W., Gu, B., Whinston, A.B.: Do online reviews matter?—An empirical investigation of panel data (2008)
3. Pang, B., Lee, L.: Opinion mining and sentiment analysis. Found. Trends®. Inf. Retr. **2**, 1–135 (2008)
4. Liu, B.: Sentiment Analysis and Opinion Mining (2012)
5. Poria, S., Cambria, E., Winterstein, G., Huang, G.: Bin: Sentic patterns: dependency-based rules for concept-level sentiment analysis. Knowl.-Based Syst. **69**, 45–63 (2014)
6. Poria, S., Cambria, E., Gelbukh, A.: Aspect extraction for opinion mining with a deep convolutional neural network. Knowl.-Based Syst. **108**, 42–49 (2016)
7. Pang, B., Lee, L., Rd, H., Jose, S.: Thumbs up? Sentiment classification using machine learning techniques. Lang. (Baltim), 79–86 (2002)
8. Aly, M., Atiya, A.: LABR: a large scale arabic book reviews dataset. In: The 51st Annual Meeting of the Association for Computational Linguistics, pp. 494–498 (2013)
9. Sarikaya, R., Kirchhoff, K., Schultz, T., Hakkani-Tur, D.: Introduction to the special issue on processing morphologically rich languages. IEEE Trans. Audio. Speech. Lang. Processing. **17**, 861–862 (2009)
10. Biltawi, M., Etaiwi, W., Tedmori, S., Hudaib, A., Awajan, A.: Sentiment classification techniques for Arabic language: a survey. In: 2016 7th International Conference on Information and Communication Systems (ICICS), pp. 339–346 (2016)
11. Al-Ayyoub, M., Essa, S.B., Alsmadi, I.: Lexicon-based sentiment analysis of Arabic tweets. Int. J. Soc. Netw. Min. (2015)
12. Nabil, M., Aly, M., Atiya, A.: ASTD: Arabic sentiment tweets dataset. In: Proceedings of the 2015 Conference on Empirical Methods in Natural Language Processing. Association for Computational Linguistics, pp. 2515–2519 (2015)
13. Al-Smadi, M., Qawasmeh, O., Talafha, B., Quwaider, M.: Human annotated Arabic dataset of book reviews for aspect based sentiment analysis. In: Proceedings of 2015 International Conference on Future Internet of Things and Cloud, FiCloud 2015 and 2015 International Conference on Open and Big Data, OBD 2015, pp. 726–730 (2015)
14. ElSahar, H., El-Beltagy, S.R.: Building large arabic multi-domain resources for sentiment analysis. In: Lecture Notes in Computer Science (including subseries Lecture Notes in Artificial Intelligence and Lecture Notes in Bioinformatics), pp. 23–34. Springer International Publishing (2015)
15. Abdul-Mageed, M., Diab, M.: SANA: a large scale multi-genre, multi-dialect Lexicon for Arabic subjectivity and sentiment analysis. In: Proceedings of the Language Resources and Evaluation Conference, pp. 1162–1169 (2014)
16. Eskander, R., Rambow, O.: SLSA: A sentiment Lexicon for standard Arabic. In: Proceedings of the 2015 Conference on Empirical Methods in Natural Language Processing. Association for Computational Linguistics, pp. 2545–2550 (2015)
17. Cherif, W., Madani, A., Kissi, M.: Towards an efficient opinion measurement in Arabic comments. In: Procedia Computer Science, pp. 122–129. Elsevier (2015)

18. Mohammad, S.M., Salameh, M., Kiritchenko, S.: How translation alters sentiment. J. Artif. Intell. Res. **55**, 95–130 (2016)
19. Abdul-Mageed, M., Diab, M.T., Korayem, M.: Subjectivity and sentiment analysis of modern standard Arabi0063. In: ACL-HLT 2011—Proceedings of 49th Annual Meet. Association for Computational Linguistics, vol. 2, pp. 587–591 (2011)
20. Al Shboul, B., Al-Ayyouby, M., Jararwehy, Y.: Multi-way sentiment classification of Arabic reviews. In: 2015 6th International Conference on Information and Communication Systems, ICICS 2015, pp. 206–211. IEEE (2015)
21. Pedregosa, F., Varoquaux, G.: Scikit-learn: machine learning in Python (2011)

Adaptation of Sentiment Analysis Techniques to Persian Language

Kia Dashtipour[1], Amir Hussain[1], and Alexander Gelbukh[2]([✉])

[1] Department of Computing Science and Mathematics, University of Stirling,
FK9 4LA Stirling, Scotland, UK
[2] CIC, Instituto Politécnico Nacional, 07738 Mexico City, Mexico
gelbukh@gelbukh.com

Abstract. In the recent years, people all around the world share their opinions about different fields with each other over Internet. Sentiment analysis techniques have been introduced to classify these rich data based on the polarity of the opinion. Sentiment analysis research has been growing rapidly; however, most of the research papers are focused on English. In this paper, we review English-based sentiment analysis approaches and discuss what adaption these approaches require to become applicable to the Persian language. The results show that approaches initially suggested for English language are competitive with those developed specifically for Persian sentiment analysis.

1 Introduction

In the recent years, web has been evolving, with the opinion of people becoming important and valuable (Cambria et al. 2017; Poria et al. 2014a). There are many unstructured data available online which can be advantageous for companies and organisation to improve their product and services; however, these data should be classified. Sentiment analysis has been used to classify these opinions into different categories (Pang and Lee 2008).

Many researchers working in the sentiment analysis field are trying to build approaches to allow computers to analyse data automatically. There are huge amounts of data available online, such as reviews, tweets, blog posts, news, etc. How to analyse these data is an interesting area for researchers, who have developed various approaches for sentiment analysis (Dashtipour et al. 2016a, 2017a, b, 2018; Adeel et al. 2019).

Sentiment analysis has been applied to different fields such as sport or news. It is also used to classify online reviews such as Twitter (Chikersal et al. 2015). Sentiment analysis became very popular in industry, because it was used to understand people's opinions towards products and services (Pang et al. 2002). Combined with author profiling (Majumder et al. 2017) and opinion extraction techniques (Poria et al. 2016), it provides solid technology underlying opinion mining applications in e-commerce and decision making.

In the recent years, much research has been done in the sentiment analysis field. The earlier research was focused on the forerunners of sentiment analysis, while later

© Springer Nature Switzerland AG 2018
A. Gelbukh (Ed.): CICLing 2017, LNCS 10762, pp. 129–140, 2018.
https://doi.org/10.1007/978-3-319-77116-8_10

research focused on an explanation of views and the evaluation of data (Dashtipour et al. 2016a). There were many works published in English-based sentiment analysis, as there were requirement to proposed approaches, tools and resources in other languages such as Persian (Kumar and Sebastian 2012).

Persian language is official language of Iran. The Persian language has more than one hundred million speakers. The main challenge for Persian sentiment analysis is lack of tools and resources (Ghassemzadeh et al. 2005).

To the best of our knowledge, there are no review papers on Persian sentiment analysis. In this paper, we discuss about current approaches in Persian sentiment analysis and about issues, challenges and available resources and tools in Persian sentiment analysis. We discuss about English sentiment analysis approaches and we implemented English approaches in using different tools and resources to make these approaches adaptable for Persian language. We discuss about our adaption to approaches and finally we evaluate the performance of these approaches using different Persian dataset.

This paper is organised as follows: Sect. 2 discusses about related work on Persian sentiment analysis, Sect. 3 discusses challenges in Persian sentiment analysis, Sect. 4 discusses English-based approaches and their adaptation, Sect. 5 gives evaluation results, and Sect. 6 concludes the paper.

2   Related Work

In this section, a brief overview about the applications of sentiment analysis, data pre-processing, available lexicon in English and Persian, features and existing approaches in Persian sentiment analysis.

2.1   Data Pre-processing

Data pre-processing is the main process in sentiment analysis for classification of text. Texts have noise, tags, scripts and advertisements. Data pre-processing can reduce the noise of texts and improve performance and accuracy of classification. Data pre-processing contains different tasks such as removing white space, stemming, tokenisation, normalisation, removing stop words, handling of negation, cleaning of text and abbreviations. These tasks are usually called transformations (Haddi et al. 2013).

- **Tokenization.** Tokenization is process of breaking the text into words, phrases, symbol or other type of meaningful elements which is called tokens. For example, "من به شام دعوت شدم" (I have been invited for dinner), it transfers into "شدم", "دعوت", "شام","به", "من" (Shukla and Kakkar 2016).
- **Normalization.** Text normalization is process of converting text into canonical form. In this stage the tokens are transferred into normal form. For example, "شام عااااالی بود" ("The dinner was greattttttttt") it becomes "شام عالی بود", ("The dinner was great") (Balahur et al. 2013).

- **Stop-word removal.** There are some words which it can be filtered before analysing data. The stop-words are most common words which it can be removed easily without effecting the performance of the sentiment analysis system. For example, "all", "the", "a", "and" etc. The list of stop-words in Persian are "و", (and), "با" (with), "به" (to), etc. (Martineau and Finin 2009).
- **Stemming.** Stemming is process of reducing inflected words and change the words to their root forms. For example, the word "stemming" will be changed into "stem" or the word "fishing" will be changed into "fish" (Savoy 2006). In Persian example, the word "کشورها" (countries) will be changed into "کشور" (country) or the word "درختان" (trees) will be changed into "درخت" (tree) (Sharifloo and Shamsfard 2008).
- **Negation.** One of the most important tasks in sentiment analysis is the control of negation. For example, the texts "I like the Rio movie" and "I do not like the Rio movie" are very similar. However, they express opposite meaning. A negation term such as "not" can change the meaning of the sentence in the above examples. Negation can be handled directly or indirectly (Zhai et al. 2010). Indirect negation can be used as a second feature and feature vector can be used as the initial presentation, whereas direct negation is encoded (Li et al. 2010; Liu 2011). For instance, assigning the word "not" to negation words or phrases such as "no" or "don't". The negation in the Persian sentences is located at the end of the sentence. For example, "من این فیلم دوست ندارم" ("I do not like this movie"). The word "ندارم" means "is not" which is located in the end of the sentence (Pang and Lee 2008).

2.2 Lexicon

Lexicons are used in many approaches to sentiment analysis to express positive and negative opinion of desired and undesired comments. Collection of terms and phrases is called lexicon. There are different lexicons available in English and Persian.

- **SentiWordNet.** One of the main lexical resources used for sentiment classification. This resource provides annotation based on three different sentiment score such as positive, negative and neutral for WordNet synset (Musto et al. 2014; Cambria et al. 2012).
- **SenticNet.** SenticNet is resources for English language which is used to recognize, interpret and process the sentiment and opinion in the English text. The SenticNet is contains more than thirty thousand concepts. The SenticNet can identify polarity, aptitude, sensitivity, pleasantness, aptitude of English concepts (Cambria et al. 2014a, 2014b; Poria et al. 2014b).
- **PerSent.** There are no well-known Persian lexicons available in this field. Thus, we have developed PerSent, a lexicon of more than thousand words along with their part-of-speech tag and their polarity. The proposed lexicon has been evaluated with two different classifiers such as SVM and Naïve Bayes and the best performance of the lexicon is 69.54% (Dashtipour et al. 2016b).

2.3   Feature Engineering

Feature selection is divided on different categories:

- **Part-of-speech tagging.** The part-of-speech tagging used in sentiment analysis and describes the type of words that can be used for disambiguation. An adjective can be effective to identify the sentiment and helpful for feature selection during sentiment analysis (Ghosh et al. 2016; Rahate and Emmanuel 2013). Very nice dinner. The POS tag, "Very" is adverb, "nice" is adjective and "dinner" is noun. For example, the Persian sentence "من به سینما رفتم" (I went to the cinema) will be POS tag as "من" is noun, "به" is determiner, "سینما" is noun and "رفتم" is verb.
- **N-grams.** The n-gram is effective feature to identify. Usually, the stop-words removed and then the unigram, bigrams are identified in the training dataset. For example, "I bought new car", the unigram is "I", "bought", "new", "car". In Persian example, "فیلم خوبی بود" (It was good movie), then it will be "فیلم", "خوبی" and "بود" (Kouloumpis et al. 2011; Xu et al. 2007).

2.4   Existing Approaches to Persian Sentiment Analysis

Various approaches exist to analyse sentiment in Persian texts.

- **Basiri et al.** (2014) proposed an approach for Persian language. The main aim of developing this approach was to use specific approach for Persian language instead of using existing English sentiment analysis approaches. The proposed approach is data pre-processing the input data, this process contains normalisation, spell correction, stemming. To evaluate the polarity of the sentence, the unigram and bigram has been selected, the SentiStrength has been used to identify the overall polarity of the sentence. The overall performance of the proposed approach is 87%. The main drawback of the proposed system is not able to detect informal words because they removed the sentences which contains informal words and phrases.
- **Dashtipour et al.** (2016b) developed a Persian lexicon of more than thousand words and phrases along with their part-of-speech tag. To evaluate the performance of the approach, the headline news dataset has been used, in the first step, the dataset has been pre-processed. The data pre-processing task contains tokenisation, normalisation, stop-words removal and stemming, then different features has been selected such frequency of positive/negative words, frequency of positive/negative adjectives, nouns, verbs and adverbs. There are two different classifiers has been used to evaluate the performance of the approach, Naïve Bayes and SVM. The SVM received better performance in compare with Naïve Bayes. The overall performance of the approach is 69.54%. The main drawback of the approach is the lexicon is very small for Persian language and it is not able to handle sarcasm.
- **Vaziripour et al.** (2016) proposed an approach for Persian twitter sentiment analysis. There are one million tweets has been collected in Persian language. There are two different Persian speakers has been invited to manually annotated the tweets. Each tweet has been annotated from 1 to 5. 1–2 is negative, 3 is neutral, 4

and 5 is positive. The stop-words has been removed and dataset has been stemmed. The SVM classifier has been used to evaluate the performance of the approach. The overall accuracy of the approach is 70%. The main drawback of the approach is there is any novelty has been proposed in this approach and most of the tools and resources has been used which has been used in English approach.

3 Challenges in Persian Sentiment Analysis

While research about Persian sentiment analysis, we understand there are various challenges in Persian sentiment analysis, in this section we will discuss these challenges.

There are different dialects in Persian language. The dialects for the Afghanistan and Iran is different, even they are used different words. For example, for the word "hospital" in Iran is "بیمارستان" but in Afghanistan is "شفاخانه". There are not any tools available to detect these dialects and because of various dialects it is difficult to build a lexicon to cover these dialects. As mentioned before, Persian is official language of Iran and Afghanistan. Even in Iran and Afghanistan, there are different dialects in south of Iran and north of Iran. These dialects have their own stop-words and negation (Windfuhr and Perry 2009).

The other challenges for Persian sentiment analysis is lack of dataset and lexicon. The Persian researchers did not address this issue and they use to translate the dataset so they able to use available lexicon in English language such as SentiWordNet and because of lack of dataset and lexicon the Persian sentiment analysis is less developed in compare with other languages (Basiri et al. 2014).

There are many sarcasm, slang and idiom expressions available in Persian language but there are not any tools in Persian language are available to address this issue. It is important to develop a tool to detect Persian slang, idiom and sarcasm because there are many sarcasms available in Persian language.

Other main challenge of Persian language is mixture of Persian and English text. For example, "WOWبود خوبی روز عجب ،". It means "WOW, what a great day". People used to comment in mixture of Persian and English, there is not any tools available in Persian to detect these words.

The other challenges of Persian sentiment analysis are use of bad words in the reviews. For example, if they did not like the movie or products, they use bad words (swear). There is an application should be developed to detect these types of words (Gholamain and Geva 1999).

4 English-Based Approaches and Their Adaptation

In this section, we will discuss about proposed approaches for English. The reason for choosing English approaches only because it is well developed and there are more tools and resources are available in English in compare with other languages.

4.1   English-Based Sentiment Analysis Approaches

Singh et al. (2013) proposed an approach which employed SentiWordNet for document-level of sentiment classification of movie and blog reviews. The Senti-WordNet approach has been used with different linguistic features and scoring schemes. The sentiment score of movie reviews and blog post reviews were calculated based on SentiWordNet. To evaluate the performance of this approach, two different datasets have been used. Dataset one contains seven hundred positive and seven hundred negative reviews, while dataset two contains one thousand positive and one thousand negative reviews. This approach used Naïve Bayes and support vector machine classifiers. The overall accuracy is 81.07%. This approach shows that using SentiWordNet does not improve performance of the classifier (Singh et al. 2013).

We used JHAZM to extract part-of-speech tag such as adjective and adverb. The JHAZM is a Java library tool, it is Persian tool is able to do different natural language processing task such as stemming, normalization, dependency parsing and part-of-speech tagging (Khallash et al. 2013). The JHAMZ is available free online to download in the following link (https://github.com/mojtaba-khallash/JHazm). They extract adjective and adverbs only and SentiWordNet has been used to assign polarity into the adjective and adverbs. We translated the SentiWordNet using Google translate into Persian and then SVM used to evaluate the performance of the combination of these two features.

Tanawongsuwan (2010) proposed a system used classify the product reviews into positive or negative. The data were collected from the Amazon website, with about 1,428 reviews being collected. After data were collected, the reviews were converted into tokens such as words, punctuation, letters, numbers, symbols, etc. Then, parts-of-speech tagging process was done on the tokens. The part-of-speech usually consist of adjectives, verbs and adverbs, this approach uses the Penn Treebank which contains more detailed part-of-speech tag set such as plural nouns, superlatives and comparative adjectives, numbers and international words. The original part-of-speech contains eleven tags; the Penn Treebank contains thirty-seven. After classifying the words using part-of-speech tagging, C4.5, Bayesian classifiers and neutral networks were used to evaluate performance. The Bayesian classifier reported an accuracy of 84.7% for the training set and 86% for the testing set (Tanawongsuwan 2010).

We did used JHAZM part-of-speech tag and there is different tag such as adjective, adverbs, noun and verb has been used. After the classify the dataset using part-of-speech tagging, we used C4.5 classifier to evaluate the performance of the approach.

Pérez-Rosas and Mihalcea (2013) proposed an approach for sentiment analysis of spoken reviews. The video reviews were collected from ExpoTV, which has a rating from 1 to 5. The rating information was used to label videos as positive, negative and neutral. Manual transcription was used for crowdsourcing and speech recognition was used for automatic transcription. The Google speech recognition was used for automatic transcription of a dataset. To evaluate the performance of the approach, support vector machine classifier was trained; the bag of words has been selected as feature, the average accuracy of classification was 75%. The weakness of this approach is using both automatic and manual transcription services, which take time. The verbal and written spoken reviews are different from each other. The verbal reviews are less informative than written reviews. This approach is used for written reviews only and cannot work consistently on the verbal reviews (Pérez-Rosas and Mihalcea 2013).

We did add "UTF-8" to the code which is available online for free. The code is available in Java.

Raina (2013) proposed an approach for sentiment analysis of news articles. This approach can categorize sentences into positive, negative and neutral. This approach employed the semantic parser, SenticNet. Sentences are passed into the semantic parser to extract features from sentences, and then sentences are sent into sentiment analyser to match the sentences with sentic vector in the SenticNet. To evaluate the performance of the approach, more than five hundred articles were collected online. The overall accuracy of the approach is 71.2%. The main weakness of this approach was that the semantic parser was not able to identify all the features and this approach requires more work to become commercially valuable (Raina 2013).

We did used JHAZM to identify the semantic parser for the sentences and translated SentiWordNet has been used to identify the polarity of extracted features.

Priyanka and Gupta (2013) proposed an approach to investigate the use of different features and to determine the best feature combination for sentiment analysis. The main components of this approach are pre-processing; feature extracting, feature combination and building classifier. In this approach, various features have been used such as n-gram and POS. The SentiWordNet has been used to calculate the positive and negative score of POS tagged sentences. For example, the positive and negative score of word "good" is 0.675 and 0.005. The following equation has been used to calculate the positive score and negative score of sentences:

$$\text{Positive Score}\,(\text{Word} = \text{Pos}) = \frac{\sum \text{Positive Score}}{\text{synsets}\,(\text{word} = \text{Pos})} \qquad (1)$$

$$\text{Negative Score}\,(\text{Word} = \text{Pos}) = \frac{\sum \text{Negative Score}}{\text{synsets}\,(\text{word} = \text{Pos})} \qquad (2)$$

The reviews of smart phones have been collected for the dataset, and it has been manually labelled. The highest accuracy obtained is 89%. However, this approach is not appropriate for large datasets (Priyanka and Gupta 2013).

We did used JHAZM part-of-speech feature such as noun, adjective, adverb and verb and also n-gram feature has been used to adapt this approach for Persian. The translated SentiWordNet has been used to assign polarity to the features.

5 Evaluation

In this section, we evaluate the performance of Persian and English approaches using different Persian dataset. Since the authors do not publish their code, we implement their approaches and we evaluate the performance of their method using different Persian dataset.

We evaluate the performance of discussed approaches in sentiment analysis using different Persian dataset. The aim was to find best approaches applied into Persian sentiment analysis and we discussed about adaption which English approaches are required to become compatible for Persian sentiment analysis and what adaptation are they required to become promising for Persian.

The Persian sentiment analysis approaches used English tools because there is not any tool and resource developed in Persian language. It is difficult and time consuming to developed Persian sentiment analysis approaches because researchers need to use English resources and to use these resources they need to translate dataset into English which it can affect the performance of the approach. We used a Persian tool called JHAZM, as we mentioned earlier it is Java library tool and contains Persian stemming, normalization, part-of-speech tagging and semantic parsing instead of English tools which it has been used in existing Persian sentiment analysis approaches such (Vaziripour et al. 2016) used Stanford part-of-tagger to identify the part-of-speech tag for sentence (Toutanova and Manning 2000), they used to translate the dataset into English and they used English tools which is time-consuming.

The main problem once the dataset has been translated into English it can cause some important part of text cannot be translated correctly especially slang and dialects.

Vaziripour et al. (2016) used twitter dataset but the dataset was not available online, we did not able to evaluate the approach with their implementation dataset.

The lesson we have learnt is some of the proposed approaches with small changes can be compatible for different languages such as Persian. However, the main challenges are lack of tools and resources. If there are enough tools and resources is available in various language some approaches which they used English dataset to report their accuracy can be compatible with various languages.

The result shows the existing English sentiment analysis with adaption which has been made to these approaches can be competitive for Persian sentiment analysis.

Dataset. To evaluate the performance of these approaches, two different datasets in Persian have been used. These datasets are reviews from different domains such as movie reviews and headline news, *First*, Persian movie review dataset which it has been collected manually from Persian movie websites such as www.caffecinema.com and cinematicket.org. The dataset contains 500 positive and 500 negative reviews. The *second* dataset is Persian Headline news for Voice of America (VOA) it contains 500 positive and 500 negative news headlines which is contains all the headline news from 2007 to 2011.

We evaluate the English approaches with different Persian dataset and we mentioned our adaptation to English approaches in order to make English approaches compatible with Persian approaches. In Table 1, we discuss about these adaptions which English approaches is required.

Table 1. Results. Lex stands for lexicon-based approach

Paper	Our adaption	Approach	Machine learning classifier	Dataset	Documented accuracy	Persian headline news	Persian movie reviews
Basiri et al. (2014)	–	Lex	SVM	Cell-phone reviews	87%	70.89%	69.76%
Dashtipour et al. (2016b)	–	Lex	SVM, Naïve Bayes	VOA Persian Headline News	69.54%	68.73%	65.24%
Vaziripour et al. (2016)	–	ML	SVM	Twitter	70%	64.41%	61.01%
Singh, et al. (2013)	SentiWordNet Translated	Lex	SVM, Naïve Bayes	Movie reviews and blog posts	79.02%.	67.85%	65.45%
Tanawongsuwan (2010)	Persian JHAZM POS Tag used	ML	Bayesian classifiers and neutral networks	Amazon Product reviews	85%	69.49%	67.34%
Pérez-Rosas and Mihalcea (2013)	Adaption "UTF_8"	SVM	SVM	Video Reviews	68%	65.85%	66.15%
Raina (2013)	Persian JHAZM used	SVM	SVM	News Articles	65%	63.49%	61.91%
Priyanka and Gupta (2013)	Persian JHAZM used	Lex	SVM	Smart phone reviews	75.33%	69.83%	68.79%

6 Conclusion

In this paper, we have reviewed current Persian sentiment analysis approaches, we discussed about tools, resources and challenges of Persian language and we review English sentiment analysis approaches and we discussed about what adaptation English approaches required to become compatible for Persian sentiment analysis.

According to our experiment, the best approach is proposed by Priyanka and Gupta (2013) which is easily compatible for Persian language and also in term of accuracy and performance received better results.

The Persian approaches used to translate their dataset to use available English tools and resources, the translated dataset is not able to detect slangs and idiom in Persian language and translation is time consuming and computationally expensive.

As future work, we evaluate the English sentiment analysis approaches with different English approaches to find best English approach which is compatible with different English dataset.

References

Adeel, A., et al.: A survey on the role of wireless sensor networks and IoT in disaster management. In: Durrani, T., Wang, W., Forbes, S. (eds.) Geological Disaster Monitoring Based on Sensor Networks, pp. 57–66. Springer Natural Hazards. Springer, Singapore (2019). https://doi.org/10.1007/978-981-13-0992-2_5

Balahur, A., et al.: Sentiment analysis in the news. ArXiv Prepr. ArXiv:13096202, pp. 2216–2220 (2013)

Basiri, M.E., Naghsh-Nilchi, A.R., Ghassem-Aghaee, N.: A framework for sentiment analysis in Persian. Open Trans. Inf. Process. 1, 14–18 (2014)

Cambria, E.: Affective computing and sentiment analysis. IEEE Intell. Syst. 31(2), 102–107 (2016)

Cambria, E., Havasi, C., Hussain, A.: SenticNet 2: a semantic and affective resource for opinion mining and sentiment analysis. In: FLAIRS Conference, pp. 202–207 (2012)

Cambria, E., Olsher, D., Rajagopal, D.: SenticNet 3: a common and common-sense knowledge base for cognition-driven sentiment analysis. In: Proceedings of the 28th AAAI Conference on Artificial Intelligence, pp. 148–152. AAAI Press (2014)

Cambria, E., Poria, S., Gelbukh, A., Kwok, K.: Sentic API. A common-sense based api for concept-level sentiment analysis. In: 4th Workshop on Making Sense of Microposts (#Microposts2014), co-located with the 23rd International World Wide Web Conference (WWW 2014). CEUR Workshop Proceedings, vol. 1141, pp. 19–24 (2014)

Cambria, E., Poria, S., Gelbukh, A., Thelwall, M.: Sentiment analysis is a big suitcase. IEEE Intell. Syst. 32(6), 74–80 (2017)

Chikersal, P., Poria, S., Cambria, E., Gelbukh, A., Siong, C.E.: Modelling public sentiment in Twitter: using linguistic patterns to enhance supervised learning. In: Gelbukh, A. (ed.) CICLing 2015. LNCS, vol. 9042, pp. 49–65. Springer, Cham (2015). https://doi.org/10.1007/978-3-319-18117-2_4

Dashtipour, K., Hussain, A., Zhou, Q., Gelbukh, A., Hawalah, A.Y.A., Cambria, E.: PerSent: a freely available Persian sentiment lexicon. In: Liu, C.-L., Hussain, A., Luo, B., Tan, K.C., Zeng, Y., Zhang, Z. (eds.) BICS 2016. LNCS (LNAI), vol. 10023, pp. 310–320. Springer, Cham (2016a). https://doi.org/10.1007/978-3-319-49685-6_28

Dashtipour, K., et al.: Multilingual sentiment analysis: state of the art and independent comparison of techniques. Cogn. Comput. **8**(4), 757–771 (2016b)

Dashtipour, K., Gogate, M., Adeel, A., Hussain, A., Alqarafi, A., Durrani, T.: A comparative study of persian sentiment analysis based on different feature combinations. In: Liang, Q., Mu, J., Jia, M., Wang, W., Feng, X., Zhang, B. (eds.) CSPS 2017. LNEE, vol. 463, pp. 2288–2294. Springer, Singapore (2017a). https://doi.org/10.1007/978-981-10-6571-2_279

Dashtipour, K., Gogate, M., Adeel, A., Algarafi, A., Howard, N., Hussain, A.: Persian named entity recognition. In: 2017 IEEE 16th International Conference on Cognitive Informatics & Cognitive Computing (ICCI* CC), pp. 79–83. IEEE (2017b)

Dashtipour, K., Gogate, M., Adeel, A., Ieracitano, C., Larijani, H., Hussain, A.: Exploiting deep learning for persian sentiment analysis. arXiv preprint arXiv:1808.05077 (2018)

Ghassemzadeh, H., Mojtabai, R., Karamghadiri, N., Ebrahimkhani, N.: Psychometric properties of a Persian-language version of the beck depression inventory-second edition: BDI-II-PERSIAN. Depress. Anxiety **21**(4), 185–192 (2005)

Gholamain, M., Geva, E.: Orthographic and cognitive factors in the concurrent development of basic reading skills in English and Persian. Lang. Learn. **49**(2), 183–217 (1999)

Ghosh, S., Ghosh, S., Das, D.: Part-of-speech tagging of code-mixed social media text. EMNLP **2016**, 90–98 (2016)

Haddi, E., Liu, X., Shi, Y.: The role of text pre-processing in sentiment analysis. Procedia Comput. Sci. **17**, 26–32 (2013)

Khallash, M., Hadian, A., Minaei-Bidgoli, B.: An empirical study on the effect of morphological and lexical features in Persian dependency parsing. In: 4th Workshop on Statistical Parsing of Morphologically Rich Languages, pp. 97–101 (2013)

Kouloumpis, E., Wilson, T., Moore, J.D.: Twitter sentiment analysis: the good the bad and the omg! Icwsm **11**(3), 538–541 (2011)

Kumar, A., Sebastian, T.M.: Sentiment analysis: a perspective on its past, present and future. Int. J. Intell. Syst. Appl. IJISA **4**, 1 (2012)

Li, S., Huang, C.-R., Zhou, G., Lee, S.Y.M.: Employing personal/impersonal views in supervised and semi-supervised sentiment classification. In: Proceedings of the 48th Annual Meeting of the Association for Computational Linguistics. Association for Computational Linguistics, pp. 414–423 (2010)

Liu, B.: Opinion mining and sentiment analysis. In: Web Data Mining. pp. 459–526. Springer (2011)

Liu, B., Zhang, L.: A survey of opinion mining and sentiment analysis. In: Mining Text Data, pp. 415–463. Springer (2012)

Majumder, N., Poria, S., Gelbukh, A., Cambria, E.: Deep learning-based document modeling for personality detection from text. IEEE Intell. Syst. **32**(2), 74–79 (2017)

Martineau, J., Finin, T.: Delta TFIDF: an improved feature space for sentiment analysis. ICWSM **9**, 106–111 (2009)

Musto, C., Semeraro, G., Polignano, M.: A comparison of Lexicon-based approaches for Sentiment Analysis of microblog posts. Inf. Filter. Retr. p. 59 (2014)

Pang, B., Lee, L.: Opinion mining and sentiment analysis. Found. Trends Inf. Retr. **2**, 128–135 (2008)

Pang, B., Lee, L., Vaithyanathan, S.: Thumbs up: sentiment classification using machine learning techniques. In: Proceedings of the ACL-02 Conference on Empirical Methods in Natural Language Processing, vol. 10. Association for Computational Linguistics, pp. 79–86 (2002)

Pérez-Rosas, V., Mihalcea, R.: Sentiment analysis of online spoken reviews. In: Inter-Speech, pp. 862–866 (2013)

Poria, S., Cambria, E., Gelbukh, A.: Aspect extraction for opinion mining with a deep convolutional neural network. Knowl.-Based Syst. **108**, 42–49 (2016)

Poria, S., Cambria, E., Ku, L.W., Gui, C., Gelbukh, A.: A rule-based approach to aspect extraction from product reviews. In: 2nd Workshop on Natural Language Processing for Social Media (SocialNLP), pp. 28–37 (2014)

Poria, S., Ofek, N., Gelbukh, A., Hussain, A., Rokach, L.: Dependency tree-based rules for concept-level aspect-based sentiment analysis. In: Presutti, V., Stankovic, M., Cambria, E., Cantador, I., Di Iorio, A., Di Noia, T., Lange, C., Reforgiato Recupero, D., Tordai, A. (eds.) SemWebEval 2014. CCIS, vol. 475, pp. 41–47. Springer, Cham (2014). https://doi.org/10.1007/978-3-319-12024-9_5

Priyanka, C., Gupta, D.: Identifying the best feature combination for sentiment analysis of customer reviews, In: 2013 International Conference on Advances in Computing, Communications and Informatics (ICACCI), pp. 102–108. IEEE (2013)

Rahate, R.S., Emmanuel, M.: Feature selection for sentiment analysis by using SVM. Int. J. Comput. Appl. **84**, 24–32 (2013)

Raina, P.: Sentiment analysis in news articles using sentic computing. In: 2013 IEEE 13th International Conference on Data Mining Workshops (ICDMW), pp. 959–962. IEEE (2013)

Savoy, J.: Light stemming approaches for the French, Portuguese, German and Hungarian languages. In: Proceedings of the 2006 ACM Symposium on Applied Computing, pp. 103–113. ACM (2006)

Singh, V.K., Piryani, R., Uddin, A., Waila, P.: Sentiment analysis of movie reviews and blog posts. In: 2013 IEEE 3rd International Advance Computing Conference (IACC), pp. 893–898. IEEE (2013)

Sharifloo, A.A., Shamsfard, M.: A bottom up approach to Persian stemming. In: IJCNLP, pp. 583–588 (2008)

Shukla, H., Kakkar, M.: Keyword extraction from educational video transcripts using NLP techniques. In: 2016 6th International Conference Cloud System and Big Data Engineering (Confluence), pp. 105–108. IEEE (2016)

Toutanova, K., Manning, C.D.: Enriching the knowledge sources used in a maximum entropy part-of-speech tagger. In: Proceedings of the 2000 Joint SIGDAT Conference on Empirical methods in Natural Language Processing and Very Large Corpora: Held in Conjunction with the 38th Annual Meeting of the Association for Computational Linguistics, vol. 13, pp. 63–70 (2000)

Tanawongsuwan, P.: Product review sentiment classification using part-of-speech. In: 2010 3rd IEEE International Conference on Computer Science and Information Technology (ICCSIT), pp. 424–427. IEEE (2010)

Vaziripour, E., Giraud-Carrier, C., Zappala, D.: Analyzing the political sentiment of Tweets in Farsi. In: 10th International AAAI Conference on Web and Social Media, pp. 669–702 (2016)

Windfuhr, G., Perry, J.R.: Persian and Tajik. The Iranian Languages, pp. 416–544 (2009)

Xu, Y., Jones, G.J., Li, J., Wang, B., Sun, C.: A study on mutual information-based feature selection for text categorization. J. Comput. Inf. Syst. **3**, 1007–1012 (2007)

Zhai, Z., Xu, H., Jia, P.: An empirical study of unsupervised sentiment classification of Chinese reviews. Tsinghua Sci. Technol. **15**, 702–708 (2010)

Verb-Mediated Composition of Attitude Relations Comprising Reader and Writer Perspective

Manfred Klenner[✉], Simon Clematide, and Don Tuggener

Institute of Computational Linguistics, Andreasstrasse 15, 8050 Zurich, Switzerland
{klenner,siclemat,tuggener}@cl.uzh.ch

Abstract. We introduce a model for attitude prediction that takes the reader and the writer perspectives into account and enables a joint reception of the attitudinal dispositions involved. For instance, a proponent of the reader might turn out to be a villain, or some moral values of his might be negatively affected. A formal model is specified that induces in a compositional, bottom-up manner informative relation tuples which indicate perspectives on attitudes. This enables the reader to focus on interesting cases, since they are directly accessible from the parts of the relation tuple.

Keywords: Attitude prediction · Sentiment implicatures
Opinion mining

1 Introduction

A sentence might express the explicit and implicit attitudes of an actor towards other actors, events or topics. From *The senator criticizes that this country supports Isis* we infer that the senator is an adversary of the addressed country and of Isis. We also conclude that the country is an advocate of Isis. Formally, these positive and negative attitudes are relations, e.g. [2] use the relations PosPair and NegPair, respectively. In the model of [6], not only the relation among the entities are captured, but also the effects on these entities. That is, Isis is – according to the writer – a beneficiary, since it actually receives support.

Clearly, not all PosPairs and NegPairs that we can retrieve from the Web on the basis of these systems are equally interesting. But what is it that makes polar facts exciting or topical? There is no prior answer to this, it depends on the reader's interests, his stances, his opinions. A reader perspective is needed. To this end, the user has to specify his profile in advance, his sympathies, stances, etc. We then might get a relation in a text collection of the following spirit: my proponent is an advocate of my opponent being a beneficiary. If we assume that Isis is an opponent of the user and if we replace *this country* by a country the user has a positive attitude towards, then we have exactly such an (in this case alarming) relation.

© Springer Nature Switzerland AG 2018
A. Gelbukh (Ed.): CICLing 2017, LNCS 10762, pp. 141–155, 2018.
https://doi.org/10.1007/978-3-319-77116-8_11

We introduce a formal model that integrates the reader and writer perspective. Verbs play a crucial role for the attitude prediction. Our model captures the interplay of truth commitment induced by verbs and event factuality. We assume the polar connotations of verbs to be shared (more or less) by the writer and reader. The reader thus is able to understand what the writer wants him to believe and to adopt. The reader however evaluates the writer perspective from his own point of view. This view includes his proponents and opponents, but also the (moral) values he shares with his peer groups. We use a polarity lexicon in the spirit of the appraisal theory [7] and an NP-level sentiment composition to identify phrases that conform with or violate values of the reader. In our model, a relation tuple is induced bottom up and it is meant to directly make overt these (sometimes antagonistic) perspectives and allow the reader to focus on interesting cases.

2   The Verb Resource

According to [2] a verb might have positive and negative effects on the filler of the direct object (patient or theme). It is, however, not only the direct object that bears a polar connotation, but also – depending on the verb – the subject (e.g. *to whitewash*), the indirect object (ditransitive *to recommend*), the PP object (*to fight for*) and the complement clause (*to criticize that*). For German, [5] have introduced a freely available verb lexicon which they call *sentiframes* (about 300 frames). It specifies for each verb, on the basis of its grammatical functions, the positive and negative effects that the affirmative (non-negated) and factual use of the verb have on the filler objects. In case that a verb subcategorizes for a complement clause, the implicature signature in the sense of [8] is specified as well. We use this resource[1] in our implementation. Note that this resource is not restricted to verbs expressing a private state of an opinion holder (sometimes called *source*) towards a target. *To survive* is clearly positive for the subject, but the subject is neither an opinion holder nor an opinion target in the classical sense (being a target of an attitude). Nevertheless, the subject is a beneficiary (i.e. has received a positive effect) and might become a target if realized as a complement clause, e.g. *I was happy that he survived*. This is to say that it makes sense to specify polar effects also for verbs that do not have opinion sources and targets. They just tell us whether an entity might be regarded as benefiting or suffering from an event. Other verbs of this class are *to profit from* and *to fall flat*.

We have specified the sources and targets of all verbs of the sentiframe resource (they are not given as part of the original release). As discussed, a source is not necessarily an opinion holder, but might just be the agent of a positive or negative event, e.g. an immoral event (*he lied*). The same applies to the target: It is either the recipient of an attitude (*he was criticized*), a positive/negative action (*he was killed*), or an experience (*he suffers*).

---

[1] Available from https://pub.cl.uzh.ch/projects/opinion/lrec_data.txt.

Formally, the roles *source* and *target* need to be qualified w.r.t. whether they bear a positive or negative effect. We thus get four roles, the superscript plus indicates a positive effect, minus indicates a negative effect:

$$\mathcal{F} = \{source^+, target^+, source^-, target^-\}$$

Let v_i denote the lemma of verb v at word position i in a given sentence and n_j a noun. The function $gf(v_i, n_j)$ returns the grammatical function of noun n_j relative to verb v_i.

We specify a verb-specific mapping, $\mathcal{P}_{role}$, from grammatical functions f to polar roles. The partial definition is given in Fig. 1.

$$\mathcal{S}_{ig}(v) =$$

$$\begin{cases} T & \text{if } v \in \{regret, \ criticize, \dots\} \\ T & \text{if } v \in \{manage, \dots\} \wedge \mathcal{A}_{f\!f}(v) = aff \\ T & \text{if } v \in \{deny, \dots\} \wedge \mathcal{A}_{f\!f}(v) = \neg aff \\ F & \text{if } v \in \{refuse, \dots\} \wedge \mathcal{A}_{f\!f}(v) = aff \\ F & \text{if } v \in \{manage, \dots\} \wedge \mathcal{A}_{f\!f}(v) = \neg aff \\ N & \text{if } v \in \{hesitate, \dots\} \wedge \mathcal{A}_{f\!f}(v) = aff \\ N & \text{if } v \in \{force, \dots\} \wedge \mathcal{A}_{f\!f}(v) = \neg aff \\ N & \text{if } v \in \{hope, fear, \dots\} \end{cases}$$

$$\mathcal{P}_{role}(v, f) =$$

$$\begin{cases} source^- & \text{if } v \in \{whitewash, \ ..\} \wedge f = subj \\ target^+ & \text{if } v \in \{whitewash, \ ..\} \wedge f = dobj \\ target^- & \text{if } v \in \{criticize, \ ..\} \wedge f = comp \\ \dots \end{cases}$$

Fig. 1. Polar roles: mapping   **Fig. 2.** Verb signatures

Given *The senator has whitewashed$_4$ an immoral affair$_7$* with $v_4 = whitewash$ and $n_7 = affair$ as its direct object (dobj), the function call $\mathcal{P}_{role}(v_4, gf(v_4, n_7))$ returns $target^+$, therefore, it is positive for the affair getting whitewashed.

3   Truth Commitment and Factuality

Verbs that subcategorize for a complement clause further receive what is called the verb's implicature signature by [8] and [3]. The implicature signature of a (matrix) verb specifies the positive or negative relative polarity of the (whole) complement clause given the affirmative status of the matrix verb. Since this usage of the term *polarity* is confusing in the context of sentiment analysis, we do not rely on it. However, the well-established term *factuality* is also inappropriate if we intend to specify implicature signatures of verbs. Factuality in the sense of [13] refers to the truth of a single event, not a whole clause as implicature signatures do. In the context of factive verbs like *regret* or *criticize*, the truth of the complement clause is implied, which does not necessarily mean that the event denoted by the verb of the complement clause is factual, it could be negated and thus counterfactual (the notion used by [13]): In *He regrets not having approved the campaign*, *approve* is counterfactual, since the complement clause is true only if *approve* is not true. Thus, factuality and verb signatures interact, but should not be confused.

We distinguish the truth commitments (T, F, N) posed by verb signatures from the factuality of events denoted by verbs. If the complement clause is implied to be true, we use T (affirmative, *manage*), F (affirmative, *forget*) if false, and N (affirmative, *refuse*) if no commitment can be made. Factive (*regret*) and non-factive (*hope*) verbs cast T and N respectively, independent of their affirmative status.

We distinguish factual (*fact*), counterfactual (*cfact*) and non-factual (*nfact*) for the factuality status. The first two follow the definition of [13], the category non-factual collapses the two values *probable* and *possible* of the certainty class of [13]. We do not need to consider these fine-grained distinctions. Take non-affirmative *force* with signature N (i.e. no commitment). If X does not force Y to lie, Y might lie anyway. While [13] would classify it as factual with certainty value *possible*, we just take it as non-factual which leaves it open whether it happens or not and what the exact epistemic value is. This is crucial. *nfact* as a factuality status blocks certain inferences. If X just hopes that Y promotes Z, then there is no relation between Y and Z we could infer, although we know that X is positive towards Z. If we replace non-factive *hope* with *approve*, then *promote* becomes factual and Y now is an advocate of Z. Still X is positive towards Z. Finally, if we negate *promote* (X *approves that Y does not promote Z*) *promote* becomes counterfactual which cancels the advocate relation, and also changes X from a proponent of Z to an opponent. Thus, all three factuality states are needed.

We follow the definition of implicature signatures from [8] except that we use T, F, N as categories as discussed above. The function $\mathcal{S}_{ig}(v)$ (see Fig. 2) specifies implicature signatures of verbs (partial definition). We do not introduce explicit verb classes, which would ease the formal definitions, but also would make it harder for the reader to follow the definitions. We rather work with illustrative (but partial) verb lists. $\mathcal{S}_{ig}(v)$ returns T (truth commited), F (falsehood commited), N (no commitment). The function $\mathcal{A}_{ff}$ returns the affirmative status of verb v, i.e. whether it is negated ($\neg aff$) or not (aff). For instance, given $v_i = deny$ and $\mathcal{A}_{ff}(v_i) = \neg aff$ then $\mathcal{S}_{ig}(v_i) = T$, indicating that the complement clause of *deny* is commited to truth.

$$\mathcal{T}_c(v) =$$

$$\begin{cases} T & if\ v^{main} = v \wedge \neg mod(v) \\ N & if\ v^{main} = v \wedge mod(v) \\ T & if\ \mathcal{T}_c(m_v(v)) = T \wedge \mathcal{S}_{ig}(m_v(v)) = T \\ F & if\ \mathcal{T}_c(m_v(v)) = T \wedge \mathcal{S}_{ig}(m_v(v)) = F \\ N & if\ \mathcal{S}_{ig}(m_v(v)) = N \wedge m_v(v) \notin \{regret, ..\} \\ T & if\ m_v(v) \in \{regret, ..\} \\ N & if\ m_v(v) \in \{hope, fear, ..\} \end{cases}$$

$$\mathcal{F}_{ac}(v) =$$

$$\begin{cases} fact & if\ \mathcal{T}_c(m_v(v)) = T \wedge \mathcal{A}_{ff}(v) = aff \\ fact & if\ \mathcal{T}_c(m_v(v)) = F \wedge \mathcal{A}_{ff}(v) = \neg aff \\ cfact & if\ \mathcal{T}_c(m_v(v)) = F \wedge \mathcal{A}_{ff}(v) = aff \\ cfact & if\ \mathcal{T}_c(m_v(v)) = T \wedge \mathcal{A}_{ff}(v) = \neg\ aff \\ nfact & otherwise \end{cases}$$

Fig. 3. Truth commitment of clauses   **Fig. 4.** Factuality determination

In a complex sentence, a verb acting as a matrix verb might itself be embedded, i.e. it might be the verb of a complement clause whose truth commitment

Table 1. R_{roles}

myValueConfirmer	myProponent	some
myValueContemner	myOpponent	
myAversions	myValues	

Table 2. W_{roles}

benefactor	villain	pos_affected	entity
beneficiary	victim	neg_affected	

needs to be taken into account. If X hopes that Y manages to win, the clause with head verb *manage* bears commitment label N (from *hope*) which absorbs the signatures of *manage* (i.e. *win* inherits N). The truth commitment of a clause needs to be determined outside-in. Figure 3 specifies this. $\mathcal{T}_c(v)$ denotes the truth commitment of a clause whose head verb is v. The main clause verb denoted as v^{main} gets T if no modality operators (function $mod(v) = false$) are present, F otherwise. Recursively, T is given if the verb is embedded into a T whose verb (the embedding verb) has signature T (row 3). The complement clauses of factive verbs like *regret* have truth commitment T in any case (row 6) while non-factive verbs like *hope* have N (row 7). Given $\mathcal{T}_c(v)$, we are able to determine event factuality, see Fig. 4.

A verb v denotes a factual (*fact*) event if the matrix verb $m_v(v)$ has truth commitment T and v is affirmative (*aff*) or if the truth commitment is F and v is negated (*forget to not inform him*). It is counterfactual (*cfact*) if F is combined with *aff* or T with ¬*aff*. Otherwise it is non-factual (*nfact*).

4 Relation Tuple Induction

Given a sentence, we extract all pairs $\langle x, y \rangle \in P$ such that x and y denote a noun position that acts as a polar role (see Fig. 1) of one or more verbs. If V is the set of verb positions of a sentence, and $\mathcal{E}$ is the set of entity positions with $x \in \mathcal{E} \wedge y \in \mathcal{E}$, then P is defined as:

$$P = \{\langle x, y \rangle | \exists r \in V : prl(x, r) \wedge \exists s \in V : prl(y, s)\}$$

The predicate $prl(x, v)$ is true if x indexes a polar role of v. We induce a polar relation tuple for each pair $\langle x, y \rangle \in P$. The general form of a relation tuple is $\langle L_r.L_w.rel.L_r.L_w \rangle$ where $L_r \in R_{roles}$ (the writer perspective roles), $L_w \in W_{roles}$ (the reader perspective roles) and $rel \in \{advocate, adversary\}$. See Tables 1 and 2 for these role sets.

L_r and L_w indicate the reader and writer (or text) perspective on the verb roles, respectively. We first introduce the reader and writer perspective before we turn to the attitude relation.

4.1 Reader Perspective

The function $\mathcal{R}_p$ (see Fig. 5) maps a word denoting an actor or an entity to a reader-specific attitude towards it (see R_{roles} from Table 1). These attitudes are based on personal preferences, political stances, moral values, etc.

myProponent and *myOpponent* indicate actors the reader likes or dislikes, be it some political party or some movie actress he finds interesting. *myValues* and *myAversion* relate to (positive or negative) concepts on the moral dimension, that is, values (*honesty*) and non-values (*lies*) that the reader more or less shares with his peer groups. This is realized by reference to our polarity lexicon which is meant to represent these common-sense polar connotations of words. It distinguishes, among others, words related to moral (judgement) from words related to emotion (affect). Sentiment composition (at the NP level only) uses this information. Finally, *myValueConfirmer* and *myValueContemner* are roles meant to capture morally positive and negative actors that are again classified (either directly or via composition) on the basis of the polarity lexicon (e.g. *honest colleague* as positive).

$$
\mathcal{R}_p(n) = \begin{cases}
myOppenent & \text{if } n \in \{Isis, \dots\} \\
myProponent & \text{if } n \in \{Red\ Cross, \dots\} \\
myValues & \text{if } pol(n) = j\_pos \wedge \neg actor(n) \\
myAversions & \text{if } pol(n) = j\_neg \wedge \neg actor(n) \\
myValueContemner & \text{if } pol(n) = j\_neg \wedge actor(n) \\
myValueConfirmer & \text{if } pol(n) = j\_pos \wedge actor(n) \\
some & otherwise
\end{cases}
$$

Fig. 5. Reader perspective

The function $pol(n)$ (not further defined in this paper) from Fig. 5 realizes rule-based sentiment composition (on the NP level) given a nominal head n. $pol(n)$ would deliver the polarity *j_neg* (judgement negative) for the phrase *immoral affair* where n would be *affair* from the sentence used earlier. $actor(n)$ is true if n is an entity that denotes a person or a group of people.

The reader perspective is independent of the factuality status, it is just a prior attitude of the reader towards actors and entities.

4.2   Writer Perspective

The writer perspective reveals to the reader what the writer wants him to believe to be true and explicates what this implies for the status of the actors and entities involved. It is the way the writer conceptualizes the world with his text and shows what the reader should adopt (which he might or might not). We use the writer perspective roles from Table 2 to characterize the status of an actor/entity. Factuality is crucial here. All roles from Table 2 except *entity* require factuality in order to get instantiated. If the event is not factual or counterfactual, *entity* is selected. For instance in *X pretends to help Y*, X is not a benefactor since *help* is counterfactual. Given *X manages to help Y*, X is a benefactor (X is the implicit subject of factual *help*).

These roles are verb-specific. *victim, villain, benefactor,* and *beneficiary* are actor roles related to the moral dimension, while *pos_affected* and *neg_affected* are used for the remaining cases. Note that in principle, very fine-grained distinctions are possible, e.g. victims of negative affect, victims of negative physical actions, victims of inhuman actions, etc. We only have to add these roles and the verbs indicative of it to the role inventory in order to cope with them.

$$
\mathcal{W}_p(r, v) : \begin{cases}
\textit{beneficiary} & \textit{if } r = target^+ \land v \in \{help, \ldots\} \land F_{ac}(v) = \textit{fact} \\
\textit{benefactor} & \textit{if } r = source^+ \land v \in \{help, \ldots\} \land F_{ac}(v) = \textit{fact} \\
\textit{villain} & \textit{if } r = source^- \land v \in \{torture, \ldots\} \land F_{ac}(v) = \textit{fact} \\
\textit{victim} & \textit{if } r = target^- \land v \in \{torture, \ldots\} \land F_{ac}(v) = \textit{fact} \\
\textit{neg\_affected} & \textit{if } r = target^- \land v \in \{criticize, \ldots\} \land F_{ac}(v) = \textit{fact} \\
\textit{pos\_affected} & \textit{if } r = target^+ \land v \in \{agree\_with, \ldots\} \land F_{ac}(v) = \textit{fact} \\
\textit{entity} & \textit{if } F_{ac}(v) = \textit{nfact} \lor F_{ac}(v) = \textit{cfact}
\end{cases}
$$

Fig. 6. Writer perspective

If *X helps Y* and Y is a negative entity, then X still is classified as a benefactor. This is intended. We could, of course, turn the benefactor into a villain given such a constellation. But the relation $\langle L_R, L_w, rel, L_r, L_w \rangle$ derived from a clause is meant to make such *odd* (and interesting) perspectives explicit in the first place and should not hide or resolve them.

Figure 6 gives the definition of the writer perspective. If the verb event is counterfactual (*cfact*), the role is neutralized (mapped to *entity*). One could argue that given counterfactuality, the roles should be inverted, i.e. *victim* to *beneficiary*. While this is appropriate sometimes, the general conditions under which it should be applied need to be further clarified.

4.3 Verb-Specific Polar Relations

The relation $\mathcal{P}_{rel}$ is verb-specific (see Fig. 7). We use *adversary* and *advocate* as roles. These roles indicate the polar relation among actors and actors/entities: negative if *adversary*, positive if *advocate*. Verb negation inverts these roles, this might not be appropriate in every case and could be made dependent on the verb.

$$
\mathcal{P}_{rel}(v) = \begin{cases}
\textit{adversary} & \textit{if } v \in \{criticize, \ldots\} \land \mathcal{F}_{ac}(v) = \textit{fact} \\
\textit{advocate} & \textit{if } v \in \{punish, \ldots\} \land \mathcal{F}_{ac}(v) = \textit{cfact} \\
\textit{advocate} & \textit{if } v \in \{approve, \ldots\} \land \mathcal{F}_{ac}(v) = \textit{fact} \\
\textit{adversary} & \textit{if } v \in \{approve, \ldots\} \land \mathcal{F}_{ac}(v) = \textit{cfact}
\end{cases}
$$

Fig. 7. Verb-specific relations of attitude

$\mathcal{P}_{rel}$ represents the verb-specific attitude that v expresses. Given a pair $\langle x, y \rangle \in P$, both might occupy polar roles of the same or of different verbs.

If x and y have different verbal heads, i.e. the verbal head of x either directly or recursively embedds a verb with y as a polar role then the relations stemming from the intermediary v_i are composed into a single relation rel. If X approves that Y criticizes Z, the relation between X and Z is that of an adversary. This depends on the advocate relation of *approve* and the adversary relation of *criticize*. To advocate that Y is an adversary of Z means to be an adversary of Z. That is, (source of) advocate combined with (target of) adversary gives adversary. In contrast, if y is the *source* of an embedded verb v instead of the target of that verb, then the polar relation of v is not needed. Take again X *approves that Y criticizes Z*. X is an advocate of Y, since X is an advocate of something Y does. That is, the relation of a source towards a source of an embedded verb is independent of the relation associated with the embedded verb, it only depends on the relation of the embedding verb (or, in case of deeper embeddings, the composition of relations originating from it).

$$
\mathcal{J}(r,s) = \begin{cases}
adversary & if\ r = advocate \wedge s = adversary \\
advocate & if\ r = adversary \wedge s = adversary \\
adversary & if\ r = adversary \wedge s = advocate \\
advocate & if\ r = advocate \wedge s = advocate \\
advocate & if\ r = advocate \wedge s = unspec \\
adversary & if\ r = adversary \wedge s = unspec
\end{cases}
$$

Fig. 8. Attitude determination

Figure 8 shows the principles behind this kind of relation composition. Note that verb negation is applied prior to a function call to $\mathcal{J}$. For instance, given that X approves that Y does not criticize Z, we first invert the relation adversary of *criticize* to advocate and then apply $\mathcal{J}\ (advocate, advocate) = advocate$ in order to infer that X is an advocate of Z. *unspec* is the identity element (see below).

Now that we have talked about relation composition, we define the function $\mathcal{P}_{rel}^+$ (see Fig. 9) used to determine the attitude relation given a pair $\langle x, y \rangle$, where x occupies a polar role of v, and y might either have the same verb v or some verb directly or recursively dependent on v. The function $h_v(x)$ returns the verb v whose polar role is occupied by x. If x and y depend on the same verb (row 1), $\mathcal{P}_{rel}(v)$ is called, i.e. the relation expressed by the verb v is returned (see Fig. 7 for the definition). The function *role* returns the type the polar role y bears given its verb – either *source* or *target*. If the type of polar role of y is source, then *unspec* is returned, since, as discussed, the relation of the verb does not play any role in the determination of the relation among x and y in this case. If the verb of y is different from v (row 3), then the relation of the current verb v is combined (by applying $\mathcal{J}$) with the result of a recursive call on $\mathcal{P}_{rel}^+$ with the verb embedded by v (denoted by $cmp_v(v)$).

$$\mathcal{P}_{rel}^{+}(x, v, y) =$$

$$\begin{cases} \mathcal{P}_{rel}(v) & \text{if } h_v(x) = h_v(y) = v \wedge role(h_v(y), y) = target \\ unspec & \text{if } h_v(y) = v \wedge role(h_v(y), y) = source \\ \mathcal{J}(\mathcal{P}_{rel}(v), \mathcal{P}_{rel}^{+}(x, cmp_v(v), y)) & \text{if } h_v(y) \neq v \end{cases}$$

Fig. 9. Attitude determination

4.4 A Complete Example

In the implementation of our model, we use a dependency parser [14] and a rule-based extractor to identify the fillers of the grammatical functions given a sentence. We also take control verbs, passive voice, etc. into account and apply rules for coreference resolution.

Take the sentence: *The right-wing politician$_3$ criticized$_4$ that the EU$_7$ helps$_8$ the refugees$_{10}$.* We get three pairs: v_4:$\langle x_3, y_7 \rangle$, v_4:$\langle x_3, y_{10} \rangle$ and v_8:$\langle x_7, y_{10} \rangle$. Let's say the reader has no prior attitudes towards right-wing politicians but refugees have his sympathy (are *myProponents* of his). We discuss the case of v_4:$\langle x_3, y_{10} \rangle$, i.e. the directed polar relation of the *right-wing politician* towards the *refugees*. We call: $\langle \mathcal{R}_p(3), \mathcal{W}_p(3), \mathcal{P}_{rel}^{+}(3, 4, 10), \mathcal{R}_p(10), \mathcal{W}_p(10) \rangle$ which returns: $\langle$some, entity, adversary, myProponent, beneficiary$\rangle$.

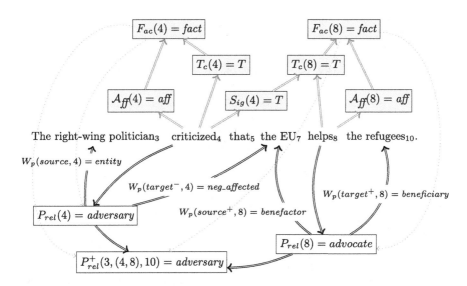

Fig. 10. Example analysis (omitting the a priori reader perspective). The part above the sentence shows the connections between the predicates that determine affirmative status, factuality, and the verb signatures. The part below the sentence depicts the predicates that determine the relations between the entities and the assigned roles. The dotted arrows between the upper and lower part indicate that the factuality determination licenses the relations found below.

Figure 10 shows a graph of the predicates in the analysis. We can paraphrase this a bit more explicitly as *some entity is an adversary of my proponent being a beneficiary*. Note that *beneficiary* as a role comes from factual *help*. This is the writer or text perspective. It tells us that the *refugees*, our proponents, are beneficiaries of some event that happened in reality. The relation tells us also that *some entity* is an adversary of this, that is, he does not approve the status of our proponents, the refugees, as beneficiaries. This immediately makes him a candidate of the class of actors that are opponents.

4.5   Notable Relation Types

We here give a couple of relations that might indicate interesting cases, see Table 3. The third column illustrates the underlying cases; US and Germany are set to be proponents of the reader (for short: *myProp*). In 1, two proponents are (surprisingly) adversaries. In 2, someone disapproves what the reader disapproves (a new proponent?). In 3, someone approves what the reader disapproves (a new opponent?). In 4, a proponent acts in a way the reader finds morally questionable (no longer a proponent?), and in 5, someone might turn out to be an opponent, since he violates the reader's values.

Table 3. Charged Relation Tuples

#	Relation Tuple
1	⟨myProp, entity, adversary, myProp, neg_affected⟩ e.g. *US refuses Germany something*
2	⟨some, entity, is, adversary, of, myAversions, neg_affected⟩ e.g. *someone condemns terror*
3	⟨some, entity, is, advocate, of, myAversions, pos_affected⟩ e.g. *someone insists on vengeance*
4	⟨myProp, benefactor, advocate, myValContemner, beneficiary⟩ e.g. *US supports dictator*
5	⟨some, villain, adversary, myValues, neg_affected ⟩ e.g. *someone ridicules human behaviour*

5   Empirical Evaluation

We carried out experiments with different data sets. Our model was developed independently from these data. The first data set consists of 80 (complex) made-up sentences (two and more subclause embeddings) and 80 real sentences. Our goal was to verify the generative capacity of our model, thus the made-up sentences. Two annotators specified advocate and adversary relations. We harmonized the annotations afterwards to get a gold standard. Our goal was to see how our lexicon including the principles of factuality determination affects the performance. Precision was 83.5% and recall was 75.2%, which gives an F measure of 79.1%. We then dropped the verb signatures from the lexicon. That is, we replaced the individual signatures by a default setting. There are three possible settings. We set the signature for the affirmative use of the verbs to 'T' (truth commitment), the signature for negated cases was set to 'F', 'N' and 'T' in turn. We got a precision of 69.06%, 75.36% and 74.88% and a recall of 69.36%, 71.62% and.75.23%. The F measure for the best setting (T,T) is 75.06% which is about 4% worse than the system's best result, 79.1%. This demonstrates that verb-specific information is crucial.

Table 4. Evaluation results on 198 extracted items independently rated by annotator A and B (Prec = Precision)

	System	A	Prec	B	Prec	A+B	Prec	Agreement
Advocate	84	64	0.76	65	0.77	61	0.73	0.95
Adversary	114	88	0.77	87	0.76	84	0.74	0.95
neg_affected	74	58	0.78	56	0.76	52	0.70	0.90
pos_affected	66	44	0.67	43	0.65	38	0.58	0.86
villain	30	24	0.80	21	0.70	21	0.70	0.88
victim	34	25	0.74	23	0.68	22	0.65	0.88
beneficiary	14	8	0.57	6	0.43	6	0.43	0.75
benefactor	34	15	0.44	16	0.47	12	0.35	0.80
myValues	20	20	1.00	19	0.95	19	0.95	0.95
myAversions	32	29	0.91	30	0.94	29	0.91	1.00
myValueConfirmer	14	11	0.79	12	0.86	10	0.71	0.91
myValueContemner	22	22	1.00	19	0.86	19	0.86	0.86

The basis of our second experiment are 3.5 million sentences taken from the German periodicals ZEIT and Spiegel. The selection criterion was that a sentence contained a verb from the sentiframe lexicon. While the first experiment was meant to prove the benefits of our lexical resource (wrt. recall and precision), the second experiment is precision oriented – we were interested in a detailed error analysis, not only for advocate and adversary relations but also for the others (e.g. victim).

We identified the most frequent named entities in these sentences in order to define *myProponent* and *myOpponent* for our experiments. We took the most frequent political parties and nations and categorized them according to the axis left- versus right-wing and west versus east as proponents and opponents of our virtual reader. We then ran our system with the restriction that only relations that carry at least one of the six core reader perspective roles (e.g. *myProponent*), i.e. not counting the neutral *entity*, are retrieved. This way, 276 relation tuple types (16,297 tokens) were induced (out of $\langle 6 * 7 * 2 * 6 * 7 \rangle = 3528$ possible types).

From the output of our system, we randomly sampled 2 relation tuples from each relation type that occurred at least 14 times, resulting in a evaluation set with 198 relations (see Table 4). Two raters independently checked whether the extracted relations were correct. The evaluation task was presented to the raters in the form of standardized verbalized textual entailments, which they could accept or reject. No false negatives had to be added, and we therefore only assess the precision of the system. For the relations *advocate* and *adversary*, the system reaches about 74%. The categories of the writer's perspective show varying results, between 35% (*benefactor*) and 70% (*villain* and *neg_affected*). The classification of the reader's perspective was in part easier for the system because of the predefined lists of named entities for *myProponent* and *myOpponent*, reaching a precision of 100% (thus omitted from the table). However, the detection of *myValueConfirmer* and *myValueContemner*, which relies on sentiment composition based on the polarity lexicon, can go wrong. An error analysis on the 198 extracted *advocate* and *adversary* relations shows that syntactic problems, that is parsing errors due to OCR errors, sentence segmentation errors, complex sentences, or predicate-argument structure extraction errors, trigger 49% of all 92 wrong relations (aggregated from both raters). Semantic errors (meaning of verbs and negation interpretation) produce 33% of the errors, factuality errors occur in 18% of all cases.

We consider the precision of our system sufficiently high for a practical extraction task, however, our evaluation says nothing about its coverage and recall.

6   Related Work

[11] determine the polarity of an event as a function of the polarity of the argument of the verb denoting the event. The polarity of an event is not what we focus on. We are interested in the various ways the participants of an event, the writer and the reader relate to each other. [11] also have no means to model positive and negative effects on objects that are neutral. Factuality is not considered at all. Work in the spirit of [11] for German is presented in [12].

The appraisal theory described in [7] classifies sentiment and opinions along the distinction between judgement (moral), affect (emotion), and appreciation (aesthetic dimension). [9] rely on this theory. They introduce a rule-based approach to sentiment inferences based on verb classes. They distinguish a subject's inner emotion from external, judgement-related attitudes. For example, *He admires a mafia leader* gives *POS affect* internally and *NEG judgement* externally. The reader is not explicitly modelled in their approach, also factuality does not play a role.

Another rule-based approach to sentiment implicatures (their term) is described in [1] and [2]. The goal is to detect entities that are in a positive (PosPair) or negative (NegPair) relation to each other. The rules are realized in the framework of Probabilistic Soft Logic, where the rule weights depend on the output of the preprocessing pipeline made out of two SVM classifiers and three existing sentiment analysis systems. The model of [2] also copes with event-level sentiment inference, however factuality is not taken into account, also the reader is not modelled.

Recently, [10] have presented an elaborate model that is meant to explicate the relations between all involved entities: the reader, the writer, and the entities referred to by a sentence. Also, the internal states of the referents and their values are part of the model. The underlying resource, called connotation frames, was created in a crowdsourcing experiment, the model parameter (e.g. values for positive and negative scores) are average values. The authors use belief propagation to induce the connotation frames of unseen verbs; they also use the connotation frames to predict entity polarities. This was applied to analyse the vocabulary of Democrats and Republicans. [10] only have two attitude relations, whereas our richer set of induced relations not only models positive and negative attitudes, but also gives some insight into the attitudinal dispositions behind them.

How Description Logics can be used to identify so-called polarity conflicts is described in [4], but attitudes are not part of that model. [6] stress the point that factuality determination is a crucial part of sentiment inferences. They introduce a rule-based system for German realized with Description Logic and SWRL, where polar roles are divided into so-called of-roles and for-roles of verbs. The rules also are taking the affirmative and factuality status of the sentence into account. The goal is to instantiate relations (con and pro) expressing the attitudes of entities towards each other. We agree that factuality is a crucial part of such a model. However, we use a tripartite distinction while their factuality labels are binary. Our aim is to have relations for attitudes that are much more fine-grained than just con and pro, indicating a reader and a writer perspective on both, the source and the target. We also utilize a polarity lexicon and sentiment composition with categories of the appraisal theory in order to define roles like myValues, which models the moral dimension of the reader. Finally, our model is not a rule-based approach, we introduce a formal description which does not presuppose a particular implementation.

7   Conclusions

We have introduced a model that integrates various notions considered relevant in the field of sentiment implicatures, a field that focuses on the explicit and implicit positive or negative attitudes between actors, but has recently started to also distinguish various perspectives: e.g. those of the reader and the writer. However, so far only plain polarities (positive or negative) are used to qualify these perspectives. In our model, more fine-grained distinctions are available: (moral) values of the reader, value confirmers, etc. Rather than to define a set of labels on the basis of predefined rules, we induce complex attitude relations bottom-up – this is also novel. Our relation tuples jointly encode the reader and the writer perspective, as well as the attitude between the source and target expressed by the verb. Such a relation expresses what the writer wants the reader to believe and how the reader given his stances, moral values, etc. might perceive this. This enables the reader to search for interesting constellations, where e.g. his proponent acts in astonishing (e.g. odd) ways.

We have integrated NP-level sentiment composition, a polarity lexicon along the lines of the appraisal theory, truth commitment, negation and factuality into a formal model. An empirical evaluation showed good precision.

References

1. Deng, L., Wiebe, J.: Sentiment propagation via implicature constraints. In: Meeting of the European Chapter of the Association for Computational Linguistics (EACL-2014) (2014)
2. Deng, L., Wiebe, J.: Joint prediction for entity/event-level sentiment analysis using probabilistic soft logic models. In: Proceedings of the 2015 Conference on Empirical Methods in Natural Language Processing, EMNLP 2015, Lisbon, Portugal, September 17–21, 2015, pp. 179–189 (2015)
3. Karttunen, L.: Simple and phrasal implicatives. In: Proceedings of the First Joint Conference on Lexical and Computational Semantics, SemEval 2012, pp. 124–131 (2012)
4. Klenner, M.: Verb-centered sentiment inference with description logics. In: 6th Workshop on Computational Approaches to Subjectivity, Sentiment & Social Media Analysis (WASSA), September 2015, pp. 134–140 (2015)
5. Klenner, M., Amsler, M.: Sentiframes: a resource for verb-centered German sentiment inference. In: Calzolari, N. (Conference Chair), Choukri, K., Declerck, T., Goggi, S., Grobelnik, M., Maegaard, B., Mariani, J., Mazo, H., Moreno, A., Odijk, J., Piperidis, S. (eds.) Proceedings of the Tenth International Conference on Language Resources and Evaluation (LREC 2016), Paris, France, May 2016. European Language Resources Association (ELRA) (2016)
6. Klenner, M., Clematide, S.: How factuality determines sentiment inferences. In: Titov, I., Gardent, C., Bernardi, R. (eds.) Proceedings of *SEM 2016: The Fith Joint Conference on Lexical and Computational Semantics, August 2016 (2016)
7. Martin, J.R., White, P.R.R.: Appraisal in English. Palgrave Macmillan, London, England (2005)
8. Nairn, R., Condoravdi, C., Karttunen, L.: Computing relative polarity for textual inference. In: Proceedings of Inference in Computational Semantics (ICoS 5), Buxton, England, pp. 67–75 (2006)
9. Neviarouskaya, Alena, Prendinger, Helmut, Ishizuka, Mitsuru: Semantically distinct verb classes involved in sentiment analysis. IADIS AC **1**, 27–35 (2009)
10. Rashkin, H., Singh, S., Choi, Y.: Connotation frames: a data-driven investigation. In: Proceedings of the 54th Annual Meeting of the Association for Computational Linguistics (ACL), Berlin, Germany, August 2016 (2016)
11. Reschke, K., Anand, P.: Extracting contextual evaluativity. In: Proceedings of the Ninth International Conference on Computational Semantics, pp. 370–374 (2011)
12. Ruppenhofer, J., Brandes, J.: Effect functors for opinion inference. In: Calzolari, N. (Conference Chair), Choukri, K., Declerck, T., Goggi, S., Grobelnik, M., Maegaard, B., Mariani, J., Mazo, H., Moreno, A., Odijk, J., Piperidis, S. (eds.) Proceedings of the Tenth International Conference on Language Resources and Evaluation (LREC 2016), Paris, France, May 2016. European Language Resources Association (ELRA) (2016)

13. Saurí, Roser, Pustejovsky, James: FactBank: a corpus annotated with event factuality. Lang. Resour. Eval. **43**(3), 227–268 (2009)
14. Sennrich, R., Schneider, G., Volk, M., Warin, M.: A new hybrid dependency parser for German. In: Proceedings of the German Society for Computational Linguistics and Language Technology, pp. 115–124 (2009)

Customer Churn Prediction Using Sentiment Analysis and Text Classification of VOC

Yiou Wang[✉], Koji Satake, Takeshi Onishi, and Hiroshi Masuichi

Fuji Xerox Co., Ltd, Yokohama, Japan
{Yiou.Wang,Koji.Satake,Takeshi.Onishi,Hiroshi.Masuichi}@fujixerox.co.jp

Abstract. In this paper, we explore the utility of sentiment analysis and text classification of voice of the customer (VOC) for improving churn prediction, which is a task to detect customers who are about to quit. Our work is motivated by the observation that *the increase of customer satisfaction will reproduce churn and the customer satisfaction can be reflected in some degree by applying NLP techniques on VOC*, the unstructured textual information which captures a view of customer's attitude and feedbacks. To the best of our knowledge, this is the first work that introduces text classification of VOC to churn prediction task. Experiments show that adding VOC analysis into a conventional churn prediction model results in a significant increase in predictive performance.

1 Introduction

It costs more than five times as much to acquire a customer than to retain a customer (Kotler and Keller 2006). Therefore, customers, who switch to competitors or move out from service providers, become critical concerns for companies. Churn prevention through churn prediction is one of the methods to ensure customer loyalty. An improvement of 5% in customer retention leads to an increase of 25% to 85% in profits (Li and Green 2011; Kerin et al. 2009; Reichheld et al. 1990). Existing research has focused on using structured data for churn prediction (Kusuma et al. 2013; Adwan et al. 2014; Huang et al. 2015; Li et al. 2016).

On the other hand, customers voice opinions and advices about some brands, companies, products or services. Today voice of the customer (VOC), capturing a view of customer's behaviors, needs, and feedbacks, can be obtained through center calls, emails, questionnaire, web reviews or SNS (Subramaniam et al. 2009; Aguwa et al. 2012; Choi et al. 2013; Saeed et al. 2013). Such unstructured textual information contains valuable information for marketing analysis. Therefore, it is attractive to consider exploiting VOC analysis in churn prediction. However, few research has been conducted in this direction and this textual information is often neglected. One exception is an investigative study showing

© Springer Nature Switzerland AG 2018
A. Gelbukh (Ed.): CICLing 2017, LNCS 10762, pp. 156–165, 2018.
https://doi.org/10.1007/978-3-319-77116-8_12

that adding the information of call center emails resulted in an increase in predictive performance (Krist 2008). They used a weighted term-by-email matrix to represent a collection of emails and used Latent Semantic Indexing (LSI) via Singular Value Decomposition (SVD) to reduce the matrix to k dimension in order to overcome disadvantages of large and sparse matrix. However, it is not possible to know what value of k will lead to an optimal solution in different business situations. They made great effort to determine the critical k for their task. Their techniques require specialized pre-processing and dimension reduction steps and are difficult to be adapted to other business situations and are limited to document-level VOC like emails.

In this paper, we present a simple and easy-to-implement approach to improve the performance of churn prediction by incorporating VOC analysis into a conventional churn prediction model. Specifically we explore the utility of sentiment analysis and text classification of VOC. We first identify sentiment polarities of VOC, categorize VOC into types, calculate VOC churn classification scores and then generate new features by these three kinds of information and train new models. We evaluate the usefulness of our approach in a series of experiments and demonstrate that VOC analysis provides substantial performance gains in churn prediction.

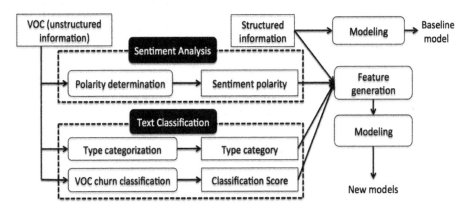

Fig. 1. Proposed method

2   The Baseline Churn Prediction Model

The churn models that exploit traditional predictors, such as demographic information (gender, birthday, career, home address, etc.), contractual details (subscribed services, etc.), customer levels (VIP or non-VIP), usage facts (customer behaviors, usage, payment staturs) or other service-related information, have been extensively studied (Kusuma et al. 2013). The same as other conventional churn models, we introduced demographic information, usage facts, other service-related information and loyalty card details as our baseline features. Our baseline features are listed as follows:

- Demographic information: age and gender.
- Usage facts: monthly, quarterly, half-yearly and yearly usage facts; customer internal comparison of quarterly, half-yearly and yearly usage facts which are used to show change and change rate of usage of a customer
- Loyalty card details: types of members, accumulation points.

In other words, our baseline features are based on the structured information. We train our baseline prediction model on random forest method, which is introduced by (Breiman 2001). It is demonstrated to be one of the most effective classification algorithms (Fernandez-Delgado et al. 2014) and is shown to perform very well compared to many other classifiers in handling imbalanced data classifications and churn prediction tasks (Xie et al. 2009; Dror et al. 2012).

3   Proposed Method

In this section, we describe our approach of effectively integrating useful information from VOC into the above baseline models through features. As we know VOC is different from profile information, which is structured information. The core problem is to extract useful information from the unstructured VOC. We first preprocess VOC by the following three steps: (i) determine the sentiment polarity of VOC; (ii) categorize VOC into types; (iii) make the model of VOC churn classification and calculate the churn classification score. We then get the sentiment polarity of VOC, type category of VOC and VOC churn classification scores from the above preprocessing. Finally, we incorporate new features by these three kinds of information into the above baseline churn prediction model and train new models. Figure 1 shows an overview of our approach. The rest of this section describes our features in detail.

3.1   Features of Sentiment Polarity

In this subsection we explore the utilization of sentiment analysis in churn prediction. A customer is likely to quit if he is unsatisfied with a service or a product. The sentiment polarity of VOC, which shows if VOC is a praise or a complaint of a service or a product, may indicate the customer satisfaction in some degree.

In order to increase the customer satisfaction, VOC Which were collected from the questionnaire in previous years were annotated manually with sentiment information and category information. For the sentiment information, trained annotators judged whether a VOC is a positive or negative opinion of a service or a product and tagged each VOC with sentiment polarity tag. There are three kinds of tags: positive $(+)$, negative $(-)$ and neutral (0). There are 32,740 VOC in the annotated corpus. In our task, we predicted the sentiment polarity of the new VOC using one-versus-rest multi-class linear kernel support vector machines (SVMs) with the features in Table 1 and built classification models on the annotated data. Here w_i, b_i, t_i, p_i and n denote the word surface form, word base form (a form of word stem), POS tag and polarity dictionary information of the i-th word, and the numbers of words in VOC, respectively. For the words in VOC, if the word surface form are matched with the polarity dictionary, p_i is the polarity tag in the dictionary. The classification model achieved 89.6% in accuracy result.

Table 1. Feature templates for sentiment analysis

Type	Feature	Description
Unigram	$w_1, b_1, t_1, ... w_i, b_i, t_i, ... w_n, b_n, t_n$	For the words in VOC, word surface form, word base form and the POS tag are added as unigram features
Bigram	$w_1 \& w_2, b_1 \& b_2, t_1 \& t_2, ...$ $w_i \& w_{i+1}, b_i \& b_{i+1},$ $t_i \& t_{i+1}, ... w_{n-1} \& w_n,$ $b_{n-1} \& b_n, t_{n-1} \& t_n$	For the words in VOC, word surface form bigram, word base form bigram and POS tag bigram features are added
Polarity dictionary	$p_1, ... p_i, ... p_n$	The polarity dictionary information of words in VOC

We added the type (a) features in Table 2 to encode the sentiment polarity information for each customer.

Table 2. Features of sentiment polarity

Type	Features	Description
(a) Sentiment polarity	$N_+/(N_+ + N_- + N_0)$	Ratio of VOC with + tag to all VOC
	$N_-/(N_+ + N_- + N_0)$	Ratio of VOC with − tag to all VOC
	$N_0/(N_+ + N_- + N_0)$	Ratio of VOC with 0 tag (neutral) to all VOC
	$R_+/(R_+ + R_- + R_0)$	Ratio of latest VOC with + tag to all tags
	$R_-/(R_+ + R_- + R_0)$	Ratio of latest VOC with − tag to all tags
	$R_0/(R_+ + R_- + R_0)$	Ratio of latest VOC with 0 tag to all tags

Here, $N_+/N_-/N_0$ is the numbers of VOC with positive/negative/neutral sentiment polarity tags. $R_+/R_-/R_0$ shows if latest VOC contain positive/negative/neutral ones. Note that one customer may have multiple VOC data and one VOC may have multiple sentiment polarity tags. We here used the last one VOC of the customers. We also tried to present the feature to show the VOC in the recent one month. Sine one customer do not have many VOC in our case, most customers do not VOC in the recent one month. In the case that customers

Table 3. Features of type category

Type	Features	Description
(b) Type category	T_i	Number of VOC in each type category i
	T_{i+}	Number of VOC in each type category i with + tag
	T_{i-}	Number of VOC in each type category i with − tag

express VOC frequently, more complicated feature representations of sequential information are necessary to show the tendency of customer sentiment.

3.2 Features of Type Category

The idea is motivated by the observation that the impact of different VOC type on customer satisfaction is different. For example, in the situation of restaurant, although a customer complained about the connection of Wi-Fi, if he was very satisfied with the food, to view the situation as a whole, he may be satisfied with the restaurant.

There are various kinds of VOC such as praises or complaints about staffs, foods, network services, environmental sanitation and so on. The types of VOC are divided into 27 categories according to evaluation targets of VOC. In detail, there are 7 categories for related staffs, 12 categories for services, 4 categories for environment or settings, 4 categories for related principles or policies. The same as sentiment analysis, the trained annotators categorized old VOC data into 27 categories (There are 32,740 VOC in the annotated corpus.) and classification models were built on the annotated data using one-versus-rest multi-class linear kernel SVMs and provided 86% in accuracy. The final feature setting is shown in Table 3.

3.3 Features of Classification Score

In this subsection, we introduce VOC churn classification models and encode classification scores of VOC as features in customer churn prediction.

The VOC from a churn customer is supposed to be a churn VOC and the VOC from a loyal customer is supposed to be a loyal VOC. In this way, we can obtain a corpus for VOC churn classification. We classified VOC into churn or loyal using polynomial kernel support vector machine (SVM). We used the features shown in Table 4 for VOC churn classification. Here w_i, b_i and n denote the word surface form, word base form (a form of word stem) of the i-th word, and the numbers of words in VOC, respectively.

We then divided the corpus into ten equal-sized sets. For each set, we used the remaining nine sets to train a VOC churn classification model and used this model to generate VOC churn classification scores for VOC from this set. In this way, for each VOC, a churn classification score was provided with the VOC classification model in this cross-validation-like techniques (Collins 2002;

Table 4. Feature templates for VOC churn classification

Type	Feature	Description
Unigram	$w_1, b_1, ...w_i, b_i, ...w_n, b_n,$	For the words in VOC, word surface form and word base form are added as unigram features
Bigram	$w_1\&w_2, b_1\&b_2, ... w_i\&w_{i+1},$ $b_i\&b_{i+1}, ...w_{n-1}\&w_n, b_{n-1}\&b_n$	For the words in VOC, word surface form bigram and word base form bigram features are added

Martins et al. 2008; Wang et al. 2011). One customer may have multiple VOC data. We added the type (c) features in Table 5 to encode the classification score information for a customer.

Table 5. Features of classification score

Type	Features	Description
(c) Classification score	$Max(S_1..S_n)$	Maximum classification score of all VOC
	$Min(S_1..S_n)$	Minimum classification score of all VOC
	$Mean(S_1..S_n)$	Mean of classification scores of all VOC
	$Meadian(S_1..S_n)$	Median of classification scores of all VOC
	$Mean(S_1..S_r)$	Mean of classification scores of all latest VOC

4   Experiments

4.1   Data Set

In this paper, we chose the customers who answered the web questionnaire as our prediction targets. VOC are collected from web questionnaires. One questionnaire, which generally contains 1 or 2 sentences in Japanese, is considered to be one VOC. We use the comparison of usage information of FY2014 and FY2013 to judge whether a customer is churn or loyal (See Fig. 2). A customer whose utilization frequencies dramatically drops is prone to churn. In particular, we define $t = 2/3$. Table 6 provides the statistics of the customers and VOC numbers.

4.2   Experimental Results

We evaluated the effectiveness of new features in a series of experiments. We used recall (R), precision (P), F and the area under the ROC curve (AUC)

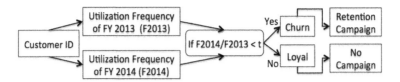

Fig. 2. Customer churn definition

Table 6. Statistics of customers and VOC data

	Number of customers	Number of VOC
Churn	1,770	4,878
Loyal	3,519	10,644
All	5,289	15,522

as evaluation metrics. Table 7 shows the final results for all experiments of our random forest models for customer churn prediction. Here P, R and F are the results with the threshold of *churn probability* = 0.5. Precision-recall curves of all the experiments are shown in Fig. 3.

Table 7. The results of churn prediction model

Methods	R	P	F	AUC
Baseline	0.662	0.447	0.534	67.4%
+ (a) Sentiment polarity	0.653	0.450	0.533	67.2%
+ (b) Type category	0.672	0.449	0.538	67.4%
+ (c) Classification score	**0.686**	**0.487**	**0.570**	**71.8%**
+ (a) +(b) +(c)	0.678	0.473	0.558	71.0%

The results of Table 7 show that sentiment polarity features and type category features did not provide additional increase in churn predictive performance. That is to say, we can not find significant relationship between sentiment polarity and customer churn. Classification score features were very effective. The AUC increased from 67.4% to 71.8% by adding classification score features. This improvement of 4.4 in AUC point is significant ($p < 0.001$). The combination of all features can not provide further improvement, since the former two kinds of features are not effective. The details of feature effect will be discussed in Sect. 5. Finally we only added the classification score features to baseline features as our final feature set. The final results show that adding VOC analysis into a conventional churn prediction model results in a significant increase in predictive performance.

5 Discussion

5.1 The Impact of Sentiment Polarity and Type Category on Churn

We analyzed the VOC data with positive and negative opinions and summarized the statistics in Table 8. Such statistics contradict our expectation. Loyal customers express more negative opinions than churn customers. This may be because Japanese customers are prone to make evaluations in a polite way, even they are not satisfied with the services or products.

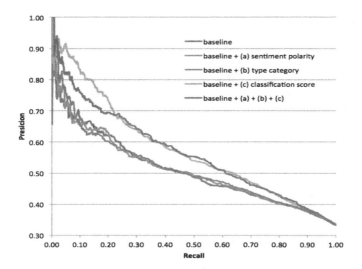

Fig. 3. Precision-recall curves

Table 8. Statistics of positive and negative VOC

	Positive VOC	Negative VOC
Churn customers	55%	45%
Loyal customers	41%	59%

We can not encode the real satisfaction information by only using the limited classification information of sentiment polarity and opinion type. We found that degrees and strengths of praises and complaints are very important for churn prediction. For example, *"the food is really amazing"* indicates a very strong positive opinion. *"the food is good"* indicates a weak positive attitude. The customer expressed VOC like the former example is likely to be a loyal customer.

5.2 The Impact of VOC Classification Score on Churn Prediction

Classification scores provide more information than binary or multi-labeled classification and are in some degree able to represent the strength of sentiment. By comparing differences of prediction results of our baseline model and the model with classification score features, the effect of VOC can be summarized in the following two directions:

(i) The churn predictions of the customers, who voiced the opinions with strong attitude or in an extreme way, achieved an improvement in performance. Such as the following VOC examples:

1. *The whole experience there was <u>extremely good</u> and we <u>really appreciate</u> it.*
2. *The staff made a <u>fatal mistake</u> and did not apologize to us.*

(ii) For the customers who have multiple VOC data, the sequential and dynamic customer satisfaction of services or products was encoded by the VOC features and provided an improvement in prediction performance, such as the following VOC examples from the same loyal customer.

1. *The food was too cold. (Jan 2013)*
2. *The dessert was very good and we ordered many for takeout. (Aug 2013)*
3. *The food was hot and very delicious. (Dec 2013)*

6   Conclusion

The main contributions of this paper are as below:

- We presented an easy-to-implement approach to improve the performance of churn prediction by incorporating VOC analysis.
- The impact of sentiment analysis and text classification of VOC on customer churn prediction was investigated.
- We found no significant relationship between the sentiment polarity of someone's VOC and his churn behavior.
- VOC classification score features provided substantial improvement over the baseline method.

This study aims to provoke increased consideration of the application of NLP techniques in churn prediction tasks.

References

Martins, A.F.T., Das, D., Smith, N.A., Xing, E.P.: Stacking dependency parsers. In: Proceedings of EMNLP-2008, pp. 513–521 (2008)

Aguwa, C.C., Monplaisir, L., Turgut, O.: Voice of the customer: customer satisfaction ratio based analysis. Expert Syst. Appl. **39**(11), 10112–10119 (2012)

Choi, C.-H., Lee, J.-E., Park, G.-S., Na, J., Wan-Sup, C.: Voice of customer analysis for internet shopping malls. Int. J. Smart Home **7**(5), 291–304 (2013)

Dror, G., Pelleg, D., Rokhlenko, O., Szpektor, I.: Churn prediction in new users of Yahoo! answers. In: Proceedings of the 21st International Conference Companion on World Wide Web (2012)

Huang, Y., Zhu, F., Yuan, M., Deng, K., Li, Y., Ni, B.: Telco Churn prediction with big data. In: ACM SIGMOD International Conference, pp. 607–618 (2015)

Coussement, K., Van den Poel, D.: Integrating the voice of customers through call center emails into a decision support system for churn prediction. Inf. Manag. **45**(3), 164–174 (2008)

Fernandez-Delgado, M., Cernadas, E., Barro, S., Amorim, D.: Do we need hundreds of classifiers to solve real world classification problems. J. Mach. Learn. Res. **15**, 3133–3181 (2014)

Breiman, L.: Random forests. Mach. Learn. **45**, 5–32 (2001)

Li, H., Yang, D., Yang, L., Yao, L., Lin, X.: Supervised massive data analysis for telecommunication customer churn prediction. In: IEEE International Conferences on Big Data and Cloud Computing, pp. 163–169 (2016)

Subramaniam, L.V., Faruquie, T.A., Ikbal, S., Godbole, S., Mohania, M.K.: Business intelligence from voice of customer. In: Proceedings of IEEE International Conference on Data Engineering (2009)

Li, M.-L., Green, R.D.: A mediating influence on customer loyalty: the role of perceived value. J. Manag. Market. Res. **7**, 1–12 (2011)

Collins, M.: Ranking algorithms for named-entity extraction: boosting and the voted perceptron. In: Proceedings of ACL-2002, pp. 489–496 (2002)

Adwan, O., Faris, H., Jaradat, K., Harfoushi, O., Ghatasheh, N.: Predicting customer churn in telecom industry using multilayer preceptron neural networks: modeling and analysis. Life Sci. J. **11**(3), 75–81 (2014)

Kerin, R.A., Hartley, S.W., Rudelius, W.: Marketing, 9th edn. McGraw-Hill Irwin, Boston (2009)

Kusuma, P.D., Radosavljevik, D., Takes, F.W.: Combining customer attribute and social network mining for prepaid mobile churn prediction. In: Proceedings of BENE-LEARN (2013)

Kotler, P., Keller, K.L.: Marketing Management. Prentice Hall, Upper Saddle River, NJ (2006)

Saeed, R., Lodhi, R.N., Munir, J., Riaz, S., Dustgeer, F., Sami, A.: The impact of voice of customer on new product development. World Appl. Sci. J. **24**(9), 1255–1260 (2013)

Reichheld, F.F., Sasser Jr., W.E.: Zero defections. Quality comes to services. Harv. Bus. Rev. **68**(5), 105–111 (1990)

Xie, Y., Li, X., Ngai, E.W.T., Ying, W.: Customer churn prediction using improved balanced random forests. Expert Syst. Appl. **36**, 5445–5449 (2009)

Wang, Y., Kazama, J., Tsuruoka, Y., Chen, W., Zhang, Y., Torisawa, K.: Improving Chinese word segmentation and POS tagging with semi-supervised methods using large auto-analyzed data. In: Proceedings of IJCNLP-2011, pp. 309–317 (2011)

Benchmarking Multimodal Sentiment Analysis

Erik Cambria[1(✉)], Devamanyu Hazarika[2], Soujanya Poria[3], Amir Hussain[4], and R. B. V. Subramanyam[2]

[1] School of Computer Science and Engineering, NTU, Singapore
cambria@ntu.edu.sg
[2] National Institute of Technology, Warangal, India
[3] Temasek Laboratories, NTU, Singapore
[4] School of Natural Sciences, University of Stirling, Stirling, UK

Abstract. We propose a deep-learning-based framework for multimodal sentiment analysis and emotion recognition. In particular, we leverage on the power of convolutional neural networks to obtain a performance improvement of 10% over the state of the art by combining visual, text and audio features. We also discuss some major issues frequently ignored in multimodal sentiment analysis research, e.g., role of speaker-independent models, importance of different modalities, and generalizability. The framework illustrates the different facets of analysis to be considered while performing multimodal sentiment analysis and, hence, serves as a new benchmark for future research in this emerging field.

Keywords: Multimodal sentiment analysis · Emotion detection
Deep learning · Convolutional neural networks

1 Introduction

Emotion recognition and sentiment analysis have become a new trend in social media analytics because of the immense opportunities they offer in terms of understanding preferences and habits of users and their contents [1]. With the advancement of communication technology, abundance of smartphones and the rapid rise of social media, a larger and larger amount of data is being uploaded in video, rather than text, format [2]. For example, consumers tend to record their reviews and opinions on products using a web camera and upload them on social media platforms such as YouTube or Facebook to inform subscribers of their views. Such videos often contain comparisons of products from competing brands, pros and cons of product specifications, and other information that can aid prospective buyers to make informed decisions.

The primary advantage of analyzing videos over mere text analysis for detecting emotions and sentiment from opinions is the surplus of behavioral cues. Video provides multimodal data in terms of vocal and visual modalities. The vocal

© Springer Nature Switzerland AG 2018
A. Gelbukh (Ed.): CICLing 2017, LNCS 10762, pp. 166–179, 2018.
https://doi.org/10.1007/978-3-319-77116-8_13

modulations and facial expressions in the visual data, along with text data, provide important cues to better identify true affective states of the opinion holder. Thus, a combination of text and video data helps to create a better emotion and sentiment analysis model.

Recently, a number of approaches to multimodal sentiment analysis producing interesting results have been proposed [3–7]. However, there are major issues that remain unaddressed in this field, such as the role of speaker-dependent and speaker-independent models, the impact of each modality across datasets, and generalization ability of a multimodal sentiment classifier. Not tackling these issues has presented difficulties in effective comparison of different multimodal sentiment analysis methods. In this paper, we address some of these issues and, in particular, propose a novel framework that outperforms the state of the art on benchmark datasets by more than 10%. We use a deep convolutional neural network (CNN) to extract features from visual and text modalities.

The paper is organized as follows: Sect. 2 provides a brief literature review on multimodal sentiment analysis; Sect. 3 presents the proposed framework; experimental results and discussion are given in Sect. 4; Sect. 5 proposes a qualitative analysis; finally, Sect. 6 concludes the paper.

2 Related Work

Text-based sentiment analysis systems can be broadly categorized into knowledge-based and statistics-based systems [8]. While the use of knowledge bases was initially more popular for the identification of emotions and polarity in text [9,10], sentiment analysis researchers have recently been using statistics-based approaches, with a special focus on supervised statistical methods [11–13].

In 1970, Ekman et al. [14] carried out extensive studies on facial expressions. Their research showed that universal facial expressions are able to provide sufficient clues to detect emotions. Recent studies on speech-based emotion analysis [15] have focused on identifying relevant acoustic features, such as fundamental frequency (pitch), intensity of utterance, bandwidth, and duration.

As to fusing audio and visual modalities for emotion recognition, two of the early works were done by De Silva et al. [16] and Chen et al. [17]. Both works showed that a bimodal system yielded a higher accuracy than any unimodal system. More recent research on audio-visual fusion for emotion recognition has been conducted at either feature level [18] or decision level [19].

While there are many research papers on audio-visual fusion for emotion recognition, only a few research works have been devoted to multimodal emotion or sentiment analysis using text clues along with visual and audio modalities. Wollmer et al. [4] and Rozgic et al. [20] fused information from audio, visual and text modalities to extract emotion and sentiment. Metallinou et al. [21] and Eyben et al. [22] fused audio and text modalities for emotion recognition. Both approaches relied on feature-level fusion. Wu et al. [23] fused audio and textual clues at decision level.

In this paper, we propose CNN-based framework for feature extraction from visual and text modality and a method for fusing them for multimodal sentiment analysis. In addition, we study the behavior of our method in the aspects rarely addressed by other authors, such as speaker independence, generalizability of the models and performance of individual modalities.

3   Method

3.1   Textual Features

For feature extraction from textual data, we used a CNN. The trained CNN features were then fed into a support vector machine (SVM) for classification, i.e., we used CNN as trainable feature extractor and SVM as a classifier (Fig. 1).

The idea behind convolution is to take the dot product of a vector of k weights w_k, known as kernel vector, with each k-gram in the sentence $s(t)$ to obtain another sequence of features $c(t) = (c_1(t), c_2(t), \ldots, c_L(t))$:

$$c_j = w_k^T \cdot \mathbf{x}_{i:i+k-1}. \tag{1}$$

We then apply a max pooling operation over the feature map and take the maximum value $\hat{c}(t) = \max\{\mathbf{c}(t)\}$ as the feature corresponding to this particular kernel vector. We used varying kernel vectors and window sizes to obtain multiple features.

For each word $x_i(t)$ in the vocabulary, a d-dimensional vector representation, called word embedding, was given in a look-up table that had been learned from the data [24]. The vector representation of a sentence was a concatenation of the vectors for individual words. The convolution kernels are then applied to word vectors instead of individual words. Similarly, one can have look-up tables for features other than words if these features are deemed helpful.

We used these features to train higher layers of the CNN to represent bigger groups of words in sentences. We denote the feature learned at a hidden neuron h in layer l as F_h^l. Multiple features are learned in parallel at the same CNN layer. The features learned at each layer are used to train the next layer:

$$F^l = \sum_{h=1}^{n_h} w_k^h * F^{l-1}, \tag{2}$$

where $*$ denotes convolution, w_k is a weight kernel for hidden neuron h and n_h is the total number of hidden neurons. The CNN sentence model preserves the order of words by adopting convolution kernels of gradually increasing sizes, which span an increasing number of words and ultimately the entire sentence.

Each word in a sentence was represented using word embeddings. We employed the publicly available word2vec vectors, which were trained on 100 billion words from Google News. The vectors were of dimensionality $d = 300$, trained using the continuous bag-of-words architecture [24]. Words not present in the set of pre-trained words were initialized randomly.

Each sentence was wrapped to a window of 50 words. Our CNN had two convolution layers. A kernel size of 3 and 4, each of them having 50 feature maps was used in the first convolution layer and a kernel size 2 and 100 feature maps in the second one. We used ReLU as the non-linear activation function of the network. The convolution layers were interleaved with pooling layers of dimension 2. We used the activation values of the 500-dimensional fully-connected layer of the network as our feature vector in the final fusion process.

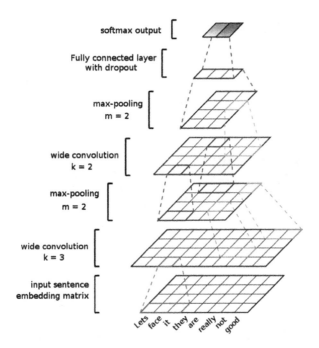

Fig. 1. CNN for feature extraction from text modality.

3.2   Audio Features

We automatically extracted audio features from each annotated segment of the videos. Audio features were also extracted in 30 Hz frame-rate; we used a sliding window of 100 ms. To compute the features, we used the open-source software openSMILE [25]. This toolkit automatically extracts pitch and voice intensity. Voice normalization was performed and voice intensity was thresholded to identify samples with and without voice. Z-standardization was used to perform voice normalization.

The features extracted by openSMILE consist of several low-level descriptors (LLD) and their statistical functionals. Some of the functionals are amplitude mean, arithmetic mean, root quadratic mean, etc. Taking into account all functionals of each LLD, we obtained 6373 features.

3.3   Visual Features

Since the video data is very large, we only considered every tenth frame in our training videos. The constrained local model (CLM) was used to find the outline of the face in each frame [26]. The cropped frame size was further reduced by scaling down to a lower resolution, thus creating our new frames for the video. In this way, we could drastically reduce the amount of training video data. The frames were then passed through a CNN architecture similar to Fig. 1.

Neuron with Highly Activated Features of Forehead and Mouth

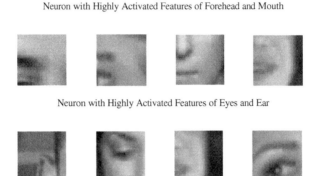

Neuron with Highly Activated Features of Eyes and Ear

Fig. 2. Top image segments activated at two feature detectors in the first layer of a deep CNN

To capture the temporal dependence of the images constituting the video, we transformed each pair of consecutive images at t and $t+1$ into a single image and provided this transformed image as input to the multilevel CNN. We used kernels of varying dimensions to learn Layer-1 2D features (shown in Fig. 2) from the transformed input. Similarly, the second layer also used kernels of varying dimensions to learn 2D features. The down-sampling layer transformed features of different kernel sizes into uniform 2D features and was then followed by a logistic layer of neurons.

Pre-processing involved scaling all video frames to half of their resolution. Each pair of consecutive video frames were converted into a single frame to achieve temporal convolution features. All frames were standardized to 250×500 pixels by padding with zeros.

The first convolution layer contained 100 kernels of size 10×20; the next convolution layer had 100 kernels of size 20×30; this layer was followed by a logistic layer of fully connected 300 neurons and a softmax layer. The convolution layers were interleaved with pooling layers of dimension 2×2. The activation of the neurons in the logistic layer were taken as the video features for the classification task.

3.4   Fusion

In order to fuse the information extracted from each modality, we concatenated feature vectors extracted from each modality and sent the combined vector to a SVM for the final decision. This scheme of fusion is called feature-level fusion.

Since the fusion involved concatenation and no overlapping merge or combination, scaling and normalization of the features were avoided. We discuss the results of this fusion in Sect. 4. The overall architecture of the proposed method can be seen in Fig. 3.

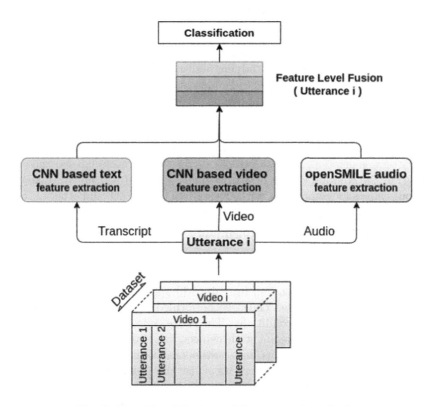

Fig. 3. Overall architecture of the proposed method.

4   Experiments and Observations

4.1   Datasets

Multimodal Sentiment Analysis Datasets. For our experiments, we used the MOUD dataset, developed by Perez-Rosas et al. [3]. They collected 80 product review and recommendation videos from YouTube. Each video was segmented into its utterances (498 in total) and each of these was categorized by a sentiment label (positive, negative and neutral). On average, each video has 6 utterances and each utterance is 5 s long. In our experiment, we did not consider neutral labels, which led to the final dataset consisting of 448 utterances. We dropped the neutral label to maintain consistency with previous work. In a similar fashion, Zadeh et al. [27] constructed a multimodal sentiment analysis dataset called multimodal opinion-level sentiment intensity (MOSI), which is bigger than MOUD, consisting of 2199 opinionated utterances, 93 videos by 89 speakers. The videos address a large array of topics, such as movies, books, and products. In the experiment to address the generalizability issues, we trained a model on MOSI and tested on MOUD.

Multimodal Emotion Recognition Dataset. The IEMOCAP database [28] was collected for the purpose of studying multimodal expressive dyadic interactions. This dataset contains 12 h of video data split into 5 min of dyadic interaction between professional male and female actors. Each interaction session was split into spoken utterances. At least 3 annotators assigned to each utterance one emotion category: *happy, sad, neutral, angry, surprised, excited, frustration, disgust, fear* and *other*. In this work, we considered only the utterances with majority agreement (i.e., at least two out of three annotators labeled the same emotion) in the emotion classes of *angry, happy, sad*, and *neutral*. We take only these four classes for comparison with the state of the art [29] and other authors.

4.2   Speaker-Independent Experiment

Most of the research in multimodal sentiment analysis is performed on datasets with speaker overlap in train and test splits. Given this overlap, however, results do not scale to true generalization. In real-world applications, the model should be robust to person variance. Thus, we performed person-independent experiments to emulate unseen conditions. This time, our train/test splits of the datasets were completely disjoint with respect to speakers. While testing, our models had to classify emotions and sentiments from utterances by speakers they have never seen before. Below, we enlist the procedure of this speaker-independent experiment:

- **IEMOCAP:** As this dataset contains 10 speakers, we performed a 10-fold speaker-independent test, where in each round, one of the speaker was in the test set. The same SVM model was used as before and macro F-score was used as a metric.
- **MOUD:** This dataset contains videos of about 80 people reviewing various products in Spanish. Each utterance in the video has been labeled as *positive, negative* or *neutral*. In our experiments, we consider only *positive* and *negative* sentiment labels. The speakers were divided into 5 groups and a 5-fold person-independent experiment was run, where in every fold one out of the five group was in the test set. Finally, we took average of the macro F-score to summarize the results (Table 1).
- **MOSI:** The MOSI dataset is a dataset rich in sentimental expressions where 93 people review topics in English. The videos are segmented with each segment's sentiment label scored between +3 to −3 by 5 annotators. We took the average of these labels as the sentiment polarity and, hence, considered only two classes (*positive* and *negative*). Like MOUD, speakers were divided into

5 groups and a 5-fold person-independent experiment was run. During each fold, about 75 people were in the training set and the remaining in the test set. The training set was further split randomly into 80%–20% and shuffled to generate train and validation splits for parameter tuning.

Comparison with the Speaker-Dependent Experiment. In comparison with the speaker-dependent experiment, the speaker-independent experiment performs poorly. This is due to the lack of knowledge about speakers in the dataset. Table 2 shows the performance obtained in the speaker-dependent experiment. It can be seen that audio modality consistently performs better than visual modality in both MOSI and IEMOCAP datasets. The text modality plays the most important role in both emotion recognition and sentiment analysis. The fusion of the modalities shows more impact for emotion recognition than for sentiment analysis. Root mean square error (RMSE) and TP-rate of the experiments using different modalities on IEMOCAP and MOSI datasets are shown in Fig. 4.

Table 1. Speaker-Independent: Macro F-score reported for speaker-independent classification. *IEMOCAP:* 10-fold speaker-independent average. *MOUD:* 5-fold speaker-independent average. *MOSI:* 5-fold speaker-independent average. *Legenda:* A stands for Audio, V for Video, T for Text.

Modality	Source	IEMOCAP	MOUD	MOSI
Unimodal	A	51.52	53.70	57.14
	V	41.79	47.68	58.46
	T	65.13	48.40	75.16
Bimodal	T + A	70.79	57.10	75.72
	T + V	68.55	49.22	75.06
	A + V	52.15	62.88	62.4
Multimodal	T + A + V	**71.59**	**67.90**	**76.66**

4.3 Contributions of the Modalities

As expected, bimodal and trimodal models have performed better than unimodal models in all experiments. Overall, audio modality has performed better than visual on all datasets. Except for MOUD dataset, the unimodal performance of text modality is notably better than other two modalities (Fig. 5). Table 2 also presents the comparison with state of the art. The present method outperformed the state of the art by 12% and 5% on the IEMOCAP and MOSI datasets, respectively.[1] The method proposed by Poria et al. is similar to ours, except for the fact they used a standard CLM-based facial feature extraction method. Hence, our proposed CNN-based visual feature extraction algorithm has helped to outperform the method by Poria et al.

---

[1] We have reimplemented the method by Poria et al. [5].

Table 2. Speaker-Dependent: Ten-fold cross-validation results on IEMOCAP dataset and 5-fold CV results (macro F-score) on MOSI dataset.

Modality	Source	IEMOCAP	MOSI
Unimodal	Audio	66.20	64.00
	Video	60.30	62.11
	Text	67.90	78.00
Bimodal	Text + Audio	78.20	76.60
	Text + Video	76.30	78.80
	Audio + Video	73.90	66.65
Multimodal	Text + Audio + Video	**81.70**	**78.80**
	Text + Audio + Video	**69.35**[a]	**73.55**[b]

[a]By [29]; [b]By [5]

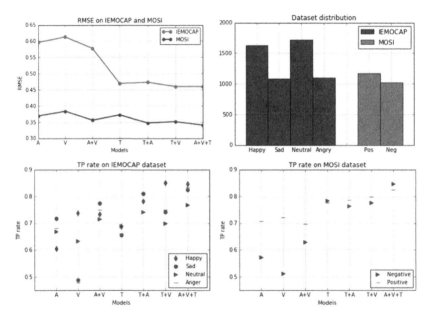

Fig. 4. Experiments on IEMOCAP and MOSI datasets. The top-left figure shows the RMSE of the models on IEMOCAP and MOSI. The top-right figure shows the dataset distribution. Bottom-left and bottom-right figures present TP-rate on of the models on IEMOCAP and MOSI dataset, respectively.

4.4   Generalizability of the Models

To test the generalization ability of the models, we have trained the framework on MOSI dataset in speaker-independent fashion and tested on MOUD dataset. From Table 3, we can see that the trained model on MOSI dataset performed poorly on MOUD dataset.

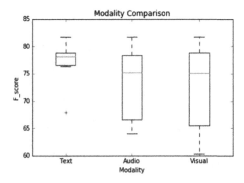

Fig. 5. Performance of the modalities on the datasets. Red line indicates the median of the F-score.

Table 3. Cross-dataset results: Model (with previous configurations) trained on MOSI dataset and tested on MOUD dataset.

Modality	Source	Macro F-score
Unimodal	Audio	41.60 %
	Video	45.50 %
	Text	50.89 %
Bimodal	Text + Audio	51.70 %
	Text + Video	52.12 %
	Audio + Video	46.35 %
Multimodal	Text + Audio + Video	52.44 %

This is mainly due to the fact that reviews in MOUD dataset had been recorded in Spanish so both audio and text modalities miserably fail in recognition, as MOSI dataset contains reviews in English. A more comprehensive study would be to perform generalizability tests on datasets in the same language. However, we were unable to do this for the lack of benchmark datasets. Also, similar experiments of cross-dataset generalization was not performed on emotion detection given the availability of only a single dataset (IEMOCAP).

4.5   Visualization of the Datasets

MOSI visualizations present information regarding dataset distribution within single and multiple modalities (Fig. 6). For the textual and audio modalities, comprehensive clustering can be seen with substantial overlap. However, this problem is reduced in the video and all modalities with structured declustering but overlap is reduced only in multimodal. This forms an intuitive explanation of the improved performance in the multimodality. IEMOCAP visualizations provide insight for the 4-class distribution for uni and multimodals, where clearly the

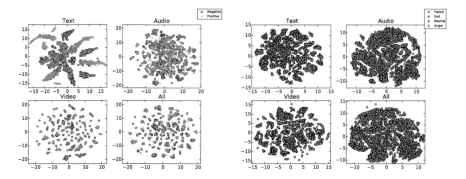

Fig. 6. T-SNE 2D visualization of MOSI and IEMOCAP datasets when unimodal features and multimodal features are used.

multimodal distribution has the least overlap (increase in red and blue visuals, apart from the rest) with sparse distribution aiding the classification process.

5   Qualitative Analysis

In order to have a better understanding of roles of modalities for the overall classification, we performed a qualitative analysis. Here, we show the cases where our model successfully comprehends the semantics of the utterances and, with aid from the multiple media, correctly classifies the emotion of the same.

While overviewing the correctly classified utterances in the validation set, we found out that text modality often helped the classification of utterances where visual and audio cues were flat with less variance. In such situations, the model gathered information from the language semantics extracted by the text modality. For example, in an utterance from the MOSI dataset "amazing special effects", there was no jest of enthusiasm in speaker's voice and face audio-visual classifier, which caused failure to identify the positivity of this utterance by the audio and video unimodal classifiers. The text classifier, instead, correctly detected the polarity as positive (given the presence of highly polar words) and, hence, helped the bimodal and multimodal classifiers to correctly classify the utterance.

The text modality also helped in situations where the face of the reviewer was not visible (which happens quite often in product reviews). Even in cases where the text modality led to a misclassification (e.g., due to the presence of misleading linguistic cues), the overall classification was correct thanks to the audio and video inputs. For example, the text classifier classified the sentence "that like to see comic book characters treated responsibly" as positive (possibly because of the presence of positive phrases such as "like to see" and "responsibly"); however, the high pitch of anger in the person's voice and the frowning face helps to identify this as a negative utterance.

The above examples demonstrate the effectiveness and robustness of our model to capture overall video semantics of the utterances for emotion and sentiment detection. They also show how bimodal and multimodal models overcome the limitations of unimodal networks, given the multiple media input.

We also explored the misclassified validation utterances and found some interesting trends. Most videos consist of a group of utterances that have contextual dependencies among them. Thus, our model failed to classify utterances whose emotional polarity was highly dependent on the context described in an earlier or later part of the video. The modeling of such an inter-dependence, however, was out of the scope of this paper and, hence, we left it to future work.

6   Conclusion

We have presented a framework (available as demo[2]) for multimodal sentiment analysis and multimodal emotion recognition, which outperforms the state of the art in both tasks by a significant margin. We also discussed some major aspects of multimodal sentiment analysis problem such as the performance of speaker-independent models and cross-dataset performance of the models.

Our future work will focus on extracting semantics from the visual features, relatedness of the cross-modal features and their fusion. We will also include contextual dependency learning in our model to overcome the limitations mentioned in the previous section.

References

1. Poria, S., Cambria, E., Bajpai, R., Hussain, A.: A review of affective computing: from unimodal analysis to multimodal fusion. Inf. Fusion **37**, 98–125 (2017)
2. Cambria, E., Hussain, A., Havasi, C., Eckl, C.: Sentic computing: exploitation of common sense for the development of emotion-sensitive systems. In: Esposito, A., Campbell, N., Vogel, C., Hussain, A., Nijholt, A. (eds.) Development of Multimodal Interfaces: Active Listening and Synchrony. LNCS, vol. 5967, pp. 148–156. Springer, Heidelberg (2010). https://doi.org/10.1007/978-3-642-12397-9_12
3. Pérez-Rosas, V., Mihalcea, R., Morency, L.P.: Utterance-level multimodal sentiment analysis. In: ACL, pp. 973–982 (2013)
4. Wollmer, M., Weninger, F., Knaup, T., Schuller, B., Sun, C., Sagae, K., Morency, L.P.: Youtube movie reviews: sentiment analysis in an audio-visual context. IEEE Intell. Syst. **28**, 46–53 (2013)
5. Poria, S., Cambria, E., Gelbukh, A.: Deep convolutional neural network textual features and multiple kernel learning for utterance-level multimodal sentiment analysis. In: Proceedings of EMNLP, pp. 2539–2544
6. Zadeh, A.: Micro-opinion sentiment intensity analysis and summarization in online videos. In: Proceedings of the 2015 ACM on International Conference on Multimodal Interaction, pp. 587–591. ACM (2015)
7. Poria, S., Chaturvedi, I., Cambria, E., Hussain, A.: Convolutional MKL based multimodal emotion recognition and sentiment analysis. In: ICDM, Barcelona, pp. 439–448 (2016)

---

[2] http://sentic.net/demo (best viewed in Mozilla Firefox).

8. Cambria, E.: Affective computing and sentiment analysis. IEEE Intell. Syst. **31**, 102–107 (2016)
9. Cambria, E., Poria, S., Hazarika, D., Kwok, K.: SenticNet 5: discovering conceptual primitives for sentiment analysis by means of context embeddings. In: AAAI, pp. 1795–1802 (2018)
10. Poria, S., Gelbukh, A., Agarwal, B., Cambria, E., Howard, N.: Advances in soft computing and its applications. In: Common Sense Knowledge Based Personality Recognition From Text, pp. 484–496. Springer (2013)
11. Pang, B., Lee, L., Vaithyanathan, S.: Thumbs up? Sentiment classification using machine learning techniques. In: Proceedings of ACL, pp. 79–86 (2002)
12. Socher, R., Perelygin, A., Wu, J.Y., Chuang, J., Manning, C.D., Ng, A.Y., Potts, C.: Recursive deep models for semantic compositionality over a sentiment treebank. In: Proceedings of EMNLP, vol. 1631, p. 1642 (2013)
13. Oneto, L., Bisio, F., Cambria, E., Anguita, D.: Statistical learning theory and ELM for big social data analysis. IEEE Comput. Intell. Mag. **11**, 45–55 (2016)
14. Ekman, P.: Universal facial expressions of emotion. Contemporary Readings/Chicago, Culture and Personality (1974)
15. Datcu, D., Rothkrantz, L.: Semantic audio-visual data fusion for automatic emotion recognition. Euromedia (2008)
16. De Silva, L.C., Miyasato, T., Nakatsu, R.: Facial emotion recognition using multimodal information. In: Proceedings of ICICS, vol. 1, pp. 397–401. IEEE (1997)
17. Chen, L.S., Huang, T.S., Miyasato, T., Nakatsu, R.: Multimodal human emotion/expression recognition. In: Proceedings of the Third IEEE International Conference on Automatic Face and Gesture Recognition, pp. 366–371. IEEE (1998)
18. Kessous, L., Castellano, G., Caridakis, G.: Multimodal emotion recognition in speech-based interaction using facial expression, body gesture and acoustic analysis. J. Multimodal User Interfaces **3**, 33–48 (2010)
19. Schuller, B.: Recognizing affect from linguistic information in 3D continuous space. IEEE Trans. Affect. Comput. **2**, 192–205 (2011)
20. Rozgic, V., Ananthakrishnan, S., Saleem, S., Kumar, R., Prasad, R.: Ensemble of SVM trees for multimodal emotion recognition. In: IEEE APSIPA, pp. 1–4 (2012)
21. Metallinou, A., Lee, S., Narayanan, S.: Audio-visual emotion recognition using Gaussian mixture models for face and voice. In: Tenth IEEE International Symposium on ISM 2008, pp. 250–257. IEEE (2008)
22. Eyben, F., Wöllmer, M., Graves, A., Schuller, B., Douglas-Cowie, E., Cowie, R.: On-line emotion recognition in a 3D activation-valence-time continuum using acoustic and linguistic cues. J. Multimodal User Interfaces **3**, 7–19 (2010)
23. Wu, C.H., Liang, W.B.: Emotion recognition of affective speech based on multiple classifiers using acoustic-prosodic information and semantic labels. IEEE Trans. Affect. Comput. **2**, 10–21 (2011)
24. Mikolov, T., Chen, K., Corrado, G., Dean, J.: Efficient estimation of word representations in vector space (2013). arXiv preprint arXiv:1301.3781
25. Eyben, F., Wöllmer, M., Schuller, B.: Opensmile: the Munich versatile and fast open-source audio feature extractor. In: Proceedings of the 18th ACM International Conference on Multimedia, pp. 1459–1462. ACM (2010)
26. Baltrušaitis, T., Robinson, P., Morency, L.P.: 3D constrained local model for rigid and non-rigid facial tracking. In: IEEE CVPR, pp. 2610–2617 (2012)
27. Zadeh, A., Zellers, R., Pincus, E., Morency, L.P.: Multimodal sentiment intensity analysis in videos: facial gestures and verbal messages. IEEE Intell. Syst. **31**, 82–88 (2016)

28. Busso, C., Bulut, M., Lee, C.C., Kazemzadeh, A., Mower, E., Kim, S., Chang, J.N., Lee, S., Narayanan, S.S.: Iemocap: interactive emotional dyadic motion capture database. Lang. Resour. Eval. **42**, 335–359 (2008)
29. Rozgić, V., Ananthakrishnan, S., Saleem, S., Kumar, R., Prasad, R.: Ensemble of SVM trees for multimodal emotion recognition. In: IEEE APSIPA, pp. 1–4 (2012)

Machine Learning Approaches for Speech Emotion Recognition: Classic and Novel Advances

Panikos Heracleous$^{(\boxtimes)}$, Akio Ishikawa, Keiji Yasuda, Hiroyuki Kawashima, Fumiaki Sugaya, and Masayuki Hashimoto

KDDI Research, Inc., 2-1-15 Ohara, Fujimino-shi, Saitama 356-8502, Japan
{pa-heracleous,ao-ishikawa,ke-yasuda,hi-kawashima,
fsugaya,masayuki}@kddi-research.jp

Abstract. Speech is the most natural form of communication for human beings, and among others, it provides information about the speaker's emotional state. The current study focuses on automatic speech emotion recognition based on classic and innovated machine learning approaches using simulated emotional speech data. Specifically, individual Gaussian mixture models (GMM) trained for each emotion, a universal background GMM model (UBM-GMM) adapted to each emotion using maximum posteriori (MAP) adaptation, and an approach based on i-vector paradigm, widely used in speaker recognition and language identification, and adapted to emotion recognition are used. When using individual GMMs, a novel technique based on multiple classifiers and late fusion is also applied. In this case, a 90.9% recognition rate is been obtained. When the state-of-the-art, i-vector paradigm based method, along with probabilistic linear discriminant analysis (PLDA) model is used, a 91.4% average rate for speaker-independent Japanese speech emotion recognition is achieved, which is a very promising result and superior to similar studies. In addition to the Japanese emotion recognition, pair-wise recognition for seven emotions in German language has also been conducted. The recognition rates obtained using the German database show the same tendency as in Japanese. In this experiment, an 89.2% average rate has been achieved.

1 Introduction

Speech is the most natural form of communication for human beings, and among others, contains also information about the speaker's emotional state. Automatic recognition of human emotions [1] is a relatively new paradigm, and is gaining high attention in research and development areas because of its high importance in real applications. Emotion recognition has an important role in human-robot communication, when robots may communicate with humans according to the detected human emotions, and can be applied at call centers to detect the callers' emotional state in case of emergency (e.g., hospitals, police stations), or to identify the level of the customer's satisfaction (providing feedback). For emotion

© Springer Nature Switzerland AG 2018
A. Gelbukh (Ed.): CICLing 2017, LNCS 10762, pp. 180–191, 2018.
https://doi.org/10.1007/978-3-319-77116-8_14

recognition, different modalities can be used such as audio, visual and tactile, or additionally a combination between them. In the current study, emotion recognition based on speech is being considered and experimentally investigated.

Automatic emotion recognition based on speech consists of feature extraction, feature selection, classification, and decision methods. Previous studies reported automatic speech emotion recognition using GMMs [2,3], hidden Markov models (HMM) [4], support vector machines (SVM) [5], neural networks (NN) [6], and deep neural network (DNN) [7]. The current study focuses on emotion recognition based on i-vector paradigm [8] along with PLDA modeling, which is a widely used approach in speaker verification and language identification frameworks, and in this study been adapted for emotion recognition. For comparison purposes, two classic GMM-based methods are also presented.

A distance measure is also calculated between pairs of GMMs. Several studies have been conducted using HMM distances to predict the performance of speech recognition systems or to select vocabularies in order to avoid confusable entries [9–11]. In the current study, distance measures between emotion-pairs are calculated to investigate the relationship between distances and recognition rates, as also to examine the variability across emotions and speakers.

What is more feature extraction and selection is an important issue in emotion recognition. Features used in emotion recognition include mel-frequency cepstral coefficients (MFCC) [12], linear predictor coefficients (LPC) [13], perceptual linear prediction (PLP) coefficients [14], pitch, energy, duration, etc.

2   Methods

2.1   Data

Four professional female actors simulate Japanese emotional speech in *neutral, joy, anger, sadness*, and *mixed* emotional states. Fifty-one utterances for each emotion are produced by each speaker. The sentences are selected from a Japanese book for kids. The data are recorded at 48 kHz and down-sampled to 16 kHz, and contain short and longer utterances of length between 1.5 s and 9 s. As for training, 28 utterances are used and as for testing 20 utterances are being used. The utterances aforementioned represent emotions such as *neutral, joy, anger* and *sadness*. Twelve PARCOR, 12 LPC, 12 PLP, and 12 MFCC plus Energy features are extracted every 10 ms by applying a window-length of 20 ms.

Additional experiments using the Berlin Emotional Speech database [15] are been conducted. Ten actors (i.e., five males and five females) simulate seven emotions (anger, happiness, sadness, neutral, boredom, disgust, anxiety) producing utterances in German. This speech material consists of 535 sentences. In this experiment, two-class recognition based on individual GMMs is conducted to investigate the discrimination of all emotion-pairs and to examine the most dominant and well recognized emotions. Also in this case, 12 PARCOR, 12 LPC, 12 PLP, and 12 MFCC plus Energy features are also extracted every 10 ms by applying a window-length of 20 ms. The German database is recorded at 16 kHz.

2.2   HMM Distance Measures Between Emotions

Juang and Rabiner [16] have addressed the problem of mathematically formulating similarity and distance measures between two HMMs. Juang and Rabiner defined the Kullback-Leibler distance rate (KLDR) as a measure of similarity between ergodic HMM, and show how to extend the KLDR to non-ergodic (i.e., left-to-right) HMMs. Euclidean and Mahalanobis distances are also often used to measure HMM similarity. However, these distances do not take into account the temporal structure represented in the Markov chain. In this study, emotion distance measures between GMMs were calculated using the Juang and Rabiner distance given in Eq. 1.

$$D(\lambda_1, \lambda_2) = \frac{1}{T_i^2}[\log P(Q_{T_i}^2|\lambda_1) - \log P(Q_{T_i}^2|\lambda_2)] \tag{1}$$

where λ_1 and λ_2 are the two GMM models, Q_T^2 is the feature sequence generated by λ_2 model, and T_i^2 is the length of feature sequence. The $D(\lambda_1, \lambda_2)$ is not symmetric and, therefore, we consider the distance of two HMMs as

$$D = \frac{D(\lambda_1, \lambda_2) + D(\lambda_2, \lambda_1)}{2} \tag{2}$$

2.3   Classification

The baseline conventional approach consists of training an individual GMM for each emotion using the corresponding training data. Given the $\mathbf{X} = (x_1, x_2..x_n)$ input feature vector, the correct emotion is detected by finding the λ_s model, which maximizes the a posterior probability, across the S emotion models

$$L = \arg \max_{1 \le k \le S} P(\mathbf{X}|\lambda_k) \tag{3}$$

In this case, a novel technique is also applied by using multiple classifiers, trained with different feature sets and late fusion. Considering the case of two classifiers (a, b) and $\mathbf{X} = (\mathbf{x}_a, \mathbf{x}_b)$ the feature vectors, the total score L is given by the following equation:

$$L = \arg \max_{1 \le k \le S} (P(\mathbf{x}_a|\lambda_k^a) + P(\mathbf{x}_b|\lambda_k^b)) \tag{4}$$

A similar approach for multi-modal emotion recognition has been reported in [17]. However, in the current study late fusion based on a single modality (i.e., audio) and using multiple classifiers trained with different feature sets is presented.

In the case of the UBM-GMM approach, a general GMM is trained using all training data, and the individual emotion GMMs are created by adapting the UBM-GMM to each emotion using MAP adaptation and the corresponding emotion training data.

In the i-vector paradigm, the limitations of high dimensional supervectors (i.e., concatenation of the means of GMMs) are overcome by modeling the variability contained in the supervectors with a small set of factors. In this case, an input utterance can be modeled as:

$$\mathbf{M} = \mathbf{m} + \mathbf{Tw} \tag{5}$$

where $\mathbf{M}$ is the emotion-dependent supervector, $\mathbf{m}$ is the emotion-independent supervector, $\mathbf{T}$ is the total variability matrix, and $\mathbf{w}$ is the i-vector. Both total variability matrix and emotion-independent supervector are estimated from the whole set of the training data.

PLDA is a popular technique for dimension reduction using Fisher criterion. Using PLDA, new axes are found, which maximize the discrimination between the different classes. PLDA is originally applied to face recognition [18], and is applied successfully to specify a generative model of the i-vector representation. Adapting to emotion recognition, for the i-th emotion, the i-vector $\mathbf{w}_{i,j}$ representing the j-th recording can be formulated as:

$$\mathbf{w}_{i,j} = \mathbf{m} + \mathbf{Sx}_i + \mathbf{e}_{i,j} \tag{6}$$

where $\mathbf{S}$ represents the between-emotion variability, and the latent variable $\mathbf{x}$ is assumed to have standard normal distribution, and represent a particular emotion and channel. The residual term $\mathbf{e}_{i,j}$ represents the within-emotion variability, and it is assumed to have a normal distribution.

3   Results

This section demonstrates the results for emotion recognition in both Japanese and German. Due to the fact that the Japanese emotional speech is recorded for speech synthesis and has not been previously evaluated with concern to emotion recognition, baseline speaker-dependent experiments are also conducted to analyze the emotion recognition rates across the speakers. In the case of the German, the experiments are conducted to cover a larger number of emotions and to compare rates and emotion recognition tendency with the Japanese. In this section, analysis based on emotion GMM distances is also reported.

3.1   Emotion HMM Distances in Japanese

For each emotion, four GMMs with 32 components are trained using the corresponding training and MFCC, PLP, PARCOR, and LPC features, respectively. Using the Juang and Rabiner formulation, the emotion distance measures are calculated for each emotion-pair and speaker separately. Figure 1a demonstrates the normalized distances when using PARCOR features. As demonstrated, the distances across the five emotions differ from each other, indicating the existence of emotion discrimination. The relationship between distances and recognition rates is also clearly shown on the figure. Figure 1b demonstrates the normalized total distances when using MFCC, PLP, PARCOR, and LPC features. The graphs show the same tendency of distance measures, except the case of using LPC features, which might indicate that LPC features are not suitable for the current task.

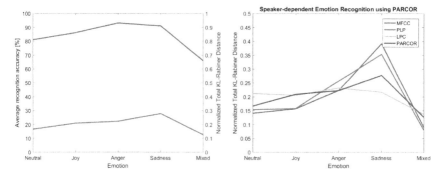

(a) Normalized KLRD distances and recognition rates

(b) Normalized KLRD distances with different features

Fig. 1. Emotion analysis results using GMM distances.

Table 1. Speaker-dependent emotion recognition rates based on individual GMMs

Classifier	Emotion				
	Neutral	Joy	Anger	Sadness	Average
MFCC	88.8	90.0	95.0	93.8	91.9
PLP	96.3	87.5	95.0	92.5	92.8
PARCOR	93.8	92.5	96.3	88.8	92.8
LPC	80.0	86.3	85.0	91.3	85.6
MFCC + PLP	98.8	88.8	95.0	93.8	94.1
MFCC + PARCOR	98.8	92.5	95.0	92.5	94.7
MFCC + LPC	92.5	95.0	96.3	91.3	93.8

3.2   Speaker-Dependent Emotion Recognition in Japanese

In order to investigate the between-emotion variability across the four speakers, analysis based on speaker-dependent emotion recognition experiments is conducted. In these experiments, four emotions are recognized using GMMs with late fusion of two classifiers. For the GMMs, 128 Gaussian components are used.

Table 1 shows the averaged recognition rates for the four emotions and also when using different kind of classifiers (i.e. single-feature, or late fusion). The results conclude, that the emotional speech simulated by the four actors could be discriminated with very high rates. The results also conclude that *anger* emotion is recognized with the highest accuracy, followed by *sadness* and *neutral* emotions. The emotion *joy* is recognized with the lowest accuracy. Concerning feature selection, PLP and PARCOR features perform better compared to MFCC and LPC features. Especially, when using LPC features alone, the recognition rate is as low as 85.6%. This result justifies the hypothesis that as GMM distances indicate, LPC features may not be suitable for the current task.

Table 2. Speaker-independent emotion recognition rates based on individual GMMs

Classifier	Emotion				
	Neutral	Joy	Anger	Sadness	Average
MFCC	87.5	75.0	73.8	97.5	80.9
PLP	87.5	77.5	95.0	96.2	89.1
PARCOR	85.0	67.5	83.8	81.3	79.4
LPC	53.8	53.8	82.5	88.8	69.7
MFCC + PLP	91.3	75.0	80.0	97.5	85.9
MFCC + PARCOR	92.5	70.0	71.3	95.0	82.2
MFCC + LPC	86.3	71.3	72.5	100.0	82.5
PLP + PARCOR	91.3	81.3	91.3	100.0	90.9
PLP + LPC	85.0	75.0	95.0	100.0	88.8
PARCOR + LPC	77.5	62.5	86.3	87.5	78.4

The results in Table 1 also clearly support the proposed novel method for using multiple classifiers with different features and fusion of the individual scores. Specifically, by applying the proposed technique, the recognition rates are significantly improved. Compared to the sole use of MFCC features in the conventional GMM-based approaches, a 28% average relative improvement is achieved. In addition to recognition rates, the Equal Error Rates (EER) (i.e., miss probability equal to false alarms) are also computed. The mean EER for the $F416$, $F418$, $F419$, and $F420$ female actors were 6.3%, 2.5%, 3.8%, and 2.5%, respectively. This result indicates, that $F418$ and $F420$ actors could simulate better the emotional speech data.

3.3   Speaker-Independent Emotion Recognition in Japanese

Speaker-independent evaluation is also conducted to recognize *neutral*, *joy*, *anger*, and *sadness* emotions. The classifiers used are an individual GMM-based using late fusion classifier, an UBM-GMM based classifier, and a PLDA with i-vector paradigm based classifier. In these experiments, emotion models using the corresponding data from all the speakers are trained. Similar to speaker dependent experiments, MFCC, PLP, PARCOR, and LPC features are used. The number of Gaussian components is 128. Table 2 demonstrates the recognition rates when using individual GMMs without and with fusion. As expected, the recognition rates are lower compared to the speaker-dependent rates. However, the results are still very promising and in some cases closely comparable with those of the speaker-dependent experiments. Specifically, using late fusion with PLP and PARCOR features, a 90.9% recognition rate is obtained. As the results show, also in these experiments PLP achieves the highest rates. The lowest rates, similar to the speaker-dependent case, are obtained when LPC features are used. The highest recognition rate is achieved from the *sadness* emotion, which is recognized with a 94.5% rate, on average. These experiments also demonstrate the improvements when using the proposed late fusion technique. Using parallel classifiers with different features and late fusion, a 25% relative improvement is obtained compared to the sole use of single classifiers.

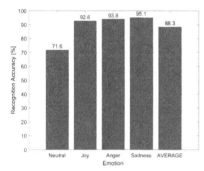
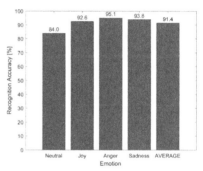

(a) Recognition rates using UBM−GMM (b) Recognition rates using i−vectors and PLDA model

Fig. 2. Results for speaker-independent speech emotion recognition in Japanese.

Figure 2a demonstrates the recognition rates when a UBM-GMM is trained using training data from all emotions, and the emotion models are created by using MAP adaptation. As shown, high recognition rates are achieved. Using this classification method, *sadness* and *anger* are recognized with the highest rates. On the other hand, the recognition rate for the *neutral* emotion is only 71.6%. The average recognition in this case is 88.3%. This rate is lower than the individual GMMs based approach. However, the computation time for training and decoding when using UBM-GMM is significantly lower.

Although, i-vector paradigm along with PLDA classifier is widely used in speaker verification, very few studies in emotion recognition have been reported. Moreover, the current study presents emotion recognition using i-vectors and PLDA model in the Japanese language, which may be very informative and helpful study for the society. Figure 2b shows the results obtained when using i-vectors and PLDA model. In this case, a 91.4% average recognition rate is obtained, which is the highest among the three different classification schemes used in this study. Similar to UBM-GMM classifier, *sadness* and *anger* emotions are recognized with the highest rates. The recognition rate for *neutral* emotion is the lowest. The results obtained are very promising, and closely comparable to the results achieved in the speaker-dependent experiments (i.e., 94.7% vs 91.4%).

Table 3. Speaker-independent two-class emotion recognition based on individual GMMs

Classifier	Emotion		
	Anger	Other	Average
MFCC	71.3	96.3	83.8
PLP	95.0	83.3	89.2
PARCOR	88.8	78.3	83.5
LPC	85.0	73.8	79.4
MFCC + PLP	82.5	92.9	87.7
MFCC + PARCOR	77.5	92.1	84.8
MFCC + LPC	75.0	91.3	83.1
PLP + PARCOR	97.5	82.9	90.2
PLP + LPC	95.0	81.7	88.3
PARCOR + LPC	88.8	77.9	83.3

3.4 Speaker-Independent Recognition of Two Emotional States in Japanese

In some real applications, only a limited number of emotions is required. A typical example is an emotion recognition system, which is implemented and operates at a call center, with task being to identify the satisfaction level of customers. In such a case, the most representative emotion that should be detected is probably the *anger* emotion. In order to address this issue, experiments are also conducted to classify the input speech to *anger* and *other* classes only. Specifically, a model for *anger* emotion is trained using the corresponding data from the *anger* emotional speech, and the *other* model is trained using the corresponding data from the *neutral*, *happiness*, and *sadness* emotional speech. In these experiments, classifiers based on individual GMMs and with/without late fusion are used. For modeling, 256 Gaussian components are used. Table 3 shows the results obtained when using different kind of classifiers. As is shown, using this classification method the variability across emotions can be captured with GMM modeling, and the two emotional states are recognized with high rates. Specifically, when using PLP and PARCOR classifiers with late fusion, a 90.2% average recognition rate is obtained (i.e., 97.5% and 82.9% recognition rates for *anger* and *other* emotions, respectively). This result is promising and shows the effectiveness of the applied method in classifying input speech into *anger* and *other* classes.

3.5   Human Evaluation Results

Human evaluation is being performed using listening tests by five native Japanese speakers. For each emotion, eight utterances randomly selected from the data of the four speakers are used. Table 4 shows the results obtained. As it is shown, the human evaluation recognition rates are similar across the speakers. The p-value using the ANOVA test is 0.977789 showing that the differences across the speakers are statistically not significant. Concerning the recognition rates across the four emotions, *joy* emotion is recognized with the lowest rate. This result shows the same tendency as in speaker-independent experiments using individual GMMs. The average recognition rate is 68.1%, almost the same as reported in [19].

Table 4. Human evaluation results for the Japanese emotional speech

Speaker	Correct samples (out of 8)				Average [%]
	Neutral	Joy	Anger	Sadness	
SP1	5	5	6	6	68.8
SP2	6	6	6	4	68.8
SP3	7	3	6	5	65.6
SP4	7	3	6	5	65.6
SP5	7	5	6	5	71.9
Average [%]	80.0	55.0	75.0	62.5	68.1

3.6   Two-Class Speaker-Independent Emotion Recognition in German

In addition to the experiments using Japanese emotional speech data, experiment using German emotional data is also conducted. This experiment aims at investigating the pair-wise classification of all emotions, and also to examine the similarity between Japanese and German emotional speech recognition. The German data is produced by ten actors and cover seven emotional states. Two-third of the data are used for training and one-third for testing. As classifier, speaker-independent individual GMMs with 32 Gaussian components are used, and PLP features are also applied. These experimental conditions were empirically adjusted to produce the highest rates.

Table 5 shows the recognition rates for all emotion-pairs. As shown, *anger*, *sadness*, and *happiness* are discriminated with the highest rates. This observation is very similar to the Japanese emotion recognition reported in the previous sections. A possible reason may be the capability of simulating these emotions due to the difference with the neutral emotion. Another possible reason lies on the production of these emotions, which may show unique spectral and temporal characteristics (e.g. speech rate, energy).

4   Discussion

Experiments on speech emotion recognition using Japanese and German emotional speech database are being conducted. Issues related to feature extraction and classification approaches are also addressed and discussed in this study. A limitation of the current study is the use of simulated emotional data in the experiments. Using emotional data obtained in real situations, more realistic results concerning speech emotion recognition may be obtained. However, the current study aims mainly at investigating the effectiveness of different classification and feature extraction techniques in speech emotion recognition. For this purpose, and similarly to other studies previously reported, emotional speech simulated by professional actors provides a reasonable and acceptable solution. To address this issue, however, the authors are currently working on obtaining/using real and spontaneous emotional speech data in order to design and implement a system for real applications. Another issue that is being discussed is the different scenarios of the Japanese and the German database. It is possible, that differences in speech production across languages and in collecting scenarios may also cause different observations in speech emotion recognition. In the current study, however, results show the same tendency in both languages, and same emotions are similarly recognized without significant differences.

Table 5. Speaker-independent pair-wise emotion recognition in German

Emotions		Rates			Emotions		Rates		
Emotion-A	Emotion-B	Rate-A	Rate-B	Average	Emotion-A	Emotion-B	Rate-A	Rate-B	Average
Anger	Happiness	100.0	20.8	60.4	Sadness	Neutral	100.0	96.3	98.2
Anger	Sadness	100.0	100.0	100.0	Sadness	Boredom	100.0	70.4	85.2
Anger	Neutral	100.0	100.0	100.0	Sadness	Disgust	100.0	100.0	100.0
Anger	Boredom	100.0	92.6	96.3	Sadness	Anxiety	90.5	95.7	93.2
Anger	Disgust	97.7	87.5	92.6	Neutral	Boredom	63.0	77.8	70.4
Anger	Anxiety	97.7	60.9	79.3	Neutral	Disgust	88.9	93.8	91.3
Happiness	Sadness	100.0	100.0	100.0	Neutral	Anxiety	88.9	100.0	94.4
Happiness	Neutral	100.0	92.6	96.3	Boredom	Disgust	70.4	100.0	85.2
Happiness	Boredom	100.0	81.5	90.8	Boredom	Anxiety	63.0	95.7	79.3
Happiness	Disgust	100.0	81.3	90.6	Disgust	Anxiety	93.8	91.3	92.5
Happiness	Anxiety	100.0	56.5	78.3					

5   Conclusion

In the current study the presentation of the analysis and recognition of emotions in Japanese and German emotional speech takes place. Speaker-dependent and speaker-independent emotion recognition experiments are applied using several classification and feature selection approaches, and simulated emotional speech. A classification method is being proposed, which uses parallel classifiers trained with different features and applying late fusion. Based on this approach, 90.9% average recognition rate for four Japanese emotions recognition is achieved. Furthermore, experiments are applied based on the state-of-the-art i-vector based paradigm along with PLDA model. In this case, a 91.4% recognition rate is achieved. These results are very promising, and show the effectiveness of our approaches in speech emotion recognition. In addition to Japanese experiments, emotion recognition using German emotional speech is also applied. The results obtained in this case show the same tendency as in Japanese, concerning the several emotions' discrimination.

References

1. Busso, C., Bulut, M., Narayanan, S.: Toward effective automatic recognition systems of emotion in speech. In: Gratch, J., Marsella, S. (eds.) Social Emotions in Nature and Artifact: Emotions in Human and Human-Computer Interaction, pp. 110–127. Oxford University Press, New York (2013)
2. Tang, H., Chu, S., Johnson, M.H.: Emotion recognition from speech via boosted Gaussian mixture models. In Proceedings of ICME, pp. 294–297 (2009)
3. Xu, S., Liu, Y., Liu, X.: Speaker recognition and speech emotion recognition based on GMM. In: 3rd International Conference on Electric and Electronics (EEIC 2013), pp. 434–436 (2013)
4. Schuller, B., Rigoll, G., Lang, M.: Hidden Markov model-based speech emotion recognition. In: Proceedings of IEEE ICASSP, vol. I, pp. 401–404 (2003)
5. Pan, Y., Shen, P., Shen, L.: Speech emotion recognition using support vector machine. Int. J. Smart Home **6**(2), 101–108 (2012)
6. Nicholson, J., Takahashi, K., Nakatsu, R.: Emotion recognition in speech using neural networks. NCA **9**(4), 290296 (2000)
7. Han, K., Yu, D., Tashev, I.: Speech emotion recognition using deep neural network and extreme learning machine. In: Proceedings of Interspeech, pp. 223–227 (2014)
8. Dehak, N., Kenny, P.J., Dehak, R., Dumouchel, P., Ouellet, P.: Front-end factor analysis for speaker verification. IEEE Trans. Audio Speech Lang. Process. **19**(4), 788–798 (2011)
9. Fosler-Lussier, E., Amdal, I., Kuo, H.: A framework for predicting speech recognition errors. Speech Commun. **46**, 153–170 (2005)
10. Silva, J., Narayanan, S.: Average divergence distance as a statistical discrimination measure for hidden Markov models. IEEE Trans. Speech Audio Process. **14**, 890–906 (2006)
11. Yamamoto, K., Nakagawa, S.: Differences of speech rate, interphoneme distance and likelihood caused by speaking style, their relationship and recognition performance. Syst. Comput. Jpn **33**(7), 50–60 (2002)
12. Sahidullah, M., Saha, G.: Design analysis and experimental evaluation of block based transformation in MFCC computation for speaker recognition. Speech Commun. **54**(4), 543565 (2012)
13. O'Shaughnessy, D.: Linear predictive coding. IEEE Potentials **7**(1), 29–32 (1988)
14. Hermansky, H.: Perceptual linear predictive (PLP) analysis of speech. J. AcousL Soc. Am. **87**(4), 1738–1752 (1990)
15. Burkhardt, F., Paeschke, A., Rolfes, M., Sendlmeier, W., Weiss, B.: A database of German emotional speech. In Proceedings of Interspeech, pp. 1517–1520 (2005)

16. Juang, B.H., Rabiner, L.: A probabilistic distance measure for hidden Markov models. AT&T Tech. J. 391–408 (1985)
17. Metallinou, A., Lee, S., Narayanan, S.: Decision level combination of multiple modalities for recognition and analysis of emotional expression. In: Proceedings of ICASSP, pp. 2462–2465 (2019)
18. Prince, S., Elder, J.: Probabilistic linear discriminant analysis for inferences about identity. In: Proceedings of International Conference on Computer Vision, pp. 1–8 (2007)
19. Lee, C.M., Yildirim, S., Bulut, M., Kazemzadeh, A., Busso, C., Deng, Z., Lee, S., Narayanan, S.: Emotion recognition based on phoneme classes. In: Proceedings of ICSLP, pp. 889–892 (2004)

Opinion Mining

Mining Aspect-Specific Opinions from Online Reviews Using a Latent Embedding Structured Topic Model

Mingyang Xu[1]([✉]), Ruixin Yang[1], Paul Jones[1], and Nagiza F. Samatova[1,2]

[1] North Carolina State University, Raleigh, NC, USA
samatova@csc.ncsu.edu
[2] Oak Ridge National Laboratory, Oak Ridge, TN, USA

Abstract. Online reviews often contain user's specific opinions on aspects (features) of items. These opinions are very useful to merchants and customers, but manually extracting them is time-consuming. Several topic models have been proposed to simultaneously extract item aspects and user's opinions on the aspects, as well as to detect sentiment associated with the opinions. However, existing models tend to find poor aspect-opinion associations when limited examples of the required word co-occurrences are available in corpus. These models often also assign incorrect sentiment to words. In this paper, we propose a Latent embedding structured Opinion mining Topic model, called the LOT, which can simultaneously discover relevant aspect-level specific opinions from small or large numbers of reviews and to assign accurate sentiment to words. Experimental results for topic coherence, document sentiment classification, and a human evaluation all show that our proposed model achieves significant improvements over several state-of-the-art baselines.

1 Introduction

There are now numerous websites and apps that offer online reviews of products, restaurants and other items. These reviews usually contain specific opinions of users and customers towards different aspects (features) of the items being reviewed. Awareness of these opinions is critical both for merchants to improve their products and for customers to make purchasing choices. However, manually extracting these opinions requires enormous human efforts. Therefore, there is an important need to automatically mine aspect-level specific opinions from online reviews. This task consists of three core sub-tasks: (1) extraction of relevant aspects and corresponding opinions, (2) sentiment classification of discovered opinions, (3) separation of general opinions and specific opinions. For example, given a review about a cellphone, *"The battery is great and durable"*, *"battery"* should be extracted as an aspect, *"great"* should be identified as a general opinion, *"durable"* should be identified as a specific opinion, and both of *"great"* and *"durable"* should be classified as positive opinions.

© Springer Nature Switzerland AG 2018
A. Gelbukh (Ed.): CICLing 2017, LNCS 10762, pp. 195–210, 2018.
https://doi.org/10.1007/978-3-319-77116-8_15

In recent years, several works [1–5] based on traditional unsupervised probabilistic topic models have been proposed to tackle some or all of the above sub-tasks simultaneously. The input to these models is typically a collection of reviews. To extract the aspects and opinions mainly discussed in the reviews, these topic models discover a set of latent topics (themes) that pervade the reviews. Each discovered topic contains a set of semantically related aspect-words and opinion-words. For sentiment classification of opinion-words, all the opinion-words in a topic are assigned a specific sentiment. The opinion-words in different topics can have different sentiments. These models have proven to be effective but they suffer from three important limitations:

1. Topics are typically discovered based on word co-occurrences - if only a small number of reviews are available or the reviews contain limited co-occurred words, the quality of the resulting topics will suffer. However, in reality, for most online items, such as those on Amazon, fewer than 100 reviews are normally available, and these tend to contain limited co-occurring words, as observed in [6].
2. With regard to sentiment classification of words, existing models typically first assign each document a distribution over sentiments, and the sentiment of each word in the document is then sampled from this distribution. This results in a tendency to assign positive sentiment to all the words in positive documents and negative sentiment to all the words in negative documents. However, this bias often results in incorrectly assigning positive sentiment to negative words in positive documents and in incorrectly assigning negative sentiment to positive words in negative documents.
3. Few existing unsupervised topic models take into account the distinction between specific opinions and general opinions. General opinion-words, such as *"great"*, can only express sentiment from the user's perspective, while specific opinions, such as *"durable"*, are more informative since they contain the reason why the user liked or disliked an item. Therefore, it is important to mine opinions that are as specific as possible.

Recently, latent embedding techniques [7,8] have gained attention, since they have proven to be effective in many NLP tasks, including topic modeling. Several topic models have been proposed to incorporate latent embeddings [9–12]. However, none of these models have attempted to mine aspect-level specific opinions. In this paper, we propose a Latent embedding structured Opinion mining Topic model, called the LOT, which is an unsupervised probabilistic topic model for mining relevant aspects and corresponding specific opinions from online reviews. Our model simultaneously addresses all of the three limitations mentioned above. The main contributions of our work are summarized as follows:

- Our model exploits two types of latent embeddings: *latent word embeddings* and *latent topic embeddings*. Latent word embeddings [7,8] have recently proven to be very effective in capturing the semantic meanings of words. Specifically, the latent word embeddings used in our model are pre-trained on a large public Amazon review dataset - these embeddings capture the semantic meanings of aspect and opinion words in a general context. Latent topic embeddings are estimated based on latent word embeddings and capture the semantic meanings of latent topics to be discovered. Informed by these two different types of embeddings, our LOT model can discover coherent topics containing relevant aspect-words and opinion-words from small numbers or large numbers of reviews without suffering from the problems caused by limited word co-occurrences.

- To assign more accurate sentiment to words, in addition to the document-sentiment distribution that existing models rely on, we also model a word-sentiment distribution, which is estimated based on the sentiment of all the instances of each word in the entire corpus. The intuition behind our approach is that document-sentiment distributions can only be used to derive the sentiment of each word based on local context, whereas word-sentiment distributions provide the sentiment of each word in a general context. By simultaneously estimating and utilizing these two distributions, our model can assign more accurate sentiment to words.

- For distinguishing between specific opinions and general opinions, we observe that words used to indicate general opinions (such as *"best"*, *"disappointed"* and *"unsatisfied"*) typically do not change with context and therefore can be pre-identified. We create and release a *general opinion lexicon* by manually selecting general opinion-words from an existing public opinion lexicon. Our proposed general opinion lexicon contains 548 of the most common general opinion-words, and each word is classified as positive or negative. To the best of our knowledge, it is the first sentiment lexicon for general opinion-words. We believe that this lexicon can assist the community with future research that requires mining specific opinions from text. We exploit this lexicon in LOT, and our experimental results show that our model can mine more specific opinions than several state-of-the-art models.

2   Related Work

Mining aspects and corresponding opinions from online reviews is an important area of current research. Topic modeling is one of the most popular approaches for this task, and various topic models have been proposed [1–3,5,13]. As our approach is itself an unsupervised topic model, we will mainly discuss the related work of the topic models that have been proposed for this task.

A topic model called JST [2] was proposed that inserted a sentiment layer into the traditional topic model, LDA [14], in order to capture the sentiment of the topics containing aspect-words and opinion-words. Another topic model called ASUM [1] added the constraint that all words in a single sentence are generated from one topic to improve the quality of the discovered topics. Subsequently, the TM model [4] was proposed to discover overall topic-sentiment correlations. However, these models discover topics based on word co-occurrences, which don't perform well when the dataset is small or few word co-occurrence examples are available. Furthermore, none of these approaches considered separating specific opinions and general opinions.

Recently, a fine-grained lifelong learning topic model [3] was proposed to mine and incorporate knowledge of word correlations for mining aspect-specific opinions. To mine reliable knowledge, they require a large number of additional datasets (e.g. 50 datasets) with relevant content to the test datasets. However, our model does not incorporate word correlation knowledge nor does our model need efforts to find additional relevant datasets for mining knowledge. A supervised topic model integrated with a discriminative maximum entropy component [13] was also proposed to mine aspect-specific opinions, but this model requires manually labeled dataset and does not consider sentiment of words. In contrast, our model is unsupervised. In addition, our model simultaneously mines aspect-specific opinions and assign sentiment to words.

Table 1. Notation used in this paper.

S, T, V^{r_i}	The number of sentiments, the number of topics, the number of words with word type r_i
$\mathbf{w}^{r_i}, \mathbf{w}^A, \mathbf{w}^E$	Pre-trained word embeddings of the words with word type r_i, aspect-words, specific opinion-words
$\boldsymbol{\lambda}^{r_i}, \boldsymbol{\lambda}^A, \boldsymbol{\lambda}^E$	Latent embeddings of the topics which contain the words with word type r_i, aspect-words, specific opinion-words
d_i, w_i, s_i, z_i	i_{th} document in corpus, i_{th} word in a document, sentiment polarity of w_i, topic of w_i
z^{-i}, s^{-i}	All the assigned topics, sentiment polarity excluding the assignment of w_i
θ	Multinomial distribution of topics for each document
π	Multinomial distribution of sentiments for each document
δ	Multinomial distribution of sentiments for each word
α, β, γ	Dirichlet prior for θ, δ, π
$N_{d,s}^{-i}$	The number of words in document d under sentiment s excluding the assignment of w_i
$N_{w,s}^{-i}$	The number of word w in corpus under sentiment s excluding the assignment of w_i
$N_{d,s,t}^{-i}$	The number of words in document d under sentiment s and topic t excluding the assignment of w_i
$N_{s,t,w}$	The number of words under topic t and sentiment s

Besides the differences discussed above, our model exploits latent embeddings and assigns the sentiment to words based on both document-sentiment distributions and word-sentiment distributions. There are existing topic models that use latent embeddings [9,11,12,15]. However, none of them were intended for the task of mining aspect-level specific opinions. For example, [9,12] incorporate word embeddings in their model, but neither of these two models consider sentiment. The model in [11] improves on [12] by considering sentiment, but it does not attempt to mine aspect-specific opinions, and also does not take into account the word-sentiment distributions considered in our model for assigning sentiment to words.

There are also other proposed topic models related to opinion mining [6,16–18]. However, these opinion-mining models do not focus on mining aspect-level specific opinions. For example, [6] focuses instead on predicting ratings of extracted aspects, [16] focuses on movie recommendation, and [18] is concerned with the task of multi-aspect sentence labeling and multi-aspect rating prediction. Finally, a model called MAS [17] was proposed for extracting topics that help to explain the ratings of aspects provided by users.

3   LOT Model

We now present our Latent embedding structured Opinion mining Topic model, LOT. We show the graphical representation of the model in Fig. 1. Each node represents a variable - the shaded variables are observed, the others are hidden and must be estimated. Our notation conventions are listed in Table 1. Our model is a generative model and the generative process for our model proceeds as follows:

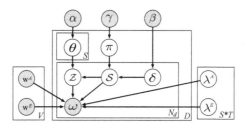

Fig. 1. Graphical model of LOT

1. For each document d, draw a sentiment distribution $\pi_d \sim Dir(\gamma)$,
2. For each sentiment s under d, draw a topic distribution $\theta_{d,s} \sim Dir(\alpha)$,
3. For each word w, draw a sentiment distribution $\delta_w \sim Dir(\beta)$,
4. For each word w_i of each document d:
   (a) choose a sentiment $s_i \sim P(s_i|\pi_d, \delta_w)$,
   (b) choose a topic $z_i \sim P(z_i|\theta_{d,s})$,
   (c) choose a word type r_i,
   (d) choose a specific opinion or aspect word $w_i \sim P(w_i|\boldsymbol{\lambda}_{z_i,s}^{r_i}, \mathbf{w}^{r_i})$.

where Dir refers to a Dirichlet distribution, and $P(w_i | \boldsymbol{\lambda}^{r_i}_{z_i,s} \mathbf{w}^{r_i})$, $P(s_i | \pi_d, \delta_w)$, $P(z_i | \theta_{d,s})$ are multinomial distributions, which are defined in Eqs. 1, 6 and 7 below.

In our approach, we model general opinion, specific opinion and aspect words separately. Specifically, each topic discovered by our model only contains specific opinion-words or aspect-words. General opinion-words are only used to determine the sentiment of the documents and assist in assigning sentiment to specific opinion-words, since general opinions typically only contain sentiment information.

To identify the word type r_i of each word, we utilize our proposed general opinion lexicon and two existing public sentiment lexicons. Specifically, to separate aspect-words and opinion-words, we identify a word as an opinion-word if it appears in the combined sentiment lexicon. As this assumption is not always correct, we also follow the approach in [3,13] to use part-of-speech (POS) tags and a maximum entropy classifier to identify the word type in a semi-supervised manner. However, based on our preliminary experiments, this approach produces worse results than solely relying on sentiment lexicon, especially when the number of reviews is small. Therefore, we solely use the combined sentiment lexicon to distinguish aspect-words and opinion-words.

For separating specific opinion-words and general opinion-words, we exploit our proposed general opinion lexicon. When an opinion-word appears in this lexicon, it is considered to be a general opinion-word, otherwise it is considered to be a specific opinion-word.

In contrast to existing models proposed for mining aspect-level specific opinions, we exploit latent word embeddings and latent topic embeddings to structure our model. The conditional probability of a word w in d generated by topic t with sentiment s and word type r_i given pre-trained word embeddings $\mathbf{w}^{r_i}$ and a topic embedding $\boldsymbol{\lambda}^{r_i}_{s,t}$ is given by the softmax function as follows:

$$\mathrm{P}(w \mid \boldsymbol{\lambda}^{r_i}_{s,t} \mathbf{w}^{r_i}) = \frac{\exp(\boldsymbol{\lambda}^{r_i}_{s,t} \cdot \mathbf{w}_w)}{\sum_{w' \in V^{r_i}} \exp(\boldsymbol{\lambda}^{r_i}_{s,t} \cdot \mathbf{w}_{w'})} \quad (1)$$

where the topic embedding $\boldsymbol{\lambda}^{r_i}_{s,t}$ is estimated using regularized maximum likelihood estimation. The negative log likelihood of the input corpus in terms of each topic t under sentiment s with L2 regularization is given by:

$$L^{r_i}_{s,t} = - \sum_{w \in V^{r_i}} N_{s,t,w} \left(\boldsymbol{\lambda}^{r_i}_{s,t} \cdot \mathbf{w}_w - \log \Big(\sum_{w' \in V^{r_i}} \exp(\boldsymbol{\lambda}^{r_i}_{s,t} \cdot \mathbf{w}_{w'}) \Big) \right) + \mu \left\| \boldsymbol{\lambda}^{r_i}_{s,t} \right\|^2_2 \quad (2)$$

where μ denotes the L_2 regularizer constant. We then obtain the estimation of $\boldsymbol{\lambda}^{r_i}_{s,t}$ by minimizing $L^{r_i}_{s,t}$. The derivative with respect to the jth element of $\boldsymbol{\lambda}^{r_i}_{s,t}$ is:

$$\frac{\partial L^{r_i}_{s,t}}{\partial \boldsymbol{\lambda}^{r_i}_{s,t,j}} = - \sum_{w \in V^{r_i}} N_{s,t,w} \left(\mathbf{w}_{w,j} - \sum_{w' \in V^{r_i}} \mathbf{w}_{w',j} \, \mathrm{P}(w' \mid \boldsymbol{\lambda}^{r_i}_{s,t} \mathbf{w}^{r_i}) \right) + 2\mu \boldsymbol{\lambda}^{r_i}_{s,t,j} \quad (3)$$

Inference: We use Gibbs Sampling [19] in our model, which is a standard inference technique for topic modeling. At each transition step of the Markov chain, the approximate probability of sentiment s in document d is defined as:

$$\pi_{d,s} = \frac{N_{d,s} + \gamma}{\sum_{s'=1}^{S}(N_{d,s'} + \gamma)} \tag{4}$$

In addition to the document-sentiment distribution π used in existing models, we estimate a word-sentiment distribution δ, which is defined as:

$$\delta_{w,s} = \frac{N_{w,s}^{-i} + \beta}{\sum_{s'=1}^{S}(N_{w,s'}^{-i} + \beta)} \tag{5}$$

We then use both δ and π for sampling sentiment to words. In particular, the approximate probability of the i_{th} word in document d under sentiment s is defined as:

$$P(s_i = s \mid \pi, \delta, \beta, \gamma) \propto \frac{N_{d,s}^{-i} + \gamma}{\sum_{s'=1}^{S}(N_{d,s'}^{-i} + \gamma)} \times \frac{N_{w,s}^{-i} + \beta}{\sum_{s'=1}^{S}(N_{w,s'}^{-i} + \beta)} \tag{6}$$

The probability of i_{th} word in document d under topic t and sentiment s is defined as:

$$P(z_i = t, s_i = m \mid \mathbf{z}^{-i}, \mathbf{s}^{-i}, \alpha, \beta, \gamma, \boldsymbol{\lambda}, \mathbf{w}) \propto$$

$$\frac{N_{d,m}^{-i} + \gamma}{\sum_{m'=1}^{S}(N_{d,m'}^{-i} + \gamma)} \times \frac{N_{w,m}^{-i} + \beta}{\sum_{m'=1}^{S}(N_{w,m'}^{-i} + \beta)} \times$$

$$\frac{N_{d,m,t}^{-i} + \alpha}{\sum_{t'=1}^{T}(N_{d,m,t'}^{-i} + \alpha)} \times \frac{\exp(\boldsymbol{\lambda}_{m,t} \cdot \mathbf{w}_{w_i})}{\sum_{w' \in V} \exp(\boldsymbol{\lambda}_{m,t} \cdot \mathbf{w}_{w'})} \tag{7}$$

The word-sentiment distribution δ and document-sentiment distribution π are initialized based on two public sentiment lexicons: *SentiWordNet* [20] and the opinion word list proposed by [21]. We combine these two lexicons into one lexicon. The processing steps are discussed in detail in Sect. 4.1. In the combined lexicon, each word is assigned a pair of values indicating the degree of positive or negative sentiment associated with the word. The word-sentiment distribution is then initialized and estimated based on the sentiment degrees of each word in the combined lexicon. Specifically, when an opinion-word w is assigned a topic t, we also add the values indicating the degree of sentiments of w to $N_{d,s}$, $N_{w,s}$, $N_{d,s,t}$ and $N_{s,t,w}$. The document-sentiment distribution is initialized to the sum of the word-sentiment distributions in each document.

4   Evaluation

In this section, we evaluate our proposed model against several related state-of-the-art baseline models, which have made their source code publicly available. We evaluate the following models:

JST [2]: Joint Sentiment/Topic Model, which can simultaneously discover aspects, opinions and their associated sentiment.

ASUM [1]: An aspect and sentiment unification model, which assumes that each sentence contains only one aspect.

ASUM+: A variant of ASUM, which uses the same sentiment lexicon as our model. The original ASUM model exploits several sentiment seed words that they define.

LOT: Our proposed opinion mining topic model, which is structured using latent embeddings and assigns sentiment to words based on document-sentiment and word-sentiment distributions. It also exploits our proposed general opinion lexicon to discover opinions that are more specific.

LOT-: A variant of LOT, which does not use our proposed general opinion lexicon.

Table 2. List of category names: 1000E and 100E (1st row), 1000NE and 100NE (2nd row), and the amazon review dataset used for training word embeddings (3rd row).

Alarm Clock, Cable Modem, Vacuum, GPS, Graphics Card, Headphone, Home Theater System, Keyboard, Projector, Rice Cooker
Baby, Bag, Dumbbell, Flashlight, Gloves, Jewelry, Movies, Magazine Subscriptions, Sandal, Video Games
Baby, Books, Beauty, Clothing, Cell Phone, Electronics, Shoes, Sports, Movies and TV, Video Games

We first evaluate the above models using the *Topic Coherence Score* proposed in [22] - topic coherence score is a very common metric to evaluate the quality of the topics discovered by a topic model. We then evaluate the models on the task of document-level sentiment classification. This allows us to compare the quality of the sentiment of words assigned by the models. Finally, we conduct a human evaluation to assess the quality of the topics, the sentiment classification of words and the specificity of discovered opinion-words.

4.1   Experimental Setup

Datasets. We use online reviews from 20 product categories in Amazon created by the authors in [23]. Each category contains 1000 online reviews. We first divide these reviews into two separate datasets: electronic (1000E) and non-electronic products (1000NE). In order to test the performance of the above models on a small number of reviews, we randomly sample 100 reviews from 1000E and 1000NE, and label these as 100E and 100NE respectively. The pre-trained word embeddings used in our model are obtained by training the classic skip-gram model [24] on related categories from a large public dataset of Amazon reviews [25]. The product categories used to evaluate the models, and those used for training word embeddings are shown in Table 2. We remove those words in our test datasets that do not have a pre-trained word embedding, and then follow the pre-processing steps in [1] to process all of the above datasets.

Lexicon. We utilize two existing public sentiment lexicons: *SentiWordNet* [20] and the opinion list created by the authors in [21]. In SentiWordNet, each word is labeled with multiple pairs of values ranging from 0.0 to 1.0 indicating the degree of positive or negative sentiment associated with the word. Different sentiment degrees of each word is based on different word contexts. We averaged these multiple sentiment degrees of each word and remove the words with low degree (less than 0.1). We also remove the words labeled as nouns.

Conversely, in the opinion list lexicon, each word is classified simply as positive or negative. As the opinion list does not contain sentiment degree information, we assign each positive and negative word in the list $(1.0, 0.0)$ and $(0.0, 1.0)$ respectively. Then we combine these two sentiment lexicons into a single lexicon. We observed that the opinion list typically contains more accurate sentiment information of words, although SentiWordNet contains many more words than the opinion list. As such, if a word appears in both lexicons, we use the sentiment degree in the opinion list in the combined lexicon.

For mining specific opinions, we rely our proposed general opinion lexicon, the words in which are manually selected from the opinion list in [21]. Our general opinion lexicon contains 548 of the most common general opinion-words, and each word is classified as positive or negative.

Parameter Setting. The parameters used in baseline models are set as described in their original papers. For each model, we set $\alpha = 0.1$, $\beta = 0.01$, $\gamma = 0.01$, which is a common setting for topic models [3,26].

4.2 Topic Coherence

We first evaluate the quality of the topics discovered by our model and baseline models using the topic coherence score proposed in [22]. This metric can accurately reflect the real semantic coherence of the topics discovered by topic models, and fits well with the goal of our model that we want to discover highly relevant aspects and corresponding opinions. We test our model and baseline models on each category of the four datasets above. Each tested model generates 10, 20 and 30 topics from each category of each dataset.

We report the average topic coherence score over 10 categories of each dataset in Fig. 2. Higher scores indicate higher coherence of a topic. It can be seen that LOT produces the best results with 10, 20 and 30 topics, and for all four datasets. This demonstrates that LOT generates more coherent topics than the baseline models based on student paired t-test ($p < 0.0001$). For the baseline models, the results show that ASUM outperforms ASUM+, which means that using the larger sentiment lexicon in ASUM is not guaranteed to improve its performance. JST also uses the sentiment lexicon used in our model, but our model still outperforms it significantly.

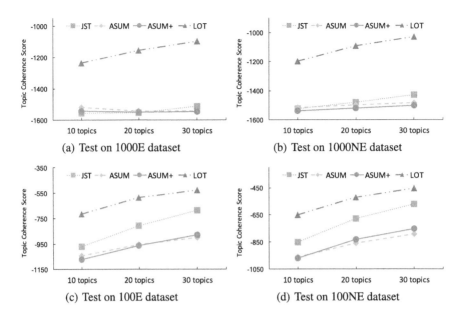

Fig. 2. Topic coherence scores when creating different numbers of topics. Each score is averaged over 10 product categories for each of the four datasets.

Table 3. Accuracy of document-sentiment classification on our four datasets (with 10 topics).

Models	100E	100NE	1000E	1000NE
LOT	**0.81**	**0.85**	**0.81**	**0.86**
ASUM+	0.70	0.71	0.72	0.74
ASUM	0.71	0.74	0.75	0.79
JST	0.69	0.68	0.65	0.67

4.3 Document Sentiment Classification

We next evaluate the models on the task of document-level sentiment classification. Although our model is not specifically designed for this task, we can assess the accuracy of the sentiment of words assigned by the models using this task. The sentiment of each document (review) for each model is determined based on the document-sentiment distribution defined in Eq. 4. Each review is assigned the sentiment with the highest probability.

For evaluation, we used the ratings already present in the datasets. Each uses a 5-star rating system - we label those reviews with 1 or 2-stars as negative, and those with 4 or 5 stars as positive. We do not consider reviews with 3-stars in this evaluation. We compare the sentiment predicted by the tested models with that from the real ratings, and then evaluate the results in terms of classification accuracy.

Table 3 shows the accuracy of the sentiment classification. We found that the accuracy is fairly insensitive to the number of topics, so we only report the

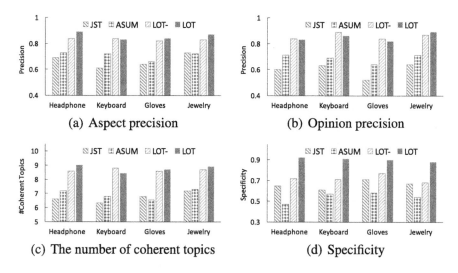

Fig. 3. Human evaluation results for the domains from 100E and 100NE

accuracy based on 10 topics. It can be seen that our LOT model produces the best results and outperforms the baseline models significantly ($p < 0.01$ based on student paired t-test) on all datasets, which shows that LOT model can assign more accurate sentiment to words. Note that ASUM+ and JST also use the sentiment lexicon used in LOT, but LOT still outperforms them significantly. The results also show that ASUM outperforms ASUM+, which indicates that using the larger sentiment lexicon in ASUM does not ensure higher accuracy of sentiment classification.

4.4   Human Evaluation

In this section, we evaluate JST, ASUM, LOT and LOT- from the perspective of human judges. ASUM+ is not evaluated since ASUM gives higher topic coherence scores and accuracy of document-sentiment classification. LOT- is a variant of LOT, which does not use our general opinion lexicon and is used to further demonstrate the effectiveness of our general opinion lexicon. Human evaluation has been widely used to evaluate the quality of the topics discovered by topic models [3, 26]. For our evaluation, we recruit five human judges who are familiar with Amazon products.

Datasets for Human Evaluation. We select four product review categories that are most familiar to human judges as the datasets. Two of the selected categories relate to electronic products, and the others are for non-electronic products. For each category, we evaluate the topics generated by the models from both of the large datasets (1000E and 1000NE) and the smaller datasets (100E and 100NE). Each topic model was configured to generate 10 topics, and the top 10 words in each topic were chosen for evaluation.

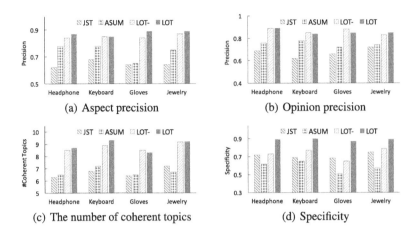

Fig. 4. Human evaluation results for the domains from 1000E and 1000NE

Topic Labeling. Each human volunteer was asked to evaluate each topic as coherent or incoherent. If more than half of the aspects and opinions in a topic were found to be related to each other, this topic was marked as coherent, otherwise it was marked as incoherent.

Aspect-Word Labeling. The human judges were first asked to identify and label each word as an aspect or opinion. Then they labeled each aspect word as correct if this aspect was relevant to the other words in the topic, otherwise the aspect word was labeled as incorrect.

Opinion-Word Labeling. For opinion words, the human judges were first asked to label each opinion-word as correct if the opinion-word was relevant to its topic and also assigned with the correct sentiment, otherwise the opinion word was labeled as incorrect. The human judges then labeled each opinion-word (that was previously marked as correct) as a specific opinion-word or a general opinion-word. Recall the example given in Sect. 1, given a sentence *"The battery is great and durable"*, *"great"* should be labelled as a general opinion and *"durable"* should be labelled as a specific opinion.

Evaluation Measure. We evaluate the results of topic labeling based on the number of coherent topics. For aspect-word labeling, we calculate the proportion of correct aspect-words in each topic. For opinion-word labeling, we calculate the proportion of correct opinion-words in each topic. Finally, we calculate the *specificity* of the opinion words, which refers to the proportion of specific opinion words in the correct opinion words in each topic.

Figure 3 shows the human evaluation results for the categories present in the small datasets (100E and 100NE). First, we can see that LOT generated more coherent topics than the baseline models (by a factor of 1.79). For the precision of aspect word labeling and opinion word labeling, LOT improves the baselines by at least 15.0% and 16.2% respectively. For the specificity of the discovered

Table 4. Example positive and negative topics of JST, ASUM and LOT from the category of Headphones. Opinion words are shown in *italics*. Incorrect aspect-words and opinion-words are underlined. Non-specific opinion-words are marked by a dashed line under them. All these errors are also marked in red.

JST		ASUM		LOT	
positive	negative	positive	negative	positive	negative
ear	music	ear	sound	headphone	headphone
pair	bass	sound	player	wear	audio
earbud	sound	headphone	quality	size	radio
hear	sony	earbud	hour	earbud	music
head	trip	*fit*	money	inside	state
around	start	*noise*	cord	*fit*	*loud*
big	noise	*good*	*unfortunately*	*comfortable*	*annoying*
better	*great*	*great*	*portable*	*durable*	*headache*
fit	*heavy*	*better*	*fall*	*light*	*loose*
comfortable	*work*	*recommend*	*right*	*decent*	*distortion*

opinion-words, we can see that LOT can mine more specific opinions than the baseline models (by at least 24.2%).

The corresponding human evaluation results from the product categories in the large datasets (1000E and 1000NE) are shown in Fig. 4. We can observe that LOT generates on average 2.13 more coherent topics than the baseline models. With regard to the precision of aspect-word labeling and opinion-word labeling, LOT improves the baselines by at least 13.5% and 10.7% respectively and can also discover at least 18.2% more specific opinion-words than the baseline models.

We can also observe from Figs. 3 and 4 that LOT- gives similar performance to LOT on aspect precision, opinion precision and the number of coherent topics. However, LOT can mine more specific opinion-words than LOT- by exploiting our proposed general opinion lexicon.

Overall, our LOT model outperforms the baseline models significantly ($p < 0.01$ for each evaluation measure based on student paired t-test). The Kappa's agreement for topic labeling, aspect-word labeling, opinion-word labeling and opinion-specificity labeling are 0.841, 0.805, 0.811 and 0.862 respectively, which shows that the human judges made similar decisions during the evaluation.

Table 4 gives some examples of positive and negative topics discovered by JST, ASUM and LOT from the reviews in the category of headphones. Each topic contains relevant aspect-words and opinion-words. Positive and negative sentiment of each topic mostly relates to these opinion-words, rather than the aspect-words; opinion-words are moved to the bottom in each topic and shown in Italics. The words not related to their topics are underlined. A dashed line is drawn under the non-specific (general) opinion-words. These errors are also marked in red. We can observe that LOT generates more coherent topics containing more relevant aspects and more specific opinions with more accurate sentiment than the baselines (less errors). For the baseline models, we can observe that ASUM can discover opinions with more accurate sentiment than JST, but JST can discover more specific opinions than ASUM.

5   Conclusions

This paper has proposed a new opinion mining topic model structured using latent embedding for mining aspect-level specific opinions, called the LOT. For discovering relevant aspects and specific opinions, we exploit latent embeddings to structure our opinion mining topic model. To improve the sentiment classification of discovered aspects and opinions, our model assigns sentiment to each word based on both document-sentiment and word-sentiment distributions. As part of this work, we created and released a general opinion lexicon for mining more specific opinions. Our experimental results show that LOT can discover more relevant and specific aspect-level opinions with more accurate sentiment classification than state-of-the-art baseline models.

Acknowledgments. This material is based upon work supported in whole or in part with funding from the Laboratory for Analytic Sciences (LAS). Any opinions, findings, conclusions, or recommendations expressed in this material are those of the authors and do not necessarily reflect the views of the LAS and/or any agency or entity of the United States Government. The authors would like to thank staff at the LAS for providing funding and inspiration for much of this work.

References

1. Jo, Y., Oh, A.H.: Aspect and sentiment unification model for online review analysis. In: Proceedings of the Fourth ACM International Conference on Web Search and Data Mining, pp. 815–824. ACM (2011)
2. Lin, C., He, Y.: Joint sentiment/topic model for sentiment analysis. In: Proceedings of the 18th ACM Conference on Information And Knowledge Management, pp. 375–384. ACM (2009)
3. Wang, S., Chen, Z., Liu, B.: Mining aspect-specific opinion using a holistic lifelong topic model. In: Proceedings of the 25th International Conference on World Wide Web, International World Wide Web Conferences Steering Committee, pp. 167–176 (2016)
4. Dermouche, M., Kouas, L., Velcin, J., Loudcher, S.: A joint model for topic-sentiment modeling from text. In: Proceedings of the 30th Annual ACM Symposium on Applied Computing, pp. 819–824. ACM (2015)
5. Mei, Q., Ling, X., Wondra, M., Su, H., Zhai, C.: Topic sentiment mixture: modeling facets and opinions in weblogs. In: Proceedings of the 16th International Conference on World Wide Web, pp. 171–180. ACM (2007)
6. Moghaddam, S., Ester, M.: The FLDA model for aspect-based opinion mining: addressing the cold start problem. In: Proceedings of the 22nd International Conference on World Wide Web, 909–918. ACM (2013)

7. Mikolov, T., Chen, K., Corrado, G., Dean, J.: Efficient estimation of word representations in vector space (2013). arXiv preprint arXiv:1301.3781
8. Liu, Y., Liu, Z., Chua, T.S., Sun, M.: Topical word embeddings. In: AAA, vol. I, pp. 2418–2424 (2015)
9. Das, R., Zaheer, M., Dyer, C.: Gaussian lda for topic models with word embeddings. In: Proceedings of the 53nd Annual Meeting of the Association for Computational Linguistics (2015)
10. Wan, L., Zhu, L., Fergus, R.: A hybrid neural network-latent topic model. AISTATS **12**, 1287–1294 (2012)
11. Fu, X., Wu, H.: Topic sentiment joint model with word embeddings. (Interactions between Data Mining and Natural Language Processing)
12. Nguyen, D.Q., Billingsley, R., Du, L., Johnson, M.: Improving topic models with latent feature word representations. Trans. Assoc. Comput. Linguist. **3**, 299–313 (2015)
13. Zhao, W.X., Jiang, J., Yan, H., Li, X.: Jointly modeling aspects and opinions with a MaxEnt-LDA hybrid. In: Proceedings of the 2010 Conference on Empirical Methods in Natural Language Processing, pp. 56–65. Association for Computational Linguistics (2010)
14. Blei, D.M., Ng, A.Y., Jordan, M.I.: Latent Dirichlet allocation. J. Mach. Learn. Res. **3**, 993–1022 (2003)
15. Xu, M., Yang, R., Harenberg, S., Samatova, N.: A lifelong learning topic model structured using latent embeddings. In: IEEE 11th International Conference on Semantic Computing. IEEE (2017)
16. Diao, Q., Qiu, M., Wu, C.Y., Smola, A.J., Jiang, J., Wang, C.: Jointly modeling aspects, ratings and sentiments for movie recommendation (JMARS). In: Proceedings of the 20th ACM SIGKDD International Conference on Knowledge Discovery and Data Mining, pp. 193–202. ACM (2014)
17. Titov, I., McDonald, R.T.: A joint model of text and aspect ratings for sentiment summarization. In: ACL, vol. 8, pp. 308–316 (2008). Citeseer
18. Lu, B., Ott, M., Cardie, C., Tsou, B.K.: Multi-aspect sentiment analysis with topic models. In: 2011 IEEE 11th International Conference on Data Mining Workshops, pp. 81–88. IEEE (2011)
19. Griffiths, T.: Gibbs sampling in the generative model of latent Dirichlet allocation (2002)
20. Baccianella, S., Esuli, A., Sebastiani, F.: Sentiwordnet 3.0: an enhanced lexical resource for sentiment analysis and opinion mining. In: LREC, vol. 10, pp. 2200–2204 (2010)
21. Hu, M., Liu, B.: Mining and summarizing customer reviews. In: Proceedings of the Tenth ACM SIGKDD International Conference on Knowledge Discovery and Data Mining, pp. 168–177. ACM (2004)
22. Mimno, D., Wallach, H.M., Talley, E., Leenders, M., McCallum, A.: Optimizing semantic coherence in topic models. In: Proceedings of the Conference on Empirical Methods in Natural Language Processing, pp. 262–272. Association for Computational Linguistics (2011)
23. Chen, Z., Ma, N., Liu, B.: Lifelong learning for sentiment classification, vol. 2, pp. 750 (2015). Short Papers
24. Mikolov, T., Sutskever, I., Chen, K., Corrado, G.S., Dean, J.: Distributed representations of words and phrases and their compositionality. In: Advances in Neural Information Processing Systems, pp. 3111–3119 (2013)

25. McAuley, J., Leskovec, J.: Hidden factors and hidden topics: understanding rating dimensions with review text. In: Proceedings of the 7th ACM Conference on Recommender Systems, pp. 165–172 ACM (2013)
26. Chen, Z., Mukherjee, A., Liu, B., Hsu, M., Castellanos, M., Ghosh, R.: Exploiting domain knowledge in aspect extraction. In: EMNLP, pp. 1655–1667 (2013)

A Comparative Study of Target-Based and Entity-Based Opinion Extraction

Joseph Lark[✉], Emmanuel Morin, and Sebastián Peña Saldarriaga

University of Nantes, LS2N CNRS, 6004 Nantes, France
{joseph.lark,emmanuel.morin}univ-nantes.fr, sebastian@dictanova.com

Abstract. Opinion target extraction is a crucial task of opinion mining, aiming to extract occurrences of the different entities of a corpus that are subjects of an opinion. In order to produce a readable and comprehensible opinion summary, which is the main application of opinion target extraction, these occurrences are consolidated at the entity level in a second task. In this paper we argue that combining the two tasks, *i.e.* extracting opinion targets using entities as labels instead of binary labels, yields better results for opinion target extraction. We compare the binary approach and the multi-class approach on available datasets in English and French, and conduct several investigation experiments to explain the promising results. Our experiment show that an entity-based labelling not only improves opinion extraction in a single domain setting, but also let us combine training data from different domains to improve the extraction, a result that has never been achieved on target-based training data.

1 Introduction

The field of sentiment analysis has attracted much interest over the recent years, and several frameworks have been proposed to tackle the challenges it presents. One of the main framework for sentiment analysis is aspect-based sentiment analysis (ABSA), which is particularly suited for the analysis of consumer reviews. This framework has been designed for summarizing points of interest (or *entities*) and causes of interest (or *aspects*) from every occurrence of opinion expression in a corpus. Consequently, the main subtasks in this framework include finding these occurrences of opinion (or *targets*) on the many subjects in the corpus and associating targets to an entity and an aspect. Initial works [1–3] on formalisation of the problem led to a binary annotation of the target extraction task, labelling as *target* a continuous span of text in a sentence representing an occurrence of a subject in an opinion expression. For instance, in the sentence *"The waiter is unfriendly but the menu is delicious"*, *waiter* and *menu* are opinion targets. Entity extraction and aspect extraction are then treated as additional classification tasks on the opinion targets. In the given example, *waiter* is an occurrence of the SERVICE entity and *menu* an occurrence of FOOD.

While this formulation of the opinion target extraction problem has helped tremendously on designing well performing systems, the binary annotation of

© Springer Nature Switzerland AG 2018
A. Gelbukh (Ed.): CICLing 2017, LNCS 10762, pp. 211–223, 2018.
https://doi.org/10.1007/978-3-319-77116-8_16

targets can seem suboptimal for the quite complex language phenomenon of opinion. A known limitation of this formulation is that opinion target extraction is very sensible to the topic of the corpus, which is often referred to as domain specificity [1,2,4,5]. We suggest that this limitation is in fact correlated to entity specificity and experiment a multi-class representation of the opinion target extraction task. To this end we use entities as target labels, in a manner similar to named entity recognition, where entities are labelled differently according to the concept they represent, but are consistent across domains.

After a brief review of related work (Sect. 2), we argue in this work in favour of a multi-class representation of opinion extraction over the current binary representation (Sect. 3). We compare extraction results on the SemEval ABSA datasets (Sect. 4), and first observe that an entity-based model improves the performance in a single domain setting. Moreover we find that in a cross-domain setting, where target-based opinion extraction learning has shown to be disadvantageous, an entity-based model improves the extraction (+1.68 points on F1 on average in English and +2.78 in French). Finally, we analyse opinion entities occurrences in the annotated data to explain these results (Sect. 5), and put forth that coherence of opinion words towards entities is critical for opinion extraction.

2   Related Work

This work is related to the formalisation of the ABSA problem, and especially to the representation of opinion target occurrences and their associated semantic category.

Definitions for ABSA tasks are fairly recent. While basic definitions such as sentiment polarity, opinion words or opinion targets remained stable since initial work on the subject [6–10], the definition of an opinion entity has been regularly revisited. The core idea of opinion entity extraction is to consolidate all occurrences of opinion targets that refer to the same object (*e.g.* a phone), object feature (*e.g.* a phone screen size), or abstract notion (*e.g.* the price or practicality of a phone) under a unique label.

Hu & Liu [7] define this task as the last step of opinion summarization, following the prior steps of entity (or *feature*) extraction, opinion word extraction and opinion orientation prediction. In their work, only explicit entities are extracted (*i.e.* opinion targets occurrences matching the entity term). Liu et al. [11] find implicit occurrences of entities by building a dictionary of variants from key entity terms, such as *weight* from *heavy* or *price* from *cost*. Kim & Hovy [12] introduce the definition of an opinion *topic* as "an object an opinion is about", which very much corresponds to what is most known in recent work as an opinion entity [2]. The approach for entity extraction presented in their work relates to ours as they first identify semantic roles, using semantic frames, to find opinion entities. However, these semantic roles are not specific to opinion targets as in our approach. In a similar manner, Mukherjee & Liu [13] suggest a topic modelling method to infer an opinion target entity (or *category*) from manually generated seed terms. Kobayashi et al. [14] formalise the opinion extraction task using a two-level hierarchy of opinion targeted objects, as in the current ABSA framework. Ding et al. [15] introduce the definition of an opinion entity as an identifiable concept (an object, person, event, etc.) in a taxonomy of components and for which a set of attributes can be defined.

SemEval ABSA workshops [3,16,17] have largely contributed to the definition of the aspect-based sentiment analysis tasks and have led to the production of labelled data. In ABSA2014, entities are associated to sentences only and not opinion targets. In ABSA2015 and ABSA2016, one of the subtasks is to find associations of two types of semantic classes for each opinion target: opinion entities and opinion categories. In this definition of the opinion extraction, a word or multi-word expression can be labelled as *target*. A target is then an occurrence of the targeted entity, which may not share the same textual form as this particular occurrence. Finally, a category of opinion describes the precise aspect that is being criticised. Intuitively, an *entity* is therefore a concept that can be the subject of an opinion or not, while an *aspect* is a subjective attribute of this concept that calls for an opinion.

3   Entity-Based Opinion Extraction

The task of opinion target extraction is to find occurrences of subjects towards which one expresses an opinion. To this end, a widely adopted approach is to consider that a subject can either be an opinion target or not. In particular, sentences such as the following are to be disambiguated:

- *We went to this restaurant based on prior internet comments.*
- *I was very disappointed with this **restaurant**.*

While both contain an occurrence of *restaurant*, only the occurrence of the second sentence is an opinion target. In order to summarize opinions of a corpus, existing works suggest to infer the opinion entity as an additional piece of information associated with the opinion target [3,7,11,14,16,17]. We differ from this approach by directly extracting entities occurrences that are subject to an opinion. We question the need for binary target extraction and argue that, in a manner similar to named entity recognition, entity labels improve opinion extraction, assuming that these are coherently defined. Using existing concepts from previous formalisation works, we suggest the labelling of targets as entities rather than as a binary information. Despite being present in the literature this formalisation has, to the best of our knowledge, not been much studied from a opinion target extraction point of view. In particular, it has never been exploited to improve opinion target extraction or to tackle the domain adaptation problem, two applications we cover in this paper.

3.1   Coherence of Opinion Entities

In addition to the coherence of targets towards opinion entities, which usually share a hyperonymy dependency, the concept of entity in opinion target extraction is strongly related to the use of opinion words. Besides some very generic adjectives such as *good* and *bad*, opinion words are associated with specific types of opinion targets, which very often coincide with opinion entities [4,5]. In the example shown in Sect. 1, *waiter* is associated with *unfriendly* and *menu* with *delicious*; the opposite associations seem highly improbable (*delicious waiter* or *unfriendly menu*). This linguistic coherence in opinion expressions towards each entity is to us a motivation to investigate entity-based opinion extraction.

3.2 Domain Adaptation Through Opinion Entities

In addition to a finer-grained extraction, we see in the annotation of opinion targets using entity labels an opportunity to tackle the domain adaptation problem in the context of opinion target extraction with a novel approach. Indeed, a domain can be defined as a set of entities, each entity being a label for opinion target extraction. Using this formalisation let us use the fact that different domains can share some entities, thus possibly sharing training data. This approach differs from existing works on domain adaptation in the sense that we do not adapt a closed and well-defined first model (specific or general) to another domain. Our hypothesis is that each domain in the context of opinion mining in user reviews is composed of several entities that can be shared across domains. When building an opinion target extraction model for a new domain, the domain adaptation task could thus be shifted to an identification task of the entities that compose the new domain. The new model would benefit in training from previously annotated data in a modular manner, as pictured on Fig. 1.

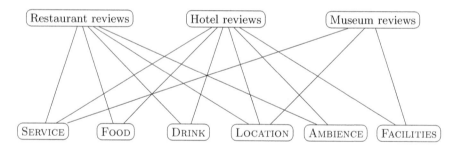

Fig. 1. Illustration of entity modularity: as several entities are shared across domains (in this example, Restaurant, Hotel and Museum), training data could be mutualised.

4 Experiments and Results

In these experiments, we compare a target-based annotation and an entity-based annotation for the task of opinion target extraction on the English and French SemEval ABSA datasets. We first describe the datasets and the extraction method we use in our experiments, and provide results for single domain and cross-domain settings.

4.1 Datasets

We conducted our study on customer reviews datasets from the SemEval ABSA workshops[1] [16,17]. These include restaurant and hotel reviews in English, and restaurant and museum reviews in French. Each dataset was annotated by a native linguist, who indicated for each sentence offsets of opinion targets and associated entities. Additional information on the datasets is shown in Table 1.

Table 1. Number of reviews, sentences and targets for each corpus..

Corpus	Reviews	Sentences	Targets
Restaurants (English)	440	2,676	2,529
Hotels (English)	30	266	264
Restaurants (French)	455	2,427	2,484
Museums (French)	162	687	582

The datasets are very relevant for our study as these cover different domains sharing some of their entities, as shown in Table 2. Indeed, this configuration let us demonstrate the usefulness of cross-domain entities in opinion target learning. Besides SemEval ABSA workshops, the datasets have been used as evaluation material for several works. However, to the best of our knowledge, there is no existing work on comparing annotations for opinion extraction.

Table 2. Opinion entities for each corpus. Shared entities are indicated in bold, and starred entities are related but don't share the same label.

Restaurants (English and French)	Hotels (English only)	Museums (French only)
AMBIENCE, DRINKS*, FOOD*, **LOCATION**, RESTAURANT, **SERVICE**	FACILITIES, FOOD AND DRINKS*, HOTEL, **LOCATION**, ROOMS, ROOMS AMENITIES, **SERVICE**	COLLECTIONS, FACILITIES, **LOCATION**, MUSEUM, **SERVICE**, TOUR GUIDING

4.2 Extraction Method

Following Jakob & Gurevych [1] and similar works that have performed well in the SemEval ABSA workshops, we train a Conditional Random Fields [18] model for target and entity extraction. Both extractions are formulated as sequence labelling tasks, and only differ from one another by the nature of the annotation: while target labels are binary, entities are annotated following a multi-class

[1] http://alt.qcri.org/semeval2016/task5/.

labelling, as shown on Fig. 2. We use the CRF++[2] toolkit for our experiments, with a segmentation on sentences. Features for each word entry include the word unigram, word bigram, part-of-speech tag and lemma of the preceding, current and following word.

Word	Lemma	POS tag	Target label	Entity label
Excellent	excellent	ADJ	0	0
atmosphere	atmosphere	NOUN	Target	AMBIENCE
,	,	.	0	0
delicious	delicious	ADJ	0	0
dishes	dish	NOUN	Target	FOOD
good	good	ADJ	0	0
and	and	CON	0	0
friendly	friendly	ADJ	0	0
service	service	NOUN	Target	SERVICE
.	.	.	0	0

Fig. 2. Annotation example on a sentence from the English train corpus. The 0 label represents the Outside class in both target-based and entity-based annotations.

We use a simple, class descriptive annotation (Target/Outside for target-based extraction and *Entity name*/Outside for entity-based extraction) instead of the often used BIO format, as early results indicated a better performance using simpler labels. Breck et al. [19] made a similar observation for opinion expression extraction, and pointed that the absence of contiguous annotations, as in our case, could explain the fact that the BIO format does not shape best the labelled data. Finally, we resolve ambiguous cases for the multi-class scenario, *i.e.* when probability of outside class is less than 0.5, by selecting the most probable entity class.

4.3   Single Domain Opinion Extraction

In order to compare a target-based and an entity-based opinion extraction, we train two distinct CRF models on the same train corpus, namely the Restaurants reviews corpus, and use the same features; only labels were replaced to compare the results from the extractions. We conduct this experiment on the two languages for which this type of annotation is available, English and French. Comparison of the results, reported in Tables 3 and 4, shows that the entity annotation enhance both precision (+0.83 percentage points in English, +0.37 pp in French) and recall (+0.65 pp in English, +0.78 pp in French) for opinion target extraction. Significance testing using a t-test showed extraction in English to be significant, but less so in French. Nonetheless, the closeness of the results between the two extractions questions the need for a target extraction step,

---

[2] https://taku910.github.io/crfpp/.

Table 3. Target-based and entity-based opinion extraction on English Restaurants reviews. The *p-value* for precision is 2.48e−3 and 3.91e−3 for recall.

Train corpus	Precision†	Recall†	F1
Targets	67.60	62.23	64.81
Entities	**68.43**	**62.88**	**65.54**

Table 4. Target-based and entity-based opinion extraction on French Restaurants reviews. The *p-value* for precision is 3.31e−1 and 3.29e−1 for recall.

Train corpus	Precision	Recall	F1
Targets	74.27	59.25	65.92
Entities	**74.64**	**60.03**	**66.55**

as the end goal. Intuitively, this result supports the hypothesis of a linguistic coherence of entities over opinion targets, in other words that entity labels help disambiguate the target extraction more than they add ambiguity. The fact that this behaviour can be observed on both languages also favours this idea.

4.4 Cross-Domain Opinion Target Learning

Using an identical framework, we now want to compare target-based and entity-based opinion extraction in a cross-domain setting. To this end, we train both models using additional out-of-domain data from the SemEval ABSA dataset. As described in Sect. 4.1, such out-of-domain data include hotel reviews for the English corpus and museum reviews for the French corpus. Results shown in Tables 5 and 6 demonstrate best the usefulness of an entity-based annotation over a target-based annotation. On one hand, we can see on line 1 that when adding target-based training data from another domain, the extraction is generally less performing.

On the other hand, adding entity-based out-of-domain training data yields opposite results, as we can see on line 2. This tends to confirm that training data on shared entities improve the extraction, or that differentiating exclusive entities help disambiguate non relevant contexts. Only precision in the case of the French corpus is lower in the cross-domain setting, which may be due to the fact that museum reviews are less similar to restaurants ones than hotels reviews.

To further investigate these results, we run an entity-by-entity opinion extraction on single and cross-domain datasets. When analysing results of this extraction, as it can be seen on Tables 7 and 8, we can observe that F1 for these entities is significantly better (+6.5 pp for LOCATION and +3.18 pp for SERVICE in English datasets, +5.36 pp for LOCATION and +1.97 pp for SERVICE in French datasets).

Evolution of entities that are exclusive to the restaurant domain is less trivial. Not only results from the cross-domain model can be better or worse than those of the single domain model, but in this case differences are not consistent across languages. For instance, F1 for the RESTAURANT entity in the English dataset has decreased (−1.05 pp), while it displays a consistent improvement (+2.6 pp) in the French dataset.

Table 5. Target and entity cross-domain opinion extraction on English Restaurants (R) and Hotels (H) datasets.

Train corpus	Precision	Recall	F1
Targets (R)	**67.60**	**62.23**	**64.81**
Targets (R+H)	67.25	61.91	64.47
Entities (R)	73.84	51.70	60.81
Entities (R+H)	**74.37**	**52.67**	**61.66**

Table 6. Target and entity cross-domain opinion extraction on French Restaurants (R) and Museums (M) datasets.

Train corpus	Precision	Recall	F1
Targets (R)	**74.27**	**59.25**	**65.92**
Targets (R+M)	72.69	58.33	64.72
Entities (R)	**74.64**	60.03	66.55
Entities (R+M)	73.34	**61.57**	**66.94**

Table 7. Single domain and cross-domain entity learning on the English Restaurants and Hotels reviews dataset.

Train corpus	Restaurants			Restaurants + Hotels			Gain (pp)
Entity	Precision	Recall	F1	Precision	Recall	F1	
AMBIENCE	76.6	61.02	67.92	80.43	62.71	70.48	+2.56
DRINKS	78.95	41.67	54.55	82.35	38.89	52.83	−1.72
FOOD	67.10	47.69	55.76	68.42	48.00	56.42	+0.66
LOCATION	100.00	40.00	57.14	58.33	70.00	63.64	+6.50
RESTAURANT	58.97	28.05	38.02	59.46	26.83	36.97	−1.05
SERVICE	78.95	69.44	73.89	81.44	73.15	77.07	+3.18

Table 8. Single domain and cross-domain entity learning on the French Restaurants and Museums reviews dataset.

Train corpus	Restaurants			Restaurants + Museums			Gain (pp)
Entity	Precision	Recall	F1	Precision	Recall	F1	
AMBIENCE	86.21	66.67	75.19	92.00	61.33	73.60	−1.59
DRINKS	73.33	32.35	44.90	72.22	38.23	50.00	+5.10
FOOD	65.37	53.00	58.54	72.96	53.62	61.81	+3.27
LOCATION	60.00	13.64	22.22	57.14	18.18	27.58	+5.36
RESTAURANT	56.45	51.47	53.85	62.50	51.47	56.45	+2.60
SERVICE	87.90	80.15	83.85	92.37	80.14	85.82	+1.97

5   Opinion Entity Analysis

In this section we conduct a study on coherence of opinion entities in the restaurants reviews datasets to best explain our results.

5.1   Target Terms and Opinion Words

We analyse the coherence of entity labels for opinion target extraction by measuring the number of target terms and opinion words by entity and across entities, as these two types of lexical elements are characteristic of the expressed opinion [2]. Coherence for each entity is here represented by the uniqueness of these elements (columns 3 and 6 of Tables 9 and 10) and coherence across entities is measured by their exclusiveness for a given entity (columns 4 and 7 of Tables 9 and 10).

Table 9. Target terms and opinion words (OW) by entity in the English dataset.

Entity	#Target terms	#Unique targets	#Exclusive targets	#OW	#Unique OW	#Exclusive OW
AMBIENCE	287	115 (40.07%)	99 (86.09%)	165	85 (51.52%)	33 (38.82%)
DRINKS	133	59 (44.36%)	58 (98.31%)	48	23 (47.92%)	4 (17.39%)
FOOD	1,311	541 (41.27%)	538 (99.45%)	776	193 (24.87%)	105 (54.40%)
LOCATION	32	16 (50.00%)	10 (62.50%)	18	15 (83.33%)	3 (20.00%)
RESTAURANT	343	119 (34.69%)	99 (83.19%)	214	90 (42.06%)	32 (35.56%)
SERVICE	432	62 (14.35%)	56 (90.32%)	250	122 (48.80%)	59 (48.36%)

Table 10. Target terms and opinion words (OW) by entity in the French dataset.

Entity	#Target terms	#Unique targets	#Exclusive targets	#OW	#Unique OW	#Exclusive OW
AMBIENCE	253	42 (16.60%)	32 (76.19%)	124	69 (55.65%)	29 (42.03%)
DRINKS	123	47 (38.21%)	46 (97.87%)	43	33 (76.74%)	4 (12.12%)
FOOD	1,248	400 (32.05%)	392 (98.00%)	540	231 (42.78%)	136 (58.87%)
LOCATION	56	28 (50.00%)	20 (71.43%)	25	19 (76.00%)	6 (31.58%)
RESTAURANT	247	42 (17.00%)	25 (59.52%)	127	75 (59.06%)	30 (40.00%)
SERVICE	536	49 (9.14%)	46 (93.88%)	321	150 (46.73%)	78 (52.00%)

The main observation here seems to be the fact that measures are consistent on both languages. Indeed, relative order of entities with regards to uniqueness and exclusivity of target terms as well as opinion words is very similar in English and French, despite the fact that entities were annotated on different datasets and by different experts. This is a strong argument towards an entity-based representation of opinion target as it tends to show a coherent conceptual coherence in entities in addition to the sheer homogeneity of target terms for each language. However, understanding the relation between coherence and extraction performance is not trivial as uniqueness and exclusivity are not correlated. For instance, the entity FOOD is represented by a large number of target terms – mainly descriptions of the different dishes – that are highly exclusive to this entity, while LOCATION is represented by a small number of target terms, including *restaurant* or *place* that are shared by other entities such as RESTAURANT or AMBIENCE. When crossing these measures with the per-entity evaluation of opinion extraction (Tables 7 and 8), we can see that entities that are best recognised, namely AMBIENCE and SERVICE, are those combining a high rate of exclusive opinion words and small number of unique target terms. In next experiments, we investigate how this coherence affects opinion extraction learning by conducting entity-by-entity active learning iterations.

5.2   Opinion Entity Learning

We analyse learning iterations from target-based and entity-based training data-sets through evaluations at the target level and at the entity level. Batches of 50 sentences are sampled from the training part of the restaurants reviews corpus in English and French. We used the uncertainty sampling strategy [20], *i.e.* we added on each iteration the 50 sentences with the lowest global output sequence probability, using annotations provided in the SemEval datasets.

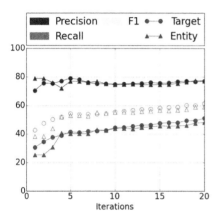
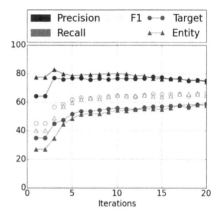

Fig. 3. Learning curves for target extraction in the English dataset using target and entity-based labelling.

Fig. 4. Learning curves for target extraction in the French dataset using target and entity-based labelling.

Although a possible drawback of a multi-class framework is an increased need for training examples, evaluation at the target level shows that the entity-based model quickly converge to a learning curve identical to the one of the target-based model. As it can be seen on Figs. 3 and 4, the entity-based model starts from a lower recall and a higher precision than the target-based model on the initial batch of reviews, and stabilises before the first five iterations.

Evaluation at the entity level, displayed on Figs. 5 and 6, shows that learning can be even faster for entities such as AMBIENCE and SERVICE, which we previously highlighted for their high coherence. Learning curves for other entities, LOCATION, DRINKS and RESTAURANT, are very chaotic. Again these results are surprisingly similar for both languages. From the observation of lexical elements associated with opinion entities shown on Tables 9 and 10, it seems that opinion word coherence impacts the most opinion entity learning. Indeed the common factor of the three entities for which learning is fast and steady (AMBIENCE, FOOD and SERVICE) is a high rate of exclusive opinion words, whereas other metrics are less conclusive. Specifically, the rate of exclusive target terms does not appear as important as we assumed. An example of this observation is the fact that the AMBIENCE entity present a stable learning curve in spite of a relatively low rate of exclusive target terms.

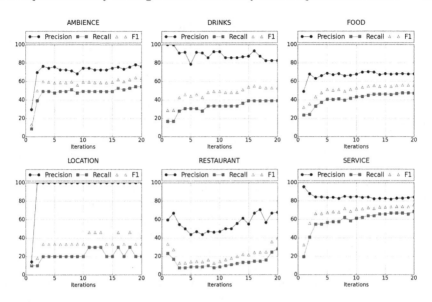

Fig. 5. Learning curves for entity extraction in the English Restaurants dataset.

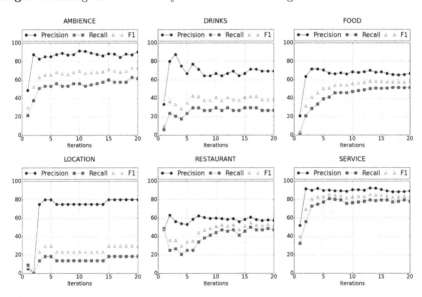

Fig. 6. Learning curves for entity extraction in the French Restaurants dataset.

6   Conclusion

In this paper we compare a target-based and an entity-based annotation for opinion extraction. From an initial intuition that the complex nature of opinion expression in language requires a fine-grained labelling, we investigate how this is depicted on real data. We use available customer reviews datasets in English and French, labelled on opinion targets and their associated entities. Our experiments show that an entity-based labelling not only improves opinion extraction in a single domain setting, but also let us combine training data from different domains to improve the extraction, a result that has never been achieved on target-based training data.

Elements as to why entity annotation improves opinion extraction are strongly related to the coherence of elements in the lexical context of opinion targets. We found that the exclusivity of opinion words towards an entity is directly linked to the capacity of the model to correctly learn to recognise its occurrences. The exclusivity of target terms representing an entity contribute to a lesser extent to the quality of the learning process. In our observations, this metric is only correlated to the convergence rate of the learning curve.

In our sense, these observations are particular signs of a need for a larger framework. In a manner similar to named entity recognition, where relevant items are defined by distinct categories (person, location, company, *etc.*), we see in entity-based opinion the opportunity to build a multi-class model for opinion extraction based on the opinion linguistic context rather than on the domain of the analysed corpus. A great advantage of a model of this kind would be to ease the domain adaptation problem. While target-based opinion extraction is very sensitive to the domain of the dataset it is trained on, we demonstrate in our experiments that an entity-based opinion extraction model could benefit from training data of multiple domains. In future works, we will investigate how multiple domains can be covered using this framework.

References

1. Jakob, N., Gurevych, I.: Extracting opinion targets in a single- and cross-domain setting with conditional random fields. In: Proceedings of EMNLP 2010, Cambridge, MA, USA, pp. 1035–1045 (2010)
2. Liu, B.: Sentiment analysis and opinion mining. In: Synthesis Lectures on Human Language Technologies. Springer, Boston (2012)
3. Pontiki, M., Galanis, D., Pavlopoulos, J., Papageorgiou, H., Androutsopoulos, I., Manandhar, S.: Semeval-2014 task 4: aspect based sentiment analysis. In: Proceedings of SemEval 2014, Dublin, Ireland (2014)
4. Li, F., Pan, S., Jin, O., Yang, Q., Zhu, X.: Cross-domain co-extraction of sentiment and topic lexicons. In: Proceedings of ACL 2010, Uppsala, Sweden, pp. 410–419 (2012)
5. Hamilton, W.L., Clark, K., Leskovec, J., Jurafsky, D.: Inducing domain-specific sentiment lexicons from unlabeled corpora (2016). arXiv preprint arXiv:1606.02820
6. Hatzivassiloglou, V., McKeown, K.R.: Predicting the semantic orientation of adjectives. In: Proceedings of ACL 1997, Madrid, Spain, pp. 174–181 (1997)
7. Hu, M., Liu, B.: Mining and summarizing customer reviews. In: Proceedings of KDD 2004, Seattle, WA, USA, pp. 68–177 (2004)
8. Kim, S.M., Hovy, E.: Determining the sentiment of opinions. In: Proceedings of COLING 2004, Geneva, Switzerland, p. 1367 (2004)
9. Bethard, S., Yu, H., Thornton, A., Hatzivassiloglou, V., Jurafsky, D.: Automatic extraction of opinion propositions and their holders. In: Proceedings of AAAI 2004, Palo Alto, CA, USA (2004)

10. Popescu, A.M., Etzioni, O.: Extracting product features and opinion from reviews. In: Proceedings of HLT-EMNLP 2005, Vancouver, Canada, pp. 339–346 (2005)
11. Liu, B., Hu, M., Cheng, J.: Opinion observer: analyzing and comparing opinions on the web. In: Proceedings of WWW 2005, Chiba, Japan, pp. 342–351 (2005)
12. Kim, S.M., Hovy, E.: Extracting opinions, opinion holders, and topics expressed in online news media text. In: Proceedings of SST 2006, Sydney, Australia (2006)
13. Mukherjee, A., Liu, B.: Aspect extraction through semi-supervised modeling. In: Proceedings of ACL 2012, Jeju Island, Korea, pp. 339–348 (2012)
14. Kobayashi, N., Inui, K., Matsumoto, Y.: Opinion mining from web documents: extraction and structurization. Trans. Jpn. Soc. Artif. Intell. **22**, 227–238 (2007)
15. Ding, X., Liu, B., Yu, P.S.: A holistic lexicon-based approach to opinion mining. In: Proceedings of WSDM 2008, Palo Alto, CA, USA, pp. 231–240 (2008)
16. Pontiki, M., Galanis, D., Papageorgiou, H., Manandhar, S., Androutsopoulos, I.: Semeval-2015 task 12: aspect based sentiment analysis. In: Proceedings of SemEval 2015, Denver, CO, USA, pp. 486–495 (2015)
17. Pontiki, M., et al.: Semeval-2016 task 5: aspect based sentiment analysis, San Diego, CA, USA (2016)
18. Lafferty, J.D., McCallum, A., Pereira, F.C.N.: Conditional random fields: probabilistic models for segmenting and labeling sequence data. In: Proceedings of ICML 2001, San Francisco, CA, USA, pp. 282–289 (2001)
19. Breck, E., Choi, Y., Cardie, C.: Identifying expressions of opinion in context. IJCAI **7**, 2683–2688 (2007)
20. Lewis, D.D., Gale, W.A.: A sequential algorithm for training text classifiers. In: Proceedings of SIGIR 1994, Dublin, Ireland, pp. 3–12 (1994)

Supervised Domain Adaptation via Label Alignment for Opinion Expression Extraction

Yifan Zhang$^{(\boxtimes)}$, Arjun Mukherjee, and Fan Yang

University of Houston, 3551 Cullen Blvd., Houston, TX 77204-3010, USA
{yzhang114,fyang11}@uh.ed, arjun@cs.uh.edu

Abstract. In this paper, we propose a supervised domain adaptation technique for opinion expression extraction task. The technique generates low dimensional projections that can improve the performance of a sequence model (e.g. CRF) in the target domain by align features with the true label sequence. We test our methods on product reviews and observe significant improvement in performance in comparison to baseline methods.

Keywords: Domain adaptation · Sequence labeling
Sequence extraction

1 Introduction

Opinion Expression Extraction (OEE) refers to the task which involves identifying segments of an input text that convey emotions, opinion, sentiments or beliefs. The OEE problem can be seen as a stem from the early works that built Subjectivity Classifiers which classify a sentence into either subjective or objective [15]. However, as pointed out by Wiebe et al. [16], a sentence can often contain more than one opinion, and it is also not unusual for a sentence to consist of both opinion and factual information. Consequently, it is better that extraction methods can be further improved to work at the sub-sentence level.

Breck et al. [3] solves the above issue by posing the identification of opinion expressions as a word sequence labeling task. More specifically, consider a sentence that consist of n words: $S = (w_1, ..., w_n)$. They propose to use conditional random fields [11] to predict a label for each of the words, $Y = (y_1, ..., y_n)$ with $y_i \in \{I, O\}$. A word will be labeled I if it is considered a part of Direct Subjective Expressions (DSE) or Expressive Subjective Expressions (ESE), and labeled O otherwise. In Yang and Cardie [17], the work was improved by using a semi-Markov Conditional Random Field instead.

Adapting these techniques into the realm of consumer reviews, we can then effectively extract opinions, compliments and complaints written by customers regarding the products. Such as in Li et al. [12], subjective expressions implying a negative opinion were discovered using sequence models and Markov networks.

© Springer Nature Switzerland AG 2018
A. Gelbukh (Ed.): CICLing 2017, LNCS 10762, pp. 224–240, 2018.
https://doi.org/10.1007/978-3-319-77116-8_17

Such system can be used to help customers making purchase decisions: instead of browsing through hundreds of reviews when researching a product, customers instead browse through pros and cons highlighted on each aspect of the product. These extracted segments are also proven to increase the accuracy of other downstream tasks such as sentiment classification.

For example, consider a Amazon review written for a earphone:

- The [*cord* is a little short] but most of the time it doesn't bother me.

Or a sentence from a flashlight reviews:

- It [does not come with an extra *bulb*] in the base of the light.

In these two examples, the phrase "cord is a little short" and "does not come with an extra bulb" is the opinion expression part of the sentence that we wish to extract, i.e., aspect specific sentiment expressions (aspect-sentiment phrases) containing the aspect and opinion expression within the sentence context. Similar to Li et al. [12] and Mukherjee [14], in this work we mainly focus on examine the performance of our proposed method on negative aspect-sentiment expressions in product reviews (as they tend to be more useful for an end-user in knowing the major issues of a product). We refer to these expression segments as *issues*. However, the proposed domain adaptation technique is generic and can be applied to any OEE or sequence modeling tasks.

Notice how this task draws in parallel with extracting DSEs and ESEs in the early literature. The main difference is an issue must includes an aspect of the product. We follow the convention in Mukherjee [14] and refer to these aspect names as Head Aspect (HA). Further more, we also limit the span of an issue should include the essence of compliment/complaint regarding the HA but not more.

As stated above, it follows that large scale labeling of such issues is both time-consuming and costly. Naturally, we would like a system to leverage small scale labeled data from one domain to discover new opinion expressions in a target domain, i.e., Domain Adaptation (DA) for Opinion Expression Extraction (DA-OEE). Both tasks OEE and DA have extensive literature. Closest related works on OEE include previously mentioned [12,14]. Also in earlier works such as Choi et al. [4] and Yang and Cardie [18] where joint models were proposed for identifying opinion holders/expression and relations among them in news articles.

In the next section, relevant literature on domain adaptation will be reviewed to establish better context for the method we will propose in Sect. 3.

2   Supervised Domain Adaptation

The goal of supervised domain adaptation can generally be described as developing techniques that allow to effectively exploit labeled data from one domain - also referred as a source domain, to enhance the performance of the model on a target domain. In our context of OEE in reviews, product reviews from two different product categories can be considered as from two different domains. We then wish to use the labeled review sentences from one of the product categories to improve the labeling performance of our algorithm in the other – target – product category.

More formally, we denote the target domain dataset as D_t and the source domain dataset as D_s. We use $\mathcal{X}$ to represent the input feature space of each dataset. In OEE problem, input features usually is a sentence representation in the form of a sequence of categorical value vectors, where each vector corresponds to a word from the sentence or features derived from that word. So a sentence consisting of n words can be denoted as $S = (v_1, ..., v_n)$, and each element v at word location i further consists of m features: $v_i = \{w_i^{f1}, w_i^{f2}, ..., w_i^{fm}\}$ (the word v_i itself is usually one of the m features). On the other hand, the output space $\mathcal{Y}$ for each input sentence in our problem is a sequence of IO (Inside-Outside) labels. So the labeling for a sentence of length n is thus $Y = (y_1, ..., y_n)$, with each label y at location i is $y_i \in \{I, O\}$.

And	the	*material*	is	plastic	.
O	O	I	I	I	O
All	wireless	*connection*	failed	.	
O	I	I	I	O	

Fig. 1. Two sample sentences.

Figure 1 shows two example sentences with their issues labeled using IO labeling scheme. Given these, a sequence labeling model M is a function that maps from space $\mathcal{X}$ to space $\mathcal{Y}$, i.e. $M : \mathcal{X} \to \mathcal{Y}$. We use M_d to represent a model that perform this mapping for domain d. The idea of supervised domain adaptation is then to use data from both the source domain D_s and the target domain D_t to improve the performance of M_s.

In the remaining of this section, three very successful ideas for performing domain adaptation will be introduced. These methods form the basis which our work builds upon.

2.1   Combining Data

The most basic yet effective domain adaptation trick is to simply combine the data set: $D_{s,t} = \{D_s \cup D_t\}$. Then a model $M_{s,t}$ is trained on the combined data. Applying $M_{s,t}$ on target domain data D_t can usually lead to a good amount of performance gain when compared to the model M_t which only trained on target domain data. The improvement can be especially pronounced when we only have access to a small amount of labeled target domain data. The improvement in performance can be attributed to the fact that increase in sample size leads to better learning of the underlying distribution. In other words, for common features that exist in both D_s and D_t, the number of training samples that contain these features were increased after combining data from both domains. Consequently, these data can more accurately depict the true underlying distribution leading to improved performance.

2.2 Easy Domain Adaptation (EasyAdapt)

In the previous Combining Data paradigm, features in both D_s and D_t are assumed to follow the exact same distribution, such that exposing more samples to the model can result in more accurate parameter estimation. However, in most of the natural language processing tasks, this assumption does not hold true for all features. This is mostly because some features – although they present in both the source domain and the target domain – can exhibit very different behavior in different domains. For these features, simply combining data will result in an averaged distribution that does not accurately reflect the true distribution in target domain, which in turn resulting in less than ideal performance. Daume in 2009 [7] offered an elegant and easy to implement solution: by simply duplicating each feature into a common feature and a domain specific feature, the technique effectively expanded the feature space to allow the learning algorithm to learn a shared weight and two domain specific weight for each feature. As observed in [7], the same feature word "the", can be a strong indicator that the following word is an entity in Broadcast News Domain. While at the same time, "the" is an indicator that the following word is not an entity in Usenet Domain.

2.3 SCL

In Blitzer et al. [2], a method called Structural Correspondence Learning (SCL) was proposed. The technique built upon the Structural Learning paradigm previously developed in [1]. The idea is to first select a set of pivot features P that exist in both domain and exhibit consistent behavior in both domains. Then, construct a total of $|P|$ linear classifiers for predicting pivots P using non-pivots as features, such that these classifiers learn the correspondence between non-pivot features and pivot features. The resulting weight values for each non-pivot feature from all pivot classifiers are then concatenated and use as a low dimensional embedding for each non-pivot feature. As shown in [2], different features that behave similarly will have similar values in this lower dimensional space (e.g., upon singular value decomposition of the feature/association matrices). For example, in the task of POS tagging, different adjective words will tend to be projected into a certain region of that space, while nouns will be projected into another.

It is worthwhile to note that domain adaptation is a broad and well developed topic that has been explored in many applications. These applications include but not limited to parsing [5], machine translation [10], and cross-lingual sentiment analysis [8]. The work in Fernández et al. [8] where Distributional Correspondence Indexing was proposed comes close to [2] and ours because all three works involve learning predictive structures of pivots that behave consistently across domains, then consequently use these predictors to generate low dimension representations. However, our focus is domain adaption for opinion expression extraction where data labeling is expensive and subjective, hence motivating us to develop approach that exploits correlation between features and actual labels.

3   Proposed Approach: Supervised Domain Adaptation via Label Alignment

3.1   Motivation

We can observe that EasyAdapt and SCL work in a rather different manner to address the task of domain adaptation - EasyAdapt works by allowing the same feature map to different behaviors in different domains; while SCL focuses on bringing different features that behave similarly into a similar projection.

We can combine the above two ideas: EasyAdapt and SCL in the hope that we can get the best of both worlds. We theorize that, for each pivot predictor, by increasing the feature space (and consequently the weight space) for each non-pivot feature during the learning stage, we can learn a more accurate projection later on. The projection will be able to demonstrate two properties: Firstly, the projection will be able to project different features that exhibit similar behavior into a similar representation, similar to what SCL was able to achieve. Secondly, the projection will also be able to capture domain specific behavior of features. That is to say, for features which behave differently in different domains, we should be able to treat them differently by using corresponding domain specific projections.

We diverge from the traditional unsupervised scheme in the SCL. Here, instead of training linear predictors for common features in the unlabeled data, we use the I/O labels as pivots in jointly learning a supervised sequence model (e.g., Conditional Random Field) from both source and target domain and then predict the actual sequence of I/O labels in $\mathcal{Y}$ corresponding to features in the input feature space $\mathcal{X}$ for the target domain's test set. While this does require a small amount of labeled data, the benefits are also obvious. First of all, our scheme eliminates the need to choose task specific pivots before applying the pipeline. This is an especially useful property for the task of opinion expression extraction, since the pivots must behave exactly the same or similar across domain. Such criteria are easier to meet in an objective task such as POS tagging in the original SCL work [2], but intrinsically difficult in subjective information extraction tasks. Secondly, using actual labels as pivot features reduces the amount of training data necessary to induce the projection space, because each pivot occurs more frequently. Lastly, since in our context the possible subsequence of output space Y is very limited, we can omit the process of applying SVD on the weight matrix W (as in traditional SCL) thereby reducing the complexity of the pipeline.

3.2   Label Alignment SCL

We call our approach Label Alignment SCL, LA-SCL for short. The naming is based on the fact our method aligns features from source and target domain based on their correlation with the ground truth labels, while also differentiates the same feature from both domain if they do not follow the same correlation.

Following the nomenclature detailed in Sect. 2, the pipeline of LA-SCL starts with generating pivot features using label sequences Y available for both domain $D_{s,t}$. For each word location i in a sentence, we can generate a number of pivot features using labels in Y according to a pivot feature scheme. We denote the set of pivot features generated at location i of sequence Y as $p_l(Y, i)$, where l represents a pivot selection scheme detailed in the next subsection. Similar to Structural Learning [1], each pivot feature in the set $p_l(Y, i)$ is a binary value in $\{0, 1\}$, indicating whether or not the label sequence conforms to that feature criteria. However, in contrast to the original SCL algorithm where pivot features are subject to the input feature X, here we generate them according to the actual labels. A pivot feature scheme is designed to capture the characteristics of the labeling segment around current word location i of the sentence.

A simple example for a pivot feature scheme in LA-SCL could be "the value of the previous label" or "the value of the next label". The second scheme for example, will generates only 1 pivot feature, $p^{right}(Y, i)$. We will say $p^{right}(Y, i) = 1$ if the next word is labeled as part of an opinion phrase, $y_{i+1} = I$. The rationale behind the above pivot selection scheme is that features which correlate similarly with the labeled pattern should be the words that articulate head aspects in similar manner. An example is shown in Fig. 1. In this example the word "plastic" and "failed" articulate their corresponding head aspects (*material* and *connection*) in a similar way. Even though one is a noun and the other is an adjective, they are both at the end of the labeled sequence and allows our model to capture this similarity.

In this manner, we can then construct l number of pivot predictors (Pivot Selection Schemes Section). Each of these predictors takes S as input and predict one corresponding pivot feature value at location i. In this work, the input of pivot predictor is limited by using a sliding window of radius 1. Such that for each training and testing instance at location i, the predictor only have access to $xp_i = (v_{i-1}, v_i, v_{i+1})$. Notice here the feature space is tripled to capture some location information: "the" being the $i - 1$th word is a different feature from "the" being the ith word, and this will affect the output of predictor differently.

Furthermore, for each of the word and its derived features $v_i = \{w_i^{f1}, \ldots, w_i^{fm}\}$, we also apply an feature space expansion by creating one common version of the feature and a domain specific one. For domain d, a mapping from the original feature space to the new one would be:

$$\Phi^d(v_i) = \{v_i^{common}, v_i^d\} \tag{1}$$

We will refer to this expanded feature space as the EA feature space. All the pivot predictors shall used non-pivot features in the EA feature space to predict pivot features, this modification allows the same non-pivot feature be projected into different latent space regions based on different domains.

Overall, in dataset $D_{s,t}$ consisting a total of k words, k training instances will be generated for each pivot predictor. In this work, each of these predictors is a linear predictor allowing bias and trained with hinge loss instead of the modified Huber loss suggested in the original Blitzer et al. [2], since it gave us better results.

After obtaining trained pivot predictors, we apply them back onto all the input features in the EA feature space from both domains. That is, at location i for pivot j, $\theta_i^j = w_j \cdot xp_i + b_j$, b is the bias term. For each pivot features, we take the real value and concatenate them to obtain a projection in $(\theta_i^1, ...\theta_i^k) \in \mathbb{R}^k$ at each location i. A sequence modeling model can then be trained on the extended features consisting of both original features, as well as the new projection features pooled from all pivots.

It is worth noting that LA-SCL does not involve transforming original features into EA feature space in the supervised sequence modeling stage - which is what the original work [7] proposed. The expanded EA feature space is only used for pivot predictor training and prediction. However, as we will later see in experiment section, combining both methods could be beneficial in certain circumstances by providing better performance.

Pivot Selection Schemes. Below are the formulations of three possible pivot schemes we developed and experimented with.

Sequence Pattern Scheme (Ptn): Sequence Pattern pivot scheme takes the form of

$$p^{ptn}(Y, i) =$$
$$\mathbb{1}_{\{y_{i-r}=\bar{y}_{i-r}\}} \cdots \mathbb{1}_{\{y_i=\bar{y}_i\}} \cdots \mathbb{1}_{\{y_{i+r}=\bar{y}_{i+r}\}} \tag{2}$$
$$\forall \bar{y}_{i-r}, \cdots, \bar{y}_{i-r} \in \mathcal{Y}$$

where r is the radius of the sliding window, and $\mathcal{Y}$ in our case is $\{I, O\}^n$. We use $\mathbb{1}$ to represent indicator function which will take value 1 if and only if, at location i, actual label y_i matches with the pivot's desired target value $\bar{y}_i$. Specifically, in this paper, we fixed $r = 1$ based on empirical results, consequently this scheme will generates a total of 8 possible pivot features: OOO, OOI, OIO, OII and so on. For example, in the first sentence of Fig. 1, the only Pattern Scheme pivot feature that take value 1 at location 1 is OOI, as it matches all of the 3 labels within the sliding window.

Distance to Sequence Head and Tail Scheme (DHT): We define Sequence Head L^h as the first I label in a segment of a labeled opinion expression. Similarly, Sequence Tail L^t is defined as the last I in a segment of a labeled opinion expression. The DHT scheme is thus defined at location i as

$$p_{i'}^{ptn\_h}(Y, i) = \mathbb{1}_{\{y_{i'}=L^h\}} \quad i' = i - r, ..., i + r$$
$$p_{i'}^{ptn\_t}(Y, i) = \mathbb{1}_{\{y_{i'}=L^t\}} \quad i' = i - r, ..., i + r \tag{3}$$
$$p^{ptn}(Y, i) = \left[p^{ptn\_h}(Y, i), p^{ptn\_t}(Y, i)\right]$$

In other words, we are predicting at each location i, for each location in sliding window, whether or not the label is a Head/Tail. We fix the window radius in this paper to 2, which consequently gives us a total of 12 possible pivot features. These consists of 5 possible sequence head location, 5 possible sequence tail location, and no Head/Tail within the sliding window. More concretely, for the first sentence in Fig. 1, at location $i = 3$, two of the DHT pivots takes value 1: $i - 1$ is a sequence head, $i + 1$ is a sequence tail.

Skip-gram Prediction Scheme (Skg): Skip-gram scheme is simply creating an independent predictor for each location within the sliding window, predicting whether or not it is a label I at that location. Consequently for a window radius of 2, this scheme will generates us 5 pivot features. Formally, the scheme is defined as:

$$p_{i'}^{skg}(Y, i) = \mathbb{1}_{\{y_{i'} = \bar{y}_{i'}\}}$$
$$i' = (i - r), ..., (i + r) \quad \forall \bar{y}_{i'} \in \mathcal{Y} \tag{4}$$

4 Data

We constructed a dataset based on the Amazon review dataset originally released in Mukherjee [14] to empirically demonstrate the effectiveness of our domain adaptation method in comparison with other baselines. The data released in [14] consists of 6 domains. Among those, 3 domains with the highest number of sentences were selected: Router (Rtr), GPS, Wireless Keyboard (Kb). In addition to these data, another 2 domains of data, Home Theatre System (HTS) and Flashlight (Fl), were collected and manually labeled following the same method and style as the original work.

Table 1. Size of domain dataset and aspects.

Domain	Aspects	Labeled Sents.
Router (Rtr)	Connection, Firmware, Signal, Wireless	1199
GPS	Screen, Voice, Software, Direction	571
Keyboard (Kb)	Spacebar, Range, Pad, Keys	570
HTS	DVD, Remote, Speaker, Subwoofer, TV	571
Flashlight (Fl)	Battery, Beam, Bulb, LED	745

As the examples shown in Sect. 1, an issue phrase was defined to be any subjective expression that captures various sentiments (evaluation, emotion, appraisal, etc.) toward the head aspect and containing the head aspect. The annotation was distributed across four human judges and cross checked to ensure consistency.

Number of labeled sentences in each domain and corresponding Head Aspects are listed in Table 1. We use these 5 domains of data to test LA-SCL and compare with other baseline methods.

4.1   Features

For each of the word in the sentences we also provide the following 5 additional feature families to help improve the labeling performance:

Head Aspect: TRUE, FALSE. The feature indicates whether the corresponding word is the aspect intended for issue extraction.

POS Tags: DT, IN, JJ, etc. Labeled using OpenNLP POS Tagger.

Syntactic Units (Chunk Tags): B-NP, I-NP, B-VP, etc. Labeled using OpenNLP POS Tagger.

Prefixes: ANTI, PRE, SUB, etc. Matched with a list of common English prefixes.

Postfixes: EST, NESS, LY, etc. Matched with a list of common English postfixes.

Sentiment Polarity: POSITIVE, NEGATIVE, NEUTRAL. Labeled by matching a list of common sentiment word using the opinion lexicon of Hu and Liu [9]

The labeled data and code of this work is provided[1] and will be released to serve as a resource and ensure reproducible research.

5   Experiments

5.1   Conditional Random Field

Similar to SCL and many other embedding methods that produce low dimensional real value projection, the projections produced using LA-SCL can be used with all feature based classifiers. In Blitzer [2], MIRA was used during supervised sequence modeling stage. However, margin based classifiers like SVM require a lot more training data to perform as well as CRF that was deployed in [3,17]. In this paper we use Conditional Random Field with High-order Dependencies (HO-CRF) developed and implemented in Cuong et al. [6] for supervised sequence labeling task in our LA-SCL pipeline.

We now describe the CRF feature functions we used during the experiments. Recall that a sentence is denoted as $S=(v_1, ..., v_n)$. For each location i, v_i consists of the original word, 5 feature types detailed in Sect. 4.1, as well as discretized projection values if the domain adaptation method being tested generates any. We use $|v|$ to denote the number of features at each word location, including any projection generated features. Thus for un-transformed data, the number of features $|v|$ at each word location is 6. Where as for LA-SCL Ptn scheme, 8 additional features are added at each word location, resulting in $|v| = 14$. And we use $\mathcal{X}_j$ to denote the feature space of jth feature. The feature function $F(X, Y, i)$ of CRF at each location i is as follow:

---

[1] https://goo.gl/fYcf98.

First-Order Edge-Observation Feature: This feature function generates all possible feature combinations using label at i, label at $i-1$ and all $|\boldsymbol{v}|$ available features at i

$$F(X, Y, i) = \{f(X, Y, i, j)\} \; for \; j = \{1 \ldots |v|\}$$
$$f(X, Y, i, j) =$$
$$\mathbb{1}_{\{y_i = \bar{y}_i\}} \mathbb{1}_{\{y_{i-1} = \bar{y}_{i-1}\}} \mathbb{1}_{\{w_i^{fj} = \bar{w}_i^{fj}\}} \tag{5}$$
$$\forall \bar{y}_i, \bar{y}_{i-1} \in \mathcal{Y} \quad \forall \bar{w}_i^{fj} \in \mathcal{X}_j$$

Label-Label Feature up to 4-Gram: This feature function generates all possible feature combinations at location i by considering label bigrams, trigrams and 4-grams, such as: OO, OI, IO, II, OOI and so on

$$F(X, Y, i) = \mathbb{1}_{\{y_i = \bar{y}_i\}} \mathbb{1}_{\{y_{i-1} = \bar{y}_{i-1}\}}$$
$$F(X, Y, i) = \mathbb{1}_{\{y_i = \bar{y}_i\}} \mathbb{1}_{\{y_{i-1} = \bar{y}_{i-1}\}} \mathbb{1}_{\{y_{i-2} = \bar{y}_{i-2}\}}$$
$$F(X, Y, i) = \tag{6}$$
$$\mathbb{1}_{\{y_i = \bar{y}_i\}} \mathbb{1}_{\{y_{i-1} = \bar{y}_{i-1}\}} \mathbb{1}_{\{y_{i-2} = \bar{y}_{i-2}\}} \mathbb{1}_{\{y_{i-3} = \bar{y}_{i-3}\}}$$
$$\forall \bar{y}_i, \bar{y}_{i-1}, \bar{y}_{i-2}, \bar{y}_{i-3} \in \mathcal{Y}$$

5.2   Baselines

Because there is relatively little research in the area of supervised domain adaptation for sequence labeling. We choose the following baseline methods to compare with LA-SCL. As previously mentioned, standard HO-CRF with feature functions detailed in Sect. 5.1 is used for sequence modeling. Additional features introduced in Sect. 4.1 are also available to all baselines listed. The following baselines either involve variate the amount of data available for the HO-CRF model during training, or rely on adding additional features/feature-space for the CRF model to work with. Our method LA-SCL falls into the second category as it also generates additional projections for features in the original data.

Source Only (SRC): Train HO-CRF model using only data from source domain, D_s.

Target Only (TGT): Train HO-CRF model with training portion of the target domain.

Combined (CMB): HO-CRF model is trained on the union of both source and target domains.

Un-supervised SCL (SCL): We implemented a version of SCL following [2]. We provide all of the unlabeled features from D_s and D_t during the process of pivot selection and pivot predictor training. A total of 70 pivot features were

Table 2. F_1 scores of baseline methods and LA-SCL tested on several domain transfer tasks

	SRC	TGR	CMB	SCL	EA-CRF	LA-SCL Ptn	LA-SCL DHT	LA-SCL Skp	EA + LA-SCL Ptn
HTS:Fl	0.7393	0.7971	0.7888	0.7935	0.7953	0.7939	**0.8006**	0.7896	0.7930
Kb:Fl	0.6761	0.7971	0.7977	0.7932	0.7961	0.7993	**0.7998**	0.7894	0.7997
Kb:GPS	0.8049	0.8377	0.8423	0.8426	0.8421	**0.8629**	0.8525	0.8516	0.8584
GPS:Rtr	0.7524	0.8748	0.8771	0.874	0.8745	0.8897	**0.8918**	0.8909	0.8839
Kb:HTS	0.6893	0.8132	0.8133	0.8068	0.8123	0.8204	0.8199	**0.8324**	0.8152
Kb:Rtr	0.7582	0.8748	0.8741	0.8730	0.8773	0.8825	**0.8833**	0.8818	0.8824
Fl:HTS	0.7374	0.8132	0.8099	0.8054	0.8089	0.8313	0.8136	0.8275	**0.8315**
HTS:GPS	0.6403	0.8377	0.8130	0.8031	0.8393	0.8374	0.8390	0.8377	**0.8516**
HTS:Rtr	0.6714	0.8748	0.8606	0.8608	0.8775	0.8873	0.8884	**0.8892**	0.8865
Average	0.7188	0.8356	0.8308	0.8281	0.8359	**0.8450**	0.8432	0.8433	0.8447

picked and following [1] we used the modified Huber loss during training process. Next, 20 columns of projection were selected after applying Singular Value Decomposition (SVD) and discretized.

Easy Adaptation CRF (EA): All features from both domains are expanded using the EasyAdapt method discussed in Sect. 2.2. A HO-CRF model is then trained on the union of modified source and target domain data.

5.3 Experiment Settings

F_1 score of 9 domain transfer tasks $(D_s:D_t)$ with one source domain (D_s) and one target domain (D_t) are considered for evaluation. These are HTS:Fl KB:Fl Kb:GPS GPS:Rtr Kb:HTS Kb:Rtr Fl:HTS HTS:GPS HTS:Rtr, where the domain short-forms are mentioned in the beginning of Data Section. Precision and recall for each of the 9 tasks is calculated using micro-average on per word label basis. For example, True positive is defined as the number of words that is correctly labeled as part of the issue sequence.

Further, to make the results more stable and statistically convincing, 5-fold cross validation was applied and all results reported in this section is the average value across 5 folds. More specifically, the both source domain data D_s and target domain data D_t are evenly split into five sets respectively. For each fold of the test, we first use 4 out of 5 sets in D_s and/or D_t for training, after which the held-out portion of D_t is used for testing. The F_1 from all 5 folds were then macro-averaged to give us the stabilized result. This is true for all five baselines as well as our methods, with the exception of pivot learning stage of un-supervised SCL - we provide the pivot predictors with all of the features $x \in \mathcal{X}$ of D_s and D_t.

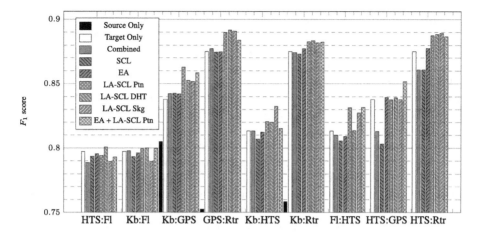

Fig. 2. Comparison of F_1 scores in 9 domain transfer tasks.

6   Results and Analysis

6.1   Results

There are several[2] observations worth noting from Fig. 2 and Table 2 which report the F_1 score of opinion expression extraction across the 9 domain transfer tasks. First, at least one of the three variations of LA-SCL is almost always performing better than the 5 baselines. And in 5 out of 9 transfer tasks, all of the 3 variations are better than the 5 baselines. Out of all paradigm tested, LA-SCL using Pattern scheme gives the highest F_1 score.

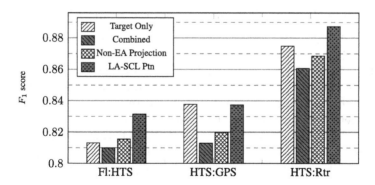

Fig. 3. Performance difference if pivot predictors do not work under EA feature space.

It is also interesting to note that while simply combining data prove to be working very well in previous work involving POS tagging and Chunking [7],

---

[2] Many of the Source Only baselines have a F_1 too low to be visible in this bar chart, please refer to Table 2 for their actual readings.

such attempt is not very fruitful in our problem. This is especially obvious in the case where source domain is HTS and target domain is GPS. We believe this phenomenon is due to a large discrepancy between distributions of the same features between source and target domains. This theory can be somewhat veri- fied by the observing the changes in performance between different methods. As shown in Fig. 2, the model trained using both HTS domain and GPS domain data without any domain adaptation applied (i.e. Combined) performs poorly when testing on GPS domain. However, simply applying EasyAdapt to expand the feature space, in this case, allows CRF to discriminate common features that occur in both domains, thereby resulting in a recovery of F_1 score. At the same time, all three LA-SCL schemes (LA-SCL Ptn, LA-SCL DHT, LA-SCL Skg) can also bridge the gap between these two domain and achieves the same F_1 perfor- mance without the need to let CRF work under EA feature space. This shows that low dimensional projection resulting from LA-SCL is consistent across two domains and is indeed useful for the sequence labeling task.

Figure 3 further compares the result if we do not allow pivot predictors to work with the EA transformed feature space as described in Eq. 1. We can see that LA-SCL performs better by allowing weights for each domain specific ver- sion of the features.

We also experimented with combining LA-SCL technique with features expanded to EA feature space. The Ptn pivot scheme is chosen because it is the best performer on average. The results are also shown in Table 2 and Fig. 2. As can be observed from the result, for majority of the domain adaptation tasks, using one of the LA-SCL schemes along will generate the best result. Using additional EA transformation is often not helpful and could potentially hurt the adaptation performance, a phenomenon we believe to be caused by overfitting.

6.2 Multiple Product Categories as Source Domain

Figures 4 and 5 show the effect of continuously adding more domains as the source domain for two target domains: HTS, GPS. We consider two models: (1) LA-SCL Ptn and (2) a standard CRF. We see that LA-SCL can leverage the additional data to improve F_1 score and performs much better than simply using more domains in training. Interesting to note is the case of HTS as target domain, where performance of a standard CRF model when training with all original data is not as good as using just the target domain data. Nonetheless the proposed LA-SCL framework is less affected and still improves performance over a standard CRF model.

6.3 Projection Space Inspection

Figure 6 shows a t-SNE graph [13] of some samples in the LA-SCL projection space. Projection vectors were selected based on original word feature at location i. About 3000 vectors from Flashlight domain that correspond to a stop word is selected as background. We then extract all the vectors that correspond to word "bulb" and "led" from Flashlight domain and plot them using Magenta and Red

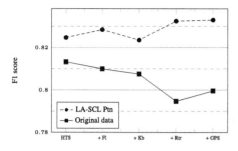

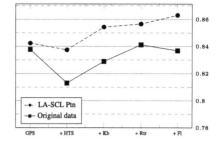

Fig. 4. F_1 score in HTS domain after each additional source domain added in training.

Fig. 5. F_1 score in GPS domain after each additional source domain added in training.

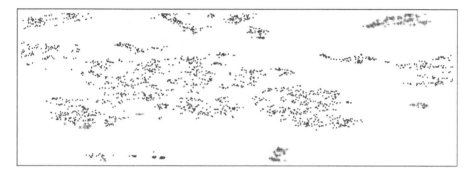

Fig. 6. t-SNE graph of "bulb" and "led" with other 3000 stop words as background blue dots. Red + is [bulb] and Magenta x is [led].

color. The distribution of projections of these two words is highly overlapped as can be observed in the graph. This shows that the low dimensional projections generated by LA-SCL preserve the original semantics of the word features.

To further validate and explain the performance improvement observed in transfer task, Table 3 shows some words and a sample of corresponding nearest neighbors in projection space. Each section is a one to one domain transfer combination previously discussed. And distance is measured using cosine distance. The list of words were all picked as a representative set out of top 20 nearest neighbors. This is mainly done to reduce the number of duplicate words for better clarity. We note from Table 3, that "brighter", "cheaper" is to Flashlight (Column FL:brighter) is projected to the same location as "great", "perfect" in Home Theatre domain. And how GPS is "weak" (Column GPS:weak) is comparable to Router being "complicated". This shows that the proposed LA-SCL scheme is also able to capture the sentiment specific semantic word associations which helps in finding the expression boundaries in the eventual downstream task of OEE.

Table 3. Words and their nearest neighbors.

Fl: battery	FL: brighter	FL: focus
HTS: speaker	FL: cheaper	FL: illuminate
HTS: subwoofer	FL: LED	FL: suit[a]
HTS: remote	HTS: great	FL: bright
HTS: DVD	FL: beam	HTS: speaker
HTS: speaker	HTS: remote	HTS: access
HTS: speaker	FL: dim	FL: term
HTS: remote	HTS: perfect	FL: filament
GPS: screen	GPS: weak	Rtr: signal
GPS: screen	GPS: archaic	Rtr: signal
GPS: screen	GPS: substandard	Rtr: connection
Kb: keys	GPS: horrible	Rtr: capability
Kb: keys	Rtr: complicated	Rtr: functionality
Kb: keys	Rtr: lost	Rtr: communication
GPS: screen	Rtr: sporadic	GPS: everything[b]

[a] "Beam can be adjust to *suit* your need"
[b] "Everything is fine except the voice *everything* sounds tinny or like it is under water"

6.4   Example Sentences

Table 4 shows few sample sentences whose opinion phrase boundaries have been corrected by applying our proposed LA-SCL domain adaptation technique (over a standard CRF model).

Table 4. Sample expressions that have been fixed after applying LA-SCL.

Wrong:	Wireless [rear speaker set-up that work ok]
Corrected:	[wireless rear speaker set-up that work ok]
Wrong:	The [rear speaker wire be not] very long
Corrected:	The [rear speaker wire be not very long]
Wrong:	I could not find the thread size for the [rear satellite speaker anywhere]
Corrected:	I [could not find the thread size for the rear satellite speaker anywhere]
Wrong:	But the [remote control do not] come with battery
Corrected:	But the [remote control do not come with battery]

7   Conclusion

In this work we proposed LA-SCL, a supervised domain transfer technique for sequence labeling task. LA-SCL generates a low dimensional projection bringing different features in source and target domains together, while at the same time differentiates the common features that behave differently in different domains. This was achieved by learning feature alignment with the actual label sequence that are available in training data. We evaluated our approach on real-world product reviews from Amazon on an actual sentiment analysis task extracting aspect-specific opinion expressions. Experimental results showed the effectiveness of the proposed technique that significantly outperformed several baselines.

References

1. Ando, R.K., Zhang, T.: A framework for learning predictive structures from multiple tasks and unlabeled data. J. Mach. Learn. Res. **6**, 1817–1853 (2005)
2. Blitzer, J., McDonald, R., Pereira, F.: Domain adaptation with structural correspondence learning. In: Proceedings of the 2006 Conference on Empirical Methods in Natural Language Processing. Association for Computational Linguistics, pp. 120–128 (2006)
3. Breck, E., Choi, Y., Cardie, C.: Identifying expressions of opinion in context. IJCAI **7**, 2683–2688 (2007)
4. Choi, Y., Breck, E., Cardie, C.: Joint extraction of entities and relations for opinion recognition. In: Proceedings of the 2006 Conference on Empirical Methods in Natural Language Processing. Association for Computational Linguistics, pp. 431–439 (2006)
5. Cohen, R., Goldberg, Y., Elhadad, M.: Domain adaptation of a dependency parser with a class-class selectional preference model. In: Proceedings of ACL 2012 Student Research Workshop. Association for Computational Linguistics, pp. 43–48 (2012)
6. Cuong, N.V., Ye, N., Lee, W.S., Chieu, H.L.: Conditional random field with high-order dependencies for sequence labeling and segmentation. J. Mach. Learn. Res. **15**(1), 981–1009 (2014)
7. Daumé III, H.: Frustratingly easy domain adaptation (2009). arXiv preprint arXiv:0907.1815
8. Fernández, A.M., Esuli, A., Sebastiani, F.: Distributional correspondence indexing for cross-lingual and cross-domain sentiment classification. J. Artif. Intell. Res. (JAIR) **55**, 131–163 (2016)
9. Hu, M., Liu, B.: Mining and summarizing customer reviews. In: Proceedings of the Tenth ACM SIGKDD International Conference on Knowledge Discovery and Data Mining, pp. 168–177. ACM (2004)
10. Koehn, P., Schroeder, J.: Experiments in domain adaptation for statistical machine translation. In: Proceedings of the Second Workshop on Statistical Machine Translation. Association for Computational Linguistics, pp. 224–227 (2007)
11. Lafferty, J., McCallum, A., Pereira, F.C.: Conditional random fields: probabilistic models for segmenting and labeling sequence data (2001)
12. Li, H., Mukherjee, A., Si, J., Liu, B.: Extracting verb expressions implying negative opinions. In: AAAI, pp. 2411–2417 (2015)

13. Van der Maaten, L., Hinton, G.: Visualizing data using t-SNE. J. Mach. Learn. Res. **9**(2579–2605), 85 (2008)
14. Mukherjee, A.: Extracting aspect specific sentiment expressions implying negative opinions. In: Proceedings of the 17th International Conference on Intelligent Text Processing and Computational Linguistics (2016)
15. Riloff, E., Wiebe, J.: Learning extraction patterns for subjective expressions. In: Proceedings of the 2003 Conference on Empirical methods in Natural Language Processing. Association for Computational Linguistics, pp. 105–112 (2003)
16. Wiebe, J., Wilson, T., Cardie, C.: Annotating expressions of opinions and emotions in language. Lang. Resour. Eval. **39**(2–3), 165–210 (2005)
17. Yang, B., Cardie, C.: Extracting opinion expressions with semi-Markov conditional random fields. In: Proceedings of the 2012 Joint Conference on Empirical Methods in Natural Language Processing and Computational Natural Language Learning. Association for Computational Linguistics, pp. 1335–1345 (2012)
18. Yang, B., Cardie, C.: Joint inference for fine-grained opinion extraction. In: ACL, vol. 1, pp. 1640–1649 (2013)

Comment Relevance Classification in Facebook

Chaya Liebeskind[(✉)], Shmuel Liebeskind,
and Yaakov HaCohen-Kerner

Department of Computer Science, Jerusalem College of Technology,
Lev Academic Center, Jerusalem, Israel
liebchaya@gmail.com, liebeskinds@gmail.com,
kerner@jct.ac.il

Abstract. Social posts and their comments are rich and interesting social data. In this study, we aim to classify comments as relevant or irrelevant to the content of their posts. Since the comments in social media are usually short, their bag-of-words (BoW) representations are highly sparse. We investigate four semantic vector representations for the relevance classification task. We investigate different types of large unlabeled data for learning the distributional representations. We also empirically demonstrate that expanding the input of the task to include the post text does not improve the classification performance over using only the comment text. We show that representing the comment in the post space is a cheap and good representation for comment relevance classification.

Keywords: Comment relevance classification · Machine learning
Semantic analysis · Social media · Supervised learning

1 Introduction

In recent years there has been increasing interest in applying various Natural Language Processing (NLP) methods to social media texts due to the increased availability and popularity of social media platforms and the significance of social data to a widespread variety of organizations. The specific linguistic features of social media texts, such as abbreviations, emojis, onomatopoeia, replicated characters, and slang words pose special challenges for NLP. Various previous NLP tasks, such as text summarization [1, 2], name entity recognition [3], and machine translation [3, 4] addressed many of these challenges.

Opinion and sentiment analysis of users' comments and reviews has received a lot of research focus as social media enable direct contact with the target public. Most previous work on sentiment analysis aimed to classify comments as positive, negative or neutral with respect to the general vibe of the blog/post [5–8]. In this study, we address the complementary preceding task of classifying the relevance of a comment to the content of its post.

The common sentiment analysis task does not depend upon a full understanding of the text's semantics because usually in the complex classification cases of messages conveying both positive and negative sentiment, the more frequent stance one is to be chosen [7]. However, in classifying the relevance of a comment to the content of its

© Springer Nature Switzerland AG 2018
A. Gelbukh (Ed.): CICLing 2017, LNCS 10762, pp. 241–254, 2018.
https://doi.org/10.1007/978-3-319-77116-8_18

post, a fuller understanding of the comment semantics and perhaps even the post's text is needed. For instance, the general attitude of the comment *John, you are the best, and above the rest*!!! is positive, but since the comment is not toward the content of the post, it is considered as irrelevant/negative in the relevance classification task.

Many text classification (TC) applications use the bag-of-words (BoW) representation [9–11], where the document is represented as a vector of words' frequencies that appear in it. In short texts (e.g., sms and tweets), usually only a few words appear in each text and the frequency of occurrence of each word is relatively low. In such cases, the BoW representation usually generates a highly sparse representation. To overcome this sparsity problem, different dimensional reduction methods for semantic analysis, such as Latent Semantic Analysis (LSA) [12, 13], Word Embedding (Word2Vec) [14], and Random Projection (RP) [15] have been explored.

In this paper, we explore the comment relevance classification task. Given only a comment text, we aim to classify its relevance to its post. To perform this task, we compare four semantic vector representations for the comment relevance classification task. Each of the representations is generated by a different dimension reduction method, which produces word vectors that collapse similar words into groups. Moreover, all of the representations are built using entirely unsupervised distributional analysis of unlabeled text.

We examine different types of large unlabeled data for learning the distributional representations, namely comment texts, post texts and both post and comment texts. Additionally, we explore whether expanding the input of the comment relevance classification task to include also the post text would increase the classification accuracy.

We demonstrate that, in the case study of Hebrew Facebook, the LSA dimensional reduction method outperforms the other methods for comment relevance classification. We show that the classification result of representing a comment in the post space is similar to that of representing it in the comment space. However, representing the comment in the post space is preferable because there are significantly less posts than comments.

The rest of this paper is organized as follows: Sect. 2 introduces relevant background about short text classification and dimensionality reduction for semantic analysis. Section 3 presents the comment relevance classification task. Section 4 introduces the experimental setting, the experimental results for four dimensionality reduction methods and a deep analysis of the best method. Finally, Sect. 5 summarizes the main findings and suggests future directions.

2   Background

During the past dozen years, short text classification is more and more popular because of the increased availability microblogs and short messages, and the usefulness of this data for various organizations, e.g., commercial companies, political organizations, and security services.

TC has been mainly addressed by supervised machine learning (ML) methods. Many of these methods use BoW representations. BoW is a high-dimensional sparse representation for documents of any length. However, the sparsity problem is much more critical for short texts. In each given short text, most words have only one occurrence. Therefore, for short texts, typical ML algorithms usually cannot be successfully applied directly to BOW's representations.

Two general approaches to address the challenges of short text classification were explored: short text expansion by external text [16–18] and dimension reduction methods [19–21], which provide a lower-dimensional representation of documents while maintaining their semantic properties. Since the adaptability of the first approach is problematic for languages that lack external resources, we focus on the second approach.

2.1   Dimensionality Reduction for Semantic Analysis

Dimensionality reduction techniques fold together terms that have the same semantics and provide a lower-dimensional representation of documents that reflects concepts instead of raw terms. As a result, terms that do not appear in a document may still be semantically related to the document, if they share common concepts.

Next, we overview four common dimensionality reduction methods for semantic analysis. One important advantage of all these methods is that the dense representations are learned unsupervisedly from a large corpus of unlabeled data.

Latent Semantic Analysis (LSA)

LSA [12, 13] constructs a semantic space from a large matrix of term-document association data via singular value decomposition (SVD). SVD is a linear algebra procedure of decomposing an arbitrary matrix into three matrices, two of which are orthonormal and the third is a diagonal matrix whose diagonal values are the singular values of the matrix.

Although LSA loses the structural information as reducing the dimension, it is effective for short text classification. It reduces the feature distribution and noise, strengthens the semantic relationship between terms and documents, and allows analyzing the similarity between term-term, document-document and term-document as terms and documents are mapped to the same k-dimensional space [19].

LSA has been widely used for short text classification [22–24]. Many classification methods that combine LSA and ML algorithms have shown to improve the accuracy of short text classification [25, 26].

Latent Dirichlet Allocation (LDA)

LDA [18, 27] is a probabilistic, generative model for discovering latent semantic topics in large corpora. Each detected topic is characterized by its own particular distribution over words. Each document is then characterized as a random mixture of topics indicating the proportion of time the document discusses each topic. The topic probabilities provide an explicit dense representation of a document. The major challenge of constructing LDA model is how to estimate the distribution information of latent topics within the document. Various algorithms, such as Expectation Maximization (EM) [28] and Gibbs sampling [29] are applied to address this challenge.

LDA can be used to find the latent structure of "topics" or "concepts" in a test corpus. It has stronger ability of describing the realistic sematic than the LSA model [19]. Thus, LDA is a common representation for short text classification [20, 30–32].

Random Projection (RP)
In RP [15], the original high-dimensional data is projected onto a lower-dimensional subspace using a random matrix whose columns have unit lengths. The key idea of random mapping arises from the Johnson-Lindenstrauss lemma [33]: if points in a vector space are projected onto a randomly selected subspace of suitably high dimension, then the distances between the points are approximately preserved.

Word Embedding
In Word Embedding [14, 21] approaches, words are embedded into a low dimensional space. A d-dimensional vector of real numbers models the contexts of each word. The vectors are meaningless on their own, but semantically similar words have similar vectors.

Inspired by the methods for learning the word vectors, document vectors are also mapped to vectors [34]. Given many contexts sampled from the document, the vector is "asked" to contribute to the prediction task of the next word.

Recently, there is a growing interest in technique to adapt unsupervised word embedding to specific applications, when only small and noisy labeled datasets are available. In sentiment classification, many previous work have incorporated the sentiment information into the neural network to learn sentiment specific word embedding. For this purpose, different neural network models with various input representations have been explored [35–38].

3 Comment Relevance Classification

Comment relevance classification is a binary classification task of determining if a given comment is relevant to the content of the post it discusses. For example, in a political domain, the comment *Each sentence is more idiotic than the other, not to mention your lies* refers to the content of the post, whereas the comment *Mark, buy a lot of books you'll have plenty of spare time in the opposition* is likely to be irrelevant to the content of the discussed post, additional detailed examples are shown in Table 1.

We adopt the supervised ML framework for the following comment relevance classification task: given only a comment as an input, we classify its relevance to its post. Since comments are usually very short texts, we represent each comment as a dense vector of d-dimensions. These dimensions are the set of features over which learning and classification are performed. The classification predicts which comments are relevant to the content of the post they refer.

An advantage of short vectors for ML systems is that they are easier to be used as features. Instead of having to learn tens of thousands of weights for each of the sparse dimensions, a classifier can just learn a few hundreds of weights to represent a document meaning. The fewer parameters in the dense representation may generalize better and help avoid overfitting.

Table 1. Comment relevance classification – examples.

#	Post	Comment	Relevance
1	"I am speaking now about the security situation in Israel. I will address the lies that the Palestinian Authority continues to tell."	"This is the truth sayings by Prime Minister of Israel…"	True
2	"Long live President Ruby Rivlin…"	"Hence, Reuven Rivlin. Enough with Ruby. The man is a President, it is not Sesame Street."	True
3	"The danger in the coming elections is the establishment of a leftist government…"	"Would love to have seen this sub-titled in english!"	False
4	"In these minutes, votes in the Knesset plenum start. Selfie from the Knesset ☺"	"I admire you!!! Be strong and of good courage!!!"	False

The dense vector representation of the input comment is learned form a large corpus of un-labeled data. We explore three different types of data that is suitable for our classification mission.

- Post data, assuming that representing the comment in the space of the posts is indicative of content overlap between them.
- Comment data, assuming that representing the comment in the comments' space is beneficial for enabling the supervised method to learn typical reference patterns or content words in comments.
- Post and comment data, assuming that correlation between post and its comment should be learned from a representation that combines the data of both. We examine two ways to combine the post and comment texts for the unsupervised dimensional reduction. In the first approach, the texts are combined through a shared BoW, the post and its comment are concatenated into a single string. The second approach assumes that the comments and posts are constructed from two different language models and combines the texts as two separated BoWs [39]. This is done by marking each of the post's words and concatenating then with the comment's words.

In addition, we explore the extension of the classification task's input, the comment text, to include both the post text and its comment text. We measure the impact of this extension on the results of the representations constructed from the different types of the unlabeled data. An example of investigation for this extension is the comparison of the classification results between representing the comment text in the space of comment + post and representing the concatenated text of the comment and the post in the same space.

4 Evaluation

4.1 Evaluation Setting

Our dataset is a large Hebrew corpus of 4.8 million comments written in Hebrew by users replying to 41,882 politicians' posts, posted on Facebook during 2014–2015. The average length of a comment is 7 words and the average length of a post is 22 words.

A sub-corpus of 1,397 comments was manually annotated for relevance classification. It contains 803 positive examples and 594 negative examples. We used three rounds to refine our coding scheme. In the first round, three judges coded 100 comments, and afterwards a discussion to find disagreements. In the second round, again each one of the three judges coded another 100 comments. Fleiss' Kappa (Fleiss and Cohen 1973) reliability of agreement measure was calculated, and appeared to be very low. This required another set of discussion, and a third round coding another 47 comments. This round resulted in a Fleiss Kappa of 0.784, which is considered as substantial agreement [40].

While the supervised classification was performed on the sub-corpus, the unsupervised learning of dense vector representations was performed on the large corpus.

For classification, Weka[1] [41, 42] data mining software was used. We used 10-fold cross-validation to estimate the classification accuracy. For dimensionality reduction, GenSim[2] python library with the default settings of 200 dimensions for the LSA, LDA and RP and 300 dimensions for Word Embedding (doc2vec) was used.

4.2 Results

In contrast to some previous works on sentiment classification [43, 44], we avoid filtering short texts of any length. Therefore, due to short length of Hebrew Facebook comments (average of 7 words per comment), we expect that the values of our evaluation measures to be lower than that of English, where the average amount of words per Facebook comment is 19 words [45]. Moreover, Hebrew is a morphologically rich language, which poses additional complexity for Natural Language Processing (NLP) tasks.

In our experiments, we used the Random Forest classification method. Table 2 presents the accuracy results of all the dimensional reduction methods, extracted by four types of unlabeled data: post data (*Post*), comment data (*Comment*), post and comment data with shared BoW (*Post_Comment*), and post and comment data as two language models (*Post_Comment_2BoWs*). The input was represented either by the comment or by the comment and the post.

In general, even though the task tests the relevance of the comment to its post, expanding the input to also include the post text did not improve the classification accuracy. This is true for all the models generated by all the explored types of

---

[1] http://www.cs.waikato.ac.nz/ml/weka/.

[2] https://radimrehurek.com/gensim/index.html.

unlabeled data except for the LDA model that was generated by the *Post_Comment_2BoWs* unlabeled data, where the difference was not statistically significant[3].

The best method for semantic analysis was LSA. LSA outperformed all the other semantic models for all types of unlabeled data. For most of the configurations the advantage of the LSA model was statistically significant.

The bolded configurations are the LSA "nature" configurations to represent the unlabeled data for a comment input and a comment + post input respectively. All the configurations that outperformed these "nature" configurations by a statistically significant amount are marked by an asterisk.

Table 3 presents the F measure results in the same manner as Table 2. Expanding the input to also include the post text did not improve the classification F measure results too. This is true for all the models generated by all the explored types of unlabeled except for the LDA model that was generated by the *Post_Comment* and the *Post_Comment_2BoWs* unlabeled data, and the Doc2Vec model that was generated by the *Post* data, where the difference was not statistically significant.

Table 2. Random forest accuracy (%) results for all the configurations.

	Input	Comment	Comment + Post
Doc2Vec	*Comment*	57.58*	54.01*
	Post_Comment_2BoWs	56.00*	54.96*
	Post_Comment	56.62*	54.22*
	Post	57.67*	57.54*
LDA	*Comment*	57.25*	53.78*
	Post_Comment_2BoWs	53.71*	54.74*
	Post_Comment	55.05*	54.25*
	Post	56.65*	54.31*
LSA	*Comment*	**64.97**	60.19
	Post_Comment_2BoWs	65.13	58.75
	Post_Comment	63.86	**60.6**
	Post	64.48	59.15
RP	*Comment*	58.12*	56.17*
	Post_Comment_2BoWs	58.01*	56.74*
	Post_Comment	57.94*	56.45*
	Post	59.45*	56.62*

The performance difference between the types of unlabeled data was not statistically significant. This implies that representing the input comment in the post space is as good as the default representation of a comment in the comment space. Since there are less posts than comments, representing the comment in the post space is cheaper. In our case study, While the matrix that the LSA reduced for the *Post* representation was

---

[3] In all the reported experiments, statistical significant was measured according to the paired t-test at the 0.05 level.

Table 3. Random forest F measure results for all the configurations

	Input	Comment	Comment + Post
Doc2Vec	*Comment*	0.69*	0.66*
	Post_Comment_2BoWs	0.68*	0.67*
	Post_Comment	0.69*	0.67*
	Post	0.70*	0.71
LDA	*Comment*	0.65*	0.65*
	Post_Comment_2BoWs	0.63*	0.66*
	Post_Comment	0.64*	0.65*
	Post	0.65*	0.66*
LSA	*Comment*	**0.74**	0.71
	Post_Comment_2BoWs	0.73	0.7
	Post_Comment	0.73	**0.71**
	Post	0.73	0.7
RP	*Comment*	0.71	0.7
	Post_Comment_2BoWs	0.71	0.7
	Post_Comment	0.71*	0.7
	Post	0.72	0.7

Table 4. Comparison of various classifiers' performances on the *Post* unlabled data

Classifier	Accuracy (%)	F measure
RandomForest	64.48	0.73
SMO	62.91	0.71
BayesNet	60.96*	0.67*
MultilayerPerceptron	57.21*	0.62*
SimpleLogistic	62.61	0.7*
Bagging	61.53*	0.69*
AdaBoostM1	60.26*	0.67*
J48	57.51*	0.63*

119,801 (dictionary size) * 41,882 (# of posts), the matrix for the *Comment* representation was much larger, 1,497,866 (dictionary size) * 4,813,013 (# of comments). The big difference in the dictionary size of the configurations is not only due to the difference in the amount of documents, but also due to a different writing style. Whereas, politicians tend to write in standard language, the comments' language is usually informal and noisy. Table 4 presents a comparison of 8 ML methods' performances on the *Post* unlabled data. The Random Forest method outperformed all the other methods.

We also compare our results to a common Tf-Idf baseline, where the input is the comment text. Tf-Idf representation, a simple transformation which takes documents represented as BoW counts and applies a weighting scheme, which discounts common terms. We obtained an accuracy of 63.92% and a F score of 0.728.

We further explore the best performing model, the LSA model. Figure 1 illustrates the effect of the dimension parameter on the accuracy of the relevance classification results for two types of unlabeled data: *Comment* and *Post*, where the input is the comment text. Similarly, Fig. 2 illustrates the effect of the dimension parameter on the F measure. Both of the figures show that the default dimension that we have selected was the optimal one.

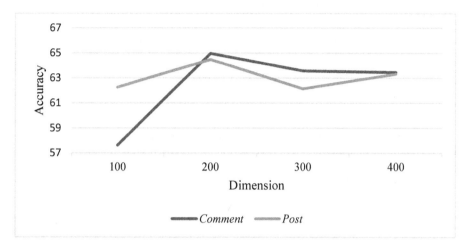

Fig. 1. LSA Accuracy (%) of the *Comment* and *Post* configurations for different dimensions.

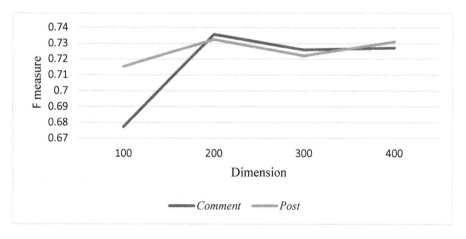

Fig. 2. LSA F measure of the *Comment* and *Post* configurations for different dimensions.

Next, we show the results of 8 classifiers on the *Post* unlabeled data using the optimized dimension parameter (200). Classification methods that the Random Forest method outperformed by a statistically significant amount are marked by an asterisk

4.3   Error Analysis

To better understand the challenges of the relevance classification task, we analyzed the classification errors of the LSA representation, constructed by the *Post* unlabeled data. First, in Table 5, we present the classification confusion matrix. Each column of the matrix represents the instances in a predicted class while each row represents the instances in an actual class. Most of the classification errors were due to incorrect classification of irrelevant comments as relevant (false positive).

Table 5. Confusion matrix for the LSA representation, constructed by the *Post* unlabeled data.

	Negative	Positive
Positive	124	669
Negative	226	360

Next, we sampled 100 misclassified examples; 50 false positive and 50 false negative and analyzed possible reasons for the two types of classification errors. We found 4 dominant types of comments which were often misclassified as relevant even though they are not (false positive):

- Greeting and swear words (40%). For example, the greeting comments (*Shabbat Shalom Bibi! Have a peaceful weekend! Always love from America!:*) and *May the Lord God be in the midst of you.*, or the rude advice *Go prepare a baguette!*.
- Comments which share an obvious context with the post (12%), such as the comment *Lots of differing stories and sites...but would hamas show the same restraint? question still is legit.* and the post *In recent days, Hamas terrorists have fired hundreds of rockets at Israel's civilians....*
- Sarcasm (12%): cynical comments like *Good luck to Tzipi on the reliability, the zigzag, sensitivity, understanding, wisdom, dedication and most of all the fitness and talent to change positions at lightening speed....* to the post *The partnership between myself and Herzog is firm and true, and nothing has changed....*
- Comments with a specific request which follow a general declarative post (8%). For example, the comment *The immediate challenge, worry that tomorrow four hundred families will not be thrown into the street from Channel 10* to the *post ... Only one force could lead the country and the people in the face of enormous challenges...*

A few of the comments refer to pictures, movies or links in the body of the post (6%). One of the comments refers to another comment (2%). For the rest of the misclassified comments (20%), we could not find any satisfying explanation.

We performed a similar analysis for the examples which were misclassified as irrelevant (false negative). We found three types of comments:

- Comments which implicitly refer to the content of the post (34%). For instance, the comment *Mr. Lapid the government is theoretically social, but apathetic and arrogant* to the post. *The government again rejected today the "law of autism."....*

- Greeting (20%). For example, the comment *A lot of health and long life* 🦵 🍎 🍖 for the post *Tonight we celebrated my mother 70 birthday with family and close friends....*
- Agreement words (20%), such as *well done!* and *Be strong and of good courage!*

Some of the comments refer to pictures, movies or links in the body of the post (14%). One of the comments is cynical (2%). For the rest of the misclassified comments (10%), we could not find any satisfying explanation.

5   Conclusions and Future Work

We introduced the task of classifying comment relevance to the content of its post. We investigated four dimensional reduction methods and different types of large unlabeled data for learning the distributional representations unsupervisingly.

We chose the best representation configuration and applied 8 ML methods for relevance classification. The Random Forest method outperformed all the other methods obtaining an accuracy of 64.48% and a F score of 0.73. The achieved results are statistically significantly better than the BoW baseline results.

We showed that expanding the input of the task to include the post text does not improve the classification performance. Additionally, we demonstrated that representing the comment in the post space is a cheap and good representation for our task.

The underlying assumption of the explored dense representations is that close vectors are semantically related. Thus, we plan to construct separate vectors for the post and the comment, calculate the distance between them by different similarity measures (e.g., Cosine and Lin similarity) and add the results as new features to our classification task. In addition, we plan to investigate the deep learning approach for our classification task as it has been shown to be effective for sentiment classification of social media short texts [35, 37, 46, 47].

References

1. Chang, Y., Wang, X., Mei, Q., Liu, Y.: Towards Twitter context summarization with user influence models. In: Proceedings of the Sixth ACM International Conference on Web Search and Data Mining, pp. 527–536. ACM (2013)
2. Raut, V.B., Londhe, D.: Survey on opinion mining and summarization of user reviews on web. Int. J. Comput. Sci. Inf. Technol. **5**, 1026–1030 (2014)
3. Ljubešić, N., Erjavec, T., Fišer, D.: Standardizing tweets with character-level machine translation. In: Gelbukh, A. (ed.) CICLing 2014. LNCS, vol. 8404, pp. 164–175. Springer, Heidelberg (2014). https://doi.org/10.1007/978-3-642-54903-8_14
4. Woodward, E.V., Yan, S.: Socially derived translation profiles to enhance translation quality of social content using a machine translation. Google Patents (2016)
5. Medhat, W., Hassan, A., Korashy, H.: Sentiment analysis algorithms and applications: a survey. Ain Shams Eng. J. **5**, 1093–1113 (2014)

6. Baldwin, T., De Marneffe, M.C., Han, B., Kim, Y.-B., Ritter, A., Xu, W.: Shared tasks of the 2015 workshop on noisy user-generated text: Twitter lexical normalization and named entity recognition. In: Proceedings of the Workshop on Noisy User-Generated Text (WNUT 2015), Beijing, China (2015)

7. Nakov, P., Rosenthal, S., Kiritchenko, S., Mohammad, S.M., Kozareva, Z., Ritter, A., Stoyanov, V., Zhu, X.: Developing a successful SemEval task in sentiment analysis of Twitter and other social media texts. Lang. Resour. Eval. **50**, 35–65 (2016)

8. Liebeskind, C., Nahon, K., Hacohen-Kerner, Y., Manor, Y.: Comparing sentiment analysis models to classify attitudes of political comments on Facebook. Polibits Res. J. Comput. Sci. Comput. Eng. Appl. (2016)

9. Manning, C.D., Raghavan, P., Schütze, H., Others.: Introduction to Information Retrieval. Cambridge University Press, Cambridge (2008)

10. Zhang, Y., Jin, R., Zhou, Z.-H.: Understanding bag-of-words model: a statistical framework. Int. J. Mach. Learn. Cybern. **1**, 43–52 (2010)

11. Wang, M., Cao, D., Li, L., Li, S., Ji, R.: Microblog sentiment analysis based on cross-media bag-of-words model. In: Proceedings of International Conference on Internet Multimedia Computing and Service, p. 76. ACM (2014)

12. Deerwester, S., Dumais, S.T., Furnas, G.W., Landauer, T.K., Harshman, R.: Indexing by latent semantic analysis. J. Am. Soc. Inf. Sci. **41**, 391 (1990)

13. Landauer, T.K., Foltz, P.W., Laham, D.: An introduction to latent semantic analysis. Discourse Process. **25**, 259–284 (1998)

14. Mikolov, T., Sutskever, I., Chen, K., Corrado, G.S., Dean, J.: Distributed representations of words and phrases and their compositionality. In: Advances in Neural Information Processing Systems, pp. 3111–3119 (2013)

15. Bingham, E., Mannila, H.: Random projection in dimensionality reduction: applications to image and text data. In: Proceedings of the Seventh ACM SIGKDD International Conference on Knowledge Discovery and Data Mining, pp. 245–250. ACM (2001)

16. Sahami, M., Heilman, T.D.: A web-based kernel function for measuring the similarity of short text snippets. In: Proceedings of the 15th International Conference on World Wide Web, pp. 377–386. ACM (2006)

17. Banerjee, S., Ramanathan, K., Gupta, A.: Clustering short texts using Wikipedia. In: Proceedings of the 30th Annual International ACM SIGIR Conference on Research and Development in Information Retrieval, pp. 787–788. ACM (2007)

18. Hu, X., Sun, N., Zhang, C., Chua, T.-S.: Exploiting internal and external semantics for the clustering of short texts using world knowledge. In: Proceedings of the 18th ACM Conference on Information and Knowledge Management, pp. 919–928. ACM (2009)

19. Song, G., Ye, Y., Du, X., Huang, X., Bie, S.: Short text classification: a survey. J. Multimed. **9**, 635–643 (2014)

20. Zhang, H., Zhong, G.: Improving short text classification by learning vector representations of both words and hidden topics. Knowl.-Based Syst. **102**, 76–86 (2016)

21. Ma, C., Xu, W., Li, P., Yan, Y.: Distributional representations of words for short text classification. In: Proceedings of NAACL-HLT, pp. 33–38 (2015)

22. Zelikovitz, S., Hirsh, H.: Transductive LSI for short text classification problems. In: FLAIRS Conference, pp. 556–561 (2004)

23. Zelikovitz, S., Marquez, F.: Transductive learning for short-text classification problems using latent semantic indexing. Int. J. Pattern Recognit Artif Intell. **19**, 143–163 (2005)

24. Pu, Q., Yang, G.-W.: Short-text classification based on ICA and LSA. In: Wang, J., Yi, Z., Zurada, Jacek M., Lu, B.-L., Yin, H. (eds.) ISNN 2006. LNCS, vol. 3972, pp. 265–270. Springer, Heidelberg (2006). https://doi.org/10.1007/11760023_39

25. Baker, L.D., McCallum, A.K.: Distributional clustering of words for text classification. In: Proceedings of the 21st Annual International ACM SIGIR Conference on Research and Development in Information Retrieval, pp. 96–103. ACM (1998)

26. Hofmann, T.: Probabilistic latent semantic indexing. In: Proceedings of the 22nd Annual International ACM SIGIR Conference on Research and Development in Information Retrieval, pp. 50–57. ACM (1999)

27. Blei, D.M., Ng, A.Y., Jordan, M.I.: Latent Dirichlet allocation. J. Mach. Learn. Res. **3**, 993–1022 (2003)

28. Minka, T., Lafferty, J.: Expectation-propagation for the generative aspect model. In: Proceedings of the Eighteenth Conference on Uncertainty in Artificial Intelligence, pp. 352–359. Morgan Kaufmann Publishers Inc. (2002)

29. Griffiths, T.L., Steyvers, M.: Finding scientific topics. Proc. Natl. Acad. Sci. **101**, 5228–5235 (2004)

30. Phan, X.-H., Nguyen, L.-M., Horiguchi, S.: Learning to classify short and sparse text & web with hidden topics from large-scale data collections. In: Proceedings of the 17th International Conference on World Wide Web, pp. 91–100. ACM (2008)

31. Chen, M., Jin, X., Shen, D.: Short text classification improved by learning multi-granularity topics. In: IJCAI, pp. 1776–1781. Citeseer (2011)

32. Wang, B., Huang, Y., Yang, W., Li, X.: Short text classification based on strong feature thesaurus. J. Zhejiang Univ. Sci. C. **13**, 649–659 (2012)

33. Johnson, W.B., Lindenstrauss, J.: Extensions of Lipschitz mappings into a Hilbert space. Contemp. Math. **26**, 1 (1984)

34. Le, Q.V., Mikolov, T.: Distributed representations of sentences and documents. In: ICML, pp. 1188–1196 (2014)

35. Tang, D., Wei, F., Yang, N., Zhou, M., Liu, T., Qin, B.: Learning sentiment-specific word embedding for Twitter sentiment classification. In: ACL, vol. 1, pp. 1555–1565 (2014)

36. Tang, D., Qin, B., Liu, T.: Document modeling with gated recurrent neural network for sentiment classification. In: EMNLP, pp. 1422–1432 (2015)

37. Wang, X., Jiang, W., Luo, Z.: Combination of convolutional and recurrent neural network for sentiment analysis of short texts. In: Proceedings of the 26th International Conference on Computational Linguistics, pp. 2428–2437 (2016)

38. Yang, Z., Yang, D., Dyer, C., He, X., Smola, A., Hovy, E.: Hierarchical attention networks for document classification. In: Proceedings of NAACL-HLT, pp. 1480–1489 (2016)

39. Dumais, S.T., Letsche, T.A., Littman, M.L., Landauer, T.K.: Automatic cross-language retrieval using latent semantic indexing. In: AAAI Spring Symposium on Cross-Language Text and Speech Retrieval, p. 21 (1997)

40. Landis, J.R., Koch, G.G.: The measurement of observer agreement for categorical data. Biometrics, 159–174 (1977)

41. Hall, M., Frank, E., Holmes, G., Pfahringer, B., Reutemann, P., Witten, I.H.: The WEKA data mining software: an update. ACM SIGKDD Explor. Newsl. **11**, 10–18 (2009)

42. Witten, I.H., Frank, E., Hall, M.A., Pal, C.J.: Data Mining: Practical Machine Learning Tools and Techniques. Morgan Kaufmann (2016)

43. Faqeeh, M., Abdulla, N., Al-Ayyoub, M., Jararweh, Y., Quwaider, M.: Cross-lingual short-text document classification for Facebook comments. In: 2014 International Conference on Future Internet of Things and Cloud (FiCloud), pp. 573–578. IEEE (2014)

44. Nakov, P., Ritter, A., Rosenthal, S., Sebastiani, F., Stoyanov, V.: SemEval-2016 task 4: sentiment analysis in Twitter. In: Proceedings of SemEval, 1–18 (2016)

45. Petz, G., Karpowicz, M., Fürschuß, H., Auinger, A., Stříteský, V., Holzinger, A.: Opinion mining on the web 2.0 – characteristics of user generated content and their impacts. In: Holzinger, A., Pasi, G. (eds.) HCI-KDD 2013. LNCS, vol. 7947, pp. 35–46. Springer, Heidelberg (2013). https://doi.org/10.1007/978-3-642-39146-0_4
46. Dos Santos, C.N., Gatti, M.: Deep convolutional neural networks for sentiment analysis of short texts. In: COLING, pp. 69–78 (2014)
47. Li, X., Pang, J., Mo, B., Rao, Y., Wang, F.L.: Deep neural network for short-text sentiment classification. In: Gao, H., Kim, J., Sakurai, Y. (eds.) DASFAA 2016. LNCS, vol. 9645, pp. 168–175. Springer, Cham (2016). https://doi.org/10.1007/978-3-319-32055-7_15

Detecting Sockpuppets in Deceptive Opinion Spam

Marjan Hosseinia$^{(\boxtimes)}$ and Arjun Mukherjee

Department of Computer Science, University of Houston, Houston, TX, USA
mhosseinia@uh.edu, arjun@cs.uh.edu

Abstract. This paper explores the problem of sockpuppet detection in deceptive opinion spam using authorship attribution and verification approaches. Two methods are explored. The first is a feature subsampling scheme that uses the KL-Divergence on stylistic language models of an author to find discriminative features. The second is a transduction scheme, spy induction that leverages the diversity of authors in the unlabeled test set by sending a set of spies (positive samples) from training set to retrieve hidden samples in the unlabeled test set using nearest and farthest neighbors. Experiments using ground truth sockpuppet data show the effectiveness of the proposed schemes.

1 Introduction

Deceptive opinion spam refers to illegitimate activities, such as writing fake reviews, giving fake ratings, etc., to mislead consumers. While the problem has been researched from both linguistic [1,2] and behavioral [3,4] aspects, the case of sockpuppets still remains unsolved. A sockpuppet refers to a physical author using multiple aliases (user-ids) to inflict opinion spam to avoid getting filtered. Sockpuppets are particularly difficult to detect by existing opinion spam detection methods as a sockpuppet invariably uses a user-id only a few times (often once) thereby limiting context per user-id. Deceptive sockpuppets may thus be considered as a new frontier of attacks in opinion spam.

However, specific behavioral techniques such as Internet Protocol (IP) and session logs based detection in [5] and group spammer detection in [6] can provide important signals to probe into few ids that form a potential sockpuppet. Particularly, some strong signals such as using same IP and session logs, abnormal keystroke similarities, etc. (all of which are almost always available to a website administrator) can render decent confidence that some reviews are written by one author masked behind a sockpuppet. This can render a form of training data for identifying that sockpuppeter; and the challenge is to find other fake reviews which are also written by the same author but using different aliases in future. Hence, the problem is reduced to an author verification problem. Given a few instances (reviews) written by a (known) sockpuppet author a, the task is to build an Author Verifier, AV_a (classifier) that can determine whether another (future) review is also written by a or not. This problem is related to authorship

© Springer Nature Switzerland AG 2018
A. Gelbukh (Ed.): CICLing 2017, LNCS 10762, pp. 255–272, 2018.
https://doi.org/10.1007/978-3-319-77116-8_19

attribution (AA) [7] where the goal is to identify the author of a given document from a closed set of authors. However, having short reviews with diverse topics render traditional AA methods, that mostly rely on content features, not very effective (see Sect. 7). While there have been works in AA for short texts such as tweets in [8] and with limited training data [9], the case for sockpuppets is different because it involves deception. Further, in reality sockpuppet detection is an open set problem (i.e., it has an infinite number of classes or authors) which makes it very difficult if not impossible to have a very good representative sample of the negative set for an author. In that regard, our problem bears resemblance with authorship verification [10].

In this work we first find that under traditional attribution setting, the precision of a verifier AV_a degrades with the increase in the diversity and size of $\neg a$, where $\neg a$ refers to the negative set authors for a given verifier AV_a. This is detailed in Sect. 4.1. This shows that the verifier struggles with higher false positive and cannot learn $\neg a$ well. It lays the ground for exploiting the unlabeled test set to improve the negative set in training. Next, we improve the performance by learning verification models in lower dimensions (Sect. 5). Particularly, we employ a feature selection scheme, ΔKL Parse Tree Features (henceforth abbreviated as ΔKL-PTFs) that exploits the KL-Divergence of the stylistic language models (computed using PTFs) of a and $\neg a$. Lastly, we address the problem by taking advantage of transduction (Sect. 6). The idea is to simply put a carefully selected subset of positive samples, reviews authored by a (referred to as a spy set) from the training set to the unlabeled test set (i.e., the test set without seeing the true labels) and extract the nearest and farthest neighbors of the members in the spy set. These extracted neighbors (i.e., samples in the unlabeled test set which are close and far from the samples in the spy set) are potentially positive and negative samples that can improve building the verifier AV_a. This process is referred to as *spy induction*. The basic rationale is that since all samples retain their identity, a good distance metric should find hidden positive and negative samples in the unlabeled test set. The technique is particularly effective for situations where training data is limited in size and diversity. Although both spy induction and traditional transduction [11] exploit the assumption of implicit clusters in the data [12], there is a major difference between these two schemes; Spy induction focuses on sub-sampling the unlabeled test set for potential positive and negative examples to grow the training set whereas traditional transduction uses the entire unlabeled test set to find the hyperplane that splits training and test sets in the same manner [13]. Our results show that for the current task, spy induction significantly outperforms traditional transduction and other baselines across a variety of classifiers and even for cross domains.

2   Related Work

Authorship Attribution (AA): AA solves the attribution problem on a closed set of authors using text categorization. Supervised multi-class classification algorithms with lexical, semantic, syntactic, stylistic, and character n-gram features have been explored in [14–16]. In [17], a tri-training method was proposed to solve AA under limited training data that extended co-training using three views: lexical, character and syntactic. The method however assumes that a large set of unlabeled documents authored by the same given closed set of authors are available which is different from our sockpuppet verification. In [18], latent topic features were used to improve attribution. This method also requires larger text collection per author to discover the latent topics for each author which is unavailable for a sockpuppet.

Authorship Verification (AV): In AV, given writings of an author, the task is to determine if a new document is written by that author or not. Koppel and Schler, (2004) [10] explored the problem on American novelists using one-class classification and "unmasking" technique. Unmasking exploits the rate of deterioration of the accuracy of learned models as the best features are iteratively dropped. In [19], the task was to determine whether a pair of blogs were written by the same author. Repeated feature sub-sampling was used to determine if one document of the pair allowed selecting the other among a background set of "imposters" reliably. Although effective unmasking requires a few hundred word texts to gain statistical robustness and was shown to be ineffective for short texts (e.g., reviews) in [20].

Sockpuppet Detection: Sockpuppets were studied in [21] for detecting fake identities in Wikipedia content providers using an SVM model with word and Part Of Speech (POS) features. In [22], a similarity space based learning method was proposed for identifying multiple userids of the same author. These methods assume reasonable context (e.g., 30 reviews per userid). These may not be realistic in opinion spamming (e.g., [6,23,24]) as the reviews per userid are far less and often only one, as shown in singleton opinion spamming [25].

3   Dataset

[26] reports that crowdsourcing is a reasonable method for soliciting ground truths for deceptive content. Crowdsourcing has been successfully used for opinion spam generation in various previous works [1,27–29]. In this work, our focus is to garner ground truth samples of multiple fake reviews written by one physical author (sockpuppet). To our knowledge, there is no existing dataset available for opinion spam sockpuppets. Hence, we used Amazon Mechanical Turk.

Participating turkers were led to a website for this experiment where responses were captured. To model a realistic scenario such as singleton opinion spamming [25], Turkers were asked to act as a sockpuppet having access to several user-ids and each user-id was to be used exactly once to write a review as if written by that alias. The core task required writing 6 positive and 6 negative deceptive reviews, each had more than 200 words, on an entity (i.e., 12 reviews per entity). Each entity belonged to one of the three domains: hotel, restaurant and product. We selected 6 entities across each domain for this task. Each turker had to complete the core task for two entities each per domain (i.e., 24 reviews per domain). The entities and domains were spread out evenly across 17 authors

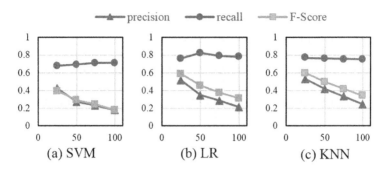

Fig. 1. Precision, recall and F-Score (y-axis) for different author diversity, $\lambda = 25\%, 50\%, 75\%, 100\%$ (x-axis) under in-training setting.

(Turkers). It took us over a month to collect all samples and the mean writing time per review was about 9 min.

To ensure original content, copy and paste was disabled in the logging website. We also followed important rubrics in [1] (e.g., restricted to US Turkers, maintaining an approval rating of at least 90%) and Turkers were briefed with the domain of deception with example fake reviews (from Yelp). All responses were evaluated manually and those not meeting the requirements (e.g., overly short, incorrect target entity, unintelligible, etc.) were discarded resulting in an average of 23 reviews per Turker per domain. The data and code of this work is available at this link[1] and will be released to serve as a resource for furthering research on opinion spam and sockpuppet detection.

Throughout the paper, for single domain experiments, we focus on the hotel domain which had the same trends to that of product and restaurant domains. However, we report results on all domains for cross domain analysis (Sect. 7.4).

4   Hardness Analysis

This section aims to understand the hardness of sockpuppet verification via two schemes.

4.1   Employing Attribution

An ideal verifier (classifier) for an author a requires a representative sample of $\neg a$. We can approximate this by assuming a pseudo author representing $\neg a$ and populating it by randomly selecting reviews of all authors except a. Under the AA paradigm, this is reduced to binary classification. We build author verifiers for each author $a_i \in A = \{a_1, ..., a_{17}\}$. As in AA paradigm, we use in-training setting, i.e., negative samples ($\neg a$) in both training and test sets are authored

---

[1] https://www.dropbox.com/sh/xybjmxffmype3u2/AAA95vdkDp6z5fnTHxqjxq5Ga?dl=0.

by the same closed set of 16 authors although the test and training sets are disjoint. Given our task, since there are not many documents per author to learn from, the effect of author diversity on problem hardness becomes relevant. Hence, we analyze the effect of the diversity and size of the negative set. Let $\lambda \in \{25\%, 50\%, 75\%, 100\%\}$ be the fraction of total authors in $\neg a$ that are used in building the verifier AV_a. Here λ refers to author diversity under in-training setting. We will later explore the effect of diversity under out-of-training setting (Sect. 5). For e.g., when $\lambda = 50\%$, we randomly choose 8 authors, 50% of total 16 authors, from $\neg a$ to define the negative set for AV_a. Note that since we have a total of 16 authors in $\neg a$ for each a and all λ values, the class distribution is imbalanced with the negative class $\neg a$ in majority. We keep the training set balanced throughout the paper as recommended in [30] to avoid learning bias due to data skewness. We use 5-fold Cross Validation (5-fold CV) so, the training fold consists of 80% of the positive (a) and equal sized negative ($\neg a$) samples. But the test fold includes the rest 20% of positive and remaining negative samples except those in training. Under this scheme, since $\neg a$ is the majority class in the test set, accuracy is not an effective metric. For each AV_a, we first compute the precision, recall and F-Score (on the positive class a) using 5-fold CV. Next, we average the results across all authors using their individual verifiers (Fig. 1). This scheme yields us a robust measure of performance of sockpuppet verification across all authors and is used throughout the paper.

We report results of Support Vector Mechine (SVM), Logistic Regression (LR) and k-Nearest Neighbor (kNN) classifiers (using the libraries LIBSVM [31] for SVM with RBF kernel, LIBLINEAR [32] for LR with L2 regularization and WEKA[2] for kNN with k = 3 whose parameters were learned via CV). The feature space consists of lexical units (word unigram) and Parse Tree Features (PTF) extracted using Stanford parser [33] with normalized term frequency for feature value assignment. Unless otherwise stated we use this feature set as well as the classifires setting for all experiments in this paper. We followed some rules from [34] in computing PTFs. The rules are generated by traversing a parse tree in three ways (i) a parent node to the combination of all its non-leaf nodes, (ii) an internal node to its grandparent, (iii) a parent to its internal child. We also add all interior nodes to the feature space (Table 1). From Fig. 1, we note:

- With increase in diversity of negative samples, λ of $\neg a$, the test set size and variety also increase and we find significant drops in precision across all classifiers. This shows a significant rise in false positives. In other words, as the approximated negative set approaches the universal negative set ($\widetilde{\neg a} \rightarrow \neg a$ with increase in diversity of $\neg a$), learning $\neg a$ becomes harder.
- Recall, however, does not experience major changes with increase in the diversity of negative set as it is concerned with retrieving the positive class (a).
- F-Score being the harmonic mean of precision and recall, aligns with the precision performance order. We also note that F-Score in SVM and LR behave similarly followed by kNN.

---

[2] http://www.cs.waikato.ac.nz/ml/weka/.

Table 1. Parse tree feature (PTF) types

Parse tree for: "The staff were friendly."		
ROOT		
S	PTF(I)	S → NP VP
NP VP	PTF(II)	JJ ˆ ADJP → VP
DT NN VBD ADJP	PTF (III)	S→ NP
The staff were JJ	Interior nodes DT, NP
friendly		

Thus, sockpuppet verification is non-trivial and the hardness increases with the increase in $\neg a$ diversity.

4.2   Employing Accuracy and F1 on Balanced Class Distribution

Under binary text classification and balanced class distribution, if accuracy or F1 are high, it shows that the two classes are well separated. This scheme was used in [10] for authorship verification. In our case, we adapt the method as follows. We consider two kinds of balanced data scenarios for a verifier for author a, AV_a: S_1 and S_2. Under S_1, we have the positive class P that consists of half of all reviews authored by a R_a, i.e., $P = \{r_i \in R_a; |P| = 1/2|R_a|\}$. The negative class N_{S_1} comprises of the other half, $N_{S_1} = \{r_i \in R_a - P; |N_{S_1}| = |P|\}$ and $S_1 = P \cup N_{S_1}$. Under S_2, we keep P intact but use a random sampling of $\neg a$ for its negative class, $N_{S_2} = \{r_i \in R_{\neg a}; |N_{S_2}| = |P|\}$ yielding us $S_2 = P \cup N_{S_2}$. Essentially, with this scheme, we wish to understand the effect of negative training set when varied from false negative (N_{S_1}) to approximated true negative (N_{S_2}). Using lexical and parse tree features and 5-fold CV we report performance under each scenario S_1 and S_2 in Table 2. We note the following:

– The precision, recall, F1 and accuracy of all models under S_2 is higher than S_1. While this is intuitive, it shows for deceptive sockpuppets, writings of an author (P) bear separation from other sockpuppeters (N_{S_2}).
– Sockpuppet verification is a difficult problem because under balanced binary classification (S_2), there is just 5–10% gain in accuracy than random

Table 2. Classification results P: Precision, R: Recall, Acc: Accuracy, F1: F-Score under two balanced data scenarios S_1 and S_2 for different classifiers.

Model	S_1				S_2			
	P	R	Acc	F1	P	R	Acc	F1
SVM	47.1	48.4	49.0	45.6	62.5	66.5	61.8	61.1
LR	47.4	46.4	49.5	44.6	63.5	67.4	61.7	62.1
kNN	41.9	57.4	49.8	44.9	51.0	68.9	56.1	53.8

(50% accuracy). Yet it does show the models are learning some linguistic knowledge that separate a and $\neg a$ and using writings of authors other than a is a reasonable approximation for universal $\neg a$.

5   Learning in Lower Dimensions

From the previous experiment, it hints that in the case of deceptive sockpuppets, only a small set of features differentiate a and $\neg a$. As explored in [34], there often exists discriminative author specific stylistic elements that can characterize an author. However, the gamut of all PTFs per author (greater than 2000 features in our data) may be overlapping across authors (e.g., due to native language styles). To mine those discriminative PTFs, we need a feature selection scheme. We build on the idea of linguistic KL-Divergence in [30] and model stylistic elements to capture *how* things are said as opposed to *what* is said. The key idea is to construct the stylistic language model for author, a and its pseudo author $\neg a$. Let A and $\neg A$ denote the stylistic language models for author a and $\neg a$ comprising the positive and negative class of AV_a respectively, where $A(t)$ and $\neg A(t)$ denote the probability of the PTF, t in the reviews of a and $\neg a$. $KL(A||\neg A) = \sum_t (A(t) \log_2 (A(t)/\neg A(t)))$ provides a quantitative measure of stylistic difference between a and $\neg a$. Based on its definition, PTF t that appears in A with higher probability than in $\neg A$, contributes most to $KL(A||\neg A)$. Being asymmetric, it also follows that PTF t' that appears in $\neg A$ more than in A contributes most to $KL(\neg A||A)$. Clearly, both of these types of PTF are useful for building AV_a. They can be combined by computing the per feature, f, ΔKL^f as follows:

$$\Delta KL_t^f = KL_t(A_t||\neg A_t) - KL_t(\neg A_t||A_t) \tag{1}$$
$$KL_t(A_t||\neg A_t) = A(t) \log_2 (A(t)/\neg A(t)) \tag{2}$$
$$KL_t(\neg A_t||A_t) = \neg A(t) \log_2 (\neg A(t)/A(t)) \tag{3}$$

Discriminative features are found by simply selecting the top PTF t based on the descending order of $|\Delta KL_t^f|$ until $|\Delta KL_t^f| < 0.01$. This is a form of subsampling the original PTF space and lowers the feature dimensionality. Intuitively, as KL_t is proportional to the relative difference between the probability of PTF t in positive (a) and negative ($\neg a$) classes, the above selection scheme provides us those PTF t that contribute most to the linguistic divergence between stylistic language models of a and $\neg a$.

To evaluate the effect of learning in lower dimensions, we consider a more realistic "out-of-training" setting instead of the in-training setting as in previous experiments. Under out-of-training setting, the classifier cannot see the writings of those authors that it may encounter in the test set. In other words test and training sets of a verifier AV_a are completely disjoint with respect to $\neg a$ which is realistic and also more difficult than in-training setting. Further, we explore the effect of author diversity under out-of-training setting, δ for the negative set (not to be confused with λ as in Sect. 4). For each experiment, the reviews from

$\delta\%$ of all authors except the intended author, $\neg a$ participate in the training of a verifier AV_a while the rest $(100-\delta\%)$ authors make the negative test set. We also consider standard lexical units (word unigram) (L), L + PTF, and top $k = 20\%$ (tuned via CV) PTF selected using χ^2 metric (L + PTF χ^2) as baselines. We examine different values of $\delta \in \{25\%, 50\%, 75\%\}$ but not $\delta = 100\%$ as that leaves no test samples due to out-of-training setting. From Table 3, we note:

- For each feature space, as the $\neg a$ diversity (δ) increases, across each classifier, we find gains in precision with reasonably lesser drops in recall resulting in overall higher F1. This shows that with increase in diversity in training, the verifiers reduced false positives improving their confidence. Note that verification gets harder for smaller δ as the size and skewness of the test set increases. This trend is different from what we saw in Fig. 1 with λ which referred to diversity under in-training setting.
- Average F1 based on three classifiers (column AVG, Table 3) improves for $\delta = 25\%, 50\%$ using L+PTF than L showing parse tree feature can capture style. However feature selection using χ^2 (L+PTF χ^2) is not doing well as for all δ values there is reduction in F1 for SVM and LR. L+ΔKL PTF feature selection performs best in AVG F1 across different classifiers. It recovers the loss of PTF χ^2 and also improves over the L+PTF space by about 2–3%.

6   Spy Induction

We recall from Sect. 1 that our problem suffers with limited training data per author as sockpuppets only use an alias few times. To improve verification, we need a way to learn from more instances. Also from Sect. 4, we know that precision drops with increase in diversity of $\neg a$. This can be addressed by leveraging the unlabeled test set to improve the $\neg a$ set in training under transduction.

Figure 2 provides an overview of the scheme. For a given training set and a test set for AV_a, spy induction has three main steps. First is spy selection where some carefully selected positive samples are sent to the unlabeled test set. The second step is to find certain Nearest and Farthest Neighbors (abbreviated NN, FN henceforth) of the positive spy samples in the unlabeled test set. As the instances retain their original identity, a good distance metric should be able to retrieve potentially hidden positive (using common NN across different positive spies) and negative (using common FN across different positive spies) samples in the unlabeled test set. These newly retrieved samples from unlabeled test set are used to grow the training set. The previous step can have some label errors in NN and FN as they may not be true positive (a) and negative ($\neg a$) samples, which can be harmful in training. These are shown in Fig. 2(B) by α_- and β_+ samples. To reduce such potential errors, a third step of label verification is employed where the labels of the newly retrieved samples from unlabeled test set are verified using agreement of classifiers on orthogonal feature spaces. with this step, we benefit from the extended training data without suffering from the possible issue of error propagation. Lastly, the verifier undergoes improved training with additional samples and optimizes the F-Score on the training set.

Table 3. P: Precision, R: Recall, F1: F-Score for out-of-training with different values of δ for three classifiers. AVG reports the average F1 across three classifiers. Feature Set: L: Lexical unit (word unigram), PTF: Parse tree feature, PTF χ^2 : PTF selected by χ^2 , ΔKL PTF: PTF selected via ΔKL

	SVM			LR			kNN			AVG
					$\delta=25\%$					
Feature Set	P	R	F1	P	R	F1	P	R	F1	F1
L	23.6	82.0	34.3	23.1	74.7	30.8	19.4	84.6	25.8	30.3
L+PTF	25.6	73.4	35.2	22.9	82.5	33.4	24.8	66.7	24.5	31.0
L+PTFχ^2	21.7	73.5	30.8	14.8	53.5	21.3	22.6	75.3	25.9	26.0
L+ΔKL PTF	25.6	79.2	36.3	21.7	80.2	32.1	22.3	81.5	27.8	**32.1**

(a)

	SVM			LR			kNN			AVG
					$\delta=50\%$					
Feature Set	P	R	F1	P	R	F1	P	R	F1	F1
L	30.7	83.6	41.8	28.7	83.1	38.7	21.1	85.1	27.1	35.9
L+PTF	33.2	73.4	42.7	30.6	78.1	40.9	28.0	73.8	28.8	37.5
L+PTFχ^2	24.8	69.2	33.7	21.0	47.8	26.9	23.4	81.6	30.2	30.3
L+ΔKL PTF	33.7	75.9	42.8	31.1	79.4	41.9	26.9	79.5	30.3	**38.3**

(b)

	SVM			LR			kNN			AVG
					$\delta=75\%$					
Feature Set	P	R	F1	P	R	F1	P	R	F1	F1
L	47.1	77.7	55.1	44.4	80.4	52.7	28.7	83.5	37.8	48.5
L+PTF	51.4	72.6	56.0	43.7	78.8	53.0	28.1	64.8	31.7	46.9
L+PTFχ^2	42.4	71.2	49.5	33.9	49.6	36.4	35.6	79.8	40.0	42.0
L+ΔKL PTF	50.5	71.9	56.2	46.3	79.4	54.9	42.2	80.8	46.1	**52.4**

(c)

6.1 Spy Selection

This first step involves sending highly representative spies that can retrieve new samples to improve training. For a given verification problem, AV_a, let $D = D.Train \cup D.Test$ denotes the whole data. Although any positive instance in $D.Train$ can be a spy sample, only few of them might satisfy the representativeness constraint. Hence, we select the spies as those positive samples that have maximum similarity with other positive instances. In other words, the selection respects class based centrality and employs minimum overall pairwise distance (OPD) as its selection criterion:

$$OPD(s) = argmin_{s \in P}(\sum_{x \in P} d(s,x)) \qquad (4)$$

where P is the positive class of training set, s denotes a potential spy sample and $d(\cdot)$ is distance function. Our spy set, $S = \{s\}$ consists of different spies that

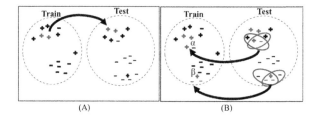

Fig. 2. Spy Induction: (A) Spies (Red plus signs) selected based on positive class centrality being put to the unlabeled test set. (B) Common nearest and farthest neighbors (Green plus and minus signs) across different spies neighborhood shown by oval boundaries found in unlabeled test set being put back in the training set.

have the least pairwise distance to all other positive samples. We also consider different sizes of the spy set $|S| = n_S$ and experiment with different values of $n_S \in N_S = \{1, 3, 5, 7\}$. The method $SelectSpy(\cdot)$ (line 4, Algorithm 1) implements this step.

6.2 New Instance Retrieval via Nearest and Farthest Neighbors

After the selected spies are put into the unlabeled test set, the goal is to find potential positive and negative samples. Intuitively, one would expect that the closest data points to positive spy samples belong to the positive class while those that are farthest are likely negative samples. For each spy, $s \in S$, we consider n_Q nearest neighbors forming the likely positive set Q_s and n_R farthest neighbors forming the likely negative set R_s specific to s. Then, we find the common neighbors across multiple spies to get confidence on the likely positive or negative samples which yields us the final set of potentially Q positive and R negative samples,

$$Q = \cap_{s \in S} Q_s; \quad R = \cap_{s \in S} R_s \tag{5}$$

This is implemented by the methods $ObtainNN(\cdot), ObtainFN(\cdot)$ (lines 5, 6, Algorithm 1). In most cases, we did not find the common neighbors Q, R to be empty, but if it is null, it implies no reliable samples were found. Further, like n_S (in Sect. 6.1), we try different values for $|Q_s| = n_Q; n_Q \in N_Q = \{1, 3\}$ and $|R_s| = n_R; n_R \in N_R = \{5, 10, 25, 40, 50, 60\}$. These values were set based on pilot experiments. The above scheme of new sample retrieval works with any distance metric.

We consider two distance metrics on the feature space L+ΔKL PTF to compute all pairwise distances in the methods $SelectSpy(\cdot)$, $ObtainNN(\cdot)$ and $ObtainFN(\cdot)$ (line 4–6, Algorithm 1): (1) Euclidean, (2) Distance metric learned from data. Specifically, we use the large margin method in [35] which learns a Mahalanobis distance metric $d_M(\cdot)$ that optimizes kNN classification in the training data using d_M. The goal is to learn $d_M(\cdot)$ such that the k-nearest neighbors (based on $d_M(\cdot)$) of each sample have the same class label as itself while different class samples are separated by a large margin.

Algorithm 1: Spy induction

$SpyInduction(D, N_S, N_Q, N_R)$
1 : $P \leftarrow \{x \in D.Train, x.label > 0\} // positive\,class$
2 : $I \leftarrow \{(n_S, n_Q, n_R) | n_S \in N_S, n_Q \in N_Q, n_R \in N_R\}$
3 : $for\ each(i = (n_S, n_Q, n_R) \in I)$
4 : $S \leftarrow SelectSpy(P, n_S)$
5 : $Q \leftarrow ObtainNN(D.Test, S, n_Q)$
6 : $R \leftarrow ObtainFN(D.Test, S, n_R)$
7 : $(Q^v, R^v) \leftarrow CoLabelingVerification(Q, R, D.Train)$
8 : $F1(i) \leftarrow CVImprovedTraining(D.Train, Q^v, R^v)$
9 : $endfor$
10 : $(n_S, n_Q, n_R)^* \leftarrow argmax\ _{i \in I}(F1(i))$
11 : $AV \leftarrow Classifier(D.Train, D.Test, (n_S, n_Q, n_R)^*)$

Fig. 3. Spy induction

6.3   Label Verification via Co-Labeling

As it is not guaranteed that the distances between samples can capture the notion of authorship, the previous step can have errors, i.e., there may be some positive samples in R and negative samples in Q. To solve this, we apply co-labeling [36] for label verification. In co-labeling, multiple views are considered for the data and classifiers are built on each view. Majority voting based on classifier agreement is used to predict labels of unlabeled instances. In our case, we consider $D.Train$ to train an SVM on five feature spaces (views): (i) unigam, (ii) unigram+bigram, (iii) PTF, (iv) POS, (v) ΔKL PTF+unigram+bigram as five different label verification classifiers. Then, the labels of samples in Q and R are verified based on agreements of majority on classifier prediction. Samples having label discrepancies are discarded to yield the verified retrieved samples, (Q^v, R^v) (line 7, Algorithm 1). The rationale here is that it is less probable for majority of classifiers (each trained on a different view) to make the same mistake in predicting the label of a data point than a single classifier.

6.4   Improved Training

The retrieved and verified samples from the previous steps are put back into the training set. However, the key lies in estimating the right balance between the amount of spies sent, and the size of the neighborhood considered for retrieving potentially positive or negative samples, which are governed by the parameters n_S, n_Q, n_R. To find the optimal parameters, we try different values of the parameter triple, $i = (n_S, n_Q, n_R) \in I$ (lines 2, 3 Algorithm 1) and record the F-Score of 5-fold CV on $D.Train \cup Q^v \cup R^v$ as $F1(i)$ (line 8, Algorithm 1). This step is carried out by the method $CVImprovedTraining(.)$. Finally, the parameters that yield the highest $F1$ in training are chosen (line 10, Algorithm 1) to yield the output spy induced verifier (line 11, Algorithm 1).

7   Experimental Evaluation

This section evaluates the proposed spy method. We keep all experiment settings same as in Sect. 5 (i.e., use out-of-training with varying author diversity δ). We fix our feature space to L+ΔKL PTF as it performed best (see Table 3). As mentioned earlier, we report average verification performance across all authors. Below we detail baselines, followed by results and sensitivity analysis.

7.1   Baselines and Systems

We consider the following systems:

MBSP runs the Memory-based shallow parsing approach [9] to authorship verification that is tailored for short text and limited training data.

Base runs classification without spy induction and dovetails with Table 3 (last row) for each δ.

TSVM uses the transductive learner of SVMLight [13] and aims to leverage the unlabeled (test) set by classifying a fraction of unlabeled samples to the positive class and optimizes the precision/recall breakeven point.

Spy (Eu.) & Spy (LM) are spy induction systems without co-labeling but use Euclidean (Eu.) and learned distance metric (LM) to compute neighbors.

Spy (EuC) & Spy (LMC) are extensions of previous models that consider label verification via co-labeling approach.

7.2   Results

Table 4 reports the results. We note the following:

- Except for two cases (F1 of SVM and kNN for Spy(LM) with $\delta = 75\%$), almost all spy models are able to achieve significantly higher F1 than base (without spy induction) and TSVM for all classifiers SVM, LR, kNN and across all diversity values δ. MBSP performs similarly as Base showing memory based learning does not yield a significant advantage in sockpuppet verification. TSVM is not doing well on F1 but improves recall. One reason could be that due to class imbalance, TSVM has some bias in classifying unlabeled examples to positive class that improves recall but suffers in precision.
- The AVG F1 column shows that on average, across three classifiers spy induction yields at least 4% gain or more. The gains in AVG F1 are pronounced for $\delta = 25\%$ with gains upto 12% with spy (EuC). For $\delta = 75\%$, we find gains of about 10% in F1 with spy (EuC). Note that we employ out-of-training setting with varying author diversity (δ) so the test set is imbalanced (i.e., the random baseline is no longer 50%). Across all classifiers, the relative gains in F1 for spy methods over base reduce with increase in author diversity δ which is due (a) better $\neg a$ samples in training that raise the base result and (b) test set size and variety reduction limiting spy induction. Nonetheless, we note that for $\delta = 25\%$ (harder case of verification), spy induction does well across all classifiers.

Table 4. P: Precision, R: Recall, F1: F-Score results for spy induction under out-of-training with different values of δ for three classifiers. AVG reports the average F1 across three classification models. Feature Set: L+ΔKL PTF. Gains in AVG F1 using spy (EuC) and (LMC) over baselines are significant at p < 0.001 using a t-test

	δ=25%									
	SVM			LR			kNN			AVG
Model	P	R	F1	P	R	F1	P	R	F1	F1
MBSP	22.9	84.1	32.0	22.1	82.1	31.1	20.7	77.5	23.5	28.9
Base	25.6	79.2	36.3	21.7	80.2	32.1	22.3	81.5	27.8	32.1
TSVM	30.6	43.9	34.3	-	-	-	-	-	-	34.3
Spy(Eu.)	39.1	51.2	40.6	51.6	42.3	43.7	43.3	52.5	39.4	41.2
Spy(LM)	42.2	49.3	42.0	44.4	49.6	43.9	34.9	62.8	33.7	39.9
Spy(EuC)	42.0	57.8	42.7	51.2	61.3	52.5	41.5	57.9	38.4	**44.5**
Spy(LMC)	38.1	60.5	40.6	42.9	64.9	47.1	35.3	68.1	36.0	41.2

(A)

	δ=50%									
	SVM			LR			kNN			AVG
Model	P	R	F1	P	R	F1	P	R	F1	F1
MBSP	31.9	85.3	42.1	25.0	81.6	34.6	21.1	84.4	28.7	35.1
Base	33.7	75.9	42.8	31.1	79.4	41.9	26.9	79.5	30.3	38.3
TSVM	20.2	83.6	31.1	-	-	-	-	-	-	31.1
Spy(Eu.)	39.1	71.2	45.9	38.1	67.9	46.2	45.1	58.7	41.7	44.6
Spy(LM)	40.1	68.5	45.9	44.7	55.5	45.6	42.7	66.2	40.2	43.9
Spy(EuC)	62.3	52.0	52.3	62.5	64.6	61.0	46.6	62.3	43.2	**52.2**
Spy(LMC)	46.8	60.9	48.0	51.5	67.4	53.7	40.7	67.6	39.2	47.0

(B)

	δ=75%									
	SVM			LR			kNN			AVG
Model	P	R	F1	P	R	F1	P	R	F1	F1
MBSP	49.9	80.4	57.2	53.9	81.9	59.1	33.8	82.2	38.6	51.6
Base	50.5	71.9	56.2	46.3	79.4	54.9	42.2	80.8	46.1	52.4
TSVM	34.4	80.4	45.8	-	-	-	-	-	-	45.8
Spy(Eu.)	55.6	70.8	58.2	50.9	77.5	57.9	57.7	57.6	50.7	55.6
Spy(LM)	53.1	62.8	54.3	51.1	69.6	56.1	51.8	57.7	45.8	52.1
Spy(EuC)	71.9	59.1	64.2	68.9	75.6	70.2	63.6	59.0	54.7	**62.4**
Spy(LMC)	55.6	72.3	58.4	60.8	68.5	61.4	53.4	60.8	48.7	56.2

(C)

- Anchoring on one distance metric (Eu./LM), we find that spy induction with co-labeling does markedly better than spy induction without co-labeling across all δ in AVG F1 across three classifiers. This shows label verification using co-labeling is helpful in filtering label noise and an essential component in spy induction.

– Between Euclidean and distance metric learned via large-margin (LM), Euclidean does better than LM in AVG F1 for both spy induction with and without co-labeling. However, using the LM metric yields higher recall than Euclidean in certain cases (underlined) which shows LM metric can yield gains in F1 beyond base with relatively lesser drops in recall which is again useful.

In summary, we can see that spy induction works in improving the F1 across different classifiers and author diversity and distance metrics. Overall, the scheme LR+Spy (EuC) does best across each δ (highlighted in gray) and is used for subsequent experiments to compare against Base.

7.3   Spy Parameter Sensitivity Analysis

To analyze the sensitivity of the parameters, we plot the range of precision, recall and F1 values as spy induction learns the optimal values in training. We focus on the variation for $\delta = 25\%, 75\%$ capturing both extremes of diversity. Figure 3 shows the performance curves for different spy parameter triples (n_S, n_Q, n_R) sorted in the increasing order of F1. We find that for both $\delta = 25\%, 75\%$, the spy induction steadily improves precision with the increase in likely $\neg a$ samples (n_R). Although the recall drops more and has more fluctuations for the harder case of $\delta = 25\%$, it stabilizes early for $\delta = 75\%$ with much lesser drop in recall. This shows that the spy induction scheme is robust in optimizing F1 with only a few (5–7) spy samples (n_S) sent to unlabeled test set.

7.4   Domain Adaptation

We now test the effective of spy induction under domain transfer. As mentioned in Sect. 3, we obtained reviews of Turkers for hotel, restaurant and product domains. Keeping all other settings same as in Tables 4 and 5 reports results for cross domain performance by training the verifiers (AV_a) using two domains and testing on the third domain. We compare sockpupet verification using LR+Spy (EuC) vs. base (LR without spy induction). We report the F1 scores as the trends of precision and recall for cross domain were similar to the trends in Table 4.

Table 5. Cross domains results of LR + Spy (EuC). Gains in F1 using spy induction over base are significant at p < 0.01 for all test domains and each δ using a t-test.

	$\delta = 25\%$		$\delta = 50\%$		$\delta = 75\%$	
	Base	Spy	Base	Spy	Base	Spy
Test domain	F1	F1	F1	F1	F1	F1
Hotel	30.3	36.4	40.0	47.0	50.3	52.6
Product	29.5	36.6	34.0	40.8	51.5	53.5
Restaurant	30.1	41.1	41.7	51.5	55.2	59.3

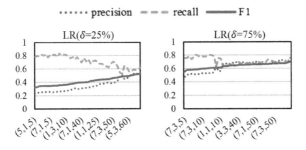

Fig. 4. Spy Parameter Sensitivity. Variation of precision, recall and F1 across different parameter triples (n_S, n_Q, n_R).

Table 6. Performance gains of Spy (EuC) in F1 over Base on Wikipedia Sockpuppet Dataset. Gains are significant ($p < 0.01$) except for LR $= 50\%$, 75%

	$\delta = 25\%$		$\delta = 50\%$		$\delta = 75\%$	
	Base	Spy	Base	Spy	Base	Spy
Classifier	F1	F1	F1	F1	F1	F1
SVM	50.7	57.6	59.6	62.9	68.0	70.3
LR	40.4	42.4	49.8	51.1	60.0	61.5
kNN	23.9	29.5	32.0	37.1	43.0	51.1

The F1 of base in cross domain (Table 5, Hotel row) is lower than corresponding LR results with base (Table 4) for all δ showing cross domain verification is harder. Nonetheless, spy induction is able to render statistically significant gains in F1 for all δ (see Table 5).

7.5 Performance on Wikipedia Sockpuppet (WikiSock) Dataset

In [37], a corpus of Wikipedia sockpuppet authors was produced. It contains 305 authors with an average of 180 documents per author and 90 words per document which we use as another benchmark for evaluating our method.

It is important to note that the base results reported in [37] are not directly comparable to this experiment (Table 6). This is because [37] used all 623 cases that were found as candidates but we focus on only 305 of them which were actually confirmed sockpuppets by Wikipedia administrators. Next, we perform experiments under realistic out-of-training setting and varying the author diversity (as in Table 4) which is different from [37]. This explains the rather lower F1 as reported in [37] for Base. We focus on F1 performance of spy (EuC) versus base (without spy) as the precision and recall trends were same as in Table 4. Compared to Table 4 base results, base does better for SVM and LR on WikiSock dataset that hints the data to be slightly easier. The relative gains of spy over base although are a bit lower than those in Table 4, spy induction consistently outperforms base.

8   Conclusion

This work performed an in-depth analysis of deceptive sockpuppet detection. We first showed that the problem is different from traditional authorship attribution or verification and gets more difficult with the increase in author diversity. Next, a feature selection scheme based on KL-Divergence of stylistic language models was explored that yielded improvements in verification beyond baseline features. Finally, a transduction scheme, spy induction, was proposed to leverage the unlabeled test set. A comprehensive set of experiments showed that the proposed approach is robust across both (1) different classifiers, (2) cross domain knowledge transfer and significantly outperforms baselines. Further, this work produced a ground truth corpus of deceptive sockpuppets across three domains.

Acknowledgments. This work is supported in part by NSF 1527364. We also thank anonymous reviewers for their helpful feedbacks.

References

1. Ott, M., Choi, Y., Cardie, C., Hancock, J.T.: Finding deceptive opinion spam by any stretch of the imagination. In: Proceedings of the 49th Annual Meeting of the Association for Computational Linguistics: Human Language Technologies, vol. 1, pp. 309–319. Association for Computational Linguistics (2011)
2. Feng, S., Banerjee, R., Choi, Y.: Syntactic stylometry for deception detection. In: Proceedings of the 50th Annual Meeting of the Association for Computational Linguistics: Short Papers, vol. 2, pp. 171–175. Association for Computational Linguistics (2012)
3. Mukherjee, A., et al.: Spotting opinion spammers using behavioral footprints. In: Proceedings of the 19th ACM SIGKDD International Conference on Knowledge Discovery and Data Mining, pp. 632–640. ACM (2013)
4. Lim, E.P., Nguyen, V.A., Jindal, N., Liu, B., Lauw, H.W.: Detecting product review spammers using rating behaviors. In: Proceedings of the 19th ACM International Conference on Information and Knowledge Management, pp. 939–948. ACM (2010)
5. Li, H., Chen, Z., Mukherjee, A., Liu, B., Shao, J.: Analyzing and detecting opinion spam on a large-scale dataset via temporal and spatial patterns. In: Ninth International AAAI Conference on Web and Social Media (2015)
6. Mukherjee, A., Liu, B., Glance, N.: Spotting fake reviewer groups in consumer reviews. In: Proceedings of the 21st International Conference on World Wide Web, pp. 191–200. ACM (2012)
7. Stamatatos, E.: A survey of modern authorship attribution methods. J. Am. Soc. Inf. Sci. Technol. **60**, 538–556 (2009)
8. Layton, R., Watters, P., Dazeley, R.: Authorship attribution for twitter in 140 characters or less. In: 2010 Second Cybercrime and Trustwor-thy Computing Workshop (CTC), pp. 1–8. IEEE (2010)
9. Luyckx, K., Daelemans, W.: Authorship attribution and verification with many authors and limited data. In: Proceedings of the 22nd International Conference on Computa-tional Linguistics, vol. 1, pp. 513–520. Association for Compu-tational Linguistics (2008)

10. Koppel, M., Schler, J.: Authorship verification as a one-class classification problem. In: Proceedings of the Twenty-first International Conference on Machine Learning, p. 62. ACM (2004)
11. Vapnik, V.: The Nature of Statistical Learning Theory. Springer, New York (2013). https://doi.org/10.1007/978-1-4757-3264-1
12. Chapelle, O., Zien, A.: Semi-supervised classification by low density separation. In: AISTATS, pp. 57–64 (2005)
13. Joachims, T.: Transductive inference for text classification using support vector machines. In: ICML, vol. 99, pp. 200–209 (1999)
14. Graham, N., Hirst, G., Marthi, B.: Segmenting documents by stylistic character. Nat. Lang. Eng. **11**, 397–415 (2005)
15. Gamon, M.: Linguistic correlates of style: authorship classification with deep linguistic analysis features. In: Proceedings of the 20th International Conference on Computational Linguistics, p. 611. Association for Computational Linguistics (2004)
16. Sapkota, U., Bethard, S., Montes-y Gómez, M., Solorio, T.: Not all character n-grams are created equal: a study in authorship attribution. In: Human Language Technologies: The 2015 Annual Conference of the North American Chapter of the ACL, pp. 93–102 (2015)
17. Qian, T., et al.: Tri-training for authorship attribution with limited training data: a comprehensive study. Neurocomputing **171**, 798–806 (2016)
18. Seroussi, Y., Bohnert, F., Zukerman, I.: Authorship attribution with author-aware topic models. In: Proceedings of the 50th Annual Meeting of the Association for Computational Linguistics: Short Papers, vol. 2, pp. 264–269. Association for Computational Linguistics (2012)
19. Koppel, M., Winter, Y.: Determining if two documents are written by the same author. J. Assoc. Inf. Sci. Technol. **65**, 178–187 (2014)
20. Sanderson, C., Guenter, S.: Short text authorship attribution via sequence kernels, Markov chains and author unmasking: an investigation. In: Proceedings of the 2006 Conference on Empirical Methods in Natural Language Processing, pp. 482–491. Association for Computational Linguistics (2006)
21. Solorio, T., Hasan, R., Mizan, M.: A case study of sockpuppet detection in wikipedia. In: Workshop on Language Analysis in Social Media (LASM) at NAACL HLT. pp. 59–68 (2013)
22. Qian, T., Liu, B.: Identifying multiple userids of the same author. In: EMNLP, pp. 1124–1135 (2013)
23. Jindal, N., Liu, B.: Opinion spam and analysis. In: Proceedings of the 2008 International Conference on Web Search and Data Mining, pp. 219–230. ACM (2008)
24. Fusilier, D.H., Montes-y-Gómez, M., Rosso, P., Cabrera, R.G.: Detection of opinion spam with character n-grams. In: Gelbukh, A. (ed.) CICLing 2015. LNCS, vol. 9042, pp. 285–294. Springer, Cham (2015). https://doi.org/10.1007/978-3-319-18117-2_21
25. Xie, S., Wang, G., Lin, S., Yu, P.S.: Review spam detection via temporal pattern discovery. In: Proceedings of the 18th ACM SIGKDD International Conference on Knowledge Discovery and Data Mining, pp. 823–831. ACM (2012)
26. Gokhman, S., Hancock, J., Prabhu, P., Ott, M., Cardie, C.: In search of a gold standard in studies of deception. In: Proceedings of the Workshop on Computational Approaches to Deception Detection, pp. 23–30. Association for Computational Linguistics (2012)
27. Li, J., Ott, M., Cardie, C., Hovy, E.H.: Towards a general rule for identifying deceptive opinion spam. In: ACL (1), Citeseer, pp. 1566–1576 (2014)

28. Li, J., Ott, M., Cardie, C.: Identifying manipulated offerings on review portals. In: EMNLP, pp. 1933–1942 (2013)
29. Banerjee, R., Feng, S., Kang, J.S., Choi, Y.: Keystroke patterns as prosody in digital writings: a case study with deceptive reviews and essays. In: Empirical Methods on Natural Language Processing (EMNLP) (2014)
30. Mukherjee, A., Venkataraman, V., Liu, B., Glance, N.S.: What yelp fake review filter might be doing? In: ICWSM (2013)
31. Chang, C.C., Lin, C.J.: LIBSVM: a library for support vector machines. ACM Trans. Intell. Syst. Technol. **2**, 27:1–27:27 (2011). http://www.csie.ntu.edu.tw/cjlin/libsvm
32. Fan, R.E., Chang, K.W., Hsieh, C.J., Wang, X.R., Lin, C.J.: LIBLINEAR: a library for large linear classification. J. Mach. Learn. Res. **9**, 1871–1874 (2008). http://www.csie.ntu.edu.tw/cjlin/liblinear/
33. Klein, D., Manning, C.D.: Accurate unlexicalized parsing. In: Proceedings of the 41st Annual Meeting on Association for Computational Linguistics, vol. 1, pp. 423–430. Association for Computational Linguistics (2003). http://nlp.stanford.edu/software/lex-parser.shtml
34. Feng, S., Banerjee, R., Choi, Y.: Characterizing stylistic elements in syntactic structure. In: Proceedings of the 2012 Joint Conference on Empirical Methods in Natural Language Processing and Computational Natural Language Learning, pp. 1522–1533. Association for Computational Linguistics (2012)
35. Weinberger, K.Q., Blitzer, J., Saul, L.K.: Distance metric learning for large margin nearest neighbor classification. In: Advances in Neural Information Processing Systems, pp. 1473–1480 (2005)
36. Xu, X., Li, W., Xu, D., Tsang, I.: Co-labeling for multi-view weakly labeled learning. IEEE Trans. Pattern Anal. Mach. Intell. **38**(6), 1113–1125 (2015)
37. Solorio, T., Hasan, R., Mizan, M.: Sockpuppet detection in wikipedia: a corpus of real-world deceptive writing for linking identities. In: Proceedings of the Ninth International Conference on Language Resources and Evaluation (LREC 2014), Reykjavik, Iceland, European Language Resources Association (ELRA) (2014)

Author Profiling and Authorship Attribution

Reading the Author and Speaker: Towards a Holistic and Deep Approach on Automatic Assessment of What is in One's Words

Björn W. Schuller[1,2(⊠)]

[1] Department of Computing, Imperial College London,
SW7 2AZ London, UK
bjoern.schuller@imperial.ac.uk
[2] University of Passau, Chair of Complex and Intelligent Systems,
94032 Passau, Germany
http://www.schuller.one

Abstract. Computational text analysis is continuously becoming richer in ways the author of a text is 'read' in terms of the states and traits of the person behind the words such as writer's age, gender, personality, emotion or sentiment to name but a few. Similarly, in the analysis of spoken language, one finds a broadening palette of such characteristics of speakers automatically analysed in recent Computational Paralinguistics research. It seems wise to assess these characteristics in one pass to understand their interrelationship rather than going one by one in isolation. As an example, it may help to estimate one's personality knowing the age, gender, and cultural background of the person. Thus, a holistic approach is advocated that aims at automatically assessing the 'larger' picture of a person that wrote or spoke words of analysis. Here, a short motivation and inspirations 'en route' to holistic author and word-based speaker profiling are given.

Keywords: Spoken language processing · Computational paralinguistics · Speaker profiling · Text analysis · Sentiment analysis Opinion mining · Affective computing

1 What One's Words Reveal

Even when speaking or writing about others, things, or when simply telling stories that appear unrelated to ourself, our words reveal an astonishing range of attributes such as states and traits on those choosing and using them. Computational linguistics since long make use of this fact to *profile* an author of written text or speaker. Over the last decades in fact, a whole range of research challenges on the topic has emerged, including major events such as the annual author profiling task at PAN within the CLEF framework, or the sentiment analysis and other tasks in SemEval. Several further signal analysis challenges include

© Springer Nature Switzerland AG 2018
A. Gelbukh (Ed.): CICLing 2017, LNCS 10762, pp. 275–288, 2018.
https://doi.org/10.1007/978-3-319-77116-8_20

at least the option of analysis of the spoken word, such as the annual Interspeech Computational Paralinguistics Challenge since its first edition in 2009 [42], or the annual Audio/Visual Emotion Challenge since 2011 [40], the Emotion in the Wild [15] challenge since 2013, the Multimodal Emotion Challenge since 2016 [20], and repeatedly tasks in the the ChaLearn Looking at People and MediaEval challenges, besides some more.

Here, I first want to give an overview on the diversity of author and speaker attributes computers can these days automatically derive from the spoken or written words. Methods may thereby reach from 'simple' bag-of-words approaches, e. g., learnt with support vector machines, looking at term frequencies or part-of-speech frequencies such as verb, adjective, and noun frequencies to more recent deep learnt word embeddings. They may, however, also go beyond and linguistically deeper by looking at contextual disambiguation, lexical repetition, and semantic priming effects [3], pragmatic language production abilities and deficits [17], or word-frequency mirror effects [2], and vocalisation composition [27,67].

Then, based on observations made in holistic analysis of spoken language, I want to emphasise on the need to model the author or speaker holistically, i. e., assess the different author and speaker attributes in parallel rather than in isolation. I will then, based on a dozen of taxonomies, show how the current space of attributes can literally be blown up for future holistic modelling aspirations. This will be exemplified based on the concrete tasks as were held over the years at the annual Interspeech conference in its above named challenge series focussed on paralinguistics which the author of this contribution co-organises.

2   Author and Speaker Attributes Mirrored in the Words

The automatic extraction from written or spoken text of attributes characterising and describing the person behind the words has been attempted in an astonishing richness over the last years, which I want to demonstrate next. To this end, I will use a first *taxonomy*, for a coarse categorisation: *states* and *traits* of an author or speaker. Later, further such taxonomies will be introduced aiming at 'blowing up' richness of ways to attribute an author or speaker.

2.1   Short-term States: Affect and Stances

The range of short-term states that can be extracted from one's words is impressively long; most frequently targeted ones from words include:

Affect, and valence [59], sentiment [6,43], basic emotions [45,60], continuous emotions such as arousal, dominance, and valence [39], irony [37], sarcasm [11,16], besides a sheer endless list of further states and stances, including awkwardness and assertiveness [30], disagreement [1], empathy [8], entrainment [4], flirting [29,30], friendliness [30], hostility [57], humour [23], interest [44,65], lying [24], nastiness [19], offensiveness [35], or politicalness [64].

2.2   Mid-term States: Health and Wellbeing, Mostly

When it comes to longer-term states, one can find mostly health and wellbeing related such attempted in the literature. Naturally, the boundaries between short, longer, and long-term temporal relation can be defined only loosely. For the sake of better structure and readability, here, I comprise all health and wellbeing related states, even if medical discussion may be ongoing whether some of these are rather traits as they are inherited and whether or not some or all of these are curable in theory or even practice. The list is again long, including in fore mostly Autism [36,67,68], ADHD [10], Alzheimer [2,3,18], cognitive impairment [18], communication abilities and disorders such as dietary influenced (e. g., by breast feeding or its absence) [56], depression [18,25,38,62] including shorter term concrete suicide risk [14], drug addiction [61], intoxication [65] such as from alcohol or drugs, Parkinson's [17], Rett Syndrome [27], or SAD [10].

Note that in the ongoing, I use the term *states* spanning across short to medium-term temporary characteristics of an author or speaker, such as affect or (non-chronical) health condition.

2.3   Traits: Identity, Age, Gender, and Whatnots

In contrast to the above discussed author or speaker states, *traits* are defined here by their long-term nature. Note that this does not necessarily require permanence, as for example, age (which obviously changes with the years) is subsumed under traits here.

In the first place, success for extracting traits of an individual from its words include the identity of the person encompassing its recognition and verification [9,12,63] next to the automatic identification of the age [31,55], gender [31,55], and personality [5,22,28,55]. Note that personality can also include the short-term perceived personality, such as featured in the MAPTRAITS challenge [7]. However, the author and speaker profiling literature is rich, touching upon a series of further traits [21,26,33,34].

3   Holism: Working from the Larger Picture

Holistic, by definition, lays weight on the whole and considers the interdependence of its parts. Likewise, rather than separating author and speaker attributes such as the above named states and traits and dealing with them in isolation – as is the largely dominating approach in the literature of word-based computational profiling – one should deal with the whole individual behind her words. In this section, I first want to motivate the need for a holistic approach to author and speaker profiling or characterisation by attributes; then, I show avenues from a computational view in practice, and how to further extend on holism in the future. Obviously, these are merely some inspirations – more powerful methods and models can be thought of and are yet to come.

3.1   Why to go Holistic?

The choice of our words and the deeper linguistic structure are impacted by the entirety of our self-characterising attributes such as all kinds of states and traits. When speaking or writing, we are not only doing this under the influence of, say, a specific emotion. Rather, at the same time, we may be suffering from drowsiness, stress, a cold, or even intoxication. In addition, a longer term mood or depression, and health condition can be influential factors. Obviously, the social role, stance, and the language we choose and our degree of nativeness will significantly impact on the linguistic aspects. Certainly, our age, personality, gender, social class, education, intellect, and many further factors will be further co-influencing this choice of words and structure. Yet, besides few examples as a study investigating the impact of emotion on author profiling [32], the vast majority of computational language processing literature ignores this co-dependence of attributes and focusses on one aspect at a time – such as emotion in this example – risking severe downgrades in accuracy once applied in real world automatic recognitions applications, where data comes from real-life highly blended out-of-the-lab situations.

3.2   How to go Holistic: Methods

A number of approaches exist in the maching learning body of literature to assess multiple attributes such as the ones described above synergistically either iteratively or in full parallel. Typical examples include neural networks with multiple output neurons for accordingly several targets such as classes or regression of several tasks. But a plethora of approaches also exists for other types of learning algorithms such as Support Vector Machines [54].

   A major bottleneck for training of such approaches is the lack of data labelled in a rich variety of attributes of authors and speakers. As re-labelling of data may be tedious and labour intensive, and partially simply not possible, as information on the ground truth states and traits may not be accessible for a broad range of these, semi automatic and fully automatic approaches were introduced to relabel databases one by another. An example is cross-task labelling (CTL) as introduced in [69]. Transfer learning can be a further alternative to learn across tasks and conditions [13] and enrich one's database in largely automated ways.

3.3   How to go Holistic: More Taxonomies

Up to this point, in Sect. 2 we were considering a single taxonomy to group or classify author and speaker attributes of interest, namely states vs traits. However, one can group these in a range of further ways [41], which will introduce also new viewing angles in terms of holism. Strictly speaking, using additional taxonomies will not increase the number of attributes one may target, but can help make their organisation easier, if we consider, e. g., acted vs spontaneous as an additional second taxonomy. As an example, we could add 'acted pain' and 'spontaneous pain' as further states of interest, but if 'pain' is handled as

Table 1. A dozen of taxonomies: frequent and important options for the grouping of author and speaker attributes with a short comment. See [41] for a detailed explanation.

Taxonomy	Comment
Trait vs state	Relates to time/permanence
Acted vs spontaneous	One may further consider masking or regulation
Complex vs simple	Blended or 'pure'
Measured vs assessed	Relates to objectivity of the 'ground truth'
Categorical vs continuous	e. g., a set of emotions vs continuous emotions
Felt vs perceived	Perceived by observers, i. e., others
Intentional vs instinctual	e. g., acting is usually intentional
Consistent vs discrepant	e. g., irony being discrepant
Private vs social	Relating to the communication intention
Prototypical vs peripheral	Salient, central example or unusual or atypical
Universal vs culture-specific	e. g., affect or acting may depend on the culture
Uni-modal vs multi-modal	Here: linguistics only or including acoustics

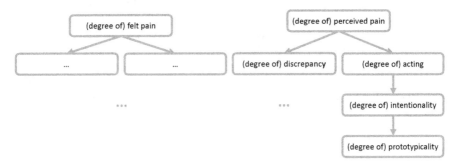

Fig. 1. An example of a semantic tree for '(degree of) pain' inspired by the taxonomies in Table 1. A higher order single root node representing the author or speaker was left out for better visibility. Explanations are given in the text.

a regression problem in the sense of 'degree of pain', it may be preferable to have a pain attribute and a related degree of acting attribute. One may now attach further related attributes and attach a property describing the cultural connection, describing to which culture the signals of pain or its acting relate. Likewise, one can think of a semantic tree structure with a root attribute – in the example '(degree of) pain' – and subsequent related attributes that describe the type of the root aspect (such as '(degree of) pain'). In Table 1, the taxonomies as were introduced in [41] – where one can find very detailed explanations of these – are given alongside a short comment for their short 'in a nutshell'-type familiarisation.

In Fig. 1, an example is given on how one may translate these taxonomies to possible ways of forming semantic trees as indicated to reach much richer descriptions of speaker attributes than current approaches in the literature target. The root nodes '(degree of) felt pain' and '(degree of) perceived pain' show, how a taxonomy was used to build two different attributes which, in a machine learning approach, could be learnt as multiple targets given their likely high correlation. One would reflect the actual pain the individual experiences (left tree), the other the pain others would perceive within the individual (right tree). Note that, one could add a new root 'author' or 'speaker' above to form a single tree, which was left out here for better visibility. On the first layer as shown in the figure, in a truly holisic approach, one would find all sorts of further attributes, e. g., grouped by states and traits, characterising the subject. Then, on the second layer in the figure, only the right hand tree is filled with exemplary further taxonomies for better visibility. Here, we find exemplary attributes that describe this perception of the author's or speaker's '(degree of) pain' by others: On the left-hand side, the '(degree of) discrepancy', on the right hand side, the '(degree of) intentionality'; up to this point, following the right hand side, we would have a description on how the portrayal of pain by a subject is perceived by others – as acted or real, and if they perceive this degree of acting as instinctual or intentional. To go one layer deeper, we could next add how prototypical this example would be by the '(degree of) prototypicality. This would give us, for example, a sample of text or speech that could be labelled as a 'typical example of intentionally acted pain in observers' eyes. Obviously, despite the figure showing dots at several places, it seems hardly reasonable to either go for full depth nor for full width, as one would need labelling for any of these aspects. However, it is interesting to see how one can likewise find a new way of describing author and speaker attributes as compared to today's dominating approach of targeting single aspects in isolation.

3.4   How to go Holistic: Adding Acoustics

Speaking of spoken language analysis, it seems wise to also consider analysis of the acoustic properties. This may also comprise adding acoustic confidences in the linguistic analysis coming from an automatic speech recognition engine. In fact, many automatic speaker attribute assessment tasks in spoken language analysis show good synergy when exploiting both acoustic and linguistic information. Examples include most notably emotion [42]. Major challenges for the exploitation of both information types thereby include the often smaller size of corpora for acoustic model training as compared to such desired and typically used in linguistic model training; also, not all corpora as used for acoustic model training can be used for linguistic analysis, as these are partially based on prompted text or even just vowel-consonant combinations. Further, fusion of the information streams can be less straight forward given the different time levels these operate on: for linguistic analyses, one mostly considers larger amounts of text than usual chunk sizes of one or a few words as are used for acoustic analysis would be – however, early fusion on the feature level has repeatedly been shown to be feasible [41]. The optimal type of fusion itself is not at all decided upon, either [41].

Table 2. Overview on the tasks of the Interspeech Computational Paralinguistics and according pre-decessor challenges centred on acoustic analysis of paralinguistic effects, while often allowing also for linguistic assessment. For simplification, a binary classification is made per taxonomy as commented upon in Table 1 and the data as was used in the according (sub-)challenge; note, however, that instead, one could also introduce a continuous dimension per taxonomy as shown in Fig. 1. In fact, this coarse discretisation of two classes renders some decisions rather ambiguous, and other classifications could partially be assigned. Also note that the assignment here does not hold for the author or speaker attribute in general (i. e., the 'phenomenon'), but is focussed on the specific task/data of the according (sub-)challenge as was held in the according year (cf. last column). '+' denotes cases where both options exist in the data. In the case of c/n (categorical or continuous), a '+' is given if the data includes continuous annotation, despite a classification task by discretisation was used in the challenge. In the case of u/m (unimodal vs multimodal), here, 'u' is given if the attributes can be inferred only by acoustics, 'm' is given, if use of linguistic analysis is reasonable, and '+' is given if u/m hold for different parts of the database. The language(s) of the content are additionally given by country code (ISO 3166 ALPHA 2). En: English (w/o specific region). Ps: Pseudo-language.

Taxonomy	trait/state	acted/spontaneous	complex/simple	measured/assessed	categorical/continuous	felt/perceived	intentional/instinctual	consistent/discrepant	private/social	prototypical/peripheral	universal/culture-specific	uni-modal/multi-modal	Lang. Year
Abbreviation	t/s	a/s	c/s	m/a	c/n	f/p	i/n	c/d	p/s	p/n	u/c	u/m	Lang. Year
Addressee	s	s	s	a	c	p	+	+	s	+	c	m	US 2017
Age	t	s	s	m	+	f	n	c	+	+	u	m	DE 2010
Autism	t	s	s	m	c	f	n	c	p	+	u	u	FR 2013
Cognitive Load	s	s	s	m	+	f	n	c	p	+	c	u	AU 2014
Cold	s	s	s	a	+	f	n	c	p	+	u	+	DE 2017
Conflict	s	s	s	a	+	p	+	+	+	+	c	m	FR 2013
Deception	s	s	s	m	c	f	+	+	p	+	c	m	US 2016
Eating	s	s	s	m	+	f	+	c	p	+	u	+	DE 2015
Emotion (acted)	s	a	s	m	c	p	i	+	s	p	c	u	Ps 2013
Emotion (spontaneous)	s	s	c	a	c	p	+	+	+	+	c	m	DE 2009
Gender	t	s	s	m	c	f	n	c	+	+	c	+	DE 2010
Pathology	t	s	s	a	+	p	n	c	p	+	u	u	NL 2012
Interest	s	s	s	a	n	p	+	+	+	+	c	m	En 2010
Intoxication	s	s	s	m	+	f	n	c	p	+	c	+	DE 2011
Likability	t	s	s	a	+	p	+	+	+	+	c	+	DE 2012
Native Language	t	s	s	m	c	f	n	c	+	+	u	+	En 2016
Nativeness	s	s	s	a	n	p	+	+	+	+	c	+	En 2015
Parkinson's	s	s	s	a	n	p	n	c	p	+	u	+	ES 2015
Personality	t	s	s	a	+	p	+	+	+	+	c	m	FR 2012
Physical Load	s	s	s	m	+	f	n	c	p	+	u	u	DE/En 2014
Sincerity	s	a	s	a	n	p	+	+	s	p	c	u	En 2016
Sleepiness	s	s	s	a	+	+	n	c	p	+	c	+	DE 2011
Snoring	s	s	s	a	c	p	n	c	p	+	u	u	– 2017
Social Signals	s	s	s	m	c	+	+	+	+	+	c	m	UK 2013

However, not all tasks have been attempted exploiting each of these two information types (acoustics and linguistics) in isolation, i.e., exclusively by acoustic, *or* linguistic information. This comes, as some speaker attributes are rather unsuited to be assessed by each of the two types of information at a similarly high precision. As an example, consider the case of speaker height recognition [66]: while height correlates in a certain age range with age, which clearly has an influence on linguistics, it will mostly manifest in acoustic features for adult speakers.

Other tasks have been attempted by both of the two types – acoustics and linguistics – also mutually exclusively such as race [58]. However, different potential can partially be observed, such as in the case of emotional speech modelling, arousal being better assessed by acoustic cues, and valence better be assessed by linguistic cues [41]. This leads to the question, when to preferably use which or both of these two types of information. As a basic assumption, one may assume that aspects which reflect physical differences reflect more or only in acoustics, whereas cognitive aspects reflect more in linguistic cues. However, in a real-world, correlation and co-effects may benefit each of the two, such as when trying to recognise if a speaker is a smoker or not. While a smoker's voice clearly differs acoustically from the one of a non-smoker, one would not assume that a smoker should linguistically differ from a non-smoker. However, depending on the test sample, or even distribution in the broader population, other effects may come into play that *correlate* with being a smoker or not, *and* impact on linguistics. For example, smokers in a test sample of a database might represent different age classes depending on when smoking was more or less popular, or represent different social classes, gender, or alike, which as a phenomenon all do impact on linguistics. Note that in the ongoing, the above introduced taxonomy 'unimodal vs multimodal' (cf. Table 1) which alludes to whether an attribute relates to one or multiple modalities will, looking at spoken language, reflect whether acoustic and linguistic phenomena can be handled only by one or by both of these types of information.

Let us now look at a concrete example of speaker attributes for further exemplification of the above principles: Based on the Interspeech Computational Paralinguistics Challenge (ComParE) series [46, 48–50, 53] (including its predecessors on emotion [42], paralinguistics [47], speaker state [52], and speaker trait [51][1], the tasks of speaker attribution as given in Table 2 were assessed by both cues (acoustics and linguistics) during the according challenge or seem to be suited to be attempted by linguistics in general.

The table provides a coarse breakdown of the specific tasks and data as were held in the challenge's sub-challenges. Likewise, note that it does not categorise the author or speaker attributes per se, but indeed the specific conditions of the data as were held. For example, in the case of acted or spontaneous, eating is marked as spontaneous. One could also act speech under eating, but during the recording of this data, actual eating was given – thus, it is marked as spontaneous in the table. A certain ambiguity here lies in the fact that, the eating

---

[1] More information can be accessed from http://www.compare.openaudio.eu.

took place in a lab with normalised intake conditions, thus, not feeling 'spontaneous', yet, clearly not being acted. Similarly, one could act a native language or sleeping, to name but two – however, in the according sub-challenge tasks, this was not the case, thus leading to the spontaneous label in the table. This is just an exemplification of the ambiguity coming with making only binary decisions. However, the main point of the table is to demonstrate the huge richness one can find in applying the dozen taxonomies shown in both tables: While the challenge dealt for example with speech under eating in 2015 as is shown in the Table 2, it specifically only dealt with non-acted eating. Accordingly, one could now consider 'speech under acted eating' as a novel attribution. Following this principle, one could now massively extend the holistic space of possible author and speaker attributes by looking at all missing combinations in the table, and – of course – by adding many more attributes in the first place.

4   Summary and Conclusion

Novel concepts of rich and structured author or speaker attributes were introduced. Repeatedly targeted dominating examples of (flat) attributes were further given that show the current state of automatic assessment from written or spoken language which is dominated by isolated handling of these. Based on these, first, a three-fold division was made into short-term and medium-term states and long-term traits to group these. Then, further taxonomies for grouping and conceptualisation were shown. Two main points were then raised:

First, author and speaker attributes should be assessed holistically in their commonality, as they largely and often co-influence each other. This is also important, as otherwise, it remains unclear what is actually recognised once going for automatic processing 'in the wild' in real-world applications. As an example, both speaking under a cold or while eating may have an impact on linguistics, as with a soar throat or while chewing one may speak in short phrases and an emphasis on specific sounds to avoid speaking too much and too strenuously. Likely, this will also influence the composition of adjectives, nouns, verbs, and other part-of-speech classes one chooses. However, a study looking only at one of these two phenomena versus 'normal speech' may be overly optimistic in its results and performance assumptions as in a real world application, both cases will happen over time, making confusions or false positives very likely. To tackle this problem, examples of ways of co-learning and multi-target classification and regression were shortly given. These included methods based on semi-supervised and transfer learning to re-label data in other databases' labels, as richly annotated let alone 'holistically' annotated data is scarce.

Second, fusion of acoustic and linguistic information in the sense of another 'holistic view' on spoken language analysis is broadly considered as synergistic, yet, for many states and traits it has hardly been attempted. A number of obstacles were named including differences in corpora usually existent in these sub-disciplines, different timing levels of analysis, and the ever-ongoing discussion of the optimal way to fuse these information sources.

To exemplify the current situation and indicate future avenues of broadening on holism, a snapshot image was given by the Interspeech Computational Paralinguistics Challenge series which is focussed on acoustic analysis, but also allows for linguistic analysis. There, one could first see the potential of not yet attempted spoken language analysis tasks by either only linguistic approaches or a combination of these with acoustic processing. Second, as these tasks were classified by a dozen of taxonomies, one could see how the kinds of attributes could be massively en-richened such as in an also introduced tree structure.

Overall, this leaves an interesting field of research efforts to mine for the future, in which intelligent systems will be able to profile authors and speakers in a broad range of attributes grouped in many ways such as by states and traits in full parallel seeing and hearing the 'larger picture' of the person behind the written and spoken words exploiting also the acoustic channel if available. One may wonder how oncoming machines equipped with such rich emotional and social intelligence may interact with us, retrieve information from spoken and written language in so far unseen richness and accuracy, but also monitor us for wellbeing, and – ultimately – change future society technology. May it be used only for society's best.

 **Acknowledgement.** The author would like to thank his colleague Anton Batliner for discussion regarding the taxonomies as discussed herein. He further acknowledges funding from the European Research Council within the European Union's 7th Framework Programme under grant agreement no. 338164 (Starting Grant Intelligent systems' Holistic Evolving Analysis of Real-life Universal speaker characteristics (iHEARu)), and the European Union's Horizon 2020 Framework Programme under grant agreement no. 645378 (Research Innovation Action Artificial Retrieval of Information Assistants – Virtual Agents with Linguistic Understanding, Social skills, and Personalised Aspects (ARIA-VALUSPA)). The responsobility lies with the author.

References

1. Abbott, R., Walker, M., Anand, P., Fox Tree, J.E., Bowmani, R., King, J.: How can you say such things? recognizing disagreement in informal political argument. In: Proceedings Workshop on Languages in Social Media, pp. 2–11. ACL, Portland, Oregon (2011)
2. Balota, D.A., Burgess, G.C., Cortese, M.J., Adams, D.R.: The word-frequency mirror effect in young, old, and early-stage alzheimer's disease: evidence for two processes in episodic recognition performance. J. Mem. Lang. **46**(1), 199–226 (2002)
3. Balota, D.A., Duchek, J.M.: Semantic priming effects, lexical repetition effects, and contextual disambiguation effects in healthy aged individuals and individuals with senile dementia of the alzheimer type. Brain Lang. **40**(2), 181–201 (1991)
4. Beňuš, Š.: Social aspects of entrainment in spoken interaction. Cogn. Comput. **6**(4), 802–813 (2014)
5. Biel, J.I., Tsiminaki, V., Dines, J., Gatica-Perez, D.: Hi youtube!: personality impressions and verbal content in social video. In: Proceedings ICMI, pp. 119–126. ACM, Sydney, Australia (2013)

6. Cambria, E., Schuller, B., Xia, Y., Havasi, C.: New avenues in opinion mining and sentiment analysis. IEEE Intell. Syst. Mag. **28**(2), 15–21 (2013)
7. Celiktutan, O., Eyben, F., Sariyanidi, E., Gunes, H., Schuller, B.: MAPTRAITS 2014: the first audio/visual mapping personality traits challenge. In: Proceedings Personality Mapping Challenge & Workshop (MAPTRAITS), Satellite of ICMI, pp. 3–9. ACM, Istanbul, Turkey (2014)
8. Chakravarthula, S.N., Xiao, B., Imel, Z.E., Atkins, D.C., Georgiou, P.G.: Assessing empathy using static and dynamic behavior models based on therapist's language in addiction counseling. In: Proceedings Interspeech, pp. 668–672. ISCA, Dresden, Germany (2015)
9. Chaski, C.E.: Empirical evaluations of language-based author identification techniques. Forensic Linguist. **8**, 1–65 (2001)
10. Coppersmith, G., Dredze, M., Harman, C., Hollingshead, K.: From adhd to sad: analyzing the language of mental health on twitter through self-reported diagnoses. In: Proceedings CLPsych at NAACL HLT, p. 1. ACL, Denver, Colorado (2015)
11. Davidov, D., Tsur, O., Rappoport, A.: Semi-supervised recognition of sarcastic sentences in twitter and amazon. In: Proceedings 14th Conference on Computational Natural Language Learning, pp. 107–116. ACL, Uppsala, Sweden (2010)
12. De Vel, O., Anderson, A., Corney, M., Mohay, G.: Mining e-mail content for author identification forensics. ACM SIGMOD Rec. **30**(4), 55–64 (2001)
13. Deng, J., Xu, X., Zhang, Z., Frühholz, S., Schuller, B.: Universum autoencoder-based domain adaptation for speech emotion recognition. IEEE Signal Process. Lett. **24**, 5 (2017)
14. Desmet, B., Pauwels, K., Hoste, V.: Online suicide risk detection using automatic text classification. In: Proceedings International Summit on Suicide Research. IASR/AFSP, New York City, 1 p. (2015)
15. Dhall, A., Goecke, R., Joshi, J., Wagner, M., Gedeon, T.: Emotion recognition in the wild challenge 2013. In: Proceedings ICMI, pp. 509–516. ACM, Sydney, Australia (2013)
16. González-Ibánez, R., Muresan, S., Wacholder, N.: Identifying sarcasm in twitter: a closer look. In: Proceedings ACL, vol. 2, pp. 581–586. ACL, Portland, Oregon (2011)
17. Holtgraves, T., Fogle, K., Marsh, L.: Pragmatic language production deficits in Parkinson's disease. In: Advances in Parkinson's Disease, vol. 2, pp. 31–36 (2013)
18. Jarrold, W.L., Peintner, B., Yeh, E., Krasnow, R., Javitz, H.S., Swan, G.E.: Language analytics for assessing brain health: cognitive impairment, depression and pre-symptomatic alzheimer's disease. In: Yao, Y., Sun, R., Poggio, T., Liu, J., Zhong, N., Huang, J. (eds.) BI 2010. LNCS (LNAI), vol. 6334, pp. 299–307. Springer, Heidelberg (2010). https://doi.org/10.1007/978-3-642-15314-3_28
19. Justo, R., Corcoran, T., Lukin, S.M., Walker, M., Torres, M.I.: Extracting relevant knowledge for the detection of sarcasm and nastiness in the social web. Knowl. Based Syst. **69**, 124–133 (2014)
20. Li, Y., Tao, J., Schuller, B., Shan, S., Jiang, D., Jia, J.: MEC 2016: the multimodal emotion recognition challenge of CCPR 2016. In: Tan, T., Li, X., Chen, X., Zhou, J., Yang, J., Cheng, H. (eds.) CCPR 2016. CCIS, vol. 663, pp. 667–678. Springer, Singapore (2016). https://doi.org/10.1007/978-981-10-3005-5_55
21. Lin, J.: Automatic author profiling of online chat logs. Ph.D. thesis, Naval Postgraduate School, Monterey, California (2007)
22. Mairesse, F., Walker, M.A., Mehl, M.R., Moore, R.K.: Using linguistic cues for the automatic recognition of personality in conversation and text. J. Artif. Intell. Res. **30**, 457–500 (2007)

23. Mihalcea, R., Strapparava, C.: Making computers laugh: investigations in automatic humor recognition. In: Proceedings Conference on Human Language Technology and Empirical Methods in Natural Language Processing, pp. 531–538. ACL, Vancouver, Canada (2005)

24. Mihalcea, R., Strapparava, C.: The lie detector: explorations in the automatic recognition of deceptive language. In: Proceedings ACL-IJCNLP Conference Short Papers, pp. 309–312. ACL, Singapore (2009)

25. Neuman, Y., Cohen, Y., Assaf, D., Kedma, G.: Proactive screening for depression through metaphorical and automatic text analysis. Artif. Intell. Med. **56**(1), 19–25 (2012)

26. Patra, B.G., Banerjee, S., Das, D., Saikh, T., Bandyopadhyay, S.: Automatic author profiling based on linguistic and stylistic features. Notebook for PAN at CLEF 1179 (2013)

27. Pokorny, F.B., Marschik, P.B., Einspieler, C., Schuller, B.W.: Does she speak RTT? Towards an earlier identification of RETT syndrome through intelligent prelinguistic vocalisation analysis. In: Proceedings Interspeech, pp. 1953–1957 (2016)

28. Poria, S., Gelbukh, A., Agarwal, B., Cambria, E., Howard, N.: Common sense knowledge based personality recognition from text. In: Castro, F., Gelbukh, A., González, M. (eds.) MICAI 2013. LNCS (LNAI), vol. 8266, pp. 484–496. Springer, Heidelberg (2013). https://doi.org/10.1007/978-3-642-45111-9_42

29. Ranganath, R., Jurafsky, D., McFarland, D.: It's not you, it's me: detecting flirting and its misperception in speed-dates. In: Proceedings Conference on Empirical Methods in Natural Language Processing, vol. 1, pp. 334–342. ACL, Singapore (2009)

30. Ranganath, R., Jurafsky, D., McFarland, D.A.: Detecting friendly, flirtatious, awkward, and assertive speech in speed-dates. Comput. Speech Lang. **27**(1), 89–115 (2013)

31. Rangel, F., Rosso, P.: Use of language and author profiling: identification of gender and age. In: Natural Language Processing and Cognitive Science, vol. 177 (2013)

32. Rangel, F., Rosso, P.: On the impact of emotions on author profiling. Inf. Process. Manag. **52**(1), 73–92 (2016)

33. Rangel, F., Rosso, P., Koppel, M.M., Stamatatos, E., Inches, G.: Overview of the author profiling task at pan 2013. In: Proceedings CLEF Conference on Multilingual and Multimodal Information Access Evaluation, pp. 352–365. CELCT, Amsterdam, The Netherlands (2013)

34. Rangel, F., Rosso, P., Potthast, M., Stein, B., Daelemans, W.: Overview of the 3rd author profiling task at pan 2015. In: Proceedings CLEF, 18 p. CLEF Association, Toulouse, France (2015)

35. Razavi, A.H., Inkpen, D., Uritsky, S., Matwin, S.: Offensive language detection using multi-level classification. In: Farzindar, A., Kešelj, V. (eds.) AI 2010. LNCS (LNAI), vol. 6085, pp. 16–27. Springer, Heidelberg (2010). https://doi.org/10.1007/978-3-642-13059-5_5

36. Regneri, M., King, D.: Automatically evaluating atypical language in narratives by children with autistic spectrum disorder. In: Sharp, B., Delmonte, R. (eds.) Natural Language Processing and Cognitive Science: Proceedings 2014, pp. 173–186. Walter de Gruyter GmbH & Co KG, Berlin (2015)

37. Reyes, A., Rosso, P., Buscaldi, D.: From humor recognition to irony detection: the figurative language of social media. Data Knowl. Eng. **74**, 1–12 (2012)

38. Rude, S., Gortner, E.M., Pennebaker, J.: Language use of depressed and depression-vulnerable college students. Cogn. Emot. **18**(8), 1121–1133 (2004)

39. Schuller, B.: Recognizing affect from linguistic information in 3D continuous space. IEEE Trans. Affect. Comput. **2**(4), 192–205 (2012)
40. Schuller, B.: The computational paralinguistics challenge. IEEE Signal Process. Mag. **29**(4), 97–101 (2012)
41. Schuller, B., Batliner, A.: Computational Paralinguistics: Emotion, Affect and Personality in Speech and Language Processing. Wiley, U.S. (2013)
42. Schuller, B., Batliner, A., Steidl, S., Seppi, D.: Recognising realistic emotions and affect in speech: state of the art and lessons learnt from the first challenge. Speech Commun. **53**(9/10), 1062–1087 (2011)
43. Schuller, B., Mousa, A.E.D., Vasileios, V.: Sentiment analysis and opinion mining: on optimal parameters and performances. WIREs Data Min. Knowl. Discov. **5**, 255–263 (2015)
44. Schuller, B., Müller, R., Hörnler, B., Höthker, A., Konosu, H., Rigoll, G.: Audiovisual recognition of spontaneous interest within conversations. In: Proceedings ICMI, pp. 30–37. ACM, Nagoya, Japan (2007)
45. Schuller, B., Rigoll, G., Lang, M.: Speech emotion recognition combining acoustic features and linguistic information in a hybrid support vector machine-belief network architecture. In: Proceedings ICASSP, vol. 1, pp. 577–580. IEEE, Montreal, Canada (2004)
46. Schuller, B., etal.: The INTERSPEECH 2017 computational paralinguistics challenge: addressee, cold & snoring. In: Proceedings Interspeech, 5 p. ISCA, Stockholm, Sweden (2017)
47. Schuller, B., et al.: Paralinguistics in speech and language—state-of-the-art and the challenge. Comput. Speech Lang. **27**(1), 4–39 (2013)
48. Schuller, B., et al.: The INTERSPEECH 2014 computational paralinguistics challenge: cognitive & physical load. In: Proceedings Interspeech. ISCA, Singapore (2014)
49. Schuller, B., et al.: The INTERSPEECH 2015 computational paralinguistics challenge: degree of nativeness, parkinson's & eating condition. In: Proceedings Interspeech, pp. 478–482. ISCA, Dresden, Germany (2015)
50. Schuller, B., et al.: The INTERSPEECH 2016 computational paralinguistics challenge: deception, sincerity & native language. In: Proceedings Interspeech, pp. 2001–2005. ISCA, San Francisco, California (2016)
51. Schuller, B., et al.: A survey on perceived speaker traits: personality, likability, pathology, and the first challenge. Comput. Speech Lang. **29**(1), 100–131 (2015)
52. Schuller, B., et al.: Medium-term speaker states—a review on intoxication, sleepiness and the first challenge. Comput. Speech Lang. **28**(2), 346–374 (2014)
53. Schuller, B., et al.: The INTERSPEECH 2013 computational paralinguistics challenge: social signals, conflict, emotion, autism. In: Proceedings Interspeech, pp. 148–152. ISCA, Lyon, France (2013)
54. Schuller, B., Zhang, Y., Eyben, F., Weninger, F.: Intelligent systems' holistic evolving analysis of real-life universal speaker characteristics. In: Proceedings 5th International Workshop on Emotion Social Signals. Sentiment & Linked Open Data (ES^{3}LOD 2014), Satellite of LREC, pp. 14–20. ELRA, Reykjavik, Iceland (2014)
55. Schwartz, H.A., et al.: Personality, gender, and age in the language of social media: the open-vocabulary approach. PloS ONE **8**(9), e73791 (2013)
56. Smith, J.M.: Breastfeeding and language outcomes: a review of the literature. J. Commun. Disord. **57**, 29–40 (2015)
57. Spertus, E.: Smokey: automatic recognition of hostile messages. In: Proceedings AAAI/IAAI, pp. 1058–1065. AAAI Press, Providence, Rhode Island (1997)

58. Squires, G.D., Chadwick, J.: Linguistic profiling a continuing tradition of discrimination in the home insurance industry? Urban Aff. Rev. **41**(3), 400–415 (2006)
59. Strapparava, C., Mihalcea, R.: Semeval-2007 task 14: affective text. In: Proceedings 4th International Workshop on Semantic Evaluations, pp. 70–74. ACL, Prague, Czech Republic (2007)
60. Strapparava, C., Mihalcea, R.: Learning to identify emotions in text. In: Proceedings ACM Symposium on Applied Computing, pp. 1556–1560. ACM, Fortaleza, Brazil (2008)
61. Strapparava, C., Mihalcea, R.: A Computational Analysis of the Language of Drug Addiction, vol. 2, pp. 136–142 (2017)
62. Valstar, M., Gratch, J., Schuller, B., Ringeval, F., Cowie, R., Pantic, M.: Summary for AVEC 2016: depression, mood, and emotion recognition workshop and challenge. In: Proceedings ACM International Conference on Multimedia, pp. 1483–1484. ACM, Amsterdam, The Netherlands (2016)
63. Van Halteren, H.: Linguistic profiling for author recognition and verification. In: Proceedings ACL, p. 199. ACL, Barcelona, Spain (2004)
64. Vijayaraghavan, P., Vosoughi, S., Roy, D.: Automatic detection and categorization of election-related tweets. In: Proceedings 10th International AAAI Conference on Web and Social Media. AAAI, Cologne, Germany (2016)
65. Wang, W.Y., Biadsy, F., Rosenberg, A., Hirschberg, J.: Automatic detection of speaker state: lexical, prosodic, and phonetic approaches to level-of-interest and intoxication classification. Comput. Speech Lang. **27**(1), 168–189 (2013)
66. Weninger, F., Wöllmer, M., Schuller, B.: Automatic assessment of singer traits in popular music: gender, age, height and race. In: Proceedings ISMIR, pp. 37–42. ISMIR, Miami, FL, USA (2011)
67. Xu, D., Gilkerson, J., Richards, J., Yapanel, U., Gray, S.: Child vocalization composition as discriminant information for automatic autism detection. In: Proceedings EMBC, pp. 2518–2522. IEEE, Minneapolis (2009)
68. Xu, D., Richards, J.A., Gilkerson, J., Yapanel, U., Gray, S., Hansen, J.: Automatic childhood autism detection by vocalization decomposition with phone-like units. In: Proceedings 2nd Workshop on Child, Computer and Interaction, Satellite of ICMI-MLMI, p. 5. ACM, Cambridge, MA, USA (2009)
69. Zhang, Y., Zhou, Y., Shen, J., Schuller, B.: Semi-autonomous data enrichment based on cross-task labelling of missing targets for holistic speech analysis. In: Proceedings ICASSP, pp. 6090–6094. IEEE, Shanghai, P.R. China (2016)

Improving Cross-Topic Authorship Attribution: The Role of Pre-Processing

Ilia Markov[1(✉)], Efstathios Stamatatos[2], and Grigori Sidorov[1]

[1] Center for Computing Research (CIC), Instituto Politécnico Nacional (IPN), Mexico City, Mexico
markovilya@yahoo.com, sidorov@cic.ipn.mx
[2] Department of Information and Communication Systems Engineering, University of the Aegean, Karlovassi, Samos, Greece
stamatatos@aegean.gr

Abstract. The effectiveness of character n-gram features for representing the stylistic properties of a text has been demonstrated in various independent Authorship Attribution (AA) studies. Moreover, it has been shown that some categories of character n-grams perform better than others both under single and cross-topic AA conditions. In this work, we present an improved algorithm for cross-topic AA. We demonstrate that the effectiveness of character n-grams representation can be significantly enhanced by performing simple pre-processing steps and appropriately tuning the number of features, especially in cross-topic conditions.

Keywords: Pre-processing · Authorship attribution · Cross-topic
Character n-grams · Machine learning

1 Introduction

Authorship Attribution (AA) is the task that aims at identifying the author of a text given a predefined set of candidate authors [1]. Practical applications of AA vary from electronic commerce and forensics, where part of the evidence refers to texts, to humanities research [2–5].

From the machine-learning perspective, AA can be viewed as a multi-class, single-label classification problem. In single-topic AA, there are no major differences in the thematic areas of training and test corpora, whereas in cross-topic AA, the thematic areas of training and test corpora are disjoint [6]. The latter better matches the requirements of a realistic scenario of forensic applications, when the available texts by the candidate authors can belong to totally different thematic areas than the texts under investigation.

Character n-grams have proved to be the best predictive feature type both under single and cross-topic AA conditions [6,7]. A reasonable explanation is that these features capture 'a bit of everything', including lexical and syntactic information, punctuation and capitalization information related with the authors' personal style. They are sensitive to both the content and form of a text [1,8,9]

© Springer Nature Switzerland AG 2018
A. Gelbukh (Ed.): CICLing 2017, LNCS 10762, pp. 289–302, 2018.
https://doi.org/10.1007/978-3-319-77116-8_21

while their higher frequency with respect to other feature types, e.g., words, make their probabilities estimation more accurate [10].

Recently, Sapkota *et al.* [11] showed that some categories of character n-grams perform better than others both for single and cross-topic settings. They claimed that a AA model trained on character n-grams that capture information about affixes and punctuation (morpho-syntactic and stylistic information) performs better than using all possible n-grams. Their results indicate that it is possible to improve basic character n-gram features without the need of extracting more complicated features.

In this paper, we present an approach that applies simple pre-processing steps, such as replacing digits, splitting punctuation marks, and replacing named entities, before extracting character n-gram features. We adopt the character n-gram categories proposed by Sapkota *et al.* [11] and examine how pre-processing steps affect their effectiveness. We evaluate the contribution of each step when applied separately and in combination. We further show that an appropriate tuning of the number of features is crucial and can further enhance AA performance, especially in cross-topic conditions.

The research questions addressed in this work are the following:

1. Can we improve the performance of AA by applying simple pre-processing steps? Which pre-processing steps are appropriate for both single and cross-topic AA settings?
2. Is it possible to enhance AA performance by selecting an appropriate feature set size using only the training corpora?
3. Is the conclusion reported in [11], that the best performing model is based solely on affix and punctuation n-grams, valid even after applying pre-processing steps? Is this conclusion valid when using different classification algorithms?

2   Related Work

Previous work in AA focuses mainly on the extraction of stylometric features that represent the personal style of authors [7,12–15]. Several studies demonstrate the effectiveness of character n-grams in AA tasks [7,16–18]. These features were also found robust in AA experiments under cross-topic conditions [6,19] despite the fact that they also capture thematic information. They are also strongly associated with compression-based models that essentially exploit common character sequences [20,21]. Character n-grams can be used either alone [18,22] or combined with other stylometric features [23].

In most previous AA studies, training and test corpora share similar thematic properties [18,20,22,24]. An early cross-topic study is described in [25] where email messages in different topic categories were used in training and test corpora. The *unmasking* method for author verification was successfully tested in cross-topic conditions [26]. A comparison of character n-grams and lexical features in cross-topic conditions is provided in [6].

Sapkota *et al.* [19] proposed to enrich the training corpus with multiple topics to enhance the performance of AA on another topic. The recent PAN evaluation campaign on author identification focused on cross-topic and cross-genre author verification [27]. As expected, the performance of AA models in cross-topic conditions is lower in comparison to single-topic conditions [6].

3 Stylometric Features

3.1 Types of n-grams

In this paper, we adopt the character n-gram types introduced by Sapkota *et al.* [11]. However, we refine the original definitions for some of the categories of character n-grams in order to make them more accurate and complete. We also follow Sapkota *et al.* [11] and focus on character 3-grams. In more detail, there are 3 main types, and each one has sub-categories as explained below:

- **Affix character 3-grams**
  prefix A 3-gram that covers the first 3 characters of a word that is at least 4 characters long.
  suffix A 3-gram that covers the last 3 characters of a word that is at least 4 characters long.
  space-prefix A 3-gram that begins with a space and does not contain punctuation.
  space-suffix A 3-gram that ends with a space, does not contain punctuation, and whose first character is not a space.
- **Word character 3-grams**
  whole-word A 3-gram that covers all characters of a word that is exactly 3 characters long.
  mid-word A 3-gram that covers 3 characters of a word that is at least 5 characters long, and that covers neither the first nor the last character of the word.
  multi-word A 3-gram that spans multiple words, identified by the presence of a space in the middle of the 3-gram.
- **Punctuation character 3-grams**
  beg-punct A 3-gram whose first character is punctuation, but the middle character is not.
  mid-punct A 3-gram whose middle character is punctuation.
  end-punct A 3-gram whose last character is punctuation, but the first and the middle characters are not.

The advantage of our modified definitions is that each occurrence of a character 3-gram is unambiguously assigned to exactly one category. For example, we directly assign the 3-gram instance '_a_' to the *space-prefix* category, excluding it from the *space-suffix* category. Note that two instances of the same 3-gram can be assigned to different categories (e.g., in phrase *the mother*, the first instance of 3-gram *the* is assigned to *whole-word* and the second instance to *mid-word*).

Moreover, when using the original definitions by Sapkota *et al.* [11], we noticed that some *n*-grams do not fall into any of the categories (e.g., when two consecutive punctuation marks are in the beginning/end of a sentence). Our refined definitions do not exclude any *n*-gram.

As an example, let us consider the following sample sentence:

(1) *John said, "Tom can repair it for 12 euros."*

The character 3-grams for the sample sentence (1) for each of the categories are following:

Table 1. Character 3-grams per category for the sample sentence (1) after applying the algorithm by Sapkota *et al.* [11].

SC	Category	*N*-grams							
affix	*prefix*	Joh	sai	rep	eur				
	suffix	ohn	aid	air	ros				
	space-prefix	_sa	_ca	_re	_it	_fo	_12	_eu	
	space-suffix	hn_	om_	an_	ir_	it_	or_	12_	
word	*whole-word*	Tom	can	for					
	mid-word	epa	pai	uro					
	multi-word	n_s	m_c	n_r	r_i	t_f	r_1	2_e	
punct	*beg-punct*	,_"	"To						
	mid-punct	d,_	_"T	s."					
	end-punct	id,	os.						

3.2   Pre-processing Steps

In this paper, we introduce simple pre-processing steps attempting to assist character *n*-gram features to capture more information related to personal style of the author and less information related to the theme of text. The pre-processing steps are applied before the extraction of *n*-grams and concern the following textual contents:

Digits (Ds) We replace each digit by 0 (e.g., 12,345 → 00,000) since the actual numbers do not carry stylistic information. However, their format (e.g., 1,000 vs. 10000 vs. 1k) reflects a stylistic choice of the author.

Punctuation marks (PMs) We split PMs in order to be able to capture their frequency separately and not just in combination with the adjacent words. For example, the character 3-grams ,_ ", "To, and _ "T in Table 1 refer to the use of a quotation mark. The use of the same PM in a different context (suffix of previous word and prefix of next word) would produce completely different 3-grams. By splitting PMs from adjacent words we allow capturing the frequency of each PM as a separate 3-gram (e.g., _ "_). We also add a space in the beginning and in the end of each line, as well as remove multiple whitespaces for the *mid-punct* category in order to be able to capture the frequency of all PMs.

Named entities (NEs) The use of NEs is strongly associated with the thematic area of texts. However, the patterns of their usage provide useful stylistic information. We replace all NE instances by the same symbol in order to keep information about their occurrence and remove information about the exact NEs.

Highly frequent words (HFWs) Usually highly frequent words are function words, e.g., prepositions, pronouns, etc. They are one of the most important stylometric features [9]. However, when a character n-gram representation is used, especially when n is low, it is not easy to capture patterns of their usage (combinations of certain HFWs with morphemes of previous or next words). To increase the ability of character 3-grams to capture such information, we replace each HFW by a distinct symbol.

As an illustrating example, the above pre-processing steps are applied to the sample sentence (1). NEs are replaced by symbol '#', HFWs *can* and *it* are replaced by symbols '%' and '$', respectively. The resulting sentence would be:

(2) *# said , " # % repair $ for 00 euros . "*

The character n-grams extracted from sample sentence (2) are shown in Table 2.

Table 2. Character 3-grams per category for the sample sentence (2) after applying our algorithm.

SC	Category	N-grams					
affix	*prefix*	sai	rep	eur			
	suffix	aid	air	ros			
	space-prefix	_sa	_#_	_%_	_re	_$_	_fo
		_00	_eu				
	space-suffix	id_	ir_	or_	00_	os_	
word	*whole-word*	for					
	mid-word	epa	pai	uro			
	multi-word	#_s	#_%	%_r	r_$	$_f	r_0
		0_e					
punct	*beg-punct*	,_"	"_#	.."			
	mid-punct	_,_	_"_	_._	" _		
	end-punct	d_,	s_.				

The proposed approach is more topic-neutral, since it does not depend on specific details that are not related to the personal style of authors. It is able to capture format of different numbers, dates, usage of NEs, the frequency of PMs and patterns of their usage, and patterns of HFWs usage. Finally, the number of features significantly decreases when applying our approach, as can be seen by comparing Tables 1 and 2 and as we show further in Sect. 5.[1]

4   Corpora and Experimental Settings

For the evaluation of our algorithm, we conducted experiments on both single-topic and cross-topic corpora. In more detail, we used CCAT_10, a subset of the Reuters Corpus Volume 1 [28], that includes 10 authors and 100 newswire stories per author on the same thematic area (corporate news). As in previous studies, we used the balanced training and test parts of this corpus [11,18].

The cross-topic corpus used in this study is composed of texts published in *The Guardian* daily newspaper. It comprises opinion articles in four thematic areas (Politics, Society, World, U.K.) written by 13 authors [6]. The distribution of texts over the authors is not balanced and, following the practice of previous studies, at most ten documents per author were considered for each of the four topic categories [6,11].

In order to be able to examine the contribution of each pre-processing step, we conducted our experiments using the same experimental settings as described in [11]. Thus, we used character 3-gram features and considered only the 3-grams that occur at least 5 times in the training corpus. We evaluate each model by measuring classification accuracy on the test corpus. For the cross-topic experiments, the results for each model correspond to the average accuracy over the 12 possible pairings of the 4 topics (training on one topic and testing on another). When the Society texts are used as training corpus, there are no training texts for one author. In that case, we removed all texts by that author from the test corpus.

To perform the pre-processing steps as described in the previous section, we used an improved version of Natural Language Toolkit[2] tokenizer, making sure that each PM is a separate token, and Stanford Named Entity Recognizer (NER) [29] in order to extract NEs, filtering out some erroneous detections. Different sets of highly frequent words were tested: 0, 50, 100, 150, and 200.

In order to examine whether different classifiers agree on the effectiveness of the proposed pre-processing steps, we compare the performance of two classifiers using their WEKA's [30] implementation: Support Vector Machines (SVM) and multinomial naive Bayes (MNB). These classification algorithms with default parameters are considered among the best for text categorization tasks [31,32].[3]

---

[1] When large sets of HFWs are replaced by distinct symbols, the size of feature set increases.

[2] http://www.nltk.org [last access: 12.01.2017].

[3] We also examined naive Bayes classifier, which produced worse results but similar behaviour (not shown).

5   Experimental Results

5.1   Contribution of Pre-processing Steps

First, we re-implemented the method of Sapkota *et al.* [11] as described in their paper and applied it to the CCAT_10 and the Guardian corpora. Although the obtained results are very similar with the ones reported in [11], we were not able to reproduce the exact results. Correspondingly, we use the results of our own implementation of the algorithm by Sapkota *et al.* [11] as baseline for the proposed method.

Moreover, following the practice of Sapkota *et al.* [11] we examine three cases according to what kind of n-gram categories are used:

(1)   **all-untyped** – where the categories of n-grams are ignored. Any distinct n-gram is a different feature.
(2)   **all-typed** – where n-grams of all available categories (**affix+punct+word**) are considered. Instances of the same n-gram may refer to different features.
(3)   **affix+punct** – where the n-grams of the **word** category are excluded.

Table 3 shows the performance of the baseline method and the contribution of each proposed pre-processing step separately, as well as their combinations on the CCAT_10 corpus. For the sake of brevity, we do not present all possible combinations, but only the most representative ones. In most cases, the pre-processing steps reduce the effectiveness of the AA models. In more detail, replacing NEs seems to be the least effective step. This can be explained by the thematic-specificity of this corpus. Each author tends to write news stories about specific topics, and this is consistent in both training and test corpora. NEs are strongly associated with thematic choices. The most useful combination of pre-processing steps for this corpus is the replacement of digits that manages to slightly improve the accuracy in most cases using either of the classification algorithms. SVM classifier seems better able to cope with this corpus.

The corresponding evaluation results on the Guardian corpus can be seen at Table 4. Here, most pre-processing steps significantly enhance the performance of AA models. In most cases, the best combination of steps is to replace digits and NEs, split PMs and not to replace HFWs. Note also, that the feature set size for this combination is significantly lower with respect to the baseline. In cross-topic conditions, the proposed approach provides a more robust reduced set of features that are not affected that much by topic shifts. Moreover, the MNB classifier provides much better results for this corpus, and it better handles the **all-untyped** features.

The main conclusion of Sapkota *et al.* [11] that models using **affix+punct** features are better than models trained on all the features is also valid in most cases of our experiments even when applying the proposed pre-processing steps. In addition, in the case of the cross-topic corpus, the highest improvement in accuracy is achieved for the **affix+punct** features when using both SVM and MNB classifiers (4.7% and 5.9% respectively).

Table 3. Accuracy results on the CCAT_10 corpus after applying the proposed pre-processing steps. Accuracy (Acc, %) and the number of features (N) are reported for each step. The "+" columns show the difference of each step and each combination with the baseline. The best accuracy and improvement for each model are in bold; in the case when the accuracies are equal, we chose the one obtained with a smaller set of features.

Approach				all-typed			affix+punct			all-untyped		
D	PM	NE	HFW	Acc	+	N	Acc	+	N	Acc	+	N
	Baseline			78.0		10,859	78.8		6,296	78.2		9,258
✓			0	77.8	−0.2	9,761	**79.6**	**0.8**	**5,503**	78.2	0.0	8,143
✓	✓		0	77.4	−0.6	8,430	77.4	−1.4	4,171	**78.2**	**0.0**	**6,648**
✓		✓	0	76.0	−2.0	7,606	75.4	−3.4	4,187	76.2	−2.0	6,364
✓	✓	✓	0	76.4	−1.6	6,651	75.8	−3.0	3,087	77.2	−1.0	5,239
✓			50	77.4	−0.6	12,457	78.4	−0.4	5,860	76.8	−1.4	10,902
✓	✓		50	76.6	−1.4	11,005	75.4	−3.4	4,416	76.4	−1.8	9,250
✓	✓	✓	50	75.2	−2.8	8,890	74.0	−4.8	3,296	75.8	−2.4	7,510
✓			100	78.0	0.0	13,687	78.6	−0.2	6,041	77.4	−0.8	12,360
✓	✓		100	77.2	−0.8	12,433	75.0	−3.8	4,570	77.4	−0.8	10,702
✓	✓	✓	100	74.6	−3.4	10,088	73.4	−5.4	3,405	74.8	−3.4	8,733
✓			150	**78.4**	**0.4**	**14,863**	78.2	−0.6	6,167	77.4	−0.8	13,931
✓	✓		150	78.0	0.0	13,520	76.4	−2.4	4,717	77.4	−0.8	11,804
✓	✓	✓	150	75.0	−3.0	11,021	73.2	−5.6	3,519	75.2	−3.0	9,682
✓			200	78.4	0.4	15,749	78.0	−0.8	6,359	78.0	−0.2	14,314
✓	✓		200	77.6	−0.4	14,260	75.2	−3.6	4,843	77.6	−0.6	12,557
✓	✓	✓	200	75.0	−3.0	11,704	72.4	−6.4	3,620	74.8	−3.4	10,382

(a) SVM classifier

Approach				all-typed			affix+punct			all-untyped		
D	PM	NE	HFW	Acc	+	N	Acc	+	N	Acc	+	N
	Baseline			73.4		10,859	**75.4**		**6,296**	74.2		9,258
✓			0	73.8	0.4	9,761	75.0	−0.4	5,503	74.4	0.2	8,143
✓	✓		0	73.6	0.2	8,430	74.0	−1.4	4,171	73.2	−1.0	6,648
✓		✓	0	71.6	−1.8	7,606	72.6	−2.8	4,187	70.8	−3.4	6,364
✓	✓	✓	0	70.2	−3.2	6,651	71.8	−3.6	3,087	70.8	−3.4	5,239
✓			50	73.2	−0.2	12,457	74.4	−1.0	5,860	74.4	0.2	10,902
✓	✓		50	73.6	0.2	11,005	74.0	−1.4	4,416	73.6	−0.6	9,250
✓	✓	✓	50	70.6	−2.8	8,890	71.4	−4.0	3,296	70.6	−3.6	7,510
✓			100	74.6	1.2	13,687	75.0	−0.4	6,041	73.6	−0.6	12,360
✓	✓		100	74.6	1.2	12,433	74.4	−1.0	4,570	73.6	−0.6	10,702
✓	✓	✓	100	70.4	−3.0	10,088	71.0	−4.4	3,405	69.8	−4.4	8,733
✓			150	**75.0**	**1.6**	**14,863**	75.2	−0.2	6,167	**74.8**	**0.6**	**13,931**
✓	✓		150	75.0	1.6	13,520	74.2	−1.2	4,717	73.8	−0.4	11,804
✓	✓	✓	150	70.4	−3.0	11,021	71.2	−4.2	3,519	69.8	−4.4	9,682
✓			200	74.6	1.2	15,749	74.4	−1.0	6,359	74.8	0.6	14,314
✓	✓		200	73.8	0.4	14,260	74.2	−1.2	4,843	74.2	0.0	12,557
✓	✓	✓	200	70.2	−3.2	11,704	71.8	−3.6	3,620	70.4	−3.8	10,382

(b) MNB classifier

To demonstrate the effectiveness of character n-gram features, we conducted experiments using the Bag-of-Words (BoW) approach, obtaining accuracy of 76.2% and 73.6% on the CCAT_10 test corpus, and 46.0% and 55.0% on the Guardian test corpus using SVM and MNB classifiers respectively. Character 3-gram features outperformed the BoW approach on both corpora for both classifiers by 1.8%–6.5%; see Tables 3 and 4.

Table 4. Accuracy results on the Guardian corpus after applying the proposed preprocessing steps (following the notations of Table 3).

Approach				all-typed			affix+punct			all-untyped		
D	PM	NE	HFW	Acc	+	N	Acc	+	N	Acc	+	N
Baseline				50.0		6,903	52.3		3,779	52.5		5,728
✓			0	50.9	0.9	6,841	52.4	0.1	3,725	52.4	−0.1	5,656
✓	✓		0	50.4	0.4	6,267	52.9	0.6	3,151	52.3	−0.2	4,985
✓		✓	0	**54.1**	**4.1**	**6,202**	56.2	3.9	3,347	54.4	1.9	5,121
✓	✓	✓	0	53.9	3.9	5,629	56.7	4.4	2,775	**55.8**	**3.3**	**4,443**
✓	✓	✓	50	52.3	2.3	7,411	56.5	4.2	2,978	52.6	0.1	6,251
✓	✓	✓	100	50.8	0.8	8,056	56.8	4.5	3,070	50.4	−2.1	6,924
✓	✓	✓	150	51.1	1.1	8,325	**57.0**	**4.7**	**3,150**	51.1	−1.4	7,210
✓	✓	✓	200	49.4	−0.6	8,451	56.1	3.8	3,219	49.7	−2.8	7,346

(a) SVM classifier

Approach				all-typed			affix+punct			all-untyped		
D	PM	NE	HFW	Acc	+	N	Acc	+	N	Acc	+	N
Baseline				56.6		6,903	58.4		3,779	56.9		5,728
✓			0	57.3	0.7	6,841	58.0	−0.4	3,725	57.1	0.2	5,656
✓	✓		0	59.5	2.9	6,267	61.6	3.2	3,151	60.2	3.3	4,985
✓		✓	0	58.0	1.4	6,202	58.9	0.5	3,347	58.5	1.6	5,121
✓	✓	✓	0	**60.8**	**4.2**	**5,629**	**64.3**	**5.9**	**2,775**	**61.9**	**5.0**	**4,443**
✓	✓	✓	50	59.1	2.5	7,411	63.8	5.4	2,978	59.5	2.6	6,251
✓	✓	✓	100	58.5	1.9	8,056	63.0	4.6	3,070	58.3	1.4	6,924
✓	✓	✓	150	58.1	1.5	8,325	62.2	3.8	3,150	57.8	0.9	7,210
✓	✓	✓	200	57.4	0.8	8,451	63.4	5.0	3,219	57.3	0.4	7,346

(b) MNB classifier

5.2 Frequency Threshold Selection

So far, all character n-grams with at least five occurrences in the training corpus were considered, similar to Sapkota et al. [11]. However, the appropriate tuning of feature set size has proved to be of great importance in cross-topic AA [6]. In this study, we attempt to select the most appropriate frequency threshold based on grid search. In more detail, we examine the following frequency threshold values: 5, 10, 20, 50, 100, 150, 200, 300, 500 and select the one that provides the best 10-fold cross-validation result on the training corpus. In the Guardian corpus, we use the average 10-fold cross-validation accuracy over the 4 training corpora.

In this experiment, we used the best combination of pre-processing steps for each corpus, as described in the previous section. For CCAT_10, the pre-processing combination was the replacement of digits. According to 10-fold cross-validation on the training corpus, the selected frequency threshold in all cases was 100 or less. This managed to slightly improve the results on the test corpus by approximately 1% with respect to a fixed frequency threshold of 5 (detailed results are omitted due to lack of space).

Table 5. Accuracy (%) variation with respect to the minimum feature frequency, where 10FCV – 10-fold cross-validation results on the training corpus; test – on the test corpus. The selected settings according to maximum 10-fold cross-validation result on the training corpus are in boldface; the top accuracies in test corpus are in italics.

min. feature frequency	all-typed			affix+punct			all-untyped		
	10FCV	test	N	10FCV	test	N	10FCV	test	N
5 (baseline)	67.9	53.9	5,629	71.7	56.7	2,775	68.4	55.8	4,443
10	69.1	55.9	4,372	73.2	59.6	2,144	71.3	57.1	3,573
20	71.5	59.8	3,249	**75.1**	**62.8**	**1,582**	73.0	60.1	2,779
50	73.1	61.5	1,956	73.4	65.2	964	72.7	62.1	1,821
100	**74.5**	**61.6**	**1,183**	74.9	*66.5*	602	71.0	61.2	1,176
150	74.1	60.9	809	74.4	65.0	436	73.2	*62.6*	856
200	74.2	62.7	604	74.2	65.9	341	**75.0**	**62.2**	**661**
300	74.4	*63.8*	386	73.5	65.2	238	73.8	62.4	437
500	67.5	60.9	205	68.9	63.3	141	70.0	60.9	227

(a) SVM classifier

min. feature frequency	all-typed			affix+punct			all-untyped		
	10FCV	test	N	10FCV	test	N	10FCV	test	N
5 (baseline)	71.7	60.8	5,629	72.6	64.3	2,775	71.6	61.9	4,443
10	73.3	63.6	4,370	74.5	67.3	2,144	73.2	64.8	3,573
20	75.6	66.4	3,249	77.6	68.6	1,582	75.1	66.7	2,779
50	76.4	66.6	1,956	77.8	70.3	964	75.4	67.1	1,821
100	**77.0**	**67.9**	**1,183**	**78.8**	**72.3**	**602**	76.2	67.6	1,176
150	76.9	68.8	809	77.1	72.3	436	77.5	69.0	856
200	76.7	*70.5*	604	78.3	*73.2*	341	**78.1**	**69.6**	**661**
300	76.4	70.4	386	76.7	72.9	238	77.4	*70.1*	437
500	73.6	69.2	205	77.4	71.3	141	73.5	68.1	227

(b) MNB classifier

For the cross-topic experiments, we applied the combination of pre-processing steps that are useful in this corpus: replacement of digits, NEs, and splitting PMs. Table 5 shows the performance results (both 10-fold cross-validation accuracy on the training corpus and the corresponding results on the test corpus) for different frequency threshold values using either SVM or MNB classifiers. We compare the obtained results with the fixed threshold of 5 used in the previous experiments, as well as by Sapkota et al. [11].

In general, any frequency threshold higher than the baseline produces better results. The best settings found by 10-fold cross-validation on the training set do not correspond to the best possible results on the test set. However, they provide a near-optimal estimation, regardless of the classifier. It is also remarkable that the settings that achieve the best performance correspond to relatively high frequency thresholds (about 100–200), much higher than the ones found for the CCAT_10 corpus. This means that low frequency features should be avoided under cross-topic conditions, since they provide confusing information to the classifiers. Note that these high values of frequency threshold drastically reduced feature set sizes (around 80% reduction in most of the cases). The selection of an appropriate frequency threshold, using only the training data, allowed us to improve the accuracy in cross-topic AA almost by around 10% for each of the models. The increase in performance is even higher if we compare the result of this experiment with the original approach of Sapkota et al. [11].

6 Conclusions

It is well-known in AA research that character n-grams provide very effective features. They are able to capture many nuances of writing style, and they are very simple to be extracted from any text in any language. However, it is not clear how thematic information can be appropriately reduced when a character n-gram representation is used. In this paper, we showed that it is possible to notably enhance the performance of AA under realistic cross-topic conditions by performing simple pre-processing steps that discard topic-dependent information from texts. It seems that the replacement of digits, punctuation marks splitting, and the replacement of named entities before the extraction of character n-grams improve the results in cross-topic AA when these steps applied separately or even better when they are combined.

On the other hand, the replacement of highly frequent words with distinct symbols does not seem to be helpful. When applied to a single-topic corpus, where authors tend to deal with specific topics, and therefore, they can be distinguished by a combination of their personal style and thematic preferences, the proposed pre-processing steps do not seem so effective.

We also showed that the appropriate selection of the dimensionality of the representation is crucial for cross-topic AA, and that it is possible to significantly improve the accuracy results by fine tuning the frequency threshold based on the training data. In cross-topic conditions, high frequency threshold values were found the most effective. It indicates that least frequent n-grams, associated with topic-specific information, should be avoided. Our approach improves the cross-topic AA accuracy by more than 10% over the baseline for the examined classifiers, while drastically reducing the size of the feature set by 80%.

Our experiments confirmed the conclusion by Sapkota *et al.* [11] that the model trained on affix and punctuation character n-grams is more effective than the models trained on all the features. This is consistent regardless of the particular learning algorithm, with or without performing pre-processing steps. It is also interesting that based on features of **affix+punct** we achieved the best increase in AA performance in cross-topic conditions.

Another interesting observation is that MNB classifier performs better than SVM under cross-topic conditions, whereas SVM is better for single-topic conditions. Further investigation is required to verify this conclusion.

One of the directions for future work would be to conduct experiments using longer character n-grams in single and cross-topic conditions and select an appropriate n-gram order. It would also be interesting to examine the effect of the proposed method to word level features, such as syntactic n-grams [33]. Moreover, the combination of different feature types should be examined since this usually improves the performance of the attribution models [23,34]. In addition, the robustness of our approach under cross-genre conditions, when training and test corpora belong to different genres (e.g., scientific papers and e-mail messages) will be tested.

Acknowledgments. This work was partially supported by the Mexican Government (CONACYT projects 240844, SNI, COFAA-IPN, SIP-IPN 20161947, 20161958, 20162204, 20162064, 20171813, 20171344, and 20172008).

References

1. Stamatatos, E.: A survey of modern authorship attribution methods. J. Am. Soc. Inf. Sci. Technol. **60**, 538–556 (2009)
2. Abbasi, A., Chen, H.: Applying authorship analysis to extremist-group Web forum messages. IEEE Intell. Syst. **20**, 67–75 (2005)
3. Chaski, C.E.: Who's at the keyboard? Authorship attribution in digital evidence investigations. Int. J. Digit. Evid. **4**, 1–13 (2005)
4. Coulthard, M.: On admissible linguistic evidence. J. Law Policy **21**, 441–466 (2013)
5. Koppel, M., Seidman, S.: Automatically identifying pseudepigraphic texts. In: Proceedings of the 2013 Conference on Empirical Methods in Natural Language Processing, (EMNLP'13), pp. 1449–1454 (2013)
6. Stamatatos, E.: On the robustness of authorship attribution based on character n-gram features. J. Law Policy **21**, 427–439 (2013)
7. Luyckx, K., Daelemans, W.: Authorship attribution and verification with many authors and limited data. In: Proceedings of the 22nd International Conference on Computational Linguistics (COLING'08), pp. 513–520 (2008)
8. Houvardas, J., Stamatatos, E.: N-gram feature selection for authorship identification. In: Proceedings of Artificial Intelligence: Methodologies, Systems, and Applications (AIMSA'06), pp. 77–86 (2006)
9. Kestemont, M.: Function words in authorship attribution. From black magic to theory? In: Proceedings of the 3rd Workshop on Computational Linguistics for Literature (EACL'14), pp. 59–66 (2014)

10. Daelemans, W.: Explanation in computational stylometry. In: Proceedings of the 14th International Conference on Intelligent Text Processing and Computational Linguistics (CICLing'13), pp. 451–462 (2013)
11. Sapkota, U., Bethard, S., Montes-y-Gómez, M., Solorio, T.: Not all character n-grams are created equal: a study in authorship attribution. In: Proceedings of the 2015 Annual Conference of the North American Chapter of the ACL: Human Language Technologies (NAACL-HLT'15), pp. 93–102 (2015)
12. Hedegaard, S., Simonsen, J.G.: Lost in translation: authorship attribution using frame semantics. In: Proceedings of the 49th Annual Meeting of the Association for Computational Linguistics: Human Language Technologies (HLT'11), pp. 65–70 (2011)
13. Schwartz, R., Tsur, O., Rappoport, A., Koppel, M.: Authorship attribution of micro-messages. In: Proceedings of the 2013 Conference on Empirical Methods in Natural Language Processing (EMNLP'13), pp. 1880–1891 (2013)
14. Sidorov, G., Velasquez, F., Stamatatos, E., Gelbukh, A., Chanona-Hernández, L.: Syntactic n-grams as machine learning features for natural language processing. Expert Syst. Appl. **41**, 853–860 (2014)
15. Gómez-Adorno, H., Sidorov, G., Pinto, D., Markov, I.: A graph based authorship identification approach. In: Working Notes Papers of the CLEF 2015 Evaluation Labs (CLEF'15), vol. 1391. CEUR (2015)
16. Grieve, J.: Quantitative authorship attribution: an evaluation of techniques. Lit. Linguist. Comput. **22**, 251–270 (2007)
17. Stamatatos, E.: Author identification using imbalanced and limited training texts. In: Proceedings of the 18th International Conference on Database and Expert Systems Applications (DEXA'07), pp. 237–241 (2007)
18. Escalante, H.J., Solorio, T., Montes-y-Gómez, M.: Local histograms of character n-grams for authorship attribution. In: Proceedings of the 49th Annual Meeting of the Association for Computational Linguistics: Human Language Technologies (HLT'11), pp. 288–298 (2011)
19. Sapkota, U., Solorio, T., Montes-y-Gómez, M., Bethard, S., Rosso, P.: Cross-topic authorship attribution: will out-of-topic data help? In: Proceedings of the 25th International Conference on Computational Linguistics (COLING'14), pp. 1228–1237 (2014)
20. Khmelev, D.V., Teahan, W.J.: A repetition based measure for verification of text collections and for text categorization. In: Proceedings of the 26th Annual International ACM SIGIR Conference on Research and Development in Information Retrieval (SIGIR'03), pp. 104–110 (2003)
21. Marton, Y., Wu, N., Hellerstein, L.: On compression-based text classification. In: Proceedings of the 27th European conference on Advances in Information Retrieval Research (ECIR'05), pp. 300–314 (2005)
22. Peng, F., Schuurmans, D., Keselj, V., Wang, S.: Language independent authorship attribution with character level n-grams. In: Proceedings of the 10th Conference of the European Chapter of the Association for Computational Linguistics (EACL'03), pp. 267–274 (2003)
23. Qian, T., Liu, B., Chen, L., Peng, Z.: Tri-training for authorship attribution with limited training data. In: Proceedings of the 52nd Annual Meeting of the Association for Computational Linguistics (ACL'14), pp. 345–351 (2014)
24. Stamatatos, E., Fakotakis, N., Kokkinakis, G.: Automatic text categorization in terms of genre and author. Comput. Linguist. **26**, 471–495 (2000)
25. de Vel, O.Y., Anderson, A., Corney, M., Mohay, G.M.: Mining email content for author identification forensics. SIGMOD Rec. **30**, 55–64 (2001)

26. Koppel, M., Schler, J., Bonchek-Dokow, E.: Measuring differentiability: unmasking pseudonymous authors. J. Mach. Learn. Res. **8**, 1261–1276 (2007)
27. Stamatatos, E., Daelemans, W., Verhoeven, B., Juola, P., López-López, A., Potthast, M., Stein, B.: Overview of the author identification task at PAN 2015. In: Working Notes of CLEF 2015-Conference and Labs of the Evaluation Forum (2015)
28. Lewis, D.D., Yang, Y., Rose, T.G., Li, F.: RCV1: a new benchmark collection for text categorization research. J. Mach. Learn. Res. **5**, 361–397 (2004)
29. Finkel, J.R., Grenager, T., Manning, C.: Incorporating non-local information into information extraction systems by Gibbs sampling. In: Proceedings of the 43rd Annual Meeting of the Association for Computational Linguistics (ACL'05), pp. 363–370 (2005)
30. Hall, M., Frank, E., Holmes, G., Pfahringer, B., Reutemann, P., Witten, I.H.: The WEKA data mining software: an update. SIGKDD Explor. **11**, 10–18 (2009)
31. Gómez-Adorno, H., Markov, I., Sidorov, G., Posadas-Durán, J., Sanchez-Perez, M.A., Chanona-Hernandez, L.: Improving feature representation based on a neural network for author profiling in social media texts. Comput. Intell. Neurosci. **2016**, 13 (2016)
32. Kibriya, A.M., Frank, E., Pfahringer, B., Holmes, G.: Multinomial naive Bayes for text categorization revisited. In: Proceedings of the 17th Australian Joint Conference on Advances in Artificial Intelligence (AI'04), pp. 488–499 (2005)
33. Sidorov, G., Gómez-Adorno, H., Markov, I., Pinto, D., Loya, N.: Computing text similarity using tree edit distance. In: Proceedings of the Annual Conference of the North American Fuzzy Information processing Society (NAFIPS'15) and 5th World Conference on Soft Computing, pp. 1–4 (2015)
34. Markov, I., Gómez-Adorno, H., Sidorov, G., Gelbukh, A.: Adapting cross-genre author profiling to language and corpus. In: Working Notes Papers of the CLEF 2016 Evaluation Labs. CEUR Workshop Proceedings, vol. 1609, pp. 947–955. CLEF and CEUR-WS.org (2016)

Author Identification Using Latent Dirichlet Allocation

Hiram Calvo[1,2(✉)], Ángel Hernández-Castañeda[1],
and Jorge García-Flores[2]

[1] Instituto Politécnico Nacional, Center for Computing Research CIC-IPN,
Av. J.D. Bátiz E/M.O. de Mendizábal, 07738 Mexico City, Mexico
{hcalvo,ahernandez}@cic.ipn.mx
[2] Laboratoire d'Informatique de Paris Nord, CNRS, UMR 7030, Université Paris
13, Sorbonne Paris Cité, 93430 Villetaneuse, France
jgflores@lipn.univ-paris13.fr

Abstract. We tackle the task of author identification at PAN 2015 through a
Latent Dirichlet Allocation (LDA) model. By using this method, we take into
account the vocabulary and context of words at the same time, and after a
statistical process find to what extent the relations between words are given in
each document; processing a set of documents by LDA returns a set of distributions of topics. Each distribution can be seen as a vector of features and a
fingerprint of each document within the collection. We used then a Naïve Bayes
classifier on the obtained patterns with different performances. We obtained
state-of-the-art performance for English, overtaking the best FS score reported in
PAN 2015, while obtaining mixed results for other languages.

1 Introduction

Author verification is an important problem to solve since many tasks require recognizing the author who wrote a specific text. For example, from knowing which author
wrote an anonymous book, up to identifying notes of a serial killer. In this paper we
deal with an author verification challenge from a more realistic approach. Specifically,
the dataset used consists of one to five documents of a known author and one document
of an unknown author. The corpus is formed by four subsets in different languages
(English, Spanish, Dutch and Greek). The aim is to identify whether a written unknown
text was written by the same author who wrote the known texts. It is important to note
that this task becomes more difficult when the dataset is composed of short documents;
since current approaches are not able to capture effective models with few amounts of
words [1]. However, on real cases as forensic field, long texts rarely exist.

Several approaches have been conducted to generate more informative features
based on text style. Nevertheless, it is also possible to generate features by extracting

The authors wish to thank the support of the Instituto Politécnico Nacional, (COFAA, SIP) and
the Mexican Government (CONACYT, SNI). The first author is currently in a research stay at
Laboratoire d'Informatique de Paris Nord, CNRS, Université Paris 13.

A. Gelbukh (Ed.): CICLing 2017, LNCS 10762, pp. 303–312, 2018.
https://doi.org/10.1007/978-3-319-77116-8_22

lexical, syntactic, semantic information among others. Lexical information is limited to word counts and occurrence of common words. On the other hand, syntactic information is able to obtain, to a certain extent, the context of the words.

In this work we use semantic information to find features that help us to discriminate texts. For this purpose, we create a model by using Latent Dirichlet Allocation (LDA). By using this method, we consider all the vocabulary from all texts at the same time, and, after a statistical process, find to which extent the relations between words are given in each document. LDA is a statistical algorithm which considers a text collection as a topics mixture; then, processing a set of documents by LDA returns a set of topic distributions. Each distribution can be seen as a vector of features and a fingerprint of each document within the collection. We use machine learning algorithms to classify the obtained patterns.

In this work we obtained the following F-measures: 85.5% for English, 76.0% for Spanish, 70.9% for Dutch and 64.0% for Greek.

2   Related Work

Several works have attempted the authorship identification challenge by generating different kinds of features [12, 14]. The nature of the dataset can determine the difficulty of the task, i.e., how hard will be to extract appropriate features [17, 18]. In [2] can be seen that, while the number of authors increases and the size of training dataset decreases, classification performance lowers. This sounds logical since, when the size of training data is lower, the identification of helpful features becomes affected.

Many works address author identification through the author's writing style [13, 16]. For instance, in [3], style-based features are compared to the BoW (Bag of Words) method. This study attempts to discriminate authors from texts in the same domain obtained from Twitter. Style markers such as characters, long words, whitespaces, punctuation, hyperlinks, parts of speech, among others, were included. The study findings showed that a style-based approach was more informative than a BoW-based method. However, their best results were obtained when considering two authors, so there was an accuracy decrease when the number of authors was increased. This suggests that, depending on how big is the training set, there will be stylistic features that help to distinguish an author from other, but not from all other authors.

Stylistic features also can be applied to other tasks. In [4], the authors combined features to address two-class problems. This work attempts to obtain style, BoW and syntax features to classify native and non-native English, texts written for conference or workshop and texts written by male or female. The dataset consists of scientific articles. This kind of texts is more extensive, compared to e-mail, tweets, or other short texts; this could have led to identify non-native written texts with promising accuracy. Nevertheless, long texts not necessarily ensure good results, since classification tasks on venue and gender obtained low accuracy.

The purpose of identifying authorship can vary. For example, Bradley et al. [5] attempt to prove that it is possible to find out which author wrote an unpublished paper (for a conference or journal); they consider only the cited works in them. By using LSA, the authors propose to create a term-document matrix wherein possible authors are considered as documents and authors who are cited are considered as terms. The results of Bradley et al. showed that the blind review system should be examined in greater detail. Another example is the Castro and Lindauer's work [15], with the task of finding out whether Twitter users identity can be uncovered by their writing style. The authors focused in features such as word shape, word length, character frequencies, stop words' frequencies, among others. With an RLSC (Regularized Least Square Classification) algorithm, the authors correctly classified 41% of the tweets.

In the work of Pimas et al. [11], the author verification task is addressed by generating three types of features. The authors extract stylometric, grammatical and statistical features. Our This study is based on PAN 2015 authorship verification challenge. In addition, Pimas et al. consider topics distribution as well, but they argue against using it, because the dataset is formed by topic mixtures. A cross validation model (10 folds) shows good performance, but, on the other hand, the model got overfitting using the training and test sets specified in the dataset.

3   Author Verification

In this section we present our method for author verification. First, in Sect. 3.1 we detail the source of features we use. Next, in Sect. 3.2 we describe the dataset used in this work for evaluation, and finally in Sect. 3.3 we give details on our feature vector construction.

3.1   Latent Dirichlet Allocation (LDA)

LDA [7] is a probabilistic generative model for discrete data collections such as texts collection. It represents documents as a mix of different *topics*. Each topic consists of a set of words that keep some link between them. Words, in its turn, can be chosen based on probability. The model assumes that each document is formed word-by-word by randomly selecting a topic and a word for this topic. As a result, each document can combine different topics. Namely, simplifying things somewhat, the generation process assumed by the LDA consists of the following steps:

1. Determine the number N of words in the document according to the Poisson distribution.
2. Choose a mix of topics for the document according to Dirichlet distribution, out of a fix set of K topics.
3. Generate each word in the document as follows:
   (a) choose a topic;
   (b) choose a word in this topic.

Assuming this generative model, LDA analyzes the set of documents to reverse-engineering this process by finding the most likely set of topics of which a document may consist. LDA generates the groups of words (topics) automatically; see Fig. 1.

Accordingly, LDA can infer, given a fixed number of topics, how likely is that each topic (set of words) appear in a specific document of a collection. For example, in a collection of documents and 5 latent topics generated with the LDA algorithm, each document would have different distributions of 5 likely topics. That also means that vectors of 5 features would be created.

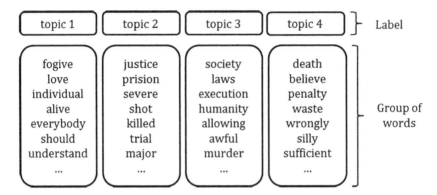

Fig. 1. Example of generated topics by using LDA.

3.2 Dataset

To conduct experiments with our approach, we use the corpus proposed in the author identification task of PAN 2015 [6]. The dataset consists of four subsets, each set written in different languages: English, Spanish, Dutch and Greek. Subsets have significant differences. The English subset consists of dialog lines from plays; the Spanish subset consists of opinion articles of online newspapers, magazines, blogs and literary essays; the Dutch subset is formed by essays and reviews; and the Greek subset is formed by opinion articles of categories as politics, health, sports among others. The corpus also has different number of documents per subset detailed in Table 1. In addition, each language consists of several problems to solve which are specifically defined below (Sect. 3.3).

Due to its nature, this dataset focused on problems which require capturing more specific information about the writing style of the author. For example, suppose we know a person who worked for a newspaper, writing articles about sports; but one day, this person decides to be independent and spend her life writing horror novels. One possible task can be to find out which articles belong to the sport ex-writer among sport articles of different authors—in this case, the vocabulary of the documents can uncover the author; for instance, by her usage rate of n-grams as features. On the other hand, another possible task is to discover whether a horror novel was written by the novelist, based on the sport articles which she wrote before., This is a drastic change in genre and topic of the documents, i.e., the intersection between vocabularies of the documents would be substantially reduced.

Table 1. Specific values for dataset of author identification task 2015

Language	Training problems			Test problems			Kind
	Items	# docs	Avg. words x doc.	Items	# docs	Avg. words x doc.	
English	100	200	366	500	452	536	Cross-topic
Spanish	100	500	954	100	1000	946	Cross-topic/genre
Dutch	100	276	354	165	380	360	Cross-genre
Greek	100	100	678	100	500	756	Cross-topic

3.3 Method

As an attempt to overcome this problem, we propose to use Latent Dirichlet Allocation (LDA) for extracting semantic information of the corpus. As mentioned before, given a collection of texts, LDA is able to find relations between words by their position in the text. Common stylistics approaches try to find discriminating symbols in the documents so they can distinguish between two documents written by different authors; however, as we stated before (Sect. 2), while texts become shorter, the amount of symbols is not enough to produce effective discriminate features. This fact becomes worse when authors number is increased. In the case of LDA we expect this issue to be less problematic.

We infer that writers have different ways to link words due to the fact that each writer makes use of favorite phrases. For example, some author usually may use the phrase "the data gathered in the study suggests that" in contrast to other author who uses "the data appears to suggest that". Thus, the words "the, in, to, that" can be included in different topics since, unlike LSA [8]. LDA can assign the same word to different topics as an attempt to better handle polysemy. As a result, to use several words at different rates shall result in different topic distributions for each document.

The task of the dataset used for this study is as follows. For each language or subset of the dataset there are specific number of problems; for each problem in turn there are from one to five documents considered as known and one document considered as unknown. These known documents are written by the same author. To solve a specific problem, we must find out whether the unknown document was written by the same author which writes the known documents.

To represent each problem, all documents in the dataset are processed with LDA. Then, we obtain vectors (with real values – probability of each topic) which represent known and unknown documents. Based on a specific problem, we do a subtraction between each known-document's vector and the unknown-document's vector (let us remember that there is only one unknown document by problem, however there are from one to five known documents). We found that converting real values to {0, 1} values slightly improved final results, so we used the arithmetic mean as threshold; 0 represents topic absence and 1 topic presence (above a certain threshold). Therefore, the subtraction between vectors can result in two possible values: 0 when topics are equal and 1 when topics are different (see Fig. 2).

Fig. 2. Example of subtraction between known-document's vector and unknown-document's vector

4 Results

In the following experiments, we use a Naïve Bayes classifier for classification. For all experiments we chose the number of topics to be 3. Therefore, patterns of three features were generated by each document. We found that varying the topics number, changed the performance classification. There is not a method for determining how many topics we should to choose for incrementing performance. Thus, we had to fix an interval until we achieved the best results. We show in Table 2 results of performance measures (explained below) regarding the number of topics selected. This table shows that the best results are around 3 topics.

Interestingly, with vectors with only a few of topics, we obtained over 64% accuracy. Actually, one might suppose that documents could have been categorized by subject; however, that assumption is unlikely because, as we showed in Sect. 3.2, the dataset used is formed of topics and genres mixtures.

Table 2. Selection of topics number based on PAN-2015's author identification task measures

No. Topics	c@1	AUC	FS
2	0.228	0.228	0.052
3	**0.856**	0.807	**0.691**
4	0.702	**0.908**	0.637
5	0.774	0.863	0.668
6	0.660	0.695	0.459
7	0.770	0.806	0.621
8	0.684	0.797	0.545
9	0.730	0.753	0.550
10	0.711	0.834	0.593
20	0.488	0.503	0.245
40	0.496	0.505	0.250
60	0.496	0.468	0.232
80	0.524	0.568	0.298
100	0.468	0.497	0.233

We conducted two experiments for knowing whether two documents written by the same author will be similar on their distribution of topics. Figure 3 shows the sum of all differences by topic in the test dataset for English. As we can see, the amount of differences is high when texts are written by different authors. In Fig. 4 is also showed that differences for Spanish language.

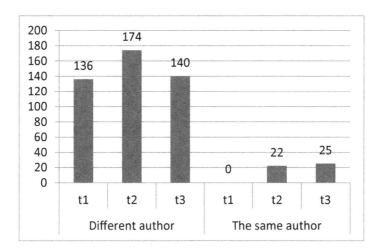

Fig. 3. Topic differences between document written either the same or different author (English subset)

We classified the dataset without pre-processing and show in Table 3 the following values: Accuracy, F-measure (F), Precision (P), Recall (R). While accuracy is a measure used in many works on deception detection and it provides us a point of comparison with other results, we also opted for showing precision, recall, and F-measure; this allows for a deeper analysis of outputs. Thus, precision shows the percentage of selected texts that are correct, while recall shows the percentage of correct texts that are selected. Finally, F-measure is the combined measure to assess the P/R trade-off.

We classified the dataset without pre-processing and show in Table 3 the following values: Accuracy, F-measure (F), Precision (P), Recall (R). While accuracy is a measure used in many works on deception detection and it provides us a point of comparison with other results, we also opted for showing precision, recall, and F-measure; this allows for a deeper analysis of outputs. Thus, precision shows the percentage of selected texts that are correct, while recall shows the percentage of correct texts that are selected. Finally, F-measure is the combined measure to assess the P/R trade-off.

We obtained the best result for English subset with 85.6% accuracy even when it has the biggest training set (500 problems) of the corpus. Spanish subset ranks second with 76.0% accuracy, Dutch subset reached 70.9% accuracy and finally Greek subset reached 64.0% accuracy. Both English and Greek subsets obtained the first and the last place of the results (Table 3) respectively; therefore, we cannot infer that the topics mixture made the difference in results since both subsets consist of themes mixture and one of them was not affected. Similarly, for both Spanish and Dutch subsets (second

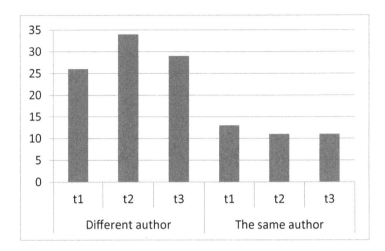

Fig. 4. Topic differences between document written either the same or different author (Spanish subset)

Table 3. Results of each subset classification

Subset	Accuracy (%)	Precision	Recall	F-measure
English	85.6	0.864	0.856	0.855
Spanish	76.0	0.760	0.760	0.760
Dutch	70.9	0.733	0.709	0.702
Greek	64.0	0.646	0.640	0.640

and third place respectively), results did not lead to conclude that the genre mixture had some correlation on it. For these reasons, we consider that the results were directly affected by the training and test set's document selection and not by the type of text.

We compare our results with those obtained in author identification task at PAN 2015 evaluation lab [6]. Therefore, we calculated, as PAN-2015 task's authors, a final score which is the product of two values: c@1 [9] and area under the ROC curve (AUC) [10]. The former is an extension of the accuracy metric and the latter is a measure of classification performance which provides more robust results than accuracy.

We show in Table 4 the better results obtained for each language subset by participants of PAN-2015 task. According to those results, our method seems to perform well for both English and Dutch languages. This work outperforms FS results with regard to English subset and had better performance than Bartoli et al. and Bagnall's result with regard to Dutch subset. On the other hand, for both Spanish and Greek subsets the proposed method did not show good performance however, ROC curve results showed that predictions are acceptable.

Table 4. Results comparison with other authors. FS = c@1*AUC.

Author	Measure	Subset			
		English	Spanish	Dutch	Greek
Bagnall 2015	c@1	0.757	0.814	0.644	0.851
	AUC	0.811	0.886	0.700	0.882
	FS	0.614	0.721	0.451	**0.750**
Bartoli et al. 2015	c@1	0.559	0.830	0.689	0.657
	AUC	0.578	0.932	0.751	0.698
	FS	0.323	**0.773**	0.518	0.458
Moreau et al. 2015	c@1	0.638	0.755	0.770	0.781
	AUC	0.709	0.853	0.825	0.887
	FS	0.453	0.661	**0.635**	0.693
This work	c@1	0.856	0.760	0.709	0.640
	AUC	0.808	0.737	0.785	0.688
	FS	**0.692**	0.560	**0.556**	0.440

5 Conclusions

A common approach to verify authorship is attempting to find the author's writing style. Therefore, the assumption is that by using that approach, it is possible to capture specific features to discriminate one author from others. This hypothesis is hard to prove; nevertheless, it is known that certain amount of data is necessary to find more appropriate features leading to high classification performance. Data is a problem, for instance, for the forensic field, since hardly there are long texts and they are in different domains. We showed in this work how LDA responds to verify authorship when there is limited data; i.e., only from one to five short texts written by a specific author to determine whether an unknown document belongs to the same author. Furthermore, the used datasets consist of topic and genre mixtures.

Basically, we used documents distributions to capture what we call the authors' fingerprint. Then, by subtraction between topic distributions, we found that documents written by different author tend to differ more than those written by the same author. This approach allowed us to achieve 74% accuracy on average.

References

1. Stamatatos, E.: A survey of modern authorship attribution methods. J. Am. Soc. Inf. Sci. Technol. **60**(3), 538–556 (2009)
2. Nirkhi, S., Dharaskar, R.V.: Comparative study of authorship identification techniques for cyber forensics analysis (2013). arXiv preprint arXiv:1401.6118
3. Layton, R., Watters, P., Dazeley, R.: Local n-grams for author identification. In: Notebook for PAN at CLEF (2013)

4. Bergsma, S., Post, M., Yarowsky, D.: Stylometric analysis of scientific articles. In: Proceedings of the 2012 Conference of the North American Chapter of the Association for Computational Linguistics: Human Language Technologies, pp. 327–337, June 2012. Association for Computational Linguistics (2012)

5. Bradley, J.K., Kelley, P.G., Roth, A.: Author identification from citations. Department of Computer Science, Carnegie Mellon University, Pittsburgh, PA, USA, Technical Report (2008)

6. Stamatatos, E., et al.: Overview of the author identification task at PAN 2015. In: CLEF (Working Notes) (2015)

7. Blei, D.M., Ng, A.Y., Jordan, M.I.: Latent dirichlet allocation. J. Mach. Learn. Res. **3**, 993–1022 (2003)

8. Dumais, S.T.: Latent semantic analysis. Annu. Rev. Inf. Sci. Technol. **38**(1), 188–230 (2004)

9. Peñas, A., Rodrigo, A.: A simple measure to assess non-response. In: Proceedings of the 49th Annual Meeting of the Association for Computational Linguistics: Human Language Technologies, vol. 1, pp. 1415–1424, June 2011. Association for Computational Linguistics (2011)

10. Fawcett, T.: An introduction to ROC analysis. Pattern Recognit. Lett. **27**(8), 861–874 (2006)

11. Pimas, O., Kröll, M., Kern, R.: Know-center at PAN 2015 author identification. In: Working Notes Papers of the CLEF (2015)

12. Narayanan, A., et al.: On the feasibility of internet-scale author identification, May 2012. In: 2012 IEEE Symposium on Security and Privacy, pp. 300–314. IEEE (2012)

13. Pateriya, P.K.: A Study on author identification through stylometry. Int. J. Comput. Sci. Commun. Netw. **2**(6), 653 (2012)

14. Madigan, D., Genkin, A., Lewis, D.D., Argamon, S., Fradkin, D., Ye, L.: Author identification on the large scale. In: Proceedings of the Meeting of the Classification Society of North America, p. 13 (2005)

15. Castro, A., Lindauer, B.: Author identification on Twitter (2012). semanticscholar.org

16. Pavelec, D., Justino, E., Oliveira, L.S.: Author identification using stylometric features. Inteligencia Artificial: Revista Iberoamericana de Inteligencia Artificial **11**(36), 59–66 (2007)

17. Green, R.M., Sheppard, J.W.: Comparing frequency-and style-based features for twitter author identification. In: FLAIRS Conference, May 2013

18. Afroz, S., Brennan, M., Greenstadt, R.: Detecting hoaxes, frauds, and deception in writing style online. In: 2012 IEEE Symposium on Security and Privacy, pp. 461–475, May 2012. IEEE (2012)

Personality Recognition Using Convolutional Neural Networks

Maite Giménez[(✉)], Roberto Paredes, and Paolo Rosso

Pattern Recognition and Human Language Technology (PRHLT) Research Center,
Universitat Politècnica de València, Camino de Vera s/n, 46022 Valencia, Spain
{mgimenez,rparedes,prosso}@dsic.upv.es

Abstract. Personality Recognition is an emerging task in Natural Language Processing due to its potential applications. However, the models which address this task rely on handcrafted resources; therefore, they are restricted by the domain of the problem and by the availability of resources. We propose a Convolutional Neural Network architecture trained using pre-trained word embeddings that is capable of learning the best features for the task at hand without any external dependence. The results show the potential of this approximation. The proposed model achieves comparable results with state-of-the-art models and is able to predict the personality traits of authors regardless of the social network and the availability of resources.

Keywords: Personality Recognition · Convolutional Neural Networks

1 Introduction

Nowadays, users share a vast amount of data in social media, and relevant information can be mined from these resources. Several Natural Language Processing (NLP) tasks are devoted to this matter like Author Profiling. Formally, Author Profiling (AP) is the task that, given a text, seeks to classify writers depending on their demographic features such as age, gender, or their personality traits. The focus of this paper will be on the Personality Recognition (PR) task.

As mentioned by Quercia et al. [23], personality is significantly correlated with different real-world behaviors. Consequently, Personality Recognition is attracting the interest of the scientific community given its breadth potential applications [7,13,14,17,20]. Even the industry is keenly committed to exploring the PR task; for instance, IBM have developed *Personality Insights*[1], an automatic service that outputs personality characteristics given at least 3500 words.

In an effort to create a shared evaluation framework, from 2013 onwards, an Author Profiling shared task has been organized at PAN lab. In 2015, PR

---

[1] http://www.ibm.com/smarterplanet/us/en/ibmwatson/developercloud/personality-insights.html.

© Springer Nature Switzerland AG 2018
A. Gelbukh (Ed.): CICLing 2017, LNCS 10762, pp. 313–323, 2018.
https://doi.org/10.1007/978-3-319-77116-8_23

was also addressed [25]. The PR task was proposed following a widely accepted model in psychology, the Big Five or Five Factor Model [6]. In this approach the personality of an author could be described in terms of five traits: Openness (O), Conscientiousness (C), Extroversion (E), Agreeableness (A), and Stability (S). The participants' models should predict a real value, in the range from -0.5 to 0.5, for each of these five personality traits. In this paper, we present a novel approach using a Convolutional Neural Network (CNN) for predicting the Big Five personality traits. The approach was tested on the PAN-AP-2015 dataset. The contributions presented in this paper could be summarized as follows:

- We used pre-trained word embeddings to develop a Personality Recognition system.
- To best of our knowledge, we have firstly used a CNN trained with word-embeddings to recognize the personality of an author of a text.
- The architecture proposed achieves comparable state-of-the-art results without using handcrafted resources.

The rest of the paper is divided in five sections. The next section gives a brief overview of the literature related to our work. Section 3 is devoted to define the architecture proposed. In Sect. 4, the dataset and the Personality Recognition task are described and the experiments that we have carried out are presented. Finally, in Sect. 5 we discussed our results and future work is proposed.

2   Related Work

2.1   Personality Recognition

The work of Argamon et al. [3] was a pioneer for the PR task; their work was focused on distinguishing two traits, neuroticism and extraversion, in authors of informal texts using Support Vector Machines (SVMs) trained with handcrafted features such as function words, conjunction words, assessment taxonomies, etc. Likewise, a different set of n-gram features, handcrafted features and resources has been employed to cluster bloggers personality [19], and to study the Big Five traits in both informal conversation and normative text [16]. With respect to social media texts, several studies have tried to predict the personality of an author using datasets extracted mainly from Twitter [8,23] and Facebook [4,26]. Furthermore, the work of Youyou et al. [29] has proven that computer-based personality judgments are more accurate than those made by humans. In addition, Farandi et al. [10] developed state-of-the-art personality prediction models and studied the variance of those models over datasets extracted from Facebook, Twitter, and Youtube. They used different content-based and context-based resources and trained several machine learning algorithms like SVMs or Decision Trees. However, they were not able to improve the performance of a model trained with data from a social media domain with data from another social media domain. Hence, there is a strong dependence on the data domain and the resources used. In summary, most approaches found in the literature used handcrafted resources which are highly dependent on the domain. Bearing in mind the amount of data generated in social media, it is likely that these resources will be outdated. Moreover, each new instance must be pre-processed to gather the relevant information from each resource which constitutes a bottle-neck decreasing the time-response of the system. The proposed model presented in this work seeks to overcome these drawbacks. As we train only with the vector representations of words, also called word embeddings, we obtained a system that is not dependent on the available resources for a language since this representation for a given word can be trained in an unsupervised fashion, and the CNN model would learn a proper representation for each tweet, the model would be feasible for languages other than English. Also, even though training a deep learning model might be more demanding, once trained the parameters of the system can be updated with new instances to improve the performance of the system or even to adapt itself to a new domain.

2.2   Convolutional Neural Networks

CNNs were proposed as a specific neural network that uses a mathematical operation called *convolution* [15]. Goodfellow et al. [11] defined Convolutional Networks as neural networks that use convolutions instead of matrix multiplication in at least one of its layers.

A convolution is a linear operation that takes a multidimensional array as an input $x \in \mathbb{R}^{m,n}$ and another multidimensional array denominated kernel $w \in \mathbb{R}^{h,k}$ where $h < m \wedge k < n$ and produce multidimensional arrays called feature maps. Each element of a feature map is obtained after applying the convolution across a window of words $\{x_{1,h}, x_{2,h-1}, \ldots x_{n-h+1:n}\}$, and it is defined as:

$$c_i = f(\mathbf{w} \cdot x_{i:i+h-1} + b_i) \tag{1}$$

where $c \in \mathbb{R}^{n-h+1}$, $b_i \in \mathbb{R}$ is a bias term, and f a linear function. The size constriction of the dimension of the the kernel has several critical characteristics that enable CNNs to learn more efficiently. Moreover, the size of a kernel enhances the generalization. A CNN model computes, during the forward phase, a convolution of the input data with a linear kernel. Then, in the backward phase, the CNN learns the values of its kernels.

CNNs have been successfully used in computer vision leading to a breakthrough in this topic. The work of Collobert et al. [9] proposed an architecture using CNNs for Natural Language Processing proving that these neural networks could also be applied competently to NLP tasks. Lately, CNNs have been applied to several NLP tasks such as: Sentence Classification [12,30] or Document Ranking [27].

Our goal is to apply CNNs to the Author Profiling task. Following a supervised approach, with the dataset gathered from the 3rd Author Profiling task at PAN, we propose a novel approximation, a CNN fed with pre-trained word embeddings, provided by GloVe [21]. Our working hypothesis was that this approach would be able to obtain comparable results to the state-of-the art models just using word embeddings rather than relying on handcrafted resources. The evaluation presented in this paper shows encouraging results, proving the validity of this hypothesis.

2.3   Text Representation

Text representation entails a challenge for Machine Learning (ML) since most of ML algorithms are trained with numerical feature vectors with a fixed size. Therefore, a method for text feature extraction is required. Within the last ten years, following the work of Bengio et al. [5], semantic vector space models have been used in different NLP tasks. These models, generally called distributed word representations models, map each word from the training vocabulary to an Euclidean space, minimizing the perplexity of the language model created, attempting to capture semantic relationships between words [1]. In this work, we have used the pre-trained word vector representations developed by Pennington et al. [21][2], using their novel logbilinear regression GloVe model, which combines global matrix factorization methods and local context window methods. This new model improves the performance across a range of tasks achieved by either family methods described.

3   Methodology

In this section, we briefly describe the model we proposed for the Personality Recognition task. The intuition behind our model is that a word can be seen as a row in an image and, by extension, the whole tweet can be represented as an image. After text is represented as an image, CNNs can be applied.

3.1   Word Embeddings

As we described in Sect. 2.3 we have fed our model using the word embeddings trained using the GloVe approach. The authors have released several word embeddings, which differ in the origin of the data used to train them in an unsupervised fashion. Thus, since our data was extracted from Twitter, we fed our system with the Twitter dataset. This dataset contains 1.2 M words, each one represented as a n-dimensional real-valued vector. For computational simplicity, we focused on the 25-dimensional and 50-dimensional word representations. Hence, a word (wr) is represented in our model following the Eq. 2:

$$wr = \begin{cases} f(w) \text{ if } w \in G \\ \mathbf{0} \quad \text{ if } w \notin G \end{cases} \tag{2}$$

being G the Glove dictionary, the function f(w) returns the Glove pre-trained vector $\in \mathbb{R}^n$ and $\mathbf{0} \in \mathbb{R}^n$, being n $\in [25, 50]$. We opt for including a vector of

---

[2] These word vector representations are available at the following URL: http://nlp.stanford.edu/projects/glove/.

zeros for those words which are not present in the GloVe dictionary, because this approximation allows us to model the relationship between seen and unseen words in the dictionary. If we would not include these vectors of zeros, the matrix that represents the tweet only will contain those words seen in the GloVe dictionary, and the order of the words would be lost. CNNs need a fixed size input matrix. Therefore, the input data must be padded. To that end, as proposed in [12], we will include as many **0** as needed at the end of a sentence to pad the data to the maximum sequence length seen in training; during the testing phase if a sentence is longer it will be trimmed.

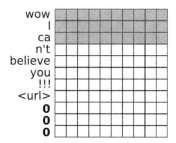

Fig. 1. Input matrix padded to feed a CNN. Shadowed rows represent the area covered by a kernel of height 3.

In Fig. 1, an example of text representation can be found. Each tweet will be represented using a matrix $\mathbf{X} \in \mathbb{R}^{m \times n}$, where m is the maximum number of words and n is the word embeddings dimension. This matrix is composed by the concatenation of the word embeddings from a tweet following the column axis.

3.2 Convolutional Neural Networks for Natural Language Processing

Several works [12,30] suggest that architectures with a convolutional layer followed by one or more fully connected layers achieve good results in NLP tasks. Therefore, we have explored these architectures that only contain a convolution layer.

Our CNN model shares some characteristics with the model proposed by Collobert et al. [9]. However, unlike the previously proposed models, our model addresses a regression problem. Consequently, the output layer will be a set of neurons with a linear activation function and it uses the Mean Square Error as a cost function.

As part of the CNN architecture definition, the kernel size must be set. A kernel, as described in Sect. 2.2, is a multidimensional array $w \in \mathbb{R}^{w \times h}$ that is applied to a window of size $w \times h$. A row represents a word, therefore, the width of the kernel is determined by the word embedding size chosen. We have tried different values for the height of the kernel.

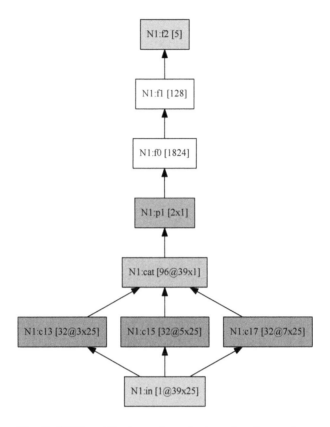

Fig. 2. CNN architecture of our best performing system.

These convolutional layers use Rectified Linear Units (ReLU) [18] as activation function. Multiple filters can be used to learn complementary features from the same regions. Thus, we have trained CNNs with 32 and 64 filters. Figure 2 shows the architecture of our best performing model. Hereafter, we will refer to this figure to describe our proposed architecture. The nodes labeled as *c13,c15,c17* represent the convolution layers using kernels of height 3, 5, and 7, respectively.

Noteworthy, we have included a concatenation layer, *cat* in Fig. 2, which concatenates the feature maps obtained after applying the different kernels in the convolution layers. This union allows the CNN to learn different representative features regardless of the kernel size.

After that, a pooling layer may be applied over the feature maps in order to reduce the size of the feature maps and to obtain local invariance to small variations of the position of the words in a tweet. We have tried models with and without max-pooling layers. However, the best performing models always had pooling layers. In Fig. 2, the node *p1* is a max-pooling layer.

Next, this pooling layer output is fed to a fully connected layer with one or more hidden units with a ReLU activation function. In the case showed in Fig. 2, this model has a reshape layer *f0* and a one hidden layer *f1* with 128 neurons. The architectures evaluated with two hidden layers had 64 neurons in the second layer. Finally, the output layer *f2* is composed of five neurons –one for each personality trait– with a linear activation function.

4   Evaluation

In order to evaluate our models we have compared them against the 2015 PAN shared task. This section describes the evaluation process that we have carried out to prove that CNNs could be an alternative approach to personality recognition.

4.1   Dataset

We have used the PAN-AP-2015 corpus which was employed in the 3rd Author Profiling shared task [25]. The organizers gathered a collection of social media interactions from Twitter. Four languages were included in this dataset. Nevertheless, we have focused our work on the English dataset. Personality traits were self-assessed with the BFI-10 online test [24][3]. Each one of the personality traits ranges between -0.5 and 0.5. The training dataset consisted of 14,166 tweets written by 152 authors, whereas the test dataset consisted of 13,172 tweets written by 142 different authors.

The evaluation of the models competing was carried out using the Root Mean Square Error (RMSE) for each trait and, the overall RMSE were calculated as the arithmetic mean of each RMSE trait.

Pre-process. Since in our work we have used GloVe vector representations, we tried to mimic their pre-process in order to maximize the number of words with a vector representation that can be found pre-trained in this resource. Firstly, we tokenized the dataset using the PTBTokenizer Stanford Tokenizer[4]. Then, we deleted those characters repeated more than three times in a token. Also, URLs, user mentions, and retweets were replaced by its corresponding GloVe token.

After this simple pre-proces, we proceeded to create the input matrix for our Convolutional Neural Network, as we described in Sect. 3.1.

4.2   Development Experiments

We instantiate different CNN architectures with the features described in Sect. 3.2. These CNN will vary in the dimension of the word embeddings (25, 50), in the number of filters (32, 64), the size of the filter (1, 3, 5, 7, 9,

---

[3] For further information about the gathering and labeling process see [25].

[4] http://nlp.stanford.edu/software/tokenizer.shtml.

12), in the number of hidden units (128 neurons in the first hidden layer, and 64 neurons in the second hidden layer when there are two hidden layers), whether or not: there is a concatenation layer, there is a pooling layer, we applied batch normalization and, we applied dropout as means of regularization.

In our approach, each tweet from each author constitutes an independent training instance. If we try to concatenate all the tweets in a single matrix, the number of parameters to fit would make unfeasible to train the model. Therefore, we have trained our model with 14,166 training instances, and it was tested with 13,172 instances. However, only a prediction for each author is required. Hence, the prediction for the author a and the trait t is the mean of the predictions obtained, defined following the Eq. 3:

$$prediction(a, t) = \frac{\sum_{i=1}^{N} p_{a,t,i}}{N} \tag{3}$$

where N is the number of tweets for each author.

4.3 Results

Table 1 shows our best five models. As can be seen, our best models were those that concatenated different convolutional filters, performed batch normalization and included max-pooling layers. Moreover, the results show that some traits accumulate a larger error consistently e.g. the trait stability.

We have trained our system *only* with word embeddings. In contrast, most of the systems have used a combination of style-based and content-based features, as well as n-gram models, and other handcrafted features that depend heavily on the domain.

Table 1. Evaluation results of our best CNN models. RMSE: the value achieved using the Root Mean Square Error, the metric used in the competition. O: Openness, C: Conscientiousness, E: Extroversion, A: Agreeableness, S: Stability, WE: Dimension of the word embedding. MP: if there is a max-pooling layer. K: height of the kernels used. N: Number of kernels used. H: the number of hidden units. DR: if dropout was applied. BN: whereas batch normalization was applied.

CNN architecture	WE	RMSE	E	S	A	C	O
K = 3, N = 32, H = 1, BN, MP	25	**0.1650**	0.1538	0.2214	0.1535	0.1457	0.1499
K = 7, N = 32, H = 1, BN, MP	25	**0.1647**	0.1584	0.2174	0.1525	0.1461	0.1491
K = 357, N=32, H = 1, BN, MP	25	**0.1625**	**0.1582**	**0.2123**	**0.1498**	**0.1438**	**0.1482**
K = 357, N = 32, H = 1, BN	50	**0.1633**	0.1564	0.2159	0.1517	0.1452	0.1473
K = 3, N = 64, H = 2, DR	50	**0.1692**	0.1587	0.2268	0.1560	0.1475	0.1569

The best result in the shared task was achieved by the team *alvarezcarmona15* (0.1442 RMSE). They have used stylistic features such as: frequency of words, contractions, words with hyphens, stop-words, punctuation marks, function words, determiners, and a set of common emoticons; combined using second order attributes [2] to build word vectors and document vectors in a space of profiles. In addition, they used the 100 most common concepts as thematic information gathered exploiting Latent Semantic Analysis. Each document is finally represented following this approach, as the union of all these features. Finally, they trained a LibLINEAR classifier. They considered each trait value in the train corpus as a class ignoring the possible personality values that were not seen in the train dataset. We have compared our results against the ones presented by *alvarezcarmona15* following the dependent t-test in order to investigate if there is a statistical difference between the results reported. The p-value obtained considering each the RMSE for each trait is 0.47. Therefore, we can accept the null hypothesis and assume that there is no statistical significance between our results and the best performing system.

5   Conclusions

In this paper we have presented a Convolutional Neural Network for predicting personality traits. Without handcrafted features or any external resource, we have trained a model able to achieve a good performance comparable withe the models of the 2015 PAN Author Profiling shared task. These models are closely dependent on the domain and on the availability of resources. In contrast, our model is not bound by any external resource, and it is domain independent. We plan to apply this model to other Personality Recognition datasets [22,28] and to investigate whether the performance of the model remains. In the future, we plan to explore deeper CNNs architectures, that have been used successfully in computer vision, but in NLP literature only shallow CNNs have been explored. In conclusion, Convolutional Neural Networks show a great potential for addressing the Personality Recognition task.

Acknowledgments. This work was developed in the framework of the TIN2015-71147-C2-1-P research project SOcial Media language understanding-EMBEDing contexts (SomEMBED), funded by the Ministry of Economy and Sustainability (MINECO). The work of the first author is financed by Grant PAID-01-2461 2015, from the Universitat Politècnica de València.

References

1. Alekseev, A., Nikolenko, S.: Predicting the age of social network users from user-generated texts with word embeddings. In: Proceeding of the AINL FRUCT 2016 Conference, pp. 3–13 (2016)
2. Álvarez-Carmona, M.A., López-Monroy, A.P., Montes-y Gómez, M., Villaseñor-Pineda, L., Escalante, H.J.: Inaoes participation at pan: author profiling task. Working Notes Papers of the CLEF, pp. 1–10 (2015)

3. Argamon, S., Koppel, M., Fine, J., Shimoni, A.R.: Gender, genre, and writing style in formal written texts. TEXT **23**(3), 321–346 (2003)
4. Bachrach, Y., Kosinski, M., Graepel, T., Kohli, P., Stillwell, D.: Personality and patterns of facebook usage. In: Proceedings of the 4th Annual ACM Web Science Conference, pp. 24–32. ACM (2012)
5. Bengio, Y., Schwenk, H., Senécal, J.S., Morin, F., Gauvain, J.L.: Neural probabilistic language models. In: Holmes, D.E., Jain, L.C. (eds.) Innovations in Machine Learning. Studies in Fuzziness and Soft Computing, vol. 194. Springer, Heidelberg (2006)
6. Boyle, G.J., Matthews, G., Saklofske, D.H.: The SAGE Handbook of Personality Theory and Assessment: Personality Measurement and Testing, vol. 2. Sage, Los Angeles (2008)
7. Cantador, I., Fernández-Tobías, I., Bellogín, A.: Relating personality types with user preferences in multiple entertainment domains. In: CEUR Workshop Proceedings. Shlomo Berkovsky (2013)
8. Celli, F., Rossi, L.: The role of emotional stability in twitter conversations. In: Proceedings of the Workshop on Semantic Analysis in Social Media, pp. 10–17. Association for Computational Linguistics (2012)
9. Collobert, R., Weston, J., Bottou, L., Karlen, M., Kavukcuoglu, K., Kuksa, P.: Natural language processing (almost) from scratch. J. Mach. Learn. Res. **12**, 2493–2537 (2011)
10. Farnadi, G., Sitaraman, G., Sushmita, S., Celli, F., Kosinski, M., Stillwell, D., Davalos, S., Moens, M.F., De Cock, M.: Computational personality recognition in social media. User Model. User Adapt. Interact. **26**(2), 109–142 (2016). https://doi.org/10.1007/s11257-016-9171-0
11. Goodfellow, I., Bengio, Y., Courville, A.: Deep Learning. MIT Press (2016). http://www.deeplearningbook.org
12. Kim, Y.: Convolutional neural networks for sentence classification, pp. 1–6 (2014). arXiv:1408.5882
13. Kosinski, M., Bachrach, Y., Kohli, P., Stillwell, D., Graepel, T.: Manifestations of user personality in website choice and behaviour on online social networks. Mach. Learn. **95**(3), 357–380 (2014)
14. Lambiotte, R., Kosinski, M.: Tracking the digital footprints of personality. Proc. IEEE **102**(12), 1934–1939 (2014)
15. LeCun, Y.: Generalization and network design strategies. In: Pfeifer, R., Schreter, Z., Fogelman, F., Steels, L. (eds.) Connectionism in Perspective. Elsevier, Zurich, Switzerland (1989)
16. Mairesse, F., Walker, M.A., Mehl, M.R., Moore, R.K.: Using linguistic cues for the automatic recognition of personality in conversation and text. J. Artif. Intell. Res. **30**, 457–500 (2007)
17. Mirkin, S., Nowson, S., Brun, C., Perez, J.: Motivating personality-aware machine translation. In: Proceedings of the 2015 Conference on Empirical Methods in Natural Language Processing, pp. 1102–1108. Association for Computational Linguistics (2015)
18. Nair, V., Hinton, G.E.: Rectified linear units improve restricted boltzmann machines. In: Proceedings of the 27th International Conference on Machine Learning (ICML-10), pp. 807–814 (2010)
19. Oberlander, J., Nowson, S.: Whose thumb is it anyway?: classifying author personality from weblog text. In: Proceedings of the COLING/ACL on Main Conference Poster Sessions, pp. 627–634. Association for Computational Linguistics (2006)

20. Paruma-Pabón, O.H., González, F.A., Aponte, J., Camargo, J.E., Restrepo-Calle, F.: Finding relationships between socio-technical aspects and personality traits by mining developer e-mails. In: Proceedings of the 9th International Workshop on Cooperative and Human Aspects of Software Engineering, pp. 8–14. ACM (2016)
21. Pennington, J., Socher, R., Manning, C.D.: Glove: global vectors for word representation. In: Empirical Methods in Natural Language Processing (EMNLP), pp. 1532–1543. http://www.aclweb.org/anthology/D14-1162 (2014)
22. Plank, B., Hovy, D.: Personality traits on twitter or how to get 1,500 personality tests in a week. In: Proceedings of the 6th Workshop on Computational Approaches to Subjectivity, Sentiment and Social Media Analysis, pp. 92–98 (2015)
23. Quercia, D., Kosinski, M., Stillwell, D., Crowcroft, J.: Our twitter profiles, our selves: predicting personality with twitter. In: 2011 IEEE Third International Conference on Privacy, Security, Risk and Trust (PASSAT) and 2011 IEEE Third Inernational Conference on Social Computing (SocialCom), pp. 180–185. IEEE (2011)
24. Rammstedt, B., John, O.P.: Measuring personality in one minute or less: a 10-item short version of the big five inventory in english and german. J. Res. Personal. **41**(1), 203–212 (2007)
25. Rangel, F., Celli, F., Rosso, P., Potthast, M., Stein, B., Daelemans, W.: Overview of the 3rd author profiling task at pan 2015. In: Proceedings of the Conference and Labs of the Evaluation Forum (Working Notes), pp. 518–538 (2015)
26. Schwartz, H.A.: Personality, gender, and age in the language of social media: the open-vocabulary approach. PloS One **8**(9), 1–16 (2013)
27. Shen, Y., He, X., Gao, J., Deng, L., Mesnil, G.: Learning semantic representations using convolutional neural networks for web search. In: Proceedings of the companion publication of the 23rd International Conference on World Wide Web, pp. 373–374. International World Wide Web Conferences Steering Committee (2014)
28. Verhoeven, B., Daelemans, W., Plank, B.: Twisty: a multilingual twitter stylometry corpus for gender and personality profiling. In: 10th International Conference on Language Resources and Evaluation (LREC 2016) (2016)
29. Youyou, W., Kosinski, M., Stillwell, D.: Computer-based personality judgments are more accurate than those made by humans. Proc. Natl Acad. Sci. **112**(4), 1036–1040 (2015)
30. Zhang, Y., Wallace, B.: A sensitivity analysis of (and practitioners' guide to) convolutional neural networks for sentence classification (2015). arXiv:1510.03820

Character-Level Dialect Identification in Arabic Using Long Short-Term Memory

Karim Sayadi[1,4(✉)], Mansour Hamidi[2], Marc Bui[1], Marcus Liwicki[2], and Andreas Fischer[2,3]

[1] CHArt Laboratory EA 4004, EPHE, PSL Research University, Paris, France
sayadi.karim@gmail.com, marc.bui@ephe.sorbonne.fr
[2] Department of Informatics, University of Fribourg, 1700 Fribourg, Switzerland
{mansour.hamidi,marcus.liwicki,andreas.fischer}@unifr.ch
[3] Institute for Complex Systems, University of Applied Sciences and Arts Western Switzerland, 1705 Fribourg, Switzerland
[4] OCTO Technology, Paris, France

Abstract. In this paper, we introduce a neural network based sequence learning approach for the task of Arabic dialect classification. Character models based on recurrent neural networks with Long Short-Term Memory (LSTM) are suggested to classify short texts, such as tweets, written in different Arabic dialects. The LSTM-based character models can handle long-term dependencies in character sequences and do not require a set of linguistic rules at word-level, which is especially useful for the rich morphology of the Arabic language and the lack of strict orthographic rules for dialects. On the Tunisian Election Twitter dataset, our system achieves a promising average accuracy of 92.2% for distinguishing Modern Standard Arabic from Tunisian dialect. On the Multidialectal Parallel Corpus of Arabic, the proposed character models can distinguish six classes, Modern Standard Arabic and five Arabic dialects, with an average accuracy of 63.4%. They clearly outperform a standard word-level approach based on statistical n-grams as well as several other existing systems.

1 Introduction

The Arabic language has different varieties that include one standard written form (*e.g.* official communications, newspapers, etc.), called the Modern Standard Arabic (MSA), and many spoken forms, each of which is a regional dialect. Arabic Dialect or the colloquial of the Arabic language is the result of a complex interaction between the Classical Arabic, ancient languages from ancient tribes and words that come from European languages in the colonization era.

In online communication, MSA and dialects are commonly used. As the Arabic world becomes more familiarized with social media (*e.g.* Facebook, Twitter, Instagram), recently people begin transcribing their respective Arabic dialects using Arabic letters on the keyboard instead of a mixture between numerical and Latin letters [1]. For example, it has been observed in the context of political

© Springer Nature Switzerland AG 2018
A. Gelbukh (Ed.): CICLing 2017, LNCS 10762, pp. 324–337, 2018.
https://doi.org/10.1007/978-3-319-77116-8_24

elections [2] that people are more willing to use their dialects written in Arabic letters on microblogging platforms like Twitter when they wanted to express important opinions and to have a broader impact on their community.

Dialect Identification refers to the task of automatically classifying text regarding its dialect. It is a challenging and an important task regarding many Natural Language Processing applications (*e.g.* Machine Translation, Sentiment Analysis, etc.). In this article, we suggest to tackle the problem of Arabic dialect identification with character models based on recurrent neural networks with Long Short-Term Memory (LSTM) [3,4]. Such character models have several promising properties. First, they are better suited than statistical n-gram models to handle long-term dependencies in sequential data, such as text.

Secondly, character-level modeling has the advantage that no knowledge about words, phrases, and sentences needs to be hard-coded into the dialect identification system, *e.g.* in form of stemming rules. Instead, the deep recurrent neural networks seem to learn such knowledge in form of syntactic and possibly semantic rules fully automatically [4]. This is especially useful for morphologically rich languages like Arabic and for dialects without strict orthographic rules, where it is difficult to hard-code knowledge at word-level. The main contributions of this paper are as follows:

- We introduce LSTM-based character models for the task of Arabic dialect identification. Text is classified based on the character perplexity.
- We provide a comprehensive experimental evaluation of the proposed approach on two publicly available datasets. One consists of MSA and Tunisian tweets and the other comprises texts in MSA and five Arabic dialects.
- We demonstrate that the suggested character models can outperform a standard word-level approach based on statistical n-grams and several other existing systems.

The remainder of this paper is organized as follows. Related work is discussed in Sect. 2. In Sect. 3, we present the proposed LSTM-based character models for dialect identification. In Sect. 4, we describe a word-level reference method used for experimental comparison. In Sect. 5, we provide details on the two datasets used in this study, *i.e.* the Tunisian Election Twitter dataset and the Multidialectal Parallel Corpus of Arabic, and present the experimental evaluation. Finally, we draw conclusions in Sect. 6.

2   Related Work

On the Dialect Identification task, several papers report different approaches with two main viewpoints on the issue: the first one consists of a split between Modern Standard Arabic and other dialects, *i.e.* a binary classification. The second is an identification of every dialect in addition to Modern Standard Arabic, which may be regarded as a fine-grained classification problem. Elfardy and Diab [5] presented a sentence-level Dialect Identification system to distinguish between MSA and Egyptian dialect. The authors used a tokenization system to extract features and trained a Naive Bayes classifier to obtain 85.5% accuracy.

[6] proposed an annotation system implemented within the Amazon's Mechanical Turk platform to identify the dialect within 100,000 sentences. The data was harvested from reader commentary on online newspaper to build a novel dataset. The authors analyzed the behavior of the Amazon's Mechanical Truk annotators and computed a human accuracy of 88%. They also trained a Dialect Identification system at word-level using n-gram models and have obtained an accuracy of 85.7% in distinguishing between MSA and dialects as two separate classes.

[7] studied the Arabic Dialect Identification at word-level and used a Support Vector Machine for classification. The authors presented the first set of experiments on the Multidialectal Parallel Corpus of Arabic (MPCA) proposed by [8] and have achieved an accuracy of 57.5% with word unigrams features on the six dialects in the MPCA dataset.

Recently, increased interest has arisen in character-level language modeling based on different kinds of Recurrent Neural Networks (RNNs) [9–11]. When modeling the language at character-level rather than word-level, the probability of a character following a sequence of previous characters can be computed. This probability can be used to generate new text, character by character, respecting the syntax and the semantics of the text corpora used for training, as shown for example by [4].

Language modeling at character-level entails many benefits like avoiding the exhaustive work of hard-coding knowledge about words, phrases, sentences or any other syntactic or semantic structures associated with a language [12]. Related works have shown that character-level RNNs can achieve state-of-the-art competitive results in language modeling [11], or even outperform word-based models on languages with rich morphology (Arabic, Czech, French, German, Spanish, Russian) [10]. However, there is no approach which can work for all kinds of datasets. To what extent a model performs in comparison depends on many factors, such as dataset size, alphabet's choice, and whether or not the texts were curated [9].

To the best of our knowledge, this paper is the first to explore the potential of LSTM-based character models for the task of Arabic dialect identification.

3   Text Classification with LSTM-Based Character Models

LSTM was proposed by [13] as a solution to the problem of explosion and vanishing of the convolved signal through RNN layers. LSTM architecture are composed of gates that retain information and thus act as memory units. LSTM combined to RNN is a robust method for sequence and contextual information modeling.

3.1   Our Approach

Our study of the Arabic dialect is based on a relatively new language model, which was thoroughly studied by [3,4]. The character-level language model consists of training an LSTM to predict, with a learned probability distribution, a sequence of characters given a sequence of previous characters. A trained LSTM would allow us to generate a new text, by sampling one character at each time step t. This output corresponds to the conditional distribution $p(x_t|x_{t-1}, \ldots, x_1)$ and can be obtained with a Softmax activation function.

$$p(x_{t,j} = 1|x_{t-1}, \cdots, x_1) = \frac{e^{w_j h_t^L}}{\sum_{j'=1}^{K} e^{w_{j'} h_t^L}} \tag{1}$$

For all possible symbols $j = 1, \ldots, K$. Where K is a fixed alphabet of characters, the input sequence of vectors $x_t, t = 1, \ldots, T$, is encoded by 1-of-K coding and given to the LSTM to obtain a sequence of D-dimensional hidden vectors $h_t^L, t = 1, \ldots, T$ at the output layer L. In Eq. 1, w_j are the rows of a $[K \times D]$ parameter matrix W. The used training criteria is a cross-entropy error which is equivalent to minimizing the negative log-likelihood of the sequence.

$$NLL = -\sum_{t=1}^{T} \log p(x_t|x_1, \cdots, x_{t-1}) \tag{2}$$

By combining these probabilities in Eq. 1, we can compute the probability of sequence x using

$$p(x) = \prod_{t=1}^{T} p(x_t|x_{t-1}, \ldots, x_1) \tag{3}$$

From the learnt distribution in Eq. 3, one can sample a new text or new sequence $x_t, t = 1, \ldots, T$ by iteratively sampling a character at each time step t. However, our goal does not consist in generating new text but rather consists in classifying short text messages into a category that corresponds to a dialect. In order to realize this task, first, we train a character-level LSTM network on each dataset. Second, we classify text with the perplexity approach.

An intuitive way to look at perplexity (PLL) is to think of it as a measure of how confused the language model is for a given sentence. The higher the perplexity is the lower the confidence of the model is, about generating the next character. The PLL of the character-level language model implemented with an LSTM network is computed over a sequence $[c_1, \ldots, c_T]$ and it is given by the following equation

$$PLL(m) = exp(\frac{NLL}{T}) \tag{4}$$

where the NLL (Negative Log Likelihood) is calculated over the test set and given by Eq. 2.

The classification task with the perplexity is carried out as follows. We divide each dialect text of different dialects into a separate dataset and train a network for each of the datasets. At testing time, we take a new unseen text m of a particular dialect and use all the trained networks to compute the perplexity of m. The network that computes the lowest perplexity, determines the class prediction of m. For example, if $PLL(m)_{tunisian} < PLL(m)_{jordanian} < PLL(m)_{syrian}$ then the network trained on Tunisian has computed the minimum perplexity and thus the message m is classified as Tunisian.

4   Reference Systems

To get some insight on the issue of dialect identification and in comparison with the character-level LSTM model, we study the effectiveness of different reference systems:

1. Word-level VS. character-level using a descriminative linear approach *i.e.* Support Vector Machine (SVM).
2. Descriminative approach using SVM VS. generative approach using traditional language model (LM) *i.e.* Bayesien estimation method both at word-level and at character-level.
3. Character-level, generative neural network model *i.e.* LSTM VS. Character-level traditional LM.
4. Word-level neural network model *i.e.* Convolutional Neural Network (CNN) over embedded word vectors VS. character-level neural network model *i.e.* LSTM.

We consider a statistical n-gram language model at word-level (we refer to the gram as a word) where the collection of texts or documents of size d is first encoded in a co-occurrence matrix $W_w \in \mathbb{R}^{d \times V}$. Each column in W_w is a vector that corresponds to the i-th word in the vocabulary V. Second, a probability distribution of the next word given a sequence of words is learned from the frequency counts provided by W_w. Finally, after representing the texts in a vector space model represented by W_w we extract the n-grams and train the classification algorithm. In order to improve the classification accuracy, we apply feature selection based on an Information Gain (IG) criterion [14] to find the most predictive n-grams. The ranked features with an IG above 0.0 are selected.

For the word-level generative neural network model, we consider a word embeddings representation where the collection of texts is encoded in a matrix where each row corresponds to one word. This matrix is given as input to the convolutional layers and it is usually refered to as the embedding layer [15]. Usually a pretrained word embeddings as word2vec [16] or GloVe [17] are used, but since we do not have a pretrained layer on dialected we initialized it from scratch and we learned the weights during training. The training is conducted with three convolutional layers, each of them is followed by a maxpooling layer. For the CNN experiments we set the dropout to 50%. The CNN network was implemented with the Tensorflow framework.

5   Experimental Evaluation

To evaluate the character-level language model approach for the task of Dialect Identification we chose to work on two different datasets. The first one contains two classes: the Modern Standard Arabic (MSA) and the Tunisian Dialect. The second contains 6 classes that include the MSA as well as 5 different Arabic dialects. The two classes of the former dataset are composed of 5778 unique sentences. The six classes of the latter dataset are composed of the same 1000 sentences which are translated for each dialect. The Dialect Identification task performed in the following experiments is divided into two subtasks: binary classification of sentences, and fine-grained classification over six classes: Modern Standard Arabic (MSA), Egyptian dialect (EGY), Syrian (SYR), Jordanian (JOR), Palestinian (PAL) and Tunisian (TUN).

Table 1. Example of annotated tweets from the tweets corpus. TUN stands for Tunisian Dialect and MSA stands for Modern Standard Arabic.

	Example	English Translation
TUN	بْرَافُو نُورَا.	Well done ! Noura.
	مَاينجمش يحكُم وحدو و لو حب.	He cannot govern alone even if he wanted to.
	كيفاش التوْانسة المقيمين بالخَارج ينتخبو ؟	How Tunisians living abroad vote ?
MSA	بعثة تصف الانتخَابَات التونسية بأنّها تعدُدية و شفَافة.	An official representation describes the elections as diverse and transparent.
	مرَاقبو الرِيَاسية يتذمرون من عدم وصول بطقَاتهم	Presidential election' watchers complain about not receiving their cards.
	التفَاصيل في مقَال الجزيرة	The details are in AlJazira' article.

5.1   Arabic Dialect Datasets

The Tunisian Election dataset [2] was collected from Tunisian users and crawled with the `Twitter Streaming API`. The tweets were published on Twitter's public message board between October 1st 2014 and December 23rd 2014. The first date is prior to the election of the 217 seat National Tunisian Assembly and the second date is posterior to the presidential elections. After separating the Arabic from non-Arabic text and further processing we obtained a total of 5778 tweets that we have manually annotated to obtain 3760 tweets written in MSA and 2018 tweets written in TUN. An example of the collected tweets is presented in Table 1.

We used the Multidialectal Parallel Corpus of Arabic (MPCA) released by [8]. The MPCA dataset is a collection of 1000 sentences in Modern Standard Arabic and translated into five regional Arabic dialects as well as in English by native speakers. The five regional dialects are Egyptian (EGY), Syrian (SYR), Jordanian (JOR), Palestinian (PAL) and Tunisian (TUN). Since identical sentences

Table 2. Example of a translation of the same sentence in different dialects. MSA stands for Modern Standard Arabic, EGY for Egyptian dialect, SYR for Syrian, JOR for Jordanian, PAL for Palestinian and TUN for Tunisian.

Dialect/Language	Example
English	Because you are a personality that I can not describe
MSA	لأنك شخصية لا أستطيع وصفها
EGY	لأنك شخصية و بجد مش هعرف أوصفها
SYR	لأنك شخصية و عنجد ما رح أعرف أوصفها
JOR	انت جد شخصية مستحيل اقدر اوصفه
PAL	عن جد ماشاء الله عليك شخصيتك ما بتنوصف
TUN	على خاطرك شخصية بلحق منجمش نوصفها

are transcribed for each dialect, the experimental comparison can be considered more balanced than the Tunisian Election dataset collected from Twitter. We divided the dataset into 80% used for a cross-validation and 20% as a holdout dataset for testing our system.

Table 3. The LSTM parameters that we optimized in the training process.

	# layers	LSTM-Size	Dropout
Net. #1	2	256	50%
Net. #2	2	512	50%
Net. #3	3	256	50%
Net. #4	3	512	50%

5.2 Experimental Setup

During the training process of LSTM, we optimized the following network parameters: The number of layers, the LSTM cells, and the dropout. We trained the LSTM networks with a base learning rate of 0.002 and a batch size of 50. We used RMSProp [18] for parameter adaptive update with 0.95 decay rate. We train each model for 50 epochs and start decaying the learning rate after 10 epochs by a factor of 0.97. We clip the gradient norm at the value of 5.

The parameter optimization process is described as follows. For the Tunisian Election Corpus, we provide an 8-fold cross-validation (CV) test, which consists of the 4k messages in the train set, divided into 3.5k CV-train set and 0.5k CV-test set for each fold. For the MPCA dataset, in order to find the optimal parameters for a successful training process, we perform a 6-fold cross-validation that is composed of our train set, which we divide into a CV-train set (650 messages from each class), and a CV-test set (130 messages from each class).

With the cross-validation, we want to figure out what parameters provide the best results regarding the classification of Arabic dialectal texts. Therefore, each fold performs 20 tests: 4 network parameter variations (see Table 3) *times* 5 different seeds (random initializations of the network weights). Preliminary tests have shown that following network parameter settings achieved the best results on the Arabic datasets: Layers 2/3; Nodes 256/512; Dropout 50%, as summarized in Table 3.

At word-level, we kept the same proportion of 20% for testing and 80% for training and optimization of the SVM with 8-fold cross-validation. We used an SVM with a linear kernel and we optimized the value of the complexity parameter C which controls the size of the hyperplane margin.

Our LSTM implementation is based on [4]'s Lua/Torch implementation. The implementation experiments were run on a configuration with 4 similar GPUs where each of them has a NVIDIA Corporation GM200 [GeForce GTX TITAN X] engine, and 3072 CUDA Cores with a VRAM of 12 GB GDDR5 and Memory Clock at 7.0 Gbps. We called this configuration `polyGPU`. We report some speed tests in the Appendix A.

Table 4. Cross-Validation results of the proposed character LSTM system and the reference word SVM system.

Tunisian election corpus						
	Net. #3	Net. #1	Net. #4	Net. #2	Mean accuracy	Variance
Net. #3		**67.7%**	79.3%	85.3%	**86.8%**	**6.2**
Net. #1	32.2%		61.3%	70.9%	86.5%	6.6
Net. #4	20.6%	38.6%		84.8%	85.9%	7.1
Net. #2	14.6%	29.0%	15.1%		85.8%	6.9
SVM					85.2%	13.6
Multidialectal parallel corpus of arabic						
	Net. #3	Net. #1	Net. #4	Net. #2	Mean accuracy	Variance
Net. #3		**63.9%**	99.5%	99.8%	**65.8%**	**3.2**
Net. #1	36.0%		96.1%	98.6%	65.2%	3.8
Net. #4	0.4%	3.8%		52.5%	64.0%	2.9
Net. #2	0.1%	1.3%	47.4%		63.6%	3.6
SVM					58.7%	24.1

5.3   Results

The results of the cross-validation for the Tunisian Election Corpus and the MPCA corpus are summarized in Table 4. The mean accuracy for the binary classification performed on the Tunisian Election Corpus is 86.8% with a variance of

6.2 and it is obtained by the parameter setting Layers 3, Nodes 256, Dropout 50%. The mean accuracy for the six-classes task performed on the MPCA corpus is 65.8% with a variance of 3.2 and it is obtained with the same parameter setting. Based on Welch's t-test, the optimal LSTM parameter setting Layers 3, Nodes 256, Dropout 50% outperforms the other settings with a minimum confidence of 67.7% for the Tunisian Election Corpus and with a minimum confidence of 63.9% for the MPCA corpus (highlighted in bold font in Table 4).

Table 5. Test results of the proposed character LSTM system.

Tunisian election corpus

#L	#N	#D	Seed. #1	Seed. #2	Seed. #3	Seed. #4	Seed. #5	Seed. #6	Seed. #7	Seed. #8	avg
3	256	50	93.0	92.0	91.4	91.0	93.8	91.4	91.2	93.6	92.2%

Multidialectal parallel corpus of arabic

#L	#N	#D	Seed. #1	Seed. #2	Seed. #3	Seed. #4	Seed. #5	Seed. #6	Seed. #7	Seed. #8	avg
3	256	50	64.2	64.2	65.3	62.4	63.0	63.2	62.3	62.9	63.4%

With these parameter values, we provide final performance results on the independent test set in Table 5. We trained 8 LSTM networks, each one with different seeds on 4000 (2000×2) text records for the Tunisian Election dataset and 4680 (780×6 dialects) text records for the MPCA. Afterwards, we test the trained networks on the test set made of 1200 records for the MPCA and 500 records for the Tunisian Election dataset. The average accuracy of the proposed LSTM-based character models for the two-classes task is 92.2% compared to 85% achieved by the SVM reference system. For the six-classes task the LSTM achieved an accuracy of 63.4% compared to 57% achieved by SVM (Fig. 1).

For a better understanding of results, we additionally provide heat maps, which illustrate the confusion matrix results with colors. While the proposed character-level approach clearly forms a colored diagonal line representing the correct classifications, the word-level approach has some trouble distinguishing between the Jordanian, Egyptian, Syrian and misclassifies them as Palestinian.

Table 6. Summary of the average test results of LSTM compared to different classification systems applied on Tunisian Election Corpus (TEC) and the Multidialectal Parallel Arabic Dataset (MPCA). The letter w stands for the word-level approach and the letter c stands for the character-level approach. The LM acronyms stands for Language Model that uses Bayesian estimation for the classification task. The CNNw stands for Convolutional Neural network applied to embedded words in a continuous vector space.

	LSTM	SVMw	SVMc	LMw	LMc	CNNw
TEC	92.2%	85%	74.5%	76.8%	71.7%	75.3%
MPCA	63.4%	57%	42.1%	51%	41.4%	44.3%

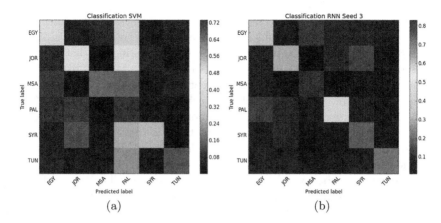

Fig. 1. (a) Confusion matrix of the SVM classifier trained on the 6 different dialect from the Multidialectal Parallel Arabic Dataset. MSA stands for Modern Standard Arabic, EGY for Egyptian dialect, SYR for Syrian, JOR for Jordanian, PAL for Palestinian and TUN for Tunisian. (b) Confusion matrix of the LSTM trained on the 6 different dialect from the Multidialectal Parallel Arabic Dataset.

We compare the results obtained by our approach to other approaches usually used by the developed identification systems in the state of art [5,7,19] as the traditional language model with Bayesian estimation or the discriminative linear approach with SVM. We also compare the results obtained by LSTM with CNN, another neural network where we used non-trained words embedding. We report in Table 6 the results obtained by all our experiments with 1-gram both at word and character level. We can clearly see the LSTM network outperform the other approaches.

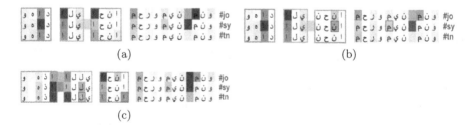

Fig. 2. Visualization of three networks outputs trained on Jordanian, Syrian and Tunisian. Arabic native speakers are invited to read from left to right.

5.4 Discussion

Previous works on Arabic Dialect Identification reveal that mainly n-gram based features at both word-level and character-level have been used for classification tasks. Distinguishing between MSA and an Arabic dialectal text, which represents a binary classification problem, has been tackled by several authors. [5] presented a sentence-level Naive Bayes classifier to distinguish between MSA and Egyptian dialect. Their supervised system which was built on using word n-gram features combined with token-based identification system, perplexity-based identification system, and meta-features. It achieved an accuracy of 85.5%. The underlying dataset was composed of 20k annotated user commentaries on news articles. A very comprehensive work on this topic is that of [6]. Since dialects do not differ equally from MSA, they provided binary classifications between MSA and Gulf, Levantine, and Egyptian dialect, using 1-gram, 2-gram, and 5-gram features at both word-level and character-level. The best performing system was the 1-gram word-level model, with an accuracy of 85.7%, followed by the 5-gram character-level model that achieved 85%. Even higher accuracies have been achieved by [19], who built a complex classification system to distinguish between MSA and Egyptian tweets. With their baseline experiments, which combines word-level with character-level n-gram features, they obtained an accuracy of 83.3% performed with a Random Forest classifier. When they additionally used a dialectal Egyptian lexicon and morphological features, the accuracy increased to over 90%. Unlike the results in the literature, they report that their character-based n-gram features significantly outperformed their word-based n-gram features.

Comparing our results with theirs, we see that our model performs slightly better (92.2% with LSTM-based character models). If we take additionally into account the fact that our model is far simpler than compared ones, we may consider our model as a progress for the task of Arabic Dialect Identification.

A related work that is based on the same dataset, is that of Malmasi et al. [7]. For the six-classes task, they trained a Support Vector Machine based on character-level and word-level n-gram features. The best achieved result in their 10-fold CV test was 65.3% when using character trigrams, and 57.5% when using word unigrams. They even could increase the accuracy to 74.3% when they used a meta-classifier performing a stacked generalization that combines all features - a method not known so far for this task. Finally, it is worthwhile noting that character n-grams outperformed word n-grams in their experiments. Related works on Arabic Dialect Identification can be found at VarDial2016[1] and VarDial2017[2]. For example, in the work of Adouane et al. [20], the authors used a dataset that they have collected and applied an SVM with 6 grams and lexicons on the Character-Level. The authors reported a macro-average F-score of 92.94% with binary classification setting each of the dialect to the other dialects. We recall that in our approach we do not use any Dialect Vocabulary neither a binary classification.

In order to have a closer look at the classification behavior of the perplexity method, we visualize the probabilities according to their occurrence likelihood. For a clear demonstration of a multi-way classification, we illustrate the

---

[1] http://ttg.uni-saarland.de/vardial2016/dsl2016.html.
[2] http://ttg.uni-saarland.de/vardial2017/sharedtask2017.html.

sentence *'and this is what we are deprived of'* translated in Jordanian, Syrian and Tunisian dialect. The colored large rectangles are those parts of the sentence, which distinguish the dialects from each other. The red small rectangles, on the other hand, designate the characters that have high probabilities and hence, probably led to the correct classification of the sentence. Considering the first illustration, Fig. 2, which is a Jordanian translation, the first row visualizes the output from the network trained on Jordanian messages, the second row visualizes the output from the network trained on Syrian messages and the third row on Tunisian messages. In this case it means coloring the probability of a letter based on the sequence of previous letters. We can observe in the first figure, which differs from the second figure by the green rectangle and from the third figure by the orange rectangle (also from the blue rectangle, but the LSTM does not make use of it), that the LSTM cells fire up with some distinctive letters (red rectangles), which in our understanding seem to be the reason that leads to the correct classification of the text. That same phenomenon is observable in the second figure, which differs from the first figure by the green rectangle, and from the third figure by the blue rectangle (also from the orange rectangle, but the LSTM does not make use of it). We can see that in the blue and green boxes the LSTM has some letters with higher probability than in the first and third row, which we highlighted with red rectangles. One can observe the same phenomenons in the third figure.

6   Conclusion

This work shows the potential of character-level language modeling with Long Short-Term Memory networks for such a complex task as Arabic Dialect Identification. We demonstrated the effectiveness of the character LSTM models by evaluating the architecture on two datasets. It outperformed a standard n-gram word language model and several other existing systems.

Future work includes the investigation of bidirectional LSTM character models [21] to improve the Dialect Identification performance. We also aim to further analyze the results to obtain a better understanding of the syntactic and possibly semantic rules that were learned by the neural network character models.

A   Speed Test

In order to get an idea about the speed of processing the text records (*e.g.* tweets) provided by the described environment based on Torch, we illustrate in the Fig. 3 the processing time per second and per tweet during the training phase.

We compared the polyGPU configuration with the macCPU configuration consists of a MacBook Air 4.2 CPU of a with a 1.7 GHz Intel Core i5 processor and 4GB of DDR3 RAM at 1333 MHz. and he monoGPU configuration consists of a single NVIDIA Corporation GK107 [GeForce GT 740] GPU deployed in a PC with CUDA Cores: 384 a VRAM: 2 GB DDR3 and Memory Clock at 1.8 Gbps.

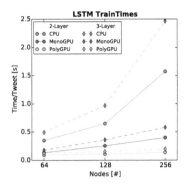

Fig. 3. Hardware comparison concerning training time

References

1. Younes, J., Achour, H., Souissi, E.: Constructing linguistic resources for the Tunisian dialect using textual user-generated contents on the social web. LNCS, pp. 3–14. Springer, Cham (2015). https://doi.org/10.1007/978-3-319-24800-4_1
2. Sayadi, K., Liwicki, M., Ingold, R., Bui, M.: Tunisian dialect and modern standard Arabic dataset for sentiment analysis: Tunisian election context. In: ACLing (2016)
3. Hermans, M., Schrauwen, B.: Training and analysing deep recurrent neural networks. In Burges, C., Bottou, L., Welling, M., Ghahramani, Z., Weinberger, K. (eds.) Advances in Neural Information Processing Systems, vol. 26, pp. 190–198 (2013)
4. Karpathy, A., Johnson, J., Fei-Fei, L.: Visualizing and understanding recurrent networks (2015). arXiv:1506.02078
5. Elfardy, H., Diab, M.T.: Sentence level dialect identification in Arabic. In: ACL (2013)
6. Zaidan, O.F., Callison-Burch, C.: Arabic dialect identification. Comput. Linguist. **40**, 171–202 (2014)
7. Malmasi, S., Refaee, E., Dras, M.: Arabic dialect identification using a parallel multidialectal corpus. In: Proceedings of the 14th Conference of the Pacific Association for Computational Linguistics (PACLING 2015), Bali, Indonesia (2015)
8. Bouamor, H., Habash, N., Oflazer, K.: A multidialectal parallel corpus of Arabic. In: LREC (2014)
9. Zhang, X., Zhao, J., LeCun, Y.: Character-level convolutional networks for text classification. In: Advances in Neural Information Processing Systems, pp. 649–657 (2015)
10. Kim, Y., Jernite, Y., Sontag, D., Rush, A.M.: Character-aware neural language models. In: Proceedings of the Thirtieth AAAI Conference on Artificial Intelligence. AAAI (2016)
11. Xiao, Y., Cho, K.: Efficient character-level document classification by combining convolution and recurrent layers (2016). arXiv:1602.00367
12. Zhang, X., LeCun, Y.: Text understanding from scratch (2015). arXiv:1502.01710
13. Hochreiter, S., Schmidhuber, J.: Long short-term memory. Neural Comput. **9**, 1735–1780 (1997)
14. Yang, Y., Pedersen, J.O.: A comparative study on feature selection in text categorization. In: ICML (1997)

15. Kim, Y.: Convolutional Neural Networks for Sentence Classification. In: EMNLP (2014)
16. Mikolov, T., Chen, K., Corrado, G., Dean, J.: Efficient estimation of word representations in vector space (2013). abs/1301.3781
17. Pennington, J., Socher, R., Manning, C.D.: Glove: global vectors for word representation. EMNLP **14**, 1532–1543 (2014)
18. Dauphin, Y.N., Vries, H.d., Bengio, Y.: Equilibrated adaptive learning rates for non-convex optimization. In: Proceedings of the 28th International Conference on Neural Information Processing Systems. NIPS'15, pp. 1504–1512. MIT Press, Cambridge, MA, USA (2015)
19. Darwish, K., Sajjad, H., Mubarak, H.: Verifiably effective Arabic dialect identification. In: EMNLP, pp. 1465–1468 (2014)
20. Adouane, W., Semmar, N., Johansson, R., Bobicev, V.: Automatic detection of arabicized berber and arabic varieties. VarDial **3**, 63 (2016)
21. Graves, A., Liwicki, M., Fernndez, S., Bertolami, R., Bunke, H., Schmidhuber, J.: A novel connectionist system for unconstrained handwriting recognition. IEEE Trans. Pattern Anal. Mach. Intell. **31**, 855–868 (2009)

A Text Semantic Similarity Approach for Arabic Paraphrase Detection

Adnen Mahmoud[1](✉), Ahmed Zrigui[2], and Mounir Zrigui[1](✉)

[1] LATICE Laboratory, University of Tunis, Tunis, Tunisia
Mahmoud.adnen@gmail.com, mounir.zrigui@fsm.rnu.tn
[2] INP Grenoble, Grenoble, France
Ahmed.zrigui@etu.univ-grenoble-alpes.fr

Abstract. The main challenge of paraphrase is how to detect the semantic relationship between the suspect text document and the source text document. Nowadays, the combination of Natural Language Processing NLP and deep learning based approaches have a booming in the field of text analysis, including: text classification, machine translation, text similarity detection, etc. In this context, we proposed a deep learning based method to detect Arabic paraphrase composed by the following phases: First, we started with a preprocessing phase by extracting the relevant information from text document. Then, word2vec algorithm was used to generate word vectors representation which they would be combined subsequently to generate a sentence vectors representation. Finally, we used a Convolutional Neural Network CNN to improve the ability to capture statistical regularities in the context of sentences which then makes it possible to facilitate the similarity measurement operation between the representations of source and suspicious sentences. The evaluation of our proposed approach gave us a promising result in term of precision.

Keywords: Plagiarism · Arabic paraphrase detection · Semantic analysis
Word2vec · Deep learning · Convolutional Neural Network CNN

1 Introduction

Plagiarism is an illegal quote from the idea, the invention, the writing, the methodology, or the design of someone. Thus, plagiarism takes place in various ways such as: copy and paste, disguised plagiarism, plagiarism by translation, shake and paste, structural plagiarism, mosaic plagiarism, metaphor plagiarism and idea plagiarism [1, 2]. However, it is often difficult to identify fragments of plagiarized text with their source [3] and particularly in the case of paraphrase which is defined as the restatement of text giving the meaning on another form [4]. In recent years, semantic text analysis is an essential problem in many Natural Language Processing NLP tasks which has drawn a considerable amount of attention by research community [5]. So, our work consists in paraphrase identification which is important for information extraction, machine translation and information retrieval trying to detect the semantic relatedness between the suspect and source text documents by combining Natural Language Processing NLP and deep learning based approaches to detect paraphrase in Arabic

© Springer Nature Switzerland AG 2018
A. Gelbukh (Ed.): CICLing 2017, LNCS 10762, pp. 338–349, 2018.
https://doi.org/10.1007/978-3-319-77116-8_25

texts: we generate word vectors representation using word2vec algorithm and subsequently apply a Convolutional Neural Networks CNN to improve the ability to capture statistical regularities in the context of sentences which then makes it possible to facilitate the similarity measurement operation between the representations of source and suspicious sentences. In this article, we start by present a state of the art in the field of Arabic paraphrase detection in Sect. 2. Thus, this section will be devoted primarily to the description of the specificities associated with Arabic language and the issues related in Arabic paraphrase detection, on the one hand; and on the works proposed in this field, on the other hand whose we will give a summary of existing works in the literature. Thereafter, to try to resolve the inherent problems in Arabic language, within the general framework of the Automatic Arabic paraphrase detection, we propose a deep learning approach for Arabic paraphrase detection in Sect. 3 which we will detail the different phases that make up our proposed approach. Finally, we present the evaluation of our proposed method in Sect. 4 as well as the result obtained and we end by a conclusion and some future works to realize in the field of Arabic paraphrase detection.

2 State of the Art

The plagiarism problem is still a challenge because of the significant technological revolution. However, it has been the biggest challenge in Arabic language. It is for this reason that several methods of detecting plagiarism have been proposed, such as: grammar based plagiarism detection, semantic based plagiarism detection, clustering based plagiarism detection, cross lingual plagiarism detection, citation based plagiarism detection, and character based plagiarism detection [2]. In this context, this section is devoted to presenting a state of the art on Arabic paraphrase detection by focusing on the specificities of the Arabic language, on the one hand, and by presenting an overview of related works in this field, on the other hand.

2.1 Arabic Language

The Arabic language is the language of the Koran and the sacred book of Muslims which is a Semitic language [6]. Thus, Arabic is the fifth most used language in the world and it is the mother tongue of over 200 million peoples and more than 450 million speakers [7]. However, Arabic is a difficult language which remains up leading to level of research and experimentation because of the great variability of morphological and typographical features of the Arabic script [8, 9]. Thus, Arabic language has many features, as follows:

- The Arabic script is written from right to left where the most letters are tied which offers Arabic script the characteristic of cursiveness [8]. Thus, Arabic language has 28 letters where three of them are vowels (such as: و ،ى ،ا) and others are consonants [1, 10].
- The diacritical marks such as diacritical dots and vowels which may be located above or below the form which they are associated. Thus, they make Arabic

language less ambiguous and more phonetic language, but the majority of Arabic texts aren't vowelized [7, 8], as shown in the following example: علم (Science) can be read in seven different ways, each having a distinct meaning; "alima", "olima", "allama", "ollima", "alamon", "eelmon" and "olim".[1]

- Arabic is highly inflected, derivational and agglutinative language which explains its sparness. Besides, its morphological and syntactic properties make hard to process when compared with other languages [6]. Thus, word can have more than one lexical category (noun, verb, adjective, etc.) in different contexts, which allows us to have different meanings of words [7].
- The root of every word in Arabic has three characters and a new word is formed by adding some suffixes (noun, verb, number…). Besides, persons and nouns have three forms (singular, dual, and plural) [1].
- The descendants (the legs) may extend horizontally below the base line which introduces descendants and the following letters, same for the ascendants (the stems) [8].

The detection of paraphrase in Arabic language based on the measurement of semantic relatedness between the suspect and source Arabic text documents is very difficult which there are limited techniques and tools in literature to detect Arabic paraphrase because of the following issues: The limited amount of annotated learning data [11] and the complexity of the nature linguistic of Arabic language [12] because of the variability of linguistic expression by the change of the word order or its meaning in the suspect sentence which causes an ambiguity during the semantic analysis between the suspect and source text documents whose a word can have more than one lexical category in different contexts which allows us to have different meanings of word what changes the meaning of sentence.

2.2   Related Works

This section provides an overview of related works that deal with Arabic paraphrase detection. Thus, among the paraphrase detection approaches existing in the literature based on semantic analysis to determine the relatedness between the suspect and source Arabic text documents, we distinguish:

In [13] it has been used Arabic Wordnet to detect sentences concepts with their synonyms, then, the medical ontologies have been used to expanding the sentences of original and suspected texts, and have been calculated the similarity between them. But, in [14] it has been shown a method for automated cross-language retrieval using Latent Semantic Indexing LSI for a new French-English collection with paraphrase detection. However, in [15] it has been analyzed a cross-lingual paraphrase based on a statistical bilingual dictionary to detect the semantic similarity. Also, in [3] it has been explored the suitability of three cross-language similarity estimation models: Cross-Language Alignment-based Similarity Analysis CL-ASA, Cross-Language Character n-Grams CL-CNG, and Translation plus Monolingual Analysis T + MA where T + MA produced the best results closely followed by CL-ASA and in [16] it has been proposed

---

[1] http://www.lexiophiles.com/english/what-is-special-about-the-arabic-language

an effective paradigm of document submission in e-learning system, plagiarism detection process and the integration of both to helping educating students about the importance of the originality by citing the original references. Its approach used Longest Common Subsequence LCS which based on the concept of similarity than distance. In [17] a Persian fuzzy paraphrase detection approach has been proposed based on the similarity among a set of synonym words which had the capacity to distinguish similar sentences. However, in [18] it has been proposed paraphrase detection tool for Arabic texts whose it built a content based method consisting mainly in fingerprinting the texts according to Arabic language specificity and comparing their logical representations by detecting synonym replacements to introduce heuristic algorithms to pass up redundant comparisons.

Many traditional unsupervised methods have been proposed to automatically infer word representations in the form of a vector to select more discriminative features such as Latent Semantic Analysis LSA which represents the meaning of words as a vector in multi-dimensional semantic space where each word has a unique vector representation and Latent Dirichlet Allocation LDA which is a probabilistic model can capture polysemy where each word have multiple meanings. We also find other traditional methods such as Vector Space Model VSM and Explicit Semantic Analysis ESA [19]. But nowadays, deep learning models have achieved better performance in Natural Language Processing NLP tasks where many works using deep learning methods have involved learning word vector representations through neural language models and performing composition over the learned word vectors for classification [20]. Thus, we will cite a few proposed approaches based on deep learning in other fields and on other languages such as the Latent or English languages: Distributed representations of words and sentences in a vector space help learning algorithms to achieve better performance in Natural Language Processing NLP by grouping similar words where the word representations using Neural Networks are very interesting because the learned vectors explicitly encode many linguistic regularities and patterns [8]. Thus, in [11, 19, 21], they have been computed word vectors representation of the sentence using word2vec or GloVe algorithms and in [19] it has been combined different word representations such as LSA, LDA and distributed vectors representation word2vec to complement the coverage of semantic aspects of a word and thus better represents the word than the individual representations. However, we use deep learning at the level of modeling the sentences and classification in order to facilitate the measurement of similarity. Originally, Convolutional Neural Networks CNN models are more useful in the literature, which they have shown to be effective for NLP tasks and have achieved excellent results in many tasks, such as: in [22, 23], they have been proposed models for text similarity; in [20], it has been proposed a model for sentence classification, in [24] it has been shown ranking in community question answering and in [25] it has been identified authorship. However, little works have been proposed in the field of Arabic paraphrase detection.

3   Proposed Approach

The detection of paraphrase based on the measurement of the relatedness similarity between source and suspected text documents is very successfully by using neural networks. In recent years, Convolutional Neural Network CNN has achieved great success in many computer visions tasks which is typically a feed forward architecture and has been characterized by local connections, shared weights among different locations, and local pooling which allows to discover local informative visual pattern with fewer adjustable parameter than Multi Layer Perceptron MLP [26], on the one hand; and CNN architecture is used to learn word embeddings and applied them of multiple Natural Language Processing NLP predictions [19], on the other hand, where CNN can learn more contextual information and represent the semantic of texts more

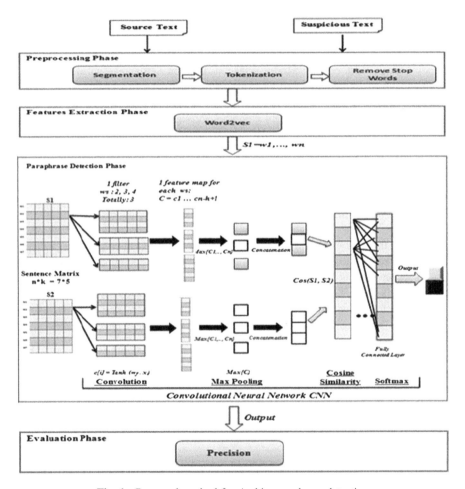

Fig. 1. Proposed method for Arabic paraphrase detection

precisely which can capture both features of n-grams [27] to extract discriminative word sequence what make CNN useful for dealing with long sentences [24]. In this context, we propose a deep learning based approach for Arabic paraphrase detection using distributed word vector representations word2vec and Convolutional Neural Network CNN composed by four phases as shown in Fig. 1 where we consider the embedding vectors of the source and suspected sentences as inputs in our CNN to modeling sentences passing by a convolution layer, a pooling layer and to determine thereafter the rate of paraphrase between them:

3.1 Preprocessing Phase

Our model begins with a phase of preprocessing to prepare text documents for subsequent processing composed by 3 steps as shown in Fig. 2: We start by a **text segmentation** phase which allows extracting the relevant information from the text. Thus, we segment the suspect document and source document into sentences. After, we proceed to the phase of **tokenization** which is the process of breaking up into words, sentences, symbols, or other meaningful elements called tokens [28]. Finally, **stop words removal** operation is applied to remove the words which aren't interest the further processing. Among the stop words in Arabic language, include for example: " في (fi: at), كل (kolla: all), لم (lam: did not), لن (lan: will not), هو (houwa: he)" [29].

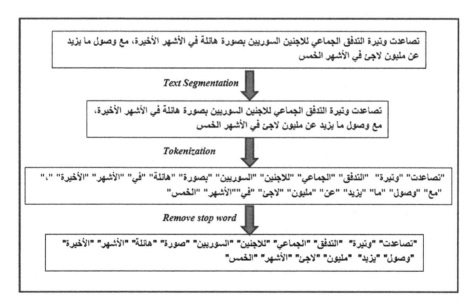

Fig. 2. An example of preprocessing steps of an Arabic text

3.2   Word Vectors Representation

In order to solve the problems inherent in the representation of words as unique or discrete identifiers, which leads to data sparsity in statistical models, we will use word vectors representation to overcome some of these obstacles such as word2vec^2 algorithm which consist of two neural network language models: Continuous Bag-Of-Words CBOW and Skip-Gram. In our work, we use the Skip-Gram model which is trained to predict the contexts of words and shows better performance in semantic analysis. Indeed, word2vec is a language modeling technique that maps words from vocabulary to continuous vectors where word embedding has been shown its ability to boost the performance in Natural Language Processing NLP and it has been trained with large datasets which it do not consume as much memory as some classic methods like Latent Semantic Analysis LSA [30, 31]. More formally, given a sentence S of length n (composed by n words), our model of sentences representation allows to extract k-dimensional word vectors representation $w_i \in R^K$ corresponding to the i-th word in the sentence S using word2vec algorithm which is an efficient model for learning word embeddings where each word is mapped into a single vector w_i represented by a column in a matrix W of size n * k. At the end of this step, a sentence will be represented by a sequence of word vectors representation (each $w_i = x_i$) as follows in Eq. 1:

$$W_{1:n} = w_1, w_2, \ldots, w_n \tag{1}$$

3.3   Convolutional Neural Network CNN

We use the word vectors representation of each sentence as inputs in our CNN which will be compared in order to determine the rate of paraphrase between source and suspected text documents. Thus, the representations of source and suspect text documents are used as entries in our CNN and processed in parallel through the following layers: Indeed, a convolution layer in order to generate new features, then, a max pooling layer to reduce their representations. Subsequently, the representations obtained are compared in a layer of similarity measure and will be transmitted later to a fully connected layer to obtain the paraphrase rate. Generally, our proposed CNN model based on a semantic analysis helps us to detect paraphrase in Arabic text documents. Given a sequence of words $(S \in R^{K*n})$ represented by a K-dimensional word embeddings, where: $x_i \in R^K$ represent the embedding of the word in the sequence and xi:j represents the concatenation of word embeddings. Our CNN model is composed by four steps as follows:

Convolution Layer. We generate a new feature S[i] from a convolution filter $w_f \in R^{j*k}$ (w_f is the weight vector of the filter) which is applied to a window (here we use different sliding window width w_s = 2, 3, 4) of j words of K-dimensional embeddings xi (the K-dimensional word vectors corresponding to the i-th word in the sentence) and a bias term $b_f \in R$ (b_f is a term of polarization) whose h_f is the activation function which is a nonlinear function (we use the Hyperbolic Tangent Tanh) of words

---

^{2} https://www.tensorflow.org/tutorials/word2vec/

$\{x_{1:j}, x_{2:j+1}, \ldots, x_{n-j+1:n}\}$ where $n - w_s + 1$ matches to the number of times the filter is moved in each sequence of words, as shown in the following Eq. 2:

$$S[i] = h_f(w_f \cdot x_{i:i+j-1} + b_f) \tag{2}$$

At the end of this step, a feature maps $\{S_1, S_2, \ldots, S_{n-j+1}\}$ will be produced as follows in Eq. 3:

$$C = [S_1, S_2, \ldots, S_{n-j+1}]; C \in R^{n-j+1} \tag{3}$$

Our model uses one filter with varying widow sizes w_s to obtain multiple features. Thus, when we increase the size of the sliding window, we will extract a longer n-gram in the input sentences.

Pooling Layer. The pooling operations sweep a rectangular window over the input vectors, computing a reduction operation for each window. Thus, we distinguish three types of pooling methods such as: average pooling, max pooling, or mean pooling.[3] In order to capture the most important feature and to reduce the representation, we apply a max pooling operation on the feature maps produced by the convolution layer to assume that the maximum value as a feature corresponding to this filter and to obtain thereafter two feature vectors that match the source and suspect sentences as shown in Eq. 4:

$$P_1 = \text{Max}\{C\} = \text{Max}\{S[i]\} \tag{4}$$

Where: i: $1, \ldots, n - j + 1$ and $1 : 1, \ldots$, number of sentences.

In our case, we will calculate the max pooling in each sentence. Afterwards, all the results obtained are concatenated to form a single feature vector for the penultimate layer as follows in Eq. 5:

$$P_{\text{vect}} = P_1, \ldots, P_n \tag{5}$$

Similarity Measurement Layer. After the feature extraction phase using word vectors representation and the phase of filtering or reducing the vectors obtained, we proceed to the similarity measurement operation between the source and suspected text documents whose sentences of suspect document is compared with all sentences of source document using as a metric of comparison the cosine similarity which had a good results in semantic analysis task [4] whose purpose is to apply a semantic comparison. More formally, given the sentences vector of suspect document S_1 and source document S_2, the semantic relation between S_1 and S_2, as follows:

$$\text{Cosine Similarity } (S_1, S_2) = \frac{S1.S2}{||S1||.||S2||}$$
$$= \frac{\sum_{i=1}^{k} S1i \, S2i}{\sqrt{\sum_{i=1}^{k} S1i^2}\sqrt{\sum_{i=1}^{k} S2i^2}} \tag{6}$$

---

[3] https://www.tensorflow.org/api_docs/python/nn/pooling

Where, K is the dimension of the vectors S_1 and S_2. At the end of this step, we obtain a vector which contains different scores of similarity according to the suspect text document size until reaching the source document size.

Fully Connected Layer. All the results of the operations we have done at each layer of our model (convolution layer, pooling layer, and similarity measurement layer) form the penultimate layer of our model and are subsequently transmitted to a fully connected layer by applying a Softmax function which convert the output score into probability as indicated in the following Eq. 7:

$$\text{Output} = \text{Softmax}(y_n) = p(y = j|x) = \frac{e^{xT\theta j}}{\sum_{k=1}^{K} e^{xT\theta k}} \tag{7}$$

Where: y is the predicted probability belonging the j-th class (where j corresponds to paraphrase class) and according to a threshold of paraphrase; x is a vector representing the different results of similarity which is a vector representing different results of similarity and Θ^K is the weight vector of the K-th class.

4 Experiments and Discussion

We used the Open Source Arabic Corpus OSAC[4] which contains about 22,429 text documents. Thus, the evaluation of our proposed approach is carried out on the category of History contains 3233 text documents which prove the relevance of our idea. The parameters we used and which made our approach efficient are: in addition, the word vectors representation using word2vec are checked in a matrix W of size n * k. In our case, we fixed k at 5 which represent the number of synonyms according to each word context. After, our CNN composed by a convolution layer which we varied window sizes ($w_s = 2, 3, 4$) to obtain multiple features and a pooling layer to calculate the maximum pooling of each sentence. Then, two sentences are considered as paraphrase, if they pass the cosine similarity threshold (α) whose the threshold was fine-tuned by several trials on the training corpus and the results achieved when $\alpha = 0.3$. The evaluation measures used on this text alignment task include: Precision and Recall.

Our proposed approach based on deep learning obtained a promising result where the detection rate of paraphrase obtained was 0.88 in term of precision in relation to other existing systems in the literature such as the system of Alzahrani [32] proposed a word correlation in N-Grams with K-overlapping approach which obtained a rate of 0.83 in term of precision and the system of Youssef et al. [12] which adapted one of famous technique that have been developed for English such as the TF × IDF using statement-based document representation to identify paraphrase in Arabic language with a rate of 0.85 in term of precision as shown in Table 1.

---

[4] http://www.academia.edu/2424592/OSAC_Open_Source_Arabic_Corpora

Table 1. Evaluation of our proposed method for paraphrase detection

Proposed approaches	Precision	Recall
Alzahrani [32]: N-Grams	0.83	0.53
Youssef et al. [12]: TF-IDF	0.85	0.88
Our approach: Word2vec+CNN	0.88	0.89

5 Conclusion

We proposed a deep learning approach for external paraphrase detection for Arabic texts. Thus, we used word2vec algorithm which is an efficient model for learning word embeddings from raw text, then, we proposed a Convolutional Neural Network CNN model to facilitate the similarity calculation between source and suspect texts. Finally, our proposed approach was evaluated on the Open Source Arabic Corpus OSAC and obtained a promising rate. Despite the promising results that we have obtained by using our proposed method, several improvements will be applied in our method later on, such as the use of recursive and recurrent connections within every convolutional layer of our CNN to improve the capability to capture statistical regularities in the context of sentences, on the one hand, and we will combine word representations for measuring word similarity such as Latent Semantic Analysis LSA, Latent Dirichlet Allocation LDA and distributed representation of words word2vec to improve the similarity measure and to improve the weakness of each method.

References

1. Abakush I.: Methods and tools for plagiarism detection in Arabic documents. In: International Scientific Conference on ICT and E-Business Related Research SINTEZA, Serbia (2016)
2. Ramesh, N.R., Landge, M.B., Namrata, M.C.: A review on plagiarism detection tools. Int. J. Comput. Appl. IJCA **125**(11), 16–22 (2015)
3. Cedeno, A.B., Gupta, P., Rosso, P.: Methods for Cross-Language Plagiarism Detection, vol. 50, pp. 211–217. Elsevier, Amsterdam (2013)
4. Samuel, F., Mark, S.: A semantic similarity approach to paraphrase detection. In: Proceedings of the Computational Linguistics UK CLUK, UK (2008)
5. Liu, Y., Sun, C., Lin, L., Zhao, Y., Wang, X.: Computing semantic text similarity using rich features. In: The 29th Pacific Asia Conference on Language, Information and Computing, PACLIC29, China (2015)
6. Ben Mohamed, M.A., Mallat, S., Nahdi, M.A., Zrigui, M.: Exploring the potential of schemes in building NLP tools for Arabic language. Int. Arab. J. Inf. Technol. IAJIT **6**(12), 13–19 (2015)
7. Zrigui, S., Zouaghi, A., Ayadi, R., Zrigui, S., Zrigui, M.: ISAO: an intelligent system of opinion analysis. Res. Comput. Sci. **110**, 21–30 (2016)
8. Meddeb, O., Maraoui, M., Aljawarneh, S.: Hybrid modeling of an offline Arabic handwriting recognition system AHRS. In: International Conference on Engineering & MIS, Maroc (2016)

9. Zouaghi, A., Zrigui, M., Antoniadis, G.: Compréhension automatique de la parole arabe spontanée. In: Traitement Automatique des Langues, Belgique (2008)
10. Saidan, T., Zrigui, M., Ahmed, M.B.: La transcription orthographique-phonetique de la langue arabe. In: RÉCITAL, Maroc (2004)
11. Le, Q., Mikolov, T.: Distributed representations of sentences and documents. In: The 31th International Conference on Machine Learning JMLR, vol. 32, pp. 1188–1196 (2014)
12. Ameer, A., Youssef, A., JuzaiddinAb, A.M.: Enhanced TF-IDF weighting scheme for plagiarism detection model for Arabic language. Aust. J. Basic Appl. Sci. AEDSI 9, **23**, 90–96 (2015)
13. Omar, K., AlKhatib, B., Dashash, M.: Plagiarism detection in Arabic using translation man medical ontology. Int. J. Curr. Med. Pharm. Res. IJCMPR **2**(9), 648–653 (2016)
14. Dumais, S.T., Letsche, T.A., Littman, M.L., Landaver, T.K.: Automatic cross-language retrieval using latent semantic indexing. In: Spring Symposium Series, Standford (1997)
15. Barron-Cedeno, A., Rosso, P., Pinto, D., Juan, A.: A cross lingual plagiarism analysis using a statistical model. In: PAN, India (2008)
16. Farhat, F., Asen, A.S., Zaher, M.A., Fahiem, A.M.: Detection plagiarism in Arabic E-learning using text mining. Britsh J. Math. Comput. Sci. BJMC **8**(4), 298–308 (2015)
17. Rakian, S., Esfahani, F.S., Rastegari, H.: A Persian fuzzy plagiarism detection approach. J. Inf. Syst. Telecommun. JIST **3**(3), 182–190 (2015)
18. Menai, M.E., Bagais, M.: A plag: a plagiarism checker for Arabic texts. In: International Conference on Computer Science & Education (ICCSE), Singapore (2011)
19. Niraula, N.B., Gantam, D., Banjadae, R., Mahayan, N., Rus, V.: Combining word representations for measuring word relatedness and similarity. In: Twenty-Eighth International Florida Artificial Intelligence Research Society Conference, Florida (2015)
20. Kin, Y.: Convolutional neural networks for sentence classification. In: Proceedings of the 2014 Conference on Empirical Methods in Natural Language Processing (EMNLP), Qatar (2014)
21. Mikolov, T., Sutskever, I., Chen, K., Corrado, G., Dean, J.: Distributed representations of words and phrases and their compositionality. In: Neural Information Processing Systems NIPS, USA (2013)
22. He, H., Gimpel, K., Lin, J.: Multi-perspective sentence similarity modeling with convolutional neural networks. In: Empirical Methods in Natural Language Processing EMNLP, Portugal (2015)
23. He, H., Wieting, J., Gimpel, K., Rao, J., Lin, J.: Attention-based multi-perspective convolutional neural networks for textual similarity measurement. In: International Workshop on Semantic Evaluation SemEval, California (2016)
24. Mohtarami, M., et al.: Neural-based approaches for ranking in community question answering. In: International Workshop on Semantic Evaluation SemEval, California (2016)
25. Zhou, L., Wang, H.: News authorship identification with deep learning. In: Conference and Labs of the Evaluation Forum, Portugal (2016)
26. Liang, M., Hu, X.: Recurrent convolutional neural network for object recognition. In: CVPR 2015, Boston (2015)
27. Lai, S., Xu, L., Liu, X., Zhao, J.: Reccurent convolutional neural networks for text classification. In: AAAI 2015 Proceedings of the Twenty-Ninth AAAI Conference on Artificial Intelligence, Texas (2015)
28. Alaa, Z., Tiun, S., Abdulameer, M.: Cross-language plagiarism of Arabi-English documents using linear logistic regression. J. Theor. Appl. Inf. Technol. **1**(83), 20–33 (2016)
29. Kahloula, B., Berri, J.: Plagiarism detection in Arabic documents: approaches, architecture and systems. J. Digit. Inf. Manag. **14**(2), 124–135 (2016)

30. Liu, Y., Sun, C., Lin, L., Zhao, Y., Wang, X.: Computing semantic text similarity using rich features. In: The 29th Pacific Asia Conference on Language, Information and Computation PACLIC29, Shanghai (2015)
31. Altszyler, E., Sigman, M., Selzak, D.F.: Comparative study of LSA vs Word2vec embeddings in small corpora: a case study in dreams database. In: Super Corr Expo Orlando, USA (2016)
32. Alzahrani, S.: Arabic plagiarism detection using word correlation in N-Grams with K-overlapping approach. In: Working Notes for PAN-ArabPlagDet at FIRE, Gandhinagar (2015)

Social Network Analysis

Curator: Enhancing Micro-Blogs Ranking by Exploiting User's Context

Hicham G. Elmongui[1,2(⊠)] and Riham Mansour[3]

[1] Computer and Systems Engineering, Alexandria University,
Alexandria 21544, Egypt
elmongui@alexu.edu.eg
[2] Umm Al-Qura University, Science and Technology Unit,
Makkah 21955, Saudi Arabia
hgmongui@uqu.edu.sa
[3] Microsoft Research Advanced Technology Lab, Cairo 11728, Egypt
rihamma@microsoft.com

Abstract. Micro-blogging services have emerged as a powerful, real-time, way to disseminate information on the web. A small fraction of the colossal volume of posts overall are relevant. We propose Curator, a micro-blogging recommendation system that ranks micro-blogs appearing on a user's timeline according to her context. Curator learns user's time variant preferences from the text of the micro-blogs the user interacts with. Furthermore, Curator infers the user's home location and the micro-blog's subject location with the help of textual features. Precisely, we analyze the user's context dynamically from the micro-blogs and rank them accordingly by using a set of machine learning and natural language processing techniques. Curators extensive performance evaluation on a publicly available dataset show that it outperforms the competitive state-of-the-art by up to 154% on NDCG@5 and 105% on NDCG@25. The results also show that location is a salient feature in Curator.

Keywords: Micro-blogs recommendation · User's context

1 Introduction

Micro-blogging services, e.g., Twitter, have emerged as a powerful real-time means of disseminating information on the web. As of January 2017, there are more than 695 M Twitter users; 342 M of them are active users posting on the average 518 M tweets every day [47]. The high volume of tweets received by the active users is continuously increasing and is reducing productivity. About 73% of companies across the United States with 100 or more employees either completely prohibited visiting social networking sites or permitted for business purposes only [8]. With 82% of the users are active on the mobile devices [48], the effect of keeping oneself "busy" skimming through the micro-blogs is becoming apparent. With many of the micro-blogs being redundant or not of interest

© Springer Nature Switzerland AG 2018
A. Gelbukh (Ed.): CICLing 2017, LNCS 10762, pp. 353–365, 2018.
https://doi.org/10.1007/978-3-319-77116-8_26

to the user, the need for ranking the micro-blogs is obvious so as to be able to show her the more relevant ones first on her timeline.

In this paper, we propose Curator, a micro-blogging recommendation system that ranks the micro-blogs by exploiting the user's context. Context is defined as "any information that can be used to characterize the situation of an entity. An entity is a person, place, or object that is considered relevant to the interaction between a user and an application, including the user and applications themselves" [2]. Main components of a user's context are her identity and her location. The former is directly reflected by her preferences, which we infer from the language used in her micro-blogs. The latter may represent the current location from which she reads or writes a micro-blog, the subject location about which she authors, or her home location which affects her culture and personality. In addition to other techniques, we use natural language techniques to infer the subject location and home location of the user. Time is an inherent component of a user's context. It reflects the evolving nature of the other context components.

Building micro-blogging recommendation systems is non-trivial. First, It needs to deal with a large, and consistently increasing, corpus of micro-blogs. Second, micro-blogs themselves lack context as they are short; users are limited to a maximum of 140 characters to post in any tweet on Twitter. Third, scarcity of author's location information is another challenge. A small percentage of micro-blogs are associated with location information for privacy purposes [39]. Fourth, with the dynamic property of real life, context changes over time, and needs to be maintained for each user.

The contributions of this paper can be summarized as follows:

- We propose Curator, a micro-blogging recommendation system that ranks the micro-blogs according to the progressing user's context.
- Curator continuously captures the user's preferences by looking at the micro-blog text and the user interaction (forwardings, replies, and likes).
- Curator infers the user's home location and the micro-blog's subject location through natural language processing on the text of the tweets.
- We perform an extensive performance evaluation of Curator on a publicly available dataset. Experimental results show that Curator outperforms the competitive state-of-the-art micro-blogging recommendation systems.

The rest of the paper is organized as follows. Section 2 summarizes the related work. Curator's details are described in Sect. 3. In Sect. 4, we evaluate Curator through a meticulous performance study. We conclude the paper in Sect. 5.

2   Related Work

The related work to Curator is two folds: micro-blogs recommendation systems and location inference techniques for micro-blogs users.

2.1   Micro-blogs Recommendation

Many systems have been propositioned as micro-blogs recommendation systems that pick which micro-blogs to show to the user. Different micro-blogs features were adopted in the recommendation; from re-tweet (i.e., forwarding) behavior as a measure of the user's interest in a tweet [15,49] to content relevance, account authority, and tweet-specific features that were used in learning-to-rank algorithm, which ranks the tweets [11].

The challenge in the personalized recommendation of micro-blogs is to learn the preference of the user. The basic solution asked the user to specify her static topics of interests [40] or to mark her tweets with pre-defined interest labels [18]. Next, this static preference was captured without user intervention either using collaborative ranking [6] or using a graph-theoretic model [53]. Nevertheless, the user interest was represented using Latent Dirichlet Allocation (LDA) [4], which is not scalable for real-time streams of micro-blogs [38].

The user's preferences naturally changes over time. This temporal dynamic property was lately accounted for in few personalized tweet recommendation systems. In [28,29], LDA was used for topic modeling and a binary "important" label is predicted for each tweet. A ranking classification of tweets is proposed in [13], which models the tweet topic detection also as a classification problem.

In contrast to all the previous work that use the dynamic user's preferences as the sole feature in the recommendation, Curator uses the dynamic user's preferences as one feature in addition to the other context features of the user. In fact, the home location of the user turns out to be a salient feature in the recommendation process as shows the thorough evaluation of Curator.

2.2   Micro-blogger's Location Prediction

Research efforts trying to infer the location of the micro-blogger can be categorized into graph-based, content-based, and hybrid techniques.

The graph-based techniques use the social graph, which connects each user with its followers and followees. The user's location was inferred from her friends' by looking at the social tie and the distance between the pairs [9,37,41], by combining weak predictors [43], or by majority voting [26]. Furthermore, the home location is inferred from landmark users who report their true locations [52] using spatial location propagation technique [14,27].

The content-based techniques get signals solely from the text of the microblogs. Signals include point of interests [32,42], local words [42], location indicative words [20], or latent topics [7] to infer the home location [5], or to infer the tweet source location [23]. Besides, statistical methods are used to infer the user current location as well as her home location [12,22,30,35]. An extensive feature selection comparison for location inference may be found in [21].

The hybrid approaches utilize both the social graph as well as the content of the micro-blog to predict the home location and visited locations of the user [14,17,33,34]. Such approaches receive added signals from both sources and therefore have improved performance over other techniques. In this work, we adopt the Injected Inferences model [14] as a building block in Curator.

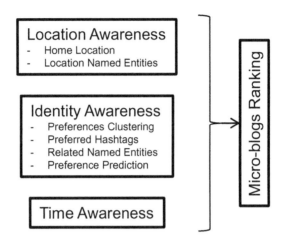

Fig. 1. Exploiting context in ranking

3 Curator: Micro-blogs Recommendation System

Curator is a context-aware micro-blogs recommendation system. When it ranks the micro-blogs on a timeline, it takes into account the context of its user. Therefore, it needs to be aware the identity, location, and time of the user as it appears in Fig. 1. In the rest of this section, we start with the pre-processing step and the feature extraction that is done on any micro-blog prior to describing how the three context components are captured by getting signals from the micro-blogs of the user and from her interaction. Next, we show how they are incorporated in the ranking model.

3.1 Micro-blogs Textual Pre-processing and Feature Extraction

Micro-blogs are to be pre-processed in Curator. This pre-processing is needed to prepare the data for the extraction of the features used in the subsequent sections. First, the text of the micro-blog is tokenized, which removes all punctuation and other white spaces. A standard list of stop words is to be used. All URLs are also removed. Tokens containing special characters are also removed except for those starting with a hash sign, '#', which denote hashtags (e.g., #cooking). Hashtags will play a role in the classification of the user's preferences are will be described later.

Micro-blogs by definition are short and lack context. Short micro-blogs make the problem worse as they do not carry enough information. Curator discards one-word-token micro-blogs.

Micro-bloggers tend to emphasize some words by repeating some letters in those words. For instance, to enthusiastically agree, one may say "yesss" instead of "yes". The #coooold shows the strong feeling of the weather being cold. For words containing excessively repeated letters (three or more occurrences), we just keep two occurrences and drop the others. Next, we use a spell checker, (e.g., GNU Aspell [16]) to detect out-of-vocabulary tokens and replace them with the best suggested replacement according to based on lexical and phonemic distance. Some out-of-vocabulary words are in fact slang. We use a slang dictionary to get their lexical meaning and use it as a substitute [25].

Named entities are to be extracted from the micro-blog text. We use a named entity recognizer to extract them [45]. Extracted named entities include, but are not limited to, locations, which will be used in Curator's location awareness (discussed next). Other named entity types will be used in Curator's identity awareness (detailed subsequently).

The last step in the pre-processing phase is representing the micro-blog tokens in a suitable representation for the machine learning techniques of Curator. We use term frequency-inverse document frequency (TF-IDF), which is a numerical statistic that reflect how important a word is to a document in a corpus [44]. Similar to the competitor state-of-the-art [13], the weights of the hashtags and named entities are doubled since micro-blogs with hashtags get two times more engagement [24].

3.2 Location Awareness in Curator

The location context of a micro-blogger is either the current location from which she reads or writes a micro-blog, the subject location about which she authors, or her home location which affects her culture and personality. These locations may or may not be the same. For instance, a French user may be traveling to India, but is micro-blogging about Wimbledon tournament in London, UK. A Londoner may be micro-blogging about the same event from his home.

The subject location of a micro-blog is inferred from textual signals in the micro-blog. In Curator, a location named entity recognizer is used to capture such signals. Upon detection, this subject location is fed into the identity awareness component as a signal of the micro-blog to be used to detect whether this location is preferred by the user.

The current location is either reported by the user's device, upon her permission, or is detected by the micro-blogging service. Only a small fraction of the users prefer to reveal their current location. However, the proposed ranking mechanism does not dependent on the current location by itself. If the user is interested about micro-blogs related to her current location, a micro-blog's subject location would be equal to the user's current location, and this subject location is already accounted for in Curator.

The home location of a user is either reported by the user on her profile, usually as a toponym, or may be predicted from the user's micro-blogs, her behavior on the micro-blogging service, or her friends. Curator infers the home location of the user by injecting the output of the Friends classifier described in [14] as an additional feature in the state-of-the-art content-based home location identification machine learning algorithm [35]. This home location is used as a feature in the proposed ranking model as will be shown later in this section.

3.3 Identity Awareness in Curator

The identity context the user is reflected by her preferences. Curator learns the user's preferences from her engagement on the micro-blogging service. If a micro-blog is replied to, forwarded, or liked by the user, it is a signal that the subject of the micro-blog lies within her preferred topics. Curator models the problem of predicting one's preferences by clustering the micro-blogs according to the topic preferences, classifying each cluster, and then detecting which cluster is closer to the micro-blogs that the user has engagement most.

The clustering phase is important to increase the context content of the micro-blogs' text that share the same topic. We use an online incremental clustering algorithm [3] on a corpus of micro-blogs. The resultant clusters have the properties that the micro-blogs of a cluster have larger cosine similarity among themselves [36], and hence share the same topic preference.

The classification phase labels each cluster with its topic by applying a set of topic-based binary SVM classifiers, hashtags classifiers, and named entities classifiers. The SVM classifiers are trained using predefined lists of keywords that are indicative of each adopted topic. The keyword lists are retrieved from web directories that are categorized by subjects. As an example, the list of *Food* retrieved from the Open Directory Project contains *drink, cheese,* and *meat* [10].

During the classification, a micro-blog may not fall in any of the existing clusters, and therefore cannot be labeled using the aforementioned SVM classifiers. For such micro-blogs, the hashtag classifiers are used to predict the topic of the micro-blog. If the micro-blog does not contain any indicative hashtags, the named entity classifiers are used for the topic prediction.

The hashtag classifier is built from the corpus used to create the clusters. Each of these hashtags are assigned a score that reflects how confident we are that the hashtag is related to the topic assigned to that cluster. Let $\mathrm{conf}(m)$ denote the SVM confidence score of the topic predicted for a micro-blog m. Let $\mathrm{tpcs}(h)$ denote the set of topics assigned of the clusters in which a hashtag h appears. Therefore, for each topic, t, each hashtag gets a score, $S(h|t)$.

$$S(h|t) = \frac{\sum\limits_{\substack{m \in t \\ h \in m}} \mathrm{conf}(m)}{|\mathrm{tpcs}(h)| + \sum\limits_{h \in m} \mathrm{conf}(m)} \tag{1}$$

where $m \in t$ denote that micro-blog m is assigned to a cluster that is labeled with topic t. From the above equation, a hashtag gets a high value when a big fraction of its micro-blogs belong to a certain topic. The number of topics in which a hashtag appears, $|\mathrm{tpcs}(h)|$, distinguishes between the heavily-used and lightly-used hashtags when such hashtags appear in a single topic as it prevents $S(h|t)$ from being 1. We would like to note that Eq. 1 looks similar but not exact to Eq. 1 in [13].

The topic with the highest score is assigned to that hashtag as shown in Eq. 2. A micro-blog is assigned to the topic of a contained hashtag if that hashtag receives a topic score above a certain threshold, $\mathbb{S} = 0.7$. We call this hashtag an indicative hashtag.

$$T(h) = \arg\max_t S(h|t) \tag{2}$$

The named entities classifiers are used when a micro-blog does not fall in any cluster and does not contain any indicative hashtag. A named entities classifier predicts the topic of a micro-blog if it contains a named entity. The different resources, i.e., canonical named entities, of Wikipedia [50] are retrieved along with their types from DBpedia [31]. An example resource type is *Musical Artist*. We project the types of the resources on the micro-blogs clusters and assign each resource type the same topic of preference of the corresponding cluster. Transitively, names entities of a certain resource type are assigned its assigned topic of preference. Also, synonyms to named entities are assigned their topic of preferences. Synonyms of canonical named entities are retrieved using WikiSynonyms service [51]. Examples of Synonyms of Elizabeth II are Queen Elizabeth II, Elizabeth II of England, and Her Majesty Queen Elizabeth II.

3.4   Time Awareness in Curator

Curator is aware of the current clock. Rankings of micro-blogs change over the time as the context itself changes over the time. The subject location changes with time as users move and talk about different places. This location variation is already accounted for as this subject location is detected separately for each arriving micro-blog in real time.

The user preferences also may change with time as situations progress. A user may be interested in micro-blogs about sports when a major tournament takes place, and then she gets interested in travel when she is arranging for an annual vacation. This is why Curator accounts for an adaptive preference detection.

The preference of a user is computed from the micro-blogs with which she engages. These contain the micro-blogs she liked, forwarded, or replied to. We denote such micro-blogs for a certain day, d, as M_d. The computation uses a $\mathrm{conf}(m)$ function, which gives Curator's confidence in its prediction of the topic t of a micro-blog m. For micro-blogs that fall in any cluster and hence take its topic, this function returns the SVM confidence of the classifier corresponding to the assigned topic. The function returns 1 if the predicted topic was using the hashtag or named entities classifiers. Otherwise, $\mathrm{conf}(m) = 0$.

Equations 3–5 give the computation for a certain user. A daily topic preference, $\mathrm{Pref}_d(t)$, is computed from that topic's micro-blogs with which that user has engaged on her timeline. A moving average on this daily topic preference is computed with a weekly window to produce the recent topic preference, $\mathrm{Pref}(t)$. The user's preference in a micro-blog is computed by multiplying the confidence in predicting its topic with that topic's recent preference as shown in Eq. 5.

The moving average definition of the topic preference enables its computation incrementally. Each day, it is updated by including a new day and removing the oldest day in the window. It is computed once a day for each topic for each user.

$$\text{Pref}_d(t) = \sum_{\substack{m \in M_d \\ m \in t}} \text{conf}(m) \tag{3}$$

$$\text{Pref}(t) = \text{MovingAverage}\big(\text{Pref}_d(t)\big) \tag{4}$$

$$\text{Pref}(m) = \text{Pref}(t) * \text{conf}(m) \qquad\qquad , \text{ where } m \text{ is of topic } t \tag{5}$$

3.5 Curator's Context Aware Micro-blogs Ranking

Curator uses a variation of the learning-to-rank model of RankSVM to rank the micro-blogs [11]. For a micro-blog m written by author a and appearing on the timeline of user u, Curator uses the following features:

- The home location of user u predicted as shown in Sect. 3.2.
- The micro-blog subject location as shown in Sect. 3.2.
- The user's adaptive topic preferences computed as described in Sect. 3.4.
- The number of forwardings and likes of that micro-blog.
- The number of the author's followers, followees, and micro-blogs.
- The number of hashtags in a micro-blog.
- Was u mentioned in the micro-blog.
- Does the micro-blog contain a hashtag that u used last week.
- The number of times u mentioned, liked, or replied to a's micro-blogs.
- The number of common users both of a and u follow.
- The number of days since the last time a and u interacted together.

RankSVM, and consequently Curator, learns the ranking function as well as the weights of the used features. The micro-blogs are shown on the user's timeline according to the learned ranking score.

4 Experimental Evaluation

We performed extensive performance evaluation of Curator against the state of the art. The machine learning algorithms were run through the WEKA suite [19]. We used a public Twitter dataset, which was used in [13,14,34] and is publicly available at [1]. This dataset contains 50 M tweets for 3 M users who have 284 M following relationships. To reproduce the results of the competitor algorithm, TRUPI, we used the same sampling algorithm as in [13], which produced 10M tweets for 20 K users who have 9.1 million following relationships. We also downloaded the user engagements from Twitter using its REST API [46].

As evaluation metrics, we use the micro-averaged F-measure (F1) and the normalized discounted cumulative gain (NDCG@k) and Mean Average Precision (MAP) for the ranked micro-blogs [36].

4.1 Evaluation of Binary Micro-blog Filtering

The binary filtering of micro-blogs refers to predicting whether or not the micro-blog is important to the user and will receive engagement from her through a reply, a like, or a forwarding [28].

The features used for this binary filtering are the same used in Sect. 3.5. The competitive baselines are the state-of-the-art binary recommendation systems that adopt a dynamic preference of the user, namely DynLDALOI and TRUPI. The major difference in both baselines is that the former uses LDA to detect the topic of interest of the user. For fairness, We compared against the J48 classifier of DynLDALOI, which gives better performance for it as shown in [28].

Table 1 shows the 10-fold cross validation for the binary micro-blog filtering. Being context-aware, Curator outperforms DynLDALOI with a relative gain of 11.3% in the micro-averaged F measure (F1). It also outperforms TRUPI with a relative gain of 6.8% on the same metric.

Table 1. 10-fold cross validation for binary micro-blog filtering

Technique	Precision	Recall	F1
DynLDALOI	74.2%	88.6%	80.7%
TRUPI	85.7%	82.7%	84.1%
Curator	93.7%	86.4%	89.9%

4.2 Evaluation of Curator Context-Aware Ranking

We performed extensive experimentation to evaluate Curator and to compare it against the state of the art recommendation systems that rank micro-blogs. We compared Curator against the 5 baselines: (1) RetweetRanker [15], whose metric of measuring user's interest is her re-tweet behavior; (2) RankSVM [11], which produces a ranking score by learning the ranking function and the weights of the input features; (3) DecisionTreeClassifier [49], which uses the tweet re-tweeting behavior to build a decision tree classifier that is used in its ranking model; (4) GraphCoRanking [53], which represents the preferences using LDA; and (5) TRUPI [13], which does not account for the home location of the author or the subject location of the micro-blog.

While comparing these techniques, the used ground truth was whether the micro-blog got any engagement from the user; i.e., whether it was replied to, forwarded, or liked by the user. Table 2 gives the evaluation of Curator and its competitor baselines using NDCG@k metric for the values of $k = 5, 10, 25$, and 50, whereas Fig. 2 gives the evaluation between the same techniques using the MAP metric. On NDCG@k, Curator consistently outperforms all other competitive baselines for all the used values of k. Specifically, Curator outperforms RetweetRanker by 154%, 117%, 105%, and 107% on NDCG@5, NDCG@10,

Table 2. Personalized ranking - NDCG@k metric

Technique	k = 5	k = 10	k = 25	k = 50
RetweetRanker	0.217	0.274	0.303	0.342
RankSVM	0.222	0.290	0.326	0.372
DecisionTreeClassifier	0.342	0.401	0.429	0.487
GraphCoRanking	0.411	0.455	0.462	0.538
TRUPI	0.508	0.543	0.577	0.615
Curator	0.551	0.595	0.622	0.706

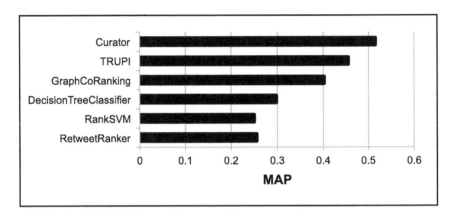

Fig. 2. Personalized ranking - MAP metric

NDCG@25, and NDCG@50 respectively. Curator outperforms the closest competitor, TRUPI, by 8%, 10%, 8%, and 15% on the same metrics. On MAP, Curator outperforms TRUPI by 13%.

4.3   Curator's Context Awareness Effect

Curator is aware of three context components, namely, time, identity, and location. From Sect. 4.2, the closest competitor was TRUPI. TRUPI already accounts for the dynamic level of interest of a user in the topic of the tweets. In this experiment, we compose a version of Curator that is not aware of the location by discarding the first two location-related features that are used in the ranking model in Sect. 3.5. We compare this version against the proposed Curator.

Table 3 and Fig. 3 give the evaluation of Curator with and without the location context using both the NDCG@k and MAP metrics. Including the location context in Curator indeed improved its performance by 12%, 12%, 10%, 18%, and 16% on NDCG@5, NDCG@10, NDCG@25, NDCG@50, and MAP, respectively. This is why we believe that the location context is a salient feature in Curator.

Table 3. Curator context awareness effect - NDCG@k metric

Context	k = 5	k = 10	k = 25	k = 50
Identity + Time	0.493	0.531	0.563	0.600
Location + Identity + Time	0.551	0.595	0.622	0.706

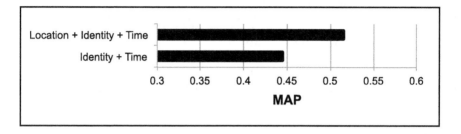

Fig. 3. Curator context awareness effect - MAP metric

5 Conclusion

In this paper, we proposed Curator, a context-aware micro-blogging recommendation system that is used to rank the micro-blogs according to the user's identity, time, and location contexts. Curator learns the user's time variant preferences from the text of the micro-blogs she engages with. Moreover, Curator infers the user's home location and the micro-blog's subject location with the help of textual features from the micro-blog. We performed an extensive performance evaluation on a publicly available dataset. Curator outperforms the competitive state-of-the-art by up to 154% on NDCG@5 and 105% on NDCG@25. The results also show that location is a salient feature in Curator.

Acknowledgement. This material is based on work supported in part by (1) Research Sponsorship from Microsoft Research, (2) the KACST National Science and Technology and Innovation Plan under grant 14-INF2461-10.

References

1. https://wiki.cites.illinois.edu/wiki/display/forward/Dataset-UDI-TwitterCrawl-Aug2012 (2012)
2. Abowd, G.D., Dey, A.K., Brown, P.J., Davies, N., Smith, M., Steggles, P.: Towards a better understanding of context and context-awareness. In: HUC (1999)
3. Becker, H., Naaman, M., Gravano, L.: Beyond trending topics: real-world event identification on Twitter. In: ICWSM (2011)
4. Blei, D.M., Ng, A.Y., Jordan, M.I.: Latent Dirichlet allocation. J. Mach. Learn. Res. **3** (2003)
5. wen Chang, H., Lee, D., Eltaher, M., Lee, J.: @Phillies tweeting from Philly? Predicting Twitter user locations with spatial word usage. In: ASONAM (2012)

6. Chen, K., Chen, T., Zheng, G., Jin, O., Yao, E., Yu, Y.: Collaborative personalized tweet recommendation. In: SIGIR (2012)
7. Cheng, Z., Caverlee, J., Lee, K.: You are where you tweet: a content-based approach to geo-locating Twitter users. In: CIKM (2010)
8. CIO survey (2009). http://rht.mediaroom.com/index.php?s=131&item=790
9. Compton, R., Jurgens, D., Allen, D.: Geotagging one hundred million Twitter accounts with total variation minimization. In: BigData (2014)
10. DMOZ - the open directory project (2014). http://www.dmoz.org/
11. Duan, Y., Jiang, L., Qin, T., Zhou, M., Shum, H.Y.: An empirical study on learning to rank of tweets. In: COLING (2010)
12. Eisenstein, J., O'Connor, B., Smith, N.A., Xing, E.P.: A latent variable model for geographic lexical variation. In: EMNLP (2010)
13. Elmongui, H.G., Mansour, R., Morsy, H., Khater, S., El-Sharkasy, A., Ibrahim, R.: TRUPI: Twitter recommendation based on users' personal interests. In: CICLing (2015)
14. Elmongui, H.G., Morsy, H., Mansour, R.: Inference models for Twitter user's home location prediction. In: AICCSA (2015)
15. Feng, W., Wang, J.: Retweet or not? Personalized tweet re-ranking. In: WSDM (2013)
16. GNU Aspell (2011). http://aspell.net/
17. Gu, H., Hang, H., Lv, Q., Grunwald, D.: Fusing text and frienships for location inference in online social networks. In: WI-IAT'12 (2012)
18. Guo, Y., Kang, L., Shi, T.: Personalized tweet ranking based on AHP: a case study of micro-blogging message ranking in T.Sina. In: WI-IAT (2012)
19. Hall, M., Frank, E., Holmes, G., Pfahringer, B., Reutemann, P., Witten, I.H.: The WEKA data mining software: an update. SIGKDD Explor. **11**(1) (2009)
20. Han, B., Cook, P., Baldwin, T.: Geo-location prediction in social media data by finding location indicative words. In: COLING (2012)
21. Han, B., Cook, P., Baldwin, T.: Text-based twitter user geolocation prediction. J. Artif. Intell. Res. **49**(1) (2014)
22. Hecht, B., Hong, L., Suh, B., Chi, E.H.: Tweets from Justin Bieber's heart: the dynamics of the location field in user profiles. In: CHI (2011)
23. Hong, L., Ahmed, A., Gurumurthy, S., Smola, A.J., Tsioutsiouliklis, K.: Discovering geographical topics in the Twitter stream. In: WWW (2012)
24. Huffington post's Twitter statistics (2013). http://www.huffingtonpost.com/belle-beth-cooper/10-surprising-new-twitter_b_4387476.html
25. Internet slang dictionary & translator (2014). http://www.noslang.com/
26. Jr., C.A.D., Pappa, G.L., de Oliveira, D.R.R., de Lima Arcanjo, F.: Inferring the location of Twitter messages based on user relationships. Trans. GIS **15**(6) (2011)
27. Jurgens, D.: That's what friends are for: inferring location in online social media platforms based on social relationships. In: ICWSM (2013)
28. Khater, S., Elmongui, H.G., Gracanin, D.: Personalized microblogs corpus recommendation based on dynamic users interests. In: SocialCom (2013)
29. Khater, S., Elmongui, H.G., Gracanin, D.: Tweets you like: personalized Tweets recommendation based on dynamic users interests. In: Social Informatics (2014)
30. Kinsella, S., Murdock, V., O'Hare, N.: "I'm eating a sandwich in glasgow": modeling locations with tweets. In: SMUC (2011)
31. Lehmann, J., et al.: DBpedia – a large-scale, multilingual knowledge base extracted from Wikipedia. Semant. Web J. (2014)
32. Li, C., Sun, A.: Fine-grained location extraction from tweets with temporal awareness. In: SIGIR (2014)

33. Li, R., Wang, S., Chang, K.C.C.: Multiple location profiling for users and relationships from social network and content. PVLDB **5**(11) (2012)
34. Li, R., Wang, S., Deng, H., Wang, R., Chang, K.C.C.: Towards social user profiling: unified and discriminative influence model for inferring home locations. In: KDD (2012)
35. Mahmud, J., Nichols, J., Drews, C.: Home location identification of Twitter users. ACM TIST **5**(3) (2014)
36. Manning, C.D., Raghavan, P., Schütze, H.: Introduction to Information Retrieval. Cambridge University Press (2008)
37. McGee, J., Caverlee, J., Cheng, Z.: Location prediction in social media based on tie strength. In: CIKM (2013)
38. Mikolov, T., Chen, K., Corrado, G., Dean, J.: Efficient estimation of word representations in vector space. In: ICLR Workshops (2013)
39. Morstatter, F., Pfeffer, J., Liu, H., Carley, K.M.: Is the sample good enough? Comparing data from Twitter's streaming API with Twitter's firehose. In: ICWSM (2013)
40. Pennacchiotti, M., Silvestri, F., Vahabi, H., Venturini, R.: Making your interests follow you on Twitter. In: CIKM (2012)
41. Rout, D., Bontcheva, K., Preoţiuc-Pietro, D., Cohn, T.: Where's @wally? A classification approach to geolocating users based on their social ties. In: HT (2013)
42. Ryoo, K., Moon, S.: Inferring Twitter user locations with 10 km accuracy. In: WWW Companion (2014)
43. Sadilek, A., Kautz, H., Bigham, J.P.: Finding your friends and following them to where you are. In: WSDM (2012)
44. Salton, G., Buckley, C.: Term-weighting approaches in automatic text retrieval. Inf. Process. Manag. **24**(5) (1988)
45. Twitter NLP tools (2011). https://github.com/aritter/twitter_nlp
46. Twitter REST API (2014). https://dev.twitter.com/docs
47. Twitter statistics (2017). http://www.statisticbrain.com/twitter-statistics/
48. Twitter usage (2014). http://about.twitter.com/company
49. Uysal, I., Croft, W.B.: User oriented tweet ranking: a filtering approach to microblogs. In: CIKM (2011)
50. Wikipedia (2001). http://www.wikipedia.org/
51. WikiSynonyms (2012). http://wikisynonyms.ipeirotis.com/
52. Yamaguchi, Y., Amagasa, T., Kitagawa, H.: Landmark-based user location inference in social media. In: COSN (2013)
53. Yan, R., Lapata, M., Li, X.: Tweet recommendation with graph co-ranking. In: ACL (2012)

A Multi-view Clustering Model for Event Detection in Twitter

Di Shang[1], Xin-Yu Dai[1(✉)], Weiyi Ge[2], Shujiang Huang[1], and Jiajun Chen[1]

[1] National Key Laboratory for Novel Software Technology, Nanjing University,
Nanjing 210023, China
{shangd,dxy,huangsj,chenjj}@nlp.nju.edu.cn
[2] Science and Technology on Information Systems Engineering Laboratory,
Nanjing 210007, China

Abstract. Event detection in Twitter is an attractive and hard task. Existing methods mainly consider words co-occurrence or topic distribution of tweets to detect the event. Few of them consider the time-series information in the text stream. In this paper, for event detection in twitter, we propose a novel multi-view clustering model which can consider both topic information and time-series information. First, we build a topic similarity matrix and a time-series similarity matrix by using the topic model and the wavelet analysis, respectively. Then, the multi-view clustering algorithms are used to group keywords. Each cluster of keywords is finally represented as an event. The experiments show that our method achieves better performance than other related works.

Keywords: Twitter event detection · Multi-view clustering
Time-series

1 Introduction

Twitter is a fast emerging social media platform. Users post short messages to share various kinds of information ranging from personal daily life to global events in real-time. Huge amounts of Twitter data contain a valuable source of timely information which covers every corner of the world. So, this raises the following question: How can we discover the interesting things from the rich Twitter data? Events detection in Twitter was defined as a task of identifying events in free text and deriving detailed and structured information about them [24].

Comparing with the traditional event detection on news articles, there are some special properties in Twitter. Firstly, under the limit to 140 characters, there is limited useful information in each message. Secondly, Twitter data is naturally real-time stream data and tweets are dynamically changing and increasing. Thirdly, topics discussed in Twitter are wide-ranging and can be very complex. It is impossible to know the event types in advance. In this case, we can not use any annotated corpora or any manually-defined linguistic patterns.

A. Gelbukh (Ed.): CICLing 2017, LNCS 10762, pp. 366–378, 2018.
https://doi.org/10.1007/978-3-319-77116-8_27

The above properties of Twitter bring new difficulties and challenges to event detection in Twitter. Considering above properties, topic and time-series information are two key factors for event detection. In most of previous studies, either topic or time-series information was used, separately. For example, ET-LDA [11] used topic model to impose topical influences and aligned tweets with events provided by traditional media. EDCoW [23] applied the word time-series similarities to group keywords together as events.

Inspired by the complementarity of the two independent information of topic and time-series, we propose a new multi-view clustering algorithm to combine both information for better event detection. The combination of topic and time-series information enforcement is non-trivial because intuitively each event has both the topic and time-series properties. After preprocessing, the classic topic model, Latent Dirichlet Allocation [6] is applied to learn the probability of words under topics. Meanwhile, a number of topic words generated from the topic model are selected as the keywords to describe events. Then, time-series similarity between these keywords is calculated by the Dynamic Time Warping (DTW) algorithm. Eventually, with the two matrices of topical similarity and time-series similarity together, our model uses a co-training spectral clustering algorithm to get the final clustering results for the keywords.

The main contributions of our method are summarized as follows: Firstly, our method does not rely on any external knowledge or labeled data. Secondly, to our knowledge, it is the first time to jointly model the topical information and temporal information together for event detection in Twitter. Thirdly, our algorithm can be extended with other more different kinds of information, such as social networking information and URLs information.

In the following we start by presenting some related work. Then we present the multi-view event detection(MVED) method, followed by experiments and concluding remarks.

2   Related Work

Previous work on event detection in Twitter can be briefly classified into three categories.

Approaches Based on Text Clustering. Text-clustering-based methods detect events by clustering documents based on the document similarity [24]. Each message from a data stream will be represented as a term vector and the similarity to the center vector of an existing event will be calculated. If the value of the similarity is larger than certain pre-determined threshold, the message will be classified into the corresponding event. If there are no existing events similar to the message, it could be set as a new event. In [1,7,19,21], this online-clustering-based method is widely applied to event detection on Twitter. However, because of the characteristics of short text, feature space is high dimensional and extremely sparse, which causes the similarity calculation can hardly be reliable. Besides, processing huge amounts of Twitter data in real time is another problem for online-clustering algorithms. Therefore, the text-clustering-based methods did not work well for event detection in Twitter data.

Approaches Based on Topic Model. Because of the fact that events discussed widely are more likely to be some hot topics, this kind of methods apply topic model to event detection. In [14], authors update the topic model in each time period by using the previously generated model. Since tweets are generated by the crowd to express their interests in the event, they are essentially influenced by the topics covered in the event in some way. But in [2], the results show that some topic models such as LDA and DTM [5] can well capture the topics happening during events with narrow topical scope. However, when many different events happen in parallel at the same time, topic models perform poorly. Also as reported in [11], topic-modeling-based methods don't work well when applied to short documents such as tweets. In general, topic-modeling-based approaches lack both stability and reliability for Twitter data.

Approaches Based on Keyword Clustering. This kind of methods study the distributions of words extracted from contents and discovers events by grouping these words together [12]. Twitter has short message with limited information. The keywords in messages might play an import role to represent an event. In recent years, the keyword-clustering-based approaches become the mainstream methods in event detection [15]. In [8], named entities in tweets are extracted as keywords. Then a graph in which each extracted name entity is represented as a node and edges represent the dynamic relationships among the entities will be built. Then, the task of word clustering is converted to a problem of graph partition. The entities divided into the same community represent an extracted event. MABED [10] takes the social aspect of tweets into account by leveraging the creation frequency of mentions that users insert in tweets to engage discussion. Based on mention anomaly, candidate events are detected preliminarily. For each event, a set of words are selected to describe it. After merging and filtering, the k most influential events are generated. However, there are a large number of non-mention-related events, which can not be detected by MABED. In addition, MABED do not make the best of temporal information. EDCoW [23] firstly introduced time-series information into event detection. They broke down the frequency of individual words into wavelets and compute the change of wavelet entropy to identify bursts. Trivial words are filtered away according to their corresponding signal's auto correlation, and the similarity between each pair of non-trivial words is measured using cross correlation. Eventually, events are represented as bags of words with high cross correlation during a predefined fixed time window, detected with modularity-based graph clustering. But the problem is, only utilizing the similarities on cross correlation may cluster several unrelated events together which happened in the same time span.

3   Multi-view Clustering Model for Event Detection

Based on the following three reasonable hypotheses, we design our method. The first is that a number of keywords can represent an event. For example, "Paris",

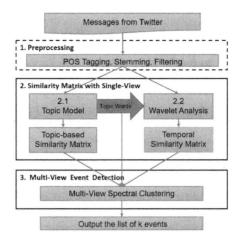

Fig. 1. Architecture of our multi-view clustering model

"attack", "explosion" and "shooting" obviously represent the event of coordinated terrorist attacks occurred in Paris on the evening of 13 November 2015. This hypothesis makes us follow the keyword-clustering research line. The second hypothesis is that topic model can well capture the topic words related to a special event, such as, "attack" "explosion" in the terrorist events. The third hypothesis is that some co-occurrence words with the same time-series tendency may represent the same event. The last two hypotheses are consistent with that the topic and time-series information are both important for event detection in twitter.

With above three hypotheses, we propose a multi-view clustering model for event detection which can use both topic and time-series information. As illustrated in Fig. 1, our proposed model consists of three main steps. We preprocess the twitter messages with POS Tagging, stemming and filtering some low frequency words firstly. Then, because our method is a keyword-clustering approach for event detection, we select some keywords with topic model (e.g. Latent Dirichlet Allocation, LDA). For the selected keywords, we construct two similarity matrices with topic information and wavelet analysis. The two similarity matrices are finally applied in our multi-view spectral clustering method to get several groups of keywords as the output events. We describe the components of our method in detail as follows.

3.1   Latent Dirichlet Allocation: Topic Information

The Latent Dirichlet Allocation (LDA) [6] model is a graph probabilistic model. It is a generative model which assumes each document is a mixture of a set of topics and the words in the document are generated given their topics. In our event detection scenario, we can model the twitter messages collection with LDA that each message can be described as a random mixture over events, and each event as a focused multinomial distribution over words. With the LDA model and the twitter data collection, we can select some keywords and get the probability of words under topics.

Keywords Selection. Given the number of topics k, the parameter φ_k, which represents the distribution of words in topic k, can be estimated through sampling in LDA. In each topic k, words with the m highest $\varphi_{k,w=1...V}$ are selected as topic words. Theoretically, up to $k \times m$ words can be selected as the keywords. In this way, the selected keywords are non-trivial words to represent events further.

Topic Words Similarity. In the related method for event detection with LDA [23], each topic is represented as a list of words. And these words in a same topic are directly used to represent an event. However, in spite of the fact that events are more likely to be topics, several events under the same topic are always extracted as one event by LDA. So, we are not going to directly get the keyword clusters from LDA as the results. We intend to apply a word clustering method to make full use of word topic distribution information. The topic vectors of keywords will be used in the clustering algorithm. For clustering we have to define a similarity measure between words.

From LDA model, we can get the word distribution on topics φ, a keyword i can be represented as a topic vectors $\varphi_{:,i}$. Then, the cosine similarity can be used to measure the similarity between two words.

$$Sim\_s(x, y) = sim\_cos(\varphi_{:,x}, \varphi_{:,y}) \tag{1}$$

3.2 Wavelet Analysis: Time-Series Information

EDCoW [23] was the first method which detected events with time-series information. It was a keyword-based approach of applying the wavelet analysis. EDCoW built signals for keywords which capture the bursts in the words' appearance. The assumption of applying wavelet analysis is that keywords related to the same event have the similar signals. Inspired by this idea, we give our similarity measure for words from the view of time-series.

Similarity Between Word Signals. The signal for each individual word is built as follows. We partition the tweet corpus C by tweet timestamps. Let's suppose that corpus covers T days. So the signal for word w can be written as a sequence:

$$S_w = [s_w(1), s_w(2), ..., s_w(T)] \tag{2}$$

As same as EDCoW, $s_w(t)$ at each time period t can be calculated in Eq. 3. It is a way like $DF - IDF$, $N_w(t)$ is the number of tweets which contain word w at time period t and $N(t)$ is the number of all the tweets at the same period.

$$s_w(t) = \frac{N_w(t)}{N(t)} \times log \frac{\sum_{i=1}^{T} N(i)}{\sum_{i=1}^{T} N_w(i)} \tag{3}$$

To make it fast and accurate, we use dynamic time warping (DTW[1]) algorithm [3] for measuring distance between two word signals S_x and S_y.

DTW is a method that calculates an optimal match between two given sequences (e.g. time series). As a dynamic programming technique, it divides the problem into several sub-problems, each of which contribute in calculating the distance cumulatively as in Eq. 4.

$$D(i,j) = d(i,j) + \min \begin{cases} D(i-1,j) \\ D(i,j-1) \\ D(i-1,j-1) \end{cases} \qquad (4)$$

And in Eq. 4, i and j are the point indices in the time series of sequence S_x and S_y. DTW distance can be calculated in Eq. 5, where m and n represent the length of time series S_x and S_y.

$$DTW(S_x, S_y) = D(m, n) \qquad (5)$$

3.3   Multi-view Clustering Model

The events in twitter have the properties of topic and time-series which are complementary to each other for better event detection. Inspired by the multi-view learning algorithm [4], we propose a new multi-view clustering event detection method. Firstly, we construct two similarity matrices of words from the view of topic and time-series, respectively. With those two matrices, we apply a co-training method to perform the multi-view clustering task for event detection.

Similarity Matrices Construction. From the topic view, we construct a similarity matrix for the N keywords selected from LDA. With the measurement of cosine similarity in Eq. 1, we can construct an $N \times N$ positive semi-definite matrix G, where G_{ij} quantifies the topic similarity between the keywords of i and j.

From the time-series view, according to the distance measurement of time sequences defined in Eq. 5, the time-series similarity between two keywords i and j is calculated as:

$$Sim\_t(i,j) = \frac{1}{1 + DTW(S_i, S_j)} \qquad (6)$$

Given Eq. 6, we build the symmetric $N \times N$ matrix T, where $T_{ij} \geq 0$ represents the time-series similarity between the keywords of i and j.

Multi-view Spectral Clustering. With the input of two similarity matrix G and T, we apply a co-training approach [13] to finish this multi-view clustering task. The main idea of this approach is that eigenvectors obtained from one view can be used to "label" the points in other view. In other words, topic matrix from semantic view can correct the clustering results generated by time-series view and vice versa.

---

[1] https://en.wikipedia.org/wiki/Dynamic_time_warping/.

Algorithm 1 Multi-view Spectral Clustering Algorithm

Require: Similarity matrix G and T for topic view and time-series view.

Ensure: Assignments N keywords to k clusters.

1: $L_1 = D_1^{-1/2} G D_1^{-1/2}$, where D_1 is a diagonal matrix with $D_1(i,i) = \sum_j G(i,j)$

2: $L_2 = D_2^{-1/2} T D_2^{-1/2}$, where D_2 is a diagonal matrix with $D_2(i,i) = \sum_j T(i,j)$

3: $U_v^0 = \arg\max_{U \in \mathbb{R}^{N \times k}} tr(U^T L_v U)$, s.t. $U^T U = I$ for $v = 1, 2$

4: **for** $i = 1$ to $iter$ **do**

5:     $S_1 = sym(U_2^{i-1} U_2^{i-1^T} G)$

6:     $S_2 = sym(U_1^{i-1} U_1^{i-1^T} T)$, where $sym()$ is defined as $sym(S) = (S + S^T)/2$.

7:     $L_v = D^{-1/2} S_v D^{-1/2}$ for v = 1,2 where D is a diagonal matrix with $D(i,i) = \sum_j S_v(i,j)$

8:     $U_v^i = \arg\max_{U \in \mathbb{R}^{N \times k}} tr(U^T L_v U)$, s.t. $U^T U = I$ for $v = 1, 2$

9: **end for**

10: Row-normalize U_1^{iter}, and form $V = U_1^{iter}$.

11: Run the k-means algorithm to cluster the row vectors of V

12: Assign example j to cluster c if the j-th row of V is assigned to cluster c by the k-means algorithm.

In Algorithm 1, we briefly outline the co-trained multi-view spectral clustering. In the process of iteration $i - 1$, spectral clustering can be solved in each individual view to get the discriminative eigenvectors, here U_1^{i-1} and U_2^{i-1}. In the next iteration i, we could have clustered points by using U_1^{i-1} and U_2^{i-1}. On the basis of the original similarity matrix G and T, we regenerate the new similarity matrix S_1, which is modified by U_2^{i-1} and vice versa for S_2. In this way, both of two views are interacted on each other. After the iteration ends, U_1^{iter} is the discriminative eigenvectors which is mainly based on topic matrix G and modified by time-series similarity T. U_2^{iter} is based on T and modified by G. On the step 10 of Algorithm 1, it should be noted that we chose $V = U_1^{iter}$. Because we believe that the topic view is the most informative view and time-series helps to correct the results from topic view. Finally, this method outputs the cluster results guided by both topic and time-series information and we name this method as MVED (Multi-view Clustering Event Detection).

4   Experiments and Results

In this section, we firstly give the experimental setup. Then, we present the experimental results in comparison with both keyword-based methods and text-based methods. Finally, we show the examples for qualitative evaluation.

4.1   Experimental Setup

Datasets. We investigate the performance of our method on two twitter datasets, FSD2011 [18] and Event2012 [16], shown as in Table 1. Both the two corpora are annotated with labels of events. There are 27 events in FSD2011.

Table 1. Dataset statistics

Dataset	Messages	Events	Vocabulary
FSD2011	2453	27	3664
Event2012	78747	500	43115
Event2012_1	6546	50	9303
Event2012_2	20486	100	19802
Event2012_3	23130	150	22412
Event2012_4	28585	200	25644

We remove 7 events which contain very few tweets and use the remaining 20 of them for the experiments. Event2012 is a very large dataset which contains 500 events. In order to observe the performance in different numbers of events, we randomly split the 500 events into 4 subdatasets, event2012_1-4, which respectively contains 50, 100, 150 and 200 events.

For these tweet message data, we perform the part-of-speech tagging with the tool ARK[2] developed by the Language Technologies Institute in CMU. After POS tagging, we just keep words with the following POS tags for analysis: 'A', 'N', '∧', '$', 'V', '#', '!', 'S', 'Z', 'G', 'R', 'E', '@'. The detailed descriptions of tags are introduced in [9].

Parameter Setting. There are some hyperparameters in our experiments. It is hard to determine these hyperparameters with grid search because of the human evaluation. Considering there are fewer events in FSD2011 dataset, we in advance conduct a series of experiments for different values of parameters on FSD2011 dataset. Fortunately, we found that the performance of our method is not very sensitive to the different parameters. For final experiments, we set the parameters α and β in LDA as 0.2 and 0.1. The number of topics k depends on the number of events in dataset. We set $m = 15$ for selecting m keywords in each topic. In building time-series signals, we partition dataset using 1-day time-slices. In multi-view clustering, the number of iterations is set as 5.

4.2 Comparing with Keywords-Clustering Methods

Our method is a kind of keyword-clustering method. We compare our method with other related keyword-clustering methods. Three comparing methods used in our paper are as follows:

- **MABED** [10]. It is a representative keyword-clustering-based studies about event detection in recent years. We use it for baseline comparison.
- **LDA** (LDA Spectral Clustering). LDA is the most typical and representative topic model. With the input of similarity matrix G from LDA, as showed in Fig. 1, we apply spectral clustering [22] for clustering words into groups.

---

[2] http://www.cs.cmu.edu/~ark/TweetNLP/.

Table 2. Comparison with keyword-clustering-based methods

Dataset	FSD2011			Event2012_1			Event2012_2		
Method	Precision	Recall	F-score	Precision	Recall	F-score	Precision	Recall	F-score
MABED	0.9000	0.5500	0.6828	0.8333	0.1800	0.2961	0.7959	0.3500	0.4862
TSC	0.7000	0.6000	0.6461	0.6000	0.5600	0.5793	0.4500	0.4100	0.4291
LDA	0.9000	0.7000	0.7875	0.7600	0.7000	0.7288	0.6600	0.6300	0.6447
MVED	0.9000	0.8000	0.8471	0.8600	0.7600	0.8069	0.7600	0.7100	0.7341

Dataset	Event2012_3			Event2012_4		
Method	Precision	Recall	F-score	Precision	Recall	F-score
MABED	0.6905	0.1867	0.2939	0.6429	0.1600	0.2562
TSC	0.3667	0.3000	0.3300	0.2350	0.2150	0.2246
LDA	0.5867	0.5667	0.5765	0.4600	0.4500	0.4549
MVED	0.6333	0.6000	0.6162	0.5050	0.4850	0.4948

And the words in a group are represented as an event. We use LDA as a comparing method, because we want to see the performance on a single view of topic information.

- **TSC** (Time-series Spectral Clustering). With the input of time-series similarity matrix T, we apply spectral clustering [22]. We use the TSC as a comparing method, because we want to see the performance on a single view of time-series information.

Evaluation Metric. Our MVED model and the three comparing methods mentioned above output a list of events, each of which is represented as a group of words. Following the evaluation used in comparing methods, we evaluate the event results manually. Two annotators who are not involved in this project are employed to do this evaluation work. Given each group of words, annotators are required to check whether it is a correct event or not. For assuring the objectiveness for human annotators, we setup three heuristic rules as follows.

(1) The group of words can represent a unique event.
(2) The unique event can be found in the event lists of dataset.
(3) The event is correct only if both of the annotators mark it as correct.

It should be noted that a cluster which does not meet these rules will be labeled as 'un' for unknown or unclear. Based on the human's annotation, we use the precision, recall and F-score to measure the performance of different methods. Precision is defined as the proportion of the correctly identified events out of the framework returned events. Recall is defined as the proportion of correctly identified true events out of the dataset contained events. And F-score is the harmonic mean of precision and recall.

Experimental Results. Table 2 shows the precision, recall and F-score of each method in different datasets. Our two single-view spectral clustering methods,

TSC and LDA, get competitive results comparing with MABED. Especially, LDA gets better performance than MABED on all datasets. Our method MVED achieves the best performance on all datasets. With the increasing number of events, the performance of all methods declines and TSC declines heavily. The biggest weakness of TSC is unable to separate unrelated events that happened in the same period of time.

In terms of MABED, it almost gets the same high precision with our method. The events generated by MABED are quite accurate. However, in the experiments, the number of event generated by MABED is much less than that the dataset contains, which results in the poor recall.

Table 3. Comparison with text-clustering-based method

Method	LSH				MVED			
Dataset	Precision	Recall	F-score	NMI	Precision	Recall	F-score	NMI
FSD2011	0.1979	0.9340	0.3265	0.7180	0.8293	0.8341	0.8317	0.8972
Event2012_1	0.1540	0.8721	0.2617	0.7127	0.6599	0.8954	0.7598	0.8909
Event2012_2	0.0479	0.8087	0.0904	0.5717	0.6377	0.8481	0.7280	0.8662
Event2012_3	0.0274	0.8635	0.0532	0.5972	0.5633	0.8585	0.6803	0.8663
Event2012_4	0.0327	0.8178	0.0628	0.5561	0.4532	0.8108	0.5814	0.8386

4.3 Comparing with Text-Clustering Methods

Because our MVED is a keyword-based clustering method, we present detailed comparison in Sect. 4.2. Here, we do a brief comparison with a text-clustering method of LSH [17] which is an outstanding work of text-clustering-based methods for event detection. This method applies document similarity and Locality Sensitive Hashing(LSH) to generate clusters of tweets to represent the events. Different from text-clustering-based methods, our method clusters words, rather than tweets. We follow the work of [20] to assign tweets to an event clusters after keyword-clusters are given. Besides precision, recall and F-score, we use another measure of NMI[3] (Normalized Mutual Information) for evaluation, which is a popular evaluation measure for clustering.

From Table 3, though LSH is found to provide a comparable performance to that of MVED with respect to Recall. Accurate clusters are not provided by LSH because of the low precision. This is a common problem in text-clustering-based methods because many noise tweets are assigned to a cluster no matter they belong to a event or not. With better precision, MVED improve the F-score significantly and also get better NMI performance than the LSH approach.

---

[3] https://en.wikipedia.org/wiki/mutual_information.

4.4   Qualitative Evaluation

We have mentioned that topic and time-series which are complementary to each other for better event detection. We perform some qualitative evaluation to demonstrate this argument. The sample results we choose are presented in Fig. 2. Words in red represent the event of "South Sudan declares independence", blue describes "the famine declared in Somalia" and green is "the first artificial organ transplant". Besides, words in gray are not related to the three events above. TSC can hardly separate unrelated events that happened at the same time, such as TSC_13. And in TSC_18, the clusters always contain some non-topic words, while MVED_15 can remove some non-topic words with the help of topic information. At the other end, LDA_4 is unable to detect different events under the similar topic. However, these two events happened in different time periods. With complementarity of time-series information, our method can detect these two different events, shown as MVED_7 and MVED_15. Overall, MVED can give us more appropriate keyword-clustering results.

Method	Cluster ID	Keywords
LDA	LDA_4	South sudan independence nation un southern officially declared help can declares famine two Somalia regions drought parts
TSC	TSC_13	South sudan independence nation fire dies Richard bowes riots ealing very London hits riot tottenham
	TSC_18	Synthetic transplant windpipe organ amazing news bbc first
MVED	MVED_7	South sudan independence nation somalia
	MVED_15	Un southern years officially help declared can declares famine two regions drought parts
	MVED_19	Synthetic organ transplant windpipe

Fig. 2. Examples of extracted events

5   Conclusion and Future Work

In this paper, we have proposed a multi-view clustering model to detect event from tweets. The contribution of our work is that we build a multi-view framework to make full use of both topic information and time-series information in tweets. The experimental results show that our method outperforms those compared methods. In future work, based on our extensible multi-view framework, we will try to further improve the performance with more kinds of information from tweets, such as social networking information.

References

1. Aggarwal, C.C., Subbian, K.: Event detection in social streams. In: SDM, vol. 12, pp. 624–635. SIAM (2012)
2. Aiello, L.M., et al.: Sensing trending topics in twitter. IEEE Trans. Multimed. **15**(6), 1268–1282 (2013)
3. Berndt, D.J., Clifford, J.: Using dynamic time warping to find patterns in time series. In: KDD Workshop, Seattle, WA, vol. 10, pp. 359–370 (1994)
4. Bickel, S., Scheffer, T.: Multi-view clustering. In: ICDM, vol. 4, pp. 19–26 (2004)
5. Blei, D.M., Lafferty, J.D.: Dynamic topic models. In: Proceedings of the 23rd International Conference on Machine Learning, pp. 113–120. ACM (2006)
6. Blei, D.M., Ng, A.Y., Jordan, M.I.: Latent dirichlet allocation. J. Mach. Learn. Res. **3**, 993–1022 (2003)
7. Brants, T., Chen, F., Farahat, A.: A system for new event detection. In: Proceedings of the 26th Annual International ACM SIGIR Conference on Research and Development in Informaion Retrieval, pp. 330–337. ACM (2003)
8. Das Sarma, A., Jain, A., Yu, C.: Dynamic relationship and event discovery. In: Proceedings of the Fourth ACM International Conference on Web Search and Data Mining, pp. 207–216. ACM (2011)
9. Gimpel, K., Schneider, N., O'Connor, B., Das, D., Mills, D., Eisenstein, J., Heilman, M., Yogatama, D., Flanigan, J., Smith, N.A.: Part-of-speech tagging for twitter: Annotation, features, and experiments. In: Proceedings of the 49th Annual Meeting of the Association for Computational Linguistics: Human Language Technologies: short papers, vol. 2, pp. 42–47. Association for Computational Linguistics (2011)
10. Guille, A., Favre, C.: Mention-anomaly-based event detection and tracking in twitter. In: 2014 IEEE/ACM International Conference on Advances in Social Networks Analysis and Mining (ASONAM), pp. 375–382. IEEE (2014)
11. Hu, Y., John, A., Seligmann, D.D., Wang, F.: What were the tweets about? Topical associations between public events and twitter feeds. In: ICWSM (2012)
12. Kleinberg, J.: Bursty and hierarchical structure in streams. Data Min. Knowl. Discov. **7**(4), 373–397 (2003)
13. Kumar, A., Daumé, H.: A co-training approach for multi-view spectral clustering. In: Proceedings of the 28th International Conference on Machine Learning (ICML 2011), pp. 393–400 (2011)
14. Lau, J.H., Collier, N., Baldwin, T.: On-line trend analysis with topic models: \# twitter trends detection topic model online. In: COLING, pp. 1519–1534 (2012)
15. Long, R., Wang, H., Chen, Y., Jin, O., Yu, Y.: Towards effective event detection, tracking and summarization on microblog data. In: Wang, H., Li, S., Oyama, S., Hu, X., Qian, T. (eds.) WAIM 2011. LNCS, vol. 6897, pp. 652–663. Springer, Heidelberg (2011). https://doi.org/10.1007/978-3-642-23535-1_55
16. McMinn, A.J., Moshfeghi, Y., Jose, J.M.: Building a large-scale corpus for evaluating event detection on twitter. In: Proceedings of the 22nd ACM International Conference on Conference on Information & Knowledge Management, pp. 409–418. ACM (2013)
17. Petrović, S., Osborne, M., Lavrenko, V.: Streaming first story detection with application to twitter. In: Human Language Technologies: The 2010 Annual Conference of the North American Chapter of the Association for Computational Linguistics, pp. 181–189. Association for Computational Linguistics (2010)

18. Petrović, S., Osborne, M., Lavrenko, V.: Using paraphrases for improving first story detection in news and twitter. In: Proceedings of the 2012 Conference of the North American Chapter of the Association for Computational Linguistics: Human Language Technologies, pp. 338–346. Association for Computational Linguistics (2012)
19. Phuvipadawat, S., Murata, T.: Breaking news detection and tracking in twitter. In: 2010 IEEE/WIC/ACM International Conference on Web Intelligence and Intelligent Agent Technology (WI-IAT), vol. 3, pp. 120–123. IEEE (2010)
20. Preotiuc-Pietro, D., Srijith, P., Hepple, M., Cohn, T.: Studying the temporal dynamics of word co-occurrences: an application to event detection (2016)
21. Sankaranarayanan, J., Samet, H., Teitler, B.E., Lieberman, M.D., Sperling, J.: Twitterstand: news in tweets. In: Proceedings of the 17th ACM Sigspatial International Conference on Advances in Geographic Information Systems, pp. 42–51. ACM (2009)
22. Von Luxburg, U.: A tutorial on spectral clustering. Stat. Comput. **17**(4), 395–416 (2007)
23. Weng, J., Lee, B.S.: Event detection in twitter. In: ICWSM, vol. 11, pp. 401–408 (2011)
24. Yang, Y., Pierce, T., Carbonell, J.: A study of retrospective and on-line event detection. In: Proceedings of the 21st Annual International ACM SIGIR Conference on Research and Development in Information Retrieval, pp. 28–36. ACM (1998)

Monitoring Geographical Entities with Temporal Awareness in Tweets

Koji Matsuda[1]([✉]), Mizuki Sango[2], Naoaki Okazaki[1], and Kentaro Inui[1]

[1] Tohoku University, Sendai, Japan
matsuda@ecei.tohoku.ac.jp
[2] Tokyo Institute of Technology, Tokyo, Japan

Abstract. To extract real-time information referring to a specific place from social network service texts such as tweets, it is necessary to analyze the temporal semantics of the reference. To solve this problem, we created a corpus with multiple annotations for more than 10,000 tweets using crowdsourcing. We constructed an automatic analysis model based on multiple neural networks and compared their characteristics. Our dataset and codes are released in our website (http://www.cl.ecei.tohoku.ac.jp/~matsuda/TA_corpus/).

1 Introduction

People often mention geographical entities (e.g., cities, tourist spots and public facilities) on social networking services (SNSs) expressing their present or past experiences and future plans for visits. Let us consider the following example.

(1) I went to Sendai yesterday, but I'm going to Nagoya today.
This example has references to two places, Sendai and Nagoya. However, the time recognition of the author for each place is different: Sendai is a *past* location, whereas Nagoya is a *present* location.

In applications such as text mining and marketing for tourism, it is crucial to distinguish such temporal references. By resolving temporal aspects, we can extract the opinions of people who actually stay(ed) at a certain location. In addition, recognizing individuals who are willing to go to a certain location facilitates targeted advertisements for potential visitors.

In this paper, we discuss tasks to monitor references to geographical entities that appear in the texts of SNSs. In particular, we work on the task of detecting users who are present in a location at the moment and detecting references including the intention to go a location. To realize this goal, we need to address at least two problems. First, we need to disambiguate references in the text into geographical entities because a reference can refer to multiple locations or to non-geographical entities. Second, we have another type of ambiguity problem concerning the time recognition of the author, i.e., whether and when the author (will) stay(ed) at the location. This paper discusses the latter problem, which recognizes the temporal relation between a geographical entity and an author given in a tweet.

© Springer Nature Switzerland AG 2018
A. Gelbukh (Ed.): CICLing 2017, LNCS 10762, pp. 379–390, 2018.
https://doi.org/10.1007/978-3-319-77116-8_28

Our task is closely related to *Temporal Awareness* [7] (*TA*, hereafter), where location references and temporal polarities in tweets are identified. However, to avoid the ambiguity effect, we assume that the detection of the location mentioned has already been completed and create an annotated corpus focusing on the classification of the temporal relation. To emulate this situation, we target tweets that contain predefined nouns that are known to be location references. Even if this is done for a limited number of targets, it is interesting to see whether linguistic features are learnable.

In addition, we created a model that automatically analyzes *TA* using this corpus. We use a model motivated by target-dependent sentiment classification [3], which is a variation of short text classification that incorporates target information.

The contributions of this paper are two folds.

- We designed an annotation scheme focusing on *TA*. We annotated more than 10,000 tweets using a crowdsourcing platform. The quality of the annotation was confirmed to be high, which indicates clearly that the task was properly designed.
- We built a model using a state-of-the-art method based on neural networks. We show its quantitative performance. In addition, we conducted experiments with cross domains to demonstrate the performance of the *TA* for unknown targets.

2   Related Work

Li et al. [7] proposed an end-to-end model to extract expressions representing places with time labels using the framework of sequence labeling.

They proposed a model that uses sequence labeling to simultaneously extract the references representing places and the time labels. The categories of location references they deal with are diverse and cover various expressions such as facility names and place names referring to unique entities. However, their model does not focus on temporal relationships, because it solves multiple tasks, such as ambiguities of references referring to places and temporal relationships.

In addition, we assume that as a practical usage scenario we should gather tweets mentioning specific places. However, it is not easy to gather tweets that refer to all entities with an open vocabulary. For example, because Twitter's streaming API can only obtain a very small sample of tweets, it is not appropriate to monitor references to specific entities with high coverage. In contrast, a search API is more realistic for entity monitoring because it provides data with a high coverage. In addition, the data created by Li et al. have not been verified by multiple annotations. Therefore, the validity of the design of their annotation scheme cannot be evaluated.

Recently, Huang et al. created a corpus annotating event information to summarize news and generate timelines [2]. In their corpus, the temporal status of an event is annotated to the a major social event (in particular, a civil unrest event) described in the news text whether the event has actually occurred or is going to occur. Their goal is similar to ours; however, we differ in that we aim to estimate the intentions of people's daily lives rather than to detect major social events. In addition, they target news text the text which is different in nature from our user generated text.

Our work is linked to the TIMEBANK [8] and TempEval [12] efforts; however, we consider lightweight corpus specifications. To scale the annotation, we created a simple annotation guideline and user interface and proposed a framework that allows annotators who are not experts to do high-quality work.

Readers may find this task is similar to Factuality analysis [9] in the task is to predict whether the events mentioned in the sentence correspond to actual events that have occurred. Typical Factuality analysis is intended for events represented explicitly in the text. However, as you can see in the example below, the fact that someone (will) stay(ed) in a certain place is not always explicitly written in the text.

(2) I lost my way in <u>Sendai station</u>.

In this sentence, the interpretation that the author visited Sendai station is reasonable; however, because it is implicitly written, it cannot be handled in the existing framework. To capture such an implicit event, we focus on the location reference rather than on the events explicitly mentioned in the text.

This task can also be seen as a short text classification problem [4,5,10]. However, it is reasonable to view our problem as a target-dependent short-document classification problem. This is because it is possible to assume that multiple targets appear in one short text, as in Example (1). In particular, our task is close to target-dependent sentiment analysis. In the target-dependent short-text classification problem, it has been reported that neural models using convolutional neural networks (CNNs) and recurrent neural networks (RNNs) show high performance [1,3,13]. Therefore, this study explores neural models in Sect. 4.

3   Dataset Construction

In this task, how to create data is also a big issue. First, an expression that points to a specific place is not limited to a proper noun. If the ambiguity can be resolved in some way, some general nouns, such as "hospital" or "city-hall", can also be considered as monitoring targets.

We focus on a realistic situation in which the entities to be monitored are known. Given this situation, we create data focusing on several pre-selected location. In general, it is necessary to create annotated data for each entity; however, the general linguistic meaning independent of entity can be learned from data for other entities. This suggests that it is possible to identify the TA with some degree of performance for both learned and unknown targets from annotated data for fixed targets. Of course, by annotating the target you want to monitor, it is possible to improve its performance. To experimentally verify this prediction, we adopted a policy to use closed vocabulary and to annotate a large number of instances for each target.

When annotating the temporal nature concurrently with picking up identifying the target reference, the attention of the annotator becomes distracted and high-quality annotation is not realistic. Therefore, we create data with closed vocabulary for nouns that are likely to be references to places. As a usage scenario, there may be situations where the target to be monitored is known and a certain amount of training data on the target can be obtained. Therefore, a closed setting is also meaningful.

Specifically, we first compiled a list of Japanese nouns representing a place (a location) and annotated only Japanese tweets containing at least one noun (one location references). This list contains five proper nouns and five common nouns that we consider equally, including the location name, facility name, and tourist spot name, and that are chosen based on the criterion that an interpretation other than a reference to a place (a personal name or organization name) rarely occurs.[1]

Table 1. Label set and description of annotating *TA*.

Label	Description	Example (Target is **bold**)
Present	The author is at or near the location represented by the target	The **Eiffel Tower** is beautiful
Past	The author was in the place represented by the target	Going to the **Eiffel Tower** gave me memories for the weekend
Future	The author is not currently in the place represented by the target word, but seems to be going there now	I am going to the **Eiffel Tower** from now
Non-Temporal	The author has never been in the place represented by the target word and has no plans to go	What time does the **Eiffel Tower** open?
Non-Mention	Author does not mention places represented by the target word	I watched the movie "**Eiffel Tower**"

For 1200 Japanese tweets for each target, seven workers annotated based on the label set in Table 1, which is based on Li et al. [7]. To minimize the annotator's load as much as possible when collecting annotations based on crowdsourcing, we tried to make the specifications in the annotation as simple as possible. In the actual annotation, we presented the tweet text, and the target noun and asked the annotator to choose one of the choices for the tweet author's time recognition from the candidate labels for the entity represented by that noun, as showed in

---

[1] We used the following 10 nouns as targets: "Akihabara","Kiyomizu-dera (Kiyomizu Temple)", "Shibuya-eki (Shibuya station)","Sky Tree", "Sendai","shiyakusyo (city hall)", "kousaten (crossing)","byouin (hospital)", "kaisatsu (ticket gate)" and"doubtsuen (zoo)".

Fig. 1. To eliminate malicious annotators, we did not collect user annotations that could not be answered correctly by mixing check questions with correct answers at a rate of 1 out of every 18 tweets. We paid approximately 0.4 JPY (approximately 0.34 US cents) to the annotators for annotating one tweet. We used the Yahoo! Crowdsourcing platform to collect the annotations. The basic statistical information of the created corpus is shown in Table 2.

To investigate the quality of the annotation, we calculated the number of annotates are consistent for each annotated tweet. This result is shown in Table 2. As a result, we found that 93% of the tweets coincided with the labels of five or more people out of seven. This indicates that a relatively stable annotation can be performed. In addition, there was no major bias in the distribution of the annotated labels. This result suggests that our label set is stable.

Total tweets		12318
Agreement	7 votes	2212(18%)
	6 votes	5452(44%)
	5 votes	3797(31%)
Label	Present	2413(20%)
	Past	2342(19%)
	Future	2134(17%)
	Non-Temporal	4416(36%)
	Non-Mention	962(8%)

Fig. 1. Example of the annotation user interface in English. (The actual work was performed in Japanese.)

Fig. 2. Dataset statistics.

Agreement Analysis. Table 2 shows the distribution of the agreement aggregated for each target. For most of the targets, it can be seen that five annotations match for more than 90% of the tweets. However, the agreement for "ticket gate" is lower than that of the other targets. Because there are relatively few people who make long-term visits to ticket gates, most tweets reported that they had simply passed the ticket gate.

Table 2. The annotation agreement rate calculated for each target. Numbers in parentheses indicate percentages. The value of the last column indicates the percentage of instances where an agreement of five votes or more was obtained.

Target (in English)	Proper?	# of tweets	Agreement			
			7 votes	6 votes	5 votes	≥5 votes
"Akihabara"	✓	1254	315 (0.25)	529 (0.42)	345 (0.28)	0.948
"city-hall"		1204	235 (0.20)	522 (0.43)	367 (0.30)	0.934
"crossing"		1233	235 (0.20)	522 (0.43)	367 (0.30)	0.926
"hospital"		1257	325 (0.26)	532 (0.42)	318 (0.25)	0.934
"Kiyomizu Temple"	✓	1199	233 (0.19)	566 (0.47)	339 (0.28)	0.949
"Sendai"	✓	1240	201 (0.16)	577 (0.47)	383 (0.31)	0.936
"Shibuya Station"	✓	1214	219 (0.18)	538 (0.44)	383 (0.32)	0.939
"Tokyo Skytree"	✓	1220	197 (0.16)	553 (0.45)	373 (0.31)	0.904
"ticket gate"		1257	150 (0.12)	496 (0.39)	469 (0.37)	0.887
"zoo"		1240	181 (0.15)	610 (0.49)	363 (0.29)	0.930
Total		12318	2212 (0.18)	5452 (0.44)	3797 (0.31)	0.930

(3) I passed by a ticket gate with a person similar to Jeson.

In Example (3), three workers annotated the reference as **Present**, but four workers annotated it as **Past**. Both interpretations are reasonable; however, we set the threshold to five votes and interpretations with values less than that ware not used in the automatic analysis experiment.

Label Distribution. The distribution of labels in the data set is shown in Table 3. It was found that the distribution of labels differed for each target.

In particular, instances that refer to sightseeing spots (e.g. "Kiyomizu Temple","Tokyo Skytree" and "zoo") were often labeled **Non-Temporal**. From observations of several instances, we see that there were many reference to sightseeing spots seen on television that ware just impressions. In addition, there were a relatively large number of instances where "Sendai" and "zoo" were labeled as **Non-Mention**; however, this was largely influenced by compound nouns, metaphorical expressions[2] and ambiguity of the semantic class (e.g., the organization or location).

(4) The representative of the Miyagi Prefecture is decided by the Sendai Ikuei High School.

In Example (4), a part of the high school name mentioned in the sentence contains the place name Sendai; however, it does not represent Sendai itself. Because it is difficult to automatically exclude such instances, we excluded instances with the **Non-Mention** label in the automatic analysis experiment in this study.

From this data, we removed the instances that were labeled **Non-mention** and divided the dataset into 700 training data, 100 development data and 100

---

[2] In Japanese, "zoo" is also used as a metaphor to indicate a lively appearance.

test data for each target. We used this split of the data to train the model and to validate its performance.

4   Automatic Analysis of *Temporal Awareness*

To analyze the *TA* automatically, we formulated the problem as a target-dependent text classification as follows. In our models, we calculate the left and right context representation $v_r \in \mathbb{R}^M, v_l \in \mathbb{R}^M$ separately. These vectors are calculated via an "Encoder" module such as CNN or BiLSTM (bidirectional long short-term memory) from a sequence of word embedding vectors. Then, the concatenation of these representations is used to calculate the label distribution $y \in \{$Present, Past, Future, Non-Temporal$\}$ based on a feed-forward neural network.

4.1   Incorporation of Target Information via the Network Structure

In this task, it was expected that a chain of words and expressions that appear in the target's neighborhood context would be a large clue. Therefore, giving clues by the position of the target is natural. A large portion of the example was expected to determined the from left or right context of the target. To incorporate the target information into this model, we separately computed the vector representation of the target-dependent sequence splitting. We considered the following two architectures in our experiment.

Table 3. Detailed distribution of the final labels of our dataset. The numbers in parentheses indicate the percentages.

Target (in English)	# of tweets	Labels				
		Present	Past	Future	Non-Temporal	Non-Mention
"Akihabara"	1254	281 (0.22)	182 (0.15)	405 (0.32)	370 (0.30)	13 (0.01)
"city-hall"	1204	159 (0.13)	169 (0.14)	291 (0.24)	495 (0.41)	87 (0.07)
"crossing"	1233	303 (0.25)	403 (0.33)	28 (0.02)	413 (0.33)	83 (0.07)
"hospital"	1257	202 (0.16)	269 (0.21)	417 (0.33)	323 (0.26)	39 (0.03)
"Kiyomizu Temple"	1199	120 (0.10)	275 (0.23)	219 (0.18)	555 (0.46)	29 (0.02)
"Sendai"	1240	177 (0.14)	108 (0.09)	231 (0.19)	440 (0.35)	276 (0.22)
"Shibuya Station"	1214	451 (0.37)	277 (0.23)	90 (0.07)	389 (0.32)	1 (0.00)
"Tokyo Skytree"	1220	212 (0.17)	125 (0.10)	215 (0.18)	532 (0.44)	130 (0.11)
"ticket gate"	1257	438 (0.35)	425 (0.34)	50 (0.04)	305 (0.24)	28 (0.02)
"zoo"	1240	70 (0.06)	109 (0.09)	188 (0.15)	594 (0.48)	276 (0.22)
Total	12318	2413 (0.20)	2342 (0.19)	2134 (0.17)	4416 (0.36)	962 (0.08)

Flat This model encodes a full of sentence at once without considering the target in context.

Target-Dependent We also tried to introduce target information following Tang et al. [11] which is a state-of-the-art model of target-dependent sentiment classification. In this model, the left context and right context of the target are encoded separately and concatenated as $v = [v_l; v_r]$.

4.2 Encoders

We used the following Encoders to compare the classification performance.

Averaging Encoder. The Averaging Encoder computes the averages of the words in a sentence. We used this model as baseline-encoding model. This model does not consider word ordering or collocation but has good performance in many tasks and can be a good baseline.

Convolutional Encoder. We also used the CNN encoder based on Kim [5]. In this model, we obtain the vector representation of the sentence v via fixed size convolution and max-pooling.

Let $\mathbf{x} \in \mathbb{R}^k$ be a k-dimensional word vector. A sentence can be expressed as follows using a concatenation of word vectors:

$$\mathbf{x}_{1:n} = \mathbf{x}_1 \oplus \mathbf{x}_2 \oplus \ldots \oplus \mathbf{x}_n \tag{1}$$

In this equation, $\oplus$ is the concatenation operator of the vector. Let $\mathbf{x}_{i:i+j}$ be the concatenation of the i th word to the $i + j$th word in the sentence. Here we introduce the filter matrix $\mathbf{W} \in \mathbb{R}^{hk \times L}$. H represents the size of the filter (which corresponds to n in the n-gram model); in this paper, we used two words. L represents the number of filters to be applied. The result of applying the lth filter (the vector of the lth row of $\mathbf{W}$) to the ith word is calculated as follows:

$$c_{i,l} = f(\mathbf{W}_{\cdot l} \cdot \mathbf{x}_{i:i+h-1} + b) \tag{2}$$

where $b \in \mathbb{R}$ is the bias term. f is a nonlinear function; we used the sigmoid function in this paper. This filter is applied to all positions $(1, 2, \ldots n - h + 1)$ in the sentence and the following feature map $\mathbf{c} \in \mathbb{R}^{n-h+1}$:

$$\mathbf{c}_l = [c_{1,l}, c_{2,l}, \ldots, c_{n-h+1,l}] \tag{3}$$

Then, the maximum value of the feature map vector is extracted via max-over-time pooling.

$$\hat{c}_l = max(\mathbf{c}_l) \tag{4}$$

This procedure is performed for each filter vector and is used as a representation of the sentence.

Bi-Directional LSTM (BiLSTM) Encoder. For the BiLSTM encoder, where u_i is a N dimensional input embedding of a word, h_{i-1} is the previous output, and s_{i-1} is the previous cell state.

$$h_i, s_i = lstm(u_i, h_{i-1}, s_{i-1}) \tag{5}$$

For given text, the LSTM encoder is applied recursively to sequence from left-to-right. Our model adopted the Bi-directional LSTM model, which is concatenate of the final output state of left-to-right encoding $\overrightarrow{h}$ and right-to-left encoding $\overleftarrow{h}$ as $v_{bilstm} = [\overrightarrow{h}; \overleftarrow{h}]$. We also tried another model that incorporated the attention mechanism; however, the performance did not improve.

5   Experiment

We experimentally examined three types of options on the sentence encoding method and two network architectures for introducing the target information. As a baseline, we trained logistic the regression model with unigram and bigram features. We experimented with the following three settings.

In-domain In this setting, training data for all target words including the target word used for the test are used for training. This setting assumes that labeled data of target word that want to monitor can be obtained.

Table 4. Overall result (accuracy) of different encoders and their composition.

Architecture	Encoder	In-domain	10%	Cross-domain
Majority baseline		0.390	0.390	0.390
MaxEnt Model (Uni+Bi)		0.673	0.639	0.593
Flat	Averaging	0.606	0.554	0.554
	BiLSTM	0.599	0.490	0.516
	CNN	0.607	0.586	0.553
Target dependent	Averaging	0.583	0.564	0.548
	BiLSTM	0.684	0.634	0.609
	CNN	0.669	0.628	0.591

Cross-domain In this setting, training data of the target word used for the test are not used. This is an experiment assuming the case where the target entity is unknown.
10% samples In addition, we considered an intermediate situation assuming that a small amount of training data could be prepared for the target. Specifically, 10% of the data (70 instances) was sampled from the training data for the target and used in addition to the training data of other targets.

In all settings, the classification accuracy was used as the evaluation metric.

5.1   Detailed Setting

For all models, we used 300 dimensional word2vec embedding, learned from the Japanese twitter corpus, as the initial value for embedding. For optimization, ADAM [6] was used with default hyper-parameters, the size of a mini batch was fixed to 100, and the dropout rate of each layer was fixed to 0.5. To optimize each model equally, we adopted "early stopping" technique to train all models. When the accuracy for the development set was not updated for more than 10 epochs, we assigned a label to the test data using the maximum accuracy model on the development set.

5.2   Result

Table 4 shows the classification performance of our models. Flat architectures were found to be ineffective. In addition, the encoder that separately encodes the right and left sides of the target shows higher performance than the flatly encoding model. Interestingly, we observed that the encoders based on BiLSTM and CNN have very different architectures, but achieved similar classification performances. In addition, even in the setting of the cross-domain, our models achieved a certain level of performance, which suggests that knowledge transfer between targets is possible. Finally, by combining Target Dependent encoding and BiLSTM encoder, we found that it is competitive or slightly better performance than the baseline maximum entropy model.

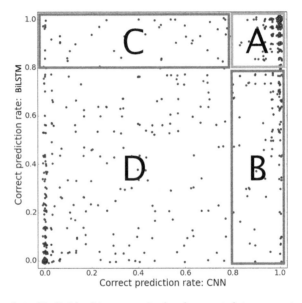

Fig. 3. Scatter plot of individual instances in development data, correct prediction rate of BiLSTM and CNN encoders over 30 different initialization.

6   Analyzing Characteristics of Different Encoders with an Initialization Test

We hypothesize that the encoders based on BiLSTM and CNN have different characteristics for the context utilization.

To compare different models, initial values in the learning stage of each model were varied randomly and each instance was analyzed to see whether it was accidentally correct or stably correct. First, we defined a metric called the correct prediction rate (CPR). This is a value defined for each instance that indicates whether the model learned from a number of initial values among the changed initial values can be corrected. We calculated the CPR for each instance for the CNN and BiLSTM models.

More specifically, the following method was used. We trained CNN-based target-dependent model and BiLSTM-based target-dependent model with 30 different initializations; we plot each example of development data in Fig. 3. The horizontal axis is the CPR of the CNN-based encoder, and the vertical axis is the CPR of the BiLSTM-based encoder. In this figure, the highlighted region in upper right means correctly predicted in most initializations, regardless of the initialization of the model. We found that a large portion of the examples are closer to both sides on the vertical axis, but not on the horizontal axis, this means that the CNN-based encoder is relatively more robust to initialization than the BiLSTM encoder.

We divided this scattergram into four regions, and investigated to which area each instances of development data belonged. We found that (A) 42.8% of examples (462/958) had over 80% CPR in both encoders; (B) 12.1% of examples (116/958) had over 80% CPR in the CNN encoder, but less than 80% CPR in the BiLSTM encoder; (C) 5.5% of examples (53/958) had over 80% CPR in the BiLSTM encoder, but less than 80% CPR in the CNN encoder; and (D) 36.8% of examples (353/958) had less than 80% CPR in both encoders.

To examine which examples are differently encoded, we employ the linguistic annotator to annotate with *TA* label and their target reference, in addition, we consulted to give label to a clue words that need to predict *TA* label. This annotation was performed on randomly shuffled (B) + (C) portion of development data, and actual *TA* label ware hide from the annotator. As a result, clues were annotated in 58% of instances included in area (B), whereas in cases included in area (C) clues were only found in 32% of instances annotated. This result suggests that CNN can successfully encode sentences with clearer clues, whereas BiLSTM is better suited to handle ambiguous clues such as chain of function words.

7   Conclusions

In this paper, we addressed the task of analyzing the Temporal Awareness for location references. As a result of crowdsourcing annotation, the agreement between annotations was high, indicating that the task was properly designed. In addition, we constructed encoders based on BiLSTM and CNN and compared them. It became clear that BiLSTM can handle blurred clues such as linkage of function words better. Investigating of a model integrated with location name extraction and disambiguation is left as a future task.

Acknowledgments. This work was partially supported by *Research and Development on Real World Big Data Integration and Analysis*, MEXT, Japan.

References

1. Dong, L., Wei, F., Tan, C., Tang, D., Zhou, M., Xu, K.: Adaptive recursive neural network for target-dependent twitter sentiment classification. In: Proceedings of the ACL (2014)
2. Huang, R., Cases, I., Jurafsky, D., Condoravdi, C., Riloff, E.: Distinguishing past, on-going, and future events: the eventstatus corpus. In: EMNLP (2016)
3. Jiang, L., Yu, M., Zhou, M., Liu, X., Zhao, T.: Target-dependent twitter sentiment classification. In: Proceedings of the ACL-HLT (2011)
4. Johnson, R., Zhang, T.: Effective use of word order for text categorization with convolutional neural networks (2014). arXiv preprint arXiv:1412.1058
5. Kim, Y.: Convolutional neural networks for sentence classification (2014). arXiv preprint arXiv:1408.5882
6. Kingma, D.P., Ba, J.: Adam: a method for stochastic optimization. In: ICLR (2014). http://arxiv.org/abs/1412.6980
7. Li, C., Sun, A.: Fine-grained location extraction from tweets with temporal awareness. In: Proceedings of the SIGIR (2014)
8. Pustejovsky et al.: The timebank corpus. In: Corpus Linguistics, vol. 2003, p. 40 (2003)
9. Saurí, R., Pustejovsky, J.: Are you sure that this happened? Assessing the factuality degree of events in text. Comput. Linguist. **38**(2), 261–299 (2012). https://doi.org/10.1162/COLI_a_00096
10. Severyn, A., Moschitti, A.: Twitter sentiment analysis with deep convolutional neural networks. In: Proceedings of the SIGIR (2015)
11. Tang, D., Qin, B., Feng, X., Liu, T.: Effective LSTMs for target-dependent sentiment classification. In: Proceedings of COLING 2016, The 26th International Conference on Computational Linguistics: Technical Papers, pp. 3298–3307. The COLING 2016 Organizing Committee, Osaka, Japan, December 2016. http://aclweb.org/anthology/C16-1311
12. Verhagen, M., Sauri, R., Caselli, T., Pustejovsky, J.: Semeval-2010 task 13: Tempeval-2. In: Proceedings of the 5th International Workshop on Semantic Evaluation, pp. 57–62. Association for Computational Linguistics (2010)
13. Vo, D.T., Zhang, Y.: Target-dependent twitter sentiment classification with rich automatic features. In: Proceedings of the IJCAI (2015)

Just the Facts: Winnowing Microblogs for Newsworthy Statements using Non-Lexical Features

Nigel Dewdney$^{(\boxtimes)}$ ⓘ and Rachel Cotterill ⓘ

Department of Computer Science, University of Sheffield, Sheffield, UK
acp08njd@sheffield.ac.uk, r.cotterill@sheffield.ac.uk

Abstract. Microblogging has become a popular method to disseminate information quickly, but also for many other dialogue acts such as expression opinion and advertising. As the volumes have risen, the task of filtering messages for wanted information has become increasingly important. In this work we examine the potential of natural language processing and machine learning to filter short messages for those that state items of news. We propose an approach that makes use of information carried at a deeper level than message's lexical surface, and show that this can be used effectively improve precision in filtering Twitter messages. Our method outperforms a baseline unigram "bag-of-words" approach to selecting news-event Tweets, yielding a 4.8% drop in false detection.

1 Introduction

Microblogging has become one of the most popular ways for people to communicate to a wide range of audiences online. As a result it has not only become a popular method for disseminating new information quickly, [1,2], but also one for marketing and entertainment purposes. For example, despite Twitter's anti-spam policies, research estimates between 10% and 46% of user accounts are automated or semi-automated [3]. Also, social media, by its very nature, is dominated by social communication, most of which is dyadic in nature and not of wider interest. It has been likened to chat rather than publication [4]. As the volume of microblog messages has risen to an estimated 400 million per day [5], the task of filtering messages for wanted information, such as items of news, has become increasingly important.

While established news events and messages on particular topics may be targeted by keywords and hashtags, breaking news stories are more problematic as the topic is unknown prior to its establishment. A mitigation for this is to detect bursts of interest in a topic within social media. However this approach has the limitation that not all topics are related to news events, and messages might not give useful information. Ideally, in filtering social media for news, one would wish to remove messages that do not assert any relevant information. A useful question to systematically answer, therefore, would be "does the Tweet make an informative statement?"

© Springer Nature Switzerland AG 2018
A. Gelbukh (Ed.): CICLing 2017, LNCS 10762, pp. 391–403, 2018.
https://doi.org/10.1007/978-3-319-77116-8_29

In this work we examine the potential for natural language processing and machine learning to filter short messages for those that state items of news. We propose an approach that makes use of information carried at a deeper level than at a message's lexical surface, and show that this can be used effectively to improve precision in filtering Twitter messages selected by an event-burst detector.

We first describe previous work related to this task before going on to describe the experiments carried out and the results obtained, concluding with a discussion on the observations made.

2   Related Work

The detection of new information in document streams is often associated with the identification of breaking news. Early work examined first story detection for newswire monitoring [6] (finding it to be a challenging task), and the detection of when new information was stated within a topic [7]. More recently focus has shifted to the detection of breaking news from social media sources as its use has become widespread. For example, Petrović et al. [8] have investigated first story detection in Twitter micro-blog feeds, while later work [2] examined breaking news in Twitter and newswire, finding that both reported the same news events (with Twitter often leading, though not with major news events). Comparing different social media platforms, Osborne and Dredze [9] found that all reported the same news events although Twitter was favoured for breaking stories. Meanwhile, Aiello et al. [10] compared the wide variety of methods used in trending topic detection. They found that while natural language processing techniques can perform well, models incorporating temporal distributions are needed where multiple stories evolve in parallel, and that use of n-gram co-occurrence and *tf-idf_t* based ranking provided the most consistent method.

Correspondingly, with the rise in microblogging popularity there has been a rise in largely unwanted content and the requirement to create models to detect spam, uninformative messages and their sources, e.g. [11–13]. Such models typically leverage many features and their construction relies on the application of machine learning techniques. The use of machine learning on Twitter data has been widely explored for various applications besides topic classification, such as political orientation [14], user demographics [15], and particularly sentiment analysis [16], which has been a popular task for researchers [17].

These applications largely make use of lexical, network and temporal features, focussing on associating vocabulary with desired classes. Of interest to us are features that capture aspects of communication *intent* rather than subject matter. These could be provided by deeper natural language analysis. For example, Higashinaka et al. [18] find that syntax-based filtering helps in selecting high quality messages for use in dialogue systems.

Filtering out unwanted message types is an alternative approach to detecting wanted ones. Spam filtering is a popularly studied example, but where this is typically an issue social dynamics may often be leveraged. For example see [19,20].

With the goal of sorting Tweets into broad categories of message purpose, Sriram et al. [21] have used the presence/non-presence of seven features (such as word-shortening, opinion words) along with the author ID. Using Naive Bayes to build classification models for Tweets into general classes of News, Events, Opinions, Deals and Private Messages, they obtained a 32% relative increase in performance over use of Bag-of-Words as features. Rather than classify individual Tweets, Zubiaga et al. [22] have developed an approach to classify trending topics as News, Ongoing Events, Memes, and Commemorative topics. Using features that characterise the social network propagating the topic rather than content features, they also achieve improvements in classification performance over Bag-of-Word models.

Another aspect to messages is the degree to which the information purported is believable. Gupta el al. [23] have proposed using a combination of Tweet metadata and content features in a trained ranking model, ordering Tweets according to credibility. They used a training set of 500 Tweets selected from six high-impact events. These were labelled as to whether the Tweet contained event information and, if so, whether it was credible, possibly credible, or incredible. The resultant model is used to re-order a feed of user's received Tweets via a browser plug-in.

The informativeness of a sentence has been considered in other contexts, notably in summarisation. Yih et al. [24] have shown that maximising the number of informative content-words (as scored by their relative frequencies) produces some of the best reported results in multi-document summarisation. In selecting sentences for opinion mining, Zhu et al. [25] form a graph representing sentences connected to some aspect of an entity, and use node centrality as a measure of sentence informativeness with respect to it. Bing et al.'s approach [26] is to both identify and extract informative noun and verb phrases using constituency trees, constructing new sentences to maximise summary content.

Selection of news bearing Tweets from opinion is similarly of growing interest. Madhawa and Atukorale [27], noting that creating domain specific hand-annotated training data is expensive, have investigated selection of newsworthy Tweets from expressions of opinion for single events (determined by query) using heuristics to obtain training data. They achieve an F1 score of 80%, although their method, which relies heavily on unigram and bigrams, does not generalise across domains.

Comparing tasks, we see that whereas others have focussed on identifying new emerging topics, message classes, credible sources, or conversely sources of spam, our focus is on detecting whether a message is explicitly informative or not, using news statements as examples.

Motivated by the need to capture some notion of informativeness, we chose to investigate similar features to those used by Bing et al. We chose to avoid lexical features, excepting our baseline, because we wished to avoid correlation with any particular topics, noting that we may not know what the topics of interest would be in the future.

News-event statement detection is complementary to other message classification tasks. The object is to refine or provide pre-filtering for other message stream classification or extraction tasks such as knowledge base population, e.g. [28]. We do not, however, consider the credibility or veracity of statements provided at this stage, focussing only on their detection.

Given the characteristic features of messages have been found to be useful in similar tasks, we hypothesise that the explicit informativeness of a message will be more associated with features derived from the syntactic role constituent words play (e.g. part-of-speech), and word frequency (or feature chunk frequency), than with the words themselves.

3   Modelling Approach

Text classification tasks are traditionally approached by creating models based on the tokenised content of the texts, and sometimes with the associated metadata as well. Machine learning techniques are used to optimise the classification performance of models. However there are many factors that may influence choice of words. Different words may perform different functions and may convey information at the syntactic, semantic, and pragmatic levels of linguistic interpretation. As we are interested in selecting Tweets containing news statements, we hypothesise that these may be discriminated from other types of message by features other than lexemes.

3.1   Features

In the news-event filtering scenario, the selection of event related Tweets has already been carried out by models built on the associated metadata. This is often carried out by detecting a burst of related Tweets. We therefore restrict features to those related to message content. Messages are tokenised by segmentation on white-space and punctuation other than '#' and '@'. We use these lexical tokens as unigram features for our baseline model.

The non-lexical features of interest break down into four categories:

- **GRP:** Orthographically identified features - Emoticons, Hashtags, URLs, and User-name Identifiers
- **NE:** Named Entities - Persons, Locations, Organisations, Dates, Sums of Money
- **SYN:** Syntactically determined features - noun-phrases, verb-phrases and pronouns
- **FRQ:** Frequency based features - total word IDF, total phrase IDF, average word IDF and average phrase IDF

The use of orthographically determined features was motivated by the expectation that statements and non-statements would either be more or less likely to contain such features. For example one might expect a Tweet expressing an opinion to be more likely to contain an emoticon than a statement would (a statement by definition must give information about something). Things are often named, hence the motivation for our Named Entity features. References may be conceptual or pronominal. We attempt to capture use of these by the syntactically determined features. Our expectation was that personal pronouns would be more prevalent in conversational messages than in news items. Our final feature subset was motivated by the idea that statements should be informative messages and therefore the words themselves should be more informative. A word's selectivity for a topic for a term can be estimated from its frequency in corpora, e.g. see [29]. Our idea was that the greater the statistical informativeness of the words in a message the more likely it would be an explicit statement.

The orthographically identified features may be readily identified by simple pattern matching. Named Entities may be detected using a named Entity extractor. We used the the TwitIE extractor [30] to detect the presence of named entity types and the set of orthographic features. It was also used to detect the presence of pronouns. The other syntactically determined features, noun and verb phrases, were detected using a simple part-of-speech parser that identified compound nouns and verb particles.

Frequency based features were based on the occurrence of the tokens or phrases across all of the Twitter messages we had available (see Sect. 3.3). Treating each message as a document, the inverse document frequency (IDF) for each word (or detected phrase) was calculated. The total and average IDF for words, and separately for detected phrases, in a message constituted the frequency feature set.

3.2   Classifiers

There are multiple machine learning methodologies. We apply four popular techniques here. Naive Bayes and Maximum Entropy classifiers build models based on the the frequency of the features in the training set. Decision trees (Quinlan's C4.5), seek to create segment classes by separating by each feature in turn. Support Vector Machines optimise a decision boundary margin between class features in a vector space. The libSVM [31] implementation was used to create the SVM models using a linear kernel. The other classifiers are provided by the Mallet toolkit [32]. All the classifiers were used "out-of-the-box" without parameter tuning as optimisation of classifier was not the objective of this investigation.

3.3   Data

We used data from the Redites project [33] which created a corpus of 1.4 million Tweets by applying a high recall, low precision, event detector on a Twitter feed. This resulted in a data set of about 489,000 events. A sample of this collection was hand annotated by the Redites project team, each event Tweet classified as either a news event or a non-news event. 2,286 Tweets were positively identified as yielding information connected to a news event, leaving approximately 1.03

million Tweets connected to non-news events. These covered a time period of 09:58 2/9/13–11:16 30/9/13.

We used the full Redites corpus to calculate IDF scores for words and our detected noun/verb phrases, and the annotated sub-corpus for our experiments.

For the first set of experiments we remove the bias towards non-news events by selecting 2,000 news-event Tweets and 2,000 non-news-events to form a sub-corpus for training and evaluating models. This sub-corpus was further divided into ten sections such that Tweets from the same event are contained within the same section. The intent here was to exercise some control for any event specific vocabulary when training and testing models using the ten-fold cross-validation methodology.

Table 1. Accuracy (Mean Precision) of models in 10-fold classification using feature set combinations with standard deviation in parentheses

	Max. Entropy	Naive Bayes	C4.5	Linear SVM
Unigrams	77.6% (26.0%)	**76.9% (34.0%)**	70.1% (31.0%)	76.9% (23.8%)
GRP	68.3% (1.7%)	68.4% (0.9%)	64.8% (6.3%)	72.6% (6.0%)
NE	70.3% (4.2%)	61.8% (4.2%)	61.8% (13.3%)	77.1% (4.0%)
SYN	50.0% (0.2%)	50.0% (0.0%)	60.2% (6.0%)	72.8% (2.4%)
FRQ	54.2% (1.2%)	61.1% (2.1%)	63.2% (14.6%)	78.6% (2.1%)
GRP+NE+SYN+FRQ	82.7% (3.5%)	73.0% (3.0%)	82.4% (3.8%)	**85.0% (2.3%)**
All-Features	86.9% (3.8%)	**92.1% (4.7%)**	85.2% (7.6%)	90.3% (8.9%)

4   Experiments

4.1   Feature Set Comparison

An initial experiment compared the classifiers with each of the feature sets using the class-balanced subset of the data comprising 2,000 news-event Tweets and 2,000 non-news event Tweets. Tweets were grouped by event and split into 10 folds such that any one event was represented in just one fold, minimising potential skew in testing from the small number of duplicates in the form of ReTweets. Standard 10-fold cross validation was used to obtain average performance figures for each of the features sets, and combinations thereof, using each of the example classifier technologies. Default hyper-parameters were used for classifiers as optimisation of the learning was not the object of the experiment.

We measured simple accuracy, i.e. percentage of correct answers, and averaged results from across the folds. This is equivalent to Mean Precision given the balanced test data. We also measured the variance in results to give an indication of how reliable the observed accuracy figures were. Results are shown in Table 1. An error margin of 1 standard deviation is shown in parenthesis, and the best results for unigrams, the non-lexical feature set and the combination of both are shown in bold.

In this experiment we observe a strong baseline performance, albeit with a high degree of variance. One reason for this could be a lack of vocabulary coverage to form reliable models that capture forms of stating news items. Another reason could be correlation between topics and news informative Tweets. (One might expect an overlap given news tends to be about particular topics.) Folds were controlled for event separation, although independence of events could not be assured.

Sole use of each non-lexical feature division resulted in lower than baseline performance but with significantly less variance across the folds. This suggests the models capture insufficient information for the task, but what information is captured is more stable than that at the lexical level. The features derived syntactically performed the worst although they did provide some classification power when used with C4.5 or linear SVM. This could be a result of the relative sparsity in the features given to the difficulty of phrase detection in informal texts. The combination of all the non-lexical features gave rise to models above baseline for three of the classifier technologies, but Naive Bayes (which assumes feature independence) failed to achieve any improvement.

Lexical and non-lexical features do capture different contributory information. This can be seen in the performance achieved by using the combination of all the features, including the Naive Bayes classifier. An average 13% improvement (18% relative) in accuracy over the baseline is achieved using all features.

The results of this experiment showed some promise that non-lexical, vocabulary independent, features could significantly help in identifying news-bearing Tweets. We next sought to further examine the contribution of our features in the classification task and model performance given new later occurring data.

Table 2. Feature contribution to classifier model accuracy (Mean Precision) on held out data

	Max. Entropy	Naive Bayes	C4.5	Linear SVM
Unigrams	52.8%	59.3%	59.8%	51.6%
NE+SYN+FRQ	72.7%	70.5%	68.4%	80.4%
GRP+SYN+FRQ	65.2%	68.0%	66.3%	82.4%
GRP+NE+FRQ	72.9%	69.8%	67.1%	83.0%
GRP+NE+SYN	56.3%	67.8%	68.9%	77.6%
GRP+NE+SYN+FRQ	71.3%	69.4%	67.3%	82.7%
Average perf.	65.2%	67.5%	66.3%	76.3%

4.2 Feature Set Contribution

The next set of experiments looked to give an indication of the portability of models and the contribution of each of the non-lexical feature types in classification models (using the "hold-out" methodology).

A further 572 Tweets from the assessed Redites corpus had been held out from the data used in the cross-fold validation experiments, equally distributed between news event and non-news event. They also had later timestamps and separate event identifiers. We used these Tweets as our test set.

Models for these experiments were built with each type of classifier using all of the 4,000 Tweets used in the first set of experiments. The non-lexical features were used with the omission of one of our feature classes, each in turn. The amount of information the omitted feature set holds can be gauged by the drop in performance that may be observed when compared with that with all features present. Again, given a balanced set of binary classes, we simply measured classification accuracy. The results are shown in Table 2.

The first notable observation is that the baseline performance (the unigram model) is significantly lower than that observed in the initial closed set experiment. Average classifier accuracy with unigrams is approximately 20% less. In comparison, the use of non-lexical features on this test set in comparison shows an average 8% drop in classifier accuracy. This lends support to the idea that apparently good results from lexical models are in some measure due to learnt vocabulary covering current news topics.

How uncommon words in a Tweet are - the informativeness as measured by their frequency in messages - though, is correlated with news-event statements. This can be seen from the 5% drop in accuracy when omitted. The presence of Named Entity features are also a factor given an average 3% drop in accuracy when omitted. Orthographically and syntactically determined features do not seem to provide any significant additional information in this particular experiment.

Having determined that more temporally stable classification models for identifying statements of news in Tweets can be learnt using non-lexical features, we finally sought to evaluate example models on a representative set of data from an event detection selection of Twitter.

4.3   Refining News-Event Tweet Detection

In the experiments above we sought to control for the prior expectation of informative event Tweets by training and testing using the same number of positive examples as negative ones. In a stream of arriving Tweets one would expect to receive many more uninformative Tweets than informative ones. The task is to filter out Tweets that do not make news-bearing statements whilst selecting those that do without missing any. In practice the filter is likely to make make false positive selections and false negative rejections. The balance between the two is something that may be controlled by the user, however, by using classifier confidence. Applying a higher threshold for accepting Tweets classified as news-bearing statements will result in fewer messages overall but those that are selected should be less likely to contain false positive examples. The Precision of the results, i.e. the number of correctly identified Tweets as a proportion of the selected Tweets, should rise. The cost however is a decline in the Recall, i.e. the number of correctly selected Tweets as a proportion of all the positive examples.

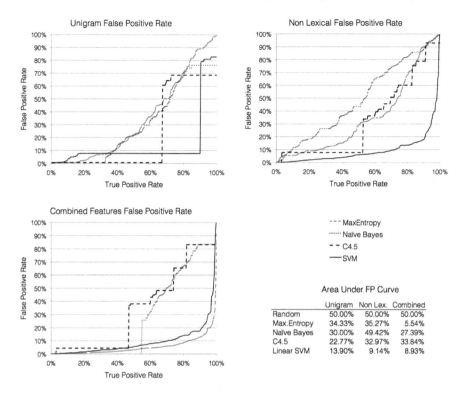

Fig. 1. ROC curves for classifiers on unseen Twitter data using unigram and non-lexical features

All of the Tweets marked as non-news event Tweets observed during the period in which the held-out news event Tweets occurred, added to the held-out data, yielded a corpus of 173,514 Tweets. We ran classifiers trained on the balanced 4,000 Tweet sub-corpus and then ranked results by classifier confidence that a Tweet was news-bearing. We measured the rate at which non-news-event messages were falsely accepted, and the corresponding proportion of the news-event Tweets correctly selected, as we reduced the threshold applied. This gave us the false/true positive receiver operator characteristics (ROC) shown in Fig. 1.

Absolute performance figures calculated at the Tweet level are naturally low. The 286 news-event Tweets in the held out data covered 187 news-events. By contrast non-news Tweets made up 84,171 non news-events. The ratio of correctly selected Tweets to incorrectly selected gives the precision in selection up to the given threshold. The area under the ROC curve gives an overall performance metric for the associated model. The false positive rate averaged over true positive rates (AUC) for the best performing models with unigram features and the non-lexical features, both using SVM, were 13.9% and 9.1% respectively. A maximum entropy model performs the best with a combination of all the features, despite having poor false positive rates for either feature set alone, with an AUC of 5.5%.

5   Error Analysis

While unigram performance within a corpus appears to be very good, we observed that it drops when the classifiers are run on later data. This does not seem to be as much as a problem for classifiers using non-lexical features. As lexical features are strongly correlated to topic (the basis for most information retrieval methods) this suggests achievement of a higher precision than should be expected through classification of informative topics. Tweets with previously unseen topics are less likely to be classed as informative because their vocabulary will not be that of the training topics. (This may be alleviated by larger training sets and regular retraining but would entail the additional associated cost.)

To gain further insight into what our models are capturing, we inspected the Tweets they most confidently classified. The false positive classification results made most confidently by the SVM classifier trained on the non-lexical features are shown below:

– 76 Regime Troops killed by #Syria free army, death toll 4,565, #Bahrain #UAE #Turkey #Iran #Russia #USA #UN cont http://t co/Pt6MbzJ2PS
– RT @———: Obama, Hollande to face off with Putin over Syria http://t co/YDbLDEX7cx #Syria #USA #UK #Iraq #Lebanon #Oman #Qatar #KSA #Y
– Allies against #Syria: US, Australia, Canada, France, Italy, Japan, South Korea, Saudi Arabia, Spain, Turkey and the United Kingdom
– RT @———: PR pros are needed for peace building project #Japan, #Thailand, #India, #Germany, #France, #UK, #USA, #Canada, #Chili
– US urges #Syria to unveil chemical weapons stockpile http://t co/2ri0165iNo #Belgium #egypt #Iraq #news #ABC #Woman #BBC #world #AP #sydney

At least three of these seem to be statements of news-related information. The others have a high number of locations or hashtags that have fooled the classifier. This suggests that the original annotation of Tweets may be debatable. Alternatively, or additionally, it may be the case that not all informative statements in Twitter are news-event related.

We then sought to better understand why statements of news items might be missed. We inspected those in our corpus most likely to result in false negatives. The least confidently predicted news-bearing examples were:

– Papers: United to bid for Coentrao http://t co/trcWis4Mlb
– #Nestl tells us it won't comment on market rumors about the sale of Powerbar gt http://t co/qEx5bHf8QK
– Kumi Naidoo on Greenpeace Activists Detained in Russia http://t co/q0LFrjDRlP

- RT @Polygon: Valve announces Steam Controller http://t co/OwOtt3p2x2
- US braces for possible shutdown http://t co/Uxp4hSQThL

It is not clear on first inspection why these examples were harder for the models to spot. URLs are present but alone seems unlikely to be the cause. Closer inspection of the associated feature vectors show that the Named Entity tagger did not recognise the entity mentions. Total IDF is also relatively low compared with those messages confidently predicted as news-bearing statements. This suggests that improved performance could be achieved with better Named Entity detection and token frequency estimates.

6   Conclusions

In this work we have proposed an approach to filtering streams of short messages for those that make informative statements. Assuming that Tweets in news events are informative we have shown that models built using a set of non-lexical features can out-perform one based on traditional unigrams ("bag-of-words"). Use of a combined feature set yields the best performance in a closed experimental setting suggesting that each feature set carries some independent information.

We have proceeded to show that when used to select Tweets from a message stream provided by a news-burst filter, non-lexical feature based models can outperform unigram based counterparts. Our best performing models were again built models built using the combination of lexical and non-lexical features.

Features making use of message token frequency, and the presence of various types of Named Entity were found to be the most useful. Although simple orthographically determined features, such as hashtags, had some discriminatory power they were not always effective. Compound phrase features which were detected using a simple part-of-speech parser were not found to be useful, although this may have been as the result of feature sparsity and low detection rates from the parser.

A Linear SVM model using our non-lexical features produces an average 4.8% drop in false news-event Tweet detection over one trained on unigrams.

Finally, manual inspection of the errors made most confidently by the classifiers suggests that although non-lexical features are useful in message type detection, they do need to be well recognised. Improved Named Entity recognition, phrase detection, and token weighting would be beneficial in identification of news-bearing Tweets by our approach. It may also be the case that not all informative statements are news-related in Twitter event bursts. Future work will examine whether or not this is the case.

References

1. Lloyd, L., Kaulgud, P., Skiena, S.: Newspapers vs. blogs: who gets the scoop. In: AAAI Spring Symposium on Computational Approaches to Analyzing Weblogs, pp. 117–124 (2006)

2. Petrovic, S., Osborne, M., McCreadie, R., Macdonald, C., Ounis, I., Shrimpton, L.: Can twitter replace newswire for breaking news? In: Seventh International AAAI Conference on Weblogs and Social Media (2013)
3. Chu, Z., Gianvecchio, S., Wang, H., Jajodia, S.: Detecting automation of twitter accounts: are you a human, bot, or cyborg? IEEE Trans. Dependable Secure Comput. **9**, 811–824 (2012)
4. Alvanaki, F., Sebastian, M., Ramamritham, K., Weikum, G.: Enblogue: emergent topic detection in web 2.0 streams. In: Proceedings of the 2011 International Conference on Management of Data. SIGMOD 2011, pp. 1271–1274. ACM, New York (2011)
5. Morstatter, F., Pfeffer, J., Liu, H., Carley, K.M.: Is the sample good enough? Comparing data from twitter's streaming API with twitter's firehose. In: Seventh International AAAI Conference on Weblogs and Social Media (2013)
6. Allan, J., Lavrenko, V., Jin, H.: First story detection in TDT is hard. In: CIKM 2000: Proceedings of the Ninth International Conference on Information and Knowledge Management, pp. 374–381. ACM, New York (2000)
7. Soboroff, I., Harman, D.: Novelty detection: the TREC experience. In: HLT 2005: Proceedings of the Conference on Human Language Technology and Empirical Methods in Natural Language Processing, Morristown, NJ, USA, pp. 105–112. Association for Computational Linguistics (2005)
8. Petrović, S., Osborne, M., Lavrenko, V.: Streaming first story detection with application to twitter. In: Human Language Technologies: The 2010 Annual Conference of the North American Chapter of the Association for Computational Linguistics, Los Angeles, CA, USA, pp. 181–189. Association for Computational Linguistics (2010)
9. Osborne, M., Dredze, M.: Facebook, twitter and google plus for breaking news: is there a winner? (2014)
10. Aiello, L.M., et al.: Sensing trending topics in twitter. IEEE Trans. Multimed. **15**, 1268–1282 (2013)
11. McCord, M., Chuah, M.: Spam detection on twitter using traditional classifiers. In: Calero, J.M.A., Yang, L.T., Mármol, F.G., García Villalba, L.J., Li, A.X., Wang, Y. (eds.) ATC 2011. LNCS, vol. 6906, pp. 175–186. Springer, Heidelberg (2011). https://doi.org/10.1007/978-3-642-23496-5_13
12. Wang, A.H.: Don't follow me: spam detection in twitter. In: Proceedings of the 2010 International Conference on Security and Cryptography (SECRYPT), pp. 1–10. IEEE (2010)
13. Haustein, S., Bowman, T.D., Holmberg, K., Tsou, A., Sugimoto, C.R., Larivière, V.: Tweets as impact indicators: examining the implications of automated "bot" accounts on twitter. J. Assoc. Inf. Sci. Technol. **67**, 232–238 (2016)
14. Cohen, R., Ruths, D.: Classifying political orientation on twitter: it's not easy! In: ICWSM (2013)
15. Pennacchiotti, M., Popescu, A.M.: A machine learning approach to twitter user classification. In: ICWSM, vol. 11, pp. 281–288 (2011)
16. Neethu, M., Rajasree, R.: Sentiment analysis in twitter using machine learning techniques. In: 2013 Fourth International Conference on Computing, Communications and Networking Technologies (ICCCNT), pp. 1–5. IEEE (2013)
17. Nakov, P., Ritter, A., Rosenthal, S., Sebastiani, F., Stoyanov, V.: Semeval-2016 task 4: sentiment analysis in twitter. In: Proceedings of the 10th International Workshop on Semantic Evaluation (SemEval 2016), San Diego, US (2016, forthcoming)

18. Higashinaka, R., Kobayashi, N., Hirano, T., Miyazaki, C., Meguro, T., Makino, T., Matsuo, Y.: Syntactic filtering and content-based retrieval of twitter sentences for the generation of system utterances in dialogue systems. In: Proceedings of the IWSDS, pp. 113–123 (2014)
19. Song, J., Lee, S., Kim, J.: Spam filtering in twitter using sender-receiver relationship. In: Sommer, R., Balzarotti, D., Maier, G. (eds.) RAID 2011. LNCS, vol. 6961, pp. 301–317. Springer, Heidelberg (2011). https://doi.org/10.1007/978-3-642-23644-0_16
20. Clark, E.M., Williams, J.R., Jones, C.A., Galbraith, R.A., Danforth, C.M., Dodds, P.S.: Sifting robotic from organic text: a natural language approach for detecting automation on twitter. J. Comput. Sci. (2015)
21. Sriram, B., Fuhry, D., Demir, E., Ferhatosmanoglu, H., Demirbas, M.: Short text classification in twitter to improve information filtering. In: Proceedings of the 33rd International ACM SIGIR Conference on Research and Development In Information Retrieval, pp. 841–842. ACM (2010)
22. Zubiaga, A., Spina, D., Martinez, R., Fresno, V.: Real-time classification of twitter trends. J. Assoc. Inf. Sci. Technol. **66**, 462–473 (2015)
23. Gupta, A., Kumaraguru, P., Castillo, C., Meier, P.: TweetCred: real-time credibility assessment of content on twitter. In: Aiello, L.M., McFarland, D. (eds.) SocInfo 2014. LNCS, vol. 8851, pp. 228–243. Springer, Cham (2014). https://doi.org/10.1007/978-3-319-13734-6_16
24. Yih, W.t., Goodman, J., Vanderwende, L., Suzuki, H.: Multi-document summarization by maximizing informative content-words. In: IJCAI, vol. 7, pp. 1776–1782 (2007)
25. Zhu, L., Gao, S., Pan, S.J., Li, H., Deng, D., Shahabi, C.: Graph-based informative-sentence selection for opinion summarization. In: 2013 IEEE/ACM International Conference on Advances in Social Networks Analysis and Mining (ASONAM), pp. 408–412. IEEE (2013)
26. Bing, L., Li, P., Liao, Y., Lam, W., Guo, W., Passonneau, R.J.: Abstractive multi-document summarization via phrase selection and merging (2015). arXiv preprint arXiv:1506.01597
27. Madhawa, P., Atukorale, A.S.: A robust algorithm for determining the newsworthiness of microblogs. In: 2015 Fifteenth International Conference on Advances in ICT for Emerging Regions (ICTer), pp. 135–139. IEEE (2015)
28. Kuzey, E., Vreeken, J., Weikum, G.: A fresh look on knowledge bases: Distilling named events from news. In: Proceedings of the 23rd ACM International Conference on Conference on Information and Knowledge Management, pp. 1689–1698. ACM (2014)
29. Joho, H., Sanderson, M.: Document frequency and term specificity. In: Large Scale Semantic Access to Content (Text, Image, Video, and Sound), Le Centre de Hautes Etudes Internationales D'informatique Documentaire, pp. 350–359 (2007)
30. Bontcheva, K., Derczynski, L., Funk, A., Greenwood, M.A., Maynard, D., Aswani, N.: TwitIE: an open-source information extraction pipeline for microblog text. In: Proceedings of the International Conference on Recent Advances in Natural Language Processing. Association for Computational Linguistics (2013)
31. Chang, C.C., Lin, C.J.: Libsvm: a library for support vector machines. ACM Trans. Intell. Syst. Technol. (TIST) **2**, 27 (2011)
32. McCallum, A.K.: Mallet: a machine learning for language toolkit (2002)
33. Osborne, M., et al.: Real-time detection, tracking, and monitoring of automatically discovered events in social media (2014)

Impact of Content Features for Automatic Online Abuse Detection

Etienne Papegnies[(✉)], Vincent Labatut, Richard Dufour, and Georges Linarès

LIA - University of Avignon, Avignon, France
{etienne.papegnies,vincent.labatut,richard.dufour,
georges.linares}@univ-avignon.fr

Abstract. Online communities have gained considerable importance in recent years due to the increasing number of people connected to the Internet. Moderating user content in online communities is mainly performed manually, and reducing the workload through automatic methods is of great financial interest for community maintainers. Often, the industry uses basic approaches such as bad words filtering and regular expression matching to assist the moderators. In this article, we consider the task of automatically determining if a message is abusive. This task is complex since messages are written in a non-standardized way, including spelling errors, abbreviations, community-specific codes, etc. First, we evaluate the system that we propose using standard features of online messages. Then, we evaluate the impact of the addition of pre-processing strategies, as well as original specific features developed for the community of an online in-browser strategy game. We finally propose to analyze the usefulness of this wide range of features using feature selection. This work can lead to two possible applications: (1) automatically flag potentially abusive messages to draw the moderator's attention on a narrow subset of messages; and (2) fully automate the moderation process by deciding whether a message is abusive without any human intervention.

1 Introduction

Among the main achievements and anonymity it brought into the way we communicate. Online communities, which are freely accessible exchange spaces on the Internet, have enjoyed a surge of users as a result. They come in many shapes and forms but they all share a common aspect: they have to be maintained by some party. Some online communities have gained considerable socio-economical importance due to their huge user base. A correct behavior in these communities is usually required to comply with a given set of rules so that users may enjoy a hospitable and productive environment. However, freedom and anonymity often give rise to *abusive behaviors*. The definition of an abusive behavior is often dependent on community rules. Almost always though, users have to show respect to one another in the way they interact, so verbal abuse as well as the expression of racist, homophobic and otherwise discriminatory views constitutes abusive behaviors. As a result, *moderation* is the task of responding

© Springer Nature Switzerland AG 2018
A. Gelbukh (Ed.): CICLing 2017, LNCS 10762, pp. 404–419, 2018.
https://doi.org/10.1007/978-3-319-77116-8_30

to abusive behaviors by sanctioning the users exceeding the rules. This moderation work is mainly done manually, which makes it very costly in terms of human and financial costs.

In this paper, we consider the problem of automatically determining if a message from a user is abusive or not. We first present our automatic abusive message classification system based on basic features. We then propose to enrich our system by considering original preprocessing approaches, as well as corpus selection and various new content features specific to the targeted community. We finally propose a qualitative study that helps to analyze the impact of each content feature on the automatic abusive message classification performance. Two types of messages are considered in this paper: one source akin to email and the other to discussions in a chatroom, both coming from a corpus of messages originating from the community of the French massively multiplayer on-line game https://play.spaceorigin.fr/ SpaceOrigin.

This paper is organized as follows. In Sect. 2, we review the main works related to abuse detection and automatic moderation. In Sect. 3, we describe the features used to automatically detect abusive messages. In Sect. 4, we present our dataset as well as the experimental protocol, and discuss the results obtained. Finally, the main points of our work are summarized in Sect. 5, which also shows how it could be extended.

2   Related Work

A number of works have tackled the problem of detecting abusive messages in on-line communities. While most of them are evaluated on English datasets, the majority of the methods used are language- and community-independent, and can therefore be applied on messages from any online community, which makes them relevant to us. This review is focused around two axes: Preprocessing techniques and Features for abuse detection. *Preprocessing* consists in taking the raw message text and attempting to alleviate most of the problems introduced by the Internet medium, such as typos, abbreviations, use of smileys and so on. The feature extraction process consists in processing a series of indicators from the raw message text, that will reflect its class.

2.1   Preprocessing Step

Preprocessing is usually a simple but important step when dealing with messages posted on-line. The Internet medium introduces specific difficulties: disregard of syntax and grammar rules, out-of-vocabulary words, heavy use of abbreviations, typos, presence of URLs, etc. The *Denoising* and *Deobfuscation* tasks both consist in mapping an unknown word back into a dictionary of known words. In the first case, a word is out of the vocabulary for unintentional reasons such as typos, e.g. "I uess so" for "I guess so". In the second case, it is due to a more deliberate attempt to conceal the word, e.g. "F8ck3r" for "fucker". Globally, mapping the word back into the dictionary increases the performance of probabilistic learning methods, since these methods need the cleanest possible text to achieve their maximum performance.

In [1], the *Levenshtein distance* (a type of *edit distance*) is used in an attempt to match unknown words against words in a crowd-sourced list of manually annotated messages containing profanity. The Levenshtein distance [2] measures the number of insertions, deletions and replacements needed to convert a string into another (*e.g.* The Levenshtein distance between "@ss" and "ass" is (1). Computing the Levenshtein distance between two words of length n and m is computationally expensive: the runtime is $\mathcal{O}(nm)$, and for this specific task each word has to be matched against each word in the dictionary, which is huge. We base some of our features on the Levenshtein Distance.

Other works, such as [3], proposed to improve on simple Levenshtein distance based denoising (for the purpose of spell checking) by considering the context of the string edition in a word (where the edited character is) and in a sentence (does the edit maps the word into a high n-gram probability?). However, this approach is based on the assumption that out-of-vocabulary words are mainly due to unintentional spelling mistakes and is therefore not applicable to deobfuscation.

In [4], the authors use a Hidden Markov Model customized with dictionary and context awareness for the purpose of deobfuscating spam messages. Those types of messages often include deliberately misspelled words in an attempt to bypass filters. Their model showed impressive results with the ability to correct both unintentional misspellings as well as deliberate obfuscation using weird characters or digits and could even map segmented words back into complete words. (i.e. "ree movee" → "remove"). This preprocessing approach, while effective, is computationally intensive and complex to implement.

Preprocessing is an important step in an automated abuse detection framework, but it should be applied with caution. The goal of preprocessing is to increase the amount of relevant information in a message, but it can have the opposite effect. For example the tendency of a user to misspell words can be viewed as an important feature to describe the user, but blind preprocessing would hide that.

2.2 Text Messaging Classification

In this section we review existing classification approaches that consider the content of the messages to detect abusive messages, and then the context of the exchanged messages.

Content-Based Approaches. The work described in [5] was one of the first to automate the detection of hostility in messages. While hostility does not imply abuse, abuse often contains hostility. The paper defined a number of rules to identify certain characteristics of the messages, such as Imperative Statement, Profanity, Condescension, Insult, Politeness and Praise. A Decision Tree classifier was then used to categorize messages into Hostile and Non-hostile classes. The setup showed good results but was limited when dealing with sarcasm, noise or innuendo. This approach is interesting but highly tuned for the grammar rules, semantics and idioms of a specific language.

In [6], the authors note that the mere presence of an offensive word in a message is not a strong enough indication that the message itself is offensive, i.e. "You are stupid." is way more offensive than "This is stupid." This is an important observation and the authors showed that the lack of context can be mitigated by looking at word n-grams instead of unigrams.

Another work [7] used features computed from tf-idf weights, a list of words reflecting negative sentiment and widely used sentences containing verbal abuse to detect cyberbullying in comments associated with Youtube videos. Their model showed good results for instances of verbal abuse and profanity but was limited with regard to sarcasm and euphemism.

Finally, in [8] the authors reviewed machine learning approaches for the classification of aggressive messages in On-line Social Networks, described a full pipeline for achieving classification of raw comments and introduced two new features: Pronoun Occurrence and Skip-Gram features. They allow for the detection of targeted phrases such as "He sucks" or "You can go die", and for the identification of long distance relationship between words, respectively.

Content-based text classification performs relatively well as a starting point. The computational cost to implement these approaches is usually reasonable. Nonetheless, these approaches have severe limitations. For instance, abuse can cross message boundaries, and therefore a message can be abusive only because of the presence of an earlier message. In other cases, messages can be abusive because they reference a shared history between two users. Therefore, studying the context of a message, its recipient, and its author might also be important.

Context-Based Approaches. To go beyond the limitations of content-based approaches, some authors proposed to take into account the context of messages, usually in addition to the textual content itself.

Some works explore the use of the content neighboring the targeted message. In [9], the authors used a supervised classifier working on n-gram features, sentence-level regular expression patterns and the features derived from the neighboring phrases of a given message to detect abuse on the Web. Their approach focused on detecting derivations in the context of a discussion around a given topic and their context features significantly improved the performance of their system. For this reason, we want to adopt a similar approach for our own method, but by focusing on derivations of users themselves from their usual patterns.

Other works have focused on modeling the users' behaviors by introducing higher-level features than the textual context. A comprehensive study of anti-social behavior in on-line discussion communities has been proposed in [10]. Their exploratory work reinforced the weight of classic features used to classify messages such as misspellings and length of words, and provided insight into the devolution of users over time in a community, regarding both the quality of their contributions and their reactions towards other members of the community. This analysis is a good step towards modeling abusive behavior. One of the essential results of the analysis is that instances of antisocial messages tend to generate a bigger response from the community compared to benign messages. The number of respondents to a given message is a feature we use in this work.

A selection of contextual features aiming at detecting abuse in on-line game communities has also been investigated in [11]. These features form a model of the users of the game by including information such as their gender, number of friends, investment in the platform, avatars, and general rankings. The goal was to help human moderators dealing with abuse reports, and the approach yielded sufficiently good results to achieve this goal. The work was however limited in applicability because of the specifics of that given community, and of the raw amount of data needed to perform similar experiments in other games.

When quantifying controversy in social media [12], the structure of the community network is exploited to identify topics that are likely to trigger strong disagreements between users. The approach relies on a network whose nodes are Twitter users and links represent communications between them. It is interesting, however hardly applicable to our case, since we cannot infer the exact network structure from our dataset unless we restrict the network to private conversations between two users.

3   Abusive Message Features

In this section, we describe the content-based features used in our automatic abusive message classification system. They can be broadly categorized as morphological, language, and context features. Some of them are quite generic, in the sense they are used for different classification tasks in the literature. The others were designed by us specifically for this experiment, and some are customized for the community where our dataset originates, but they can sometimes be generalized to other communities (we reflect on this in Sect. 5). The features we developed are denoted with a star (*) preceding their name and description.

Some features require the data to be preprocessed before being extracted, so we start with the description of our preprocessing approach first.

3.1   Preprocessing

We distinguish two preprocessing phases. In the *basic phase*, we first lower-case the raw text and tokenize it using spaces. Each token in the list is then stripped of punctuation before the message is reassembled.

In the *advanced phase*, the data undergo some additional preparation steps. First, we revert elision. *Elision* refers to the suppression of a final unstressed vowel immediately before another word beginning with a vowel. For the French language, we therefore replace instances of "j", "t" by their respective long forms "je", "te", so that, for instance, "j'arrive" becomes "je arrive". Second, we run a deobfuscation pass by mapping hexadecimal or binary encoded text in the message back to ASCII. This is highly specific to the considered online community, because users sometimes encode part of their messages in that way. Third, we convert each URL into a sequence of tokens. The first describes whether this URL is an internal link (to a server hosting the community) or an external one. The rest are words that could possibly be extracted from the name of the web page. For instance: http://edition.cnn.com/2017/01/31/politics/donald-trump-immigration-white-house/index.html is mapped to: **__url_external cnn com politics donald trump immigration white house index**. Finally, we use the FrenchStemmer from the Natural Language Toolkit [13] to convert words into their stem.

3.2   Morphological Features

Message Length. This feature corresponds to the length of a message, expressed in number of characters, before any pre-processing. The intuition is that abusive messages are usually either kept short (*e.g.* "Go die."), or extremely long, which is symptomatic of a massive copy/paste.

We also consider the length expressed in terms of words. In conjunction with the character count, it can match certain overly emphasized messages (e.g.: "Shuuuuuuuuuuuuuuuuuuuuuuut uuuuuuuuuuuuuuuuuuuup!").

Character Classes. We split characters into 5 classes: *Letters, Digits, Punctuation, Spaces* and *Others*. We keep track of the number of characters in those classes and the ratio of those classes in the message. This is done on the raw message, before any preprocessing.

We selected these features based on several observations we made on the abusive messages. First, some of them have an unusual number of characters in the "Other" class, e.g.: "8===================D". Second, some use digits to obfuscate their meaning, in violation of the game rules. For instance, "0100011101101111001000000110010001101001011001010010100101110" and "476F206469652E" are obfuscated versions of the text "Go die.": the first one is coded in binary, and the second in hexadecimal.

Abusive users also commonly "yell" insults using capital letters, which is why we keep track of both the number of caps and the corresponding ratio of caps in the message.

Compression Ratio. This feature is defined as the ratio of the length of the compressed message to that of the original message, both expressed in characters. It is based on the observation that certain users tend to repeat *exactly* and many times the same text in their abusive messages. We use the Lempel–Ziv–Welch (LZW) compression algorithm [14], and the feature therefore directly relates to the number of copy/pastes made in the same message.

Unique Characters. By counting the number of distinct characters in the message, we can detect the use of binary or hexadecimal obfuscation, as well as the overuse of punctuation. For instance, for the message "01000111011011110111011110", this feature has only a value of 2. This value is also computed before any preprocessing.

Collapsed Characters. This feature is computed after the message is lowercased. When three or more identical consecutive characters are found in the message, they are collapsed down to two characters. For instance, "looooooool" would be collapsed to "lool". The feature is the difference between the length of the original message and that of the collapsed one. This preprocessing step has been widely used in the classification of Tweets, for instance in [15].

3.3 Language Features

Bag of Words. We transform the message into a Bag of Words (BoW). This is a sparse binary vector that has the same dimension as the known vocabulary of the corpus and where each component corresponds to one word. The component has value 1 if the word is present in the message and 0 if it is not. We use the output of a Naive Bayes classifier for a given BoW as an input feature into our larger system.

Word Length. This feature is a component of the Automated Readability Index (ARI) [16]. It measures how proficient someone is at creating text documents. While abusive messages are sometimes surprisingly well written, this remains rare.

Unique Words. We consider the number of unique words in the message. The intuition is that messages with more words are likely to be more constructive in terms of their content. Moreover, we observed that people are generally straightforward when verbally abusing others, and rarely take the time to elaborate.

tf-idf Scores. Those features are the sums of the *tf-idf* scores of each individual word in the message. We use two distinct scores: the so-called *non-abuse score* is processed relatively to the non-abuse class (randomly chosen messages that have not been flagged as abusive), whereas the *abuse score* is processed over the abuse class.

If the considered word is unknown, in the sense it does not appear in the training set, we process an approximation of these scores. For this purpose, we first search for known words located within a given Levenshtein distance from the word of interest, and average their own scores.

Computing the full Levenshtein Distance between two words is computationally expensive. For this reason, solutions proposed in this paper never compute the full Levenshtein Distance between two words. Instead, a specialized tree-based index data-structure with a search function that yields all words in the tree within a given maximum edit distance is used. We use 2 as the maximum edit distance. This is considerably faster because branches of the tree are pruned as soon as we reach a state where the maximum edit distance is exceeded. It is still the second most computationally intensive operation in our experiments.

Sentiment Scores. These features are numeric values derived from the number of words in the message that have a sentiment weight. It is based on the sentiment corpus presented in [17], which was automatically generated for the French language. We augmented it manually by selecting words from a large list of insults.

Bad Words. This feature corresponds to the number of words in the message that appear in a manually crafted list of bad words. The list of words was created from a list of insults in French and then augmented with common Internet shorthand and symbolism. (*e.g*: 'connard', 'fdp', 'stfu' but also '..—..', '8==D' etc.)

When we cannot match the considered word to any of the bad words in our list, we try to perform a fuzzy match using the Levenshtein distance. This is supposed to allow us picking up some obfuscated bad words.

We also perform the same tests using the collapsed version of the message (as described for the *Collapsed Characters* feature).

Business Score*. We first mine messages for patterns specific to the community. In the targeted context of this particular online strategy game, we chose to focus on: names of buildings and military units, war vocabulary and other game-specific jargon, sets of Coordinates, and report links. The latter refers to internal URLs generated by some actions, and pointing towards summaries that the players can share. For each pattern, we manually developed a regular expression and used it to find the number of non-overlapping occurrences of the pattern in the message. We then produced the Business Score by combining the individual scores obtained for each pattern. This is a measure of how the message relates to the focus of the community. By observing the corpus we noticed that abusive messages tend to be strictly personal attacks with no pretense of roleplay and no mention of game jargon.

3.4   User Behavior Features

Number of Respondents. Given a fixed size window after a target message, this feature tracks how many distinct users replied to this message. This feature is likely to be relevant, because it has been shown that abusive comments tend to trigger big responses from the community.

Probability of n-Gram Emission (PNE)*. We investigate if the abusiveness of a user's message can be detected by considering the effect it has on the other users participating to the same conversation. To do this, we compare the writing behavior of the other users before and after the apparition of the targeted message. We model this behavior through a user-specific Markov chain, which we use to compute how likely some text is to have been generated by the considered user.

We first sort the messages in chronological order. For each participant other than the user who wrote the targeted message, we build a word n-gram Markov chain using all but the last W n-grams in the messages posted before the target message.

The Markov chain is a convenient way to store the transition probabilities for all couples of n-grams in a participant's history. We compute two values: the average emission probabilities of the n-grams in the W-length window before and after the target message, as represented in Fig. 1.

Let $P_{i,i+1}$ be the emission probability of a transition between the ith and $i+1^{th}$ n-grams in the window of length W. Then we define the average emission

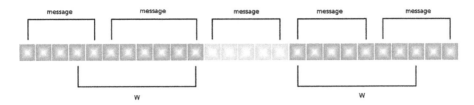

Fig. 1. A sequence of messages broken down into n-grams. Each square represents an n-gram: red for the targeted message, blue for the surrounding messages written by other users.

probability S over the set of W n-grams as:

$$S = \frac{\sum_{i=0}^{W-1} P_{i,i+1}}{W} \tag{1}$$

We note $S_B(u)$ and $S_A(u)$, respectively, the average probabilities processed before and after the targeted message, for the same user u. The final score $S(u)$ for user u corresponds to their difference:

$$S(u) = S_A(u) - S_B(u) \tag{2}$$

This score is processed for every respondent to a message in a window of fixed length after the message. We then compute our feature by averaging this score over all the responding users.

Applicability Criterion for PNE*. The previous feature requires averaging scores, which makes sense only if the considered users have sufficient history: we define a limit of at least 300 bigrams. This feature reflects the fulfillment of this constraint.

4   Experiments

In this section, we describe the data used in our experiments (Subsect. 4.1) as well as the experimental protocol (Subsect. 4.2). We then evaluate the proposed system, including the various features and original preprocessing approaches (Subsect. 4.3).

4.1   Dataset

We have access to a database of users' in-game interactions for the considered MMO. This user-generated content was manually verified, in the sense the game users had the ability to flag parts of the content as inappropriate (*i.e. abusive*). There are many types of reportable contents, but, in this paper, we focus on two of them: ingame-messages (iM) and chat messages (cM), collectively referred to as messages.

Ingame-messages (iM) are on-line messages with a clearly defined reach. They are the equivalent of e-mails and can be sent to specific users or groups of users. They can be edited *a posteriori* by moderators when an abuse case is reported. The reach of *chat messages (cM)* is loosely defined because it is limited to users currently active in a chatroom. However, there is no way to determine which user has actually seen a specific chat-message based on the available data. Users are fed recent scroll history for a chatroom upon joining, but it is not possible to reliably determine who has joined when from the chat logs. Chat-messages cannot be edited by moderators afterwards.

The database contains 474, 599 in-game messages and 3, 554, 744 chat-messages. We extract 779 *abusive* messages (0.02%), which constitute what we call the *Positive Class* (Class 1) of messages. These messages were first flagged by the game users using a built-in reporting tool, and then confirmed as being abuse cases by the game community moderators. Of these 779 abusive messages, 14% are ingame-messages, and the rest are chat-messages. We then extract *non-abuse* messages from the database, in order to constitute the so-called *Negative Class* (Class 0). They are chosen at random from a pool of messages written by users which have *never* been flagged by a confirmed abuse report. For each message, we also gather context data: a window of messages occurring before and after each message.

We run the experiments with different versions of the corpus: in-game messages only *(iM)*, chat-messages only *(cM)* and messages of both types combined *(iM+cM)*. Sizes of each considered corpus configuration are reported in Table 1. These configurations are considered as "unbalanced" (U), since there are twice as many non-abusive messages as abusive messages. As a result, we also experiment with the use of "balanced" data (B), where the number of abusive messages is equals to the non-abusive ones.

Table 1. Corpus sizes depending on the considered experimental setup (unbalanced data)

Configuration	Abusive messages	Non-abusive messages
iM+cM	779	1558
iM	111	222
cM	668	1336

4.2   Experimental Setup

Our experiment is designed as a multi-stage classifier pipeline, as described in Fig. 2 (each box corresponds to a stage). The first stage (*Raw Messages*) consists in building the corpus. Messages from both the Abuse and Non-abuse classes are extracted from the database as explained in Subsect. 4.1. The corpus is then split

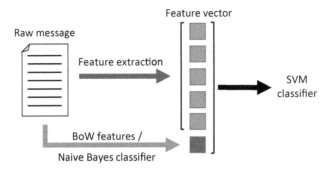

Fig. 2. The full experimental setup

into a *Train* set containing 70% of the messages, and a *Test* set containing the remaining 30%.

In the second stage (*Bag of Words Features*), messages are normalized, tokenized and converted into Bag Of Words. In the third stage (*Naive Bayes Classifier*), the Bag Of Words representations of the *Train* messages are used to build a Naive Bayes classifier. This classifier is then used to generate predictions for the class of the *Test* messages.

In the fourth stage (*Feature Extraction*), we extract the features described in Sect. 3 from the messages. As explained before, some of these are derived from the messages before any normalization or preprocessing, whereas some others require a specific preprocessing. We then use another classifier, a Support Vector Machine (SVM). We could directly feed the Bag Of Words to the SVM. However, given the size of the vocabulary in our experiments, this would lead to a dimensionality issue, with a number of features greatly exceeding the number of instances in the corpus. Therefore, we prefer to consider the decision from the Naive Bayes classifier (third stage) as an additional feature given to the SVM. We get a total of 67 distinct features, including the Naive Bayes decision, which are all gathered into an array.

The fifth stage (*SVM Classifier*) is the final classification: the feature arrays from the Train set are fetched to an SVM classifier, and the resulting model is then used to generate class predictions for the Test set.

4.3   Results

We evaluate the performance of our proposed abusive message detection system in terms of the traditional Recall, Precision and *F*-Measure. Given the relatively low number of abusive samples of the targeted corpus, the whole dataset was split into 10 parts and every result given in this section is the average value over a 10-fold cross validation. In order to show the contribution of the features as well as pre-processing approach proposed, three system configurations are studied. The first is the *baseline*, which relies on the classic feature set and the basic preprocessing, as previously described. The two others are our contributions: on the one hand the full feature set with basic preprocessing, and on the other hand the full feature set with advanced preprocessing.

Table 2 presents the performance obtained by the proposed system for all the studied configurations, using *unbalanced* data. We can firstly see that, no matter the considered message type (iM only, cM only, or iM+cM), improvements in terms of Precision, Recall and *F*-measure are observed when completing the baseline system (classic features and basic preprocessing) with our new features. This gain is even more important when using our advanced preprocessing, with *F*-measure increases of 3.1 points (iM only), 3.3 points (cM only) and 3.2 points (iM+cM) compared to the baseline system. The same observations can be made for the results obtained on the *balanced* data, displayed in Table 3, but with smaller gains (3.3, 1.4 and 1.3 points, respectively).

Table 2. Classification results (in %) of the automatic abusive message classification system, obtained by applying different feature sets and preprocessing configurations to the *unbalanced* data.

Data	Features	Preprocessing	Precision	Recall	*F*-measure
iM only	Classic set	Basic	66.9	72.8	69.7
	Full set	Basic	67.2	73.4	70.2
	Full set	Advanced	**69.6**	**76.2**	**72.8**
cM only	Classic set	Basic	65.2	71.6	68.2
	Full set	Basic	65.5	72.2	68.7
	Full set	Advanced	**67.6**	**75.9**	**71.5**
iM+cM	Classic set	Basic	65.7	72.3	68.9
	Full set	Basic	65.9	73.2	69.3
	Full set	Advanced	**68.3**	**76.4**	**72.1**

Let us now compare the results obtained for the different types of messages. When considering the unbalanced data (Table 2), iM and cM only lead to globally similar performances for all three considered measures. Combining them (iM+cM) does not bring any significant change. However, this is not the case for the balanced data (Table 3): the performance obtained for cM only is quite different, with a much higher Precision (+7.6 points on the advanced setup) and a lower Recall (−2.9 points). This pulls up the overall performance when using both message types (iM+cM), leading to a 76.5 *F*-Measure for the advanced setup, which is 4.4 points higher than with the unbalanced data.

Our experiments show that, even if acceptable results could be obtained with our abusive message detection system (best *F*-measure of more than 70%), performance is still not good enough to be directly used as a fully automatic system that replaces human moderation. Nonetheless, we think that this system could be useful to help moderators focus on messages considered as potentially abusive, instead of having to analyze all messages. This is illustrated by the left

Table 3. Classification results (in %) of the automatic abusive message classification system, obtained by applying different feature sets and preprocessing configurations to the *balanced* data.

Data	Features	Preprocessing	Precision	Recall	F-measure
iM only	Classic set	Basic	67.2	70.4	68.7
	Full set	Basic	67.6	70.8	69.2
	Full set	Advanced	**70.8**	**73.3**	**72.0**
cM only	Classic set	Basic	77.2	67.5	72.0
	Full set	Basic	77.1	67.4	71.9
	Full set	Advanced	**76.8**	**70.3**	**73.4**
iM+cM	Classic set	Basic	76.9	73.6	75.2
	Full set	Basic	77.3	74.5	75.9
	Full set	Advanced	**76.1**	**76.9**	**76.5**

plot in Fig. 3, which represents the Precision-Recall curve (traditionally obtained by varying the decision threshold on the SVM posterior probability obtained by applying the Platt Scaling implementation of the Scikit-Learn Library [18]). For a fully automatic system, requiring to be very precise on the decision to take (*i.e.* be sure that the message is abusive), a higher threshold should be used, with a loss in terms of number of detected abusive messages (*i.e.* lower Recall). On the contrary, for a software assisting a moderator, needing to recover as many abusive messages as possible, a lower threshold should be used, resulting in a higher recall (more abusive messages are retrieved) associated to a lower precision (more non-abusive messages be wrongly returned by the system). The plot shows a short plateau in the middle, which means it would be possible to increase the Recall without losing much Precision. However, estimating the exact optimal decision threshold will require more data. We now take a look at how the features are contributing to the result. We use a tree-based estimator from the Scikit-Learn library to estimate the importance of the features for our classification problem. This tool is stochastic, so the score measuring this importance can vary from one run to the other. Thus, we ran it 200 times to get stable results. The right plot in Fig. 3 shows all of these runs as well as the average curve. It displays how the F-Measure evolves as the features are removed one by one, by increasing order of estimated importance. Our SVM classifier is trained and evaluated at each feature removal. Despite the stochastic nature of the process, the last removed (and therefore most important) feature is always the Naive Bayes decision: this makes sense, since it is already the output of a full-fledged classifier. This is confirmed by the tree-based estimator, which gives an importance score of 42.5%.

Each of the 200 runs shows a sharp drop at the end. We detected that this drop is due to the removal of any feature in the following group: Number of bad words in the collapsed comment, Average word length, PNE and Applicability criterion for PNE. We therefore conclude that these features are com-

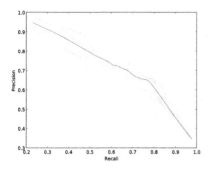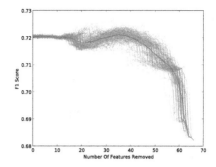

Fig. 3. Left: Precision-Recall curve of the SVM classifier. Right: Evolution of the classifier performance when sequentially dropping all features but one.

plementary, and result in a strong classifier when combined. According to the tree-based estimator, these four features have a combined importance score of 15.9%. So, our results show that a small group of 5 features account for 58.4% of the classifier performance. The rest of the features improve the performance only marginally. Other noteworthy features include the ratio of letters and other characters (5%), the ratio of capitalized letters (2.1%), and the positive and negative scores (4.23%). The Business Score feature, defined by us specifically for the targeted online community, has only an importance of 1.13%: it accounts for a small part of the classifier decision, but on the positive side it is fast to compute. This is not the case for the PNE feature: computing it is expensive both in terms of CPU time and memory since we need to build and store a complete model of multiple user speech patterns.

5   Conclusion

In this paper, we developed a system to classify abusive messages from an online community. It is developed on top of a first-stage Naive Bayes classifier and relies on multiple types of features: morphological, language- and context-based features, that have proven their usefulness in previous research. We added several features that we derived directly from observations of our corpus, and developed a context-based feature that aims to capture abnormal reactions from users caused by an abusive message. Our goal here was to explore a large number of features to identify the most relevant one for the problem at hand.

Our results show that abusive messages have characteristics that can be caught by an automatic system, our proposed system achieving a Recall and a Precision of more than 76% on our dataset. While the performance of the system is not good enough yet to be deployed as fully automatic moderation tool, this can already help moderators focus on messages being identified as abusive, before a manual verification is made. However, because some features used in the system are specific to the community in which it is meant to operate, care must be taken when adapting the system to work on a different dataset.

Our results also show that a small number of features, including both generic and problem-specific ones, account for most of the classifier decision.

We now plan to pursue our work in several ways. First, because preprocessing has been shown to have an important effect on overall performance, we will experiment with computationally more demanding preprocessing methods, such as the HMM-based preprocessor from [4], and evaluate their contribution to the classifier performance. Second, we want to derive variants of our PNE feature, and assess which one is the most appropriate in our situation. More generally, we plan to propose other context-based features, especially ones based on the network of user interactions.

References

1. Sood, S.O., Antin, J., Churchill, E.F.: Using crowdsourcing to improve profanity detection. In: AAAI Spring Symposium: Wisdom of the Crowd (2012)
2. Levenshtein, V.I.: Binary codes capable of correcting deletions, insertions and reversals. In: Soviet Physics Doklady, vol. 10, p. 707 (1966)
3. Brill, E., Moore, R.C.: An improved error model for noisy channel spelling correction. In: 38th Annual Meeting on Association for Computational Linguistics, pp. 286–293 (2000)
4. Lee, H., Ng, A.Y.: Spam deobfuscation using a hidden Markov model. In: 2nd Conference on Email and Anti-Spam (2005)
5. Spertus, E.: Smokey: automatic recognition of hostile messages. In: 14th National Conference on Artificial Intelligence and 9th Conference on Innovative Applications of Artificial Intelligence, pp. 1058–1065 (1997)
6. Chen, Y., Zhou, Y., Zhu, S., Xu, H.: Detecting offensive language in social media to protect adolescent online safety. In: International Conference on Privacy, Security, Risk and Trust and International Conference on Social Computing, pp. 71–80 (2012)
7. Dinakar, K., Reichart, R., Lieberman, H.: Modeling the detection of textual cyberbullying. Soc. Mob. Web **11**, 02 (2011)
8. Chavan, V.S., Shylaja, S.S.: Machine learning approach for detection of cyber-aggressive comments by peers on social media network. In: International Conference on Advances in Computing, Communications and Informatics, pp. 2354–2358 (2015)
9. Yin, D., Xue, Z., Hong, L., Davison, B.D., Kontostathis, A., Edwards, L.: Detection of harassment on web 2.0. In: WWW Workshop: Content Analysis in the WEB 2.0, pp. 1–7 (2009)
10. Cheng, J., Danescu-Niculescu-Mizil, C., Leskovec, J.: Antisocial behavior in online discussion communities (2015). Preprint arXiv:1504.00680
11. Balci, K., Salah, A.A.: Automatic analysis and identification of verbal aggression and abusive behaviors for online social games. Comput. Hum. Behav. **53**, 517–526 (2015)
12. Garimella, K., De Francisci Morales, G., Gionis, A., Mathioudakis, M.: Quantifying controversy in social media. In: 9th ACM International Conference on Web Search and Data Mining, pp. 33–42 (2016)
13. Bird, S.: NLTK: the natural language toolkit. In: Proceedings of the COLING/ACL on Interactive presentation sessions, Association for Computational Linguistics, pp. 69–72 (2006)

14. Batista, L.V., Meira, M.M.: Texture classification using the Lempel-Ziv-Welch algorithm. In: Brazilian Symposium on Artificial Intelligence, pp. 444–453 (2004)
15. Roy, S., Dhar, S., Bhattacharjee, S., Das, A.: A lexicon based algorithm for noisy text normalization as pre processing for sentiment analysis. Int. J. Res. Eng. Technol. **2**, 67–70 (2013)
16. Senter, R.J., Smith, E.A.: Automated readability index. Technical Report AMRL-TR-6620, Wright-Patterson Air Force Base (1967)
17. Chen, Y., Skiena, S.: Building sentiment lexicons for all major languages. In: 52nd Annual Meeting of the Association for Computational Linguistics, pp. 383–389 (2014)
18. Pedregosa, F., et al.: Scikit-learn: machine learning in python. J. Mach. Learn. Res. **12**, 2825–2830 (2011)

Detecting Aggressive Behavior
in Discussion Threads Using Text Mining

Filippos Karolos Ventirozos[1], Iraklis Varlamis[1], and George Tsatsaronis[2(✉)]

[1] Department of Informatics and Telematics, Harokopio University of Athens,
Athens, Greece
{it21105,varlamis}@hua.gr
[2] Content and Innovation, Elsevier B.V., Amsterdam, The Netherlands
g.tsatsaronis@elsevier.com
https://www.dit.hua.gr/~varlamis/

Abstract. The detection of aggressive behavior in online discussion communities is of great interest, due to the large number of users, especially of young age, who are frequently exposed to such behaviors in social networks. Research on cyberbullying prevention focuses on the detection of potentially harmful messages and the development of intelligent systems for the identification of verbal aggressiveness expressed with insults and threats. Text mining techniques are among the most promising tools used so far in the field of aggressive sentiments detection in short texts, such as comments, reviews, tweets etc. This article presents a novel approach which employs sentiment analysis at message level, but considers the whole communication thread (i.e., users discussions) as the context of the aggressive behavior. The suggested approach is able to detect aggressive, inappropriate or antisocial behavior, under the prism of the discussion context. Key aspects of the approach are the monitoring and analysis of the most recently published comments, and the application of text classification techniques for detecting whether an aggressive action actually emerges in a discussion thread. Thorough experimental validation of the suggested approach in a dataset for cyberbullying detection tasks demonstrates its applicability and advantages compared to other approaches.

Keywords: Aggressive behavior · Cyberbullying
Sentiment analysis · Thread classification

1 Introduction

Sentiment analysis methods aim at identifying the sentiment orientation of a piece of text (e.g., sentence, paragraph, snippet) by analyzing lexical features at word or term level. The problem is either handled as a binary classification problem [1] where only positive and negative sentiments are considered, or as a multi-class classification problem when a fine-grained list of sentiments is used (e.g., anger, disgust, fear, guilt, interest, joy, sadness, shame, surprise).

© Springer Nature Switzerland AG 2018
A. Gelbukh (Ed.): CICLing 2017, LNCS 10762, pp. 420–431, 2018.
https://doi.org/10.1007/978-3-319-77116-8_31

Despite the large number of works on sentiment analysis [2] and cyberbullying detection [3], text classification methods have focused only on single posts and not yet on the complete discussion thread. Such methods have several disadvantages; for instance, they can be misled by attackers who intentionally misspell words to prevent detection [4], or they may falsely categorize the responses of the victims or their defenders as aggresive behavior. Existing methods actually neglect the fact that the inherent characteristics of bullying are *repetitiveness*, *intentionality* and *imbalance of power* between the harasser and the victim [5].

In an attempt to address these limitations and omissions we present in this work for the first time, to the best of our knowledge, a supervised learning model that detects aggressive behavior events by considering the whole thread in order to extract features which relate to changes in sentiment between consecutive messages. In order to validate experimentally the suggested approach we compared its performance in terms of accuracy in a benchmark set against a previously published state-of-the-art method which has been applied in the same set, and we also experimented with different variations of the method. The benchmark set is publicly available and comprises 139 discussion threads from *MySpace*. Results suggest that the presented method offers a more accurate predictor of aggressive behaviors in discussion threads.

The remainder of the article is organized as follows: Sect. 2 provides an overview of related literature. Section 3 summarizes the steps of the processing pipeline and highlights the novelties of the proposed methodology. Section 4 presents and discusses the experimental results, and, finally, Sect. 5 concludes and gives pointers to future work.

2   Related Work

The problem of textual harassment or aggressive behavior detection in text has been tackled by researchers as a classification problem. In [6], authors applied a supervised machine learning approach for detecting cyber-harassment, in which posts are represented using word frequency features, sentiment features and features that capture the similarity to neighboring posts. In [7] a rule-based model using a number of lexical features (e.g. bad words) outperformed the baseline bag-of-words (BOW) model. In [8] authors applied a range of binary and multi-class classifiers on a corpus of comments from YouTube videos in various topics. The findings show that topic-sensitive binary classifiers improved the performance of generic multi-class classifiers.

The authors in [9] compared an rule-based expert system, a supervised machine learning model, and a hybrid approach and showed that the latter outperformed the other two. In [10] a fuzzy support vector machine classifier using lexical features, sentiment features and user metadata was employed.

In [11] authors developed and applied a classification scheme for cyberbullying, which may detect cyberbullying presence, the judgment of its severity, and the role of the posts' authors (i.e., harasser, victim or bystander). Authors focused on specific cyberbullying-related text categories such as *threat/blackmail*, *insult*, and *curse/exclusion*, and the experimental results demonstrated the feasibility of fine-grained cyberbullying detection. Character and word n-grams as well as lexicon based sentiment features were used.

All of the research works referenced so far approached the problem as a binary classification problem of single messages, without considering analyzing the entire thread. In addition, almost all of these approaches employed a very similar text pre-processing pipeline comprising stop-word removal, tokenization, POS tagging, emoticon detection, stemming, etc., and a typical text feature extraction step which resulted in bag-of-words, or, bag-of-stems representations that employ words, word and character n-grams, sentiment lexicon or even emoticon-related features, used for classifying texts at post level.

As a result, it is likely that these automated post labeling techniques may be inaccurate when an aggressive post does not contain bad words, when profanity or pronouns are misspelled, or when the posts are not in the language matching the aggressive words. In fact some works attempt to overcome these limitations by employing user-based features [10, 12], thus taking into account the history of users' activities. However, such features, or history, are not widely available, limiting from another angle this time, as opposed to the aforementioned approaches, their application at large scale and big heterogeneity of fora.

It is only recently that researchers focused on thread-level comment analysis to address such limitations. In [13], authors used thread-level features in a classification task, which exposed paid opinion manipulation trolls. One such feature was the number of times a certain users comments were among the top-k most loved/hated comments in some thread. However, the instances of the classification task were the users and not the thread messages. In [14], authors analyzed a question-answering community and used the whole communication thread as content, in a different application than the one discussed here, namely that of answer selection, and of evaluating the quality (good or bad) of given answers.

Perhaps closer to our work, in [15] authors focus on whole threads of comments. The authors acknowledge the fact that cyberbullying can take place even without the use of profane words. However, it is the occurrence of profane words in one or more comments in the same thread that is leveraged to decide whether cyberbullying is committed. The same authors in a more recent work [16], use an incremental classifier which sums the polarity of comments posted in a thread and decides when the thread must be blocked because of potential aggressiveness. Motivated by similar ideas, authors in [17–19] agree that aggressive posts can be persistent and not single acts, thus highlight the need for whole thread processing.

The current work is distinguished, however, from the works in this latter category, which embed the notion of thread analysis, in the actual way the thread is used in the method. More precisely, in our work the thread is used to generate "sentiment n-grams" which represent the sequence of sentiments expressed within a thread, by the same, or different users that participate in the thread. We demonstrate experimentally that the consideration of this sequence reduces the effect of misclassifications at the comment-level and improves the performance of the aggressive behavior detection methodology overall.

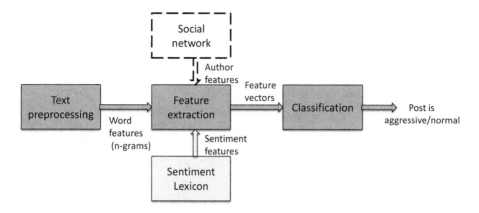

Fig. 1. Pipeline for detecting an aggressive behavior in text threads

3 Cyber-Bullying Detection in Text

3.1 Sentiment Detection and Cyberbullying

The two main approaches for extracting sentiment from text are lexicon- and machine learning-based. Lexicon-based approaches first calculate the semantic orientation of words or phrases in the text using one or more pre-compiled lexicons [20] (e.g., *SentiWordnet*[1], *Sentiful*[2], *ANEW*[3] *LIWC*[4], *WordNetAffect*[5], *SenticNet*[6]) and then decide on the document orientation and strength by adding up individual sentiment scores [21]. On the other side, machine learning methods build classifiers from labeled instances of texts or sentences and use a wide range of features in order to capture the orientation and strength of a sentiment in the text [22]. Support Vector Machine classifiers and Deep Learning approaches that use features such as word n-grams, with or without part-of-speech labels perform very well in this task [23].

The typical cyberbullying detection methodology in social media [24] as depicted in Fig. 1 has two phases: First it extracts general keyword features, features for sentiments that are rare in other contexts but frequently expressed in bullying posts, and, possibly features that draw the author profile of each message. Then it classifies the message as aggressive or not.

3.2 The Proposed Method

The first step of the proposed methodology for aggressive behavior detection, as depicted in Fig. 2, is the selection of a set of sentiments that will be used as

---

[1] http://sentiwordnet.isti.cnr.it/.

[2] https://sites.google.com/site/alenaneviarouskaya/research-1/sentiful.

[3] http://csea.phhp.ufl.edu/media/anewmessage.html.

[4] http://liwc.wpengine.com/.

[5] http://wndomains.fbk.eu/wnaffect.html.

[6] http://sentic.net/downloads/.

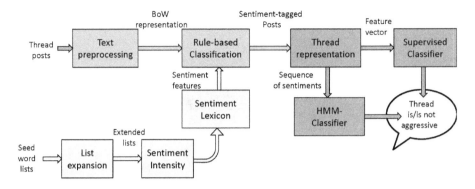

Fig. 2. The proposed pipeline for detecting an aggressive behavior in text threads

features in the classification model. The proposed method considers 7 possible sentiments for a message: anger, disgust, fear, happiness, sadness, surprise and trust. These 7 have been characterised as basic emotions, clearly distinguished from other affective phenomena [25]. For these sentiments a lexicon is compiled, which contains related terms to each sentiment. Starting from a set of seed words for each sentiment, extracted from related sentiment lists on the web, we retrieve more synonym terms from *Wordnet* and expand the seed words to the final sentiment lexicon. The lexicon also contains a degree of strength for each sentiment word using information from ANEW, SentiWordnet and SenticNet 2 lexicons.

In the next step, using a rule based classifier[7] that takes into account the occurrence of sentiment words in the text, each message is tagged with one or more sentiments depending on our strategy, which will be explained in the following. The output of this step for a discussion thread is a sequence of sentiments, expressed by different users, which interchange during the discussion.

The last step comprises the thread classification algorithm, which decides whether an aggressive behavior is expressed within the thread. We evaluate different types of classifiers, that fall into two main categories depending on how the thread is represented: (*i*) when the thread is represented as a feature vector, then sentiment unigrams and bigrams are the features and any supervised classifier can be applied, and, (*ii*) when the thread is represented as an independent sequence of sentiments, then a Hidden Markov Model classifier is applied. The comparison of these two representations gives insight as to whether the order of the messages in the thread is important for the task.

---

[7] Rule-based classifier, called *BullyTracer*, was used in [26] in the same dataset that we use in this work. However, any other classification method can be applied in this step.

4   Experimental Evaluation

For the evaluation of the methodology, the (*Original*) dataset provided by the authors in [26] is used. The set comprises 139 discussion threads from *MySpace* forums, each containing 7 to 48 consecutive posts[8]. In the original work, authors acknowledge the interactive nature of cyberbullying, and process the conversations using a moving window of 10 posts to capture context (referenced as *window* dataset in the following). However, they never use sentiment n-grams as features as we do in our work. From the 139 threads, 39 discussions have been characterized for aggressive behavior (binary classification) and from the 2,062 windows (each having 10 consecutive posts length), 425 have been marked for cyberbullying (binary classification), which creates an imbalance in the dataset. Our results are directly comparable to the original results. In the original results, accuracy ranges from 32% to 84%, the average overall accuracy is 58.63%, and the true positive ratio is 85.30%.

In our experiments, we evaluated the two representations and the classification alternatives presented in Fig. 2 and compared our results against the original method and a random classifier. After pre-processing and rule-based classification, each post in the thread was tagged with the sentiments it contains, each one with a score, which corresponds to the total occurrences of the related sentiment words in the post.

4.1   Feature Vector Representation

In this alternative, each post is either classified to a single sentiment (the prevailing sentiment) or is tagged with multiple sentiment tags, depending on the sentiment words it contains. The post information is summed up at thread level and populates a feature vector that comprises as features:

- sentiment uni-grams (*Unig*), the degree of a sentiment in the thread (i.e., anger, disgust, fear, happiness, sadness, surprise, trust and neutral when no sentiment is expressed at all).
- sentiment bi-grams (*Big*), the occurrences of sentiment changes, among consecutive posts. The features are the 64 ordered sentiment combinations.
- personal pronouns(*PP*) e.g. I, me, you, him, it, they, etc., in consecutive posts, which frequently denote an aggressive stance towards another user.
- bullying bi-grams (*BBig*). When a post contains a word from the list of BullyTracer lexicon, then it is characterized as bullying (*b*) and neutral (*n*) in the opposite case. The feature counts the occurrences of different pairwise combinations (i.e. *nn, nb, bn, bb*) within the thread.
- bullying tri-grams (*BTrig*). The number of different triple combinations (*nnn,nnb*, etc.) of bullying or neutral posts within the thread.

---

[8] The original dataset and the datasets we used in the current research can be downloaded from: https://goo.gl/wPrU2n.

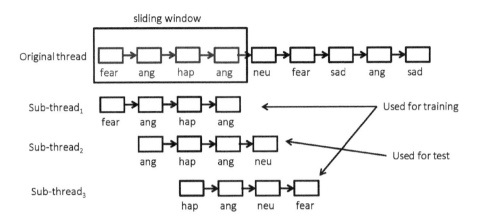

Fig. 3. An example of the dataset creation process and the possible similarities between training and test samples due to overlapping posts

Sub-thread Classification. We evaluated several supervised classification algorithms on different combinations of features using a 10-fold cross validation technique. All experiments were performed both in the *Original* and the *Window* dataset. Since the *Window* dataset was created using a sliding window over each thread it will not be fair to evaluate the algorithms using a completely random 10-fold split of the dataset, because in this case several highly overlapping sub-sequences of the same thread may split between the training and the test dataset as shown in Fig. 3. It is indicative that a lazy 1-Nearest neighbor classifier achieved 95.43% accuracy, in such a type of split.

For this reason, in the *first set of experiments* we experimented with a 90%–10% training-test split of the dataset, taking care that no sub-threads of the same thread occur both in the training and test dataset. We repeated the experiment 10 times with an 40%–60% split of positive and negative samples in average in the test set and a respective split of 20%–80% in the training set. Using only combinations of uni-gram, bi-gram and personal pronoun features and a Radial Basis Function (RBF) classifier[9] an overall accuracy of 67.11% was achieved, whereas when an oversampling technique (*SMOTE*) was used to balance the number of positive an negative samples in the training dataset the overall accuracy dropped to 65.79%. Since the employed datasets are inbalanced, we also report the ROC area in each experiment in order to compare with a random classifier (ROC Area = 0.5).

In the second experiment of this set, the same training-test splits were used but this time each post is tagged with a single sentiment. The performance of the *RBF* classifier dropped for most of the setups except for one that used the *SMOTE* over-sampled training dataset and all the features (*Unig* + *Big* + *PP* feature set), which achieved an accuracy of 72.80%.

---

[9] The *RBFClassifier* implementation for *Weka* has been used.

Table 1. Sub-thread classification (*Window* dataset) using single and multi-class classification and sentiment uni-grams and bi-grams as features

Features	Post class	Balanced training	Accuracy	ROC area
Unig	Multiple	No	60.96	0.529
Big	Multiple	No	61.40	0.532
Unig + Big	Multiple	No	61.84	0.569
Unig + Big	Multiple	Yes	64.91	0.595
Unig + Big + PP	Multiple	No	67.11	0.682
Unig + Big + PP	Multiple	Yes	65.79	0.621
Unig	Single	No	61.84	0.517
Big	Single	No	60.09	0.523
Unig + Big	Single	No	61.84	0.545
Unig + Big	Single	Yes	56.58	0.571
Unig + Big + PP	Single	No	63.60	0.754
Unig + Big + PP	Single	Yes	**72.81**	**0.777**

The results of this evaluation, which are summarized in Table 1 show that the combination of sentiment unigrams with bigrams and personal pronoun usage in consecutive posts can improve the overall performance. Also, the classification of each post to a single class (sentiment) in combination with the personal pronoun usage feature balances the impact of feature values to the final decision and achieves the best prediction performance so far.

Whole Thread Classification. In these experiments whole threads (*original* dataset) were classified in order to avoid the bias of fragmenting a conversation. The 10-fold cross validation strategy was directly applicable, since threads were not overlapping. The class distribution in the original dataset was 30%–70% with the majority being non-aggressive threads. In order to balance this ratio, a *SMOTE* filter was applied. Both types of post classification (single class and multi class) have been tested and the same feature set combinations have been evaluated (Uni-grams only, Bi-grams only, their combination and their combination plus the personal pronouns count). We also considered the user that posts each comment and merge any consecutive comments by the same user to a single comment. The only change in the results of this experiment is that the *RBF* classifier was outperformed by an *SVM* classifier[10] using a radial based kernel function. Results are depicted in Fig. 2.

Subthread Classification Based on Post Type Changes. The last set of experiments that represented threads as feature vectors was based on the

---

[10] The *LibSVM* implementation of *Weka* with default parameters.

Table 2. Whole thread classification (*Original* dataset) using single and multi-class classification of posts and 10-fold cross validation

Features	Post class	Balanced Training	Accuracy	ROC Area
Unig	Multiple	No	73.38	0.526
Big	Multiple	No	73.38	0.526
Unig + Big	Multiple	No	73.38	0.526
Unig + Big	Multiple	Yes	73.50	0.735
Unig + Big + PP	Multiple	No	74.45	0.542
Unig + Big + PP	Multiple	Yes	75.12	0.762
Unig	Single	No	73.38	0.500
Big	Single	No	71.94	0.500
Unig + Big	Single	No	72.66	0.521
Unig + Big	Single	Yes	72.50	0.725
Unig + Big + PP	Single	No	73.38	0.533
Unig + Big + PP	Single	Yes	**77.50**	**0.775**

conversion of threads into a sequence of bullying or neutral posts. This binary classification of a post was performed using BullyTracer's lexicon and significantly reduces the number of features. More specifically when bullying bi-grams are employed then we have only four features and in the case of tri-grams only eight features. We count the number of occurrences of each double or triple combination as well as the total number of bullying or non-bullying posts in the thread.

Experiments were performed on the *Window* dataset following the same test-training split as before. The algorithm that outperformed all others was the *RBF* classifier with a *PCA* attribute selection filter applied in a first step[11]. The results presented in Table 3 show an improved performance when compared to the respective results on the *Window* dataset with the sentiment features. They also show that using tri-grams instead of bi-grams in combination with simple counts of aggressive posts (both using BullyTracer lexicon and personal pronouns) gives a better performance. The above hold both when the training dataset is balanced or imbalanced, but the overall performance was better in the former case as expected.

4.2 Sequence Representation

An alternative representation for a discussion thread is as an independent sequence of sentiments. In this case, only the prevailing sentiment is used for each post and a *Hidden Markov Model* classifier is applied. Since the *SMOTE*

---

[11] Attribute Selected Classifier with *PCA* as attribute selection method and *RBF* classifier as classification method, was used.

Table 3. Sub-thread classification threads using single-class classification of posts, a training-test split and bullying/neutral bi-grams and tri-grams as features

Features	Balanced training	Accuracy	ROC area
Bcount	No	69.74	0.808
Bcount	Yes	75.00	0.808
Bbig + Bcount	No	67.98	0.803
Bbig + Bcount	Yes	72.81	0.801
Bbig + Bcount + PP	No	71.05	0.879
Bbig + Bcount + PP	Yes	76.75	0.871
Btrig + Bcount	No	69.74	0.797
Btrig + Bcount	Yes	76.32	0.817
Btrig + Bcount + PP	No	69.30	0.870
Btrig + Bcount + PP	Yes	**79.39**	**0.867**

Table 4. Classification of threads and sub-threads as sequences of sentiments or aggressive/neutral posts. An HMM classifier was used in all cases

Features	Split	Dataset	Balanced training	Accuracy	ROC area	TP ratio
Sentiments	90-10	Window	No	51.32	0.532	**94.6**
Sentiments	90-10	Window	Yes	53.07	0.517	79.3
Sentiments	10 fold	Original	No	49.64	0.457	48.7
Sentiments	10 fold	Original	Yes	56.41	0.504	61.5

over-sampling method cannot be applied in the sequence attribute, we apply a down-sampling technique (remove majority class samples without replacement) in order to balance the training dataset. From the results in Table 4 we see that the *HMM* classifier does not perform well with the sentiment sequences, probably because a larger training set is needed. The performance of the classifier for the sequence of aggressive/neutral posts is still high and comparable to the classifiers that use feature representations. Although the accuracy scores are not high, the reported true positive rate in the *Window* dataset, when the original data without sampling are employed is really high (94.6%), higher than that report in the original work.

5   Conclusions

This article presents a novel approach for the detection of aggressive and anti-social behavior in discussion threads, using text mining. The proposed method processes the thread of messages as a whole and captures the changes in sentiment between consecutive posts, which are used in turn for classifying the whole thread as aggressive or neutral. This reduces the effect of misclassifications in message-level and improves the performance of the aggressive behavior detection methodology. A set of n-gram like features, that capture the change of sentiment in consecutive posts or the interchange between bullying and neutral posts, as well as the use of personal pronouns in consecutive posts, are combined and evaluated. Experimental evaluation on a publicly available dataset shows that the proposed method outperforms a related, state-of-the-art method applied in the same dataset. As our future work, we plan to investigate the representation of sentiments as word embeddings directly learned from deep neural network architectures, such as *long short-term memory recurrent neural networks*.

References

1. Katakis, I.M., Varlamis, I., Tsatsaronis, G.: PYTHIA: employing lexical and semantic features for sentiment analysis. In: Calders, T., Esposito, F., Hüllermeier, E., Meo, R. (eds.) ECML PKDD 2014. LNCS (LNAI), vol. 8726, pp. 448–451. Springer, Heidelberg (2014). https://doi.org/10.1007/978-3-662-44845-8_32
2. Medhat, W., Hassan, A., Korashy, H.: Sentiment analysis algorithms and applications: a survey. Ain Shams Eng. J. **5**(4), 1093–1113 (2014)
3. Van Royen, K., Poels, K., Daelemans, W., Vandebosch, H.: Automatic monitoring of cyberbullying on social networking sites: from technological feasibility to desirability. Telemat. Inform. **32**(1), 89–97 (2015)
4. Moore, M.J., Nakano, T., Enomoto, A., Suda, T.: Anonymity and roles associated with aggressive posts in an online forum. Comput. Hum. Behav. **28**(3), 861–867 (2012)
5. Olweus, D.: Bullying at school: knowledge base and an effective intervention programa. Ann. N. Y. Acad. Sci. **794**(1), 265–276 (1996)
6. Yin, D., Xue, Z., Hong, L., Davison, B.D., Kontostathis, A., Edwards, L.: Detection of harassment on web 2.0. In: Proceedings of Content Analysis in the WEB 2.0 (CAW2.0) Workshop at WWW (2009)
7. Reynolds, K., Kontostathis, A., Edwards, L.: Using machine learning to detect cyberbullying. In: Proceedings of Machine Learning and Applications and Workshop (ICMLA), vol. 2, pp. 241–244. IEEE (2011)
8. Dinakar, K., Reichart, R., Lieberman, H.: Modeling the detection of textual cyberbullying. In: Proceedings of the Workshop on the Social Mobile Web, at the International AAAI Conference on Weblogs and Social Media (2011)
9. Dadvar, M., Trieschnigg, D., de Jong, F.: Experts and machines against bullies: a hybrid approach to detect cyberbullies. In: Sokolova, M., van Beek, P. (eds.) AI 2014. LNCS (LNAI), vol. 8436, pp. 275–281. Springer, Cham (2014). https://doi.org/10.1007/978-3-319-06483-3_25
10. Nahar, V., Al-Maskari, S., Li, X., Pang, C.: Semi-supervised learning for cyberbullying detection in social networks. In: Wang, H., Sharaf, M.A. (eds.) ADC 2014. LNCS, vol. 8506, pp. 160–171. Springer, Cham (2014). https://doi.org/10.1007/978-3-319-08608-8_14
11. Van Hee, C., et al.: Detection and fine-grained classification of cyberbullying events. In: Proceedings of the Recent Advances in NLP Conference (RANLP), pp. 672–680 (2015)
12. Dadvar, M., de Jong, F., Ordelman, R., Trieschnigg, D.: Improved cyberbullying detection using gender information. In: Proceedings of the Dutch-Belgian Information Retrieval Workshop (DIR), pp. 23–25 (2012)

13. Mihaylov, T., Nakov, P.: Hunting for troll comments in news community forums. In: Proceedings of the 54th Annual Meeting of the Association for Computational Linguistics (ACL), vol. 16, pp. 399–405 (2016)
14. Joty, S., et al.: Global thread-level inference for comment classification in community question answering. In: Proceedings of the Conference on Empirical Methods in Natural Language Processing (EMNLP), vol. 15 (2015)
15. Rafiq, R.I., Han, H.H.R., Lv, Q., Mishra, S., Mattso, S.A.: Careful what you share in six seconds: Detecting cyberbullying instances in vine. In: Proceedings of the International Conference on Advances in Social Networks Analysis and Mining (ASONAM), pp. 617–622. IEEE (2015)
16. Rafiq, R.I., Hosseinmardi, H., Han, R., Lv, Q., Mishra, S.: Investigating factors influencing the latency of cyberbullying detection (2016). arXiv preprint arXiv:1611.05419
17. Smith, P.K., Mahdavi, J., Carvalho, M., Fisher, S., Russell, S., Tippett, N.: Cyberbullying: its nature and impact in secondary school pupils. J. Child Psychol. Psychiatry **49**(4), 376–385 (2008)
18. Dooley, J.J., Pyżalski, J., Cross, D.: Cyberbullying versus face-to-face bullying: a theoretical and conceptual review. J. Psychol. **217**(4), 182–188 (2009)
19. Grigg, D.W.: Cyber-aggression: definition and concept of cyberbullying. Aust. J. Guid. Couns. **20**(02), 143–156 (2010)
20. Poria, S., Gelbukh, A., Cambria, E., Yang, P., Hussain, A., Durrani, T.: Merging senticnet and wordnet-affect emotion lists for sentiment analysis. In: Proceedings of the 11th International Conference on Signal Processing (ICSP), vol. 2, pp. 1251–1255. IEEE (2012)
21. Taboada, M., Brooke, J., Tofiloski, M., Voll, K., Stede, M.: Lexicon-based methods for sentiment analysis. Comput. Linguist. **37**(2), 267–307 (2011)
22. Thelwall, M., Buckley, K., Paltoglou, G., Cai, D., Kappas, A.: Sentiment strength detection in short informal text. J. Am. Soc. Inf. Sci. Technol. **61**(12), 2544–2558 (2010)
23. Glorot, X., Bordes, A., Bengio, Y.: Domain adaptation for large-scale sentiment classification: a deep learning approach. In: Proceedings of the 28th International Conference on Machine Learning (ICML), pp. 513–520 (2011)
24. Xu, J.M., Jun, K.S., Zhu, X., Bellmore, A.: Learning from bullying traces in social media. In: Proceedings of the North American Chapter of the Association for Computational Linguistics (NAACL), pp. 656–666. Association for Computational Linguistics (2012)
25. Ekman, P.: An argument for basic emotions. Cogn. Emot. **6**(3–4), 169–200 (1992)
26. Bayzick, J., Kontostathis, A., Edwards, L.: Detecting the presence of cyberbullying using computer software. In: Proceedings of the 3rd International Conference on Web Science (WebSci) (2011)

Machine Translation

Combining Machine Translation Systems with Quality Estimation

László János Laki[1,2] and Zijian Győző Yang[1(✉)]

[1] MTA-PPKE Hungarian Language Technology Research Group,
Pázmány Péter Catholic University, Faculty of Information Technology and Bionics,
Práter str. 50/A, Budapest 1083, Hungary
{laki.laszlo,yang.zijian.gyozo}@itk.ppke.hu
[2] MorphoLogic Lokalizáció Kft., Logodi str. 54, Budapest 1012, Hungary

Abstract. Improving the quality of Machine Translation (MT) systems is an important task not only for researchers but it is a substantial need for translating companies to create translations in a quicker and cheaper way. Combining the outputs of more than one machine translation systems is a common technique to get better translation quality because the strengths of the different systems could be utilized. The main question is to find the best method for the combination. In this paper, we used the Quality Estimation (QE) technique to combine a phrase-based and a hierarchical-based machine translation systems. The composite system was tested on several language combinations. The QE module was used to compare the outputs of the different MT systems and gave the best one as the result translation of the composite system. The composite system gained better translation quality than the separated systems.

1 Introduction

In the past few years Machine Translation (MT) systems have undergone significant changes. The goal of the researchers was to create better and better translations, which lead them to implement numerous MT systems based on the actual technological brands (e.g. rule-based MT, statistical MT, syntactical MT, neural MT). These methods have different advantages, therefore these systems could be used for different problems with high efficiency. For example, the rule-based MT system is more efficient when translating between inflected languages, because it is able to generate inflected word forms based on language specific grammar rules. However it is not able to translate unknown words effectively, which is the strength of SMT systems. Combining different kinds of MT systems can join their advantages and reduce the deficiency of the systems. The combined result can achieve better quality than the original MT systems.

Recently, there have been several experiments in combining outputs from different MT systems [1,10,13,16]. These combinations can be realized in many ways. One of these methods is to build a word-level confusion network from

© Springer Nature Switzerland AG 2018
A. Gelbukh (Ed.): CICLing 2017, LNCS 10762, pp. 435–444, 2018.
https://doi.org/10.1007/978-3-319-77116-8_32

hypotheses translations [12], others work on sentence-level [10]. There are differences in the alignment methods used for the generation of the confusion network [6,15,17,18,20].

The novelty of our system is that we used sentence level quality estimation (QE) calculation for the phrase-based (PB) and hierarchical-based (HB) MT system outputs to choose the best translation. The QE (see Fig. 1) estimates the quality of the MT translated segment without reference translation. It is based on a statistical model trained by regression analysis. Quality indicators are extracted based on the source segment, the machine translated segment and inner parameters of the MT system. The QE model is trained based on these indicators and on human or automatic evaluation scores. After all, using the trained statistical model, we can predict the quality of the new unseen segments. In our research we translated the source segments with two different MT systems, then using a QE model we chose the better quality translation as the final MT output.

The structure of this paper is as follows: First, we discuss the related work. Then, we shortly introduce the quality estimation approach. After this, we explain our methods and experiments in detail. Finally, our results and conclusions are described.

2   Related Work

The most common combination method is using confusion network decoding method to combine and choose the best translation outputs from multiple MT systems. A word-level alignment method is to select a monolingual reference against a hypothesis. Then, using this alignment, a confusion network is built from the hypotheses [13,19]. For this task, Rosti et al. [18] and Heafield et al. [6] used the TER [20] algorithm, Okita et al. [15] used the BLEU [17] method. Rosti et al. used an incremental word-based alignment method to build a confusion network. The alignment is based on the TER algorithm. This incremental alignment uses a pair-wise hypothesis. All alternative translation hypotheses are aligned against a "skeleton" hypothesis independently. Then, using the incremental alignment decoding, confusion networks are built from the union of hypotheses alignments. Heafield et al. [6] improved the confusion network decoding with phrase-level alignment. Huang and Papineni [7] created a hierarchical system combination strategy. This approach can combine MT hypotheses on word, phrase and sentence levels.

In our research, we used the QE approach to combine the MT outputs.

3   Quality Estimation

In the QE [21] task (see Fig. 1), we extract different kinds of quality indicators from the source and translated sentences without using reference translations. Following the research of Specia et al. [21], we can separate different feature categories. The first class contains the complexity features (e.g. number of tokens in the source segment), which are extracted from the source sentences. To the second class we extract fluency features (e.g. percentage of verbs in the target sentences) from the translated sentences. The third category includes the adequacy features (e.g. ratio of percentage of nouns in the source and target) extracted from the comparison of the source and the translated sentence. Finally, we can also extract features from the decoder of the MT system, namely confidence features (e.g. features and global score of the SMT system).

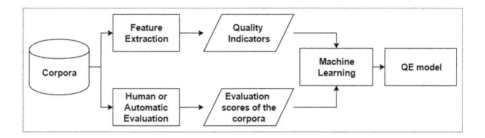

Fig. 1. Quality estimation model

From another perspective the features could be divided into two main groups: "black-box" and "glass-box" features. The features extracted from the inner parameters of the MT system are called "glass-box" features. Definition of these features requires access to the MT system, but in most cases the internal accessibility of the systems is not allowed. The features which do not depend on the MT inner parameters are the "black-box" features. These features make decisions based on the source and the hypothesis translation only. Since in our experiments we have translations from different MT systems, we use only the "black-box" features.

Thanks to the regression analysis, the QE system is able to predict the quality of the MT output compared to any measure score using the extracted quality indicators and features. Ideally, this measure would be human evaluation, but it is expensive and time-consuming. That is why standard automatic metrics (e.g. BLEU [17], OrthoBLEU [5], TER [20], etc.) are used to determine the quality of a translation. One of the advantages of the QE model is that it does not require reference translations for quality prediction, therefore it is an applicable and fully automatic solution to combine different kind of MT systems.

4   Introducing the Used MT Systems

Thanks to the QE technique, it is possible to create any number and any types of multiple MT systems. In this paper we used phrase-based (PB) and hierarchical-based (HB) statistical machine translation systems.

PBMT [11] systems rely on statistical observations derived from those phrase alignments which are automatically extracted from parallel bilingual corpora.

The main reason to use SMT is its language independent behavior, which can be used successfully in the case of language pairs with similar syntactic structure and word order. PBMT is a solution to handle local reordering, but it has difficulties with long distant ones. HBMT [3] tries to solve this reordering issue.

HBMT is the extension of the PBMT. While PBMT uses a phrase-to-phrase stack decoder, HBMT uses context free grammar based chart decoder. This technique helps HBMT to learn reordering patterns in a partially lexicalized form. For example the English-French negation is stored as *don't X → ne X pas*, where X can be replaced by any verb phrase.

If we compare the PB an HB systems, we can observe that their performance highly depends on the language pairs and the domain of the corpus. This explains why a HB system is not able to outperform the PB system in all cases (shown in Table 2). Consequently more robust MT systems can be created by choosing the better translation from the outputs of these two MT systems.

5 Methods and Experiments

In this section we will describe our experimental setups. First of all we will show the settings of the translation systems and after that the settings of the quality estimation system. Finally the system combination will be presented.

5.1 Dataset

We performed experiments on four language pairs, where English was the source language and the target languages were Hungarian, German, Italian and Japanese. These language combinations gave us a wide overview of the performance of the system, since the structure of these languages is very different. The wide overview was also supported by the used corpus, which was built from four domains (car industry documentation, European Parliament documentations, IT and industrial product documentation). The IT or the industrial documents contained mostly short segments, while the segments in law texts usually included more than 20 words. The biggest parts of the corpora were given by a translation company, therefore this part is not open-source. In order to make our experiments reproducible, one of our corpora was an open-source one: The Acquis Communautaire multilingual parallel corpus[1] used in the case of English-German translation. The exact size of the resources are shown in Table 1. The segments in each text are unique without repetitions. Also there are no overlapped segments between the train and the test sets.

5.2 Machine Translation System Setup

First of all, preprocessing was made on the training set, which included tokenization and truecaseing. The word alignment was created with GIZA++ [14].

---

[1] https://ec.europa.eu/jrc/en/language-technologies/jrc-acquis.

Table 1. Corpora used in the experiment

Language pair	Domain	# of MT training segments	# of QE training segments	# of QE test segments
English-Hungarian (en-hu)	Car industry	240,000	6,000	1,500
English-German (en-de)	Law	1,000,000	5,250	1,300
English-Italian (en-it)	Product description	800,000	3,143	785
English-Japanese (en-ja)	IT	1,000,000	3,169	790

Both phrase-based and hierarchical machine translation systems were realized by Moses [9]. This system was used for building 5-gram translation models as well. The 3-gram language models were built by the IRSTLM [4] tool. Our translation system could handle XML markups correctly, thanks to the *m4loc* architecture [8]

5.3   Quality Estimation System Setup

In our research, we used automatic evaluation metrics for building QE models, i.e. BLEU [17], OrthoBLEU [5] and OrthoTER scores. It means that our QE systems will predict a float number between 0 and 1 based on these metrics. The OrthoBLEU and OrthoTER methods work on character level, therefore these perform better than BLEU in the case of agglutinating and compounding languages like Hungarian. If the translation fails only at inflections, the BLEU gives a low-score, even if the stem is translated correctly. In such cases OrthoBLEU scores are more accurate.

As it was presented in Sect. 3, feature extraction is needed to build the QE model. For this task we applied the QuEst [21] system. We used only "black box" and language independent features to create the proper QE models for both MT systems for all four languages. In our research 67 features were applied, which were developed by Specia et al. [21]. These 67 features contain adequacy and fluency features (number of tokens in source and target segments, percentage of source 1–3-grams observed in different frequency quartiles of the source side of the MT training corpus, average number of translations per source word in the segment as given by IBM-1 model [2], etc.).

For QE model training, we tried several regression methods, for example support vector regression, decision trees and rules, Gaussian process etc. The Gaussian process (GP) with RBF Kernel gained the highest performance, thus in the results section we show only the GP scores.

For Hungarian, an optimization task (referred to as *en-hu-opt*) has been applied. According to the research of Yang et al. [22,23] an additional 60 features were added, from which 53 were developed by the authors of this paper. These features were language specific features (ratio of percentage of verbs and nouns in the target, percentage of nouns in the target, etc.), n-gram features (perplexity of the target, log probability of the target, etc.), error features (percentage of unknown words in the target, percentage of XML tags in the target, etc.) and semantic features (WordNet features, dictionary features, etc.). In the evaluation section we will compare the results of the basic and the optimized systems.

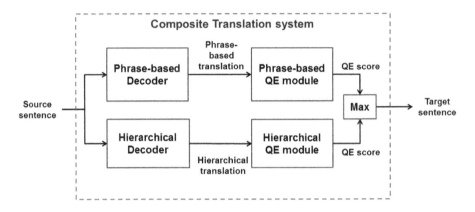

Fig. 2. Composite decoder architecture

5.4 Composite Translation System

The composite translation system was built from the PBMT and HBMT systems. The architecture of the system is shown in Fig. 2. First of all, the input segment translation is processed by the PBMT and the HBMT modules, then the appropriate QE modules predict an evaluation score for these translations. Finally, the system chooses the better translation based on the estimated rate, which will be the final output of the composite system.

6 Results and Evaluation

In our experiments, the QE train and test sets were translated with both MT systems. Based on the source and the translated sentences, the QE features have been extracted, then the GP regression was trained on one of the automatic evaluation metrics. Finally, the QE scores of the test sets were predicted. These steps were performed on the four language pairs and the three evaluation metrics. In Table 2, we can see the BLEU, OrthoBLEU and OrthoTER scores of the PBMT and the HBMT systems separately and the scores of the composite MT (Co MT). One more row is shown, namely the Max MT, which means the theoretical maximum score for Co MT translation, if the QE model would be perfect. Co MT system outperforms PBMT and HBMT systems in every cases.

During the deeper evaluation of the composite system, we took a closer look at the performance of the translation selection module. The precision, the recall and the F-measure were calculated both for the PB and for the HB selection cases

Table 2. Evaluation scores of the combined MT systems

		en-hu	en-hu-opt	en-de	en-it	en-ja
BLEU mean score ↑	PB MT	0.5156	0.5156	0.6288	0.7513	0.5945
	HB MT	0.6157	0.6157	0.4808	0.6998	0.6044
	Co MT	**0.6360**	**0.6375**	**0.6302**	**0.7525**	**0.6057**
	max MT	0.6702	0.6702	0.6475	0.7660	0.6458
OrthoBLEU mean score ↑	PB MT	0.7381	0.7381	0.6757	0.8202	0.5361
	HB MT	0.7679	0.7679	0.6221	0.7993	0.5536
	Co MT	**0.7795**	**0.7788**	**0.6788**	**0.8246**	**0.5553**
	max MT	0.8023	0.8023	0.6979	0.8374	0.5832
OrthoTER mean score ↓	PB MT	0.2903	0.2903	0.3574	0.1669	0.4281
	HB MT	0.2193	0.2193	0.4170	0.1995	0.4075
	Co MT	**0.2085**	**0.2108**	**0.3540**	**0.1662**	**0.4055**
	max MT	0.1848	0.1848	0.3349	0.1542	0.3769

as well. With these statistics, the performance of the quality predication could be measured. The last row contains the system level accuracy of the selection method, which means how many times the system chose the right one of the compared translations. For the calculation of F-measure, we counted as a positive predication when the PB and the HB translations were identical. This is the reason why F-measure could be higher than accuracy in the case of both systems.

These results are shown in Table 3. In most cases, our QE models are able to estimate with high accuracy. It is interesting to see that the precision of the selector metric is above 80% in the case of word-level metric and it is above ~72% in the case of character level measures. From these numbers we could see that the problem is with the recall of the MT system which has the lower quality. In most cases recall is near 65%, which could be an answer for the decrease of the accuracy of the Co MT system.

In the case of the comparison of the basic Hungarian model to the optimized Hungarian model, we used the statistical correlation, the MAE (Mean absolute error) and the RMSE (Root mean-squared error) evaluation metrics. The correlation ranges are from −1 to +1; the correlation is better if it is closer to the edge of the range. In the case of MAE and RMSE metrics, closer values to 0 mean a better QE model.

In Table 2 we can see that in the case of the BLEU metric ~2% higher correlation was reached with the optimized feature set. It could also be noticed, that in Table 4 the optimized Hungarian BLEU model could make higher accuracy prediction than the basic Hungarian BLEU model. The features we used for optimization were word-based features, hence only the optimized BLEU model could gain higher accuracy.

Table 3. Evaluation of the performance of the composite system

			en-hu	en-hu-opt	en-de	en-it	en-ja
BLEU	Precision	PB	88.224%	85.662%	85.229%	93.808%	90.846%
		HB	81.707%	82.979%	96.790%	97.513%	86.483%
	Recall	PB	64.809%	68.328%	98.810%	98.072%	84.976%
		HB	94.783%	93.103%	67.703%	92.114%	91.821%
	F-measure	PB	74.725%	76.020%	91.518%	95.892%	87.813%
		HB	87.761%	87.750%	79.675%	94.737%	89.072%
	System accuracy		80.067%	80.400%	84.615%	92.229%	81.519%
OrthoBLEU	Precision	PB	81.261%	78.547%	71.736%	90.997%	92.647%
		HB	79.834%	80.394%	97.324%	94.679%	85.894%
	Recall	PB	64.986%	67.003%	98.794%	95.773%	83.306%
		HB	90.244%	88.086%	52.980%	88.812%	93.893%
	F-measure	PB	72.218%	72.317%	83.118%	93.324%	87.728%
		HB	84.720%	84.064%	68.611%	91.652%	89.716%
	System accuracy		76.867%	76.267%	71.846%	88.025%	82.152%
OrthoTER	Precison	PB	85.714%	82.305%	73.132%	91.081%	94.559%
		HB	80.168%	80.833%	97.066%	95.644%	86.915%
	Recall	PB	59.627%	62.112%	98.712%	96.700%	84.140%
		HB	94.260%	92.287%	54.014%	88.441%	95.606%
	F-measure	PB	70.330%	70.796%	84.018%	93.807%	89.046%
		HB	86.645%	86.181%	69.406%	91.902%	91.053%
	System accuracy		78.400%	78.000%	73.077%	88.662%	84.304%

Table 4. Optimized BLEU Hungarian model

			en-hu	en-hu-opt
BLEU	Correlation	PB	0.6667	**0.6884**
		HB	0.5926	**0.6199**
	MAE	PB	0.1809	**0.1730**
		HB	0.1953	**0.1888**
	RMSE	PB	0.2266	**0.2196**
		HB	0.2402	**0.2341**

7 Conclusion

We created a composite machine translation system, which combines the output of multiple machine translation systems based on sentence-level quality estimation. In our experiments, phrase-based and hierarchical-based MT systems were combined, but with this technique any kind and number of systems could be combined. Quality estimation method with "black-box" features was used for the combination. The composite system was tested on four different language pairs. Results showed that our Co MT system gained better final translation quality compared to PBMT and HBMT systems in any experiments. In the case of English-Hungarian language pairs, some language dependent QE features were integrated to the Hungarian QE model, which led us to better prediction.

Acknowledgement. We would like to thank MorphoLogic Lokalizáció Kft. for the data sets which were used in our research.

References

1. Bangalore, S., Bordel, G., Riccardi, G.: Computing consensus translation from multiple machine translation systems. In: Proceedings of the IEEE Automatic Speech Recognition and Understanding Workshop (ASRU). Madonna di Campiglio, Italy, pp. 350–354 (2001)
2. Brown, P.F., Pietra, V.J.D., Pietra, S.A.D., Mercer, R.L.: The mathematics of statistical machine translation: Parameter estimation. Comput. Linguist. **19**(2), 263–311 (1993)
3. Chiang, D.: A hierarchical phrase-based model for statistical machine translation. In: Proceedings of the 43rd Annual Meeting on Association for Computational Linguistics, ACL 2005, pp. 263–270. Association for Computational Linguistics, Stroudsburg (2005)
4. Federico, M., Bertoldi, N., Cettolo, M.: IRSTLM: an open source toolkit for handling large scale language models. In: INTERSPEECH, pp. 1618–1621 (2008)
5. FTSK: Orthobleu mt evalution based on orthographic similarities @ONLINE (May 2014)
6. Heafield, K., Hanneman, G., Lavie, A.: Machine translation system combination with flexible word ordering. In: Proceedings of the Fourth Workshop on Statistical Machine Translation, pp. 56–60. Association for Computational Linguistics, Athens, Greece (March 2009)
7. Huang, F., Papineni, K.: Hierarchical system combination for machine translation. In: EMNLP-CoNLL, pp. 277–286 (2007)
8. Hudk, T., Ruopp, A.: The integration of Moses into localization industry. In: Proceedings of the 15th International Conference of the European Association for Machine Translation, Leuven, Belgium (2011)
9. Koehn, P., et al.: Moses: open source toolkit for statistical machine translation. In: Proceedings of the 45th Annual Meeting of the ACL, pp. 177–180 (2007)
10. Kumar, S., Byrne, W.: Minimum bayes-risk decoding for statistical machine translation. In: Susan Dumais, D.M., Roukos, S. (eds.) HLT-NAACL 2004: Main Proceedings, pp. 169–176. Association for Computational Linguistics, Boston (2004)
11. Lopez, A.: Statistical machine translation. ACM Comput. Surv. **40**(3), 8:1–8:49 (2008)
12. Mangu, L., Brill, E., Stolcke, A.: Finding consensus among words: lattice-based word error minimization. In: Sixth European Conference on Speech Communication and Technology, EUROSPEECH 1999, Budapest, Hungary, 5–9 Sept 1999, pp. 495–498 (1999)

13. Matusov, E., Ueffing, N., Ney, H.: Computing consensus translation for multiple machine translation systems using enhanced hypothesis alignment. In: McCarthy, D., Wintner, S. (eds.) EACL 2006, 11st Conference of the European Chapter of the Association for Computational Linguistics, Proceedings of the Conference, Trento, Italy, 3–7 April 2006. The Association for Computer Linguistics (2006)

14. Och, F.J., Ney, H.: A systematic comparison of various statistical alignment models. Comput. Linguist. **29**(1), 19–51 (2003)

15. Okita, T., van Genabith, J.: Minimum bayes risk decoding with enlarged hypothesis space in system combination. In: Gelbukh, A. (ed.) CICLing 2012. LNCS, vol. 7182, pp. 40–51. Springer, Heidelberg (2012). https://doi.org/10.1007/978-3-642-28601-8_4

16. Okita, T., Rubino, R., Genabith, J.v.: Sentence-level quality estimation for mt system combination. In: Proceedings of the Second Workshop on Applying Machine Learning Techniques to Optimise the Division of Labour in Hybrid MT, pp. 55–64. The COLING 2012 Organizing Committee, Mumbai, India (December 2012)

17. Papineni, K., Roukos, S., Ward, T., Zhu, W.J.: Bleu: A method for automatic evaluation of machine translation. In: Proceedings of the 40th Annual Meeting on Association for Computational Linguistics, ACL 2002, pp. 311–318. Association for Computational Linguistics, Stroudsburg (2002)

18. Rosti, A.V., Zhang, B., Matsoukas, S., Schwartz, R.: Incremental hypothesis alignment for building confusion networks with application to machine translation system combination. In: Proceedings of the Third Workshop on Statistical Machine Translation, pp. 183–186. Association for Computational Linguistics, Columbus (June 2008)

19. Sim, K.C., Byrne, W.J., Gales, M.J.F., Sahbi, H., Woodland, P.C.: Consensus network decoding for statistical machine translation system combination. In: 2007 IEEE International Conference on Acoustics, Speech and Signal Processing-ICASSP 2007, vol. 4, pp. IV-105–IV-108 (April 2007)

20. Snover, M., Dorr, B., Schwartz, R., Micciulla, L., Makhoul, J.: A study of translation edit rate with targeted human annotation. In: Proceedings of Association for Machine Translation in the Americas, pp. 223–231 (2006)

21. Specia, L., Shah, K., de Souza, J.G., Cohn, T.: Quest-a translation quality estimation framework. In: Proceedings of the 51st Annual Meeting of the Association for Computational Linguistics: System Demonstrations, Sofia, Bulgaria, pp. 79–84 (2013)

22. Yang, Z.G., Laki, L.J.: Minőségbecslő rendszer egynyelvű természetes nyelvi elemzőhöz. In: XIII. Magyar Számítógépes Nyelvészeti Konferencia. pp. 37–49. Szegedi Tudományegyetem, Informatikai Tanszékcsoport, Szeged, Hungary (2017)

23. Yang, Z.G., Laki, L.J., Siklsi, B.: Quality estimation for English-Hungarian with optimized semantic features. In: Computational Linguistics and Intelligent Text Processing, Konya, Turkey (2016)

Evaluation of Neural Machine Translation for Highly Inflected and Small Languages

Mārcis Pinnis[✉], Rihards Krišlauks, Daiga Deksne, and Toms Miks

Tilde, Vienibas gatve 75A, Riga, Latvia
{marcis.pinnis,rihards.krislauks,daiga.deksne,
toms.miks}@tilde.lv

Abstract. The paper evaluates neural machine translation systems and phrase-based machine translation systems for highly inflected and small languages. It analyses two translation scenarios: (1) when translating broad domain data from a morphologically rich language into a morphologically rich language or English (and vice versa), and (2) when translating narrow domain data and there are limited amounts of training data available for training machine translation systems. The paper reports on experiments for English (Germanic), Estonian (Finno-Ugric), Latvian (Baltic), and Russian (Slavic) languages. The scenarios are evaluated using automatic and manual – system comparative and error analysis-based – evaluation methods. The paper also analyses the aspects where neural systems are superior to statistical (phrase-based) systems and vice versa.

Keywords: Neural machine translation · Small languages
Highly inflected languages · System evaluation

1 Introduction

During the past three years, the research field of machine translation (MT) has experienced a paradigm shift from traditional statistical phrase-based machine translation (SMT) technologies (e.g., Koehn et al. 2003, 2007) to neural machine translation (NMT) technologies (Bahdanau et al. 2015; Devlin et al. 2014; Jean et al. 2015; Luong et al. 2015, etc.). Just recently in 2016, neural machine translation systems showed to achieve significantly better results than statistical systems for multiple language pairs including English-German, English-Czech, and English-Romanian (Sennrich et al. 2016a; Bojar et al. 2016), thereby paving the way for neural machine translation as the potential future technology for state-of-the-art machine translation system development. The first production level NMT system was introduced by Google at the end of September 2016 for Chinese-English (Le and Schuster 2016), followed by support for additional language pairs in November (Turovsky 2016). Recent research in NMT involves analysis of factored input support (Sennrich and Haddow 2016), character level NMT (in order to remove the necessity of pre-processing and post-processing data; e.g., Lee et al. 2016), methods for improved attention mechanisms (Meng et al. 2016), multi-language NMT (Firat et al. 2016), and multi-modal NMT (Caglayan, et al. 2016). However, the research and evaluation efforts have mainly focussed on large well-resourced languages with limited morphological complexity or just automatic

© Springer Nature Switzerland AG 2018
A. Gelbukh (Ed.): CICLing 2017, LNCS 10762, pp. 445–456, 2018.
https://doi.org/10.1007/978-3-319-77116-8_33

evaluation (Junczys-Dowmunt et al. 2016; Toral and Sánchez-Cartagena 2017), which as shown by our results and as previously indicated by Pinnis (2016) can have conflicting tendencies to manual evaluation results.

We believe that it is important to validate whether NMT technologies can be applied to more difficult language pairs with less resources available for training NMT systems. Therefore, in this paper we compare SMT and NMT systems trained for highly inflected small languages in two scenarios: (1) when translating broad domain data from a morphologically rich language into a morphologically rich language and English (and vice versa), and (2) when translating narrow domain data and there are limited amounts of training data available for training machine translation systems. More specifically, we analyse Estonian as a Finno-Ugric language, Latvian as a Baltic language and Russian as a Slavic language. We do not rely solely on automatic methods to compare systems trained on two different paradigms (i.e., SMT and NMT), but we report also results of two manual evaluation tasks: (1) system comparative evaluation according to the methodology by Skadiņš et al. (2010) and (2) error analysis of SMT and NMT translations.

The paper is further structured as follows: Sect. 2 describes the data used in the experiments, Sect. 3 describes system training, Sect. 4 discusses both automatic and manual evaluation efforts, and we conclude the paper in Sect. 5.

2 Data

In our experiments, we analyse the translation quality of SMT and NMT systems trained on relatively large and relatively small corpora. The large systems were trained on corpora covering texts from a broad domain, for instance, texts from legal, news, information technology, medicine, mechanical engineering, tourism and other sources. The large systems were trained for six language pairs including Latvian-English, Estonian-English, Estonian-Russian, and vice versa.

For the small systems, we were interested in identifying: (1) whether NMT systems can reach the SMT system quality when trained on narrow domain data and (2) what the optimal configuration parameters for training NMT systems on small data sets are. Therefore, the small systems were trained in multiple configurations for one language pair (English-Latvian). For training, we used the parallel corpus of the European Medicines Agency (EMEA). The corpus consists of two parts: (1) the OPUS EMEA corpus (Tiedemann 2009) and (2) the latest documents from EMEA's website[1] (years 2009–2014). The EMEA corpus is a narrow domain corpus in the medical domain.

Prior to training both SMT and NMT systems, we pre-processed the training data using the standard data pre-processing workflow of the *TildeMT* (Vasiļjevs et al. 2012) platform. At first, the data were cleaned (e.g., by normalising punctuation, whitespace, removing control symbols, decoding XML and HTML entities, etc.) and filtered (e.g., by deleting duplicates, sentences with word count differences and alphanumeric symbol and other symbol ratio differences higher than a standard threshold of the platform).

---

[1] The website of the European Medicines Agency can be found at: http://www.ema.europa.eu/.

Then non-translatable tokens were identified (e.g., e-mail addresses, file addresses, codes, etc.) to allow creating more generalised models (and reduce data sparsity). Finally, the data were tokenised and truecased. The statistics of the corpora after pre-processing are given in Table 1.

Table 1. Statistics of training data used for training SMT and NMT systems

Language pair	Unique sentence pairs (running tokens in source/target) in parallel corpora	Unique sentences in monolingual corpora
Broad domain systems (large data set systems)		
en-et	21,900,622 (368.7 M/288.8 M)	48,567,363
et-en	21,900,794 (288.2 M/368.6 M)	217,724,716
ru-et	4,179,198 (41.4 M/38.7 M)	48,606,392
et-ru	4,179,153 (38.7 M/41.4 M)	138,001,100
en-lv	7,300,666 (162.8 M/140.3 M)	74,741,452
lv-en	7,300,666 (140.3 M/162.8 M)	95,259,699
Narrow domain systems (small data set systems)		
en-lv	316,443 (6.0 M/5.3 M)	309,182

The English-Estonian and English-Latvian (and vice versa) broad domain systems were tuned using the ACCURAT balanced tuning corpus (Skadiņš et al. 2010) of 1,000 sentence pairs. The Estonian-Russian (and vice versa) systems were tuned using a held-out data set of 2,000 sentences from the training corpus. All broad domain systems were evaluated using the ACCURAT balanced evaluation corpus consisting of 512 sentences. The English-Latvian narrow domain systems were tuned and evaluated using held-out data sets from the narrow domain system training data.

3 System Training

3.1 Phrase-Based Statistical Machine Translation Systems

We started by training baseline SMT systems. All systems were trained in the cloud-based MT system development platform *TildeMT* using the *Moses* SMT system (Koehn et al. 2007). Default training set-up was used for all systems. For language model training, the *KenLM* (Heafield 2011) toolkit was used. All SMT systems were tuned using *MERT* (Bertoldi et al. 2009).

3.2 Neural Machine Translation Systems

NMT systems were trained using the sub-word neural machine translation toolkit *Nematus*[2] (Sennrich et al. 2016a) that is based on the toolkit *dl4mt-tutorial*[3] (Bahdanau et al. 2015). The toolkit supports training attention-based encoder-decoder

---

[2] https://github.com/rsennrich/nematus.

[3] https://github.com/nyu-dl/dl4mt-tutorial.

models with gated recurrent units. The architecture of the NMT systems is depicted in Fig. 1. For word segmentation in sub-word units, we use the byte pair encoding (BPE) tool from the *subword-nmt*[4] (Sennrich et al. 2016b) toolkit.

The broad domain NMT models were trained using a vocabulary size of 100,000 segments (99,500 for byte pair encoding). All other parameters were set to the default values used by the developers of *Nematus* for their WMT 2016 submissions,[5] namely, a projection (embedding) layer of 500 dimensions, recurrent units of 1024 dimensions, a batch size of 80, etc.

The narrow-domain NMT models were trained using multiple configurations to identify the best performing configuration. We analysed different vocabulary sizes, training with or without dropout and reduced hidden layer dimensionality. The configurations are as follows (other parameters were set to default settings):

- vocabulary size of 40,000 segments, batch size of 20 (*40K*);
- vocabulary size of 40,000 segments, dropout enabled, batch size of 20 (*40K+D*);
- vocabulary size of 40,000 segments, projection layer with just 250 dimensions, recurrent layers with 512 dimensions, batch size of 40 (*40K+R*);
- reduced vocabulary size of 8,000 segments, batch size of 40 (*8K*);
- reduced vocabulary size of 8,000 segments, dropout enabled, batch size of 40 (*8K+D*).

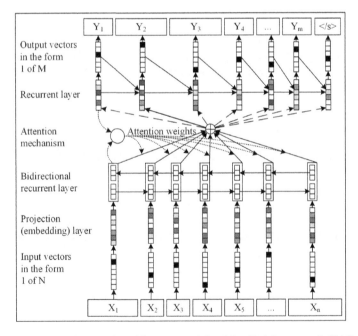

Fig. 1. Neural network architecture as defined by (Bahdanau et al. 2015)

---

[4] https://github.com/rsennrich/subword-nmt.

[5] https://github.com/rsennrich/wmt16-scripts/blob/master/sample/config.py.

4 Evaluation

We evaluated MT systems using both automated and manual evaluation methods. For automated evaluation, we report BLEU (Papineni et al. 2002), NIST (Doddington 2002), and ChrF2 (Popovic 2015). Two different manual evaluation methods were used – system comparative evaluation according to the methodology by Skadiņš et al. (2010), and error analysis of SMT and NMT translations. In comparative evaluation, NMT systems were compared to SMT and to Google Translate for all language pairs except English and Latvian for which only NMT and SMT systems were compared. Error analysis was performed for English-Latvian.

4.1 Automatic Evaluation

The results of the broad domain system automatic evaluation (see Table 2) show that for English-Estonian the NMT system achieves better results than both the baseline system and the Google Translate system. However, for the remaining translation directions Google Translate shows significantly better results than both the baseline SMT system and the NMT system. Note that the automatic evaluation results differ from the manual evaluation results (see Sect. 4.2) where the NMT system translations are found to be better than the SMT system translations for almost all language pairs.

Table 2. Translation system automatic evaluation scores

Dir.	System	BLEU	NIST	ChrF2
en-et	SMT	22.53 (20.39–24.95)	5.78 (5.53–6.04)	0.624
	Google Translate	19.80 (18.00–21.60)	5.51 (5.28–5.77)	0.621
	NMT	24.64 (22.76–26.54)	6.25 (6.01–6.49)	0.660
et-en	SMT	32.52 (30.55–34.53)	7.84 (7.61–8.06)	0.673
	Google Translate	40.57 (38.48–42.84)	8.56 (8.31–8.76)	0.712
	NMT	31.74 (29.91–33.45)	7.63 (7.40–7.86)	0.645
ru-et	SMT	09.87 (08.73–11.01)	3.79 (3.61–3.96)	0.512
	Google Translate	12.52 (11.03–14.01)	4.16 (3.94–4.37)	0.546
	NMT	09.02 (08.02–10.00)	3.72 (3.55–3.88)	0.495
et-ru	SMT	07.94 (07.07–08.82)	3.77 (3.63–3.91)	0.458
	Google Translate	14.74 (13.18–16.15)	4.56 (4.36–4.75)	0.530
	NMT	09.39 (08.33–10.46)	3.90 (3.73–4.06)	0.456
en-lv	SMT	32.57 (29.96–35.33)	6.96 (6.66–7.29)	0.647
	Production (SMT)	37.54 (34.65–40.50)	7.48 (7.14–7.81)	0.671
	NMT	24.77 (22.94–26.72)	6.22 (5.99–6.45)	0.615
lv-en	SMT	28.79 (26.84–30.82)	7.07 (6.84–7.31)	0.646
	Production (SMT)	43.76 (41.25–46.45)	8.35 (8.06–8.66)	0.714
	NMT	29.62 (27.62–31.44)	7.12 (6.89–7.35)	0.649

The automatic evaluation for the narrow domain systems (see Table 3) shows that the SMT system achieves better results than all NMT systems (the results are statistically significant with p equal to 0.01). From NMT systems the best result is achieved by *40K+D*. NMT systems with the reduced vocabulary size generally performed worse than their counterparts, but the results are still sufficiently good – the difference is around 2 BLEU points. Our observations align with the results by Wu et al. (2016).

Table 3. Narrow-domain NMT system automatic evaluation results (the underlined systems were compared in the human comparative evaluation)

System	BLEU	NIST	ChrF2
SMT	49.67 (47.57–51.73)	9.3723 (9.1531–9.5856)	0.776
40K voc., without dropout (*40K*)	43.15 (40.73–45.58)	8.6666 (8.4712–8.8736)	0.737
40K voc., with dropout (*40K+D*)	44.34 (42.33–46.43)	9.0127 (8.8150–9.2119)	0.762
40K voc., without dropout + reduced hidden layers (*40K+R*)	43.97 (42.00–45.65)	8.7630 (8.5586–8.9654)	0.736
8K voc., no dropout (*8K*)	40.95 (39.09–42.78)	8.4511 (8.2499–8.6494)	0.719
8K voc., dropout enabled (*8K+D*)	43.56 (41.10–45.41)	8.9045 (8.7162–9.1162)	0.749

Automatic evaluation results during system training (measured on the validation dataset) in Fig. 2 (BLEU) and Fig. 3 (loss) reveal a more detailed view on the training dynamics. For both vocabulary sizes, NMT systems with dropout eventually surpass their counterparts without dropout despite the slower learning in the first half of the experiments.

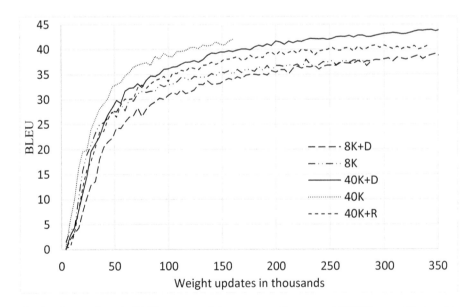

Fig. 2. Narrow-domain NMT system training progress (in terms of BLEU) on validation data

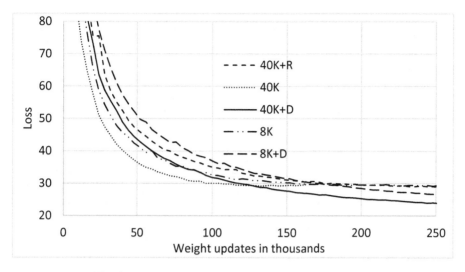

Fig. 3. Mean binary cross-entropy loss on validation data

4.2 Human Comparative Evaluation

To compare the translation quality of traditional SMT engines (Moses and Google Translate systems) with NMT engines, we enlisted professional translators (5–8 depending on the language pair) to perform blind comparative evaluation of translations from the evaluation sets. When comparing two translations produced by systems A and B for a given source sentence the translators were presented with three options: (1) A is better than B; (2) B is better than A; (3) neither A nor B is better or both are equally good. The results are summarised in Figs. 4, 5, and 6. The chart area marked with a pattern represents the lower bound for a 95% confidence interval. We show both the relative preference of translators for all three choices (left) and only for the choices (1) and (2) (right).

The broad domain system comparative evaluation results show that the translations of the NMT systems are preferred more by professional translators than the translations of the baseline SMT systems for all language directions except Latvian-English for which the systems performed similarly (see Fig. 4). The results are strongly sufficient for English-Estonian and weakly sufficient for all other language directions except Latvian-English, according to the methodology by Skadiņš et al. (2010).

When comparing NMT systems to Google Translate (see Fig. 5), the translations of the NMT system are preferred to Google Translate for English-Estonian and Estonian-English. For Estonian-Russian, translations of Google Translate are preferred instead. For Russian-Estonian, the systems performed similarly. The results are strongly sufficient for English-Estonian and weakly sufficient for Estonian-English and Estonian-Russian.

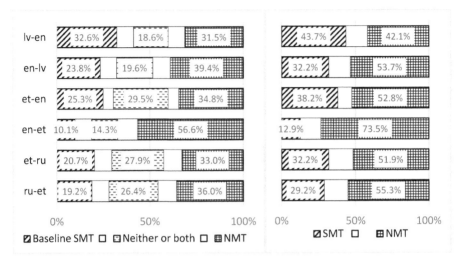

Fig. 4. Baseline SMT and NMT comparative evaluation results

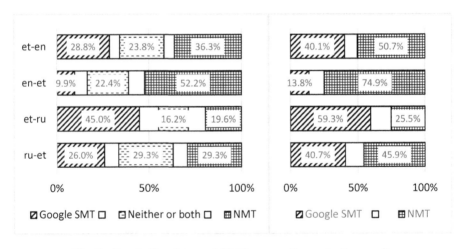

Fig. 5. Google Translator and NMT comparative evaluation results

The system selection for the narrow domain manual evaluation task was guided by the automatic evaluation results during system training. For this reason, *40K+D* was selected for comparison with the SMT system. Although the results in Fig. 6 show an advantage of the baseline SMT system over the NMT system, the results are insufficient according to the methodology by Skadiņš et al. (2010) to decide that any of the systems performs better than the other system.

Fig. 6. Comparative evaluation results for Baseline SMT and 40K+D NMT system

4.3 Translation Error Analysis

Human comparative evaluation in an essence allows identifying, which system is more preferred by the evaluators. It does not tell what kind of errors are produced more by one or the other system and where one of the systems is stronger than the other system. Therefore, we performed an error analysis task for 196 sentences from the English-Latvian broad domain systems in order to identify the strengths and weaknesses of the NMT technology in comparison to the SMT technology. The following errors were asked to be identified in MT system translations:

- word order errors;
- morphology (i.e., incorrect surface form selection), syntax (i.e., incorrect syntactic structures) and agreement (i.e., morphological agreement between words is broken) errors;
- non-translated or missing phrases in translations;
- additional phrases in translations (i.e., phrases appearing in the translation that are not present in the source sentence);
- wrong lexical choice errors (i.e., selection of a translation candidate that does not correspond to the context, including terminology errors).

The results of the error analysis (see Fig. 7) show that the NMT system handles (1) word ordering and (2) morphology, syntax and agreements (including long distance agreements) up to respectively five and three times better than the SMT system. This means that the NMT system produces more fluent translations than the SMT system.

The analysis also shows that the NMT system produces almost twice as many wrong lexical choice errors. However, in overall the results of the error analysis show that the NMT systems produce translations of higher quality. An additional positive result is that the percentage of completely correctly (without a single error) translated sentences for the balanced evaluation set is increased from 25% (for the SMT system) up to 35% (for the NMT system).

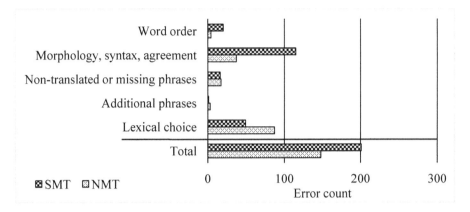

Fig. 7. Error analysis of English-Latvian SMT and NMT translations for 196 sentences from the balanced evaluation set

5   Conclusion

In the paper, we evaluated NMT and SMT systems for highly inflected and small languages using automatic and manual evaluation methods. Human comparative evaluation showed that for five out of six language pairs broad domain NMT systems achieved better results than SMT systems, compared to only three out of six language pairs in automatic evaluation. We see that the largest difference between automatic and manual evaluation results is evident for English-Latvian for which the automatic evaluation shows that the SMT system is by 7.8 BLEU points better than the NMT system. However, the comparative evaluation shows that professional translators prefer the translations of the NMT system more. The error analysis of the translations of the two systems further strengthens the findings of the comparative evaluation by showing that the NMT system produces significantly less errors than the SMT system. This means that it is not sufficient to rely on automatic evaluation results when comparing MT systems that are developed using two different MT paradigms.

The narrow domain experiments showed that in terms of automatic evaluation the SMT system produced significantly better translations than all NMT systems. The comparative evaluation showed that there is no significant difference between the quality of the SMT system and the best performing NMT system. This means that we have yet to identify a good NMT method for narrow domain and/or limited data NMT system development that could surpass the quality of SMT systems.

The error analysis for English-Latvian showed (as has also been shown by related research) that the NMT system produces more fluent translations. However, the SMT system is better at lexical choice, thereby producing translations that are more accurate.

We believe (after preliminary analysis of translation errors and training data) that a part of the lexical choice errors is caused by the level of noise in the parallel data (i.e., non-parallel segments). However, this has to be investigated further in future research.

Acknowledgements. The research has been supported by the European Regional Development Fund within the project "Neural Network Modelling for Inflected Natural Languages" No. 1.1.1.1/16/A/215. We would like to thank the Tilde Localization team for work on evaluating the translations.

References

Bahdanau, D., Cho, K., Bengio, Y.: Neural machine translation by jointly learning to align and translate. In: Proceedings of the International Conference on Learning Representations (ICLR) (2015)

Bertoldi, N., Haddow, B., Fouet, J.-B.: Improved minimum error rate training in moses. Prague Bull. Math. Linguis. **91**(1), 7–16 (2009)

Bojar, O., et al.: Findings of the 2016 conference on machine translation. In: Proceedings of the First Conference on Machine Translation. Shared Task Papers, vol. 2, pp. 131–198 (2016)

Caglayan, O., Barrault, L., Bougares, F.: Multimodal attention for neural machine translation (2016). arXiv preprint arXiv:1609.03976

Devlin, J., Zbib, R., Huang, Z., Lamar, T., Schwartz, R.M., Makhoul, J.: Fast and robust neural network joint models for statistical machine translation. In: Proceedings of the 52nd Annual Meeting of the Association for Computational Linguistics, pp. 1370–1380 (2014)

Doddington, G.: Automatic evaluation of machine translation quality using N-gram Co-occurrence statistics. In: Proceedings of the Second International Conference on Human Language Technology Research, pp. 138–145. Morgan Kaufmann Publishers Inc. (2002)

Firat, O., Cho, K., Bengio, Y.: Multi-way, multilingual neural machine translation with a shared attention mechanism. In: NAACL-HLT 2016, pp. 866–875 (2016)

Heafield, K.: KenLM: faster and smaller language model queries. In: Proceedings of the sixth workshop on statistical machine translation, pp. 187–197. Association for Computational Linguistics (2011)

Jean, S., Firat, O., Cho, K., Memisevic, R., Bengio, Y.: Montreal neural machine translation systems for WMT15. In: Proceedings of the Tenth Workshop on Statistical Machine Translation, pp. 134–140 (2015)

Junczys-Dowmunt, M., Dwojak, T., Hoang, H.: Is neural machine translation ready for deployment? A case study on 30 translation directions (2016). Arxiv. Retrieved from http://arxiv.org/abs/1610.01108

Koehn, P., et al.: Moses: open source toolkit for statistical machine translation. In: Proceedings of the 45th Annual Meeting of the Association for Computational Linguistics on Interactive Poster and Demonstration Sessions, pp. 177–180 (2007)

Koehn, P., Och, F.J., Marcu, D.: Statistical phrase-based translation. In: Proceedings of the 2003 Conference of the North American Chapter of the Association for Computational Linguistics on Human Language Technology, vol. 1, pp. 48–54 (2003)

Le, Q.V., Schuster, M.: A neural network for machine translation, at production scale (2016). Retrieved from https://research.googleblog.com/2016/09/a-neural-network-for-machine.html, 19 Oct 2016

Lee, J., Cho, K., Hofmann, T.: Fully character-level neural machine translation without explicit segmentation (2016). Retrieved from http://arxiv.org/abs/1610.03017

Luong, M.-T., Pham, H., Manning, C.D.: Effective approaches to attention-based neural machine translation. In: Proceedings of the 2015 Conference on Empirical Methods in Natural Language Processing, pp. 1412–1421 (2015)

Meng, F., Lu, Z., Li, H., Liu, Q.: Interactive attention for neural machine translation. In: Proceedings of the 26th International Conference on Computational Linguistics, Osaka, Japan, pp. 2174–2185 (2016)

Papineni, K., Roukos, S., Ward, T., Zhu, W.-J.: BLEU: a method for automatic evaluation of machine translation. In: Proceedings of the 40th Annual Meeting on Association for Computational Linguistics, pp. 311–318 (2002)

Pinnis, M.: Towards hybrid neural machine translation for English-Latvian. In: Human Language Technologies—The Baltic Perspective—Proceedings of the Seventh International Conference Baltic HLT 2016, pp. 84–91. IOS Press, Riga, Latvia (2016)

Popovic, M.: chrF: character n-gram F-score for automatic MT evaluation. In: Proceedings of the 10th Workshop on Statistical Machine Translation (WMT-15), pp. 392–395 (2015)

Sennrich, R., Haddow, B.: Linguistic input features improve neural machine translation. In: Proceedings of the First Conference on Machine Translation (WMT 2016). Research Papers, vol. 1, pp. 83–91 (2016)

Sennrich, R., Haddow, B., Birch, A.: Edinburgh neural machine translation systems for WMT 16. In: Proceedings of the First Conference on Machine Translation (WMT 2016). Shared Task Papers, vol. 2 (2016a)

Sennrich, R., Haddow, B., Birch, A.: Neural machine translation of rare words with subword units. In: Proceedings of the 54th Annual Meeting of the Association for Computational Linguistics (ACL 2016) (2016b)

Skadiņš, R., Goba, K., Šics, V.: Improving SMT for baltic languages with factored models. In: Human Language Technologies: The Baltic Perspective: Proceedings of the Fourth International Conference, Baltic HLT 2010, vol. 219, pp. 125–132. IOS Press (2010)

Tiedemann, J.: News from OPUS—a collection of multilingual parallel corpora with tools and interfaces. In: Proceedings of Recent Advances in Natural Language Processing, vol. 5, pp. 237–248 (2009)

Toral, A., Sánchez-Cartagena, V.M.: A multifaceted evaluation of neural versus phrase-based machine translation for 9 language directions (2017). Retrieved from http://arxiv.org/abs/1701.02901

Turovsky, B.: Found in translation: more accurate, fluent sentences in google translate (2016). Retrieved from https://blog.google/products/translate/found-translation-more-accurate-fluent-sentences-google-translate/, 26 Jan. 2017

Vasiļjevs, A., Skadiņš, R., Tiedemann, J.: LetsMT! a cloud-based platform for do-it-yourself machine translation. In: Proceedings of ACL 2012 System Demonstrations, pp. 43–48 (2012)

Wu, Y., et al.: Google's neural machine translation system: bridging the gap between human and machine translation (2016). arXiv preprint arXiv:1609.08144

Towards Translating Mixed-Code Comments from Social Media

Thoudam Doren Singh[1,2(✉)] and Thamar Solorio[2]

[1] Department of Computer Science and Engineering, IIIT Manipur,
Imphal, India
thoudam.doren@gmail.com
[2] Department of Computer Science, University of Houston, Houston, TX, USA
solorio@cs.uh.edu

Abstract. The translation task of social media comments has attracted researchers in recent times because of the challenges to understand the nature of the comments and its representation and the need of its translation into other target languages. In the present work, we attempt two approaches of translating the Facebook comments – one using a language identifier and other without using the language identifier. We also attempt to handle some form of spelling variation of these comments towards improving the translation quality with the help of state-of-the-art statistical machine translation techniques. Our approach employs n-best-list generation of the source language of the training dataset to address the spelling variation in the comments and also enrich the resource for translation. A small in-domain dataset could further boost the performance of the translation system. Our translation task focuses on Hindi-English mixed comments collected from Facebook and our systems show improvement of translation quality over the baseline system in terms of automatic evaluation scores.

Keywords: Code-Switching · Social media · Transliteration
Statistical machine translation

1 Introduction

The present day social media contents of Facebook, Twitter etc. are filled with several types of comments/tweets with different languages. One of the very common phenomenona is code switching by way of using two or more languages in the same post or comment. To the best of our knowledge, there is no report on the translation of Hindi and English code-switched comments from Facebook into English. There are two main reasons to work on this technology:

– The presence of large amount of Hindi-English code-switched (CS) comments in social media in general and Facebook in particular.

© Springer Nature Switzerland AG 2018
A. Gelbukh (Ed.): CICLing 2017, LNCS 10762, pp. 457–468, 2018.
https://doi.org/10.1007/978-3-319-77116-8_34

– Hindi is spoken mainly in the northern and western belts of India. According to census 2001, India claims to be the world's second-largest English-speaking country, second only to the US and expected to quadruple in the next decade. The impact of English on the native languages of other countries is on the rise.

Translating these comments into a target language would make a significantly larger access to other users who are monolingual in nature or media persons and researchers as well. From the research point of view, multiple interesting issues are to be addressed. First of all, it is important to identify the languages of comment, the domain and context in which the comments are passed from one user to other users. The comments in Facebook are often misspelled or filled with typos and not-so-well formed sentences. There are a lot of code-switched comments with different word forms (abbreviated, incorrect spelling, several smileys and emoticons etc.). Dealing with such kind of communication for natural language processing tasks is a big challenge, be it information retrieval or machine translation apart from developing basic NLP toolkits. The present paper attempts to address two important aspects:

– Preliminary analysis of the Facebook comments from the perspective of machine translation and collect the code-switched comments
– Attempt to translate comments with typos, misspellings and out-of-vocabulary words and handle code-switched comments with in-domain dataset built through selective approach of comments from different Facebook Ids with and without language identifier.

CS data faces several challenges for language computing technologies at syntactic and semantic levels and related applications of automatic speech recognition, information retrieval and extraction etc. Previous work of [24] reported the prediction of code switching points for English and Spanish language pair. Our attempt of translation chooses statistical machine translation (SMT) system. The present day SMT systems rely on the existence of two important resources, i.e., a parallel corpus between the source and target language and a monolingual corpus for the target language model. In the process, a very careful approach to collect data is carried out in such a way that we can build small in-domain parallel corpora which consists of common occurrences of the code-switched comments. This parallel corpus is used to build a translation model together with the baseline parallel corpus as detailed in Sect. 5. Most of the present day parallel corpus contains bit of noisy data and handling them is altogether a different challenge.

2   Related Work

The work of [21] presented a mechanism for machine translation systems of Hinglish (mixed code of English and Hindi) to pure (standard) Hindi and pure English. In this translation system, an additional layer is incorporated to the existing English to Hindi translation (AnglaBharti-II) which is basically a rule-based system and Hindi to English translation (AnuBharti-II) which is an EBMT (Example Based MT) system and both the systems are hybridized with varying degree of different paradigms developed by [20].

Machine translation experiments of [25] using the harvested data in two domains: edited news and microblog are reported with two test sets (1) news test set from Chinese-English documents on the project Syndicate for tuning and testing and using the full NIST dataset as the language model training data (2) manually indentified Weibo parallel segments used for development and testing tweets from Twitter for English language model and tweets from Weibo for the Chinese language model. They presented an efficient method to detect naturally occurring parallel text from the microblog for over 1 million Chinese-English parallel segments from Sina Weibo (Chinese couterpart of Twitter). Using these parallel segments as supplement to existing parallel training data, a substantial translation quality improvement is yielded in translating microblog text and modest improvement in translating edited news commentary. Impact of misspelled words in SMT and an enhancement of the system by decoding a word-based confusion network representing spelling variations of the input text is reported by [15]. A mixed-script query expansion [17] for information retrieval to handle the mixed-script term matching and spelling variation where the terms across the scripts are modeled jointly in a deep-learning architecture comparable in a low-dimensional abstract space.

3 Nature of Facebook Comments

Facebook comments are found in several scripts. The Hindi-English mixed Facebook comments are largely in Roman script than the Devanagari script. No specific standard rules are followed to spell a Hindi word in a non-native script. As a common feature, the Facebook comments consist of several code-switched texts either at sentence level or word level with several scripts for representation. Analysis of Facebook data generated by English – Hindi bilingual users on the significance of identification of borrowing and mixing between the two languages are presented by [9]. In the present task, we consider Facebook comments written only in Roman scripts for both English and Hindi or mixed comments. There are several conversational comments. These comments contain several slangs, code switch comments, exclamation and new terminology of other origin or language. There are diverse vocabulary and the spelling variation in these comments and no spelling rules are followed. Some of the comments collected from the Facebook are given below along with their translation.

i. *Source Comment*: plzzzz aap cptn mt chhodna ...... u r a best cptn of team india ......

   Translated Comment: please don't leave your captainship ... you are the best captain of team India ...

ii. *Source Comment*: Chinta mat karo, u have done it., u will see change in 6 month max.

   Translated Comment: Don't worry, you have done it .., you will see change in 6 month maximum.

iii. *Source Comment*: no wonder! ab King ke liye thora hee hotee hain Deadlines??

   Translated Comment: no wonder! Are there deadlines for king??

iv. *Source Comment*: Bhai mere 1st day se aaj 5th day tak every ball dekh raha hu,
Best of luck. Dil maange Ind Vs Eng 4-0.
Translated Comment: Brother I have been watching every ball from first day till
today the fifth day, Best of luck. Heart wants Ind Vs Eng 4-0.

Some comments consist of word-level code switch as given by example (i).
Example (ii) shows sentence level code-switching. In most of the comments, there are
non-standard spellings of words both for Hindi and English words.

4 Our Approach

Translation of social media comments is a challenging task. We consider English-Hindi
code mixed comments from Facebook to translate into English output. A comparative
study of the translation systems are carried out between two approaches – (1) transliterate the Hindi words of the comments and then translate the whole comment into
English (2) transliterate the training Hindi sentences of training data and translate the
Hinglish comments using the system trained using the transliterated Hindi sentences
and the corresponding English sentence of the Hindi-English parallel corpus.
Translating them into well formed sentences depends on the availability of a good
translation system. While statistical machine translation system works well for many
major language pairs, the quality of translation depends on the availability of large
good quality parallel corpus between the source and target language. We make use of a
small freely available Hindi-English parallel corpus. We also propose a simple method
to handle spelling variation of Hindi words of the input comments to be translated by
means of n-best-list generation as detailed in Sect. 6.

In the experiment, the English words and out of vocabulary words in the comments
are passed through the statistical machine translation system and Hindi words are
translated into English to achieve translated comments in English. The SMT systems
are built to translate Hindi to English using freely available Hindi-English parallel
corpus and a target language monolingual corpus. The Hindi sentences of the original
parallel corpus are in Devanagari script and English sentences are in Roman. In the first
approach, a language identification module is plugged-into identify Hindi words found
in the comments. The Hindi words found in the comments are transliterated using a
Roman to Devanagari statistical transliteration system. In this approach, the Hindi-
English SMT system is built using the original Hindi-English parallel corpus which
uses Devanagari script for Hindi and Roman script for English. So, the Hindi words
found in Devanagari scripts are translated using the SMT system and other English
words and out of vocabulary words are passed through.

The second approach of the experiment uses a Devanagari script to Roman script
transliteration system to transliterate Hindi sentences of the Hindi-English parallel
corpus. Additional experiments to handle spelling variation of the Hindi words in the
comments are carried out using n-best-list generated from a Devanagari script to
Roman script statistical transliteration system of the Hindi sentences of the Hindi-
English parallel corpus. Experiment to introduce domain adaptation is carried out by
introducing a small in-domain dataset.

We attempt to translate Facebook comments which consist of several spelling variants and code mixed comments with and without language identification. We attempt two approaches. Our approach attempts to narrow down the problem-solving from three angles (1) use of comments written only with Roman script (2) taking advantage of the n-best-list spelling variants of a transliteration system from Devanagari to Roman scripts and (3) using small in-domain parallel dataset in Roman script.

Domain adaptation is important in statistical machine translation; the translation performance degrades when translating an input which is out-of-domain. The larger part of training dataset that we use in the present work is from different sources and they are in different domain. We add a small in-domain dataset which is hand-aligned in the training dataset. We, also, tune the system with an in-domain dataset collected from the Facebook and translated into English with gain in BLEU score.

Challenges to handle new terminology and word spellings are some of the most important problem to be tackled to translate Facebook comments. Spelling variation leads to misspelling. These misspelled words are formed by incorrect sequence of characters. Several social media comments consists of intentional short form of a word, or unintentional typing errors. A challenging issue is to resolve the confusion between the spellings of a word with the spelling of another word with similar sound. Thus the social media notation for these words may lead to ambiguous words. Consider a Hindi word *aapka*, this could be written *apka, aapaka, aapka, aapk, apaka, apaaka, apakaa*. Dealing with unknown words is very common for resource scarce language pairs. There are large numbers of out-of-vocabulary (OOV) words in the social media comments, specially in Facebook comments. We do exercise passing-through the OOV by keeping unknown words at the output of the SMT system.

5 Corpus Collection and Preparation

The Hindi-English parallel corpus[1] (HindEnCorp 0.5) of 273.9 k (3.76 M English tokens and 3.88 M Hindi tokens) is based on Devanagari and Roman scripts respectively. We use this parallel corpus after passing through a filtering and cleaning pipeline. The Hindi-English parallel corpora [16] using Devanagari and Roman scripts were collected from different web sources and preprocessed primarily for building Hindi-English SMT system. The main sources are Tides, Commentaries by Daniel Pipes, EMILLE and the extended Indic multi-parallel corpus [13, 14], Launchpad.net, TED talks, Intercorp [6] etc. Some of the noise content and some of the character level inconsistencies and errors are resolved through simple filtering.

Additional small in-domain development dataset in Roman script is also added to the training dataset for additional model. This dataset is the selective collection of Facebook comments from three different categories. We collected comments from two Indian Bollywood actors, two Indian sportsperson and one Indian politician. The public Facebook IDs are AmitabhBachchan, IamSRK, SachinTendulkar, Circle of Cricket. MSDhoni and narendramodi. We choose these IDs so that we get comments written

---

[1] https://ufal.mff.cuni.cz/hindencorp.

mainly either in English or Hindi or in Hinglish from different domains. This dataset consists of several spelling variants and good amount of code-switched comments. The comments are filtered from the raw data setting criteria for certain amount of English words and Hindi words in the comments based on CMU dictionary and a collection of high frequency Hindi words that are found in the code-mixed data.

The CMU lexicon[2] consists of 133,771 entries. All the entries are lowercased. The filtering is done in such a way that at least 20% but less than or equal to 80% of the comments should match from the CMU lexicon. There are named entities in the lexicon. This dataset consists of code mixing at word level as well as sentence level. After the filtering with CMU lexicon, the 2-3-4 g characters of frequently used Hindi words [2] in the Hinglish comments are again applied to filter in order to ensure that there are Hindi words as well. The Hindi 2-3-4 character grams and words consist of 156 entries. Additional collection can be made to improve the quality. One of the most important steps in translation from the Facebook comments to English is to make correct spelling at the target side. Certain punctuation marks such as long trail of dots (….) are replaced with a three dot (…). While the guideline for translating from the Facebook comments to English follows certain rules, there are still issues to get the right translation from the comments. One such rule is to remove extra characters, such as *speechlesssss* is corrected as *speechless*.

6 Statistical Transliteration and N-Best-List Generation

The transliteration work to represent texts from a source script into a target script conforming to the target language phonology and orthographic conventions for major Indian languages pairs are reported by [3]. They used only the CFILT, IITB transliteration corpus.[3] We start our work with corpus collection and preparation followed by building statistical transliteration system for Devanagari to Roman script and Roman to Devanagari transliteration system. In the parallel corpus we used, the source side of the Hindi-English parallel dataset is available in Devanagari script. The Romanization of Devanagari script based text must account for the language phonology and orthographic conventions. We employ a Devanagari to Roman statistical transliteration system for the generation of the n-best-list for the top 3, 5 and 7 transliterations in Roman script. The n-best list is a by-product of log-linear based decoding process in the phrase based machine translation system generated as a full hypothesis from a word graph for additional re-ranking. There may be several look-alike hypotheses with different scores in the n-best list based on the mapping of two words by a single phrase or two individual phrases by a single word. These top 3, 5 and 7 transliterated Roman script based Hindi source sentences are included in the training process to increase the coverage of the spelling variants. The training dataset of the transliteration system consist of the Devanagari-Roman datasets from CFILT, IIT Bombay[4] transliteration

---

[2] http://svn.code.sf.net/p/cmusphinx/code/trunk/cmudict/cmudict-0.7b.

[3] http://www.cfilt.iitb.ac.in/brahminet/static/register.html.

[4] http://www.cfilt.iitb.ac.in/brahminet/static/register.html.

corpus and FIRE 2013[5] shared task [10] datasets. The tune and test dataset [22] are part of the dictation experiment.[6] The performance of the transliteration systems are measured in terms of acceptability by subjective evaluation approach. This transliteration is a phrase based approach of SMT using Moses [18]. MGIZA [19] a multithreaded word-to-word aligner version of GIZA++ [8] is used with grow-diag-final-and symmetrization technique. The training dataset pairs are split into character levels. The target language model is built using SRILM [1] using character level split dataset using Good-Turing discounting. The tuning is carried out using MERT [7] with Batch MIRA [5]. Table 1 gives the transliteration training corpus statistics and Table 2 gives the tune and test sets statistics of transliteration system.

We also built a transliteration system from Roman to Devanagari using the same dataset as given by Table 1 and toolkits mentioned above followed by evaluation. This system is used to transliterate the code-switched test datasets for the SMT systems which use Hindi-English parallel corpora in Devanagari and Roman scripts respectively.

Table 1. Transliteration training corpus break-up statistics

Training dataset source	Number of tokens	Number of characters	
		Devanagari	Roman
CFILT IIT Bombay	21,031	136,861	150559
FIRE 2013	30,823	158526	205161

Table 2. Tune and Test sets statistics of transliteration

Datasets	Number of tokens	Number of characters	
		Devanagari	Roman
Total training datasets	51854	295387	355720
Tuning	519	2071	2428
Testing	512	2161	2475

7 Experimental Setup

We use Moses [18], an open-source phrase-based statistical machine translation system for all the models built in the following experiments. KenLM [11], which is fast and memory efficient is used for target language modelling of order 5. The language model

---

[5] http://cse.iitkgp.ac.in/resgrp/cnerg/qa/fire13translit/Hindi%20-%20Word%20Transliteration%20Pairs%201.txt.

[6] http://cse.iitkgp.ac.in/resgrp/cnerg/qa/fire13translit/Hindi%20-%20Word%20transliteration%20pairs%202.zip.

uses modified Kneyser-Ney smoothing [23] without pruning. The MGIZA [19], a multi-threaded and drop-in replacement version of GIZA++ [8], which is a toolkit of the original IBM alignment models implementation is used for word alignment with grow-diag-final symmetrization technique. The hier-mslr-bidirectional-fe is used for reordering. BLEU [12] is used for automatic scoring, MERT [7] with Batch-MIRA [5] is used for tuning the system. The target monolingual corpus for language modeling is built using the Xinhua part of English Gigaword with additional 2.47 million comments which is by and large in English from the comments collected from the Facebook IDs mentioned earlier after setting criteria for filtering. There are approximately 17.11 million tokens in the comments collected from the Facebook.

8   Our Translation Models of Facebook Comments

We carry out experiments with different datasets for two approaches.

First Approach: We transliterate the Hindi words of the code-switched Hinglish comments using a Roman to Devanagari transliteration system then translate these mixed script comments using the SMT model built using Hindi-English parallel corpora which are in Devanagari script for Hindi and Roman script for English respectively.

This model is tuned and tested using the datasets given by Table 4. For this approach, we plug in a language identifier to identify the Hindi words for transliteration.

We use a weakly supervised method of identifying words in mixed-language document [4] with CRF option as a sequence labeling problem. The monolingual English and Hindi corpus given by Table 3 is used by the language identifier. Subjective evaluation on a testset of 8978 tokens of Hinglish comments gives an accuracy of 89.34%.

Table 3. Language Identifier Statistics

Languages	Number of Comments	Number of Tokens
English monolingual	811,085	4,063,764
Hindi monolingual	37121	456,057

Table 4. Tune and test datasets statistics

Datasets	Number of comments	Number of tokens	
		Hinglish	English
Tune dataset	542	10243	9939
Test dataset	500	7677	7506

Second Approach: We translate the code-switched Hinglish comments in Roman script using the SMT model built using the existing Hindi-English parallel corpus after transliterating the Hindi sentences of training corpora which are in Devanagari into Roman. So, both the Hindi and English sentences of the parallel training corpus are in Roman scripts. Table 4 gives the tune and test dataset statistics. Table 5 gives statistics of parallel corpora used in this approach. Hinglish Facebook comments collected are in Roman script. Under this approach, three categories (I, II and III) of phrase based statistical translation models are trained to translate Hinglish comments to English.

Table 5. Roman-roman parallel corpus

Datasets	Number of sentences	Number of tokens	
		Hindi	English
Model-I	269,541	3,252,034	2,970,423
Model-II	808,623	9,643,037	8,860,830
Model-III	1,347,705	16,071,742	14,768,050
Model-IV	1,886,787	22,705,879	20,779,600
Model-V	1,078,706	12,873,695	11,831,635
Model-VI	1,617,788	19,302,400	17,738,855
Model-VII	2,156,870	25,968,156	23,759,962

Category I: This is the baseline model. The Hindi sentences of the original parallel corpus in Devanagari script are transliterated into Roman script using the Devanagari to Roman transliteration system for training a translation model. The phrase based translation model is trained based on the parallel corpus (Hindi in Roman script and English in Roman scripts) using MGIZA word aligner and Moses SMT system. In the decoding process, the Hindi words in the Hinglish comments will get translated to English if it is found in the translation model (Phrase Table) and other out of vocabulary (OOV) words including the English words in the Hinglish comments get passed through at the output. Model I (Baseline) is under this category.

Category II: There are large numbers of spelling variation for the words in the Facebook comments. This model attempts to address some of the spelling variation found in the Hinglish comments. In order to handle this spelling variation, we generate some of the spelling variation of Hindi sentences using the n-best-list generation feature of a statistical transliteration system. We picked up 3, 5 and 7-best list transliterations of the Hindi sentence using the Devanagari to Roman transliteration system. These translation models are trained using the 3, 5 and 7-best-list transliterated Hindi sentences in Roman script and corresponding English sentences. Models II (3-Best-List), III (5-Best-List), IV (7-Best-List) are under this category.

Category III: This model is trained on the combination of training dataset of Model I, Model-II and in-domain dataset (development set) between Hinglish and English. Models V (3-Best+Baseline, DevSet), Model-VI (5-Best+Baseline+DevSet), Model-VII (7-Best+Baseline+DevSet) are under this category.

9   Evaluation Result

We carried out the automatic scoring for evaluation with a single reference translation. The automatics scores of the SMT systems are given by Tables 6 and 7.

Table 6. BLEU and NIST scores of first approach

Datasets	BLEU score	NIST score
Baseline	12.17	4.03

Table 7. BLEU and NIST scores of second approach

Datasets	BLEU score	NIST score
Model-I	23.19	5.03
Model-II	24.42	5.04
Model-III	22.91	5.05
Model-IV	24.25	5.06
Model-V	28.98	5.90
Model-VI	29.13	5.77
Model-VII	27.17	5.52

We see a significant drop in the BLEU scores in the first approach as shown by Table 6 due to the data sparsity problem as compared to the SMT models which use Hindi-English parallel corpora in Roman scripts for both the source and target languages of second approach. The data sparsity is due to lack of following standard spelling rules in the code-switched comments, not-so-well formed sentences and the transliteration system used.

10   Discussion and Conclusion

Translating code-switched comments to another target language often fails to work or results in a poor translated output for multiple reasons over and above the challenges in machine translation tasks. The present task attempts to address some of the issues and challenges while translating a code-switched data to a target language from the perspective of statistical machine translation systems. We consider translating from Hindi-English code-switched data to English as our task for the comments collected from public Facebook Ids. The SMT-based system performance degrades while translating out of domain source sentences. Another important issue is handling unknown and out-of-vocabulary words. The present task also attempts to build small in-domain parallel dataset between the code-switched data and the target language.

Considering English and Hindi at the source side in case of mixed-code and English at the target side, the MT performance is affected due to the richer morphology of Hindi compared to English and sparse data problem due to the orthographic variation. Using standard tools would be hard for these kinds of comments and there are abbreviations or exclamations that may not have been seen in the training data which further affect the quality of translation. One of the important issues handled is the length of the comments. This is an important issue in case of translation as a trade-off between the length and quality of translation.

The current work also tries to rectify any not-so-well formed English part of the comments into well formed English apart from translating Hindi part of the comments into a well formed English output. The misspelling and wrong syntactic structure of the Facebook comment can be further improved with additional in-domain parallel corpora between English-Hindi mixed comments and the target language, i.e. English. The automatic score shows that there is a big jump of 5.94 absolute score from the baseline system by adding in-domain dataset and 5-best list. Thus, the models of second approach outperform the models of first approach. While the quality of transliteration used is paramount, improving the n-best-list quality is an important step ahead.

Acknowledgements. We acknowledge the HPCC of UH for providing facility to carry out experiments. We thank Kunal Parmar for his help in the data preparation. Thanks to Paolo Rosso, UPV, Valencia, Spain for his valuable feedback and Heba Elfardy of George Washington University for helping during the data collection process.

References

1. Stolcke, A.: SRILM - an extensible language modeling toolkit. In: Proceedings of the International Conference Spoken Language Processing, Denver, Colorado (2002)
2. Dey, A., Fung, P.: A Hindi-English code-switching corpus. In: The 9th International Conference on Language Resources and Evaluation (LREC) (2014)
3. Kunchukuttan, A., Puduppully, R., Bhattacharyya, P.: Brahmi-Net: a transliteration and script conversion system for languages of the Indian subcontinent. In: Conference of the North American Chapter of the Association for Computational Linguistics—Human Language Technologies: System Demonstrations (2015)
4. King, B., Abney, S.: Labelling the languages of words in mixed-language documents using weakly supervised methods. In: Proceedings of NAACL-HLT 2013, Atlanta, Georgia, 9–14 June, pp. 1110–1119 (2013)
5. Cherry, C., Foster, G.: Batch tuning strategies for statistical machine translation. In: Proceedings of the NAACL-HLT, 2012, Montreal, Canada, June 3–8, pp. 427–436 (2012)
6. Čermák, F., Rosen, A.: The case of InterCorp, a multilingual parallel corpus. Int. J. Corpus Linguis. **13**(3), 411–427 (2012)
7. Och, F.J.: Minimum error rate training in statistical machine translation. In: Proceedings of the 41st Annual Meeting on Association for Computational Linguistics, ACL 2003, Stroudsburg, PA, USA, vol. 1, pp. 160–167. Association for Computational Linguistics (2003)
8. Franz Josef Och and Herman Ney: A systematic comparison of various statistical alignment models. Comput. Linguis. **29**(1), 19–51 (2003)
9. Bali, K., Sharma, J., Choudhury, M., Vyas, Y.: "I am borrowing ya mixing ?" An analysis of English-Hindi code mixing in Facebook. In: Proceedings of the First Workshop on Computational Approaches to Code Switching, Oct 25, 2014, Doha, Qatar, pp. 116–126 (2014)

10. Gupta, K., Choudhury, M., Bali, K.: Mining Hindi-English transliteration pairs from online Hindi lyrics. In: Proceedings of the Eight International Conference on Language Resources and Evaluation (LREC 2012), 23–25 May, Istanbul, Turkey, pp. 2459–2465 (2012)
11. Heafield, K.: KenLM: faster and smaller language model queries. In: Proceedings of the EMNLP 2011 Sixth Workshop on Statistical Machine Translation, Edinburgh, Scotland, United Kingdom, pp. 187–197 (2011)
12. Papineni, K., Roukos, S., Ward, T., Zhu, W.J.: BLEU: a method for automatic evaluation of machine translation. In: Proceedings of 40th ACL, Philadelphia, PA (2002)
13. Birch, L., Callison-Burch, C., Osborne, M., Post, M.: The Indic multi-parallel corpus (2011). http://homepages.inf.ed.uk/miles/babel.html
14. Post, M., Callison-Burch, C., Osborne, M.: Constructing parallel corpora for six Indian languages via crowdsourcing. In: Proceedings of the Seventh Workshop on Statistical Machine Translation, Montreal, Canada, pp. 401–409. Association of Computational Linguistics (2012)
15. Bertoldi, N., Cettolo, M., Federico, M.: Statistical machine translation of texts with misspelled words. In: Human Language Technologies: The 2010 Annual Conference of the North American Chapter of the Association for Computational Linguistics, June 02–04, Los Angeles, California, pp. 412–419 (2010)
16. Bojar, O., Diatka, V., Rychlý, P., Stranak, P., Suchomel, V., Tamchyna, A., Zeman, D.: HindEnCorp—Hindi-English and Hindi-only corpus for machine translation. In: Proceedings of the Ninth International Conference on Language Resources and Evaluation (LREC 2014), Reykjavik, Iceland (2014)
17. Gupta, P., Bali, K., Banchs, R.E., Choudhury, M., Rosso, P.: Query expansion for mixed-script information retrieval. In: SIGIR 2014, Gold Coast, Queensland, Australia, July 6–11, 2014 (2014)
18. Koehn, P., Hoang, H., Birch, A., Callison-Burch, C., Federico, M., Bertoldi, N., Cowan, B., Shen, W., Moran, C., Zens, R., Dyer, C., Bojar, O., Constantin, A., Herbst, E.: Moses: open source toolkit for statistical machine translation. In: Annual Meeting of the Association for Computational Linguistics (ACL), demonstration session, Prague, Czech Republic (2007)
19. Gao, Q., Vogel, S.: Parallel implementations of word alignment tool. In: Software Engineering, Testing, and Quality Assurance for Natural Language Processing, pp. 49–57 (2008)
20. Sinha, R.M.K.: An engineering perspective of machine translation: AnglaBharti-II and AnuBharti-II architectures. In: Invited Paper, Proceedings of International Symposium on Machine Translation, NLP and Translation Support System (iSTRANS-2004), November 17–19, Tata Mc Graw Hill, New Delhi (2004)
21. Sinha, R.M.K., Thakur, A.: Machine translation of bi-lingual Hindi-English (Hinglish) text. In: Proceedings of the 10th Conference on Machine Translation (2005)
22. Sowmya, V.B., Choudhury, M., Bali, K., Dasgupta, T., Basu, A.: Resource creation for training and testing of transliteration systems for Indian languages. In: LREC 2010 (2010)
23. Stanley, F.C., Goodman, J.: An empirical study of smoothing techniques for language modeling. Technical Report TR-10–98, Harvard, Cambridge, MA, USA (1998)
24. Solorio, T., Liu, Y.: Learning to predict code-switching points. In: Proceedings of the 2008 Conference on Empirical Methods in Natural Language Processing, Honolulu, Hawaii, pp. 973–981 (2008)
25. Ling, W., Xiang, G., Dyer, C., Black, A., Trancoso, I.: Microblogs as parallel corpora. In: Proceedings of the 51st Annual Meeting of the Association for Computational Linguistics, Sofia, Bulgaria. Long Papers, vol. 1, pp. 176–186 (2013)

Multiple System Combination for PersoArabic-Latin Transliteration

Nima Hemmati[1(✉)], Heshaam Faili[1], and Jalal Maleki[2]

[1] Department of ECE, University of Tehran, Tehran, Iran
{nimahemmati,hfaili}@ut.ac.ir
[2] Natural Language Processing, Department Computer and Information Science,
Linköpings Universitet, 581 83 Linköping, Sweden
jalal.malek@liu.se

Abstract. In this paper, we model a PersoArabic to Latin transliteration system as grapheme-to-phoneme (G2P) and word lattice methods combined with statistical machine translation (SMT). Persian is an Indo-Iranian branch of the Indo-European family of languages belonging to Arabic script-based languages. Our transliteration model is induced from a parallel corpus containing the PersoArabic script of a Persian book together with its Romanized transcription in Dabire. We manually aligned the sentences of this book in both scripts and used it as a parallel corpus. Our results indicate that the performance of the system is improved by adding grapheme-to-phoneme and word lattice methods for out-of-vocabulary handling task into the monotonic statistical machine transliteration system. In addition, the final performance on the test corpus shows that our system achieves comparable results with other state-of-the-art systems.

1 Introduction

Transliteration is a task for converting words in the source language using the approximate phonetic or spelling equivalents into ones in the target language (Kirschenbaum and Wintner 2009). Transliteration and translation belong to two different categories. For instance the English transliteation of the Persian word "كتاب/ book/ketâb" is "Book", whereas its transliteration is "ketâb".

Machine transliteration has proven to be an important and useful research area in the field of natural language processing (NLP). One of the main uses of transliteration schemes is in Machine Translations (MT). Despite the large amount of data available for machine translations, the MT systems are still suffering from the presence of the out-of-vocabulary (OOV) words. OOV words are mostly Named Entities such as people, company and place names, technical terms and foreign words which usually do not appear in the dictionaries. Transliterating the source languages and using it directly in the target language, is a solution for OOV words.

In addition to the MT systems, there are many challenging tasks in NLP dealing with machine transliteration:

- Cross-Lingual Information Retrieval.
- Real-time translation for emails, blogs, etc.

© Springer Nature Switzerland AG 2018
A. Gelbukh (Ed.): CICLing 2017, LNCS 10762, pp. 469–481, 2018.
https://doi.org/10.1007/978-3-319-77116-8_35

- Multilingual chat applications.
- Cross-Lingual Question Answering Systems.
- Text to speech (TTS) systems.

In this paper, we discuss a statistical machine translation (SMT) based model that uses Persian-Latin parallel corpus to mitigate the OOV words problem. In the first step, we used a simple phrase based SMT model. Analyzing the errors indicated that the Ezafe markers in the Persian texts have a considerable contribution to the errors. Ezafe used to link two words in some contexts and it is an unstressed short vowel/-e/ (or/-ye/after vowels) (Asghari et al. 2014). Therefore, we used a CRF method for Ezafe recognition system to determine Ezafe markers in Persian text. Then, to deal with the OOV words problem we trained two grapheme-to-phoneme (G2P) conversion and word lattice models which were integrated into the SMT system.

Unfortunately, there is no standard way of writing Persian using the Latin alphabet and words are Romanized in various ways. In order to avoid ambiguities, we chose to use Dabire, a romanized transcription scheme (Maleki 2008) and created our PersoArabic-Dabire parallel corpus (see Sect. 5.1).

The contributions of this research are as follows:

- Development of a clear and relatively large Dabire-PersoArabic parallel corpus as training data, so the results are considerably reliable.
- Modeling the transliteration task as a statistical translation system which can be handled with SMT techniques.
- Using a grapheme-based method to deal with the OOV words problem.
- Using word lattice method to transliterate OOV words.
- Using the CRF approach for Ezafe recognition system to determine Ezafe markers in Persian text which have a considerable contribution to the errors.

The paper is organized as follows. In Sect. 2, an overview of Persian language and its characteristics are introduced. Section 2 also includes a brief overview of Dabire writing system. Section 3 provides a list of previous works in machine transliteration. Section 4 describes our approach. Finally, experiments and results are provided in Sect. 5 including experimental setup, evaluation measures, and the implementation of our baseline and proposed method. Conclusion will be discussed in the final section.

2 Overview

In this section, some specific features of Persian language are briefly mentioned. Furthermore, Dabire (Maleki 2008), which serves as our standard transcription scheme is described.

2.1 An Overview of Persian Language

Persian language belongs to Indo-Iranian branch of the Indo-European family of languages. Persian is the formal language of Iran, Tajikistan and Afghanistan. Furthermore, it is an Iranian language belonging to Arabic script-based languages and unlike many of the Indo-European languages it is not written in the Latin alphabet. This category of languages includes Arabic, Pashtu, Urdu, Kurdish and Persian. Some characteristics of the Arabic-based writing systems are as follows (Farghaly and Shaalan 2009):

- Absence of capitalization
- Non-representation of short vowels in writing that makes high degree of ambiguity.
- Lack of clear word boundaries

Persian is a rich morphology language. Many methods including rule based, semantic, statistical and also hybrid methods usually used for Persian processing tasks.

Consequently, we propose grapheme-based and word lattice methods that exploits the statistical machine translation (SMT) techniques in order to deal with the challenges above-mentioned in machine transliteration system.

2.2 Dabire Writing System

The writing system of Persian in Iran, is the PersoArabic script (PA-Script) (Neysari 1996). Due to the technological limitations in deploying PA-Script in many software systems, the Latin script is very common in blogs, email, SMS and other online chat services and there are many Latin-based scripts for Persian. The main goal of Dabire initiated in (Maleki 2008) is bridging the gap between Latin-based scripts and PA-scripts by developing algorithms and rules for converting back and forth between these writing systems (Maleki and Ahrenberg 2008).

3 Related Works

In a research accomplished by (Masmoudi et al. 2015) a rule-base method has been presented for Tunisian Dialect text that is written in Latin script (also known as Arabizi) into Arabic script. The input of this transliteration system consists of communication on social networks in form of messages, chat, SMS etc. To perform the transliteration, a rule-based approach generates all possible transliterations for the Latin script input. Then, an annotator is instructed to select from a pool of given choices.

In another approach to Arabic transliteration, Sellami et al. develop a cross-linguistic method for the recognition of Arabic name entities and their transliteration into French using a well-resourced language (English) as the pivot (Sellami et al. 2015). First, English-Arabic and English-French parallel corpora were used to extract English-Arabic and English-French lexicons. Then, these terms were merged using the pivot English language and some transliteration rules.

Mathur et al. propose a hybrid approach to transliteration of name entities in English to Indian languages (Mathur and Parakash Saxena 2014). They have applied a rule based approach to extract individual phonemes and then a statistical approach to convert them into their equivalent Hindi phonemes. In another attempt Kaur et al. describe a hybrid method to transliterate proper nouns from Hindi to English. They combined direct mapping, rules and statistical machine translation approach to present a novel and effective method (Kaur et al. 2014). Durrani et al. developed an unsupervised transliteration model which was integrated in a SMT system. With this approach they generated a transliteration model from parallel data and used it to translate OOV words (Durrani et al. 2014). In yet another work in this domain, Balabantary et al. used phrase-based SMT techniques for the task of machine transliteration for Odia-English and Odia-Hindi language pairs (Balabantaray and Sahoo 2014). A rule based method has been presented to transliterate English-Punjabi language pair (Bhalla et al. 2013). Rules can extract or separate the syllables from the words and translate them using *MOSES* SMT toolkit (Koehn et al. 2007). Transliteration has been shown to improve MT quality, especially when it comes to named entities (Al-Onaizan and Knight 2002; Habash 2009; Kashani et al. 2007; Azab et al. 2013).

For the English to Persian transliteration task Mousavi Nejad et al. (2011) have proposed three systems: a maximum entropy model, a grapheme to phoneme convertor and a phrase based statistical machine translation system.

Karimi (2008) has proposed a combined transliteration method in a black-box framework for OOV words in both English-Persian and Persian-English transliteration task. In this approach, multiple spelling-based transliteration systems were aggregated into one system M with the combination method being a mixture of a Naïve Bayes classifier and a majority voting scheme. Persian-English task was evaluated on a corpus of 2,010 Persian person names. Experiments using ten-fold cross-validation of these corpora led to 85.5% and 69.5% word accuracies for English-Persian and Persian-English cases, respectively.

4 Transliteration Models

In the following we describe our proposed method for transliterating PersoArabic into Dabire. In our proposed method we use IBM source-channel model, joint source-channel model and word lattice methods. So, at the first in the following we explain these models and then describe proposed method in Sect. 4.4.

4.1 IBM Source-Channel Model

One of the most widely studied machine translation models is the IBM source-channel model (Brown et al. 1993). In this paper we used the IBM source-channel model as a baseline method. In other words, we model the transliteration task as a machine translation task. For this purpose, we use Moses (Koehn and Hoang, Factored Translation Models, 2007), a phrase-based SMT as a generation transliteration system and align the word pairs with GIZA++ (Och and Ney 2003). As this induced alignment is monotonic and there are no distortions in it, the reordering parameter is set to zero in GIZA++. It is worth mentioning that the monotonicity nature of our task appears in the transliteration system as well. Also, we train a 3-gram language model on the target side of parallel training data with KenLM (Heafield 2011).

4.2   Joint Source-Channel Model

In contrast to the IBM source-channel model, the joint source-channel estimates the most optimal transliteration string without decomposing the joint probability. It means that, the joint probability can be estimated through an N-gram transliteration model (Bisani and Ney 2008).

In this work we use SEQUITUR (Sequitur G2P 2008) tool to train an N-gram model iteratively: in the first step, a unigram model is created; next, a bigram model is trained with this unigram model, which in turn is used to train a trigram model, and so on.

4.3   Word Lattices

A word lattice is a directed acyclic graph with a single start point and weighted edges labelled with a word. In a word lattice the end point is unique and has no outgoing edge. Formally, a word lattice is a finite state automata (FSA) that can represent any finite set of strings (Dyer et al. 2008). Figure 1 illustrates an example of non-linear word lattice.

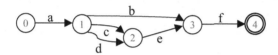

Fig. 1. Example of non-linear word lattice

For transliteration, we find it useful to encode input as a word lattice and maximize the transliteration probability along any path in the input.

4.4   Proposed Method

An overview of the proposed method for PersoArabic to Dabire transliteration task is shown in Fig. 2. Our system consists of the following steps: (1) preprocessing, (2) word level monotone statistical machine translation (SMT) and (3) handling the OOV words.

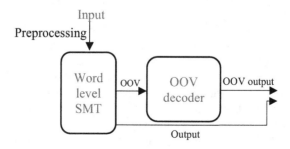

Fig. 2. Overview of the proposed method

Preprocessing. In many NLP applications, the preprocessing plays an important role in the performance improvement. Preprocessing of Persian language can be of different types and resolves many issues (Mousavi Nejad et al. 2011). Some of these issues are listed below:

- *Multiple Unicode*: in PersoArabic some letters have more than one Unicode (for instance "ی / i" or "ک / k").
- *Different spelling*: some words in PersoArabic may be written with different letters. For instance "Tehran" can be written either "طهران / *Tehrân*" or "تهران / *Tehrân*". Different spelling such as example, can be confusing for the machine.
- *Ezafe construction*: in Persian, the elements within a noun phrase are linked by the enclitic particle called Ezafe. Ezafe (written as *e* or *ye*) is a vowel that is unstressed and usually not written but pronounced. Adding Ezafe to the end of a noun indicates that it is modified by either *(i)* another noun, *(ii)* an adjective, or *(iii)* a pronoun, which follows the noun. Some examples of Ezafe are:
  - /dust e man/ (my friend)
  - /sib e sorx/ (red apple)
  - /xâne ye zibâ/ (beautiful house)
- *Imported letters from Arabic*: some Arabic diacritics or letters such as "Tanwin" or "Hamza" used in Persian words do not always appear in text. For instance, the word "معمولاً /ma'mulan/ usually" which is marked by "Tanwin" diacritic often appears without it.
- *Different spacing*: in Persian writing, space is not exclusively a word-delimiter, it may also appear within a word. For example "میگفت / *migoft* / was saying", "می‌گفت / *migoft* / was saying", and "می گفت / *migoft* / was saying" are all possible forms of "was saying". Intra-word space should be written using the zero width space, but for various reasons it is written as the normal inter-word space.

To overcome issues mentioned above, a normalization tool is required to be run on the training and the test sets.

Word Level Monotone Statistical Machine Translation (SMT). After the preprocessing step, our statistical model that is used to transliterate the PersoArabic text into Dabire. To do so, we used a uningual, sentence-aligned training corpus for the training system. This model was implemented using the Moses statistical machine translation system. The SMT method were performed on the whole unilingual corpus using 10-fold cross-validation method. The results of this approach appear in Sect. 6 under the 'PB-SMT' row of Table 1.

Our error analysis of the SMT results shows that around 30% of the total errors are related to Ezafe marker. Consequently, we tried to build an Ezafe recognition system in order to reduce the corresponding error as much as possible.

For Ezafe recognition task we used the CRF method (Asghari et al. 2014) with an accuracy of about 98.04%. After that, we re-performed the SMT method to the whole corpus, this time with the Ezafe marker. In Fig. 3 the word level monotone SMT method combined with the Ezafe recognition system has been shown. The results of this approach are shown in Sect. 6 under the 'SMT + CRF Ezafe recognition' row of Table 1.

Handling OOV Words. Despite the availability of large amount of data, machine translation systems suffer from the presence of the out-of-vocabulary (OOV) words. We present two methods for transliteration of OOV words:

- Joint source-channel model
- Word lattice

In order to compare with previous work, we use manually created test corpus that consists of 2500 word pairs selected randomly from (Karimi et al. 2007) test data. These words were manually converted from English to Dabire scheme and our methods were applied to the resulting data. In the rest of this paper, this test set is referred to as Karimi dataset.

Joint Source-Channel Model. As the first step of handling OOV words, we apply word origin recognition to grapheme-to-phoneme (G2P) conversion, the task of predicting the phonemic representation of a word given its written form. We specifically study G2P conversion on Karimi dataset and OOV.

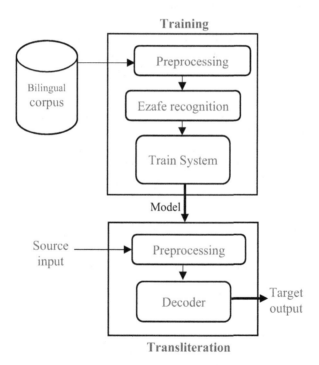

Fig. 3. Overview of SMT method combined with Ezafe recognition

In G2P conversion we assumed syllables of English words (with Dabire scheme) as output and Persian words as input. To train this system, we need a parallel corpus containing of Persian words and their syllables in Dabire scheme.

So, the first step in the G2P conversion is to syllabify Dabire words. We used syllabification method introduced in (Maleki and Ahrenberg 2008). This syllabification transducer ensures that the number of consonants after the process is maximized. The following examples show this syllabification:

- مهمان/ mehmân/ guest ➔ مه/meh+مان/ mân
- برو/ boro / go ➔ ب/bo + رو/ro

After the syllabification, Sequitur G2P conversion system (Bisani and Ney 2008) was implemented. The results of this approach on our complete parallel corpus are shown in Sect. 6 under the 'SMT + G2P (OOV)' row of Table 1. Tables 2 and 3 (Sect. 6) contain the results of G2P conversion on only OOV words of our parallel corpus and on the Karimi dataset.

Word Lattice. Overview of the word lattice method for PersoArabic to Dabire task is shown in Fig. 4. At the beginning we used phrase table of character level SMT to create phrase-based corpus. In this corpus, each PersoArabic text is aligned with all possible Dabire alternatives. For example, Persian character "ن / N"/N" is aligned with "ne","na","no" in Dabire.

For translation, we find it useful to encode input as a word lattice and maximize the transliteration probability along any path in the input. Therefore, for an input word we created the word lattice with any possible path by paraphrase corpus. So, the lattice of the input word contain of any possible path would be created. Then, the decoder maximizes the probability of transliteration along any path in the input lattice. The results of word lattice on our whole parallel corpus are shown in Table 1 of Sect. 6 marked as 'SMT + Word lattice' (4th row). Furthermore, the result of this method on only OOV words in our parallel corpus are shown in Table 2 and in Table 3 where comparison of word lattice with other methods on Karimi dataset are expressed.

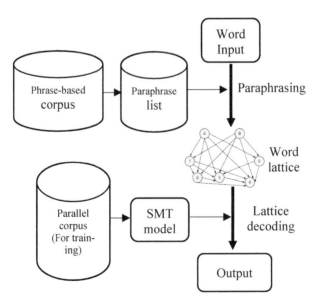

Fig. 4. Overview of word lattice method

5 Experiments

5.1 Data Sets

In the first step of this research we used a phrase based SMT model, trained on a sentence aligned PersoArabic-Dabire parallel corpus. This parallel corpus extracted from a Persian biography book written in both PersoArabic and Dabire. According to the first and last words of each sentence, we manually aligned the sentences of this book. Then for checking and eliminating the errors, the length of the sentences on both sides was compared. Finally, our parallel corpus with 13933 sentence pairs, 155623 words in PersoArabic text and 170702 Dabire words was created[1].

In the next step, we used a joint source-channel model, trained on a word aligned parallel Persian- syllables of Dabire words. We performed the syllabification method (see Sect. 5) to syllabify the Dabire words of the above corpus. So, our parallel corpus for G2P conversion was created with 136364 pair words of which 12644 were unique.

5.2 Evaluation Measures

In order to accurately evaluate the results, we should consider effective evaluation measures. Some of the evaluations measures used to evaluate the results are described below.

Precision or transliteration accuracy measures the proportion of transliterations that are correct:

$$accuracy = \frac{number\ of\ correct\ transliterations}{total\ number\ of\ test\ words}$$

Recall is the fraction of correct pairs that are retrieved:

$$recall = \frac{number\ of\ correct\ pairs\ extracted}{total\ number\ of\ test\ words}$$

Finally, F-measure is calculated as the harmonic mean of precision (P) and recall (R):

$$F = \frac{2P \times R}{P + R}$$

In this paper, we assumed that the precision is equal to the accuracy.

5.3 Experimental Setup

In this research, the experiments were performed on the whole PersoArabic-Dabire corpus using 10-fold cross-validation method. We use GIZA++ (Och and Ney 2003) to

---

[1] This dataset is free available and can be obtain by contacting the corresponding authors.

align source and target words with the grow-diag-and-final heuristic, while other parameters are set to default. The KenLM (Heafield 2011) toolkit is used to count n-gram on the target of the training set. Here we use a 3-gram language model.

The IBM Source-channel model was implemented with Moses statistical machine translation system (Koehn et al. 2007). Sequitur G2P was used to implement the Joint source-channel model (Sequitur G2P 2008).

6 Results

Table 1 gives a comprehensive evaluation of methods which were implemented in this study. We reported accuracy, recall, F-measure and BLEU gains (Papineni et al. 2002) obtained by each approach. The results of our PB-SMT method show a precision of 92.63%. With adding CRF Ezafe recognition to our system precision is increased to 94.55%. A precision improvement of 95.81% is observed when word lattice method is added to CRF Ezafe recognition. In this method a recall improvement of 97.06% is observed. As shown in Table 1, adding G2P method to CRF Ezafe recognition has resulted in the best performance with respect to other approaches. The results show a precision of 96.24% and recall of 97.21%, leading to an F-measure of 96.72 with BLEU score of 81.59.

Table 1. Transliteration performance on PersoArabic-Dabire corpus

Experiments	Performance measure			
	Accuracy	Recall	F-measure	BLEU
PB-SMT	92.63	93.27	92.95	75.83
SMT + CRF Ezafe recognition	94.55	93.49	93.76	81.05
SMT + G2P (OOV)	96.24	97.21	96.72	81.59
SMT + Word lattice (OOV)	95.81	97.06	96.43	81.32

From Table 2, we can see the result of G2P and word lattice methods on only OOV word in PersoArabic-Dabire parallel corpus. Both G2P and word lattice methods are shown to have acceptable performance. A closer look at this table reveals that the G2P method is slightly better compared to word lattice especially in the resulting accuracy and BLEU measures.

Table 2. Transliteration performance on only OOV words in PersoArabic-Dabire corpus

Experiments	Performance measure			
	Accuracy	Recall	F-measure	BLEU
G2P	85.61	67.13	72.25	58.69
Word lattice	76.41	64.99	70.23	51.22

Table 3 gives a comprehensive results for the accuracy of methods which were implemented on Karimi et al. (2007) dataset which manually created from 2500 Persian to English OOV word pairs. As can be seen in Table 3, both of G2P and word lattice methods have a better performance compared to the Karimi et al's method.

Table 3. Transliteration performance on Karimi dataset

Experiments	Performance measure
	Accuracy
Karimi et al.	69.5
PB-SMT	16.1
G2P	81.3
Word lattice	73.9

7 Conclusion

In this study, we presented grapheme-to-phoneme (G2P) conversion and word lattice methods as a novel approach to PersoArabic to Dabire transliteration task and also conducted some experiments on it. In the first step, we used a simple SMT model. The results show a precision of 92.63%. Due to the high contribution of the Ezafe marker to the errors, we proposed the use of a CRF method for Ezafe recognition system to determine Ezafe markers in Persian text. The resulting approach increases the precision to 94.55%.

As the next experiment we trained a G2P conversion model that was integrated into the SMT model. The best results on the PersoArabic-Dabire parallel corpus were due to this method with a precision and recall of 96.24% and 97.21% respectively. We have also tested a word lattice method integrated with Ezafe recognition on PersoArabic-Dabire parallel corpus that resulted in a precision of 95.81%.

Furthermore, we trained G2P conversion and word lattice method on only OOV words in PersoArabic-Dabire parallel corpus. The experiments showed that G2P conversion method has the best result on OOV words with accuracy of 85.61%.

Also, for comparison with previous works, we manually created a test corpus that consists of 2500 Persian to English word pairs selected randomly from (Karimi et al. 2007). These words were manually converted from English to Dabire scheme. We tested SMT, G2P conversion and word lattice methods on this test data and the results show that both of G2P and word lattice methods have better performance compared to the Karimi et al. method.

References

Al-Onaizan, Y., Knight, K.: Translating named entities using monolingual and bilingual resources. In: 40th Annual Meeting of the Association for Computational Linguistics (2002)

Asghari, H., Maleki, J., Faili, H.: A probabilistic approach to Persian Ezafe recognition. In: 14th Conference of the European Chapter of the Association for Computational Linguistics (EACL 2014). Gothenburg, Sweden (2014)

Azab, M., Bouamor, H., Mohit, B., Oflazer, K.: Dudley north visits north London: learning when to transliterate to Arabic. In: Proceedings of the 2013 Conference of the North American Chapter of the Association for Computational Linguistics: Human Language Technologies, pp 439–444. Association for Computational Linguistics, Atlanta, Georgia (2013)

Balabantaray, R.C., Sahoo, D.: Odia transliteration engine using moses. In: Business and Information Management (ICBIM 2014) (2014)

Bhalla, D., Joshi, N., Mathue, I.: Rule based transliteration scheme for english to punjabi. Int. J. Nat. Lang. Comput. (IJNLC) **2**(2), 67–73 (2013)

Bisani, M., Ney, H.: Joint-sequence models for grapheme-to-phoneme conversion. Speech Commun. **5**, 435–451 (2008)

Brown, F.P., Della, S.A., Pietra, V.J., Pietra, D., Robert Mercer, L.: The mathematics of statistical machine translation: parameter estimation. Comput. Linguist. **19**(2), 263–311 (1993)

Durrani, N., Sajjad, H., Hoang, H., Koehn, P.: Integrating an unsupervised transliteration model into statistical machine translation. In: 14th Conference of the European Chapter of Association for Computational Linguistics (EACL 2014), Gothenburg, Sweden (2014)

Dyer, C., Muresan, S., Resnik, P.: Generalizing word lattice translation. In: Annual Meeting of the Association for Computational Linguistics (ACL) (2008)

Farghaly, A., Shaalan, K.: Arabic natural language processing: Challenges and solutions. ACM Trans. Asian Lang. Inf. Process. (TALIP) **8**(4), 14 (2009)

Habash, N.: REMOOV: a tool for online handling of out-of-vocabulary words in machine translation. In: Proceedings of the Second International Conference on Arabic Language Resources and Tools. The MEDAR Consortium, Cario, Egypt (2009)

Heafield, K.: KenLM: faster and smaller language model queries. In: Proceedings of the Sixth Workshop on Statistical Machine Translation, pp 187–197. Edinburgh, Scotland, United Kingdom (2011)

Karimi, S., Scholer, F., Turpin, A.: Collapsed consonant and vowel models: new approaches for English-Persian transliteration and back-transliteration. In: 45th Annual Meeting of the Association of Computational Linguistics, Prague, Czech Republic (2007)

Karimi, S.: Machine transliteration of proper names between English and Persian. Ph.D. dissertation, RMIT University, Melbourne (2008)

Kashani, M.M., Joanis, E., Kuhn, R., Foster, G., Popwich, F.: Integration of an Arabic transliteration module into a statistical machine translation system. In: Proceedings of the Second Workshop on Statistical Machine Translation, Prague, Czech Republic (2007)

Kaur, V., Kaur Sarao, A., Singh, J.: Hybrid approach for Hindi to English transliteration system for proper nouns. Int. J. Comput. Sci. Inf. Technol. **5**(5), 6361–6366 (2014)

Kirschenbaum, A., Wintner, S.: Lightly supervised transliteration for machine translation. In: 12th Conference of the European Chapter of the Association for Computational Linguistics (EACL), Athens, pp 433–441 (2009)

Koehn, P., Hoang, H.: Factored translation models. In: EMNLP (2007)

Koehn, P., et al.: Moses: open source toolkit for statistical machine translation. In: Human Language Technology Conference of the NAACL, Main Conference. Association for Computational Linguistics, New York, USA (2007)

Maleki, J.: A romanized transcription for persian. In: Natural Language Processing Track (INFOS2008). Cario (2008)

Maleki, J., Ahrenberg, L.: Converting romanized persian to the arabic writing systems. In: Language Resources and Evaluation Conference (2008)

Masmoudi, A., Habash, N., Ellouze, M., Estève, Y., Belguith, L.H.: Arabic transliteration of romanized tunisian dialect text: a preliminary investigation. In: Gelbukh, A. (ed.) CICLing

2015. LNCS, vol. 9041, pp. 608–619. Springer, Cham (2015). https://doi.org/10.1007/978-3-319-18111-0_46

Mathur, S., Parakash Saxena, V.: Hybrid appraoch to English-Hindi name entity transliteration. In: Electrical, Electronics and Computer Science (SCEECS) IEEE Students' Conference (2014)

Mousavi Nejad, N., Khadivi, S., Taghipour, K.: The Amirkabir machine transliteration system for NEWS 2011. In: Named Entities Workshop (2011)

Neysari, S.: A Study on Persian Orthography (in Persian). Sazmane Cap o Entesarat, Tehran (1996)

Och, F.J., Ney, H.: A systematic comparison of various statistical alignment models. Comput. Linguist. **29**(1), 19–51 (2003)

Papineni, K., Roukos, S., Ward, T., Zhu, W.: BLEU: a method for automatic evaluation of machine translation. In: Proceedings of the 40th Annual Meeting on Association for Computational Linguistics, ACL. Morristown, NJ, USA, pp 311–318 (2002)

Sellami, R., Deffaf, F., Sadat, F., Belguith, L.H.: Improved statistical machine translation by cross-linguistic projection of named entities recognition and translation. Computación y Sistemas **19**(4), 701–711 (2015)

Sequitur   G2P,   https://www-i6.informatik.rwth-aachen.de/web/Software/g2p.html   (2008). Accessed 1 Apr 2016

Building a Location Dependent Dictionary for Speech Translation Systems

Keiji Yasuda[(⊠)], Panikos Heracleous, Akio Ishikawa, Masayuki Hashimoto,
Kazunori Matsumoto, and Fumiaki Sugaya

KDDI Research, Inc., Garden Air Tower, 3-10-10,
Iidabashi, Chiyoda-ku, Tokyo 102-8460, Japan
{ke-yasuda,pa-heracleous,ao-ishikawa,
masayuki,matsu,fsugaya}@kddi-research.jp

Abstract. Mis-translation or dropping of proper nouns reduces the
quality of machine translation output or speech recognition output as
input of a dialog system. In this paper, we propose an automatic method
of building a location dependent dictionary for speech recognition and
speech translation systems. The method consists of two parts: location
dependent word extraction and word classification. The first part extracts
the word by using micro blog data based on Akaike's information criteria. The second part classifies the words by using the Convolutional Neural Net (CNN) trained on crawled data. According to the experimental
results, the method extracted around 2,000 location dependent words in
the Tokyo area with 75% accuracy.

1 Introduction

As a result of drastic advances in technical innovations of speech processing and
natural language processing, a speech-to-speech translation system is becoming
a realistic tool for travelers, internet users and others. Especially for travel, the
coverage of proper nouns for tourist spots, landmarks, restaurants, and accom-
modations highly influences system performance. Okuma et al. proposed the
class based method to install hand-crafted bilingual dictionaries into a Statistical Machine Translation (SMT) framework to improve the proper noun coverage
of Machine Translation (MT).

However, there are two remaining problems for practical usage. The first
problem is the cost of building the dictionary. Since the number of proper noun
suchs as restaurant names or product names, is increasing daily, the cost to manually maintain a dictionary is high. The second problem is system performance
degradation. Especially for Automatic Speech Recognition (ASR), too large dictionary causes gushing-out errors. It also has an adverse effect on the class based
SMT by class collision of polysemic words.

In this paper, we propose a method to automatically build a location dependent dictionary. Since this method extracts dictionary entries related to the
targeted location, the method yields an adequate size dictionary.

© Springer Nature Switzerland AG 2018
A. Gelbukh (Ed.): CICLing 2017, LNCS 10762, pp. 482–491, 2018.
https://doi.org/10.1007/978-3-319-77116-8_36

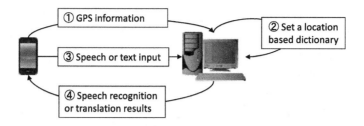

Fig. 1. Practical use of location dependent dictionary.

Section 2 describes related work in this research. Section 3 explains the proposed method. Section 4 details the experiments using the crawled data. Section 5 concludes the paper and presents some directions for future work.

2   Related Work

There are two conventional approaches for Out Of Vocabulary (OOV) treatment. One is to use bilingual dictionary [1]. The other is the transliteration approach [2–4]. The bilingual dictionary approach estimates translation of the OOV in MT input, using several language resources.

With the transliteration approach, since a typical system uses phoneme or grapheme mapping rules to produce transliterations, the systems sometimes yields non-word output or incorrect word output for the English translation.

Since these methods only assume text input, the methods are not suitable for speech input.The proposed method, however, extracts source or target language side words frequently used in the target location. Since the proposed method also classifies the category of the extracted word, it is easy to introduce into an Automatic Speech Recognition (ASR) system and MT that have a class-based language model.

Figure 1 shows the practical use case of a location dependent dictionary. Although, the proposed method does not have a mechanism to infer the translation of the extracted words, conventional OOV treatment methods can complement the proposed method. For actual use, we also need an additional model, such as the Grapheme to Phoneme(G2P) model [5] to infer the phoneme expression of the words.

3   Proposed Method

As shown in the Fig. 2, the proposed method consists of a location dependent word extraction part and a word classification part. This section explains details of two parts. Each parts uses different corpus. Details of these two parts are explained in this section.

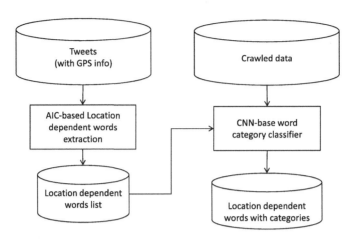

Fig. 2. Framework of the proposed system.

3.1 Location Dependent Word Extraction

To extract location dependent words from tweets, we introduce Akaike's Information Criteria (AIC) [6,7] which is defined by the following formula:

$$AIC = -2 \log \hat{L} + 2k \tag{1}$$

where $\hat{L}$ and k are the maximum likelihood of the given model and number of free parameters in the model, respectively. AIC is the popular criteria to evaluate the statistical model. In our case, AIC is used to detect location independent words. First, we collect tweets with GPS information. Second, we divide the tweets into two groups as follows. Tweets that are tweeted at the target location (l) are $\mathcal{T}_l$. And the other tweets are $\mathcal{T}_{\neg l}$. Then, by analyzing $\mathcal{T}_l$, we extract proper noun ($w \in W$) from $\mathcal{T}_l$.

Finally, for $\mathcal{T}_l$ and $\mathcal{T}_{\neg l}$, we count the number of tweets that contain w and not contain w. Table 1 shows the notation of the counted numbers of tweets.

To express the four joint probabilities, the location dependent model introduces three parameters that are $p(w,l)$, $p(w,\neg l)$ and $p(\neg w,l)$, where

$$p(\neg w, \neg l) = 1 - p(w,l) - p(\neg w,l) - p(w,\neg l) \tag{2}$$

Meanwhile, the location independent model expresses the four joint probabilities by two parameters that are $p(w)$ and $p(l)$ as follows

$$p(w,l) = p(w) \times p(l) \tag{3}$$

$$p(\neg w, l) = (1 - p(w)) \times p(l) \tag{4}$$

$$p(w, \neg l) = p(w) \times (1 - p(l)) \tag{5}$$

$$p(\neg w, \neg l) = (1 - p(w)) \times (1 - p(l)) \tag{6}$$

Using the likelihoods calculated based on numbers of tweets shown in Table 1, we calculate AIC for the location independent model (AIC_i) and location dependent model (AIC_d) for each word (w) as follows:

$$AIC_i = -2(N_{1.} \ln \frac{N_{1.}}{N_{total}} + N_{.1} \ln \frac{N_{.1}}{N_{total}} + N_{2.} \ln \frac{N_{2.}}{N_{total}} + N_{.2} \ln \frac{N_{.2}}{N_{total}}) + 2 \times 2 \quad (7)$$

$$AIC_d = -2(N_{11} \ln \frac{N_{11}}{N_{total}} + N_{21} \ln \frac{N_{21}}{N_{total}} + N_{21} \ln \frac{N_{21}}{N_{total}} + N_{22} \ln \frac{N_{22}}{N_{total}}) + 2 \times 3 \quad (8)$$

For word extraction, the proposed method adds the target word to the dictionary only if $(AIC_i - AIC_d)$ is greater than threshold.

Table 1. Tweet count for AIC calculation.

	w is observed	w is not observed	Total
$\mathcal{T}_l$	N_{11}	N_{12}	$N_{1.}$
$\mathcal{T}_{\neg l}$	N_{21}	N_{22}	$N_{2.}$
Total	$N_{.1}$	$N_{.2}$	N_{total}

3.2 Proper Noun Classification

This subsection explains the method of deciding class of proper nouns extracted by the AIC based method as explained in the previous subsection. For proper noun classification, we train classifiers that are configured as the Convolutional Neural Network (CNN). CNN gave superior performance in the research fields of image processing and speech recognition [8,9]. Currently, CNN has also outperformed the Natural Language Processing (NLP) task such as text classification [10,11] by incorporating word-embedding [12] in the input layer.

The network configuration for our proper noun categorization is shown in Fig. 3.[1]

Here, $\mathbf{x}_i (\in \mathbb{R}^k)$ is the word-embedding vector of i-th word in a given sentence. By concatenate word-embedding vectors, a sentence whose length is n is expressed as the following formula:

$$\mathbf{x}_{1:n} = \mathbf{x}_1 \cdots \oplus \mathbf{x}_i \cdots \oplus \mathbf{x}_n \quad (9)$$

The convolutional layer maps the n-gram features whose length (or filter window size) is h to the j-th feature map by using the following formula:

$$c_{h,j,i} = \tanh(\mathbf{w}_{h,j} \cdot \mathbf{x}_{i:i+h-1} + b_{h,j}) \quad (10)$$

---

[1] The actual hyper parameters' setting is different from the example shown in the figure. Detail setting will be explained in Sect. 4.

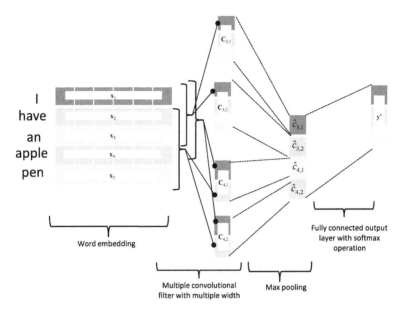

Fig. 3. Convolutional neural network for proper noun classification.

where $w_{h,j}$ and $b_{h,j}$ are weights for filtering and bias terms, respectively. For each n-gram length and feature map number, concatenate the results of Eq. 10 as follows

$$\mathbf{c}_{h,j} = [c_{h,j,1}, c_{h,j,2}, \cdots, c_{h,j,n-h+1}] \tag{11}$$

The max pooling layer chooses the element that has the largest value from all elements in $\mathbf{c}_{h,j}$ as follows

$$\hat{c}_{h,j} = \max_{i=1}^{n-h+1} \mathbf{c}_{h,j} \tag{12}$$

Fully connected output layer takes the softmax operation to yield the probability distribution of the categories ($\hat{\mathbf{y}} \in \mathbb{R}^{n_c}$) as follows

$$\hat{\mathbf{y}} = \frac{\exp(z_q)}{\sum_{p=1}^{n_c} \exp(z_p)}, q = 1, \cdots, n_c \tag{13}$$

where n_c is the total number of categories. And, $\mathbf{z}$ ($\in \mathbb{R}^{n_c}$) are the raw output values from output layer.

4   Experiments

4.1   Experimental Settings

For the extraction of proper nouns, we use Japanese tweets which posted in Japan between April 2015 to June 2016. All of the tweets have GPS information. We used the other corpus collected by web crawling for the category classification. The crawler collected data from review sites that have place names, reviews were written in natural language, and the category of the place included restaurant, accommodation, building, transportation, and so on. The statistics for this corpus are shown in Table 2. The crawled corpus is used to train the CNN-base classifier. We used reviews in natural language as input for CNN and categories as the label to be inferred.

For the actual classification of the proper nouns extracted from tweets, the classifier reads the sentence of the tweets containing the target proper nouns and outputs the category of the proper noun. Table 3 shows the detailed parameter setting of the CNN-based classifier. As shown in the table, the CNN has five output units, and each of them outputs the probability of the category, which is restaurant, accommodation, building, transportation or other. The details of the crawled corpus are shown in Table 4. As shown in the table, data size varies depending on the category. To reduce the adverse effect of data imbalances, we sampled training data to be balanced while mini batch training.

Table 2. Statistic of training corpora used for the experiments.

Corpus type	# of tweets or reviews	# of words	Lexicon size
Tweet corpus	13,161,098	436,002,321	9,037,274
Crawled corpus	20,786	437,383	23,694

Table 3. CNN parameter setting

Parameters	Setting
Maximum length of input sentence	150 words
Mini batch size	65
Dimension of word-embedding vector (k)	400
Filter window size (n-gram length)	3–5-gram
Number of filters for each window size	128
Drop out rate for fully connected layer	0.5
Optimizer	Adam optimizer
# of output units	5

4.2 Experimental Results

Table 5 shows the subjective evaluation results of location dependent proper noun extraction in the Tokyo area. We randomly sampled 20 words for three results in different threshold settings. Then, an annotator evaluated whether the words were related to Tokyo area or not. As shown in the table, the result with threshold 100 gives decent performance. However, accuracy drops when the threshold is small as 50.

Table 4. Detail of the crawled corpus.

Category	# of Venue	# Reviews
Accommodation	72 (5)	389 (100)
Building	1,857 (83)	7,878 (1,660)
Restaurant	3,197 (174)	16,065 (3,480)
Transportation	195 (50)	2,273 (1,000)
Others	142 (4)	501 (80)
All	5,463 (316)	27,106 (6,320)

The numbers in parentheses are the held out data for evaluation.

Table 5. Evaluation results of the keyword extraction

Threshold $(AIC_i - AIC_d)$	Accuracy	# of Extracted words
10	30%	5,366
50	55%	2,738
100	75%	1,931

Table 6 shows the evaluation results of the proper noun classifier. Here, we used the held out data of the crawled corpus as the test set. The classifier infers the words category by using a single review in this condition. The table shows the baseline SVM result and three results by CNNs with three different conditions of word-embedding training as follows:

Condition 1 Randomly initialize word-embedding matrix.
Condition 2 Train the word-embedding matrix using Word2Vec [12] on tweets and crawled corpora. The word-embedding matrix is fix through training.

Table 6. Accuracy of the proper noun classification on held out set of crawled corpus

Classifier	WE pretraining	WE training	Accuracy
SVM	N/A (Bag of words)	N/A (Bag of words)	76.38%
CNN	Random initialization	Tuned while CNN training	78.54%
CNN	W2V(Tweet + Crawled corpora)	Fixed	80.24%
CNN	W2V(Tweet + Crawled corpora)	Tuned while CNN training	**81.55%**

WE: Word Embedding, W2V: Word 2 Vec

Condition 3 Train Word2Vec model using tweets and crawled corpora. Word2Vec word-embedding matrix is used as an initial value of the CNN input layer. The word-embedding matrix is tuned while CNN training.

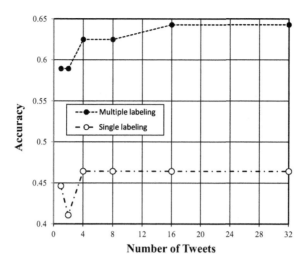

Fig. 4. Relationship between number of tweets and classification accuracy (CNN with W2V(Tweet + Crawled)

As shown in the table, Condition 3 gives the best performance. Thus, we set the experimental condition to Condition 3 for the following experiments.

To make Twitter test set for word classification, we sampled 100 words out of the 1,931 proper nouns shown in Table 5. First, we pick up all accommodation words from the full set, since the occurrence of words in the accommodation category is very low. Then, we randomly sample the words from the rest of the full set.(Details of the Twitter word set will be shown in Table 7.) The Twitter test set is manually labeled. Since some of the words are ambiguous, we allow an annotator to tag primary and secondary labels.

For the actual use of the proper noun classification, we can use one or more tweets for each extracted word. To see the effect of the use of multiple tweets, Fig. 4 shows the relationship between classification performance and the number of tweets. The vertical and horizontal axes represent the accuracy of the classification and the number of tweets used for each classification, respectively. Here, we use the top 50 most frequent words from Twitter test set. Then, for each word, randomly extract 32 tweets that contain the target word. For the classification using multiple tweets, we simply calculate the average of the category probabilities over all outputs from CNN and choose the most probable category. The figure shows two evaluation results. One is a strict evaluation that allows primary labels. The other evaluation allows system output to be either primary or secondary labels. As shown in the figure, classification accuracy is saturated at 16 tweets for both evaluation conditions. Thus, we use at most 16 tweets for the following word categorization experiments.

Table 7 shows the evaluation results using the full Twitter test set. As shown in the table, the CNN based method gives better results than the SVM-based

Table 7. Accuracy of the proper noun classification on proper nouns extracted from tweet corpus

Category	# test words	Accuracy			
		Single labeling		Multiple labeling	
		CNN	SVM	CNN	SVM
Accommodation	20	14.29%	9.52%	85.71%	71.43%
Building	18 (+20)	94.44%	83.33%	94.74%	84.21%
Restaurant	19	73.68%	94.74%	73.68%	94.74%
Transportation	10	60.00%	0.00%	60.00%	0.00%
Others	33	0.00%	0.00%	12.50%	9.38%
Total	100	40.00 %	35.00%	59%	51%

At most 16 tweets are used to classify one word.
The number in parentheses is the word number of secondary label

method except for the restaurant category. The accuracy for "the other" category is very low for both classifiers. One of the reasons is the insufficient size of the training set. The other reason is the way we create data. We merged several low frequent categories, including natural park, hospital, and other terms. So, there are no common features that can characterize the other category.

5   Conclusions and Future Work

We proposed a method to build a location dependent dictionary for a speech translation system. We carried out experiments for dictionary extraction and word classification using several corpora. According to the experiments, the method can extract around 2,000 words for the Tokyo area with 75% dictionary extraction accuracy. For word categorization, we compared the SVM and CNN-based classifiers using several parameter settings. The word categorization accuracy is highly depended on the training corpus size. However, CNN-based classifier with W2V pre-training setting gave the best results for most of cases.

As future work, we will increase the crawled corpus for improvement of classification accuracy for transportation category and the other category. We will also carry out a speech translation field experiments to evaluate effects location dependent dictionary in the near future.

Acknowledgments. This research is supported by Japanese Ministry of Internal Affairs and Communications as a Global Communication Project.

References

1. Tonoike, M., Kida, M., Takagi, T., Sasaki, Y., Utsuro, T., Sato, S.: Translation estimation for technical terms using corpus collected from the web. In: Proceedings of the Pacific Association for Computational Linguistics, pp. 325–331 (2005)

2. Al-Onaizan, Y., Knight, K.: Translating named entities using monolingual and bilingual resources. In: Proceedings of the 40th Annual Meeting of the Association for Computational Linguistics (ACL), pp. 400–408 (2002)
3. Sato, S.: Web-based transliteration of person names. In: Proceedings of IEEE/WIC/ACM International Conference on Web Intelligence and Intelligent Agent Technology, pp. 273–278 (2009)
4. Finch, A., Dixon, P., Sumita, E.: Integrating a joint source channel model into a phrase-based transliteration system. In: Proceedings of the NEWS, vol. 2011, pp. 23–27 (2011)
5. Rao, K., Peng, F., Sak, H., Beaufays, F.: Grapheme-to-phoneme conversion using long short-term memory recurrent neural networks. In: 2015 IEEE International Conference on Acoustics, Speech and Signal Processing (ICASSP), pp. 4225–4229. IEEE (2015)
6. Akaike, H.: A new look at the stastical mpdel identification. IEEE Trans. Autom. Control **19**, 716–723 (1974)
7. Ikeda, K., Hattori, G., Ono, C., Asoh, H., Higashino, T.: Twitter user profiling based on text and community mining for market analysis. Knowl. Based Syst. **51**, 35–47 (2013)
8. Krizhevsky, A., Sutskever, I., Hinton, G.E.: Imagenet classification with deep convolutional neural networks. In Pereira, F., Burges, C.J.C., Bottou, L., Weinberger, K.Q. (eds.) Advances in Neural Information Processing Systems, vol. 25, pp. 1097–1105. Curran Associates, Inc., New York (2012)
9. Abdel-Hamid, O., Mohamed, A., Jiang, H., Deng, L., Penn, G., Yu, D.: Convolutional neural networks for speech recognition. IEEE/ACM Trans. Audio Speech Lang. Process. **22**, 1533–1545 (2014)
10. Kim, Y.: Convolutional neural networks for sentence classification. In: Proceedings of the 2014 Conference on Empirical Methods in Natural Language Processing, pp. 1746–1751 (2014)
11. Kalchbrenner, N., Grefenstette, E., Blunsom, P.: Convolutional neural networks for modeling sentences. In: Proceedings of the 52nd Annual Meeting for Computational Linguistics, pp. 655–665 (2014)
12. Mikolov, T., Sutskever, I., Chen, K., Corrado, G.S., Dean, J.: Distributed representations of words and phrases and their compositionality. In Burges, C.J.C., Bottou, L., Welling, M., Ghahramani, Z., Weinberger, K.Q. (eds.) Advances in Neural Information Processing Systems, vol. 26, pp. 3111–3119. Curran Associates, Inc. (2013)

Text Summarization

Gold Standard Online Debates Summaries and First Experiments Towards Automatic Summarization of Online Debate Data

Nattapong Sanchan[1(✉)], Ahmet Aker[2(✉)], and Kalina Bontcheva[2(✉)]

[1] School of Information Technology and Innovation, Bangkok University, 9/1 Moo 5 Phaholyothin Road, Klong 1, Klong Luang 12120, Pathumthani, Thailand
nattapong.sa@bu.ac.th
[2] Natural Language Processing Group, Department of Computer Science, The University of Sheffield, 211 Portobello, Sheffield, UK
{ahmet.aker,k.bontcheva}@sheffield.ac.uk
https://www.sheffield.ac.uk/dcs

Abstract. Usage of online textual media is steadily increasing. Daily, more and more news stories, blog posts and scientific articles are added to the online volumes. These are all freely accessible and have been employed extensively in multiple research areas, e.g. automatic text summarization, information retrieval, information extraction, etc. Meanwhile, online debate forums have recently become popular, but have remained largely unexplored. For this reason, there are no sufficient resources of annotated debate data available for conducting research in this genre. In this paper, we collected and annotated debate data for an automatic summarization task. Similar to extractive gold standard summary generation our data contains sentences worthy to include into a summary. Five human annotators performed this task. Inter-annotator agreement, based on semantic similarity, is 36% for Cohen's kappa and 48% for Krippendorff's alpha. Moreover, we also implement an extractive summarization system for online debates and discuss prominent features for the task of summarizing online debate data automatically.

Keywords: Online debate summarization · Text summarization
Semantic similarity · Information extraction · Sentence extraction

1 Introduction

With the exponential growth of Internet usage, online users massively publish textual content on online media. For instance, a micro-blogging website, Twitter, allows users to post their content in 280-characters length. A popular social media like Facebook allows users to interact and share content in their communities, as known as "Friends". An electronic commercial website, Amazon, allows

© Springer Nature Switzerland AG 2018
A. Gelbukh (Ed.): CICLing 2017, LNCS 10762, pp. 495–505, 2018.
https://doi.org/10.1007/978-3-319-77116-8_37

users to ask questions on their interested items and give reviews on their purchased products. While these textual data have been broadly studied in various research areas (e.g. automatic text summarization, information retrieval, information extraction, etc.), online debate domain, which recently becomes popular among Internet users, has not yet largely explored. For this reason, there are no sufficient resources of annotated debate data available for conducting research in this genre. This motivates us to explore online debate data.

In this paper, we collected and annotated debate data for an automatic summarization task. There are 11 debate topics collected. Each topic consists of different number of debate comments. In total, there are 341 debate comments collected, accounting for 2518 sentences. In order to annotate online debate data, we developed a web-based system which simply runs on web browsers. We designed the user interface for non-technical users. When participants logged into the system, a debate topic and a comment which is split to a list of consecutive sentences were shown at a time. The annotators were asked to select salient sentences from each comment which summarize it. The number of salient sentences chosen from each comment is controlled by a compression rate of 20% which is automatically calculated by the web-based system. For instance, Table 1 shows a debate comment to be annotated by an annotator. Based on the compression rate of 20%, the annotator needs to choose 1 sentence that summarizes the comment. This compression rate was also used in [1,11]. In total, we obtained 5 sets of annotated debate data. Each set of data consists of 341 comments with total 519 annotated salient sentences.

Inter-annotator agreement in terms of Cohen's Kappa and Krippendorff's alpha are 0.28 and 0.27 respectively. For social media data such low agreements have been also reported by related work. For instance, [13] reports Kappa scores between 0.20 and 0.50 for human constructed newswire summaries. [7] reports again Kappa scores between 0.10 and 0.35 for the conversation transcripts. Our agreement scores are based on strict conditions where agreement is achieved when annotators have selected exact the same sentences. However, such condition does not consider syntactically different sentences bearing the same semantic meaning. Thus we also experimented with a more relaxed version that is based on semantic similarity between sentences. We regard two sentences as identical when their semantic similarity is above a threshold. Our results revealed that after applying such an approach the averaged Cohen's Kappa and Krippendorff's alpha increase to 35.71% and 48.15% respectively.

Finally we report our results of automatic debate data summarization. We implemented an extractive text summarization system that extracts salience sentences from user comments. Among the features the most contributing ones are sentence position, debate titles, and cosine similarity of the debate title words and sentences.

The paper is structured as follows. First we describe the nature of our online debate data. In Sect. 3 we discuss the procedures of data annotation and discuss our experiments with semantic similarity applied on inter-annotator agreement

Table 1. Examples of the debate data to be annotated.

Task 02: Is global warming fictitious?
[1] I do not think global warming is fictitious
[2] I understand a lot of people do not trust every source and they need solid proof
[3] However, if you look around us the proof is everywhere
[4] It began when the seasons started getting harsh and the water levels were rising
[5] I do not need to go and see the ice caps melting to know the water levels are rising and the weather is changing
[6] I believe global warming is true, and we should try and preserve as much of the Earth as possible

Table 2. Examples of Paraphrased Arguments.

Example 1: Propositions from the proponents
– Global warming is real
– Global warming is an undisputed scientific fact
– Global warming is most definitely not a figment of anyone's imagination, because the proof is all around us
– I believe that global warming is not fictitious, based on the observational and comparative evidence that is currently presented to us
Example 2: Propositions from the opponents
– Global warming is bull crap
– Global Warming isn't a problem at all
– Just a way for the government to tax people on more things by saying their trying to save energy
– Yes, global warming is a myth, because they have not really proven the science behind it

computation. In Sect. 4, we present our first results on automatically performing debate data summarization. We conclude in Sect. 5.

2   Online Debate Data and Their Nature

The nature of online debate is different from other domains. It gives opportunities to users to discuss ideological debates in which users can choose a stance of a debate, express their opinions to support their stance, and oppose other stances. To conduct our experiments we collected debate data from the Debate discussion forum.[1] The data are related to an issue of the existence of global warming. In the data, there are two main opposing sides of the arguments. A side of proponents believes in the existence of global warming and the other side, the opponents,

---
[1] http://www.debate.org.

says that global warming is not true. When the proponents and the opponents express their sentiments, opinions, and evidences to support their propositions, the arguments between them arise. Moreover, when the arguments are referred across the conversation in the forum, they are frequently paraphrased. Table 2 illustrates examples of the arguments being paraphrased. Sentences expressing related meaning are written in different context.

3   Annotation Procedures

In this paper, we collected and annotated debate data for an automatic summarization task. There are 11 debate topics collected. Each topic consists of a different number of debate comments as shown in Table 3. The annotation was guided through a web-based application. The application was designed for non-technical users. When participants logged in to the system, a debate topic and a comment which is split to a list of sentences were shown at a time. The annotators were given a guideline to read and select salient sentences that summarize the comments. From each comment we allowed the participants to select only 20% of the comment sentences. These 20% of the sentences are treated as the summary of the shown comment. In the annotation task, all comments in the 11 debate topics were annotated. We recruited 22 participants: 10 males and 12 participants to annotate salient sentences. The participants' backgrounds were those who are fluent in English and aged above 18 years old. We aimed to have 5 annotations sets for each debate topic. Due to a limited number of annotators

Table 3. Statistical information of the online debate corpus.

Topic ID	Debate topics	Comments	Sentences	Words
01	Is global warming a myth?	18	128	2701
02	Is global warming fictitious?	28	173	3346
03	Is the global climate change man made?	10	47	1112
04	Is global climate change man-made?	103	665	12054
05	Is climate change man-made?	9	46	773
06	Do you believe in global warming?	21	224	3538
07	Does global warming exist?	68	534	9178
08	Can someone prove that climate change is real (yes) or fake (no)?	8	49	1127
09	Is global warming real?	51	434	6749
10	Is global warming true?	5	26	375
11	Is global warming real (yes) or just a bunch of scientist going to extremes (no)?	20	192	2988
	Average	31	229	3995
	Total	341	2518	43941

and a long list of comments to be annotated in each debate topic, 11 participants were asked to complete more than one debate topic, but were not allowed to annotate the same debate topics in which they had done before. In total, 55 annotation sets were derived: 11 debate topics and each with 5 annotation sets. Each annotation set consists of 341 comments with total 519 annotated salient sentences.[2]

3.1 Inter-Annotator Agreement

In order to compute inter-annotator agreement between the annotators we calculated the averaged Cohen's Kappa and Krippendorff's alpha with a distant metric, Measuring Agreement on Set-valued Items metric (MASI). The scores of averaged Cohen's Kappa and Krippendorff's alpha are 0.28 and 0.27 respectively. According to the scale of [12], our alpha did neither accomplish the reliability scale of 0.80, nor the marginal scales between 0.667 and 0.80. Likewise, our Cohen's Kappa only achieved the agreement level of *fair agreement*, as defined by [9]. However, such low agreement scores are also reported by others who aimed creating gold standard summaries from news texts or conversational data [7,13].

Our analysis shows that the low agreement is caused by different preferences of annotators in the selection of salient sentences. As shown in Table 2 the sentences are syntactically different but bear the same semantic meaning. In a summarization task with a compression threshold, such situation causes the annotators to select one of the sentences but not all. Depending on each annotator's preference the selection leads to different set of salient sentences. To address this we relaxed the agreement computation by treating sentences equal when they are semantically similar. We outline details in the following section.

3.2 Relaxed Inter-Annotator Agreement

When an annotator selects a sentence, other annotators might select other sentences expressing similar meaning. In this experiment, we aim to detect sentences that are semantically similar by applying Doc2Vec from the Gensim package [16]. Doc2Vec model simultaneously learns the representation of words in sentences and the labels of the sentences. The labels are numbers or chunks of text which are used to uniquely identify each sentence. We used the debate data and a richer collections of sentences related to climate change to train the Doc2Vec model. In total, there are 10,920 sentences used as the training set.

To measure how two sentences are semantically referring to the same content, we used a function provided in the package to calculate cosine similarity scores among sentences. A cosine similarity score of 1 means that the two sentences are semantically equal and 0 is when it is opposite the case. In the experiment, we manually investigated pairs of sentences at different threshold values and found that the approach is stable at the threshold level above 0.44. The example below

---

[2] This dataset can be downloaded at https://goo.gl/3aicDN.

shows a pair of sentences obtained at 0.44 level.

S1: *Humans are emitting carbon from our cars, planes and factories, which is a heat trapping particle.*

S2: *So there is no doubt that carbon is a heat trapping particle, there is no doubt that our actions are emitting carbon into the air, and there is no doubt that the amount of carbon is increasing.*

In the pair, the two sentences mention the same topic (i.e. *carbon emission*) and express the idea in the same context. We used the threshold 0.44 to re-compute the agreement scores. By applying the semantic approach, the inter-annotator agreement scores of Cohen's Kappa and Krippendorff's alpha increase from 0.28 to 35.71% and from 0.27 to 48.15% respectively. The inter-annotator agreement results are illustrated in Table 4. Note that, in the calculation of the agreement, we incremented the threshold by 0.02. Only particular thresholds are shown in the table due to the limited space.

Table 4. Inter-Annotator Agreement before and after applying the semantic similarity approach.

Trial	Threshold ($\geq$)	κ	α
Before		0.28	0.27
After	0.00	0.81	0.83
	0.10	0.62	0.65
	0.20	0.46	0.50
	0.30	0.40	0.43
	0.40	0.39	0.41
	0.42	0.38	0.41
	0.44	**0.38**	**0.40**
	0.46	0.38	0.40
	0.48	0.38	0.40
	0.50	0.38	0.40
	0.60	0.38	0.40
	0.70	0.38	0.40
	0.80	0.38	0.40
	0.90	0.38	0.40
	1.00	0.38	0.40

4   Automatic Salient Sentence Selection

4.1   Support Vector Regression Model

In this experiment, we work on extractive summarization problem and aim to select sentences that are deemed important or that summarize the information mentioned in debate comments. Additionally, we aim to investigate the keys features which play the important roles in the summarization of the debate data. We view this salient sentence selection as a regression task. A regression score for each sentence is ranged between 1 to 5. It is derived by the number annotators selected that sentence divided by the number of all annotators. In this experiment, a popular machine learning package which is available in Python, called Scikit-learn [6] is used to build a support vector regression model. We defined 8 different features and the support vector regression model combines the features for scoring sentences in each debate comment. From each comment, sentences with the highest regression scores are considered the most salient ones.

4.2   Feature Definition

1. **Sentence Position (SP).** Sentence position correlates with the important information in text [2,5,8]. In general, humans are likely to mention the first topic in the earlier sentence and they express more information about it in the later sentences. We prove this claim by conducting a small experiment to investigate which sentence positions frequently contain salient sentences. From our annotated data, the majority votes of the sentences are significantly at the first three positions (approximately 60%), shaping the assumption that the first three sentences are considered as containing salient pieces of information. Equation 1 shows the calculation of the score obtained by the sentence position feature.

$$SP = \begin{cases} \frac{1}{sentence\ position}, & \text{if } position < 4 \\ 0, & \text{otherwise} \end{cases} \tag{1}$$

2. **Debate Titles (TT).** In writing, a writer tends to repeat the title words in a document. For this reason, a sentence containing title words is likely to contain important information. We collected 11 debate titles as shown in Table 3. In our experiment, a sentence is considered as important when it contains mutual words as in debate titles. Equation 2 shows the calculation of the score by this feature.

$$TT = \frac{number\ of\ title\ words\ in\ sentence}{number\ of\ words\ in\ debate\ titles} \tag{2}$$

3. **Sentence Length (SL).** Sentence length also indicates the importance of sentence based on the assumption that either very short or very long sentences are unlikely to be included in the summary. Equation 3 is used in the process of extracting salient sentences from debate comments.

$$SL = \frac{number\ of\ words\ in\ a\ sentence}{number\ of\ words\ in\ the\ longest\ sentence} \tag{3}$$

4. **Conjunctive Adverbs (CJ).** One possible feature that helps identify salient sentence is to determine conjunctive adverbs in sentences. Conjunctive adverbs were proved that they support cohesive structure of writing.

For instance, "the conjunctive adverb *moreover* has been used mostly in the essays which lead to a conclusion that it is one of the best accepted linkers in the academic writing process." [10]. The NLTK POS Tagger[3] was used to determine conjunctive adverbs in our data.

5. **Cosine Similarity.** Cosine similarity has been used extensively in Information Retrieval, especially in the vector space model. Documents will be ranked according to the similarity of the given query. Equation 4 illustrates the equation of cosine similarity, where: q and d are n-dimensional vectors [14]. Cosine similarity is one of our features that is used to find similarity between two textual units. The following features are computed by applying cosine similarity.

$$cos(q, d) = \frac{\sum_{i=1}^{n} q_i d_i}{\sqrt{\sum_{i=1}^{n} q_i^2} \sqrt{\sum_{i=1}^{n} d_i^2}} \qquad (4)$$

(a) **Cosine similarity of debate title words and sentences (COS_TTS).** For each sentence in debate comments we compute its cosine similarity score with the title words. This is based on the assumption that a sentence containing title words is deemed as important.

(b) **Cosine similarity of climate change terms and sentences (COS_CCTS).** The climate change terms were collected from news media about climate change. We calculate cosine similarity between the terms and sentences. In total, there are 300 most frequent terms relating to location, person, organization, and chemical compounds.

(c) **Cosine similarity of topic signatures and sentences (COS_TPS).** Topic signatures play an important role in automatic text summarization and information retrieval. It helps identify the presence of complex concepts or the importance in text. In a process of determining topic signatures, words appearing occasionally in the input text but rarely in other text are considered as topic signatures. They are determined by an automatic predefined threshold which indicates descriptive information. Topic signatures are generated by comparing with pre-classified text on the same topic using a concept of likelihood ratio [3,15], λ presented by [4]. It is a statistical approach which calculates a likelihood of a word. For each word in the input, the likelihood of word occurrence is calculated in pre-classified text collection. Another likelihood values of the same word is calculated and compared in another out-of-topic collection. The word, on the topic-text collection that has higher likelihood value than the out-of-topic collection, is regarded as topic signature of a topic. Otherwise the word is ignored.

6. **Semantic Similarity of Sentence and Debate Titles (COS_STT).** Since the aforementioned features do not semantically capture the meaning

---

[3] http://www.nltk.org/api/nltk.tag.html.

of context, we create this feature for such purpose. We compare each sentence to the list of debate titles based on the assumption that forum users are likely to repeat debate titles in their comments. Thus, we compare each sentence to the titles and then calculate the semantic similarity score by using Doc2Vec [16].

4.3 Results

In order to evaluate the system summaries against the reference summaries, we apply ROUGE-N evaluation metrics. We report ROUGE-1 (unigram), ROUGE-2 (bi-grams) and ROUGE-SU4 (skip-bigram with maximum gap length of 4). The ROUGE scores as shown in Table 5 indicate that sentence position feature outperforms other features. The least performing feature is the cosine similarity of climate change terms and sentences feature.

Table 5. ROUGE scores after applying Doc2Vec to the salient sentence selection.

ROUGE-N	CB	CJ	COS_CCT	COS_TTS	COS_TPS	SL	SP	COS_STT	TT
R-1	0.4773	0.4988	0.3389	0.5630	0.3907	0.4307	**0.6124**	0.4304	0.5407
R-2	0.3981	0.4346	0.2558	0.5076	0.2986	0.3550	**0.5375**	0.3561	0.4693
R-SU4	0.3783	0.4147	0.2340	0.4780	0.2699	0.3335	**0.4871**	0.3340	0.4303

Table 6. The statistical information of comparing sentence position and other features after applying Doc2Vec.

Comparison pairs	ROUGE-1		ROUGE-2		ROUGE SU4	
	Z	Asymp. Sig. (2-tailed)	Z	Asymp. Sig. (2-tailed)	Z	Asymp. Sig. (2-tailed)
SP VS CB	-4.246^a	0*	-3.962^a	0*	-3.044^a	0.002
SP VS CJ	-3.570^a	0*	-3.090^a	0.002	-2.192^a	0.028
SP VS COS_CCTS	-6.792^a	0*	-6.511^a	0*	-6.117^a	0*
SP VS COS_TTS	-1.307^a	0.191	$-.789^a$	0.43	$-.215^a$	0.83
SP VS COS_TPS	-6.728^a	0*	-6.663^a	0*	-6.384^a	0*
SP VS SL	-4.958^a	0*	-4.789^a	0*	-4.110^a	0*
SP VS COS_STT	-4.546^b	0*	-4.322^b	0*	-3.671^b	0*
SP VS TT	-3.360^b	0.001*	-2.744^b	0.006	-2.641^b	0.008

[a] Based on negative ranks.
[b] Based on positive ranks.

To measure the statistical significance of the ROUGE scores generated by the features, we calculated a pairwise Wilcoxon signed-rank test with Bonferroni correction. We report the significance p = .0013 level of significance after the correct is applied. Our results indicate that there is statistically significance among the features. Table 6 illustrates the statistical information of comparing sentence position and other features. The asterisk indicates that there is a statistical significant difference between each comparison pair.

5   Conclusion

In this paper we worked on an annotation task for a new annotated dataset, online debate data. We have manually collected reference summaries for comments given to global warming topics. The data consists of 341 comments with total 519 annotated salient sentences. We have performed five annotation sets on this data so that in total we have 5 × 519 annotated salient sentences. We also implemented an extractive text summarization system on this debate data. Our results revealed that the key feature that plays the most important role in the selection salient sentences is sentence position. Other useful features are debate title words feature, and cosine similarity of debate title words and sentences feature.

In future work, we aim to investigate further features for the summarization purposes. We also plan to integrate stance information so that summaries with pro-contra sides can be generated.

Acknowledgments. This work was partially supported by the UK EPSRC Grant No. EP/I004327/1, the European Union under Grant Agreements No. 611233 PHEME, and the authors would like to thank Bankok University of their support.

References

1. Morris, A.H., Kasper, G.M., Adams, D.A.: The effects and limitations of automated text condensing on reading comprehension performance. Inf. Syst. Res. **3**(1), 17–35 (1992)
2. Baxendale, P.B.: Machine-made index for technical literature: an experiment. IBM J. Res. Dev. **2**(4), 354–361 (1958)
3. Lin, C.-Y., Hovy, E.: The automated acquisition of topic signatures for text summarization. In: Proceedings of the 18th Conference on Computational Linguistics, COLING 2000, vol. 1, pp. 495–501. Association for Computational Linguistics, Stroudsburg (2000)
4. Dunning, T.: Accurate methods for the statistics of surprise and coincidence. Comput. Linguist. **19**(1), 61–74 (1993)
5. Edmundson, H.P.: New methods in automatic extracting. J. ACM **16**(2), 264–285 (1969)
6. Pedregosa, F., et al.: Scikit-learn: machine learning in python. J. Mach. Learn. Res. **12**, 2825–2830 (2011)
7. Liu, F., Liu Y.: Correlation between rouge and human evaluation of extractive meeting summaries. In: Proceedings of the 46th Annual Meeting of the Association for Computational Linguistics on Human Language Technologies: Short Papers, HLT-Short 2008, pp. 201–204. Association for Computational Linguistics, Stroudsburg (2008)
8. Goldstein, J., Kantrowitz, M., Mittal, V., Carbonell, J.: Summarizing text documents: sentence selection and evaluation metrics. In: Proceedings of the 22nd Annual International ACM SIGIR Conference on Research and Development in Information Retrieval, SIGIR 1999, pp. 121–128. ACM, New York (1999)
9. Landis, R.J., Koch, G.G.: The measurement of observer agreement for categorical data. Biometrics **33**(1), 159–174 (1977)

10. Januliene, A., Dziedraviius, J.: On the use of conjunctive adverbs in learners' academic essays. Verbum **6**, 69–83 (2015)
11. Neto, J.L., Freitas, A.A., Kaestner, C.A.A.: Automatic text summarization using a machine learning approach. In: Bittencourt, G., Ramalho, G.L. (eds.) SBIA 2002. LNCS (LNAI), vol. 2507, pp. 205–215. Springer, Heidelberg (2002). https://doi.org/10.1007/3-540-36127-8_20
12. Krippendorff, K.: Content Analysis: An Introduction to its Methodology, 2nd edn. Sage Publications Inc., Thousand Oaks (2004)
13. Mitrat, M., Singhal, A., Buckleytt, C.: Automatic text summarization by paragraph extraction. In: Intelligent Scalable Text Summarization, pp. 39–46 (1997)
14. Manning, C.D., Schtze, H.: Foundations of Statistical Natural Language Processing. MIT Press, Cambridge (1999)
15. Nenkova, A., McKeown, K.: Automatic summarization. Found. Trends Inf. Retr. **5**(2), 103–233 (2011)
16. Řehůřek, R., Sojka, P.: Software framework for topic modelling with large corpora. In: Proceedings of the LREC 2010 Workshop on New Challenges for NLP Frameworks, pp. 45–50. ELRA, Valletta (2010). http://is.muni.cz/publication/884893/en

Optimization in Extractive Summarization Processes Through Automatic Classification

Angel Luis Garrido[1(✉)], Carlos Bobed[1,2], Oscar Cardiel[3],
Andrea Aleyxendri[3], and Ruben Quilez[3]

[1] Department of Computer Science and Systems Engineering, University of Zaragoza,
Zaragoza, Spain
{garrido,cbobed}@unizar.es
[2] Aragon Institute of Engineering Research (I3A), Zaragoza, Spain
[3] IT Department, Grupo Heraldo, Zaragoza, Spain
{ocardiel,aaleyxendri,rquilez}@heraldo.es

Abstract. The results of an extractive automatic summarization task depends to a great extend on the nature of the processed texts (e.g., news, medicine, or literature). In fact, general-purpose methods usually need to be adhoc modified to improve their performance when dealing with a particular application context. However, this customization requires a lot of effort from domain experts and application developers, which makes it not always possible nor appropriate. In this paper, we propose a multi-language approach to extractive summarization which adapts itself to different text domains in order to improve its performance. In a training step, our approach leverages the features of the text documents in order to classify them by using machine learning techniques. Then, once the text typology of each text is identified, it tunes the different parameters of the extraction mechanism solving an optimization problem for each of the text document classes. This classifier along with the learned optimizations associated with each document class allows our system to adapt to each of the input texts automatically. The proposed method has been applied in a real environment of a media company with promising results.

Keywords: Extractive summarization · Optimization
Machine learning · Automatic classification

1 Introduction

Automatic text summarization consists of decreasing the size of a given text while retaining its most relevant information. It is a challenge task that requires an extensive knowledge of the context of the text, its structure, and its writing style. Automatic text summarization is increasingly used both in research and industry, in areas such as information retrieval [1], question answering [2], or data

© Springer Nature Switzerland AG 2018
A. Gelbukh (Ed.): CICLing 2017, LNCS 10762, pp. 506–521, 2018.
https://doi.org/10.1007/978-3-319-77116-8_38

mining [3]. Besides, the existence of the World Wide Web has caused an explosion in the amount of textual information in all areas, so automatic summarization becomes a great tool for accessing information in a consistent and summarized way.

These techniques are divided into two categories: *extractive summarization*, and *abstract summarization*. The former ones are produced by concatenating several sentences literally, while the latter ones require transformation of the sentences by deleting, substituting, and rearranging them. Summarization algorithms based on extractive techniques are generally simpler to implement, and they are mainly based on statistical or linguistic approaches to find the most relevant sentences to be included in the final summary [4]. This approach allows a simple way of working with any kind of text in any language, regardless of the textual context. In recent years new approaches have been studied in order to improve the outcomes: detection of co-ocurrence [5], simplification of sentences [6], use of named entities [7], etc. However, one of the problems with such algorithms is that a general approach prevents optimal results when it is applied in a particular domain, which leads us to the importance of the application context [8].

Hence, when an extractive summarization algorithm is used in a professional context, it is very common to perform tasks that improve the efficiency of the algorithm according to the context in which it works. For example, in the news context, it may be very interesting to take advantage of the existence of a title, a subtitle, or even the caption of the photographs accompanying the news, in order to modify the sentence selection procedure for generating the final summary. Of course, this method can not be used if what we are doing is to summarize another type of text, such as a judicial sentence, which lacks all of these elements. Again in the news domain, we can find that it is not the same to summarize a teletype, or a musical review, or an interview. All of these formats can have large structural differences that reduce the performance of generalist algorithms, but such characteristics can be exploited by customizable expert systems.

In this paper, we propose a supervised learning methodology to automatic extractive summarization, which is based on leveraging the features of the documents in order to classify them *before* the summarizing process. Firstly, we train a model to classify the documents we are working with. Then, we solve an optimization problem for each of the text document classes to learn the optimum parameters that guide the extraction of sentences in each of the cases. The combination of both methods in the system allows us to achieve better results.

The main contribution of this work is to improve the generation process of a extractive summary from a single-source document with a multi-language and general focus. This is achieved by combining an automatic categorization of texts with performing a personalized adjustment of the summarization process according to this categorization. We have implemented this methodology in the development of a system devoted to summarizing news according to their typology. We have tested such system using a real dataset, which we make available for other researchers. The experiments performed show the multilingual capabilities of our approach, as well as a good outcome on a real working environment.

This paper is organized as follows: Sect. 2 studies the state of the art related to summarization close to this context. Our methodology is detailed in Sect. 3, and its application to a working system is presented in Sect. 4. Section 5 explains the different experiments we have performed and interprets the outcomes. Finally, Sect. 6 summarizes the key points of this work, provides conclusions, and explores future work.

2   Related Work

Interest in automatic summaries appeared back in the 50's. Luhn suggested in [9] to weight the sentences of a document as a function of high frequency words and disregarding the very high frequency common words. Apart from such approach, other methods valued the use of certain words (*cue words*), the headers of the document, and the structure [10] in order to determine the weight of each sentence.

In the 1990s, machine learning techniques started to be used in Natural Language Processing (NLP) and, therefore, also in text summarization. At the beginning, most systems relied on naive-Bayes methods [11], but others focused on learning algorithms that make no independence assumptions [12]. More recently, some works have used hidden Markov models [13], log-linear models [14], and even neural networks [15] to improve extractive summarization.

Some recent works have leveraged the context regarding summarization. For example, in [16] the authors suggest and bring experimental evidence about that the effectiveness of sentence scoring methods for automatic extractive text summarization algorithms depends on certain features of each document typology, working with news, blogs, and articles. Another inspiring work is [17], where authors face the problem of Twitter context summarization by adapting certain environment signals in the context of the tweet. Both approaches are interesting but lack the generality necessary to be applied in very different contexts.

As seen, there are a variety of approaches for generic summarization applicable when the purpose of the reader is unknown. But the main drawback of a generic summarization is the difficulty of getting precise results. However, this performance is a strong requirement in real environments. Hence, new proposals with the aim of covering this need are required.

3   Methodology

This section describes a working methodology applicable to any system dedicated to produce extractive summaries from single-source text documents.

Our method customizes and enhances the process for each existing document typology. The key for improving is to identify the typology of the source documents, and then, to automatically obtain the most suitable parameters of the summarization process with the aim of improving the outcomes for each type. This is done in an off-line step over a corpus which represents the documents the system is going to process in the production stage. Then, in the summarization stage, the system's input is a text that will go through a set of specialized treatments until the generation of the output, the final summary. The size of that summary is established by the *compression rate (CR)*, given by a specific number of words.

In the following, we focus on explaining the training stage of our methodology. Firstly, we introduce the definitions of the different elements in the problem. Then, we move onto the details of both training steps, the automatic classification of documents, and the optimization of the summarization parameters.

3.1 Definitions

Before starting, it is important to give a formal definition of the different elements that take part in the problem. In this paper, we consider that a document D is defined as a tuple $< T, A >$, where:

- T is the text to be summarized. We consider it as being formed by the ordered set $S = \{s_1, \ldots, s_n\}$, with s_i being each of the sentences in order of appearance in the text.
- A is a set $\{a_1, \ldots, a_n\}$, with a_i being attributes of the text T which can be interesting for the elaboration of its summary.

Each of these attributes a_i is formed by a tuple $< name_i, \{value_i\} >$, as well. Examples of attributes could be: title, subtitle, author, place, etc. Note how an attribute can be multivalued in our setting; for example, several captions may correspond to a same text T.

Thus, given a document D, our goal is to obtain a set $R \subset S$, which contains the most relevant sentences, keeping their order. To assess the relevance of a particular sentence within such document, we need a function $ValF$, such as:

$$ValF : String \times \{D\} \rightarrow \mathcal{R}^+$$

with *String* being the set of all possible text strings, and $\{D\}$ the set of all possible documents. Besides, as above mentioned, we have to bear in mind the size constraint imposed by the *compression rate* (CR), which can be given by a specific number of words.

So, in general, given a document D a summarization system would return a set of sentences R such as:

$$R = \{r_i | r_i \in S \ \wedge \ \nexists s_j \in S.ValF(s_j, D) > ValF(r_i)\}$$

satisfying

$$\sum_{i=1..|R|} size(r_i) < CR$$

with S being the set of sentences in D, and *size* a function that gives the word count of a given sentence. It can be considered that, given a set of texts, the optimal way to make an extractive summary of each is to devise a particular

ValF function which provides an optimal result over the corpus. The evaluation of the result is usually be done by comparing each of the summaries obtained with a set of model summaries, for which a pre-established comparison function *(CompF)* is used.

However, this *ValF* function might be optimal for the corpus globally, but not for each of the different text typologies that might be included in the corpus. Thus, we advocate for finding first a categorization of the different texts and styles that might be the target of our system, and, for each of the categories, obtain an optimized *ValF* function in an automated way.

3.2 Automatic Classification of Input Documents

Many works can be found focused on finding an optimal function *ValF* in order to obtain optimal results. In this work, as above mentioned, we propose an additional consideration that improves the application of these techniques to real-world scenarios. Our proposal is that, given a set of texts, *there may be a set of ValF functions that provide an optimal result, corresponding to the different typologies of texts existing in the set*. That is, if we are able to apply a specific *ValF* function on each type of text, we will achieve better results with the application of a generic *ValF* function.

Therefore, we consider obtaining the set of text typologies within the set of documents as the starting part of the methodology. We denote the set of such typologies as $P = \{p_1, \ldots, p_x\}$. To perform this categorization, both the text T of each document and its set of attributes A can be taken into account. Note that the typologies can be established *a priori* by domain experts adopting a supervised approach, or they could be obtained by applying an unsupervised clustering algorithm (e.g., K-clustering, or hierarchical clustering algorithms). The particular technique to detect the different underlying text typologies is out of the scope of this work.

Although categorization of each input document could be manually performed, it would be very expensive and time-consuming. Therefore, once the categories have been defined and assuming that there is an acceptable set of hand-made examples of summaries for each category, we advocate for using an automatic classification using a supervised machine learning method, for example naive bayes, support vector machines, random forests, or artificial neural networks. This input categorization guides which *valF* is going to be used in the summarization process, adapting the system automatically to the input presented to it.

3.3 Summarization Process

The next step is to find out the most suitable *ValF* for each of document typology present in our system. Again, the collaboration of experts and custom development would be the most valuable method in order to design optimal *ValF* functions in real contexts for each of them.

Nevertheless, it is clear that this ideal situation is not always feasible. Therefore, it is necessary to find a method as automatic as possible. Nowadays, we can find a great deal of off-the-self generic extractive summarization approaches which work directly on the text of the document, so we can assume that we can always select one of the relevance evaluation functions they use as baseline, which we will call *BaseVal*. As we want to adapt its evaluation to the different typologies, we propose to extend it externally by adding different terms, so we can build a family of *ValF* functions defined by:

$$ValF_\alpha^{\{\beta_i\}}(s, D) = \alpha * BaseVal(s, T) + \sum_{i=1..n} \beta_i * attrVal(s, D)$$

with s being the sentence to be evaluated, and *attrVal* being functions that evaluate s according to different attributes of the document[1].

In this way, we can automatize the adaptation of the extraction to the different typologies by tuning the α and $\{\beta_i\}$ parameters that weight the generic approach and the extensions over the attributes we use to extend the function[2]. As it happened with the classification processes, we can also assume that at least an acceptable set of model summaries are available for each of the document classes. For each of the defined categories, the summaries belonging to them will be used to optimize and adapt the weight values, and, doing so, to obtain a particular *ValF* function for such document typology.

In our work, we have devised and applied some methods we have found useful for improving the performance of different summarization processes. In particular, we suggest:

1. *Exploring Attributes:* Firstly, for each attribute of a document D, the substantives in the attribute value are obtained. A relevant presence of these words in the training summaries may indicate that the sentences containing such words should be more relevant.
2. *Vocabulary Analysis:* The frequencies of the words used in the summaries are obtained. After studying them, we have found that there is a correlation among the types and quantities of words of each type, and the type and style of the text to be summarized. So, we gradually increase the relevance of the sentence containing them in each type of document summarization.
3. *Sentence Order:* First, the order of the sentences in the examples summaries is checked. If a significant percentage of sentences are located in a certain area of the documents, we leverage that circumstance by increasing the scores of these sentences gradually.

If the three methods are combined along with a selected *baseline* function, solving a multiple optimization problem [18] using separately each of the subsets of documents of each document typology give us the values for the parameters to

---

[1] In fact, they belong to *ValF* family of functions as well, but for the sake's of readability we have decided to change their name.

[2] We are aware we could get rid of the baseline term, but it is useful for the sake of comparing our approach with generic approaches.

maximize the result in each situation. Once we have calculated all the parameter values for each of the document typologies, we just have to classify the documents using the trained text classifier, and then apply the optimized $ValF$ for such typology.

4   Applying the Methodology

In order to validate our methodology, we have applied it to a summarization system in a real-world application. In particular, we have chosen the context of the news produced by a media company, as it is appropriate to evaluate the performance of a system of these characteristics. We have designed a system called NESSY (NEws Summarizer SYstem), a news extraction-based summarization system customized for the specific treatment of news, interviews, briefs, editorials, letter to the editors, and reviews. The system is devoted to perform the task of creating a single summary from a single-source document customizing the process for each typology.

In this section, firstly, the theoretical context of the news text is introduced. Secondly, the application of the methodology to develop the system is detailed.

4.1   News Genres and Structure

The purpose of news is to report on events and topics of general interest. Journalistic genres are ways of written communication that differ according to the needs or objectives of who write it. Overall, experts generally agree that there are three main journalistic genres: Informational, Interpretative, and Borderline [19–21].

1. *Informational:* They aim to narrate the news with an objective and direct language. The person writing the text does not appear explicitly. The texts are informative when transmitting data and facts of interest to the public, whether new or known in advance. The information does not allow personal opinions, much less judgmental.
2. *Interpretative:* They are intended to express the point of view of who writes. The author interprets and discusses reality, evaluates the circumstances in which the incident occurred, and he/she expresses judgments on the reasons and the consequences that may arise from them.
3. *Borderline:* Those in which, in addition to report an occurrence or event, the journalist expresses his opinion. Its purpose is to relate the event to the temporal and spatial context in which it occurs.

These genres are further subdivided into different types, which can be appreciated in Table 1. Each typology has its own characteristics that identify it: the text structure, linguistic aspects, use of verbal forms, explanation of technical terms, syntax, quotes, signatures, and the use of rhetorical figures, are examples of features.

Table 1. Types of news for each of the journalistic genres.

Informational	Interpretative	Borderline
Report	Review	Chronicle
Journalistic report	Editorial	Opinion
Biography	Newspaper column	Letter to the editors
Press review	Obituary/Farewell	Journalistic interview
Interview	Journalistic article	Debate
Documentary report		Talk show
Brief news		

Furthermore, news presents a number of common structural elements [22]:

- *Headline:* Short set of sentences, including the most relevant information, which intend to capture the attention of readers.
- *Caption:* One or few sentences located below the photos or pictures accompanying text.
- *Body:* Set of paragraphs that make up the text and details on the topic. It is the longest section, and its structure is an inverted pyramid regarding the importance of information, whose top is the first paragraph (called *lead*). The lead is situated at the beginning of the body in order to catch the reader's attention.
- *Layout Resources:* Mass media use strategies and resources to capture the reader's attention, which provide extra information or highlight the most important aspects. Examples can be: quotes, documentary data, tables, etc.

Summaries of each type of news must leverage different attributes in different ways in order to achieve good results. It is mainly for this reason that the context of the news has been chosen as an appropriate use case for the evaluation of our methodology.

4.2 The NESSY System

The aim of this system is the individual summarization a piece of news (document) which belong to one of the types of news journalistic genre previously mentioned (see Sect. 4.1). Each document type has unique features that can be used for classification purposes: writing style, structure, presentation, predominant vocabulary, linguistic resources, etc.

Following our proposed methodology, the system is divided into two steps: Text Classification, and Text Summarization. They are directly the application of the steps of our proposal.

Text Classification. First, the system categorizes the input news by using support vector machines (SVM), a well known supervising learning model [23]. The reasons for choosing this methodology are several [24]: (1) SVMs are able to extract an optimal solution with a very small training set size; (2) as SVMs use the feature space images provided by the kernel function, SVMs are applicable in such circumstances that have proved difficult or impossible for other methodologies like bayes or back-propagation neural networks (for example, when data is randomly scattered, and when the density of the distribution of the data is not even well defined); and (3) last but not least, for its simplicity and its speed. In addition, our previous experience with this tool has demonstrated several times [25–28] its good performance in this type of scenarios.

The correct categorization of the input document determines the following stage. In NESSY, the following types of news have been considered: analysis, editorial, interview, letter, opinion, piece of news, report, review, short piece of news, and documentation.

Text Summarization. This second step is in charge of applying the actual summarization process. As stated before, the system has been tuned offline by optimization techniques with the objective of applying the most suitable *ValF* function to each of the categories.

If we analyze the different categories from the point of view of the methodology proposed in Sect. 3, we find that the summaries of each category really can be optimized if the system leveraging the relevant features from each of them. In particular, we present here some relevant examples that have been considered in NESSY successfully:

– *Standard Report:* Journalistic texts of this type are clear and concise, and consist of a recent event that has an interest or curiosity for readers. The title, the caption, and the first paragraph are elements that usually contain the most relevant words, and therefore sentences with these words most likely should appear in the summary.
– *Interview:* An interview is composed by a series of questions, and their answers. The percentage of question marks and interrogative words (*who, when, where,* etc.) compared with the total amount of words is frequently higher than other types of news. The interviewee's name and his/her main quote is usually found in the headline, along (typically) with his/her profession. If an introduction paragraph exists, it usually contains relevant keywords and named entities. The first questions and the included ones into the Layout Resources are also typically the most important.
– *Review:* In this kind of text, in which the writer tries to explain his/her opinion about artistic productions, the main characteristic is that vocabulary is quite repetitive. In particular, if there exists any caption in, it is usually very important.
– *Brief News:* The brief news are a set of short texts that are characterized by their brevity and conciseness. They are summarized news which kept only the most relevant data. The title and the very first sentence are the most relevant elements.

The summaries corresponding to the different categories can be improved if these types of modifications are taken into account when establishing the weights of the sentences to be extracted. We capture these different aspects in NESSY thanks to the use of the proposed *extensions* (*attrVal*) to improve the final *ValF* functions.

Once the text has been classified, NESSY only has to apply the appropriate *ValF* function with the off-line calculated *attrVal* functions, and retrieve the most relevant sentences according to the definition presented in Sect. 3.1.

5 Evaluation

To evaluate our approach, we have performed two different sets of experiments aimed at evaluating the performance of the complete setting, considering the quality of the resulting summaries. We first present the experimental settings and the datasets we have used, and then we detail and discuss the results for each of the experiments.

5.1 Experimental Settings

Datasets. In our experiments, we have used two datasets:

- *DSHA-1* is a corpus composed by 14,000 news taken from *Heraldo de Aragón*[3], a major Spanish media. These news were previously categorized by the Documentation Department of the company with one of the next 10 types: Analysis, Editorial, Interview, Letter, Opinion, Piece of News, Report, Review, Shorts Piece of News, and Documentation. There are 1,400 news of each type. This dataset is used in both experiments for training and evaluating the text classifier precision, which is evaluated applying k-fold cross validation.
- *DSHA-2* is a smaller corpus of 400 news (40 of each aforementioned type) also taken from *Heraldo de Aragón*. Each of the news has an associated summary which has been made by professional documentalists. This dataset is used to test the precision and the recall of the summarization task.

Both datasets are available upon request to the authors exclusively for research purposes, subject to confidentiality agreements due to copyright issues.

Classifier Features. As features for the classifier, the frequency of the each word in the text is used. To achieve the value of those features, stop words[4] are firstly removed from the text. Then, a lemmatization process is applied in order to extract the lemma[5] of each remaining word. The lemmatization process is usually useful on languages with declensions and a lot of verbal forms, such as French, German or Spanish, because it reduces the frequencies catalog. Finally, TF-IDF [29] algorithm is calculated for each lemma obtained from the text.

---

[3] http://www.heraldo.es.

[4] Stop words are common words without relevant information (e.g. articles or conjunctions).

[5] A lemma is the canonical form of a word. For example, in English, sing, sings, sang, sung, and singing are different forms of the same verb, with "sing" as their common lemma.

Results Comparison. For comparison, we have used several on-line summarizers: SWESUM[6] (Sw), Tools4noobs[7] (T4n), Autosummarizer[8] (AS), and the Mashape Tools[9] (MT). We have configured all the summarizers to get a compression rate of 20%. This rate is easily translated to the required number of words, which is the CR we work with. Finally, we have used ROUGE-L to compare the automatic summaries obtained with the summary models in the dataset. ROUGE is a recall-based metric for fixed-length summaries which is based on n-gram co-occurrence, and ROUGE-L is one of the five evaluation metrics available, and it is based on finding the longest common subsequence. It takes into account sentence level structure similarity naturally and identifies longest co-occurrences in sequence n-grams automatically.

5.2   Experiment 1

For this experiment 4,800 news from the dataset *DSHA-1* are selected, and 6 typologies are considered: Editorial, Interview, Letter, Piece of News, Review, and Short Piece of News. So, we have used 800 news of each type: 600 texts are used to train the model and 200 are used to test it. In this experiment different kernels and different types of multiclassifiers are employed. The techniques used are: SVM Multiclass with linear Kernel, SVM Multiclass with radial basis function (RBF) kernel, SVM Multiclass 4th degree polynomial kernel and SVM Binary Tree with RBF kernel. Table 2 shows the experimental results.

Table 2. Accuracy results for 4-fold validation categorization test with 6 categories

	Accuracy
SVM Multiclass linear kernel	87.45%
SVM Binary Tree RBF kernel	90%
SVM Multiclass RBF kernel	92.45%
SVM Binary Tree 4th degree polynomial kernel	92.59%

We selected the best classifier, in this case SVM Binary Tree 4th degree polynomial kernel (92.59% of accuracy), and we used it in the second task to sort a set of 240 news items (40 of each type mentioned before) from the *DSHA-2* dataset, corresponding to the six aforementioned types. After solving the multiple optimization problem we obtained six different *ValF* functions, which are used to customize the summarization process of each type of news. The results are shown in Table 3, where it can be seen how in those more specific categories the most significant improvements are achieved.

---

[6] http://swesum.nada.kth.se/index-eng.html.
[7] https://www.tools4noobs.com/summarize/.
[8] http://autosummarizer.com/.
[9] http://textsummarization.net/.

Table 3. F-measure results regarding a subset of the *DSHA-2* dataset, composed by 240 news, six types, and 40 news of each type. The ROUGE-L algorithm has been used to compare the summaries with the models.

	Sw	T4n	AS	MT	Nessy
Editorial	0.55	0.36	0.38	0.41	0.64
Interview	0.51	0.46	0.45	0.52	0.65
Letter	0.49	0.21	0.29	0.48	0.55
Piece of news	0.42	0.33	0.34	0.33	0.44
Review	0.36	0.37	0.33	0.26	0.46
Short piece of news	0.51	0.46	0.45	0.52	0.65
Average	0.47	0.37	0.37	0.42	0.57

5.3 Experiment 2

For this experiment, the whole dataset (14,000 news) is used and the 10 categories are considered. In this case, 1,200 text of each genre are employed to train the model and 200 to test it. The techniques are the same as the used in the Experiment 1. Table 4 shows the experimental results.

Table 4. Accuracy results for 7-fold validation categorization test with 10 categories.

	Accuracy
SVM Multiclass RBF kernel	77.51%
SVM Multiclass linear kernel	77.73%
SVM Binary Tree RBF kernel	83.59%
SVM Binary Tree 4th degree polynomial kernel	37.90%

We selected the best classifier, in this case SVM Binary Tree RBF kernel (83.59% of accuracy), and we used it in the second task to sort the complete dataset *DSHA-2*, composed of 400 news items (40 of each type mentioned before). After solving the multiple optimization problem we obtain ten different *ValF* functions, which are used to customize the summarization process of each type of news. The results are shown in Table 5.

5.4 Discussion

As it can be seen in the previous tests, we have obtained satisfactory results, especially in Experiment 1 with more than 92% using the SVM Binary Tree with RBF kernel and the SVM Multiclass 4th degree polynomial kernel. However, when it has been included more categories the accuracy decreases to 83.59%.

Table 5. F-measure results regarding the whole *DSHA-2* dataset, composed by 400 news, 10 types, and 40 news of each type using ROUGE-L for comparing.

	Sw	T4n	AS	MT	Nessy
Analysis	0.38	0.29	0.31	0.52	0.47
Documentation	0.31	0.24	0.27	0.28	0.39
Editorial	0.55	0.36	0.38	0.41	0.61
Interview	0.51	0.46	0.45	0.52	0.62
Letter	0.49	0.21	0.29	0.48	0.53
Opinion	0.41	0.35	0.36	0.39	0.51
Piece of news	0.42	0.33	0.34	0.33	0.44
Report	0.31	0.28	0.29	0.30	0.41
Review	0.36	0.37	0.33	0.26	0.45
Short piece of news	0.51	0.46	0.45	0.52	0.62
Average	0.43	0.34	0.35	0.40	0.51

That is because it is difficult to distinguish between some categories with similar linguistics contexts, such as reporting, opinion or short. It is remarkable that the use of SVM with Binary Tree 4th grade polynomial kernel, the best in the Experiment 1, becomes the worst in the Experiment 2, where SVM Binary Tree RBF kernel is the best technique. We wanted to delve into this behaviour and, in Fig. 1, it can be seen the relation between the number of categories and the accuracy of these techniques. We observe that as the categories increase, the

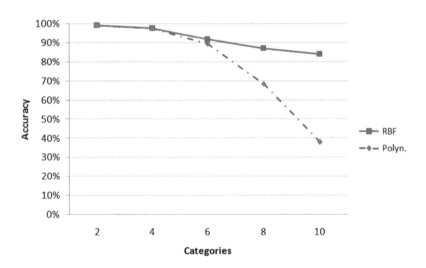

Fig. 1. Relation between number of categories and accuracy comparing SVM Binary Tree RBF versus SVM Binary Tree 4th degree Polynomial.

performance of SVM techniques is getting worse, but not in the same way. It is therefore very important to select a suitable kernel if the number of categories is high.

It is noteworthy also to point out that the mistakes in the classification stage negatively affect the preparation of the summaries, since a news classified in a wrong way will be summarized in the second stage by means of an inadequate $ValF$ function. That is why it is important to classify as best as possible. Even so, in both experiments the improvement that is obtained in the summary process is quite significant, so we can conclude that the applied methodology optimizes the process.

6 Conclusions and Future Work

In this work, we have presented a multilingual supervised learning methodology to generate automatic extractive summaries. Our work focuses on the single-document general purpose extractive summaries, but with a significant difference: whereas other approaches considered a homogeneous corpus, we think that this aspect does not fit well to real scenarios, since within a set of documents, it is very usual to see different subsets with very different characteristics.

Our methodology can be applied in multiple working environments, with the advantage that the system, from a sample, is able to adapt to that context and specialize its way of making summaries. One of the key elements to make this work is the realization of an automatic categorization of source documents, to then, by solving an optimization problem, perform the adaptation of personalized summaries on each of the subsets resulting from such classifying them. The main contribution of this work is to improve the generation process of extractive summaries, in a general case, combining an automatic categorization of texts with performing a personalized adjustment of the summarization process according to this categorization. The utility of these kind of systems is clear for example for enhancing any generic documentation system, as we proposed in our initial works [30], or even for improving automatic infoboxes generation [31].

To evaluate our approach, we have applied this methodology in the field of news by developing a system specialized in summarizing news from media. We have performed the experiments over a real dataset, which is made available to other researchers on demand. The promising outcomes suggest that the methodology can be very useful in multiple scenarios and languages, an aspect that will be verified exhaustively in our following works. Also, our plans are to expand the text features to be considered, including more linguistic and semantic issues, to enrich in that way the work done so far.

Acknowledgments. This research work has been supported by the CICYT project TIN2013-46238-C4-4-R, TIN2016-78011-C4-3-R (AEI/FEDER, UE), and DGA/FEDER. We want to thank Grupo Heraldo for their collaboration, and specially to Domingo Tardos and Susana Sangiao.

References

1. Brandow, R., Mitze, K., Rau, L.F.: Automatic condensation of electronic publications by sentence selection. Inf. Process. Manag. **31**(5), 675–685 (1995)
2. Liu, Y., Li, S., Cao, Y., Lin, C.-Y., Han, D., Yu, Y.: Understanding and summarizing answers in community-based question answering services. In: Proceedings of the 22nd International Conference on Computational Linguistics (COLING 2008), pp. 497–504. Association for Computational Linguistics (2008)
3. Padhy, N., Mishra, P., Panigrahi, R.: The survey of data mining applications and feature scope. Int. J. Comput. Sci. Eng. Inf. Technol. **2**(3), 43–58 (2012)
4. Gupta, V., Lehal, G.S.: A survey of text summarization extractive techniques. J. Emerg. Technol. Web Intell. **2**(3), 258–268 (2010)
5. Lin, C.-Y., Hovy, E.: Automatic evaluation of summaries using n-gram co-occurrence statistics. In: Proceedings of the 2003 Conference of the North American Chapter of the Association for Computational Linguistics on Human Language Technology (NAACL 2003), pp. 71–78. Association for Computational Linguistics (2003)
6. Lal, P., Ruger, S.: Extract-based summarization with simplification. In: Proceedings of the 2002 Workshop on Text Summarization (DUC 2002), pp. 1–8, NIST (2002)
7. Li, W., Wu, M., Lu, Q., Xu, W., Yuan, C.: Extractive summarization using inter- and intra-event relevance. In: Proceedings of the 21st International Conference on Computational Linguistics and the 44th Annual Meeting of the Association for Computational Linguistics (COLING ACL 2006), pp. 369–376. Association for Computational Linguistics (2006)
8. Nenkova, A., McKeown, K.: A survey of text summarization techniques. In: Mining Text Data, pp. 43–76. Springer, Boston (2012)
9. Luhn, H.P.: The automatic creation of literature abstracts. IBM J. Res. Dev. **2**(2), 159–165 (1958)
10. Edmundson, H.P.: New methods in automatic extracting. J. ACM (JACM) **16**(2), 264–285 (1969)
11. Kupiec, J., Pedersen, J., Chen, F.: A trainable document summarizer. In: Proceedings of the 18th International ACM SIGIR Conference on Research and Development in Information Retrieval (SIGIR 1995), pp. 68–73. ACM (1995)
12. Lin, C.-Y.: Training a selection function for extraction. In: Proceedings of the 8th International Conference on Information and Knowledge Management (CIKM 1999), pp. 55–62. ACM (1999)
13. Conroy, J.M., O'leary, D.P.: Text summarization via hidden Markov models. In: Proceedings of the 24th International ACM SIGIR Conference on Research and Development in Information Retrieval (SIGIR 2001), pp. 406–407. ACM (2001)
14. Osborne, M.: Using maximum entropy for sentence extraction. In: Proceedings of the ACL-02 Workshop on Automatic Summarization (AS 2002), pp. 1–8. Association for Computational Linguistics (2002)
15. Svore, K.M., Vanderwende, L., Burges, C.J.: Enhancing single-document summarization by combining ranknet and third-party sources. In: Proceedings of the 2007 Joing Conference on Empirical Methods in Natural Language Processing and Computational Natural Language Learning (EMNLP-CoNLL 2007), pp. 448–457. Association for Computational Linguistics (2007)

16. Ferreira, R., Freitas, F., de Souza Cabral, L., Lins, R.D., Lima, R., Franca, G., Simske, S.J., Favaro, L.: A context based text summarization system. In: Proceedings of the 11th IAPR International Workshop on Document Analysis Systems (DAS 2014), pp. 66–70. IEEE Xplore (2014)

17. Chang, Y., Wang, X., Mei, Q., Liu, Y.: Towards Twitter context summarization with user influence models. In: Proceedings of the 6th ACM International Conference on Web Search and Data Mining (WSDM 2013), pp. 527–536. ACM (2013)

18. Hwang, C.-L., Yoon, K.: Multiple attribute decision making: methods and applications a state-of-the-art survey, vol. 186. Springer Science & Business Media (2012)

19. Bond, F.F.: An Introduction to Journalism: A Survey of the Fourth Estate in all its Forms. Macmillan, New York (1954)

20. MacQuail, D.: Mass Communication Theory: An Introduction. Sage Publications, London (1983)

21. Wolny-Zmorzyński, K., Kozieł, A.: Journalistic genology. Media Stud. **54**, 1–16 (2013)

22. Bell, A.: The discourse structure of news stories. In: Approaches to Media Discourse, pp. 64–104 (1998)

23. Joachims, T.: Text categorization with support vector machines: learning with many relevant features. In: Nédellec, C., Rouveirol, C. (eds.) ECML 1998. LNCS, vol. 1398, pp. 137–142. Springer, Heidelberg (1998). https://doi.org/10.1007/BFb0026683

24. Shin, K.-S., Lee, T.S., Jung Kim, H.: An application of support vector machines in bankruptcy prediction model. Expert. Syst. Appl. **28**(1), 127–135 (2005)

25. Garrido, A.L., Gomez, O., Ilarri, S., Mena, E.: NASS: news annotation semantic system. In: Proceedings of the 23rd IEEE International Conference on Tools with Artificial Intelligence (ICTAI 2011), pp. 904–905. IEEE (2011)

26. Garrido, A.L., Gómez, O., Ilarri, S., Mena, E.: An experience developing a semantic annotation system in a media group. In: Bouma, G., Ittoo, A., Métais, E., Wortmann, H. (eds.) NLDB 2012. LNCS, vol. 7337, pp. 333–338. Springer, Heidelberg (2012). https://doi.org/10.1007/978-3-642-31178-9_43

27. Garrido, A.L., Buey, M.G., Ilarri, S., Mena, E.: GEO-NASS: a semantic tagging experience from geographical data on the media. In: Catania, B., Guerrini, G., Pokorný, J. (eds.) ADBIS 2013. LNCS, vol. 8133, pp. 56–69. Springer, Heidelberg (2013). https://doi.org/10.1007/978-3-642-40683-6_5

28. Garrido, A.L., Buey, M.G., Escudero, S., Peiro, A., Ilarri, S., Mena, E.: The GENIE project-a semantic pipeline for automatic document categorisation. In: Proceedings of the 10th International Conference on Web Information Systems and Technologies (WEBIST 2014), pp. 161–171, SCITEPRESS (2014)

29. Salton, G., Buckley, C.: Term-weighting approaches in automatic text retrieval. Inf. Process. Manag. **24**(5), 513–523 (1988)

30. Garrido, A.L., Peiro, A., Ilarri, S.: Hypatia: An expert system proposal for documentation departments. In: Proceedings of the 12th International Symposium on Intelligent Systems and Informatics (SISY 2014), pp. 315–320. IEEE (2014)

31. Garrido, A.L., Ilarri, S., Sangiao, S., Gañan, A., Bean, A., Cardiel, O.: NEREA: named entity recognition and disambiguation exploiting local document repositories. In: Proceedings of the 28th IEEE International Conference on Tools with Artificial Intelligence (ICTAI 2016), pp. 1035–1042. IEEE (2016)

Summarizing Weibo with Topics Compression

Marina Litvak[1], Natalia Vanetik[1(✉)], and Lei Li[2]

[1] Department of Software Engineering, Shamoon Engineering College,
Beer Sheva, Israel
{marinal,natalyav}@sce.ac.il
[2] Department of Computer Science, Beijing University of Posts and
Telecommunications, Beijing, China
leili@bupt.edu.cn

Abstract. Extractive text summarization aims at selecting a small subset of sentences so that the contents and meaning of the original document are best preserved. In this paper we describe an unsupervised approach to extractive summarization. It combines hierarchical topic modeling (TM) with the Minimal Description Length (MDL) principle and applies them to Chinese language. Our summarizer strives to extract information that provides the best description of text topics in terms of MDL. This model is applied to the NLPCC 2015 Shared Task of Weibo-Oriented Chinese News Summarization [1], where Chinese texts from news articles were summarized with the goal of creating short meaningful messages for Weibo (Sina Weibo is a Chinese microblogging website, one of the most popular sites in China.) [2]. The experimental results disclose superiority of our approach over other summarizers from the NLPCC 2015 competition.

1 Introduction

The MDL principle is widely used in compression techniques of non-textual data, such as summarization of query results for OLAP applications [3,4]. However, only a few works on text summarization based on MDL can be found in the literature. According to MDL, the best summary is the one that leads to the *best compression* of the text by providing its *shortest and most concise description*. Nomoto and Matsumoto [5] used K-means clustering extended with the MDL principle for finding diverse topics in the summarized text. [6] also extended the C4.5 classifier with MDL for learning rhetorical relations. [7] formulated the problem of micro-review summarization within the MDL framework. The authors viewed the tips as having been encoded by snippets, and sought to find a collection of snippets that could produce the encoding with the minimum number of bits.

All summarization approaches must refer to *text informativeness* when extracting or generating short informative summaries. However, how to measure it best remains an open question, with multiple choices available in the

© Springer Nature Switzerland AG 2018
A. Gelbukh (Ed.): CICLing 2017, LNCS 10762, pp. 522–534, 2018.
https://doi.org/10.1007/978-3-319-77116-8_39

literature. It is quite common to measure text informativeness by the frequency of its components–words, phrases, concepts, and so on. In addition to operating term frequency, some works applied the data mining technique for calculating frequent itemsets to the text summarization task, where text is represented as a transactional data. For example, ItemSum [8] represents sentences in a transactional data format and measures their relevancy by coverage of frequent items. SciSumm [9] uses frequent term-based clustering for summarization of scientific papers. The opinion summarizer introduced in [10] extracts a frequent opinion feature set from review texts through the multiword approach, which uses an ordered sequence of words. An algorithm introduced in [11] extracts fuzzy association rules between weighted key phrases in collections of text documents. The summarization approach from [12] integrated frequent sequence mining with the MDL principle and used frequent word sequences as a description model. That approach represents documents as a sequential transactional dataset and then compresses it by replacing frequent sequences of words by codes. The summary is then compiled from sentences that best compress (or describe) the document content, ranked by their coverage of best compressing frequent word sequences.

Nevertheless, as recent research shows, frequency analysis of a document vocabulary is not enough to express its informativeness. Extracting central topics of a document or a document set by topic modeling (TM) can guide sentence selection from the perspective of document-level knowledge, instead of analyzing isolated sentences. Most works in TM use Latent Dirichlet Allocation (LDA) to generate topic words. Hierarchical LDA (hLDA) [13] is an extension of LDA that can model a tree of topics, instead of a flat topic structure. hLDA is an unsupervised method in which topic numbers could grow automatically with the data set. At present, there are many works where topic models have been applied for better summarization of text documents [14–16].

In this work, we propose a new approach to unsupervised text summarization, one that combines hierarchical TM and text encoding, based on the MDL principle that is suitable for Chinese Hanzi script and short Weibo texts. The intuition behind this approach suggests that a summary that best describes the original text should cover its main topics in the form of word sequences with the most important topic words. As such, the problem of summarization is very naturally reduced to the maximal coverage problem [17,18], where the extract must maximally cover the information contained in the source text. We apply the greedy approximation method that ranks sentences by their coverage of best compressing (frequent and representing main topics) word sequences and then selects the top-ranked sentences to a summary. Our results show the significant superiority of the proposed approach over other summarizers.

2   Text Preprocessing

There are several text preprocessing steps that we perform in our work: (1) sentence splitting; (2) word segmentation (or tokenization); and (3) stop-word removal. Word segmentation is a non-trivial step in Chinese because Hanzi words are not separated by spaces. We used the ICTCLAS tool[1] that is based on a dictionary and a machine learning algorithm. The detailed description of the algorithm for Chinese word segmentation can be found in [19].

3   Topic Modeling

The hLDA method of [13] implements a TM that finds a hierarchy of topics forming a tree. The structure of the tree is determined by the input data. A node in this tree structure indicates a topic, with words of the document being probabilistically assigned to this topic. An algorithm of collapsed Gibbs sampling is used for approximating the posterior distribution for hLDA. A topic is defined as a probability distribution across vocabulary words. Given an input document consisting of a sequence of words, hierarchical TM finds useful sets of topics and learns to organize the topics according to a hierarchy. There are no limitations such as a maximum depth or maximum branching factor. Multiple parameters can change the structure of the resulting topics. Their adjustment can be found in [16].

In classic hLDA modeling, every document is allocated to a path from the root to the leaf in the tree. Each node is associated with a topic, which is a distribution across words. Documents sharing the same path should be similar to each other. All documents share the topic distribution associated with the root node.

Because we are interested in detecting important sentences for a single-document summarization, we strive to get a topic distribution over sentences in a document, instead of distribution over documents in a corpus. Therefore, we applied the TM model on single documents as corpora, where the model treats every single sentence as a document. Despite the fact that a single sentence contains much fewer words than a document, TM provided us a considerable discrimination among sentences, that helped us at better sentence selection. We set the depth of the tree as 3. Because a sentence of 2 levels seems too simple, while a tree model with a depth of 4 or greater is too complex for computing and understanding. There are several parameters in hLDA, including ETA, GAM, GEM_MEAN, GEM_SCALE, SCALING_SHAPE, SCALING_SCALE. Different parameters lead to different trees, while the efficacy of a tree can be evaluated by human checking. The parameter settings in our system are as defined in Table 1, and explained and justified in [20].

Figure 1 shows an example hLDA hierarchy built from the M004 Chinese corpus of MultiLing 2015. Figure 2 depicts its translation to English, for non-Chinese readers.

We analyzed both the word distribution between topic nodes and levels of the hierarchy generated by hLDA on Chinese texts, as well as how this information can be used for better summarization. The focus of this study was to provide better sentence scoring for extractive summarization, using hLDA.

---

[1] http://ictclas.nlpir.org/.

Table 1. hLDA parameters

Parameter	Setting
ETA	1.2, 0.5, 0.05
GAM	1.0,1.0
GEM_MEAN	0.5
GEM_SCALE	100
SCALING_SHAPE	1
SCALING_SCALE	0.5

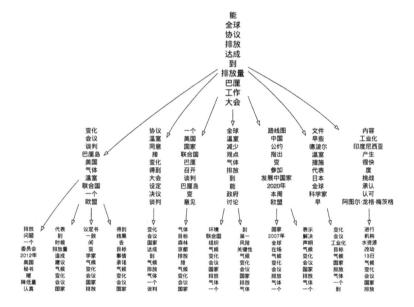

Fig. 1. An example of hLDA hierarchy.

The following observations were adapted to our summarization model (see Sect. 4.3 for details):

1. There are mainly two kinds of words in the root level (0): highly frequent words and stop words with low frequency. The latter occur in cases of minimizing the stopwords list to avoid losing useful information. In general, most non-frequent words appear in level 0, and their number decreases as the levels increase. As such, increasing the frequency threshold when analyzing hLDA results should help ignore stopwords or unimportant words in sentence scoring.

2. Words appearing in top levels are more general. In this way, words that appear in level two are more specific. If we wish to hold specific information in a summary, we must give a higher impact to these words.

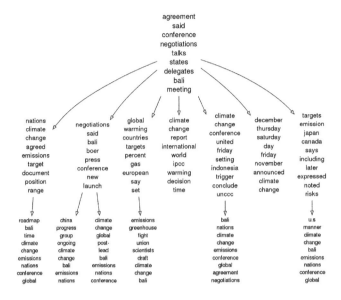

Fig. 2. An example of hLDA hierarchy, translated to English.

3. There are non-frequent words appearing only in level 2 that are also non-frequent in human summaries, and these words can be ignored in scoring sentences for summarization.
4. Most of the frequent words in human gold-standard summaries appear in *multiple levels* of hLDA hierarchy. Therefore, frequent words that are assigned to multiple nodes in different levels, should play the *primary role* in assigning sentence scores, as words carrying various contexts.
5. Some not very frequently used words that are assigned to multiple levels in hLDA hierarchy also appear in human summaries. We decided to consider these kinds of words with secondary priority when assigning scores to sentences.
6. Applying hLDA on texts without removing stopwords results in their high appearance in all levels of hierarchy, especially the root level. As a conclusion, the stopwords must be removed before hLDA modeling if we do not want them to influence sentence scores.

4   MDL-based Approach

4.1   MDL Principle

The proposed summarization algorithm is based on the MDL principle that Mitchell [21] formally defined as follows: Given a set of models $\mathcal{M}$, a model $M \in \mathcal{M}$ is considered the *best* if it minimizes $L(M) + L(D|M)$, where $L(M)$ is the bit length of the description of M and $L(D|M)$ is the bit length of the dataset D encoded with M.

In our approach, we represent an input text as a transactional dataset. Then, using the Krimp dataset compression algorithm presented by [22], we build the MDL for this dataset using its frequent sequences (sequences of words after word segmentation) and TM output. The sentences that cover most frequent word sequences are chosen to a summary.

4.2   Frequent Sequences

We represent a text as a *transactional dataset*. Such a dataset consists of *transactions* (sentences) that are denoted by $T_1, \ldots, T_n$, and *items* (words following segmentation and stop-word removal) that are denoted by $I_1, \ldots, I_m$ and that are unique across the entire dataset. The *size* of a dataset is determined by the number of sentences it contains. Transaction T_i is composed of items $I_1, \ldots, I_m$. Support $supp(s)$ of an itemset s in the dataset is the number of transactions containing it as a subset. Given a support bound S, a set s is called *frequent* if $supp(s) \geq S$.

We generate *sequences* of words as itemsets, where the order of items is meaningful and where items may not be unique; an example of such a sequence is (I_2, I_1, I_2).

A paper by [23] has proposed two algorithms–Apriori and Apriori-TID–for mining frequent itemsets in large databases in efficient time. The Apriori algorithm makes multiple passes over the database. In the first pass frequent individual items are determined. Then, for each pass $k > 1$, the frequent itemsets from the previous pass, $k - 1$, are grouped in sets of k items and form candidate itemsets. The support for each candidate is then counted by passing the database, and those with lower than minimum support are filtered out. This process continues until the set of frequent itemsets in a particular pass is an empty set.

Unlike the Apriori algorithm, the Apriori-TID algorithm uses the database only once, in the first pass. In each consecutive pass $k > 1$ it uses a storage set C^{k-1} of pairs $< TID, X_i >$, where X_i is a candidate itemset of $k - 1$ items in transaction TID. C^1 is built during the first pass. If C^k fits in memory (as it happens in processing our textual data), Apriori-TID is much faster than Apriori. In this work, we use the Apriori-TID algorithm that is adapted for frequent sequence mining.

4.3   The Algorithm

The purpose of the Krimp algorithm [22] is to use frequent sets to compress a transactional database in order to achieve MDL for that database. Let FS be the set of all frequent sequences in the database. A collection CT of sequences from FS, called the coding table, is called *best* when it minimizes the size of $L(CT) + L(D|CT)$. In order to encode the database, every itemset $s \in CT$ is associated with its binary code; any type of code may be used as long as its size logarithmically depends on its itemset size. Because our summarization algorithm never uses the codes, themselves, we only assume that the length of a code for item s_i is bound by $log(i)$ ($|code(s_i)| = log(i)$), where i is the index of itemset s_i in the coding table.

We adapt Krimp to TM knowledge by using the following output of TM (hLDA):

1. Word vocabulary, i.e., words sorted by their number of appearances in a document;
2. Word hLDA level data for every appearance of a word in a sentence: we collected, for every word, the number of times it was classified by hLDA to be at each level K ($K = 0, 1, 2$). Intuitively (according to our observations), words appearing at different levels many times have greater importance than words appearing at only one level, notwithstanding their having a high frequency count in the text.

As we use TM, we define *topical term importance* for term t as the sum of importance of each appearance of t in the text:

$$imp(t) := \sum_{\text{level } K} \sum_{t \text{ appears at level } K} imp(K)$$

The intuition is that a term appearing more times at more hLDA levels across the document receives a higher score.

Here, $imp(K)$ measures *level importance* for level K, and is measured as:

$$imp(K) := \frac{K + 1}{\max + 1}$$

and max $= 2$ in our case; adding 1 is done in order to include all level data.

In this way, lower level results are of greater importance because these words are less general, and a word that appears more times at more levels is of greater importance.

We calculate the importance of a word sequence *seq* as the sum of importance scores for its terms:

$$imp(seq) := \sum_{t \in seq} imp(t)$$

We use TM results in two stages of our algorithm–initial candidate ordering and final sentence ranking–as follows:

1. We find all frequent sequences of terms in the document using the Apriori-TID algorithm of [23] for the given support value *supp*, and store them in set *FS*. This set is kept in the Initial Candidate Ordering according to the topical sequence importance $imp(seq)$.
2. The coding table CT is initialized to contain all single normalized terms and their frequencies. CT is always kept in TM Cover Order in order to give preference to sequences with more topical importance as follows:
   - If topic information is not available (a document is too short), CT is sorted by first decreasing sequence length, then by decreasing support, and finally, in lexicographical order (this is Standard Cover Order).

- If topic information is available, CT is sorted by topical importance $imp(t)$ of sequence I, first, and then by Standard Cover Order.
3. We repeatedly choose frequent sequences from the set FS so that the size of the encoded dataset is minimal, with every selected sequence replaced by its code. Selection is done by computing the decrease in size of the encoding when each one of the sequences is considered to be a candidate to be added to CT, and choosing the best candidate sequence whose replacement *maximally decreases* encoding size.
4. The size of CT should be limited by required summary size in order to obtain a better distinction between encoded and non-encoded parts of the document. The best (most informative) parts must be encoded.
5. The summary is constructed by incrementally adding sentences with the highest weighted coverage rate of codes in CT. This score is computed as the number of common terms in the sentence and CT divided by the size of CT and multiplied by the coefficient measuring the sentence's distance from the beginning of a document (also according to our observations regarding importance of leading sentences). Note that our system very rarely selected just the leading sentences; thus CT coverage rate proved to be more important than sentence index.
6. Sentences are selected in the greedy manner as long as the summary word limit L is not exceeded.

Our approach takes into account the observations from Sect. 3 as follows:

- We give a non-zero weight to the root level, considering word frequency by generating only frequent sequences in CT, which is compatible with observation one; where only frequent words from the root level are considered. Using only frequent sequences by encoding is also covers observation three.
- Following observation two, we give different weights to different levels so that specific words get higher importance than general ones.
- Following observation four, we count every word appearance in all levels. As such, we prefer words that appear many times in many levels over frequent words from one single level. However, our scheme does not match both observation two and observation four perfectly. For example, it will prefer word X appearing twice in level two (getting the highest weight) over word Y appearing once in level one and once in level two.
- Not very frequent words (having support value close to the threshold) may enter CT if there is space left there. After that, their weights correspond to their multiple level appearances. In this way, these words have a chance to influence the sentence scores but not the first priority. This behaviour matches observation five.
- We use texts and hLDA results post stopwords removal, according to observation six.

Our approach is illustrated in Fig. 3.

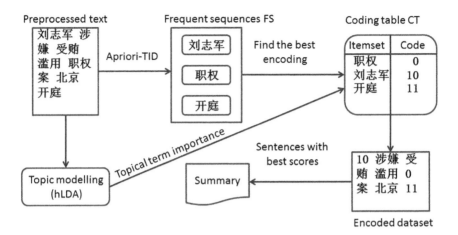

Fig. 3. Encoding approach to summarization.

5   Experimental Results

5.1   The Dataset

We performed our experiments on Weibo-Oriented Chinese News Summarization test set data from NLPCC 2015. This corpus contains 250 news documents and is described in [1]; average document size is 2.96 KB. The dataset was constructed in an automatic way by collecting messages from a few news accounts on Sina Weibo, such as Renminwang, Beijingdaily, SouthernMetropolisWeekly, Breakingnews, etc. According to NLPCC 2015 competition organizers [1], all messages with a URL link to the full Chinese news article were kept, and the news URLs that corresponded to two different Weibo [2] messages were stored. The web pages were downloaded via the URL links and the news articles were extracted from the web pages. Each Weibo message was written and posted by a human editor; it was therefore considered to be a human-written model summary for the associated news article. As a result, two human gold standard summaries are associated with every such document.

This dataset was used for a summarization contest organized in the 4th CCF Conference on Natural Language Processing & Chinese Computing, where 16 systems participated. The maximal length of a summary was limited by 140 characters because the target task was to generate short messages for Sina Weibo.

5.2   Evaluation Setup

We used the ROUGE-1.5.5 toolkit [24] adapted to Chinese, with its command line adopted from the NLPCC 2015 competition: -c 95 -2 4 -U -r 1000 -n 4 -w 1.2 -a -l 140.

The NLPCC 2015 competition [1] used character-based evaluation for evaluating Chinese summary, meaning that no Chinese word segmentation needs to be performed when running the ROUGE toolkit. Instead, all Chinese characters were separated with blank spaces. We used the same approach in our experiments and separated Chinese characters in both gold standard summaries and system summaries in order to measure ROUGE scores correctly. Slight differences in ROUGE scores for participating systems, from those reported at the competition, are due to the fact that we implemented simple Chinese character splitting ourselves.[2]

5.3   Summarizer Systems

We compared our approach (denoted by SkrimpTM) with:

- Two runs of the top system[3] from the NLPCC 2015 competition, denoted by WUST-1 and WUST-2;
- MDL-based approach without TM (denoted by Skrimp), to see what difference TM makes;
- TM-based system without MDL (denoted by TMSumm), to see what difference MDL makes;
- Top-K system that just takes 140 top characters, as a baseline; and
- FWCov that selects sentences ranked by coverage of singular frequent words, to see the difference between the MDL principle and simply covering frequent words.

It is worth noting that the length of summaries generated by WUST-1 and WUST-2 is not consistent; sometimes the summaries are very short (minimal length is about 20 characters). The balance of the systems produced summaries of exact 140-character length.

5.4   Evaluation Results

Table 2 contains the comparative results in terms of Rouge-1, 2, 3, and 4, recall, precision, and F-measure, with confidence intervals (95%). WUST-1 and WUST-2 – are two runs of the same system, supplied by Wuhan University of Science and Technology and denoted in [1] by NLP@WUST. It sometimes selects parts of sentences instead of whole sentences.

For FWCov, Skrimp, and SkrimpTM, a minimum support was set to 2, meaning that word/sequence must appear at least in two sentences in a document to be considered frequent. No gaps are allowed, meaning that sequence words must appear in sentences *together*. For example, sequence XY appears in the sentence (represented by a transaction as a sequence of sentence words) XYZ, but not in sentence XZY. The maximal size of a coding table was set to 50 in all cases, because the average word in our systems contains 2–3 Hanzi characters, and the target summary length is 140 characters. Note that this limitation was used

---

[2] Because our system did not participate in the NLPCC competition, all experiments were re-run by ourselves.

[3] ranked first by Rouge, F-measure.

Table 2. Experimental results: Rouge-1,2,3,4

System	R1-R	R1-P	R1-F	R2-R	R2-P	R2-F	R3-R	R3-P	R3-F	R4-R	R4-P	R4-F
WUST-1	0.586	0.378	0.449	0.395	0.258	0.304	0.275	0.182	0.213	0.206	0.136	0.160
WUST-2	0.599	0.378	0.455	0.409	0.261	0.312	0.290	0.187	0.222	0.218	0.142	0.168
Top-K	0.558	**0.430**	**0.484**	0.414	**0.319**	0.359	0.334	**0.257**	0.290	0.288	**0.222**	0.250
FWCov	0.559	0.347	0.425	0.382	0.238	0.291	0.291	0.182	0.222	0.246	0.154	0.188
SkrimpTM	**0.615**	0.389	0.473	**0.472**	0.298	**0.363**	**0.396**	0.250	**0.304**	**0.353**	**0.222**	**0.270**
Skrimp	0.563	0.352	0.430	0.392	0.246	0.299	0.303	0.190	0.231	0.256	0.160	0.195
TMSumm	0.562	0.413	0.474	0.408	0.299	0.344	0.326	0.238	0.274	0.283	0.206	0.238

solely to speed up the computation process without affecting the algorithm in any way.

Skrimp does not use TM results. Sentences were ranked by their coverage of the coding table–the first sentence was selected as the one that covers the maximum number of codes (term sequences) in the coding table, the second sentence was selected as the one that covers the maximum number of codes uncovered by the first sentence, and so on.

Experimental results show that MDL with topical knowledge provides an excellent basis for extracting valuable information for generating short messages. SkrimpTM has the best results in terms of most Rouge metrics. It is also a very interesting (however, not surprising) fact that the Top-K approach performs extremely well, and even best, in some cases.

According to the Wilcoxon statistical test, FWCov, Skrimp, and SkrimpTM significantly outperform WUST-2 (P values are 0.0237, 0.0221, and 0.0160, respectively) in terms of Rouge-1, Recall. Also, SkrimpTM is significantly better than Top-K, with P value < 0.0001.

According to the same test performed on Rouge-1 Precision scores, FWCov, Skrimp, and SkrimpTM significantly outperform WUST-2 (P values are 0.0156, 0.0425, and 0.0036, respectively). However, Top-K is significantly outperforms both WUST-2 and SkrimpTM (with $P < 0.0001$). Confidence intervals (95%) for FWCov, Skrimp, WUST-2, Top-K, and SkrimpTM are: 0.348 ± 0.017, 0.352 ± 0.016, 0.378 ± 0.019, 0.431 ± 0.018, and 0.389 ± 0.017, respectively.

SkrimpTM is significantly better than Skrimp in terms of all Rouge metrics ($P < 0.0001$). This outcome confirms a claim from [12] that the MDL approach by itself is inappropriate for summarizing single short documents that do not contain enough content for providing a high-quality description.

Given TM results, practical average running time for SkrimpTM per file is: 22.728 ms on a platform with dual-core I7 CPU, 8GB RAM and 64-bit OS.

6   Conclusions

This paper introduces a new method for generating short summaries, using TM and the MDL principle. The method represents a document as a transactional database and uses frequent sequences of meaningful words as a description model for that database. We adapted this method for Chinese Hanzi script and short Weibo texts. The results of experiments performed on the NLPCC 2015 dataset shows superiority of the proposed approach when compared to other systems that participated in the NLPCC 2015 competition.

In the future, we intend to try this method on other languages and domains. Also, adapting the proposed MDL principle for other summarization tasks, such as query-based or updated summarization, should be very challenging, while promising intriguing results.

References

1. Wan, X., Zhang, J., Wen, S., Tan, J.: Overview of the NLPCC 2015 shared task: weibo-oriented chinese news summarization. In: Li, J., Ji, H., Zhao, D., Feng, Y. (eds.) NLPCC 2015. LNCS (LNAI), vol. 9362, pp. 557–561. Springer, Cham (2015). https://doi.org/10.1007/978-3-319-25207-0_52
2. SINA, C.: Sina weibo (2009). www.weibo.com
3. Lakshmanan, L.V.S., Ng, R.T., Wang, C.X., Zhou, X., Johnson, T.J.: The generalized mdl approach for summarization. In: Proceedings of the 28th International Conference on Very Large Data Bases, VLDB 2002, pp. 766–777 (2002)
4. Bu, S., Lakshmanan, L.V.S., Ng, R.T.: Mdl summarization with holes. In: Proceedings of the 31st International Conference on Very Large Data Bases, VLDB 2005 pp. 433–444 (2005)
5. Nomoto, T., Matsumoto, Y.: A new approach to unsupervised text summarization. In: Proceedings of the 24th Annual International ACM SIGIR Conference on Research and Development in Information Retrieval. SIGIR '01, pp. 26–34 (2001)
6. Nomoto, T.: Machine learning approaches to rhetorical parsing and open-domain text summarization. Ph.D. thesis, Nara Institute of Science and Technology (2004)
7. Nguyen, T.S., Lauw, H.W., Tsaparas, P.: Review synthesis for micro-review summarization. In: Proceedings of the Eighth ACM International Conference on Web Search and Data Mining, WSDM 2015, pp. 169–178 (2015)
8. Baralis, E., Cagliero, L., Jabeen, S., Fiori, A.: Multi-document summarization exploiting frequent itemsets. In: Proceedings of the 27th Annual ACM Symposium on Applied Computing, SAC 2012 pp. 782–786 (2012)
9. Agarwal, N., Gvr, K., Reddy, R.S., Ros, C.P.: Scisumm: a multi-document summarization system for scientific articles. In: Proceedings of the ACL-HLT 2011 System Demonstrations, pp. 115–120 (2011)
10. Dalal, M.K., Zaveri, M.A.: Semisupervised learning based opinion summarization and classification for online product reviews. Appl. Comput. Intell. Soft Comput. **2013**,10 (2013)
11. Danon, G., Schneider, M., Last, M., Litvak, M., Kandel, A.: An apriori-like algorithm for extracting fuzzy association rules between keyphrases in text documents. In: Proceedings of the 11th International Conference on Information Processing and Management of Uncertainty in Knowledge-Based Systems (IPMU 2006), Special Session on Fuzzy Sets in Probability and Statistics, pp. 731–738 (2006)
12. Litvak, M., Vanetik, N., Last, M.: Krimping texts for better summarization. In: Conference on Empirical Methods in Natural Language Processing, pp. 1931–1935 (2015)
13. Blei, D.M., Ng, A.Y., Jordan, M.I.: Latent dirichlet allocation. J. Mach. Learn. Res. **3**, 993–1022 (2003)

14. Wang, D., Zhu, S., Li, T., Gong, Y.: Multi-document summarization using sentence-based topic models. In: Proceedings of the ACL-IJCNLP 2009 Conference Short Papers, pp. 297–300. Association for Computational Linguistics (2009)
15. Lee, S., Belkasim, S., Zhang, Y.: Multi-document text summarization using topic model and fuzzy logic. In: Perner, P. (ed.) MLDM 2013. LNCS (LNAI), vol. 7988, pp. 159–168. Springer, Heidelberg (2013). https://doi.org/10.1007/978-3-642-39712-7_12
16. Li, L., Heng, W., Yu, J., Liu, Y., Wan, S.: Cist system report for ACL multiling 2013-track 1: multilingual multi-document summarization. MultiLing **2013**, 39 (2013)
17. Takamura, H., Okumura, M.: Text summarization model based on maximum coverage problem and its variant. In: EACL '09: Proceedings of the 12th Conference of the European Chapter of the Association for Computational Linguistics, pp. 781–789 (2009)
18. Gillick, D., Favre, B.: A scalable global model for summarization. In: Proceedings of the NAACL HLT Workshop on Integer Linear Programming for Natural Language Processing, pp. 10–18 (2009)
19. Zhang, H.P., Yu, H.K., Xiong, D.Y., Liu, Q.: HHMM-based Chinese lexical analyzer ICTCLAS. In: Second SIGHAN Workshop Affiliated with 41th ACL, pp. 184–187 (2003)
20. Wei, H., Jia, Y., Lei, L., Yongbin, L.: Research on key factors in multi-document topic modeling application with HLDA. J. Chin. Inf. Process. **27**(6), 117–127 (2013)
21. Mitchell, T.M.: Machine Learning, 1st edn. McGraw-Hill Inc., New York, NY, USA (1997)
22. Vreeken, J., Leeuwen, M., Siebes, A.: Krimp: mining itemsets that compress. Data Min. Knowl. Discov. **23**, 169–214 (2011)
23. Agrawal, R., Srikant, R.: Fast algorithms for mining association rules. In: 20th International Conference on Very Large Databases, pp. 487–499 (1994)
24. Lin, C.Y.: ROUGE: a package for automatic evaluation of summaries. In: Proceedings of the Workshop on Text Summarization Branches Out (WAS 2004), pp. 25–26 (2004)

Timeline Generation Based on a Two-Stage Event-Time Anchoring Model

Tomohiro Sakaguchi[(⊠)] and Sadao Kurohashi

Graduate School of Informatics, Kyoto University, Kyoto, Japan
sakaguchi@nlp.ist.i.kyoto-u.ac.jp, kuro@i.kyoto-u.ac.jp

Abstract. Timeline construction task has become popular as a way of multi-document summarization. Dealing with such a problem, it is essential to anchor each event to an appropriate time expression in a document. In this paper, we present a supervised machine learning model, two-stage event-time anchoring model. In the first stage, our system estimates event-time relations using local features. In the second stage, the system re-estimates them using the result of first stage and global features. Our experimental results show that the proposed method surpasses the state-of-the-art system by 3.5 F-score points in the TimeLine shared task of SemEval 2015.

1 Introduction

In understanding text, it is essential to understand temporal information in texts correctly. There have been a lot of studies to understand temporal expressions and event ordering [1–3].

Furthermore, as a way of multi-document summarization, timeline construction has become popular recently [4–6]. In SemEval 2015, a shared task, *TimeLine: Cross-Document Event Ordering*, was proposed to create a timeline in which events related to a given target entity are extracted from a set of news articles, and they are ordered along the time axis [6]. For example in Fig. 1, a timeline of the target entity "iPhone 4" is generated from articles related to the topic "Apple Inc.". A timeline consists of an ordered list of <time value, event> pairs, and the finest granularity of time values is day.

The timeline generation task consists of two subtasks: extraction of events related to a target entity, and anchoring those events to appropriate time values. The severe problem lies in the latter subtask. In anchoring events to time values in a document, easy cases and difficult cases are mixed up. In some cases, an event expression is explicitly modified by a time expression; in other cases, the time value cannot be estimated without understanding the context.

For example, "introduced" in the second sentence in Fig. 1 is a relatively easy case, since it has a dependency relation with the corresponding temporal expression "7 June". On the other hand, the event "announced" in the first

© Springer Nature Switzerland AG 2018
A. Gelbukh (Ed.): CICLing 2017, LNCS 10762, pp. 535–545, 2018.
https://doi.org/10.1007/978-3-319-77116-8_40

DocID: 39896,   Document Creation Date: 2010-06-16

1. **Pre-orders** of the recently **announced** iPhone 4 began **Tuesday**.
   2010-06-15
2. The newest iPhone, **iPhone 4** was **introduced** by Apple CEO
   Steve Jobs at the company's 2010 Worldwide Developer's
   Conference on 7 **June**.
   2010-06-07
3. He **praised** it as "the biggest **leap** we've taken since the original
   iPhone."

Timeline of target entity *"iPhone 4"*

\<time value\>	\<event\>
2010-06-07	39896-1-announced
2010-06-07	39896-2-introduced
2010-06-07	39896-3-praised
2010-06-15	39896-1-pre-orders

Fig. 1. An example of timeline construction. Underline denotes events, red denotes events related to the target entity "iPhone 4", blue denotes phrases which corefer the target entity and green denotes temporal expressions.

sentence cannot be anchored correctly to 2010-06-07 without using the event-time anchoring result of the same event, "introduced" in the next sentence.

The contribution of our work is to propose a two-stage event-time anchoring model which enables us to consider wider context than previous work.

The TimeLine task of SemEval 2015 has two tracks: Track A and Track B. In Track A, raw texts are given as input; in Track B, texts with gold event mentions are given. Since we focus on the two problems, namely, extracting target-entity-related events and anchoring events to time values, Track B setting is used. Our experimental results show that the proposed method surpasses the state-of-the-art system by 3.5 F-score points.

2   Related Work

Several works tackled the TimeLine task of SemEval 2015.

HeidelToul team (Moulahi et al. [7]) proposed a rule based approach. They first extract sentences and events which are relevant to the target entity. They apply string matching using cosine similarity matching function with a threshold, and also apply entity coreference resolution using Stanford CoreNLP [8] to extract terms which refer to the target entity. Then, temporal expressions are extracted and normalized by HeidelTime [9], and associated with events in the same sentence. Finally, the events are pruned using the token distance between event and the closest term which refers the target entity.

GPLSIUA team (Navarro and Saquete [10]) proposed another rule based method using two clustering processes. They first extract events which are relevant to the target entity. They resolve the named entity recognition and corefer-ence resolution using OPENER web service[1] and extract sentences which include the target entity or its coreference entity. Events in the sentences are selected as relevant events of the target entity. Next, they apply two clustering processes in sequential order: temporal clustering and lemma clustering. The idea of the clustering is that events which occur in the same date and refer to the same fact

---
[1] http://www.opener-project.eu/webservices/.

are regarded as coreferent events. In temporal clustering, they extract temporal expressions, events and links between them using TIPSem [11], and group events which occurred in the same date. Lemma clustering groups events which have the same head word lemma, the same date and the same target entity.

Navarro and Saquete [12] improved the GPLSIUA system. In extracting target entity related events, they additionally consider whether the event and the target entity have a *has_participant* relation with the semantic role ARG0 or ARG1 in the Propbank Project [13]. In the clustering processes, they expand the lemma clustering by using synonymy relations, and added distributional clustering after the lemma clustering. Distributional clustering groups semantically compatible events which do not have same lemma or synonyms.

Cornegruta and Vlachos [14] first introduced supervised approach to this TimeLine task. They estimated <event, target entity> and <event, temporal expression> anchoring by machine learning. Since the gold timelines consist of <time value, event> pairs, they first generate pseudo training data using distant supervision method. They recognize entities by approximate string matching with the Stanford Coreference Resolution System [15], and extract temporal expressions using UWTime temporal parser [16]. Correct <event, target entity> labels are generated by associating each event to the nearest mention of the target entity in the same sentence. Similarly, each event is associated to the nearest temporal expression which has consistency in the <event, time value> pair in the gold timeline. After that, they train each anchoring using alignment model at the document level with global information. The difference with our event-time anchoring is that they anchor events to temporal expressions, though we anchor events to time value. Another difference is that they imposed a first order Markov assumption and use only preceding information, though we use wider context information.

Laparra et al. [17] proposed rule based method using tense information in the Track A of the TimeLine task. They extracted events and temporal expressions by a semantic role labelling tool, MATE Tools [18] and TextPro suite [19] respectively. They first expand the target entity using DBpedia and extract events which have the target entity as their ARG0 or ARG1. Events are anchored to corresponding time values by a rule-based strategy which uses tense information.

3 Our Approach

The proposed method first anchors all the events in a document to time values. Then, events related to a target entity are extracted by matching the event specification phrases in a document to various expressions denoting a target entity. A timeline of a target entity is generated by ordering relevant events according to their anchored time values. Figure 2 exemplifies the proposed method.

We describe these steps in detail in the following subsections.

Fig. 2. Process of timeline construction: anchoring events to appropriate time values and extraction of target-entity-related events.

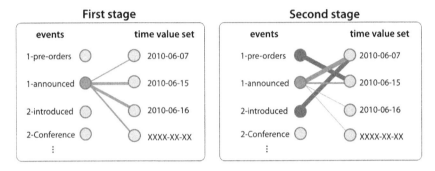

Fig. 3. Outline of the two-stage event-time anchoring method. In the first stage, each event estimates the probabilities of associating time values. In the second stage, each event updates the probabilities considering its neighbour events (blue events), and is associated to the time value which has highest probability.

3.1 Anchoring Events to Appropriate Time Values

We start with the task of anchoring events in a document to appropriate time values. We assume that all events in a document are given as a gold data, since we use the setting of Track B of the TimeLine task. First, extraction of time values in a document is explained. Then, the two-stage event-time anchoring method is described.

Time Value Set. Each event in a document corresponds to either the time value represented by a time expression in a document, or the document creation date, or uncertain time value[2]. We use UWTime temporal parser [16], the state-of-the-art temporal expression analysis system, to detect and normalize temporal expressions in documents. Note that temporal expressions which do not represent

---

[2] When the time value of an event is uncertain, it is treated as corresponding to the special time value "XXXX-XX-XX" in the TimeLine task.

dates but periods like "six months" are removed. We call the set of time values in a document as *time value set*.

Training Data for Event-Time Anchoring. Though it is desirable that all events in a document are anchored to appropriate time values in event-time anchoring training data, the annotated data of the TimeLine task provides only the event-time correspondences related to specific target entities. We use the annotated data as pseudo training data by ignoring the unanchored events.

Learning to Rank in Two Stages. The selection of the most relevant time value for an event among time value set is relative, and it is appropriate to use the framework of learning to rank. Learning to rank is performed so as to make *score of selecting correct time value* is larger than *score of selecting other time values* for each event.

As described in the Introduction section, in anchoring events to time values in a document, easy cases and difficult cases are mixed up. To cope with such a mixed problem, we considered a two-stage method: the first stage estimates event-time relations using local features, and the second stage estimates event-time relations again using global features including the first stage estimation results (Fig. 3).

The local features are extracted from the event expression and a time value/expression. They are all binary and are classified into the following three types:

1. Features of the event expression:
   - tense, aspect and POS tag of the event expression.
   - the event expression is a communication event such as "say" and "announce" or not.
   - the event expression is included in the headline of the document.
   - the event expression has a direct dependency relation with any temporal expression.
2. Features of a time value/expression:
   - a time value is the document creation date (DCD), next day of DCD, future from DCD, or uncertain.
   - the granularity of a time value is day or larger.
   - a time expression depends on the dependency root of the sentence.
3. Features concerning a pair of the event expression and a time value/expression:
   - the event expression is before or after a temporal expression.
   - the event expression and a temporal expression are in the same sentence or not; they have a direct dependency relation or not.

The global features represent the relation between the event under consideration, E_c, and its four types of neighbour events: the preceding event, the following event, the nearest event that has the same stem with E_c, and the nearest event that has the same tense with E_c.

For each of these four events, E_x, we use the following global features:

- E_x exists or not.
- E_x is a communication event or not.
- E_c and E_x are in the same sentence or not.
- the sentence distance between E_c and E_x.
- E_x's time value estimated in the first stage and its confidence score.

3.2   Extraction of Target-Entity-Related Events

Next, events related to a given target entity are extracted in a document. We realize a flexible extraction both by expanding a target entity expression and by collecting event-related phrases in a document.

A target entity can be expressed in various expressions in a text. For example, "Toyota Motor" can appear in a text as "Toyota" or "Toyota Company". Therefore, we expand a target entity by using two external knowledge.

- DBpedia
  Paraphrases of proper nouns are acquired from DBpedia[3], using redirect links [17].
- PPDB
  A target entity is sometimes not a named entity, but an ordinary expression like "stock markets worldwide". Since most entries of DBpedia are named entities, we employed The Paraphrase Database (PPDB) [20], to obtain paraphrases of ordinary expressions. By using PPDB, for example, we can obtain "stock markets around the world" as a paraphrase of "stock markets worldwide".

For each event, we need to extract what is the event about from the context. For example, the event "introduced" in the second sentence in Fig. 1 is about "iPhone 4" and "Steve Jobs". We call them *event specification phrases* (ESPs in short).

First, we apply dependency parsing (Turbo Parser [21]), and for each event expression, we extract phrases in sub-trees under its children and its siblings, as ESPs.

Furthermore, when a phrase in ESPs corefers other expressions in a document, they are added to ESPs. BART [22] is used for coreference resolution. For example, in the case of the event "praised" in Fig. 1, since "it" corefers "iPhone 4", "iPhone 4" is also included in ESPs.

As a final result, if there is any exact match between ESPs of the event and the paraphrases of the target entity, the event is judged to be related to the target entity.

4   Experiments and Results

The dataset used in the SemEval 2015 TimeLine task is composed of articles from Wikinews. The development dataset consists of timelines for six target

---

[3] http://dbpedia.org/.

Table 1. Results on SemEval 2015 task-4 Track B

System	Airbus	GM	Stock	Total		
	F1	F1	F1	P	R	F1
HeidelToul	16.50	10.82	25.89	13.58	28.23	18.34
GPLSIUA	22.35	19.28	33.59	21.73	30.46	25.36
(Navarro+, 2016)	26.21	21.08	31.58	23.68	30.37	26.61
(Cornegruta+, 2016)	25.65	26.64	32.35	29.05	28.12	28.58
One stage	28.32	27.49	16.49	30.60	20.54	24.58
One stage+DBpedia+PPDB	27.46	27.42	31.83	30.07	28.58	29.31
Two stages	**31.06**	**29.52**	18.71	32.42	22.57	26.94
Two stages+DBpedia+PPDB	29.63	29.44	**36.34**	**32.50**	**31.64**	**32.06**

Table 2. Accuracy of event-time anchoring.

	Airbus	GM	Stock
One stage	37.46	28.29	42.21
Two stages	41.09	31.58	55.89

entities (e.g. "Steve Jobs", "iPhone 4"), generated from 30 documents related to "Apple Inc.". We used the development dataset for training. The test dataset consists of three documents set, each of which related to "Airbus and Boeing", "General Motors, Chrysler and Ford" and "Stock Market", and each set has 30 documents. In the TimeLine task, participants generate timelines of dozen entities for each topic. Output timelines are evaluated by the time value of event and the order of event, and Precision, Recall and F-score are calculated.

In the experiment, we utilized SVM-rank [23] as a learning to rank tool. We compared our results with four systems. HeidelToul and GPLSIUA are systems participating in SemEval 2015 task-4 TrackB, and the rest are systems constructed after that. The system by Cornegruta et al. uses development dataset for training as our system do, though the others do not use.

The results of the experiment are shown in Table 1. The proposed method surpasses the state-of-the-art by 3.5 points in F-score. Looking at the results of the proposed method in detail, the two-stage model is 2.7 points better than the one-stage model which just utilizes local features. Expansion of target entity expressions using DBpedia, PPDB improved the recall scores significantly.

5 Discussion

In this section, we discuss the results from the viewpoints of event-time anchoring and extraction of target-entity-related events.

5.1   Anchoring Events to Time Values

Table 2 shows the evaluation results of event-time anchoring. The second stage improved the result of first stage in every corpus.

In the first stage, events which have dependency relation with temporal expressions tend to be correctly associated to the corresponding time value. For example, the event "entered into" in the following sentence is correctly associated to the time value 2007-08-10 ("Friday", DCD).

Table 3. Evaluation of the selection of events which related to target entities.

	Airbus			GM			Stock		
	P	R	F1	P	R	F1	P	R	F1
Two stages	62	52	56	73	66	69	76	26	39
Two stages+DBpedia+PPDB	52	59	55	71	79	75	75	57	65

(1) [Document Creation Date: 2007-08-10]
    On Friday, the Fed <u>entered into</u> a \$38 billion repurchase agreement of mortgage-backed securities, easing stockholder worries.

In the second stage, events which are refered in other sentences are modified. The following example consists of the two consecutive sentences.

(2) [Document Creation Date: 2005-06-13]
    (a) Ryanair <u>exercises</u> options on five Boeing 737s.
    (b) Irish low cost airline, Ryanair, announced today that it is <u>exercising</u> its options with Boeing to purchase five new 737 aircraft.

In the first stage, while the system correctly associated the event "exercising" in the second sentence to 2005-06-13 ("today", DCD), the event "exercises" in the first sentence was wrongly associated to XXXX-XX-XX. However, in the second stage, the anchoring is modified to 2005-06-13 by using the information of "exercising" in the next sentence.

The majority of errors are due to our not considering event-event temporal and semantic relations. For example, there is an implicit temporal relation between "purchase" and "deliver". Since the amount of training data is not enough for acquiring these relations, using distant supervision or external knowledge would be needed. Some errors are related to the temporality of events. For example, a verb "plan" tends to represent events in the future. There are also errors related to the event-time features. Especially in complex sentences, not only the information of direct dependency relations but also the structure of sentences and semantic roles are essential to identify the event-time relations.

5.2 Extraction of Target-Entity-Related Events

Table 3 shows the evaluation results of extracting target-entity-related events. In every corpus, the expansion of target entities improved the recall score significantly.

On the other hand, the precision scores are decreased as its trade-off. The main reason of the decrease is the acquisition of terms which are relative but not paraphrase. For example, as a result of expanding the target entity "EADS" (the predecessor of Airbus group), "Airbus group" and "Airbus company" are extracted as the same entity.

In GM corpus, though the F-score of extracting target-entity-related events is improved by the expansion, the F-score of generated timelines is slightly decreased. Most of the improvement in GM corpus is due to the target entity "Frederick Henderson", and the average F-score of extracting target-entity-related events without it is 73.13 and 74.00 for with and without the expansion. Since 95% of the events in the timeline of "Frederick Henderson" are extracted from one document and the event-time anchoring model could not work well in the document, the advantage of expansion did not lead to improvement in the final result.

In Stock Market corpus, greater improvement is achieved than the other corpora. This is due to the category of target entities. In Airbus and GM corpora, most of the target entities are company name, product name and person name (e.g. "China Eastern Airlines", "Barack Obama"). Since these entities are often written without abbreviated at their first appearances in document, many target-entity-related events can be extracted by just using string matching and coreference resolution. On the other hand, in the Stock Market corpus, most of the target entities are indexes and money expressions (e.g. "Dow Jones Industrial Average", "FTSE 100 index", "US Dollar") which are usually abbreviated or paraphrased. For example, "Dow Jones Industrial Average" is usually written as "the Dow Jones" or "the Dow Industrials". In these cases, much more related events can be extracted by using knowledge of paraphrases.

6 Conclusion and Future Work

The timeline generation problem consists of two subtasks: extraction of events related to a target entity, and anchoring those events to appropriate time values.

We proposed a two-stage event-time anchoring model which can consider wide context information. The evaluation of our proposed method showed that it surpasses the state-of-the-art system by 3.5 F-score points in the TimeLine task of SemEval 2015.

In this work, we focused on event-time anchoring in a document. In the future work, we are going to tackle with cross-document event-time anchoring and event coreference to construct a consistent timeline of multiple documents.

Acknowledgments. This work was partially supported by JST CREST Grant Number JPMJCR1301 including AIP challenge program, Japan.

References

1. Verhagen, M., Gaizauskas, R., Schilder, F., Hepple, M., Katz, G., Pustejovsky, J.: Semeval-2007 task 15: tempeval temporal relation identification. In: Proceedings of the Fourth International Workshop on Semantic Evaluations (SemEval-2007), pp. 75–80. Association for Computational Linguistics (2007)
2. Verhagen, M., Sauri, R., Caselli, T., Pustejovsky, J.: Semeval-2010 task 13: Tempeval-2. In: Proceedings of the 5th International Workshop on Semantic Evaluation, pp. 57–62. Association for Computational Linguistics (2010)
3. UzZaman, N., Llorens, H., Derczynski, L., Allen, J., Verhagen, M., Pustejovsky, J.: Semeval-2013 task 1: Tempeval-3: evaluating time expressions, events, and temporal relations. In: Second Joint Conference on Lexical and Computational Semantics (*SEM), Volume 2: Proceedings of the Seventh International Workshop on Semantic Evaluation (SemEval 2013), pp. 1–9. Association for Computational Linguistics (2013)
4. Swan, R.C., Allan, J.: Timemine: visualizing automatically constructed timelines. In: SIGIR, p. 393 (2000)
5. Tran, G., Alrifai, M., Herder, E.: Timeline summarization from relevant headlines. In: ECIR (2015)
6. Minard, A.L. et al.: Semeval-2015 task 4: Timeline: Cross-document event ordering. In: Proceedings of the 9th International Workshop on Semantic Evaluation (SemEval 2015), Denver, Colorado, pp. 778–786. Association for Computational Linguistics (2015)
7. Moulahi, B., Strötgen, J., Gertz, M., Tamine, L.: Heideltoul: a baseline approach for cross-document event ordering. In: Proceedings of the 9th International Workshop on Semantic Evaluation (SemEval 2015), pp. 825–829. Association for Computational Linguistics (2015)
8. Manning, C.D., Surdeanu, M., Bauer, J., Finkel, J., Bethard, S.J., McClosky, D.: The Stanford CoreNLP natural language processing toolkit. In: Association for Computational Linguistics (ACL) System Demonstrations, pp. 55–60 (2014)
9. Strötgen, J., Gertz, M.: Multilingual and cross-domain temporal tagging. Lang. Resour. Eval. **47**, 269–298 (2013)
10. Navarro, B., Saquete, E.: Gplsiua: combining temporal information and topic modeling for cross-document event ordering. In: Proceedings of the 9th International Workshop on Semantic Evaluation (SemEval 2015), pp. 820–824. Association for Computational Linguistics (2015)
11. Llorens, H., Saquete, E., Navarro, B.: Tipsem (english and spanish): evaluating crfs and semantic roles in tempeval-2. In: Proceedings of the 5th International Workshop on Semantic Evaluation, Uppsala, Sweden, pp. 284–291. Association for Computational Linguistics (2010)
12. Navarro-Colorado, B., Saquete, E.: Cross-document event ordering through temporal, lexical and distributional knowledge. Know. Based Syst. **110**, 244–254 (2016)
13. Palmer, M., Kingsbury, P., Gildea, D.: The proposition bank: An annotated corpus of semantic roles. Comput. Linguist. **31**, 71–106 (2005)
14. Cornegruta, S., Vlachos, A.: Timeline extraction using distant supervision and joint inference. In: Proceedings of the 2016 Conference on Empirical Methods in Natural Language Processing, pp 1936–1942. Association for Computational Linguistics (2016)
15. Lee, H., Chang, A., Peirsman, Y., Chambers, N., Surdeanu, M., Jurafsky, D.: Deterministic coreference resolution based on entity-centric, precision-ranked rules. Comput. Linguist. **39** (2013)

16. Lee, K., Artzi, Y., Dodge, J., Zettlemoyer, L.: Context-dependent semantic parsing for time expressions. In: Proceedings of the 52nd Annual Meeting of the Association for Computational Linguistics (Volume 1: Long Papers), Baltimore, Maryland, pp. 1437–1447. Association for Computational Linguistics (2014)
17. Laparra, E., Aldabe, I., Rigau, G.: Document level time-anchoring for timeline extraction. In: Proceedings of the 53rd Annual Meeting of the Association for Computational Linguistics and the 7th International Joint Conference on Natural Language Processing (Vol. 2: Short Papers), pp. 358–364. Association for Computational Linguistics (2015)
18. Björkelund, A., Hafdell, L., Nugues, P.: Multilingual semantic role labeling. In: Proceedings of the Thirteenth Conference on Computational Natural Language Learning (CoNLL 2009): Shared Task, pp. 43–48. Association for Computational Linguistics (2009)
19. Emanuele Pianta, C.G., Zanoli, R.: The textpro tool suite. In Nicoletta Calzolari (Conference Chair) et al. (eds.) Proceedings of the Sixth International Conference on Language Resources and Evaluation (LREC 2008), Marrakech, Morocco, European Language Resources Association (ELRA) (2008). http://www.lrec-conf.org/proceedings/lrec2008/
20. Ganitkevitch, J., Van Durme, B., Callison-Burch, C.: PPDB: the paraphrase database. In: Proceedings of NAACL-HLT, Atlanta, Georgia, pp. 758–764. Association for Computational Linguistics (2013)
21. Martins, A.F.T., Smith, N.A., Xing, E.P., Aguiar, P.M.Q., Figueiredo, M.A.T.: Turbo parsers: dependency parsing by approximate variational inference. In: Proceedings of the 2010 Conference on Empirical Methods in Natural Language Processing, EMNLP 2010, 9–11 October 2010, , pp. 34–44. MIT Stata Center, Massachusetts, USA, A meeting of SIGDAT, a Special Interest Group of the ACL(2010)
22. Versley, Y. et al.: Bart: a modular toolkit for coreference resolution. In: Proceedings of the 46th Annual Meeting of the Association for Computational Linguistics on Human Language Technologies: Demo Session. HLT-Demonstrations 2008, Stroudsburg, pp. 9–12. Association for Computational Linguistics, PA, USA (2008)
23. Joachims, T.: Making large-scale svm learning practical. LS8-Report 24, Universität Dortmund, LS VIII-Report (1998)

Information Retrieval and Text Classification

Efficient Semantic Search Over Structured Web Data: A GPU Approach

Ha-Nguyen Tran[(✉)], Erik Cambria, and Hoang Giang Do

School of Computer Science and Engineering, Nanyang Technological University,
Singapore, Singapore
{hntran,cambria,do0008ng}@ntu.edu.sg

Abstract. Semantic search is an advanced topic in information retrieval which has attracted increasing attention in recent years. The growing availability of structured semantic data offers opportunities for semantic search engines, which can support more expressive queries able to address complex information needs. However, due to the fact that many new concepts (mined from the Web or learned through crowd-sourcing) are continuously integrated into knowledge bases, those search engines face the challenging performance issue of scalability. In this paper, we present a parallel method, termed *gSparql*, which utilizes the massive computation power of general-purpose GPUs to accelerate the performance of query processing and inference. Our method is based on the backward-chaining approach which makes inferences at query time. Experimental results show that gSparql outperforms the state-of-the-art algorithm and efficiently answers structured queries on large datasets.

1 Introduction

In recent years, the Resource Description Framework (RDF) and the Web Ontology Language (OWL) are widely applied to model knowledge bases such as DBPedia [2], Yago [17], or SenticNet [6]. Searching and retrieving the complete set of information on such knowledge bases is a complicated and time-consuming task. The crucial issue is that the searching process requires a deep understanding of the semantic relations between concepts. In other words, semantic search engines generally need to integrate an inference layer which derives implicit relations from the explicit ones based on a set of rules.

With the immense volume of daily crawled data from the Internet sources, the sizes of many knowledge bases have exceeded millions of concepts and relations [7]. Real-time inference on such huge datasets with various user-defined rulesets is a non-trivial task which faces challenging issues in term of system performance. As a consequence, efficiently searching and retrieving information on large-scale semantic systems have attracted increasing interest from researchers recently. Query engines such as OWLIM [3], Sesame [4], and Jena [8] integrate an inference layer on top of the query layer to perform the reasoning process and

© Springer Nature Switzerland AG 2018
A. Gelbukh (Ed.): CICLing 2017, LNCS 10762, pp. 549–562, 2018.
https://doi.org/10.1007/978-3-319-77116-8_41

retrieve the complete set of results. This approach is termed backward-chaining reasoning.

Those methods are also designed to support in-memory execution. Most of them, however, are facing the problems of scalability and execution time. As a result, it has until now been limited to either small datasets or weak logics. Distributed computing methods [21] have been introduced to deal with large graphs by utilizing parallelism, yet there remains the open problem of high communication costs between the participating machines.

Recently, Graphic Processing Units (GPUs) with massively parallel processing architectures have been successfully leveraged for query processing [14,18–20]. GPUs have also been utilized to enhance the performance of forward-chaining reasoning [12,15] which makes explicit all implicit facts in the pre-processing phase. The benefits of the forward-chaining inference scheme are (1) the time-consuming materialization is an off-line computation; (2) the inferred facts can be consumed as explicit ones without integrating the inference engine with the runtime query engine. However, the drawback of this approach is that we can only reason and query on the knowledge bases with a pre-processed rule-set. In addition, the amount of inferred facts could be very large in comparison with the original dataset.

To address the requirements of scalability and execution time for semantic search engines over large-scale knowledge bases with custom rules, in this paper we introduce a parallel method, termed *gSparql*, which utilizes the massive computation power of general-purpose GPUs. Our method accepts different rulesets and executes the reasoning process at query time when the inferred triples are determined by the set of triple patterns defined in the query. To answer SPARQL queries in parallel, we convert the execution plan into a series of primitives such as sort, merge, prefix scan, and compaction which can be efficiently done on GPU devices. We also present optimization techniques to improve the performance of reasoning and query processing. To highlight the efficiency of our solution, we perform an extensive evaluation of gSparql against the state-of-the-art semantic query engine Jena. Experiment results on LUBM [10] show that our solution outperforms the existing method on large datasets.

The rest of the paper is structured as follows: Sect. 2 provide formal definitions of our problem; Sect. 3 introduces the GPU-based approach to accelerate semantic search engines; Sect. 4 presents the GPU implementation of the inference engine; experiment results are shown in Sect. 5; finally, Sect. 6 concludes the paper.

2   Semantic Search on Web Data

RDF[1] is a W3C recommendation that is used for representing information about Web resources. Resources can be anything, including documents, people, physical objects, and abstract concepts. The RDF data model enables the encoding,

---

[1] https://www.w3.org/TR/rdf-primer/.

exchange, and reuse of structured data. It also provides the means for publishing both human-readable and machine-processable vocabularies.

RDF data is represented as a set of triples $< S, P, O >$, as in Table 1, where each triple $< s, p, o >$ consists of three components, namely *subject*, *predicate*, and *object*. Each component of the RDF triple can be represented in either as a universal resource identifier (URI) or in literal form.

Subject (s)	Predicate (p)	Object (o)
x:Alice	y:isSisterOf	x:Andy
x:Bob	y:isSiblingOf	x:Andy
x:Bob	y:hasParent	x:Brad
x:Alice	y:liveIn	x:London
x:Bob	y:liveIn	x:Paris
x:Andy	y:liveIn	x:England
x:Alice	rdf:type	x:Female
x:Alice	y:age	23
x:Andy	y:age	17
x:Bob	y:age	25
x:London	y:isPartOf	x:England
y:hasParent	rdfs:domain	x:Person
x:Female	rdfs:subClassOf	x:Person
x:isSiblingOf	rdf:type	owl:SymProp

Fig. 1. RDF triples

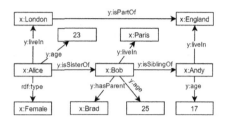

Fig. 2. RDF knowledge graph

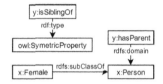

Fig. 3. RDF schema graph

RDF data is also represented as a directed labeled graph. The nodes of such a graph represent the subjects and objects, while the labeled edges are the predicates. Such a representation is believed to be cognitive inspired and it is adopted in many areas of artificial intelligence [5,16,23]. We give the formal definition of an RDF graph as follow:

Definition 1. An RDF graph is a finite set of triples (subject, predicate, object) from the set $T = U \times U \times (U \cup L)$, where U and L are disjoint, U is the set of URIs, and L the set of literals.

For example, Fig. 2 illustrates the RDF graph based on the RDF triples in Fig. 1. RDF graphs are further classified into two sub-types, namely *RDF knowedge graph* and *RDF schema graph*. The set of nodes in an RDF knowledge graph includes entities, concepts, and literals, as can be seen in Fig. 2. On the other hand, the RDF schema graph describes the relationships between types/predicates. Each edge connects two types or predicates (Fig. 3).

Similar to a RDF graph, a SPARQL query[2] also contains a set of triple patterns. The subject, predicate and object of a triple pattern, however, could be a variable, whose bindings are to be found in the RDF data.

---

[2] https://www.w3.org/TR/rdf-sparql-query/.

Definition 2. A SPARQL triple pattern is any element of the set $T = (U \cup V) \times (U \cup V) \times (U \cup L \cup V)$, where V is the variable set.

A SPARQL triple pattern can also be recursively defined as follows:

(1) If P_1 and P_2 are SPARQL triple patterns, then expressions with the forms of $P_1 . P_2$, P_1 *OPTIMAL* P_2, and P_1 *UNION* P_2 are also SPARQL triple patterns.

(2) If P is a SPARQL triple pattern and C is a supported condition, then P *FILTER* C is also a SPARQL triple pattern.

In a SPARQL query, the *SELECT* keyword is used to identify the variables which appear in the result set. For example, one wants to list all people whose parent is Brad and whose ages are greater than 20. The SPARQL query for this question is illustrated below:

```
SELECT ?a ?b
FROM {
    ?a rdf:type x:Person.
    ?a y:hasParent x:Brad.
    ?a y:age ?b
    FILTER (?b > 20)
}
```

This query returns an empty result set because we cannot find any matches in the RDF data triples in Table 1. However, if we consider the semantic relations by using rules $R = \{R_1, R_2, R_3, R_4, R_5, R_6, R_7\}$, which are given below, the result set of the search query is (?a, ?b) = {(x:Alice, 23), (x:Bob, 25)}.

R_1 : (?x y:isSisterOf ?y) $\rightarrow$ (?x y:isSiblingOf ?y)

R_2 : (?x y:isSiblingOf ?y) (?y y:isSiblingOf ?z) $\rightarrow$ (?x y:isSiblingOf ?z)

R_3 : (?x rdf:type ?y) (?y rdfs:subClassOf ?z) $\rightarrow$ (?x rdf:type ?z)

R_4 : (?x y:isSiblingOf ?y) (?y y:hasParent ?z) $\rightarrow$ (?x y:hasParent ?z)

R_5 : (?x ?p ?y) (?p rdfs:subPropertyOf ?q) $\rightarrow$ (?x ?q ?y)

R_6 : (?x ?p ?y) (?p rdf:type owl:SymmetricProperty) $\rightarrow$ (?y ?p ?x)

R_7 : (?x ?p ?y) (?p rdfs:domain ?z) $\rightarrow$ (?x rdf:type ?z)

These results can be explained as follows: based on rules R_1, R_2, and R_6, we can infer a triple (x:Alice y:isSiblingOf x:Bob); then, we obtain a triple (x:Alice y:hasParent x:Brad) by applying R_4 to that triple; finally, R_7 generates two other triples relevant to the query, i.e., (x:Alice rdf:type x:Person) and (x:Bob rdf:type x:Person).

3   GPU-Accelerated Semantic Search

In this section, we first present our schema to index the RDF data. Then, we introduce an overview of our rule-based reasoning and query processing system, termed *gSparql*. Our method is based on backward-chaining reasoning and is accelerated by the massive parallel computing power of GPUs.

3.1   RDF Index

In gSparql, the triplestore layout is based on the *property tables* approach in which all triples with the same predicate name are stored in the same Table [1]. In the traditional method, a property table consists of two columns $< s, o >$ and is sorted by *subject*. In order to support efficient merging and joining operations during reasoning and query processing, we maintain another $< o, s >$ column table, which is sorted by *object*, for each predicate name. Our method uses a dictionary to encode URI and literal RDF terms into numeric values (Fig. 4). This encoding is commonly used in large-scale triplestores such as RDF-3X [13] and Hexastore [22] to reduce tuple space and faster comparison. The numeric values are stored in their native formats.

URI/Literal	Numeric ID
y:liveIn	0
y:age	1
rdfs:subClassOf	2
...	...
x:Alice	10
x:Andy	11
x:Bob	12
x:England	13
x:London	14
x:Paris	15
...	...

Fig. 4. URI/Literal dictionary

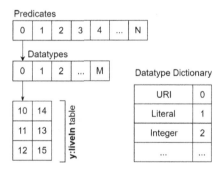

Fig. 5. RDF index

In practice, the objects of the same predicate name might have different datatypes. The objects in the predicate related to *Born* in Wikipedia, for example, consist of Literal values (e.g., "179-176 BC"), Integer values (e.g., 1678), and Datetime values (e.g., 06 June 1986). For each predicate name, we further divide the column tables $< s, o >$ and $< o, p >$ into smaller ones based on the object datatypes. Figure 5 illustrates the triplestore layout used in our method.

The advantages of our RDF data index are: (1) gSparql is able to immediately make comparisons between numeric data in the FILTER clauses without further requesting actual values from the dictionary; (2) Our method can directly return the results of unary functions on RDF terms such as *isIRI*, *isLiteral*, or *isNumeric*; (3) We only need to execute joining operations on columns with the same datatype. Thus, unnecessary joins can be pruned out; (4) The vertical partitioning approach enables GPU kernels to retrieve the table content in a coalesced memory access fashion which significantly improves the performance of GPU-based systems.

3.2   Overview of gSparql

In this section, we present an overview of our GPU-based semantic search engine over structured Web data in RDF/OWL format, termed *gSparql*. Our method integrates an inference layer to make explicit semantic relations between concepts at query time. As can be seen in Fig. 6, the input of our system is a SPARQL query and the output is a result set which contains a collection of relevant tuples in which the selected variables are bound by RDF terms.

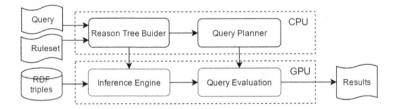

Fig. 6. An overview of gSparql

The engine module first parses the input query Q into a set of single triple patterns. For each triple pattern p, the Reason Tree Builder module generates a reasoning tree which is in fact a rooted DAG (directed acyclic graph) with the root of p based on the predefined ruleset. Figure 7 shows a part of the reasoning tree of the *(?a y:hasParent x:Brad)* triple pattern. A reasoning tree

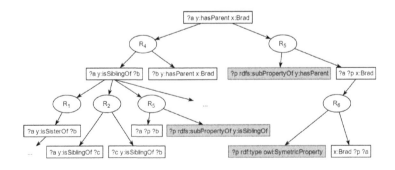

Fig. 7. Reasoning tree of (?a y:hasParent x:Brad)

comprises two types of nodes, namely pattern nodes and rule nodes. A pattern node is established by connecting rule nodes using *or* operations. A rule node, in contrast, is created by applying *and* operations between pattern nodes. General speaking, the parent of a rule node R_i is the consequent of the rule and its child nodes are the antecedents of R_i. The module builds those trees by recursively applying rules to pattern nodes in DFS fashions. The reasoning tree construction terminates when no more rules can be further applied. The Inference engine, then, searches for all RDF triples which can match p using the reasoning tree. The matching process takes the majority of the querying time due to time-consuming operations such as joining, filtering, and duplication elimination. Thus, our method takes advantage of massively parallel computing ability of GPUs to accelerate such operations. In particular, we employ an efficient parallel scheme which combines GPU-friendly primitives such as sort, merge, prefix scan.

The Query Evaluation module then executes joining the resulting triples of all single triple patterns in the search query. Similar to the Inference Engine, the matching process takes the majority of the querying time and thus is performed on the GPU. A data buffer in the device memory is utilized to temporarily maintain the required data and intermediate results during execution. After performing the query evaluation phase, the final results are transferred back to the main memory.

4   Inference Engine

In comparison to materialization-based method, the reasoning and query processing system based on backward-chaining is usually required to perform more computation at the query time. The real-time inference is considered as the bottleneck of this approach which decreases the overall response time. Thus, the objective of gSparql is to enhance the performance of the backward-chaining reasoning process.

4.1   Optimized Backward-Chaining

The main routine of Inference Engine execution path is illustrated in Algorithm 1. The input of the algorithm is a single triple pattern p in the SPARQL query. To find the results of p, we first identify the set of rules R_p which can be applied to generate p (Line 2). For each rule $r \in R_p$, we recursively find the matches of triple patterns in the left hand side (LHS) of r, then apply the rule to generate the matches of the right hand side (RHS) pattern. The results of all rules are then merged to obtain the inferred triples of p (Lines 3–14). The inferred results are finally combined with the matches of p in the triplestore to produce the final solutions (Lines 16–17).

The real-time backward-chaining inference is executed by matching and joining RDF triples based on the reasoning trees using bottom-up approaches. To match a rule node, we first search for the matches of its child nodes in the RDF

Algorithm 1. bwc-reasoning: Backward-Chaining reasoning

Input: triple pattern p, ruleset R, triplestore D
Output: set of triples T

```
1  R_p := find_rule(p, R);
2  T := {};
3  foreach rule r ∈ R_p do
4      LP := get_lhs(r, p);
5      LP_Result := {};
6      foreach pattern i ∈ LP do
7          if is_computed(i) then
8              if can_terminate(i) then
9                  break;
10             else
11                 LP_Result[i] = get_result(i);
12         else
13             LP_Result[i] := bwc-reasoning(i, R, D);
14     T_r := make_inference(r, LP_Result);
15     T := merge(T, T_r);
16 T_p = match_pattern(p, D);
17 T := merge(T, T_p);
18 return T
```

---

triplestore. Then, joining operations are applied to return the rule node's results. This is the common procedure to make inferences on *general rules*. The matches of a pattern node are obtained by merging the results of its child rule nodes. In rules R_2 and R_4 of the example ruleset, the triple patterns related to column tables *y:isSiblingOf* and *y:hasParent* appear on both sides of the rules respectively. In these cases, new triples are potentially generated when we continuously apply the same rules to the derived triples. We call such rules *recursive rules*.

Optimizations: In order to reduce the number of rule nodes in the reasoning tree and enhance the performance of real-time inference, we apply the following optimization techniques:

- *Pre-compute the RDF schema graph:* Based on the observation that the triple pattern related to the RDF schema graph such as *(?a rdfs:subPropertyOf y:hasParent)* is more generic than *(?a ?p x:Brad)*, since the latter pattern depends on the input query while the former does not. Such schema triple patterns, however, appear frequently in the reasoning trees. In order to reduce the execution time spending on searching the matches of those schema triple patterns, we perform the materialization process on the RDF schema graph in the pre-processing step [21]. As a result, our system only needs to perform reasoning on the non-schema triple patterns (Line 13).
- *Memorize intermediate results:* An optimization technique to reduce the processing time is memorizing the reasoning results of triple patterns which

appear many times in all reasoning trees. In other words, we only perform the backward-chaining reasoning on the subtree of such triple patterns once, then we maintain results for the next calls. The two techniques are applied to the *is_compute* function at Line 7.

– *Prune out unnecessary tree branches:* The pre-calculation of the triple patterns allows us to prune out the reasoning branches which cannot contribute to the final solutions. For example, assume that we have $B \cap C \rightarrow A$ and $D \cap E \rightarrow C$. This implies that $B \cap (D \cap E) \rightarrow A$, or $(B \cap D) \cap E \rightarrow C$ in a different expression. If $B \cap D = \emptyset$, the conclusion A cannot be derived from D and E. By applying this property, we can avoid the unnecessary expensive join operation between D and E [21]. In the reasoning tree described in Fig. 7, we can consider the triple pattern *(?p rdf:type owl:SymmetricProperty)* as D and the triple pattern *(?p rdfs:subPropertyOf y:hasParent)* as D. In this case, Rule R_6 will fire only if some of the subjects of the triples that are matched to *(?p rdfs:subPropertyOf y:hasParent)* is also the subject of triples part of *(?p rdf:type owl:SymmetricProperty)*.Since *(?p rdfs:subPropertyOf y:hasParent)* and *(?p rdf:type owl:SymmetricProperty)* are schema triple patterns which are computed in the pre-processing step, we are able to check the condition $B \cap D = \emptyset$ very fast and efficiently using the GPU.

– *Remove redundant triple patterns:* This technique is based on the consistency property of a semantic knowledge base and the search query. The triple pattern *(?a rdf:type x:Person)* can be derived from R_7: *(?a ?p ?b) (?p rdfs:domain x:Person)* $\rightarrow$ *(?a rdf:type x:Person)*. In the sample SPARQL query, we need to perform the joining operation between the results of *(?a y:hasParent x:Brad)* and *(?a rdf:type x:Person)*. Due to the pre-computation of the schema triple *(?p rdfs:domain x:Person)*, we can understand that the results contain *(y:hasParent rdfs:domain x:Person)*. Therefore, the result set of *(?a rdf:type x:Person)* is the superset of the matched solution of *(?a y:hasParent x:Brad)*. As a consequence, the joining operation between the result sets of the two single triple patterns is redundant. Thus, the time-consuming reasoning process for the triple pattern *(?a rdf:type x:Person)* can be ignored.

4.2 GPU Implementation

This subsection discusses about the GPU implementation of the inference engine. Initially, we give brief descriptions of some important GPU primitives which significantly outperform the CPU-based counterparts. Then, we discuss how to map these primitives to different groups of inference rules.

Prefix scan: A prefix scan (in short, *scan*) employs a binary operator to the input array of size N and generates an output of the same size. An important example of prefix scan is prefix sum which is commonly used in database operations. In gSparql, we apply the GPU implementation from [11].

Sort: Our system employs Bitonic Sort algorithm for sorting relations in parallel. The bitonic sort merges bitonic sequences, which are in monotonic ascending or descending orders, in multiple stages. We adapt the standard bitonic sort algorithm provided by NVIDIA library.

Merge: Merging two sorted triples is a useful primitive and is a basic building block for many applications. To perform the operation on GPUs, we apply an efficient GPU Merge Path algorithm [9].

Sort-Merge Join: Following the same procedure as the traditional sort-merge joins, we execute sorting algorithms on two relations and after that merge the two sorted relations. Due to the fact that our triplestore layout is based on vertical partitioning approach, the sort-merge join is well-suited for reasoning and query processing execution.

Next, we discuss how to apply these operations on various groups of rules. Our considered inference rules can be divided into some major groups, namely copy rules, subject/object join rules, and predicate join rules.

Copy Rules: For this group of rules, the joining operations are not required. We simply copy the whole column table into a new one. The rule R_1 is an example of the rule group. At implementation level, we do not perform actual copy operations.

Subject/Object Join Rules: Performing this rule group requires joining two predicate tables in the positions of subject or object (e.g, rules R_2, R_3, R_4). Since our triplestore maintains both sorted predicate tables $< s, o >$ and $< o, s >$, these join rules are straightforwardly executed by the standard sort-merge join.

Predicate Join Rules: This kind of rules joins two triple patterns in which one join attribute is located at the predicate position of a triple pattern and the other attribute is in the subject or object position of the remain triple pattern (e.g, rules R_5, R_6, R_7). We reduce the joining operation of the rules to *scan* and *merge*. First, we scan the triplestore to collect the predicate names of the join attribute in the first triple pattern. Then, we merge column tables of the obtained predicate names.

Recursive Rules: The main routine of making inferences on recursive rules is illustrated in Algorithm 2. The general idea of the algorithm is to recursively apply the rule R to derived triples until no new triple is found.

Algorithm 2. Reasoning procedure for recursive rules

Input: set of triples T, rule R
Output: set of triples T

1 NewT := T;
2 **while** *NewT not empty* **do**
3 | InferT := `apply_rule`(*NewT, R*);
4 | NewT := T \ InferT;
5 | T := T ∪ NewT;
6 **return** *T*

The algorithm often generates a large number of duplicated triples in each iteration. The first type of data duplication is witnessed in the inferred triples (i.e., *InferT* set) which are produced by applying rules on the input triple sets

(Line 3). To remove such duplicated triples, we implement the sort-based approach [12] on the GPU. First, the derived triples are sorted by using the GPU Sort primitive. Duplicated triples are then pruned out by applying the compaction algorithm. In this method, we identify the valid triples which are different from their next ones. We then employ the Prefix Scan operation to calculate the output locations of these triples. Finally, we write the output triples to the *InferT* array in parallel. The second type of triple duplication is observed when the derived triples in *InferT* have already existed in *T* (Lines 4–5). Since *InferT* and *T* are two sorted triple sets, we resolve the duplication problem by modifying the GPU-based Merge operation. In this modification, we detect the new triples and maintain them in the *NewT* array. After that, we merge the remaining triples in *InferT* to the *T* set.

The approach, however, needs to request all existing triples of the related column table to identify triple duplications and new derived triples from the inferred set. Unlike in-memory systems which are able to maintain all data in the main memory, the storage capability of a typical GPU device is very limited. For the property table whose size cannot fit into the global memory, we must frequently transfer data between GPU and CPU memory during execution. This might become the bottleneck which significantly reduces the overall reasoning performance. To achieve the high performance in such cases, we implement a GPU-based Bloom Filter algorithm to resolve the problem of triple duplication.

5 Performance Evaluation

We evaluate the performance of gSparql in comparison with the state-of-the-art reasoning and query answering system based on backward-chaining inference, named Jena [8]. We perform the experiments using a set of 14 queries taken from LUBM [10] (as shown in the Appendix section). These queries involve

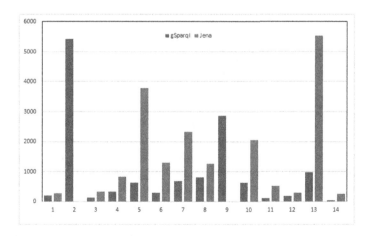

Fig. 8. Comparison with Jena on LUBM10

properties associated with the LUBM university-world ontology, with none of the custom properties/rules whose support is actually our end goal.

The runtime of the CPU-based algorithms is measured using an Intel Xeon E5-1650 v2 3.50GHz CPU with 16GB of memory. Our GPU algorithms are tested using the CUDA Toolkit 6.0 running on NVIDIA Tesla K20c GPU with 5 GB global memory and 48 KB shared memory per Stream Multiprocessor. For each of those tests, we execute 100 different queries and record the average elapsed time. Figure 8 and Table 1 compare our gSparql system with the state-of-the-art Jena using backward-chaining reasoner. We perform our experiments on two different LUBM settings. The first dataset consists of 1.3 million triples generated from 10 universities (Fig. 8). The second one has 13.5 million triples (i.e., 100 universities) (Table 1). The whole datasets are maintained in the main memory. As for the ruleset, we utilize the standard RDFS ruleset which is supported by Jena.

Table 1. Comparison with Jena on LUBM100

	gSparql (ms)	Jena (ms)
Q1	245	285
Q2	26657	>15m
Q3	178	336
Q4	1765	5454
Q5	3315	50571
Q6	679	9691
Q7	4572	21712
Q8	5830	9533
Q9	17945	>15m
Q10	2144	19735
Q11	154	1360
Q12	211	294
Q13	4899	67630
Q14	54	267

As can be seen in the Fig. 8 and Table 1, the response time of gSparql is faster than that of Jena up to several orders of magnitude. When the size of the LUBM dataset increases, the response time of the Jena system rises significantly. In contrast, the processing time of our method increases at a much slower rate. In our gSparql system, we take advantage of a large number of parallel threads, can efficiently handle operations such as joining, merging and sorting in large-scale, thus its performance remains stable. We take the Query 2 and 9 for examples, the Jena system cannot handle a large amount of data in the joining operations while our system still answers those questions efficiently. As can be seen in Query

13, by taking advantage of the fourth optimization technique, we can ignore the reasoning process on the triple pattern *(?X rdf:type ub:Person)*. As a result, gSparql can decrease the overall processing time.

6 Conclusion

In this paper, we introduce gSparql, a fast and scalable inference and querying method on mass-storage RDF data with custom rules to retrieve information on semantic knowledge bases. Our method focuses on dealing with backward-chaining reasoning, which makes inferences at query time when the inferred triples are determined by the set of triple patterns defined in the query.

To efficiently answer SPARQL queries in parallel, we first build reasoning trees for all triple patterns in the query and then execute those trees on GPUs in a bottom-up fashion. In particular, we convert the execution tree into a series of primitives such as sort, merge, prefix scan, and compaction which can be efficiently done on GPU devices. We also utilize a GPU-based Bloom Filter method and sorting algorithms to overcome the triple duplication. Extensive experimental evaluations show that our implementation scales in a linear way and outperforms current optimized CPU-based competitors.

References

1. Abadi, D.J., Marcus, A., Madden, S.R., Hollenbach, K.: Scalable semantic web data management using vertical partitioning. Proc. VLDB Endow. **1**(1), 411–422 (2007)
2. Auer, S., Bizer, C., Kobilarov, G., Lehmann, J., Cyganiak, R., Ives, Z.: DBpedia: a nucleus for a web of open data. In: Aberer, K., et al. (eds.) ASWC/ISWC -2007. LNCS, vol. 4825, pp. 722–735. Springer, Heidelberg (2007). https://doi.org/10.1007/978-3-540-76298-0_52
3. Bishop, B., Kiryakov, A., Ognyanoff, D., Peikov, I., Tashev, Z., Velkov, R.: Owlim: a family of scalable semantic repositories. Semant. Web **2**(1), 33–42 (2011)
4. Broekstra, J., Kampman, A., van Harmelen, F.: Sesame: a generic architecture for storing and querying RDF and RDF schema. In: Horrocks, I., Hendler, J. (eds.) ISWC 2002. LNCS, vol. 2342, pp. 54–68. Springer, Heidelberg (2002). https://doi.org/10.1007/3-540-48005-6_7
5. Cambria, E., Hussain, A.: Sentic Computing: A Common-Sense-Based Framework for Concept-Level Sentiment Analysis. Springer, Cham, Switzerland (2015)
6. Cambria, E., Poria, S., Bajpai, R., Schuller, B.: SenticNet 4: a semantic resource for sentiment analysis based on conceptual primitives. In: the 26th International Conference on Computational Linguistics, pp. 2666–2677 (2016)
7. Cambria, E., Wang, H., White, B.: Guest editorial: Big social data analysis. Knowl. Based Syst. **69**, 1–2 (2014)
8. Carroll, J.J., Dickinson, I., Dollin, C., Reynolds, D., Seaborne, A., Wilkinson, K.: Jena: implementing the semantic web recommendations. In: World Wide Web Conference, pp. 74–83. ACM (2004)
9. Green, O., McColl, R., Bader, D.A.: Gpu merge path: a gpu merging algorithm. In: Proceedings of the 26th ACM International Conference on Supercomputing, pp. 331–340. ACM (2012)

10. Guo, Y., Pan, Z., Heflin, J.: Lubm: a benchmark for owl knowledge base systems. Web Semant. Sci. Serv. Agents World Wide Web **3**(2), 158–182 (2005)
11. Harris, M., Sengupta, S., Owens, J.D.: Gpu gems 3, chapter parallel prefix sum (scan) with cuda (2007)
12. Heino, N., Pan, J.Z.: RDFS reasoning on massively parallel hardware. In: Cudré-Mauroux, P., et al. (eds.) ISWC 2012. LNCS, vol. 7649, pp. 133–148. Springer, Heidelberg (2012). https://doi.org/10.1007/978-3-642-35176-1_9
13. Neumann, T., Weikum, G.: Rdf-3x: a risc-style engine for rdf. Proc. VLDB Endow. **1**(1), 647–659 (2008)
14. Paul, J., He, J., He, B.: Gpl: a gpu-based pipelined query processing engine. In: Proceedings of the 2016 International Conference on Management of Data, pp. 1935–1950. ACM (2016)
15. Peters, M., Brink, C., Sachweh, S., Zündorf, A.: Scaling parallel rule-based reasoning. In: Presutti, V., d'Amato, C., Gandon, F., d'Aquin, M., Staab, S., Tordai, A. (eds.) ESWC 2014. LNCS, vol. 8465, pp. 270–285. Springer, Cham (2014). https://doi.org/10.1007/978-3-319-07443-6_19
16. Rajagopal, D., Cambria, E., Olsher, D., Kwok, D.: A graph-based approach to commonsense concept extraction and semantic similarity detection. In: WWW, pp. 565–570. Rio De Janeiro (2013)
17. Suchanek, F.M., Kasneci, G., Weikum, G.: Yago: a core of semantic knowledge. In: Proceedings of the 16th International Conference on World Wide Web, pp. 697–706. ACM (2007)
18. Tran, H.-N., Cambria, E.: GpSense: A GPU-friendly method for common-sense subgraph matching in massively parallel architectures. In: CICLing, Konya (2016)
19. Tran, H.-N., Cambria, E., Hussain, A.: Towards gpu-based common-sense reasoning: using fast subgraph matching. Cognit. Comput. **8**(6), 1074–1086 (2016)
20. Tran, Ha-Nguyen, Kim, Jung-jae, He, Bingsheng: Fast subgraph matching on large graphs using graphics processors. In: Renz, Matthias, Shahabi, Cyrus, Zhou, Xiaofang, Cheema, Muhammad Aamir (eds.) DASFAA 2015. LNCS, vol. 9049, pp. 299–315. Springer, Cham (2015). https://doi.org/10.1007/978-3-319-18120-2_18
21. Urbani, Jacopo, van Harmelen, Frank, Schlobach, Stefan, Bal, Henri: QueryPIE: backward reasoning for OWL horst over very large knowledge bases. In: Aroyo, L., et al. (eds.) ISWC 2011. LNCS, vol. 7031, pp. 730–745. Springer, Heidelberg (2011). https://doi.org/10.1007/978-3-642-25073-6_46
22. Weiss, C., Karras, P., Bernstein, A.: Hexastore: sextuple indexing for semantic web data management. Proc. VLDB Endow. **1**(1), 1008–1019 (2008)
23. Zheng, V., Cavallari, S., Cai, H., Chang, K., Cambria, E.: From node embedding to community embedding (2017). https://arxiv.org/abs/1610.09950

Efficient Association Rules Selecting for Automatic Query Expansion

Ahlem Bouziri[1,2(✉)], Chiraz Latiri[1,2], and Eric Gaussier[3]

[1] ISAMM, Manouba University, 2010 Tunis, Tunisia
ahlembou@yahoo.com
[2] LIPAH Research Laboratory, FST, Tunis EL Manar University, Tunis, Tunisia
[3] LIG Research Laboratory, Joseph Fourier University (Grenoble I),
Grenoble, France

Abstract. Query expansion approaches based on term correlation such as association rules (ARs) have proved significant improvement in the performance of the information retrieval task. However, the highly sized set of generated ARs is considered as a real hamper to select only most interesting ones for query expansion. In this respect, we propose a new learning automatic query expansion approach using ARs between terms. The main idea of our proposal is to rank candidate ARs in order to select the most relevant rules tSo be used in the query expansion process. Thus, a pairwise learning to rank ARs model is developed in order to generate relevant expansion terms. Experimental results on TREC-Robust and CLEF test collections highlight that the retrieval performance can be improved when ARs ranking method is used.

Keywords: Query expansion · Association rules · Learning to rank

1 Introduction and Motivations

In Information Retrieval (IR), Automatic Query Expansion (AQE) refers to techniques, algorithms or methodologies used to reformulate an original query by adding new terms to it, in order to achieve better retrieval results. Many query expansion techniques and algorithms were developed in the last decades. An interesting survey on AQE is given in [5].

As detailed in [5], the expanded query is the result of the four sequential steps, namely: preprocessing of data source, generation and ranking of candidate expansion features, selection of expansion features and query expansion. In this paper, we focus on techniques used to generate candidate expansion features, i.e., *terms*, and their ranking according to the type of their relationship with the original query terms. In this respect, we argue that a conjunction between machine learning methods and some advanced text mining methods, especially Association Rules (ARs) between terms extraction [1], is particularly appropriate and should outperforms classical AQE methods. Interestingly enough, ARs between terms have been explicitly used in previous works for finding expansion

© Springer Nature Switzerland AG 2018
A. Gelbukh (Ed.): CICLing 2017, LNCS 10762, pp. 563–574, 2018.
https://doi.org/10.1007/978-3-319-77116-8_42

features correlated with the query terms like [3,7,9,11,12,14]. The advantage of the insight gained through ARs is in the contextual nature of the discovered inter-term correlations. Indeed, more than a simple assessment of pair-wise term occurrences, an AR binds two sets of terms, which respectively constitute its premise (X) and conclusion (Y) parts. Thus, a rule approximates the probability of having the terms of the conclusion in a document, given that those of the premise are already there. So, ARs reflect more thoroughly terms use context in the document collection.

In this paper, we propose to revisit the AQE approach based on non-redundant ARs introduced in [9] which has a main drawback, namely: the candidate terms for expansion are selected from a large set of valid ARs with respect to their confidence value and based only to the correlation between original terms and the candidate ones. This blind query expansion approach leads to add several terms to the original query, without however being assessed as the most relevant ones. The driving idea behind our proposal in this paper is to enhance AQE based on ARs by means of a machine learning approach to rank candidate ARs for expansion, in order to select the best candidate features for expansion. Although the motivation is similar, our work differs from previous "learning to rank" based IR approaches [10]. These approaches rank documents according to their relevance with respect to the query, in that we propose to rank a set of candidate ARs for expansion which are mined from the documents collection.

The rest of the paper is organized as follow: Sect. 2 presents a brief literature review on AQE. In Sect. 3, theoretical foundation of mining ARs is given as the conducted experiments on two test collections of documents. Then, in Sect. 4, we describe our learning to rank ARs approach for query expansion. Experimental validation is detailed and discussed in Sect. 5. Finally, we conclude in Sect. 6.

2   Related Work

The principle of query expansion is to add new query terms to the initial query in order to enhance its formulation. Candidate terms for expansion are either extracted from external resources such as WordNet or from the documents themselves; based on their links with the initial query terms. In the latter types of methods, the most popular one is the pseudo-relevance feedback [4]. Our work focuses on document analysis methods. This analysis may be either (i) *global* (corpus analysis to detect word relationships) or (ii) *local* feedback (analysis of documents retrieved by the initial query) [5].

Local AQE methods use retrieved documents produced by the unmodified query. It uses usually pseudo relevance feedback [4] approach to reformulate the query. These methods use top-ranked documents retrieved by the unmodified original query. However, the top retrieved documents may not always provide good terms for expansion, particularly for difficult queries with few relevant documents in the collection that do not share the relevant terms. These methods lead to topic drift and negatively impact the results.

To overcome this drawback, Global methods work alike but in that case candidate terms come from the entire document collection rather than just (pseudo-)

relevant documents. In [16], authors proved that using global analysis techniques produces results that are both more effective and more predictable than simple local feedback. Such AQE approaches are generally based on extraction of relationships between terms among the whole document collection and based on their co-occurrences where the window size used is a document.

Among global AQE techniques, AR mining targets to retrieving correlated patterns [1]. An AR binds two sets of terms namely *a premise* and *a conclusion*. This means that the conclusion occurs whenever the premise is observed in the set of documents. To each AR, a confidence value is assigned to measure the likelihood of the association. It is proved in the literature that the use of such dependencies for query expansion could significantly increase the retrieval effectiveness [15].

Authors in [14] performed a small improvement when using the APRIORI algorithm [1] with a high confidence threshold (more than 50%) that generated a small amount ARs. Using a lower confidence threshold (10%), authors performed better results [14]. Haddad et al. [7] proposed the same approach performing improvement when using the APRIORI algorithm to extract ARs. The best improvements were performed with low confidence values. The main limitation of this approach consists in the huge number of generated ARs while a large part of them are redundant in the sense that several rules convey the same information. A more adapted mining algorithm to text that avoids redundancy is proposed in [9]. A Minimal Generic Basis ($\mathcal{MGB}$) of non-redundant ARs between terms is first derived from the document collection. This compact basis is then used to blindly expand the user query considering all terms that appear in the conclusions of the non-redundant ARs whose premise is contained by the original query. Experimental evaluation of this approach shows an improvement of the IR task. In this paper, we propose to enhance the proposed query expansion approach in [9] by learning to rank ARs.

3   Mining Association Rules Between Terms for Query Expansion

In text mining field, an *extraction context* is a triplet $\mathcal{K} = (\mathcal{D}, \mathcal{T}, \mathcal{R})$ where $\mathcal{D}$ represents a finite set of documents, $\mathcal{T}$ is a finite set of terms and $\mathcal{R}$ a binary relation (i.e., $\mathcal{R} \subseteq \mathcal{D} \times \mathcal{T}$). Each couple $(d, t) \in \mathcal{R}$ means that the document $d \in \mathcal{D}$ contains the term $t \in \mathcal{T}$.

ARs techniques start with finding out the frequent sets of terms called *termsets*[1] from the textual context. These termsets must occur more than a user-defined threshold, denoted *minsupp*. Many representations of frequent termsets were proposed in the literature [1] where terms are characterized by the frequency of their co-occurrence. The ones based on *closed termsets* and *minimal generators* [13] result from the mathematical bases of the Formal Concept Analysis (FCA) [6]. Indeed, the mining process heavily relies on the *Galois closure operator* [6].

---

[1] By analogy to the *itemset* terminology used in data mining.

The Galois closure operator splits the set of frequent termsets into *equivalence classes*. Each class contains termsets characterizing the same set of documents. These termsets share the same closure which is obtained by intersecting the associated documents. A *closed termset* represents a maximal group of terms sharing the same documents. While, often several *minimal generators* constitute the *minimal* incomparable elements within each equivalence class. Intuitively, we can say that a closed termset includes the most general terms, while a minimal generator includes one of the most specific terms describing the set of documents. Therefore, the derivation of ARs is achieved starting from the set of frequent closed termsets extracted from the textual context $\mathfrak{K} = (\mathcal{D}, \mathcal{T}, \mathcal{R})$.

Let $T \subseteq \mathcal{T}$, be a set of distinct terms and D. The support of T in $\mathfrak{K}$ is equal to the number of documents in $\mathcal{D}$ containing all the term of T. The support is formally defined as follows[2]:

$$Supp(T) = |\{d|d \in \mathcal{D} \land \forall t \in T : (d, t) \in \mathcal{R}\}| \tag{1}$$

$Supp(\text{T})$ is called the *absolute* support of T in $\mathfrak{K}$. The *relative* support (*aka* frequency) of $T \in \mathfrak{K}$ is equal to $\frac{Supp(T)}{|\mathcal{D}|}$.

A termset is said *frequent* (or *covering*) if its terms co-occur in the collection a number of times greater than or equal to a user-defined support threshold, denoted *minsupp*.

An AR R is a correlation of the form $R: T_1 \Rightarrow T_2$, where T_1 and T_2 are subsets of $\mathcal{T}$, and $T_1 \cap T_2 = \emptyset$. The termsets T_1 and T_2 are, respectively, called the *premise* and the *conclusion* of R. The rule R is said to be based on the termset T equal to $T_1 \cup T_2$. The *support* of a rule $R: T_1 \Rightarrow T_2$ is then defined as:

$$Supp(R) = Supp(T) \tag{2}$$

while its *confidence* is computed as:

$$Conf(R) = \frac{Supp(T)}{Supp(T_1)}. \tag{3}$$

An association R is said to be *valid* if its confidence value, *i.e.*, $Conf(R)$, is greater than or equal to a user-defined threshold denoted *minconf*. This confidence threshold is used to exclude non valid rules. Also, the given support threshold *minsupp* is used to remove rules based on termsets T that do not occur often enough, *i.e.*, rules having $Supp(T) < minsupp$.

3.1   Conducted Experiments for Mining Association Rules

In order to generate efficient non-redundant ARs between terms for learning and ranking candidate terms for AQE, we propose to use the Minimal Generic Basis $\mathcal{MGB}$ of ARs presented in [9] where authors proved that $\mathcal{MGB}$ generic basis was suitable for query expansion based on ARs.

---

[2] In this paper, we denote by $|X|$ the cardinality of the set X.

Document Collections Description. We use two documents collections described in Table 1. In order to extract the most representative terms, a linguistic preprocessing is performed on the corpora by using a part-of-speech tagger TREETAGGER.[3] In this work, we focus only on terms related to three grammatical categories, namely: common nouns, proper nouns and adjectives, since they are the most informative grammatical categories and are most likely to represent the content of documents [2]. A stoplist is used to discard functional French and English terms that are very common.

The minimal threshold of confidence is set to 20% and 50% respectively for CLEF2003 collections and Robust TREC 2004 collections. These values of $minconf$ threshold are determined empirically so as to obtain a number ARs allowing to apply our learning to rank model. We also varied the minimum threshold of the support, $i.e.$, $minsupp$ $w.r.t.$ the document collection size and to term distributions. While considering the $Zipf$ distribution of every collection, the minimum threshold of the support $minsupp$ values is experimentally set in order to eliminate marginal terms which occur in few documents, and are then not statistically important when occurring in a rule.

Table 1. Statistical analysis on generic basis of ARs $\mathcal{MGB}$ derived from CLEF 2003 french collections and TREC 2004 Robust collections

Collection	#docs	Minsupp	Minconf	#Closed termsets	#Minimal generators	#$\mathcal{MGB}$
CLEF 2003 French collections (Nbre of queries = 60)						
Le Monde 94	44,013	700	0.2	446870	446870	1 404 933
SDA 94	43,178	200	0.2	328969	332143	591564
SDA 95	42,615	200	0.2	352759	357368	627518
TREC 2004 Robust track (Nbre of queries = 250)						
FBIS	130,471	1700	0.5	463323	463323	770 359
Federal Register 94	55,63	1500	0.5	186703	190332	211 759
LA Times	131,896	2000	0.5	538323	538323	514 039
Financial Times	210,158	2500	0.5	437099	437099	379 248

Experiments and Results. Table 1 summarizes the number of mined ARs between terms as well as the number of closed termsets and minimal generators for the different $minsupp$ threshold values, respectively to the considered collections. We observe that for "Le monde 94 collection", an important number of ARs between terms were discovered for an absolute minimal support equal to 700 documents. This result is not in contradiction with "Le Monde 94" terms distribution, where the support of terms is important according to the zipfien distribution, between 200 and 700 documents. Moreover, the collections

---

[3] http://www.cis.uni-muenchen.de/~schmid/tools/TreeTagger/.

"Le Monde94", "FBIS", "LA Times" and "Financial Times" behave as "a worst case" context, *w.r.t.* the Galois closure operator, where each closed termset is exactly equal to its minimal generator. This arises for almost all tested *minsupp* values, and even for very low ones as depicted in Table 1. As a consequence and contrary to "SDA94", "SDA95" and "Federal Register 94" document collections, a very few number of exact ARs (i.e., with confidence equal to 1) with is generated starting from these collections since each frequent minimal generator is itself a closed termset.

4 Learning to Rank Association Rules for Query Expansion

4.1 General Layout of the Proposed Model

Efficiency of query expansion is highly dependent on the terms selected to expand the query. In this work, expansion terms are extracted from ARs conclusions. It is then required to proceed to a selection of the RAs to be used in the expansion process to find out the more efficient ones to improve quality of information retrieval. We propose in this paper, an approach that learns to rank ARs according to their efficiency to query expansion. Figure 1 presents the main components and steps of our AQE approach, as follows:

1. Mining ARs as mentioned in Sect. 3.
2. Building a training data set of ranked queries according to results of an algorithm that explores RAs in order to optimize MAP of expanded queries.
3. Producing a ranking Model based on the data set and using a pair-wise learning to rank algorithm.
4. Predicting the ranking of the corresponding ARs of a new query to expand then proceed to expansion using the top k ARs.

4.2 Paire-Wise Learning to Rank Approach

Many learning-to-rank algorithms are proposed in the litterature [10]. In particular, we are interested in the pairwise approach.[4] In IR field, this latter does not focus on accurately predicting the relevance degree of each document; instead, it cares about the relative order between two documents. With a pairwise approach, the ranking problem of ARs for AQE is reduced to a classification problem on ARs pairs. That is, the goal of learning is to minimize the number of missclassified ARs pairs (*i.e.*, make positive predictions on those pairs whose first rule is more relevant than the second rules for the expansion, and make negative predictions on other pairs). In the extreme case, if all the AR pairs are correctly classified, all the rules will be correctly ranked [10].

---

[4] Also referred to as preference learning in the literature.

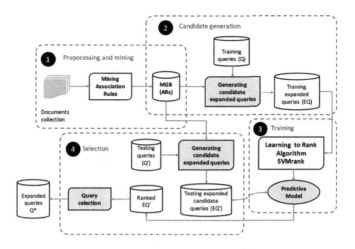

Fig. 1. General layout of learning to rank RAs for query expansion.

4.3   Association Rules Ranking Model

In this section, we introduce a global formalization of learning to rank a set of expansion ARs according to their performance in expanding a given query. We assume that, given a set of queries $q = \{q_1, \ldots, q_m\}$, each query q_i is associated with a set of candidate ARs to expansion $\mathcal{R}_i = \{R_{i,1}, \ldots, R_{i,n_i}\}$, where $R_{i,j}$ denotes the jth AR having all its premise terms in i^{th} query , thus $prem(R_{i,j}) \subseteq q_i$ and n_i denotes the size of $\mathcal{R}_i$. Subsequently, each q_i is associated with a set of possible expanded queries $E_i = \{eq_{i,1}, \ldots, eq_i, n_i\}$ where $eq_{i,j} = q_i \cup conclusion(R_{i,j}.$

Each set $\mathcal{R}_i$ of ARs is associated with a set of relevance judgement $y_i = \{y_{i,1}, \ldots, y_{i,n_i}\}$, where y_{ij} denotes the extent to which the candidate AR $R_{i,j}$ is relevant to query q_i. This degree of relevance is evaluated using the performance improvement on MAP measur as follows:

$$y_{ij} = \frac{MAP(eq_{ij}) - MAP(q_i)}{MAP(q_i)} \tag{4}$$

The assumption is that the higher performance improvement is observed for combination of R_{ij} and q_i, the stronger relevance exists between them.

The training is denoted as $\Gamma = \{(x_i, y_i)\}$ $i = 1, \ldots, m$ where $x_i = \{x_{i,1}, \ldots, x_{i,n_i}\}$ is the list of feature vectors and $y_i = \{y_{i,1}, \ldots, y_{i,n_i}\}$ is the list of the corresponding scores. The feature vector $x_{i,j}$ is created from each query-AR pair $(q_i, R_{i,j}), i = 1, \ldots, m$ and $j = 1, \ldots, n_i$. For each feature vector $x_{i,j}$, the ranking function f outputs a score $f(x_{i,j})$. We will obtain a list of scores $z_i = (f(x_{i,1}), , \ldots, f(x_{i,n_i}))$ for the list of vectors x_i. Details about features we adopt are given in Sect. 4.4.

The objective of learning is formalized as minimization of the total losses with respect to the training data, given by:

$$\sum_{i=1}^{m} L(y_i, z_i) \tag{5}$$

where L is a loss function.

Given a new query q' and the set of its ARs R', we construct feature vectors x' and use the trained ranking function to assign scores to the ARs in R'. We then select the top k ARs and expand the original query q using conclusion terms of the selected ARs.

4.4 Features for Association Rules Ranking

Feature vectors are computed based on statistical distribution measures of terms in the query's text and in AR conclusion. These features are of three categories and are explained in the following paragraphs.

Document Frequency Based Features. The document frequency (DF) is a statistical predictor that measures whether a term is rare or common in the corpus. Its value for a query represents the average of the DF for all query terms. The $DF(q)$ of a query q is given by:

$$DF(q) = \frac{1}{|q|} \sum_{i=1}^{|q|} \frac{|\{d_j, t_i \in d_j\}|}{|D|} \tag{6}$$

We compute also the inverted document frequency (iDF) for a query as follows:

$$iDF(q) = \frac{1}{|q|} \sum_{i=1}^{|q|} \log \frac{|D|}{|\{d_j, t_i \in d_j\}|} \tag{7}$$

Both DF and iDF are also computed for an AR, they represent the average of the DF or iDF for all the terms in the conclusion of the AR.

Term Frequency Based Features. We include features dealing with term frequency in the documents of the collection. The term frequency ($TF(t,d)$) is simply the number of occurrences of term t in document d. For each term of the query or of the AR conclusion we use an average of its frequencies in all documents and we note it $ATF(t_i)$ for term t_i. For a query, the term frequency is considered to be the average of the ATF of al the terms in the query and is calculated as in Eq. 8.

$$TF(q) = \frac{1}{|q|} \sum_{i=1}^{|q|} ATF(t_i) \tag{8}$$

$$ATF(t_i) = \frac{1}{|D|} \sum_{j=1}^{|D|} TF(t, d_j)$$

When dealing with an average, it is important to know how the values are distributed around it. We introduce a variance measure to evaluate the variation of term frequencies. For a query we compute the average of theses variations for all query terms as given in Eq. 9

$$VTF(q) = \frac{1}{|q|}\sum_{i=1}^{|q|}\frac{1}{|D|-1}\sum_{j=1}^{|D|}(TF(t_i, d_j) - ATF(t_i, d_j))^2 \tag{9}$$

Association Rule Based Features. In addition to the features based on term's distribution measures, we use 6 features to characterise an AR:

- the number of terms in the AR premise,
- the number of terms in the AR conclusion,
- the AR confidence,
- the AR support,
- a matching factor representing the number of terms in the AR premise by the number of terms in the query.
- a relevance factor computed as the proportion of terms in the AR conclusion that are present in the other ARs.
- the rank of the AR with respect to the query

5 Experimental Validation

Experiments are conducted on two test document collections described in Table 1. We used title, description and narration fields of the topics for query creation. For the task of pairwise learning to rank, we use a Ranking SVM based approach, i.e., SVM^{Rank} algorithm [8]. The parameter k of the number of the top ranked AR to consider for expansion is chosen while optimising LTR parameters using a 5-fold nested cross validation. We vary k values in the interval [1..20] with a step of 1. The optimal value of k depends on the tested collection and it is set to 56 for CLEF collection to 23 for Robust collection. We compared the effectiveness of our query expansion methods, denoted Blind-MGB-QE (with all $\mathcal{MGB}$ rules) and LTR-MGB-QE (with leraning to rank AR model) (*cf.* Table 2), to two baselines: (1) the Okapi probabilistic model (BM25) which was parametrized as recommended in the literature: $k_1 = 1.2$, $k_3 = 7$ and $b = 0.75$; and (2) the Pseudo-Relevance-Feedback (PRF). For the purpose of evaluating and comparing retrieval effectiveness, experiments are conducted under version 4.0 of the Terrier search engine,[5] using the Okapi probabilistic retrieval model (BM25) and considering as performance measures The Mean Average Precision (MAP) measure and the exact precision @5, @10 documents and the NDCG@5.

In order to highlight the interest and the efficiency of ranking ARs for AQE, we first propose to perform a $\mathcal{MGB}$-based blind query expansion. It means that we consider the whole set of rules in the generic basis $\mathcal{MGB}$ without any selection or ranking on these rules (Blind-MGB-QE in Table 2). Hence, the proposed blind

---

[5] http://www.terrier.org.

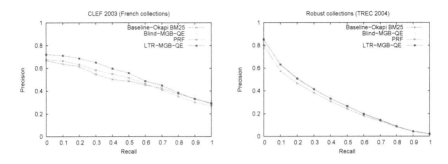

Fig. 2. MAP improvement for the tested document collections of CLEF 2003 (*left*) and Robust collections (*right*) under the OKAPI BM25 model.

query expansion process based on ARs between terms consists of expanding each query by all terms that appear in the conclusions of the non-redundant ARs in $\mathcal{MGB}$ whose premise is contained by the original query. Each term of the query is handled individually. Given an original query q, the basic idea of our blind query expansion process for obtaining the associated expanded query eq is defined as follows:

$$\forall R: T_1 \Rightarrow T_2, \text{an AR} \in \mathcal{MGB}: \; if \; T_1 \subseteq q, \text{then } eq := q \cup T_2. \qquad (10)$$

Table 2 highlights significant and important improvements for all the evaluation measures over the baseline and the Blind-MGB-QE model, except for the PRF model concerning CLEF 2003 collection where the improvement is low and not significant. We can note that the expansion model using learning to rank ARs of the $\mathcal{MGB}$ generic basis (*i.e.*, LTR-MGB-QE) leads to an increase in the exact precision at low recall (*i.e.*, P@5 and P@10 documents) for all collections. This means an increase in the number of retrieved relevant documents put in the head of the top ranked documents list. Moreover, as depicted in Table 2, the performance of the proposed learning to rank AR based AQE model was assessed by evaluating the NDCG@5 measure which is often used to evaluate search engine algorithms and other techniques whose goal is to order a subset of items in such a way that highly relevant documents are placed on top of the list, while less important ones are moved further down. We achieve an interesting improvement of the NDCG@5 with the LTR-MGB-QE model for both test collections (+5.63 for CLEF collection test and +5.13 for TREC-Robust). It is worth noting that higher values of NDCG mean that the system output gets closer to the ideally ranked output.

From Fig. 2, we notice that the curves associated to the CLEF2003 French test collection show a better improvement compared to the Robust test collection. This can be explained by the fact that a part of the Robust vocabulary is not used, since the considered values of *minsupp* are more greater than those considered for CLEF2003 test collection, owing to the high size of Robust corpora. Moreover, this Fig. 2 sheds light that our LTR-MGB-QE model gives an improvement greater than the PRF model. Meanwhile, considering MAP measure, we can observe that the LTR-MGB-QE significantly overpasses the standard BM25 model with a gain equal to 10.41% and 08.28% respectively for CLEF 2003 French test collection and TREC 2004 Robust test collection.

Table 2. Comparative evaluation of retrieval effectiveness. *% Chg.* indicates the model improvements in terms of MAP, P@*n* and NDCG@5. The symbols †, ‡ and †‡ denote the student *t*-test significance: † : $0.01 < t \leq 0.05$; ‡ : $0.001 < t \leq 0.01$; †‡ : $t \leq 0.001$.l

Evaluation	MAP	% Chg.	P@5	% Chg.	P@10	% Chg.	NDCG@5	% Chg.
CLEF 2003 French collections								
Baseline	47.26	–	43.00	–	33,83	–	63.86	–
PRF	48.64	+2.92	44.33	+3.09	36.50	+7.89	64.67	+1.26
Blind-MGB-QE	48.11	+1.79†‡	47.33	+10.06	37.67	+11.35	63,36	−0.78
LTR-MGB-QE	**52.17**	**+10.38‡**	**49.67**	**+15.51**	**38.50**	**+13.8**	**67.46**	**+5.63**
TREC 2004 Robust track								
Baseline	27.28	–	55.82	–	47.91		55.86	–
PRF	29.45	+7.95†‡	56.40	+1.03	50.28	+4.94	58.06	+3.93
Blind-MGB-QE	29.55	+8.32†‡	60.56	+8.49	51.77	+8.05	58.20	+4.18
LTR-MGB-QE	**29.54**	**++8.28†‡**	**61.04**	**+9.35**	**51.77**	**+8.05**	58.73	**++5.13**

6 Conclusion

This paper presents a new automatic query expansion approach based on ARs between terms. A blind use of ARs shows that these latter are an interesting source of information for query expansion. The main idea of our proposal is to rank candidate ARs in order to select the more efficient ones to be used in the query expansion process. Experiments are conducted on the French test collection of CLEF 2003 campaign and Robust track test collection of TREC 2004 campaign. Results highlight significant improvements for all the evaluation measures over the baseline BM25 and the Blind-expansion-$\mathcal{MGB}$.

References

1. Agrawal, R., Skirant, R.: Fast algorithms for mining association rules. In: Proceedings of the 20th International Conference on Very Large Databases, VLDB 1994. Santiago, Chile, pp. 478–499, September 1994
2. Barker, K., Cornacchia, N.: Using noun phrase heads to extract document keyphrases. In: Hamilton, H.J. (ed.) AI 2000. LNCS (LNAI), vol. 1822, pp. 40–52. Springer, Heidelberg (2000). https://doi.org/10.1007/3-540-45486-1_4
3. Belalem, G., Abbache, A., Belkredim, F.Z., Meziane, F.: Arabic query expansion using wordnet and association rules. Int. J. Intell. Inf. Technol. **12**(3), 51–64 (2016)

4. Buckley, C., Salton, G., Allan, J., Singhal, A.: Automatic query expansion using SMART: TREC-3. In: Proceedings of the 3rd Text REtrieval Conference, pp. 69–80 (1995)

5. Carpineto, C., Romano, G.: A survey of automatic query expansion in information retrieval. ACM Comput. Surv. (CSUR) **44**(1), 1 (2012)

6. Ganter, B., Wille, R.: Formal Concept Analysis. Springer, Heidelberg (1999)

7. Haddad, H., Chevallet, J.P., Bruandet, M.F.: Relations between terms discovered by association rules. In: Proceedings of the Workshop on Machine Learning and Textual Information Access in Conjunction with PKDD 2000, Lyon, France, September 2000

8. Joachims, T.: Training linear svms in linear time. In: Proceedings of the 12th ACM SIGKDD International Conference on Knowledge Discovery and Data Mining, KDD '06, pp. 217–226. ACM, New York, NY, USA (2006)

9. Latiri, C., Haddad, H., Hamrouni, T.: Towards an effective automatic query expansion process using an association rule mining approach. J. Intell. Inf. Syst. **39**(1), 209–247 (2012)

10. Li, H.: Learning to rank for information retrieval and natural language processing. Synth. Lectures Hum. Lang. Technol. **7**(3), 1–121 (2014)

11. Lin, H.C., Wang, L.H., Chen, S.M.: Query expansion for document retrieval by mining additional query terms. Inf. Manag. Sci. **19**(1), 17–30 (2008)

12. Liu, C., Qi, R., Liu, Q.: Query expansion terms based on positive and negative association rules. In: 2013 IEEE Third International Conference on Information Science and Technology (ICIST), pp. 802–808, March 2013

13. Pasquier, N., Bastide, Y., Taouil, R., Stumme, G., Lakhal, L.: Generating a condensed representation for association rules. J. Intell. Inf. Syst. **24**(1), 25–60 (2005)

14. Tangpong, A., Rungsawang, A.: Applying association rules discovery in query expansion process. In: Proceedings of the 4th World Multi-conference on Systemics, Cybernetics and Informatics, SCI 2000, Orlando, Florida, USA, July 2000

15. Wei, J., Bressan, S., Ooi, B.C.: Mining term association rules for automatic global query expansion: methodology and preliminary results. In: Proceedings of the First International Conference on Web Information Systems Engineering (WISE 2000) (2000)

16. Xu, J., Croft, W.B.: Query expansion using local and global document analysis. In: Proceedings of the 19th Annual International ACM SIGIR Conference, pp. 4–11. ACM Press, Zurich, Switzerland, August 1996

Text-to-Concept: A Semantic Indexing Framework for Arabic News Videos

Sadek Mansouri[1($\boxtimes$)], Chahira Lhioui[1], Mbarek Charhad[2],
and Mounir Zrigui[1]

[1] LATICE Laboratry, Tunis, Tunisia
mansouri_sadek@hotmail.fr, chahira_ml983@yahoo.fr,
mounir.zrigui@fsm.rnu.tn
[2] Taibah University, Madina, Saudi Arabia
mbarek.charhad@gmail.com

Abstract. In these last years, many works have been published in the video indexing and retrieval field. However, few are the methods that have been designed to Arabic video. This paper's aim is to achieve a new approach for Arabic news video indexing based on embedded text as the information source and Knowledge extraction techniques to provide a conceptual description of video content. Firstly, we applied a low level processing in order to detect and recognize the video texts. Then, we extract the conceptual information including name of person, Organization and location using local grammars that have been implemented with the linguistic platform NooJ. Our proposed approach was tested on a large collection of Arabic TV news and experimental results were satisfactory.

Keywords: Arabic text detection · Concept extraction · NooJ
Arabic news indexing

1 Introduction

Due to the increasing number of TV channels and the technological advances in the field of computer science, the amount of news video is growing rapidly especially on the World Wide Web (WWW). This diversity as well as the amount of these collections make access to useful information a complex task. Hence, there is a huge demand for efficient tools that enable users to find the required information in large news videos archives. To this end, efficient efficient handling of the video data relies on data analysis and the extraction of semantics that help users searching video sequences. The first proposed approaches for video annotation are key-word based. This can work for small databases, but when it deals with important ones it will be difficult and time-consuming. The manual annotation is not only daunting in terms of time, but also for its subjectivity. Indeed, for a person to annotate a video, he must show the tendency to use keywords that reflect what he understands when viewing this video. The same video may be annotated differently by another person. All these problems have pushed researchers to focus more on content based video retrieval. In fact, Video news is rich in semantics and information but the problem is how to extract from this signal

© Springer Nature Switzerland AG 2018
A. Gelbukh (Ed.): CICLing 2017, LNCS 10762, pp. 575–584, 2018.
https://doi.org/10.1007/978-3-319-77116-8_43

semantics information. Indeed, when analyzing a video sequence we have at our disposal pixels and audio signals and the challenge then is how we can recognize real events and concepts from these signals.

Therefore, automatic video semantic annotation has become a hot research area in recent years and is an important task in TRECVID benchmark[1]. Various approaches have been proposed for concept detection and semantic annotation of news video [1, 2]. The basic idea of these approaches is to use low-level features and rule-based or statistical learning algorithms to detect visual concepts in video content. However, these methods face a challenge called "semantic gap". This semantic gap represents the distance between low level video features and high level visual concepts. To overcome this problem, many approaches have been proposed for semantic news videos indexing using external knowledge.

One of the external knowledge is text embedded in video frames. In fact, this type of text is artificially added to the video at the time of editing providing high-level information of video content that seems to be a utile clue in the multimedia indexing system such as name of the speaker, headlines summarizing the reports in news video, event place and the score or name of the player. Nevertheless, semantic video annotation based on text information will encounter two difficulties:

– The first problem is how we can detect and recognize text information from video frame taking into account the variety of text features (size, color and style), presence of complex background and conditions of video acquisition.
– The second problem concerns conceptual knowledge extraction such as name of person and specific event from text information in order to provide high-level annotation and help user to search and comment the video content.

To treat these various problems, we propose in this article an approach of Arabic news video indexing using the semantic content. Our approach is based on a conceptual description of the contents by using list of concepts (person, localities, Organization). First of all, we propose an original approach for Arabic-text detection and localization from news videos. The main idea of this approach consists of detecting candidate text regions and then validates the effective ones by a filtering process based on specific features of Arabic text. Then, we suggest a symbolic approach for concepts extraction using local grammars which have been implemented with the linguistic platform NooJ.

The rest of the paper is organized as follows: In Sect. 2, we discuss works related Knowledge Extraction for semantic news video indexing. Section 3 introduces our proposed approach and its different stages. Section 4 exposes experiments results and Sect. 5 states the conclusion.

---

[1] http://trecvid.nist.gov/.

2   Related Works

Knowledge Extraction is considered as an important preprocessing step for many tasks such as document classification or clustering, machine translation (MT), information retrieval (IR), automatic language comprehension and other text processing applications.

In this section, we present a literature review of the different Knowledge extraction techniques of semantic news video indexing. Among these, in [3], the authors propose a multilingual information extraction (IE) system for annotating sports videos in English, German, and Dutch using ASR (Automatic speech recognition) tools. The IE components of this system include tools for tokenizing, part-of-speech tagging, knowledge extraction, and co-reference resolution.

In [4], the system's aim is to perform automatic knowledge extraction from Italian TV news. This system also utilizes an ASR tool to obtain the video texts and IE techniques (named entities recognition).

Another semantic video annotation application called Rich News has been described in [5], where the authors make use of the resources on the web to enhance the indexing process. The overall system contains the following modules: automatic speech recognition, key-phrase extraction from the speech transcripts and searching the video using key phrases. Moreover, the proposed system allows also manual annotation to ameliorate segmentation results.

In [6] a system has been implemented to annotate Turkish news video using video text as a source of information and IE techniques including named entity recognition, person entity extraction, co-reference resolution, and semantic event interpretation.

For better knowledge, our work presents the first attempt for semantic Arabic news video indexing based on text analysis and information extraction (IE) techniques that subsume low and conceptual features of video content.

3   Proposed System

In this part, we present an overview of our semantic video indexing system. Figure 1 introduces the main system components which are based on two levels of analysis. The first level puts a focus on low-level processing such as video segmentation, text detection and recognition. The second level seeks for extracting the semantic concepts including named entity such as name of person, organization, and location. The extracted semantic information is used to build a data structure to annotate the news video and facilitates the search using metadata.

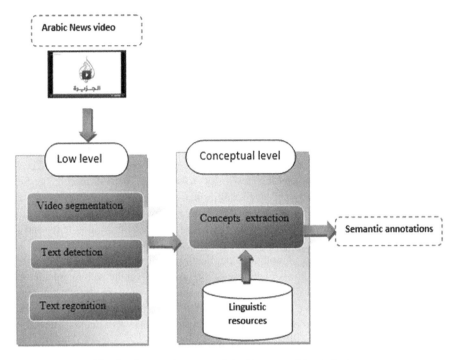

Fig. 1. Proposed system of Arabic news video indexing.

3.1 Low Level

Video segmentation

Techniques at this level tend to model the apparent characteristics inside video (color, texture, shape) [7]. The video data is continuous and unstructured. To analyze and understand its contents, the video needs to be parsed into segments. Most existing video database systems start with temporal segmentation of video into a hierarchical model of frames, shots and scenes. Temporal segmentation is then followed by the representation and modeling of contents inside each shot using key-frames and objects. From a visual point of view, the low-level description consists of a series of shots or a single shot that takes place in a single location and deals with a single action. Transitions or boundaries [8] between shots can be abrupt (Cut) or they can be gradual. Most of the existing techniques reported in literature detect shot boundary by extracting some form of feature for each frame in the video sequence, then evaluating a similarity measure on features extracted from successive pairs of frames in the video sequence. Eventually, a shot boundary is declared if the difference exceeds a fixed global threshold.

In this work, we have applied a temporal segmentation based on the following assumption "the text in the image requires at least 2 s to be readable by the user", to generate shots. Then for each video shot, the middle image will be selected as a key-frame.

Text detection and localization

After video segmentation into shots, text information are detected and extracted from each key-frame. Our text detection method relies on two necessary steps: text detection and text validation. The first step detects connected components (CC) using a hybrid method which combines MSER and edge information. These CC are then grouped by mathematical morphology operators to reassemble candidate text regions. The second stage aims to remove non-text region using geometric constraints and specific signature of Arabic script called baseline (Fig. 2).

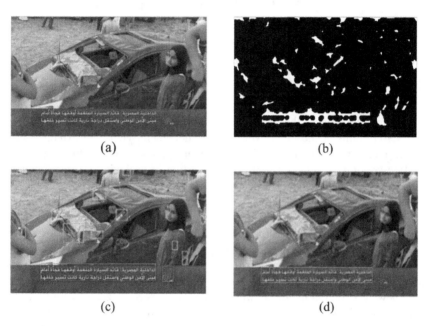

(a)   (b)

(c)   (d)

Fig. 2. Step of text detection: (a) original image, (b) CC extraction, (c) candidates text regions and (d) final detection.

Text recognition

After text detection in video frame, the next step target is to segment and binarize text region in order to separate it from the rest of the frame using Otsu's global thresholding method as described in [9]. An optimal threshold is calculated on the basis of the grey level histogram by assuming Gaussian distributions of text pixels and non-text pixels. The method aims to maximize the interclass variance. Figures 3 and 4 illustrate the results attained.

قصف تركي بري وجوي علــى مواقــع حزب العمال
الكردستاني وتنظيم الدولة شمالي العراق وسوريـــا

Fig. 3. Text region.

قصف تركي بري وجوي علـــى مواقـــع حزب العمال
الكردستاني وتنظيم الدولة شمالي العراق وسوريـــا

Fig. 4. Result of binarization step.

In the last stage, commercial OCR engine ABBYY FineReader[2] has been applied for the recognition of the video news text.

3.2 Conceptual Level

In this section, we propose the implementation of an approach that makes it possible to specify descriptions at the high-level of video document. Here, we exploit concepts notion to represent symbolic data elements. Our approach intends to represent conceptual description of video content. It is a question of designing a generic representation. This representation is independent of video contents. However, it allows to incorporate the various data elements and to give more information about semantic content of the video text in indexing process. To this end, we propose a symbolic approach for concepts extraction (person, location and organization) from Arabic news video using linguistic resources. Firstly, we parse transcriptions files to detect symbolic information by comparing their items to elements in the three concept classes (person identity, the name of a city and organization). This procedure is based on the projection of each news text on the list of keywords called gazetteers. Gazetteers are of a varied nature: lists of first names for the recognition of persons names, cities names for the detection of location, etc. Each list is associated with a semantic label which shall be the type of annotation.

Due to Arabic language complexity and specific characteristics, we also exploit a set of Lexical triggers to extract the name of person, location and organization not covered by the gazetteer resources as shown in Table 1. We have used the three kinds of lexical triggers to detect the name of person.

Lexical trigger with right context:

US Vice President Joe Biden

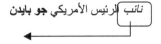

نائب الرئيس الأمريكي **جو بايدن**

---

2 https://www.abbyycom/fr-fr/finereader/.

Name of person: جو بايدن

Lexical trigger with left context:

Ban Ki-moon, Secretary General of the United Nations

Name of person: بان غي مون
Lexical trigger with both left and right context:
Dr. Moncef Marzouki, President of the Conference Party

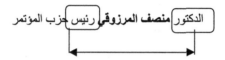

Name of person: منصف المرزوقي

The Fig. 5 illustrates an implementation grammar in NooJ linguistic platform. This grammar summarizes all possible cases of persons names use.

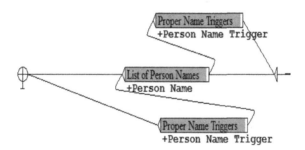

Fig. 5. Person names linguistic grammar implementation.

To extract the concept of location and organization, we have only used the lexical trigger with right context:

Exemple 1: Regime forces renewed shelling of Qaboun neighborhood in Damascus

Location: ريف دمشق, حي القابون

Exemple 2: Tunisian Parliament approves budget

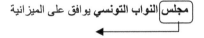

Organization: مجلس النواب التونسي

The Fig. 6 reveals the grammar used for organization recognition.

Fig. 6. Organization linguistic grammar implementation.

Table 1. List of some lexical triggers

Lexical triggers of person	Lexical triggers of location	Lexical triggers of organization
السيد،،رئيس ،العقيد ،وزير ،قائد ،نائب،طالب،مستشار ،أستاذ،	قرب،في،منطقة،مدينة ،قرية،ريف،أحياء،حي،بلدة ،قطاع	كلية،جامعة،منظمة،حزب، مؤسسة، جمعية، وزارة ،تنظيم شركة

The extracted conceptual information (name of person, location, and organization) is used to build a data structure to annotate the video text and facilitate the searching using metadata. The original descriptions are attached to the news video as xml file.

4 Experimentation and Result Evaluation

4.1 Corpus

In order to evaluate the performance of our proposed system in terms of robustness and effectiveness, we used a set of 20 video news (10,000 images) that have been collected from different Arabic TV channels: Aljazeera, Alarabiya, Wataniya 1, Elmayadeen, RT-arabe over the period of September 15, 2016 until the 5th of December, 2016 and they have a total duration of about 2 h. The videos have been automatically transcribed leading to a transcription text of 9704 words. Besides, the concepts extraction phase is done with NooJ using Gazetteers and lexical triggers as linguistic resources as shown in Tables 2.

Table 2. Linguistic resources.

Concept	Gazetteer	Lexical triggers
Person	3000	200
Location	2500	200
Organization	1000	100

4.2 Experiment and Results

The evaluation results of the system on these data sets are presented in Table 3 in terms of precision, recall, and f-measure.

Table 3. Annotation results

Concepts	Precision	Recall	F-measure
Person	83.02%	79.56%	81.25%
Location	80.23%	77.62%	78.90%
Organization	82.5%	80.35%	81.41%
Overall			**80.52%**

From Table 3 above, the results may be satisfactory achieving 80.52% as overall of F-measure. The main reason for these results is the use of local grammars, which permit the detection of semantic concepts despite the small size of Gazetteer.

5 Conclusion

Video texts tend to be useful information sources for semantic indexing with the noticeable increase of large scale videos news collection. NLP approaches such as knowledge extraction have been widely emphasized to improve multimedia indexing by providing high level information such as semantic concepts. In this paper, we have introduced a conceptual approach for Arabic videos news based on text analysis process and concepts extraction techniques. The experimentation and the evaluation results are promising.

In future work, we will try to improve our concept extraction tool by implementing other local grammars that cover all structure of Arabic text. In addition, our focus will also be on developing a retrieval interface that allows user to search the desired video.

References

1. Ye, Y., Rong, X., Yang, X., Tian, Y.: Region trajectories for video semantic concept detection. In: ICMR, pp. 255–259 (2016)
2. Markatopoulou, F., Mezaris, V., Patras, I.: Online multi-task learning for semantic concept detection in video. In: 2016 IEEE International Conference on Image Processing (ICIP), Phoenix, AZ, pp. 186–190 (2016)
3. Saggion, H., Cunningham, H., Bontcheva, K., Maynard, D., Hamza, O., Wilks, Y.: Multimedia indexing through multi-source and multi-language information extraction: the MUMIS project. Data Knowl. Eng. **48**, 247–264 (2004)
4. Basili, R., Cammisa, M., Donati, E.: RitroveRAI: a web application for semantic indexing and hyperlinking of multimedia news. In: Proceedings of the International Semantic Web Conference (ISWC) (2005)
5. Dowman, M., Tablan, V., Cunningham, H., Popov, B.: Web-assisted annotation, semantic indexing and search of television and radio news. In: Proceedings of the International Conference on World Wide Web (WWW) (2005)
6. Küçük, D., Yazıcı, A.: A semi-automatic text-based semantic video annotation system for Turkish facilitating multilingual retrieval. Expert Syst. Appl. **40**(9), 3398–3411 (2013)
7. Lezama, J., Alahari, K., Sivic, J., Laptev, I.: Track to the future: spatiotemporal video segmentation with long-range motion cues. In: Computer Vision and Pattern Recognition (CVPR), pp. 3369–3376, 20–25 June 2011
8. Li, H., Ngan, K.N.: Saliency model-based face segmentation and tracking in head-and-shoulder video sequences. J. Vis. Commun. Image Represent. **19**, 320–333 (2008)
9. Otsu, N.: A threshold selection method from gray-level histograms. IEEE Trans. Syst. Man Cybern. **9**(1), 62–66 (1979)

Approximating Multi-class Text Classification Via Automatic Generation of Training Examples

Filippo Geraci[1(✉)] and Tiziano Papini[2]

[1] Istituto di Informatica e Telematica, CNR, Via G. Moruzzi, 1, 56124 Pisa, Italy
`filippo.geraci@iit.cnr.it`
[2] Quest-it.com, Via Firenze 33, 53048 Sinalunga, Italy

Abstract. Text classification is among the most broadly used machine learning tools in computational linguistic. Web information retrieval is one of the most important sectors that took advantage from this technique. Applications range from page classification, used by search engines, to URL classification used for focus crawling and on-line time-sensitive applications [2]. Due to the pressing need for the highest possible accuracy, a supervised learning approach is always preferred when an adequately large set of training examples is available. Nonetheless, since building such an accurate and representative training set often becomes impractical when the number of classes increases over a few units, alternative unsupervised or semi-supervised approaches have come out. The use of standard web directories as a source of examples can be prone to undesired effects due, for example, to the presence of maliciously mis-classified web pages. In addition, this option is subjected to the existence of all the desired classes in the directory hierarchy.

Taking as input a textual description of each class and a set of URLs, in this paper we propose a new framework to automatically build a representative training set able to reasonably approximate the classification accuracy obtained by means of a manually-curated training set. Our approach leverages on the observation that a not negligible fraction of website names is the result of the juxtaposition of few keywords. Yet, the entire URL can often be converted into a meaningful text snippet. When this happens, we can label the URL by measuring its degree of similarity with each class description. The text contained in the pages corresponding to labelled URLs can be used as a training set for any subsequent classification task (not necessarily on the web). Experiments on a set of 20 thousand web pages belonging to 9 categories have shown that our auto-labelling framework is able to attain an approximation factor over 88% of the accuracy of a pure supervised classification trained with manually-curated examples.

© Springer Nature Switzerland AG 2018
A. Gelbukh (Ed.): CICLing 2017, LNCS 10762, pp. 585–601, 2018.
https://doi.org/10.1007/978-3-319-77116-8_44

1 Introduction

Text classification is the activity of assigning a document to a category. This machine learning tool has become a key component for a plethora of applications in computational linguistic and information retrieval. In the context of the web, in particular, the range of applications that benefit from classification is incredibly wide. Search engines exploit the membership class of web pages as a filter to remove spurious results unrelated to the query class. Web directories can take advantage of classification in a variety of ways: from the automatic update of the directory content to the verification of the adequateness of the user classification. Ad targeting can benefit from text classification improving the pertinence of ads to the website content.

Nonetheless, training a classifier requires a large number of labelled examples that may be not available. In this case, building an accurate and adequately large training dataset can quickly become impractical, in particular when the number of classes exceeds few units. The dual approach to supervised clussification is that of clustering. However, the lack of domain knowledge makes clustering not accurate enough for business quality applications.

The need of large sets of examples has driven researchers to use data coming from open source collaborative directory projects like DMOZ. However, the lack of a central supervision on the quality of the classification done by the volunteers exposes these directories to the risk of misclassification or even spamming. As a result, taking the examples from these directories without any control could affect the overall classification accuracy. In addition, this approach may have copyright restrictions and it is possible only if the desired categories are well represented in the directory hierarchy.

Although some semi-supervised approaches that mix together labelled and unlabelled examples have been proposed, to the best of our knowledge the problem of automatically building an entire set of high quality examples to be used for training was not addressed yet in previous research.

The main contribution of this paper is a novel semi-supervised algorithm that, exploiting the strong correlation between the textual representation of certain URLs and the corresponding page content, automatically builds a set of web pages for each class that have a very high chance to be good positive examples. The texts of these automatically labelled pages can be used as a training set for any subsequent classification task even outsite the web context.

The ultimate aim of our auto-labelling approach is that of approximating the accuracy of a standard supervised classification even in absence of a predetermined set of training example. In order to achieve this goal our method exploits a learning to rank approach where each class is represented as a document consisting of a set of words describing its content and a URL is a text snippet obtained by means of a dictionary-driven tokenization algorithm. The ranking of the classes is based on the degree of relatedness with the URL text snippet. A web page is assigned to the top rank class.

Since the relationship between a URL and the corresponding web page is not necessarily revealing, we introduced in our auto-labelling framework a filtering step aimed at excluding from the labelling phase those URLs that do not show an evident correlation. This filtering phase is possible because the final goal of our framework is not that of the exhaustive classification of all the input pages, but only the creation of a high quality subset of labelled examples.

We evaluated our algorithm using a set of about 20 thousand web pages belonging to 9 categories extracted from the DMOZ web directory. Our experiments have shown that a training set created with our method contains over 85% of correctly labelled web pages. This accuracy is consistently higher than that of three tested state of the art semi-supervised approaches. We also evaluated the effect of using our auto-labelled examples as a training set for two standard applications in web information retrieval: web page classification and class size quantification (i.e. the estimation of the number of documents belonging to a given category). We report that, when a classifier is trained using our training set, its accuracy is never lower than the 88% of the accuracy attained with a manually curated training set. We believe that this result opens to the possibility of using classification even in those situations where a high quality set of examples is not available.

The remainder of this paper is organized as follows: Sect. 2 provides a brief overview of the main classification tasks and algorithms in web information retrieval, Sect. 3 describes in detail our approach. In Sect. 4 we report the outcome of our experiments, while we draw conclusions in Sect. 5.

2   Related Work

Classification is a common task in web information retrieval. According to the availability of the page content or only the URL, different sub-problems and approaches have been proposed. In particular: web page classification is the problem of assigning a web page to a predetermined category according to its content, URL classification is a similar problem based only on the analysis of the URL structure. The growth in size of the web has suggested the large-scale use of machine learning techniques for about two decades. In [16] the authors propose one of the first approaches in this sense.

The range of applications that benefit from classification is incredibly wide. Just to cite few examples: focus crawlers exploit classification for: automatic service discovery [12], language classification [17], relevancy prediction [19,38]. Search engines take advantage of classification for: query results filtering [33], spam detection [14], advertisement relevance refinement [8]. Among the other possible applications we also mention: parental control [13] and business intelligence [10].

With the purpose of increasing classification accuracy, special purpose features and algorithms have been studied [30]. In [28] the authors introduce the use of genetic algorithms (GA) trained on both the structure and the text of pages. In [41] the authors introduce a web page classification algorithm based on the k-nearest neighbour that combines link and textual information. In [42] a SVM based classification strategy is introduced. In their paper, the authors discuss the problem of the availability of negative examples and show how is it possible to obtain state of the art accuracy even using only positive examples for training. In [29] the authors face the web page classification problem exploiting a naive Bayes classifier. A similar approach restricted only to the URL structure has been proposed in [31]. A survey of the algorithms for URL-based topic classification can be found in [3], while URL-based language classification is reviewed in [4].

In [25] the authors use Ward's minimum variance as a distance to cluster together web pages and after that remove redundant features. The reduced set is later used for content based classification. In [22] the authors exploit the semi structured nature of web sites to represent them as disjoint feature vector sets. These sets are used as input for k-NN based classification. A similar approach is used in [43] where the authors used a pool of support vector machines. Heuristic filters to enhance classification by removing potentially misclassified examples are discussed in [7,34,36].

Few unsupervised approaches have been proposed in the literature and they are typically devoted to specific tasks. For example in [21] the authors propose a clustering-based approach to build intra-website topic-oriented models of URL structures. A similar task is done in [20] where the authors exploit a statistical model. In [39] the authors leverage on the social tag information of web pages for clustering with the purpose of showing the benefit of using this data for web mining. Other authors have used clustering algorithms for the automatic categorization of web search results (see [9] for a survey).

Some authors have studied the problem of automatically building or enhancing web directories via clustering or classification. In [1] the authors propose a hierarchical clustering approach based on self organizing maps aimed at aiding directory editors in building an initial categorization of the entire directory hierarchy. In [26] the authors investigate the problem of large scale content based classification of imbalanced hierarchical data providing ad-hoc solutions for different class sizes. To show the effectiveness of their approach they classified the entire Yahoo^{TM} web directory using DMOZ for validation.

3   Our Approach

The goal of our algorithm is that of automatically building a set of examples to use as a training set during the learning phase of classification. To do this, we use a learning to rank approach [23] getting a collection of URLs as well as the textual description of each class as input, and returning a labelled subset of the input URLs.

As URL classification methods, our algorithm leverages on the fact that a domain name (or the entire URL of a page) is often the result of the juxtaposition of few representative words predictive of the website content [3] and thus it can be used to measure its degree of membership with a class. When this happens, the text snippet extracted from the URL tends to be more focused on the page/website topic than the entire text of the page. In fact, due to the limited number of available characters, publishers tend to introduce mostly topic specific keywords in the page URL.

Once we succeed to convert a URL into a meaningful text, we can use it to measure the degree of membership of the page with each class. For example, the presence of the word *hotel* in the domain *hotel-sweet-dreams.tld* is a clue of the fact that the page topic is hotel and thus the membership class is travel.

Since we are not interested in labelling all the pages, but we can limit only to that fraction of pages showing a revealing degree of membership with only one class, when a URL does not follow our assumption we can safely ignore it.

Our algorithm works in five main phases: tokenization, URL/page relevance filtering, classification, ranking, and quality filtering. In the first step we analyze each URL and attempt to extract from it a possible corresponding text snippet. The second phase is aimed at filtering out from the set of tokenized URLs those elements for which the correlation with the corresponding web page is not evident enough. In the third step we build a vector of features for each class/URL pair in order to measure the degree of relevance of the class description for the URL textual representation. These vectors are later used in the forth phase to derive a labelling for a given URL by ranking the classes. In the last phase the algorithm filters out those URLs for which the association with the top rank class is not strong enough to be considered reliable. Figure 1 graphically summarizes our pipeline.

Once all the URLs have been analysed, the subset of labelled URLs can be used as a training set for any subsequent supervised classification task.

3.1 Tokenization Algorithm

Although URL tokenization is not the main focus of our work, the outcome of this step could impact on the classification accuracy of the overall system. In fact, URLs not appropriately tokenized will probably not be assigned to any class with a high enough confidence, but they will be filtered out from the training set reducing the number of available examples for the subsequent learning process, and in turn affecting the classification accuracy.

Our tokenization algorithm follows a greedy approach and it is based on the recognition of words inside the URL or the domain name. Since domains appropriate for our purposes are supposed to consist of a small phrase, we attempt to recognize words from left to right. A URL is firstly divided into base units by extracting subdomains and splitting the directory path, then each base unit is analyzed. The algorithm scans each base unit searching for the smallest possible word. Once a word is found the procedure is recursively repeated on the remaining part of the base unit. If all the characters of the base unit are assigned to a word the tokenization succeeds and it is returned as part of the textual representation of the URL, otherwise the procedure backtracks and repeats the tokenization with a longer word.

If the procedure fails in splitting the domain name into words, the URL is discarded.

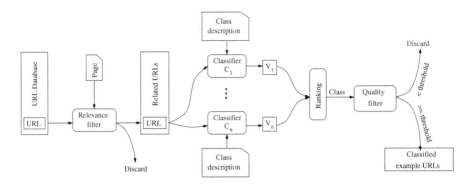

Fig. 1. Visual representation of the auto-labelling pipeline.

3.2   Relevance Filtering

There can be several reasons for a URL to be unrelated with the corresponding page content. Automatic server responses, spam, and evocative names are among the major reasons, but there can be other possibilities. A close examination of them is beyond the purposes of this paper. However, a URL whose textual representation does not reflect the content of the corresponding web page is more likely to produce an incorrect labelling and thus it should not be included in a training set. In the relevance filtering phase we verify that such a relationship between the textual representation of a URL and the content of the corresponding page exists. In particular, we suppose that a web page is unrelated to its URL if they do not have any words in common. We do not constraint the web page to contain all the URL' words because URLs are likely to contain also irrelevant or off topic words. In the previous example, the evocative domain name *hotel-sweet-dreams.tld* is tokenized into three words, but only the first one is connected to the commodity sector of the web site. As a measure of URL/document correlation we used the well-known cosine similarity. In order to prevent naive or misleading word matches, we pre-processed URLs by removing stopwords, but we did not apply stemming. Since the goal of this step is only that of excluding evidently inappropriate URLs from the labelling phase, we set a low cut-off threshold (as low as 0.2) for this filter.

3.3   Feature Extraction

The main step of our example extraction pipeline consists in representing each URL/class pair by means of a vector of features able to capture the appropriateness of the class for the URL. We selected standard features available in the literature trying to highlight different aspects.

Let $C = \{C_1, \ldots, C_m\}$ be the set of classes where, for the sake of simplicity, we denote with the same symbol both the class and its textual description; and let $U_j = \{w_1, \ldots, w_k\}$ be the outcome of the tokenization of the URL u_j. We denote with $b(C_i)$ the vectorial representation of the bag-of-words of the class C_i

and with $b(U_j)$ the same representation for URL u_j. In this latter case the IDF component is computed on C.

Our feature vector consist of the seven following similarity measures:

1. **Cosine similarity.** It is defined as the cosine of the angle between $b(U_j)$ and $b(C_i)$.

$$s(b(U_j), b(C_i)) = \frac{b(U_j) \cdot b(C_i)}{\|b(U_j)\| \cdot \|b(C_i)\|}$$

2. **Generalized Jaccard Coefficient** [11]. It is defined as the ratio of the weights of each term in common between $b(U_j)$ and $b(C_i)$.

$$GJC(b(U_j), b(C_i)) = \sum_{k=1}^{n} \frac{\min(b(U_j)[k], b(C_i)[k])}{\max(b(U_j)[k], b(C_i)[k])}$$

GJC is proven to be a metric in [11].

3. **Dice Coefficient.** Provides an estimation of the relative importance of the terms in common between U_j and C_i.

$$DC(b(U_j), b(C_i)) = \frac{2 * b(U_j) \cdot b(C_i)}{\|b(U_j)\| + \|b(C_i)\|}$$

4. **Overlap Coefficient.** It is a measure of the importance of the portion of the URL's text matched by the class.

$$OC(b(U_j), b(C_i)) = \frac{\sum_{k=1}^{n} \min(b(U_j)[k], b(C_i)[k])}{\min(\|b(U_j)\|, \|b(C_i)\|)}$$

5. **Weighted Coordination Match Similarity** [40]. Provides a measure of the distribution of matched words among the classes. It is computed as follows:

$$WCM(U_j, C_i) = \sum_{w \in [U_j \cap C_i]} \log\left(\frac{|C|}{DF(w)}\right)$$

6. **Okapi BM25** [32]. It is calculated as follows:

$$BM25(U_j, C_i) =$$

$$= \sum_{x=1}^{k} \cdot IDF_C(w_x) \frac{TF(w_x, C_i) \cdot (k_1 + 1)}{TF(w_x C_i) + k_1 \cdot (1 - b + b \cdot \frac{|C_i|}{avg(|C|)})}$$

where $avg(C)$ denotes the average number of words per class, k_1, and b are free parameters. In our case we set these free parameters according to LETOR [24] thus $k_1 = 2.5$ and $b = 0.8$.

7. $LMIR_{DIR}$ [5]. It is a language model for information retrieval computed by the Dirichlet smoothing method and calculated as follows:

$$LMIR_{DIR}(U_i, C_i) = \prod_{x=1}^{k} \frac{TF(w_x, C_i) + \mu P(w_x/C)}{|C_i| + \mu}$$

where $P(w_x/C)$ is the probability of finding w_x in the entire collection and μ is a free parameter.

All the above features have non-negative values and act as similarity measures, thus to higher values correspond a higher degree of similarity.

3.4 Ranking

Once all the feature vectors related to a given URL are built, they are used to obtain a score for each class and consequently a ranking of their relevance.

The heterogeneity of the ranges of the features requires a range normalization step to enable their direct comparison and, in turn, achieve a proportional contribution of each feature to the class score. To this end, we applied the following feature scaling normalization:

$$\hat{f_i} = \frac{f_i - \min(f(U_j))}{\max(f(U_j)) - \min(f(U_j))}$$

where f_i is the value of the feature f for the i-th class and $f(U_j)$ is the vector of all the values of the feature f for the URL u_j.

Subsequently, we used the average of the normalized vector as URL/class relevance score.

The final labelling is derived from the highest score class. In this phase we did not consider the case of score tie because URLs without a predominant class will be removed from the training set in the last phase of the algorithm.

3.5 Quality Filtering

The goal of the last phase of our procedure is that of retaining in the training set only those web sites whose class label assignment is not ambiguous. We accomplished this task by simply setting a cut-off threshold so that a URL is kept in the training set only if the top rank class score exceeds the threshold, it is discarded otherwise. Consider again the example of the URL *hotel-sweet-dreams.tld*. In this case the word *hotel* induces the correct classification, but the word *sweet* could produce a misclassification because of its membership with the category *food*. Because of the absence of a clearly predominant class this URL is discarded. Setting the threshold parameter can influence the accuracy of the target classification application. In fact, the higher the threshold, the more accurate is the final set of examples. In contrast, a higher value of the threshold reduces the number of available labelled examples and, consequently, the learning accuracy. Although the dependence on the size of the document corpus makes fixing a default threshold impractical, the strategy of using the highest value that maintain the size of the resulting training set reasonably large often leads to an easy choice of the threshold value. As shown in Sect. 4.2 in our experiments we empirically set this parameter to 0.7 obtaining a valuable boost of labelling accuracy at the cost of a reasonable reduction of number of labelled examples.

4 Results and Discussion

We conducted an extensive experimentation on two common classification tasks in web information retrieval using an adequately large dataset to obtain a realistic representation of our auto-labelling accuracy. We measured how well our learning framework is able to approximate the standard supervised classification and compared it with a semi-supervised learning approach.

To test our approach we implemented a prototype in Java and performed experiments using a single workstation endowed with a 2.6 GHz Intel Core i7, 16 GB of RAM and Mac OS X operating system.

4.1 Dataset

We built a collection of web pages from the Italian section of the well-known DMOZ[1] web directory. Since we did not use any language specific pre-processing tool, the choice of Italian instead of English had no practical effects in the outcome of our experimentation. The Italian DMOZ consists in 13 categories and over 150 thousand pages. We removed categories largely overlapping with most of the other classes (i.e. business, online shopping) as well as extremely underrepresented categories (house, news). As a result, we obtained a sample of 20,016 pages divided into 9 non-overlapping classes. We subsequently partitioned each class into training and test sets. Since our auto-labelling algorithm autonomously computes the size of the training set, we decided to use the same size to train supervised algorithms. This choice allowed us to measure how our method approximates the supervised classification accuracy.

Table 1 summarizes the distribution of the number of pages per class and the bipartition into training and test sets.

Table 1. Summary of the number of documents per class and relative distribution into training and test sets.

Category	# doc	# train	# test
Shopping	2,359	465	1,894
Art	5,303	1,702	3,601
Computer	1,197	309	888
Games	1,058	206	853
Wellness	1,513	424	1,089
Science	1,357	283	1,074
Society	3,328	865	2,463
Sport	1,642	434	1,208
Free time	2,259	583	1,676
# docs	20,016	5,270	14,746

---

[1] http://www.dmoz.org/World/Italiano/.

We used the DMOZ directory also as source to build a textual description of each class. Similarly to web search results, to each listed web site in DMOZ a small textual description is associated. These descriptions are manually curated by the editors who added the webpages to the directory, thus they are often quite accurate. For this reason we used as textual representation of a class the union of all the descriptions of the URLs belonging to it.

4.2   Auto-labelling Evaluation

The main purpose of our approach is that of assigning a label to an adequately large set of URLs of the data corpus. Both the number of labelled web pages and the classification accuracy will affect the outcome of the subsequent classification task. In this section we show that our algorithm is able to provide a reasonable amount of examples per class regardless to its size and topic. In addition, we show the classification accuracy of our labelling algorithm and compare it with three standard classification algorithms: k-NN, SVM and the naive Bayesian classifier (NBC). All the tested algorithms receive as input the textual description of the classes and a list of URL to label.

Table 2. Distribution as percentage of the amount of URLs filtered out in each phase of our pipeline. *Token.* reports the number of URLs for which the tokenization failed, *Relevnce* reports the amount of URLs with cosine similarity equal to 0 or under the threshold. *QF* reports the URLs involved in the quality filtering. *# Lab.* counts the labelled URLs.

Category	Token	Relevnce		QF	# Lab
		$S = 0$	$S < 0.2$		
Shopping	39.93	12.76	22.89	4.71	19.71
Art	33.08	11.41	15.41	8.01	32.00
Computer	43.27	11.53	11.78	7.60	25.82
Games	48.02	8.21	15.88	8.51	19.38
Wellness	23.13	20.56	20.95	7.34	28.02
Science	25.79	26.09	19.75	7.52	20.85
Society	29.06	14.93	21.91	8.11	25.99
Sport	27.04	12.91	22.17	11.45	26.43
Free time	28.51	14.52	21.07	10.09	25.81
Average	32.36	14.15	19.09	8.07	26.33

In Table 2 we summarize the amount of web pages discarded in each phase of our labelling pipeline and the amount of labelled examples. For the sake of comparison among different classes we expressed data as percentages.

Overall, our pipeline does not assign a label to the 75, 67% of URLs. This result is consistent with the intuition that only for a small fraction of the URLs a text snippet can be derived and, among them, only a portion shows a revealing relationship with a class. As shown in Table 2 a notable part of the unlabelled URLs derive from the fact that we could not find an appropriate tokenization in the first phase of the pipeline (32.36%). We observed that the distribution of the amount of untokenized URLs depends on the class topic. In particular, the abundance of proper names or brands in the classes *Games*, *Shopping*, and *Computer* produces an increase in the number of untokenized URLs. We reported also the effect of the relevance filtering phase on the number of unlabelled URLs. We separated the case where the cosine similarity between the URL and the web page is 0 from the case in which this similarity is under the cut-off threshold. The first case is more often due to technological reasons. For example: it might not be possible to extract the text of the page because it was dynamically generated via web 2.0 technologies, the corresponding page is empty or no longer exists. Instead, web pages discarded because of a cosine similarity under the cut-off threshold are more likely to contain spam. Although at first sight the amount of these pages could seem high, their number is coherent with a recent estimation made on the .com and .biz gTLDs [18]. A relatively modest amount of web pages (8.07% on average) are discarded after labelling because of the quality filtering (see column QF in Table 2). Although this filtering could be avoided, it enables a valuable boost of accuracy (as shown in Table 3) at the cost of a small reduction in the amount of labelled examples. Finally, the last column of Table 2 shows the percentage of labelled pages per class. This value can be read as a measure of the ability of our method to retrieve examples for each class. As shown in Table 2, the least represented class can rely on a 19.38% of training examples enabling a fair learning. We also observe that the ability of our method to retrieve examples for a class is not subjected to the class size.

Table 3. Accuracy and number of examples of: our method (*ALP*), our method without the final quality filtering (*ALP (nt)*), the semi-supervised naive Bayesian classifier (*NBC*), the semi-supervised k-NN and the semi-supervised SVM

	ALP	ALP (nt)	NBC	k-NN	SVM
Shopping	0.847	0.788	0.719	0.780	0.563
Art	0.894	0.865	0.631	0.588	0.405
Computer	0.864	0.780	0.810	0.533	0.835
Games	0.849	0.725	0.793	0.515	0.742
Wellness	0.811	0.744	0.750	0.473	0.736
Science	0.760	0.647	0.675	0.470	0.709
Society	0.867	0.821	0.604	0.377	0.630
Sport	0.873	0.818	0.730	0.299	0.693
Free time	0.816	0.742	0.649	0.374	0.695
Accuracy	0.858	0.800	0.674	0.496	0.598
# docs	5,270	6,886	6,886	6,886	6,886

If fact, although the two classes *Shopping* and *Sport* have different sizes, they can rely approximately on the same number of examples.

In Table 3 we report in detail the classification accuracy of our labelling pipeline. Results show that, even without quality filtering, our method achieves a consistently higher accuracy than other methods (as high as at least 13%). Considering that without applying the quality filtering all the compared algorithms return an equal number of labelled documents, we can state that our method always enables a more accurate learning of the final target application than the tested semi-supervised state of the art solutions.

According to the results shown in Table 3, using our quality filter we obtain a further accuracy increase of the 5.8% at the cost of a smaller number of labelled pages. Although a general evaluation of the possible negative effects due to the lower number of examples is impractical, some considerations are still possible. In principle, even if the contribution quickly tends to become marginal, extending a training set adding further correct examples should yield a more accurate classification. In contrast, the introduction of a significant number of misclassified examples can cause an unpredictable behaviour of the classification with a consequent degradation of the classification accuracy. The accuracy increase due to the introduction of the quality filter is a clue to the inappropriate labelling of the withdrawn examples. In fact, in our tests the accuracy of the set of removed examples is 0.661. This low value suggests that, in general, the accuracy of our pipeline without the use of the quality filter is expected to be lower regardless the higher number of examples.

4.3   Web Page Classification

Aimed at evaluating the feasibility of using our approach to automatically build a set of examples for the task of web page classification, we tested our method training three broadly used classifiers: the Support Vector Machine (SVM), the Naive Bayes Classifier (NBC), and the k-Nearest Neighbours (k-NN). As a baseline for the comparison we used a set of manually curated examples to train the same classifiers.

We also investigated the behaviour of the same classifiers used in a semi-supervised setting. The most natural alternative to our approach is a k-NN where the textual representations of classes are used as pivot points. We also included SVM and NBC trained only with the class description. However, in this latter case the class' prior cannot be estimated during the training phase because each class contains exactly one element. Thus classes are assumed to be equally probable.

In Table 4 we report the comparison of our approach with the semi-supervised framework. As evaluation measures we used: accuracy, precision, recall and f-measure. Being interested in how well each method approximates the performance of the supervised approach, for the sake of comparison we report results as percentage of the corresponding value of the baseline.

As shown in Table 4, except for a single case our auto-labelling method outperforms the semi-supervised approach independently of the chosen classifier.

Table 4. Approximation (as percentage) of the accuracy, precision, recall and f-measure of supervised classification using our method (ALP) or the semi-supervised classifiers (S.S.)

Algorithm		Accuracy	Precision	Recall	f-measure
SVM	ALP	90.45	90.77	88.92	89.78
	S.S.	59.54	77.74	77.23	77.41
NBC	ALP	89.61	98.74	92.59	95.49
	S.S.	49.23	105.82	60.19	76.16
k-NN	ALP	95.48	90.07	96.39	93.50
	S.S	39.54	48.76	62.55	55.37

In particular, comparing results obtained with the same classifier, we measured an improvement in term of accuracy never lower than 30% while the increment of f-measure is always higher than 10%. An interesting exception is the precision of *S.S.* obtained with NBC. In fact, in this case the classifier achieve a precision even higher than that of a pure supervised approach. Nonetheless, this result is balanced with a consistently lower recall.

Table 4 also shows that, independently from the chosen classification algorithm and evaluation measure, the set of examples labelled with our method allows an approximation factor of over 88% of the performances of standard supervised classification. This result suggests that achieving a high level of classification accuracy is still possible even in absence of a manually labelled training set.

4.4 Quantification

The new trend in social science of inferring socio-economic information from the web has raised the need to use web classification to quantify the diffusion of certain phenomena [6]. In this application the focus is not the exact labelling of each web site, but the quantitative evaluation of the number of pages belonging to each class. For example, we could want to compare the importance of certain commodity sectors by quantifying the number of web sites belonging to the corresponding classes. As shown in [35], in this scenario, achieving a high quantification accuracy requires not only a classifier to be accurated, but also to uniformly spread errors across all the classes thus avoiding to concentrate them into a single or a few classes.

In this section we experimentally evaluated how our approach approximates the quantification accuracy of the standard supervised classification and compared it with a semi-supervised approach. Different measures have been proposed in the literature to evaluate quantification [37]. Following the same approach of [15,27] we used the Kullback-Leibler divergence (KLD) that evaluates the loss of information when a classifier is used to estimate the real frequency of an event.

For a single class classifier the KL divergence is defined as follows:

$$KLD(c) = p(c)\ln\frac{p(c)}{\hat{p}(c)} + p(1-c)\ln\frac{p(1-c)}{\hat{p}(1-c)}$$

where $p(c)$ is the real probability of belonging to the class c and $\hat{p}$ is the approximation returned by the classifier. The KL divergence has values in the range $[0, +\infty)$ and holds 0 when $\hat{p} = p$. Since KLD is additive for independent probability distributions, it can be naturally extended to multi class classification by mean of the following:

$$KLD(C) = \sum_{c \in C} KLD(c)$$

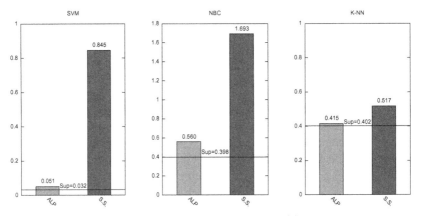

(a) Support Vector Machine   (b) Naive Bayesian Classifier   (c) K-Nearest Neighbour

Fig. 2. Kullback-Leibler divergence (KLD) of the quantification task for: our method (ALP), and the semi-supervised classifiers (S.S.). The figure reports also the KLD of the baseline supervised classification (SUP).

Here we report in detail the KLD for the three classifiers already tested in Sect. 4.3. As Fig. 2 shows, our approach is always able to obtain a tight approximation (within a factor 2) of the canonical supervised approach regardless of the chosen classification algorithm. In contrast, the semi-supervised approach achieves a KLD score an order of magnitude higher than our method in two out of three cases (see Fig. 2(a) and (b)).

Besides the application-specific importance, this result witnesses the stability of our approach. In fact, a low KLD score is the symptom of the fact that our algorithm does not concentrate errors in one or few classes and, consequently, it has been able to label enough examples to enable a correct learning of each of the considered classes.

5   Conclusions

Text classification is a common tool in information retrieval. The supervised learning approach has demonstrated to be effective and accurate enough to be successfully employed in industrial applications. Nonetheless, the need of classifiers to be trained with thousands of examples per class together with the cost of manual labelling can make its use impractical. Relying on semi-supervised or unsupervised alternatives can help to reduce costs at the price of a consistent drop of accuracy.

In this paper we showed how it is possible to automatically extract from the web an high quality training set by exploiting the strong relationship between certain web pages and the text contained in their URLs. Experiments on a set of 20 thousand web pages divided in 9 classes have demonstrated that our framework is able to create a reasonably accurate training set and, consequently, achieve a tight approximation of the accuracy of a supervised learning approach independently from the specific task or from the choice of the classifier. As a result, our method has demonstrated to be a cost effective alternative to supervised classification in those cases where examples are not available.

Funding. This work was supported by: the Regione Toscana of Italy under the grant POR CRO 2007/2013 Asse IV Capitale Umano; the Italian Registry of the ccTLD "it" Registro.it.

References

1. Adami, G., Avesani, P., Sona, D.: Clustering documents into a web directory for bootstrapping a supervised classification. Data Knowl. Eng. **54**(3), 301–325 (2005). https://doi.org/10.1016/j.datak.2004.11.003
2. Baykan, E., Henzinger, M., Marian, L., Weber, I.: Purely URL-based topic classification. In: Proceedings of the 18th International Conference on World Wide Web, pp. 1109–1110. ACM (2009)
3. Baykan, E., Henzinger, M., Marian, L., Weber, I.: A comprehensive study of features and algorithms for URL-based topic classification. ACM Trans. Web **5**(3), 15:1–15:29 (2011). https://doi.org/10.1145/1993053.1993057
4. Baykan, E., Henzinger, M., Weber, I.: A comprehensive study of techniques for URL-based web page language classification. ACM Trans. Web **7**(1), 3:1–3:37 (2013). https://doi.org/10.1145/2435215.2435218
5. Bennett, G., Scholer, F., Uitdenbogerd, A.: A comparative study of probabilistic and language models for information retrieval. In: Proceedings of the Nineteenth Conference on Australasian Database, vol. 75, pp. 65–74. Australian Computer Society Inc. (2008)
6. Boyd, D., Crawford, K.: Critical questions for big data: provocations for a cultural, technological, and scholarly phenomenon. Inf. Commun. Soc. **15**(5), 662–679 (2012)
7. Breunig, M.M., Kriegel, H.P., Ng, R.T., Sander, J.: LOF: identifying density-based local outliers. In: ACM Sigmod Record, vol. 29, pp. 93–104. ACM (2000)
8. Broder, A.Z., Ciccolo, P., Fontoura, M., Gabrilovich, E., Josifovski, V., Riedel, L.: Search advertising using web relevance feedback. In: Proceedings of the 17th ACM Conference on Information and Knowledge Management, CIKM 2008, pp. 1013–1022. ACM, New York (2008). https://doi.org/10.1145/1458082.1458217

9. Carpineto, C., Osiński, S., Romano, G., Weiss, D.: A survey of web clustering engines. ACM Comput. Surv. **41**(3), 17:1–17:38 (2009)

10. Castellanos, M., Daniel, F., Garrigós, I., Mazón, J.N.: Business intelligence and the web. Inf. Syst. Front. **15**(3), 307–309 (2013)

11. Charikar, M.S.: Similarity estimation techniques from rounding algorithms. In: Proceedings of STOC-02, 34th Annual ACM Symposium on the Theory of Computing, , Montreal, CA, pp. 380–388 (2002)

12. Dong, H., Hussain, F.: Focused crawling for automatic service discovery, annotation, and classification in industrial digital ecosystems. IEEE Trans. Ind. Electron. **58**(6), 2106–2116 (2011)

13. Eickhoff, C., Serdyukov, P., de Vries, A.P.: Web page classification on child suitability. In: Proceedings of the 19th ACM International Conference on Information and Knowledge Management, CIKM 2010, pp. 1425–1428. ACM, New York (2010). https://doi.org/10.1145/1871437.1871638

14. Erdélyi, M., Garzó, A., Benczúr, A.A.: Web spam classification: a few features worth more. In: Proceedings of the 2011 Joint WICOW/AIRWeb Workshop on Web Quality, WebQuality 2011, pp. 27–34. ACM, New York (2011). https://doi.org/10.1145/1964114.1964121

15. Forman, G.: Quantifying counts and costs via classification. Data Min. Knowl. Discov. **17**(2), 164–206 (2008)

16. Fürnkranz, J.: Exploiting structural information for text classification on the WWW. In: Hand, D.J., Kok, J.N., Berthold, M.R. (eds.) IDA 1999. LNCS, vol. 1642, pp. 487–497. Springer, Heidelberg (1999). https://doi.org/10.1007/3-540-48412-4_41

17. de Groc, C.: Babouk: Focused web crawling for corpus compilation and automatic terminology extraction. In: 2011 IEEE/WIC/ACM International Conference on Web Intelligence and Intelligent Agent Technology (WI-IAT), vol. 1, pp. 497–498 (2011)

18. Halvorson, T., et al.: The BIZ top-level domain: ten years later. In: Taft, N., Ricciato, F. (eds.) PAM 2012. LNCS, vol. 7192, pp. 221–230. Springer, Heidelberg (2012). https://doi.org/10.1007/978-3-642-28537-0_22

19. Hao, H.W., Mu, C.X., Yin, X.C., Li, S., Wang, Z.B.: An improved topic relevance algorithm for focused crawling. In: 2011 IEEE International Conference on Systems, Man, and Cybernetics (SMC), pp. 850–855, October 2011

20. Hernández, I., Rivero, C.R., Ruiz, D., Corchuelo, R.: A statistical approach to URL-based web page clustering. In: Proceedings of the 21st International Conference Companion on World Wide Web, WWW 2012 Companion, pp. 525–526. ACM, New York (2012). https://doi.org/10.1145/2187980.2188109

21. Hernández, I., Rivero, C.R., Ruiz, D., Corchuelo, R.: CALA: an unsupervised URL-based web page classification system. Knowl.-Based Syst. **57**, 168–180 (2014). http://www.sciencedirect.com/science/article/pii/S0950705113003997

22. Kriegel, H.P., Schubert, M.: Classification of websites as sets of feature vectors. In: Databases and Applications, pp. 127–132 (2004)

23. Liu, T.Y.: Learning to rank for information retrieval. Found. Trends Inf. Retr. **3**(3), 225–331 (2009). https://doi.org/10.1561/1500000016

24. Liu, T.Y., Xu, J., Qin, T., Xiong, W., Li, H.: LETOR: Benchmark dataset for research on learning to rank for information retrieval. In: Proceedings of SIGIR 2007 Workshop on Learning to Rank for Information Retrieval, pp. 3–10 (2007)

25. Mangai, J.A., Kumar, V.S., Alias Balamurugan, S.A.: A novel feature selection framework for automatic web page classification. Int. J. Autom. Comput. **9**(4), 442–448 (2012)

26. Marath, S.T., Shepherd, M., Milios, E., Duffy, J.: Large-scale web page classification. In: 2014 47th Hawaii International Conference on System Sciences (HICSS), pp. 1813–1822. IEEE (2014)
27. Milli, L., Monreale, A., Rossetti, G., Giannotti, F., Pedreschi, D., Sebastiani, F.: Quantification trees. In: 2013 IEEE 13th International Conference on Data Mining (ICDM), pp. 528–536. IEEE (2013)
28. Özel, S.A.: A web page classification system based on a genetic algorithm using tagged-terms as features. Expert Syst. Appl. **38**(4), 3407–3415 (2011). https://doi.org/10.1016/j.eswa.2010.08.126
29. Patil, A.S., Pawar, B.: Automated classification of web sites using naive Bayesian algorithm. In: Proceedings of the International Multi-Conference of Engineers and Computer Scientists, vol. 1, pp. 14–16 (2012)
30. Qi, X., Davison, B.D.: Web page classification: features and algorithms. ACM Comput. Surv. (CSUR) **41**(2), 12 (2009)
31. Rajalakshmi, R., Aravindan, C.: Naive Bayes approach for website classification. In: Das, V.V., Thomas, G., Lumban Gaol, F. (eds.) AIM 2011. CCIS, vol. 147, pp. 323–326. Springer, Heidelberg (2011). https://doi.org/10.1007/978-3-642-20573-6_55
32. Robertson, S.E.: Overview of the okapi projects. J. Doc. **53**(1), 3–7 (1997)
33. Rose, D.E., Levinson, D.: Understanding user goals in web search. In: Proceedings of the 13th International Conference on World Wide Web, pp. 13–19. ACM (2004)
34. Saad, M.K., Hewahi, N.M.: A comparative study of outlier mining and class outlier mining. Comput. Sci. Lett. **1**(1) (2009)
35. Sebastiani, F.: Text quantification. In: de Rijke, M. (ed.) ECIR 2014. LNCS, vol. 8416, pp. 819–822. Springer, Cham (2014). https://doi.org/10.1007/978-3-319-06028-6_104
36. Smith, M., Martinez, T.: Improving classification accuracy by identifying and removing instances that should be misclassified. In: The 2011 International Joint Conference on Neural Networks (IJCNN), pp. 2690–2697, July 2011
37. Tang, L., Gao, H., Liu, H.: Network quantification despite biased labels. In: Proceedings of the Eighth Workshop on Mining and Learning with Graphs, pp. 147–154. ACM (2010)
38. Taylan, D., Poyraz, M., Akyokus, S., Ganiz, M.: Intelligent focused crawler: learning which links to crawl. In: 2011 International Symposium on Innovations in Intelligent Systems and Applications (INISTA), pp. 504–508, June 2011
39. Trivedi, A., Rai, P., Daumé III, H., DuVall, S.L.: Leveraging social bookmarks from partially tagged corpus for improved web page clustering. ACM Trans. Intell. Syst. Technol. (TIST) **3**(4), 67 (2012)
40. Wilkinson, R., Zobel, J., Sacks-davis, R.: Similarity measures for short queries. In: Fourth text Retrieval Conference (TREC-4), pp. 277–285 (1995)
41. Xu, Z., Yan, F., Qin, J., Zhu, H.: A web page classification algorithm based on link information. In: 2011 Tenth International Symposium on Distributed Computing and Applications to Business, Engineering and Science (DCABES), pp. 82–86, October 2011
42. Yu, H., Han, J., Chang, K.C.: PEBL: web page classification without negative examples. IEEE Trans. Knowl. Data Eng. **16**(1), 70–81 (2004)
43. Zhong, S., Zou, D.: Web page classification using an ensemble of support vector machine classifiers. J. Netw. **6**(11), 1625–1630 (2011)

Practical Applications

Generating Appealing Brand Names

Gaurush Hiranandani[1]([✉]), Pranav Maneriker[1], and Harsh Jhamtani[2]

[1] Adobe Research, Bangalore, India
[2] Language Technology Institute, Carnegie Mellon University,
Pittsburgh, PA, USA
{ghiranan,pmanerik}@adobe.com, jharsh@cs.cmu.edu

Abstract. Providing appealing brand names to newly launched products, newly formed companies or for renaming existing companies is highly important as it can play a crucial role in deciding its success or failure. In this work, we propose a computational method to generate appealing brand names based on the description of such entities. We use quantitative scores for readability, pronounceability, memorability and uniqueness of the generated names to rank order them. A set of diverse appealing names is recommended to the user for the brand naming task. Experimental results show that the names generated by our approach are more appealing than names which prior approaches and recruited humans could come up.

1 Introduction

Choosing right brand names for newly launched products, newly formed companies and entities like social media campaigns, apps, websites etc. is critical. In the context of creating brands, it is believed that such a naming decision may well be "the most important marketing decision one can make" [1]. A marketer may often spend a lot of time in coming up with an appealing name which can achieve favorable outcomes on various key performance indicators (KPIs) like website visits, number of customer acquisitions, etc. This becomes critical in scenarios like quickly planned campaigns, where there is not enough time for marketers or authors to come up with an appealing name. However, prior technologies are insufficient to computationally come up with appealing names for such entities based on a provided description. Moreover, rarely is the management provided with interpretable objective criteria upon which a brand name is suggested [1]. This creates a need for an algorithm which automatically generates appealing names from the description of an entity in a justified manner.

This paper has the following contributions. Firstly, we define and infer the importance of various linguistic and statistical features for the task of suggesting names for brands, products or other such entities. Secondly, we propose computational methods to generate brand names given the description of the entity in question. Though coming up with names for entities like brands is considered mostly a creative task, our MTurk based evaluation study determining the

© Springer Nature Switzerland AG 2018
A. Gelbukh (Ed.): CICLing 2017, LNCS 10762, pp. 605–616, 2018.
https://doi.org/10.1007/978-3-319-77116-8_45

appeal of the recommended names shows that names from our method obtain ratings comparable with human-provided names.

Rest of the paper is organized as follows. Section 2 reviews the existing relevant works. Section 3 explains methodology behind generating and ranking names. Section 4 explains the conducted experimental studies and evaluation. In Sect. 5 we discuss some of the limitations and future work. Lastly in Sect. 6, we provide conclusions.

2   Related Work

Robertson [2] showed characteristics of a 'good' name which include short, easy to say, spell, read, understand and easily retrievable from memory. Yorkston and Menon [3] showed consumers use the information they gather from phonemes in brand names to infer product's attributes. Little et al. [4] suggested that a recommendation tool for performing the naming task better should aid in ideation as well.

From the perspective of word-generation, there are prior works on password and domain name generation. Some studies have focused on memorability of passwords [5], while tools like *PWGEN* [6] and *Kwyjibo* [7] generate pronounceable passwords and domain names respectively. However, using such tools and being limited to one attribute is not useful for brand-naming generation given the description.

Bauer [8] treated blending of words[1] as a process to create neologisms. Such blending can be based on various phonetic and syllable alignment techniques, such as those used by Kondrak [9] and Hedlund et al. [10]. Özbal and Strapparava [11] proposed a computational approach to generate neologisms consisting of homophonic puns and metaphors based on the category and properties of the entity. However, for recommendation purposes, it is important to define a ranking based on appeal of a name given the properties or description of the entity. Özbal and Strapparava [11] carry out filtering/ranking during the evaluation by combining the phonetic structure and language model with equal weights. However, ranking based on the appeal of a name is not motivated by the crucial metrics mentioned above. This becomes critical when one needs to recommend a few appealing names rather than generate a large number of names. Further, some online tools[2] provide names by concatenating random strings and some consulting companies[3] are engaged in brand naming but do not use any automated processes.

---

[1] Forming a word by combining sounds from two or more distinct words - e.g. *Wikipedia* by blending "Wiki" and "encyclopedia".

[2] www.online-generator.com, www.namegenerator.biz.

[3] ABC Namebank, A Hundred Monkeys.

3   Methodology

Figure 1 provides an outline of our proposed solution. We do some basic preprocessing on the words in description, followed by expansion of the set of words using an external ontology. We blend the words to generate candidate names. Then, we score and rank the names based on readability, memorability, pronounceability and uniqueness. Finally we postprocess the ranked list of names to provide a diverse set of suggestions.

As mentioned earlier, we generate candidate names based on blending of syllables present in the word set. The choice of following this approach is based on prior works. Özbal et al. [12] provided an annotated dataset of 1000 brand names to understand linguistic creativity involved in naming. In the data, around 20% of the names are created either by juxtaposition or clipping which are morphological mechanisms similar to blending.[4] Further, Bauer [8] treated blending of words as a process to create neologisms. This convinced us to generate candidate names based on blending of words.

3.1   Generate Names

The method takes the description of an entity as input. For example, an input can be '**Creating** an **application** to **split expense wisely**.' We use words other than stop-words [13] while generating names and call them *root* words. Thereafter, our method looks for part-of-speech (POS) tags [14] of the root words. POS tags are important as words can have a different semantic orientation based on usage and thus have different set of synonyms (used later). Let D denote the set of words along with their POS tags. For example,

$D = \{$(Creating, Verb), (Application, Noun), (Split, Verb), (Expense, Noun), (Wisely, Adverb)$\}$.

Next, we use *Wordnet* [15] to obtain synonyms of root words based on POS. Further, we obtain synonyms and metaphors by applying a strategy similar to Özbal and Strapparava [11]. We call these words *related words* and attribute the

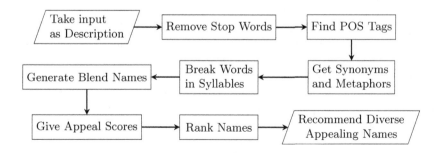

Fig. 1. Overview of the algorithm

---

[4] e.g. *DocuSign* from "Document" and "Signature".

Table 1. Percentage of blending rules

Rule	%	Rule	%
Noun-Adjective	40.10	Adjective-Adverb	3.2
Noun-Verb	8.02	Noun-Noun	36.36
Noun-Adverb	4.81	Verb-Verb	0.00
Verb-Adjective	0.53	Adjective-Adjective	3.28
Verb-Adverb	3.7	Adverb-Adverb	0.00

same POS tag as their root word. Let C_1, C_2,..., A_1, A_2,.., S_1, S_2,.., E_2, E_2,.., W_1, W_2,.. be the related words obtained for *Creating, Application, Split, Expense* and *Wisely* respectively. Further let R denote the list of related words with POS tags. Then,

$R = \{$*(Creating, Verb)*, *(C_1, Verb)*, *(C_2, Verb)*,...., *(Application, Noun)*, *(A_1, Noun)*, *(A_2, Noun)*,..., *(Split, Verb)*, *(S_1, Verb)*, *(S_2, Verb)*,..., *(Expense, Noun)*, *(E_1, Noun)*, *(E_2, Noun)*,..., *(Wisely, Adverb)*, *(W_1, Adverb)*, *(W_2, Adverb)*,...$\}$

Each word in R is split into syllables using PyHyphen.[5] We attach the same POS tag for each syllable as its parent word. Let us denote the set of syllables of the root and related words along with their POS tags by L. For example,

$L = \{$(Cre, Verb), (At, Verb), (Ing, Verb), (App, Noun), (Li, Noun), (Ca, Noun), (Tion, Noun),...$\}$.

We observed certain *rules* used in blended names from the description provided by the annotators in Özbal et al. [12]. A *rule* is a combination of unordered POS tags of the blended words (syllables). For example, *SplitWise* and *WiseSplit* are created from two syllables whose POS tags are verb and adverb. Hence, they follow the *Verb-Adverb* rule. Table 1 presents the percentages of each rule used in blended names. Note that the percentage of names created using a certain rule are subject to discretionary choice based on the annotator's bias. As in, for the same entity having similar properties, there can be two different descriptions from two different annotators leading to different empirical estimates. For example, *data platform enables solving problems quickly* or *data platform is quick in problem solving*. Here, we will have the same syllable *quick* in both sentences. However, in the former case, the POS tag is Adverb whereas, in the latter, it is Adjective. For the results shown in Table 1, we have used the POS tags obtained from the description provided by the annotators in Özbal et al. [12]. We observed that some of the blending rules were less frequent. Hence, we omitted the rules having less than 1% of the names under their category and called the rest as *allowed rules*. For example, a name using two syllables coming from words having verb as their POS tag is usually not created. An example can be *SplitBreak* formed by the syllables *Split* and *Break*. With respect to the data, this essentially removed three rules but had a significant reduction in the number of names generated.

---

[5] https://pypi.python.org/pypi/PyHyphen/.

Finally, the method creates new blended names by joining two or three syllables at a time taken from the permutation set of L, given that the blending is within allowed rules. Let N denote the set of generated names. For the given example, some names generated by two syllables are *SplitWise*, *BudSplit* and *BreakOwl*. Note that *break* is a synonym of *split* and *owl* is a metaphor for *wise* obtained from the idiom - *as wise as an owl*. One of the names generated by three syllables is *ExPenseBreak*. We prefer syllables over morphemes as the combination of syllables can generate any name that a combination of morphemes can. The increased number of names generated due to this choice are handled by ranking and selecting the top candidate names. This is explained in the next section.

3.2 Ranking Names

Scores Formulation. Every description can potentially generate thousands of names using the above method. However, it is crucial to rank them for recommendation purposes. Therefore, each name is given a score based on the mathematical formulations of 4 features: readability, pronounceability, memorability, and uniqueness. The scores are normalized in the range of $[0, 1]$ over the English dictionary [16], with 0 and 1 representing the least and the maximum score respectively. Let n and $|n|$ denote a name and its length respectively.

Readability: A good name should be easy to read. Hence, to assign readability score, we use Flesch—Kincaid Reading-Ease Score [17]. Since each name is a single word, readability (denoted by $R(.)$) becomes:

$$R(n) = 205.82 - 84.6 * |syllables(n)| \tag{1}$$

Where, $|syllables(n)|$ denotes the number of syllables in n. Later in this section, we will observe formulation in Eq. 1 being reduced to number of syllables due to a linear model computing appeal of a name.

Pronounceability: The more permissible the combinations of phonemes is, the more pronounceable the word becomes. We adapt the concept from Schiavoni et al. [18] with some refitting to measure the extent with which a string adheres to the phonotactics of the language. By taking substrings (n-grams) of n of length $l \in \{2, 3, 4\}$ with frequencies from the dictionary [16], we compute certain features as follows:

$$S_l^n = \frac{\sum\limits_{t \in n-grams(n)} freq(t)}{|n| - l + 1} \tag{2}$$

Here, $freq(t)$ is the frequency of the n-gram t in the dictionary. For example,

$$S_2(facebook) = \frac{fa_{109} + ac_{343} + ce_{438} + eb_{29} + bo_{118} + oo_{114} + ok_{109}}{8 - 2 + 1} = 170.8$$

Feature values for smaller l will be higher than larger l, but feature values for larger l will be more important, since there are around 4000 meaningful 4-letter words in comparison to around 1000 meaningful 3-letter words in English [19]. Hence, the probability of a 4-letter word being meaningfully used will be higher than a 3-letter word. Therefore, weights for S_l^n are assumed as $w_l = \frac{l}{2+3+4}$ for $l = \{2, 3, 4\}$. Finally, we define pronounceability $P(.)$ as:

$$P(n) = w_2 S_2^n + w_3 S_3^n + w_4 S_4^n \tag{3}$$

This formulation is a simple back off model [20], where we always back off to a lower order n-gram with fixed probability.

Memorability: Danescu-Niculescu-Mizil et al. [21] claim that memorable quotes use rare word choices, but at the same time are built upon a scaffolding of common syntactic patterns. Adapting it to blended syllables, we use Meaningful Characters Ratio defined by Schiavoni et al. [18] to capture memorability. It models the ratio of characters of n that comprise a meaningful word. A word is said to be meaningful if it occurs in [19]. We define memorability $M(.)$ as follows:

$$M(n) = \max_{\text{all splits of n}} \frac{\sum_{i=1}^{k} |s_i|}{|n|} \tag{4}$$

Here, s_i are the meaningful substrings of length ≥ 3 obtained by splitting n and k is their number. For example, If $n = facebook$,

$$M(facebook) = \frac{(|face| + |book|)}{8} = 1$$

Uniqueness: This feature prefers names having low usage and non-dictionary words. Consider a time series (V, T) for n such that $V = \{v_1, ..., v_T\}$ and $T = \{t_1, ..., t_T\}$. $t_i's$ represent consecutive years and v_i represents the normalized usage of n in year t_i as provided by GoogleNgrams [22]. Then, uniqueness $U(.)$ is defined as:

$$U(n) = \frac{\sum_{k=1}^{T_c} v_k * (t_k - t_1)}{\sum_{k=1}^{T_c} (t_k - t_1)} \tag{5}$$

where T_c represents the latest year. The intuition is that less usage in recent years is more important. Further, if GoogleNGrams fails to produce any time series for n, then $U(n)$ is taken to be 1.

Combining Scores. We needed to model the appeal of a brand name depending on the quantitative definitions of the mentioned linguistic features. We asked few annotators to provide descriptions of the entities they want to name. Among the descriptions provided, we randomly chose three of them. Thereafter, we conducted a survey of 20 participants who were shown 3 lists of 15 names generated by our method, one list for each of the three descriptions. They were asked to rank the names in a list from 1 to 15. On an average, the Kendall-Tau correlation between the "average ranks" and the "individual ranks" came out to be 0.66, 0.68 and 0.62 for the 3 lists. High correlation suggested that people give similar rankings of names if shown a description. Hence, we took average ranks to be the ground truth rankings for the 3 lists.

Next we test if the four feature scores described earlier were correlated. In other words, we wanted to test if the individual features provide information not covered by other features. While the pairwise Pearson's Correlation among the 4 scores based on 45 names ranged from -0.48 to 0.23, the Kendall Tau correlation among ranking from each individual score ranged from -0.48 to 0.21. This implied lack of correlation amongst the scores. Therefore, we defined appeal $A(.)$ of a name to be weighted linear combination of the mentioned scores. That is,

$$A(n) = a_r R(n) + a_p P(n) + a_m M(n) + a_u U(n) \tag{6}$$

Where, $a = (a_r, a_p, a_m, a_u)$ is the weight vector. Then we applied rank-svm [23] which used the obtained 315 pairwise comparisons to learn the weights showing importance for different features. The learned weights were as follows: $a = (2.18, 1.63, 0.91, 1.05)$. Interestingly participants indicated *readability* as the most crucial factor for comparing names amongst the four measures. For $n = SplitWise$, we obtained: $A(n) = 3.71$, $R(n) = 0.77$, $P(n) = 0.04$, $M(n) = 1.0$, $U(n) = 1.0$.

3.3   Recommendation

The user can be recommended with top names as per their appeal scores but since a single syllable may appear in many top names (eg. *fur*, *con* etc. which are meaningful and easy to read), this set of recommendation may not help ideation. Hence, we diversify [24] the set of names to aid ideation by an update rule [25]. The intuition is that after we choose the top candidate name based on appeal score, we update the appeal scores of the names formed by the same syllables as that of the top candidate name. And then choose the top name from the rest of the names having updated appeal scores. Suppose one chooses n' in the first iteration, then the update for diversity is defined as:

$$A(n) \leftarrow \frac{1}{|m| * |k|} * A(n) \quad \text{for} \quad n \in N \setminus [\{n'\} \cup N'] \tag{7}$$

Here, $|m|$, k, and N' denote the number of common syllables in n and n', number of syllables in n, and names sharing no common syllable with n respectively. We iteratively choose the best candidate (n') and then update appeal of other names (n) by using Eq. 7. The names chosen after some iterations (say 30) are recommended. This ensures that the recommended set of names as a whole become useful for the naming task. For the given example, the top 5 names are *ConTear*, *BreakWise*, *BudSplit*, *BreakOwl* and *DisCleave*.

4   Evaluation

To estimate the quality of the generated names from our method, we conducted an MTurk study and compared our names with two baselines. We created descriptions for 10 entities which usually require brand names. For example, one of the descriptions was *Light-weight software to locate virus on computer.* Our method took 4 min on our machine to generate 984 ranked names on an average for a description. We took the top 10 names for each description, names generated from the two baselines (described below) and compared the approaches through an MTurk study. The code, data, examples and results are available at this link.[6]

4.1   Baselines

Prior Art: Özbal and Strapparava [11] describe a method to generate names based on homophonic puns and metaphors by combining natural language processing techniques with various linguistic resources available online. We replicated the work of Özbal and Strapparava [11] to generate names. Adapting it our case, the category and properties were provided manually from the descriptions until it output atleast 10 names for the description. After following the original specification, if the number of names generated by this algorithm were fewer than the number required for our experimental setup, we added related properties to the set of properties taken as input. For more details about the approach see [11]. The output generated from this system was used for further experiments.

Human: 10 participants were recruited to give 10 names for one of the descriptions in 4 min (time taken by our method). The participants were given information about the criteria being used for creating new names, i.e. unique and appealing names. This experiment gave 10 human generated names for each description.

4.2   MTurk Survey: Results and Observations

For each description, we created 2 lists of 15 names, each containing 5 names randomly picked from the list of 10 names generated by the three approaches. Table 2 shows a few example inputs to the three approaches and the names generated by them. Then, 100 recruited judges from Amazon Mechanical Turk were shown one of the 20 lists and asked to rate each of the 15 names in it as *Good*, *Fair* or *Bad* based on their relevance to the description and uniqueness. Some of the participants of human experiment in Sect. 4.1 provided names of currently existing companies. Therefore, the latter instruction was added explicitly to avoid participants rating irrelevant existing names as *Good*. Each list was annotated by 5 judges resulting in 1500 responses.

---

[6] http://www.cicling.org/2017/data/326.

Table 2. Input and Output by the three approaches

Approach	Input	Output
Our method	Fabulous furniture to decorate your home	FabFur, MythRate, DressHouse HomeDec, FurDeck
Our method	Light weight software to locate virus on computer	FeatherTor, PingWare, Clean-Den, FaintCate, ClearSet
Prior art	Category: furniture; Properties: fabulous, decorative, attractive, homely, comfortable	Woodroom, Houly, Flooroom, Dinnel, Bedroose
Prior Art	Category: software; Properties: light, locate computable, buggy, safety	Luggyte, Cebuter, Safetyre, Locatr, Coftwarele
Human	Fabulous furniture to decorate your home	Decorature, FabHomes, Home-Decor FabulousHomes, FabFurnish
Human	Light weight software to locate virus on computer	Ubuntu, Nortun, Ad-Blocker, Windows, Web-sites

Table 3. Ratings for generated names

Approach	Good	Fair	Bad
Our method	16.6%	41.8%	41.6%
Human	20.4%	32.8%	46.8%
Prior art	13.2%	38.8%	48%

Table 3 shows percentages of ratings received by names generated by the three approaches considering all 10 descriptions. Our method outperforms the prior art. 16.6% of the names generated by our method received *Good* rating in comparison to 13.2% of the prior art. Similar is the case in *Fair* rating as well. Humans outperform both the automated approaches considering the *Good* rating. However, our method has significantly fewer *Bad* ratings when compared to humans.

We believe that 4 min constraint on humans is harsh. They can think of better names if given sufficient time. As a useful observation, one of the participants seemed to be following our approach to generate names using only the root words. The given description was *Showroom of Fabulous Furniture for Decorating Home*. Our method output names like *HomeDec* and *FabFurNi* which were rated mostly as *Fair* whereas the participant generated names like *HomeDecor* and *FabFurnish* which were rated mostly as *Good*. This tells us that the method described in this work is indeed a mechanism that humans use to generate names and further, it can also be used for ideation purposes.

Additionally, we calculated nDCG [26] to know whether our method's rankings match *Good/Fair/Bad* ratings by human judges. In order to define relevance of name n in nDCG formulation, we used 1, 0.5 and 0 as weights for number of *Good*, *Fair* and *Bad* ratings respectively. The nDCG averaged over 10 descriptions was 0.78 indicating that the ranking generated by our method indeed concurs with human rankings.

5   Limitations and Future Work

Our methodology will generate homogeneous names given the same description. Hence, there are opportunities for leveraging enterprise based personalization. Further, we agree that there are brand names like Apple, Fox, etc. which cannot be generated by our approach. However, examples like CarMax, DocuSign, etc. and aforementioned online generators, led us to believe that our approach is one of the ways by which humans create names. In future, we plan to investigate abbreviations, reduplications, and modifications over blended syllables to generate better names.

6   Conclusions

Our work is one of the first approaches to algorithmically generate appealing brand names from description. In addition to being directly used, the recommended names can also aid in ideation. Quantitative definitions of pronounciability, memorability, and uniqueness have been proposed. Further, the set of names generated by us is diverse. The inclusion of diversity aids in the ideation process, providing a rich set of names to any user of the system. Achieving near human results certainly opens the door for automation in this human dominated domain.

Acknowledgments. We thank Dr. Niloy Ganguly for providing valuable comments and feedback.

References

1. Ries, A., Trout, J.: Positioning: The Battle for Your Mind (1981)
2. Robertson, K.: Strategically desirable brand name characteristics. J. Consum. Mark. **6**, 61–71 (1989)
3. Yorkston, E., Menon, G.: A sound idea: phonetic effects of brand names on consumer judgments. J. Consum. Res. **31**, 43–51 (2004)
4. Little, G., Chilton, L.B., Goldman, M., Miller, R.C.: Exploring iterative and parallel human computation processes. In: Proceedings of the ACM SIGKDD Workshop on Human Computation, pp. 68–76. ACM (2010)
5. Clements, J.: Generating 56-bit passwords using markov models (and charles dickens). arXiv preprint arXiv:1502.07786 (2015)
6. Allbery, B.: PWGEN-random but pronounceable password generator. USENET posting in comp. sources. misc (1988)
7. Crawford, H., Aycock, J.: Kwyjibo: automatic domain name generation. Softw. Pract. Exp. **38**, 1561–1567 (2008)

8. Bauer, L.: English Word-Formation. Cambridge University Press, Cambridge (1983)
9. Kondrak, G.: Phonetic alignment and similarity. Comput. Humanit. **37**, 273–291 (2003)
10. Hedlund, G.J., Maddocks, K., Rose, Y., Wareham, T.: Natural language syllable alignment: from conception to implementation. In: Proceedings of the Fifteenth Annual Newfoundland Electrical and Computer Engineering Conference (2005)
11. Özbal, G., Strapparava, C.: A computational approach to the automation of creative naming. In: Proceedings of the 50th Annual Meeting of the Association for Computational Linguistics: Long Papers-Volume 1, pp. 703–711. Association for Computational Linguistics (2012)
12. Özbal, G., Strapparava, C., Guerini, M.: Brand pitt: a corpus to explore the art of naming. In: Proceedings of the Eighth International Conference on Language Resources and Evaluation (LREC-2012), Istanbul, Turkey, May, Citeseer (2012)
13. Bird, S., Klein, E., Loper, E.: Natural Language Processing with Python. O'Reilly Media, Inc., Beijing (2009)
14. Toutanova, K., Klein, D., Manning, C.D., Singer, Y.: Feature-rich part-of-speech tagging with a cyclic dependency network. In: Proceedings of the 2003 Conference of the North American Chapter of the Association for Computational Linguistics on Human Language Technology-Volume 1, pp. 173–180. Association for Computational Linguistics (2003)
15. Miller, G.A.: Wordnet: a lexical database for english. Commun. ACM **38**, 39–41 (1995)
16. Goldhahn, D., Eckart, T., Quasthoff, U.: Building large monolingual dictionaries at the leipzig corpora collection: from 100 to 200 languages. In: LREC, pp. 759–765 (2012)
17. Kincaid, Jr., J.P., Rogers, R.L., Chissom, B.S.: Derivation of new readability formulas (automated readability index, fog count and flesch reading ease formula) for navy enlisted personnel. Technical report, DTIC Document (1975)
18. Schiavoni, S., Maggi, F., Cavallaro, L., Zanero, S.: Phoenix: DGA-based botnet tracking and intelligence. In: Dietrich, S. (ed.) DIMVA 2014. LNCS, vol. 8550, pp. 192–211. Springer, Cham (2014). https://doi.org/10.1007/978-3-319-08509-8_11
19. Webster, M.: The Official Scrabble Players Dictionary. Springfield, USA (2005)
20. Chen, S.F., Goodman, J.: An empirical study of smoothing techniques for language modeling. In: Proceedings of the 34th Annual Meeting on Association for Computational Linguistics, pp. 310–318. Association for Computational Linguistics (1996)
21. Danescu-Niculescu-Mizil, C., Cheng, J., Kleinberg, J., Lee, L.: You had me at hello: how phrasing affects memorability. In: Proceedings of the 50th Annual Meeting of the Association for Computational Linguistics: Long Papers-Volume 1, pp. 892–901. Association for Computational Linguistics (2012)
22. Michel, J.B., et al.: Quantitative analysis of culture using millions of digitized books. Science **331**, 176–182 (2011)
23. Joachims, T.: Optimizing search engines using click through data. In: Proceedings of the Eighth ACM SIGKDD International Conference on Knowledge Discovery and Data Mining, pp. 133–142. ACM (2002)
24. Carbonell, J., Goldstein, J.: The use of MMR, diversity-based reranking for reordering documents and producing summaries. In: Proceedings of the 21st Annual International ACM SIGIR Conference on Research and Developement in Information Retrieval, pp. 335–336. ACM (1998)

25. Modani, N., Khabiri, E., Srinivasan, H., Caverlee, J.: Creating diverse product review summaries: a graph approach. In: Wang, J., Cellary, W., Wang, D., Wang, H., Chen, S.-C., Li, T., Zhang, Y. (eds.) WISE 2015. LNCS, vol. 9418, pp. 169–184. Springer, Cham (2015). https://doi.org/10.1007/978-3-319-26190-4_12
26. Järvelin, K., Kekäläinen, J.: Cumulated gain-based evaluation of IR techniques. ACM Trans. Inf. Syst. (TOIS) **20**, 422–446 (2002)

Radiological Text Simplification
Using a General Knowledge Base

Lionel Ramadier[1][(✉)] and Mathieu Lafourcade[2]

[1] Department of Radiology, University Hospital of Montpellier, Montpellier, France
l-ramadier@chu-montpellier.fr
[2] LIRMM, University of Montpellier, Montpellier, France
mathieu.lafourcade@lirmm.fr

Abstract. In the medical domain, text simplification is both a desirable and a challenging natural language processing task. Indeed, first, medical texts can be difficult to understand for patient, because of the presence of specialized medical terms. Replacing these difficult terms with easier words can lead to improve patient's understanding. In this paper, we present a lexical network based method to simplify health information in French language. We deal with semantic difficulty by replacement difficult term with supposedly easier synonyms or by using semantically related term with the help of a French lexical semantic network. We extract semantic and lexical information present in the network. In this paper, we present such a method for text simplification along with its qualitative evaluation.

Keywords: NLP · BioNLP · Text simplification

1 Introduction

Text simplification (TS) is a challenging natural language processing (NLP) task. It is an operation to simplify an existing corpus, texts or sentences while the underlying meaning and information remain the same. The main goal of TS is to make information more accessible to the large numbers of people with reduced literacy. TS can be viewed as an example of a monolingual translation task, where the source language needs to be translated into a simplified version of the same language.

Its application to the medical domain is of special importance. Understanding medical text might be particularly challenging for laymen readers who are not used to looking up unknown terms while reading. So, making record information available to the patients is a prioritized goal for many countries. It is crucial for patients to understand texts from the medical domain.

Medical texts are difficult to understand for non expert [1] because doctors often write with specialized terms (*ataxia*) and abbreviations (*HIV* for *human immunodeficiency virus*) which may require advance knowledge of medicine or

© Springer Nature Switzerland AG 2018
A. Gelbukh (Ed.): CICLing 2017, LNCS 10762, pp. 617–627, 2018.
https://doi.org/10.1007/978-3-319-77116-8_46

biology. There is a mismatch between the content delivered by medical practitioners and the consumers who have a limited health knowledge. Medical terms have been shown to be obstacles for patients [2,3]. Moreover, medical reports are often written under time pressure by professionals for professionals. This results in a telegraphic style, with omissions, abbreviations, and sometimes misspellings [4].

However, there has been relatively limited prior research on tools to automate the simplification medical texts [5]. One tool that address this problem is a system built by Elhadad [6]. The author identifies difficult terms and retrieves definitions thanks to Google search engine. The tool improve reader's comprehension by an average 1.5 points on a 5 point scale. An other method [7] identified difficult terms in the document and try to simplify by replacing them with synonyms or by explaining them using easier words. The results reported correct simplification in 68% of identified difficult word. For the French language, few studies deal with the issue of simplification of medical texts. A study [8] try to simplify a dialogue task between a virtual patient and a doctor.

The aim of our study is to simplify French radiology reports thanks to a general knowledge base that contains both general and specialized knowledge. For this issue, we use not only synonyms but also other hierarchically and/or semantically relates terms. In this paper, after a presentation of related work (Sect. 2) we present our method (Sect. 3) of semantic simplification thanks to a French lexical network (JeuxDeMots (JDM)), then we discuss experiments and analyze the results (Sects. 4 and 5).

2   Related Work

The level of difficulty can vary between kind of medical texts [9], and even brochures for patients can be difficult to understand [10]. Medical texts, such as radiology reports, are characterized by sentences containing a lot of medical terms and a frequent use of abbreviation form. Previous studies [11] have shown that replacing difficult words with easier synonyms can reduce the level of difficult in a medical text. This synonym replacement method has been evaluated on medical English text [12,15] and also on Swedish medical Text [11]. Semi-automatic adaption of word choice has been evaluated on English medical text [12] and automatic adaption on Swedish non-medical text [13]. Study used synonym lexicons and replaced difficult word with easier synonyms. The level of difficulty of a word was determined by measuring its frequency in a general corpus. In [7], the author used two sources of vocabulary knowledge: Unified Medical Language System[1] and the open-access collaborative (OAC) consumer health vocabulary (CHV). They employ two strategies to reduce the vocabulary difficulty of medical reports:

– synonym replacement
– explanation insertion

---

[1] https://www.nlm.nih.gov/research/umls/.

Leroy et al. [12] developed an algorithm that uses term familiarity to identify difficult text and select easier alternatives from lexical resources such as WordNet, UMLS and Wiktionary. Their results show that term familiarity is a valuable component in simplifying text in an efficient manner.

For synonym replacement to be a meaningful method for text simplification, there have to exist synonyms that are near enough not to change the content of what is written. For describing medical concepts, there is, however, often one set of terms that are used by medical professionals, whereas another set of easier terms are used by patients [14]. This means that synonym replacement could have a large potential for simplifying medical text. For English, there is a consumer health vocabulary initiative connecting laymen's expressions to technical terminology [5] as well as several medical terminology containing synonyms like MeSH[2] and SNOMED CT[3]. MeSH (Medical Subject Headings) is the National Library of Medicine's controlled vocabulary thesaurus, used for indexing articles for the MEDLINE database and SNOMED CT is one of a suite of designated standards for use in U.S. Federal Government systems for the electronic exchange of clinical health information.

In the radiology domain, several studies have shown that radiology reports are among the most difficult form of clinical text to understand [16]. The aim of a Swedish study [17] is to be able to develop a text simplification tool enabling patients to better understand text for a large corpus of Swedish radiology reports.

3 Our Approach

We study simplification of one medical text genre, radiology reports. We use a replacement method by synonym but also by hierarchical relations. This latter are very useful because a term can be explained as a specific incidence of its parents. For example *hepatocellular carcinoma* is a *tumor of liver* or *pulmonary embolism* is a *lung disease*. The knowledge base on which our radiology reports simplification relies is the French lexical network JDM[4] [18].

3.1 Resources

The JeuxDeMots Lexical Network. JDM network is a lexical-semantic graph for the French language whose lexical relations are generated both through GWAP (Games With A Purpose) and via a contributory tool called Diko (manual insertion and automatic inferences with validations) [19]. At the time of this writing (February, 2017), the JDM network contains over 67 millions of relations between around one million of terms. The following table provides an order of size of the amount of information we have at our disposal about the radiology areas (Table 1).

---

[2] www.nlm.nih.gov/mesh/.

[3] http://www.snomed.org/snomed-ct.

[4] http://www.jeuxdemots.org/.

Table 1. Number of relations of some key terms within the JDM lexical network.

Terms	Outgoing links	Incoming links
Medicine	22 108	24 100
Anatomy	10 477	11 453
Radiology	382	502
Medical imaging	541	556

It exists 80 lexico-semantic relations into the network but in this work we use only three different relations in addition to lexical information (Fig. 1).

$r\_synonym$ synonyms or quasi-synonyms
r_syn_strict strict synonyms
r_isa generic term
r_equiv acronym or abbreviation

Fig. 1. The relations used for medical simplification text

We use this network because it combines weight and annotations [20] on typed relations between terms. In the network JDM, the relations are weighted, the weight reflects the strength of association between terms. In specialized knowledge, the correlation between the weight of the relation and its importance is not strict. This is why it appears interesting to use annotations for some relations as they can be of a great help in the medical area. This annotations could help us in the task of text simplification (thanks to the annotation *ordinary language*) (Fig. 3). A given relation to be annotated is reified (represented by a specific node) and this node is associated to various annotations and any other regular terms. The annotation relation type is a kind of relation among others Fig. 2.

The Corpus of Radiology Reports. The corpus contains 30 000 radiology reports, from different institutions, concerning the different medical imaging techniques (MRI, scanner, medical ultrasound, X-ray radiology, vascular radiology, etc). These reports are written in semi-structured way. They are generally divided into four parts. Each part is written by the radiologist in a very free style, often with a profusion of acronyms and specialized terms. The records are deidentified with anonymized serial numbers for individual patients. The reports are examinations of all patients for this period i.e. both genders and all ages from babies to a 98 year-old.

3.2 Method

To support patient radiology report comprehension, it is important to identify words that matter most to patients in their reports. The identification of the

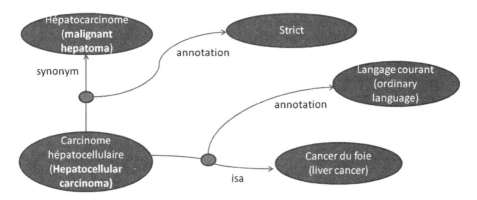

Fig. 2. A given relation to be annotated is reified (represented by a specific node, here with green circles) and this node is associated to various annotations and any other regular terms. (Color figure online)

compound terms is made compared to the content of JDM, in a first step. We use the underscore to separate the two parts of a compound word so that it is considered as an entity at the time of the extraction (tibia_fracture).

In a second step, we extract difficult term by using traditional methods i.e. term and document frequency (TF and DF) to calculate the IDF (Inverse Document Frequency). For each difficult term, we look at to the content of JDM for the synonym or hyperonym relation. A difficult word can have several syn-

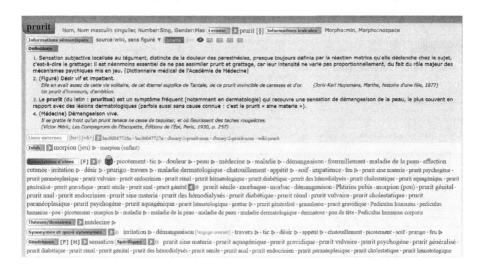

Fig. 3. Example of term *prurit* (*pruritus*) with annotations between brackets. Several annotations are possible for a given relation like *ordinary language* (*langage courant* in French)

le patient a une *aphasie* depuis deux jours. (The patient has an *aphasia* for two days)
Le patient a un **mutisme** depuis deux jours(The patient has a **mutism** for two days)

Le patient se plaint de *céphalée* (The patient complains of cephalgia).
Le patient se plaint de **maux de tête** (The patient complains of *headache*).

Symptôme : *hématurie* (symptom:*hematuria*)
Symptôme : **sang dans les urines** (symptom:**blood in the urine**)

Fig. 4. Example of sentence translation. The replaced terms are on bold

onyms and sometimes this latter are not easier. For instance, in French language
the term *carcinome hepatocellulaire* (*Hepatocellular carcinoma*) has for synonym
hepatocarcinome. In this case, we can look at the hyperonym relation. To choose
the right easier term, we use relation annotations [21]. If the relation has an
annotation (*langage courant* in French or *ordinary language*), then the original words are replaced by pertinent synonym or hyperonym. For instance, the
term *aphasia* will be replaced by *mutism* because in the network JDM *aphasia*
r_synonym *mutism* (ordinary language) (Fig. 4). The multi-word *hepatocellular
carcinoma* will be replaced by the hyperonym *liver cancer* thanks to the network
JDM (hepatocellular carcinoma r_isa *liver cancer* (*ordinary language*)). In this
case, we have chosen the hyperonym relation because the different synonym of
hepatocellular carcinoma are not easier to understand. We systematically replace
the term *anterior* by *front of* and *posterior* by *behind*. Some abbreviations (e.g.
"MVC") will be replaced with their more understandable full names (e.g. "motor

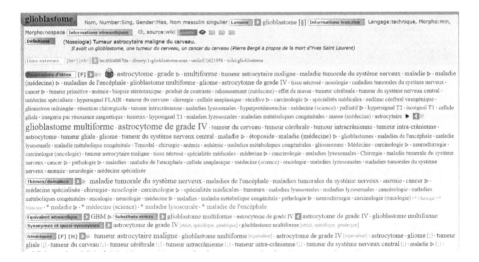

Fig. 5. Example of term *glioblastome* (*glioblastoma*) with annotations between brackets. Several annotations are possible for a given relation.

vehicle collision"), but the abbreviations AIDS will not be replaced by their full names because the patient know and understand this abbreviation. We can distinguish the two case, because in the network the relation annotation allows to distinguish both cases.

If the difficult term has not a relation annotation, we apply another approach. If the synonym or hyperonym are compound terms like *tumeur du cerveau* (*brain tumor*) do not have relation annotations, we extract semantic information for each word that makes up the compound terms from the JDM network. Indeed, lexical information indicates whether a word is part of the common language or not. For instance, glioblastoma (Fig. 5) is a brain tumor (hyperonym relation). As this hyperonym relation has not an annotation, we extract for *brain* and *tumor*, semantic information from JDM. Each word belongs to the ordinary language, then we replace *glioblastoma* by *brain tumor*.

4   Experiment and Results

4.1   Experiment

We use a corpus subset (200 radiology) reports, and we simplified them using our method (thanks to the network JDM).

For the manual evaluation, 250 sentences were randomly selected for human review and cloze testing (a standard comprehension test procedure) [7]. An expert reviewed the translations for corrections. According to the standard cloze procedure, every 5th word of each report was replaced with a blank space.

We have recruited 4 persons, who were not doctors but highly educated (1 at undergraduate school level and 3 at graduate school level), to evaluate the system. Each subject has evaluated the original and simplified reports. They were asked to fill in the blank spaces.

We calculate for each report a cloze score which is the percentage of answers that matched with the deleted word. We compared the average cloze scores of the original and translated radiology reports.

4.2   Results

On average, 10.6 terms were simplified in reports. Most of the simplifications (75%) were deemed correct by an expert reviewer. For 12% of the sentences, the replaced word has a slightly different meaning for the original word. This errors can explain because sometimes the synonym was not strict. For instance a cyst (kyste in French) and abscess (abcès in French) are synonym or quasi-synonym in the network but in the field on medicine, the meaning are different. We show some words typical for a professional language that have been replaced with every day French words, or abbreviations that have been replaced by an expanded form (Table 2).

If the cloze score is between 50–60%, then the document should be readable. In Tables 3 and 4, we show the results for original and simplified reports.

The cloze score of the original radiology reports (18%) indicate that these documents are difficult for lay people to understand.

Table 2. Examples of replaced terms.

Original terms	Replaced with
aphasie (aphasia)	mutisme (mutism)
céphalée (cephalgia)	maux de tête (headache)
prurit (pruritus)	démangeaison (itch)
dyspnée (dyspnea)	difficulté à respirer (shortness of breath)
glioblastome (glioblastoma)	tumeur maligne du cerveau (brain tumor)
CHC (hepatocellular carcinoma)	cancer du foie (liver cancer)
arthrite (arthritis)	inflammation des articulations (joint-inflammation)
TS (SA)	tentative de suicide (suicide attempt)

Table 3. Cloze score for original and simplified reports using only annotation relation.

Original reports	Simplified reports
18%	48%

Table 4. Cloze score for original and simplified reports using annotation and semantic information.

Original reports	Simplified reports
18%	57%

5   Discussion

We describe a text simplification system for a French radiology corpus. The need to improve the understanding of medical reports for patient is important. The patient want more and more to understand the different medical records. The radiology reports are the most difficult to understand for the consumers. The cloze score for the original report is lower than other study [7] that deal with various medical reports (discharge summaries, surgery report, only one radiology reports). We have implemented a prototype to improve the readability for lay readers. This study focused on vocabulary difficulty. Our method relies on the JDM network to try to simplify difficult words. Indeed, to choose an easier term, we make use of relation annotations present in the lexical semantic network. This method allows us to choose the right term easily. 80% of replaced term seem helpful with the same meaning. If we use only the relation annotations for the task of simplification, we get a cloze score of 48%. If we use the second approach based on semantic information we improve our results and we reach a cloze score of 57%.

But 35% of difficult terms are not replaced because they have not annotations (ordinary language) in the network. It needs to improve the coverage of the annotations inside the network to reach better results. The manual evaluation also showed that the original semantic meaning had been slightly altered in some sentences. In some case, some words are not strictly synonym and the replacement involve a slightly change of meaning. For instance, the replacement oedema by swelling entail a change of meaning. Moreover, in order to include abbreviations and acronyms in the synonym replacement method studied here, an abbreviation disambiguation needs to be carried out first. An acronym or an abbreviations can have two different meanings in the field of medicine.

The average cloze test of the simplified reports are high, it reach score 50–60% to be fairly readable. Our results are close to those [7] although our corpus is larger and contains only radiology reports.

Our system needs much improvement. We intend to simplify the syntax. Another task is to improve the coverage of the annotation relations (*ordinary language*) inside the network.

6   Conclusion

We have developed a system which goal is to improve patient comprehension. The results presented here are preliminary but are very promising. In this work, we have used the JeuxDeMots lexical-semantic network as a support of knowledge. Although this network is general, it contains many specialty data, including medicine/radiology that may helpful in the simplification task framework.

The difficulty of a word was assessed by the presence or not of relation annotations in the network JDM. It seems a good way to evaluate the difficulty of a word. The replacement was mainly evaluated by the cloze test. Studies on a larger reader group are required to draw any conclusions on the effect of our method for assessment of simplification. We have to recognize errors in order to eliminate them. An another future improvement is to use the definitions present in the network in order to generate explanations.

In a future work, another challenge is to simplify the syntax of radiology reports. A previous study [22] showed significant differences in syntactic content and complexity between medical discharge summaries and everyday English papers. An other survey emphasized the difficulty of syntactic text simplification [23]. For this task, we would be able to realize a grammar simplification (for instance, long sentences were broke down into two or more shorter sentences).

We also plan to test our approach in other medical domains, such as for example the oncology, because JDM contains data about this domain.

References

1. Keselman, A., Smith, C.A.: A classification of errors in lay comprehension of medical documents. J. Biomed. Inf. **45**(6), 1151–1163 (2012)
2. Chapman, K., Abraham, C., Jenkins, V., Fallowfield, L.: Lay understanding of terms used in cancer consultations. PsychoOncology **12**(6), 557–566 (2003)

3. Lerner, E.B., Jehle, D.V., Janicke, D.M., Moscati, R.M.: Medical communication: do our patients understand? Am. J. Emerg. Med. **18**(7), 764–766 (2000)
4. Hagège, Caroline, Marchal, Pierre, Gicquel, Quentin, Darmoni, Stefan, Pereira, Suzanne, Metzger, Marie-Hélène: Linguistic and Temporal Processing for Discovering Hospital Acquired Infection from Patient Records. In: Riaño, David, ten Teije, Annette, Miksch, Silvia, Peleg, Mor (eds.) KR4HC 2010. LNCS (LNAI), vol. 6512, pp. 70–84. Springer, Heidelberg (2011). https://doi.org/10.1007/978-3-642-18050-7_6
5. Keselman, A., Logan, R., Smith, C.A., Leroy, G., Zeng-Treitler, Q.: Developing informatics tools and strategies for consumer-centered health communication. J. Am. Med. Inf. Assoc. **15**(4), 473–483 (2008)
6. Comprehending Technical Texts: Predicting and Defining Unfamiliar Terms. AMIA, Maryland (2006)
7. Zeng-Treitler, Q., Goryachev, S., Kim, H., Keselman, A., Rosendale, D.: Making Texts in Electronic Health Records Comprehensible to Consumers: a Prototype Translator, pp. 846–850. AMIA, Maryland (2007)
8. Pierre, L.C.L.D.B., Rosset, Z.S.: Managing linguistic and terminological variation in a medical dialogue system. In: LREC, pp. 3167–3173. Portoroz (2016)
9. Leroy, G., Helmreich, S., Cowie, J.R.: The influence of text characteristics on perceived and actual difficulty of health information. Int. J. Med. Inf. **79**(6), 438–449 (2010)
10. Kokkinakis, Dimitrios, Forsberg, Markus, Johansson Kokkinakis, Sofie, Smith, Frida, Öhlen, Joakim: Literacy Demands and Information to Cancer Patients. In: Sojka, Petr, Horák, Aleš, Kopeček, Ivan, Pala, Karel (eds.) TSD 2012. LNCS (LNAI), vol. 7499, pp. 64–71. Springer, Heidelberg (2012). https://doi.org/10.1007/978-3-642-32790-2_7
11. Skeppstedt, E.A.T.F.M., Kvist, M.: Medical text simplification using synonym replacement: adapting assessment of word difficulty to a compounding language. In: Proceedings of the 3rd Workshop on Predicting and Improving Text Readability for Target Reader Populations (PITR)@ EACL, pp. 57–65 (2014)
12. Leroy, G., Endicott, J.E., Mouradi, O., Kauchak, D., Just, M.: Improving Perceived and Actual Text Difficulty for Health Information Consumers Using Semi-automated Methods. AMIA, Maryland (2012)
13. Keskisarkka, R.: Automatic text simplification via synonym replacement. In: Proceedings of Swedish Language Technology Conference (2012)
14. Kokkinakis, D., Gronostaj, M.T.: Lay language versus professional language within the cardiovascular subdomain a contrastive study. Proceedings of BIO'06 (2006)
15. Slaughter, L., Keselman, A., Kushniruk, A., Patel, V.L.: A framework for capturing the interactions between laypersons' understanding of disease, information gathering behaviors, and actions taken during an epidemic. J. Biomed. Inf. **38**(4), 298–313 (2005)
16. Keselman, A., Slaughter, L., Arnott-Smith, C., Kim, H., Divita, G., Browne, A., Zeng-Treitler, Q.: Towards consumer-friendly PHRs: patients' experience with reviewing their health records. In: AMIA Annual Symposium Proceedings, vol. 2007, p. 399. American Medical Informatics Association, Maryland (2007)
17. Kvist, M., Velupillai, S.: Professional language in swedish radiology reports characterization for patient-adapted text simplification. In: Scandinavian Conference on Health Informatics 2013, pp. 55–59. Linköping University Electronic Press, Denmark, 20 Aug 2013

18. Lafourcade, M.: Making people play for lexical acquisition with the JeuxDeMots prototype. In: SNLP'07: 7th International Symposium on Natural Language Processing, p. 7 (2007)

19. Lafourcade, M., Joubert, A., Le Brun, N.: Games with a Purpose (GWAPS). Wiley, New York (2015). ISBN: 978-1-84821-803-1

20. Ramadier, Lionel, Zarrouk, Manel, Lafourcade, Mathieu, Micheau, Antoine: **Spreading Relation Annotations in a Lexical Semantic Network Applied to Radiology**. In: Gelbukh, Alexander (ed.) CICLing 2014. LNCS, vol. 8403, pp. 40–51. Springer, Heidelberg (2014). https://doi.org/10.1007/978-3-642-54906-9_4

21. Ramadier, L., Zarrouk, M., Lafourcade, M., Micheau, A.: Inferring relations and annotations in semantic network: application to radiology. Comput. Sist. **18**(3), 455–466 (2014)

22. Campbell, D.A., Johnson, S.B.: Comparing syntactic complexity in medical and non-medical corpora. In: Proceedings of the AMIA Symposium, p. 90. American Medical Informatics Association, Maryland (2001)

23. Kandula, S., Curtis, D., Zeng-Treitler, Q.: A semantic and syntactic text simplification tool for health content. AMIA Annu. Symp. Proc. **2010**, 366–370 (2010)

Mining Supervisor Evaluation and Peer Feedback in Performance Appraisals

Girish Keshav Palshikar, Sachin Pawar[(✉)], Saheb Chourasia,
and Nitin Ramrakhiyani

TCS Research, Tata Consultancy Services Limited, 54B Hadapsar Industrial Estate,
Pune 411013, India
{gk.palshikar,sachin7.p,saheb.c,nitin.ramrakhiyani}@tcs.com

Abstract. Performance appraisal (PA) is an important HR process to periodically measure and evaluate every employee's performance vis-a-vis the goals established by the organization. A PA process involves purposeful multi-step multi-modal communication between employees, their supervisors and their peers, such as self-appraisal, supervisor assessment and peer feedback. Analysis of the structured data and text produced in PA is crucial for measuring the quality of appraisals and tracking actual improvements. In this paper, we apply text mining techniques to produce insights from PA text. First, we perform sentence classification to identify strengths, weaknesses and suggestions of improvements found in the supervisor assessments and then use clustering to discover broad categories among them. Next we use multi-class multi-label classification techniques to match supervisor assessments to predefined broad perspectives on performance. Finally, we propose a short-text summarization technique to produce a summary of peer feedback comments for a given employee and compare it with manual summaries. All techniques are illustrated using a real-life dataset of supervisor assessment and peer feedback text produced during the PA of 4528 employees in a large multi-national IT company.

1 Introduction

Performance appraisal (PA) is an important HR process, particularly for modern organizations that crucially depend on the skills and expertise of their workforce. The PA process enables an organization to periodically measure and evaluate every employee's performance. It also provides a mechanism to link the goals established by the organization to its each employee's day-to-day activities and performance. Design and analysis of PA processes is a lively area of research within the HR community [10,13,20,22].

The PA process in any modern organization is nowadays implemented and tracked through an IT system (the *PA system*) that records the interactions that happen in various steps. Availability of this data in a computer-readable database opens up opportunities to analyze it using automated statistical, data-mining and text-mining techniques, to generate novel and actionable

© Springer Nature Switzerland AG 2018
A. Gelbukh (Ed.): CICLing 2017, LNCS 10762, pp. 628–641, 2018.
https://doi.org/10.1007/978-3-319-77116-8_47

insights/patterns and to help in improving the quality and effectiveness of the PA process [1,15,19]. Automated analysis of large-scale PA data is now facilitated by technological and algorithmic advances, and is becoming essential for large organizations containing thousands of geographically distributed employees handling a wide variety of roles and tasks.

A typical PA process involves purposeful multi-step multi-modal communication between employees, their supervisors and their peers. In most PA processes, the communication includes the following steps: (i) in *self-appraisal*, an employee records his/her achievements, activities, tasks handled etc.; (ii) in *supervisor assessment*, the supervisor provides the criticism, evaluation and suggestions for improvement of performance etc.; and (iii) in *peer feedback* (aka 360° *view*), the peers of the employee provide their feedback. There are several business questions that managers are interested in. Examples:

1. For my workforce, what are the broad categories of strengths, weaknesses and suggestions of improvements found in the supervisor assessments?
2. For my workforce, how many supervisor comments are present for each of a given fixed set of perspectives (which we call *attributes*), such as FUNCTIONAL_EXCELLENCE, CUSTOMER_FOCUS, BUILDING_EFFECTIVE_TEAMS etc.?
3. What is the summary of the peer feedback for a given employee?

In this paper, we develop text mining techniques that can automatically produce answers to these questions. Since the intended users are HR executives, ideally, the techniques should work with minimum training data and experimentation with parameter setting. These techniques have been implemented and are being used in a PA system in a large multi-national IT company.

The rest of the paper is organized as follows. Section 2 summarizes related work. Section 3 summarizes the PA dataset used in this paper. Section 4 applies sentence classification algorithms to automatically discover three important classes of sentences in the PA corpus viz., sentences that discuss strengths, weaknesses of employees and contain suggestions for improving her performance. Section 5 considers the problem of mapping the actual targets mentioned in strengths, weaknesses and suggestions to a fixed set of attributes. In Sect. 6, we discuss how the feedback from peers for a particular employee can be summarized. In Sect. 7 we draw conclusions and identify some further work.

2   Related Work

We first review some work related to sentence classification. Semantically classifying sentences (based on the sentence's purpose) is a much harder task, and is gaining increasing attention from linguists and NLP researchers. McKnight and Srinivasan [12] and Yamamoto and Takagi [23] used SVM to classify sentences in biomedical abstracts into classes such as INTRODUCTION, BACKGROUND, PURPOSE, METHOD, RESULT, CONCLUSION. Cohen et al. [3] applied SVM and other techniques to learn classifiers for sentences in emails into classes, which are speech acts defined by a verb-noun pair, with verbs such as `request, propose, amend, commit, deliver` and nouns such as `meeting, document, committee`; see also [2]. Khoo et al. [9] uses various classifiers to classify sentences in emails into classes such as APOLOGY, INSTRUCTION, QUESTION, REQUEST, SALUTATION, STATE-MENT, SUGGESTION, THANKING etc. Qadir and Riloff [17] proposes several filters and classifiers to classify sentences on message boards (community QA systems) into 4 speech acts: COMMISSIVE (speaker commits to a future action), DIRECTIVE (speaker expects listener to take some action), EXPRESSIVE (speaker expresses his or her psychological state to the listener), REPRESENTATIVE (represents the speaker's belief of something). Hachey and Grover [7] used SVM and maximum entropy classifiers to classify sentences in legal documents into classes such as FACT, PROCEEDINGS, BACKGROUND, FRAMING, DISPOSAL; see also [18]. Deshpande et al. [5] proposes unsupervised linguistic patterns to classify sentences into classes SUGGESTION, COMPLAINT.

There is much work on a closely related problem viz., classifying sentences in dialogues through dialogue-specific categories called *dialogue acts* [21], which we will not review here. Just as one example, Cotterill [4] classifies questions in emails into the dialogue acts of YES_NO_QUESTION, WH_QUESTION, ACTION_REQUEST, RHETORICAL, MULTIPLE_CHOICE etc.

We could not find much work related to mining of performance appraisals data. Pawar et al. [16] uses kernel-based classification to classify sentences in both performance appraisal text and product reviews into classes SUGGESTION, APPRECIATION, COMPLAINT. Apte et al. [1] provides two algorithms for matching the descriptions of goals or tasks assigned to employees to a standard template of model goals. One algorithm is based on the co-training framework and uses goal descriptions and self-appraisal comments as two separate perspectives. The second approach uses semantic similarity under a weak supervision framework. Ramrakhiyani et al. [19] proposes label propagation algorithms to discover aspects in supervisor assessments in performance appraisals, where an aspect is modelled as a verb-noun pair (e.g. `conduct training, improve coding`).

3   Dataset

In this paper, we used the supervisor assessment and peer feedback text produced during the performance appraisal of 4528 employees in a large multi-national IT company. The corpus of supervisor assessment has 26972 sentences. The summary statistics about the number of words in a sentence is: min:4 max:217 average:15.5 STDEV:9.2 Q1:9 Q2:14 Q3:19.

4   Sentence Classification

The PA corpus contains several classes of sentences that are of interest. In this paper, we focus on three important classes of sentences viz., sentences that discuss strengths (class STRENGTH), weaknesses of employees (class WEAK-NESS) and suggestions for improving her performance (class SUGGESTION). The strengths or weaknesses are mostly about the performance in work carried out, but sometimes they can be about the working style or other personal qualities. The classes WEAKNESS and SUGGESTION are somewhat overlapping; e.g., a suggestion may address a perceived weakness. Following are two example sentences in each class.

STRENGTH:

- `Excellent technology leadership and delivery capabilities along with ability to groom technology champions within the team.`
- `He can drive team to achieve results and can take pressure.`

WEAKNESS:

- `Sometimes exhibits the quality that he knows more than the others in the room which puts off others.`
- `Tends to stretch himself and team a bit too hard.`

SUGGESTION:

- `X has to attune himself to the vision of the business unit and its goals a little more than what is being currently exhibited.`
- `Need to improve on business development skills, articulation of business and solution benefits.`

Several linguistic aspects of these classes of sentences are apparent. The subject is implicit in many sentences. The strengths are often mentioned as either noun phrases (NP) with positive adjectives (`Excellent technology leadership`) or positive nouns (`engineering strength`) or through verbs with positive polarity (`dedicated`) or as verb phrases containing positive adjectives (`delivers innovative solutions`). Similarly for weaknesses, where negation is more frequently used (`presentations are not his forte`), or alternatively, the polarities of verbs (`avoid`) or adjectives (`poor`) tend to be negative. However, sometimes the form of both the strengths and weaknesses is the same, typically a stand-alone sentiment-neutral NP, making it difficult to distinguish between them; e.g., `adherence to timing` or `timely closure`. Suggestions often have an imperative mood and contain secondary verbs such as `need to, should, has to`. Suggestions are sometimes expressed using comparatives (`better process compliance`). We built a simple set of patterns for each of the 3 classes on the POS-tagged form of the sentences. We use each set of these patterns as an unsupervised sentence classifier for that class. If a particular sentence matched with patterns for multiple classes, then we have simple tie-breaking rules for picking the final class. The pattern for the STRENGTH class looks for the presence of positive words/phrases like `takes ownership, excellent, hard working, commitment`, etc. Similarly, the pattern for the WEAKNESS class looks for the presence of negative words/phrases like `lacking, diffident, slow learner, less focused`, etc. The SUGGESTION pattern not only looks for keywords like `should, needs to` but also for POS based pattern like "a verb in the base form (VB) in the beginning of a sentence".

We randomly selected 2000 sentences from the supervisor assessment corpus and manually tagged them (dataset D1). This labelled dataset contained 705, 103, 822 and 370 sentences having the class labels STRENGTH, WEAKNESS, SUGGESTION or OTHER respectively. We trained several multi-class classifiers on this dataset. Table 1 shows the results of 5-fold cross-validation experiments on dataset D1. For the first 5 classifiers, we used their implementation from the SciKit Learn library in Python (scikit-learn.org). The features used for these classifiers were simply the sentence words along with their frequencies. For the last 2 classifiers (in Table 1), we used our own implementation. The overall *accuracy* for a classifier is defined as $A = \frac{\#correct\_predictions}{\#data\_points}$, where the denominator is 2000 for dataset D1. Note that the pattern-based approach is unsupervised i.e., it did not use any training data. Hence, the results shown for it are for the entire dataset and not based on cross-validation.

Table 1. Results of 5-fold cross validation for sentence classification on dataset D1.

Classifier	STRENGTH			WEAKNESS			SUGGESTION			
	P	R	F	P	R	F	P	R	F	A
Logistic Regression	0.715	0.759	0.736	0.309	0.204	0.246	0.788	0.749	0.768	0.674
Multinomial Naive Bayes	0.719	0.723	0.721	0.246	0.155	0.190	0.672	0.790	0.723	0.646
Random Forest	0.681	0.688	0.685	0.286	0.039	0.068	0.730	0.734	0.732	0.638
AdaBoost	0.522	0.888	0.657	0.265	0.087	0.131	0.825	0.618	0.707	0.604
Linear SVM	0.718	0.698	0.708	0.357	0.194	0.252	0.744	0.759	0.751	0.651
SVM with ADWSK [16]	0.789	0.847	**0.817**	0.491	0.262	0.342	0.844	0.871	**0.857**	**0.771**
Pattern-based	0.825	0.687	0.749	0.976	0.494	**0.656**	0.835	0.828	0.832	0.698

4.1   Comparison with Sentiment Analyzer

We also explored whether a sentiment analyzer can be used as a baseline for identifying the class labels STRENGTH and WEAKNESS. We used an implementation of sentiment analyzer from TextBlob[1] to get a polarity score for each sentence. Table 2 shows the distribution of positive, negative and neutral sentiments across the 3 class labels STRENGTH, WEAKNESS and SUGGESTION. It can be observed that distribution of positive and negative sentiments is almost similar in STRENGTH as well as SUGGESTION sentences, hence we can conclude that the information about sentiments is not much useful for our classification problem.

4.2   Discovering Clusters Within Sentence Classes

After identifying sentences in each class, we can now answer question (1) in Sect. 1. From 12742 sentences predicted to have label STRENGTH, we extract

---

[1] https://textblob.readthedocs.io/en/dev/.

Table 2. Results of TextBlob sentiment analyzer on the dataset D1

Sentence Class	Positive	Negative	Neutral
STRENGTH	544	44	117
WEAKNESS	44	24	35
SUGGESTION	430	52	340

Table 3. 5 representative clusters in strengths.

Strength cluster	Count
motivation expertise knowledge talent skill	1851
coaching team coach	1787
professional career job work working training practice	1531
opportunity focus attention success future potential impact result change	1431
sales retail company business industry marketing product	1251

nouns that indicate the actual strength, and cluster them using a simple clustering algorithm which uses the cosine similarity between word embeddings[2] of these nouns. We repeat this for the 9160 sentences with predicted label WEAKNESS or SUGGESTION as a single class. Tables 3 and 4 show a few representative clusters in strengths and in weaknesses, respectively. We also explored clustering 12742 STRENGTH sentences directly using CLUTO [8] and Carrot2 Lingo [14] clustering algorithms. Carrot2 Lingo[3] discovered 167 clusters and also assigned labels to these clusters. We then generated 167 clusters using CLUTO as well. CLUTO does not generate cluster labels automatically, hence we used 5 most frequent words within the cluster as its labels. Table 5 shows the largest

Table 4. 5 representative clusters in weaknesses and suggestions.

Weakness cluster	Count
motivation expertise knowledge talent skill	1308
market sales retail corporate marketing commercial industry business	1165
awareness emphasis focus	1165
coaching team coach	1149
job work working task planning	1074

---

[2] We used 100 dimensional word vectors trained on Wikipedia 2014 and Gigaword 5 corpus, available at: https://nlp.stanford.edu/projects/glove/.

[3] We used the default parameter settings for Carrot2 Lingo algorithm as mentioned at: http://download.carrot2.org/head/manual/index.html.

5 clusters by both the algorithms. It was observed that the clusters created by CLUTO were more meaningful and informative as compared to those by Carrot2 Lingo. Also, it was observed that there is some correspondence between noun clusters and sentence clusters. E.g. the nouns cluster `motivation expertise knowledge talent skill` (Table 3) corresponds to the CLUTO sentence cluster `skill customer management knowledge team` (Table 5). But overall, users found the nouns clusters to be more meaningful than the sentence clusters.

Table 5. Largest 5 sentence clusters within 12742 STRENGTH sentences

Algorithm	Cluster	#Sentences
CLUTO	`performance performer perform years team`	510
	`skill customer management knowledge team`	325
	`role delivery work place show`	289
	`delivery manage management manager customer`	259
	`knowledge customer business experience work`	250
Carrot2	`manager manage`	1824
	`team team`	1756
	`delivery management`	451
	`manage team`	376
	`customer management`	321

5 PA Along Attributes

In many organizations, PA is done from a predefined set of perspectives, which we call *attributes*. Each attribute covers one specific aspect of the work done by the employees. This has the advantage that we can easily compare the performance of any two employees (or groups of employees) along any given attribute. We can correlate various performance attributes and find dependencies among them. We can also cluster employees in the workforce using their supervisor ratings for each attribute to discover interesting insights into the workforce. The HR managers in the organization considered in this paper have defined 15 attributes (Table 6). Each attribute is essentially a work item or work category described at an abstract level. For example, FUNCTIONAL_EXCELLENCE covers any tasks, goals or activities related to the software engineering life-cycle (e.g., requirements analysis, design, coding, testing etc.) as well as technologies such as databases, web services and GUI.

In the example in Sect. 4, the first sentence (which has class STRENGTH) can be mapped to two attributes: FUNCTIONAL_EXCELLENCE and BUILD-ING_EFFECTIVE_TEAMS. Similarly, the third sentence (which has class WEAK-NESS) can be mapped to the attribute INTERPERSONAL_EFFECTIVENESS and so forth. Thus, in order to answer the second question in Sect. 1, we need to map each sentence in each of the 3 classes to zero, one, two or more attributes, which is a multi-class multi-label classification problem.

We manually tagged the same 2000 sentences in Dataset D1 with attributes, where each sentence may get 0, 1, 2, etc. up to 15 class labels (this is dataset D2). This labelled dataset contained 749, 206, 289, 207, 91, 223, 191, 144, 103, 80, 82, 42, 29, 15, 24 sentences having the class labels listed in Table 6 in the same order. The number of sentences having 0, 1, 2, or more than 2 attributes are: 321, 1070, 470 and 139 respectively. We trained several multi-class multi-label classifiers on this dataset. Table 7 shows the results of 5-fold cross-validation experiments on dataset D2.

Table 6. Strengths, weaknesses and suggestions along performance attributes

Performance Attributes	#Strengths	#Weaknesses	#Suggestions
FUNCTIONAL_EXCELLENCE	321	26	284
BUILDING_EFFECTIVE_TEAMS	80	6	89
INTERPERSONAL_EFFECTIVENESS	151	16	97
CUSTOMER_FOCUS	100	5	76
INNOVATION_MANAGEMENT	22	4	53
EFFECTIVE_COMMUNICATION	53	17	124
BUSINESS_ACUMEN	39	10	103
TAKING_OWNERSHIP	47	3	81
PEOPLE_DEVELOPMENT	31	8	57
DRIVE_FOR_RESULTS	37	4	30
STRATEGIC_CAPABILITY	8	4	51
WITHSTANDING_PRESSURE	16	6	16
DEALING_WITH_AMBIGUITIES	4	8	12
MANAGING_VISION_AND_PURPOSE	3	0	9
TIMELY_DECISION_MAKING	6	2	10

Table 7. Results of 5-fold cross validation for multi-class multi-label classification on dataset D2.

Classifier	Precision P	Recall R	F
Logistic Regression	0.715	0.711	**0.713**
Multinomial Naive Bayes	0.664	0.588	0.624
Random Forest	0.837	0.441	0.578
AdaBoost	0.794	0.595	0.680
Linear SVM	0.722	0.672	0.696
Pattern-based	0.750	0.679	**0.713**

Precision, Recall and F-measure for this multi-label classification are computed using a strategy similar to the one described in [6]. Let P_i be the set of predicted labels and A_i be the set of actual labels for the ith instance. Precision and recall for this instance are computed as follows:

$$Precision_i = \frac{|P_i \cap A_i|}{|P_i|}, \;\; Recall_i = \frac{|P_i \cap A_i|}{|A_i|}$$

It can be observed that $Precision_i$ would be undefined if P_i is empty and similarly $Recall_i$ would be undefined when A_i is empty. Hence, overall precision and recall are computed by averaging over all the instances except where they are undefined. Instance-level F-measure can not be computed for instances where either precision or recall are undefined. Therefore, overall F-measure is computed using the overall precision and recall.

6   Summarization of Peer Feedback Using ILP

The PA system includes a set of peer feedback comments for each employee. To answer the third question in Sect. 1, we need to create a summary of all the peer feedback comments about a given employee. As an example, following are the feedback comments from 5 peers of an employee.

1. vast knowledge on different technologies
2. His experience and wast knowledge mixed with his positive attitude, willingness to teach and listen and his humble nature.
3. Approachable, Knowledgeable and is of helping nature.
4. Dedication, Technical expertise and always supportive
5. Effective communication and team player

The individual sentences in the comments written by each peer are first identified and then POS tags are assigned to each sentence. We hypothesize that a good summary of these multiple comments can be constructed by identifying a set of *important* text fragments or phrases. Initially, a set of candidate phrases is extracted from these comments and a subset of these candidate phrases is chosen as the final summary, using Integer Linear Programming (ILP). The details of the ILP formulation are shown in Table 8. As an example, following is the summary generated for the above 5 peer comments.

humble nature, effective communication, technical expertise, always
supportive, vast knowledge

Following rules are used to identify candidate phrases:

- An adjective followed by in which is followed by a noun phrase (e.g. good in customer relationship)
- A verb followed by a noun phrase (e.g. maintains work life balance)
- A verb followed by a preposition which is followed by a noun phrase (e.g. engage in discussion)

- Only a noun phrase (e.g. `excellent listener`)
- Only an adjective (e.g. `supportive`)

Various parameters are used to evaluate a candidate phrase for its *importance*. A candidate phrase is more important:

- if it contains an adjective or a verb or its headword is a noun having WordNet lexical category *noun.attribute* (e.g. nouns such as `dedication`, `sincerity`)
- if it contains more number of words
- if it is included in comments of multiple peers
- if it represents any of the performance attributes such as *Innovation, Customer, Strategy* etc.

A complete list of parameters is described in detail in Table 8.

There is a trivial constraint C_0 which makes sure that only K out of N candidate phrases are chosen. A suitable value of K is used for each employee depending on number of candidate phrases identified across all peers (see Algorithm 1). Another set of constraints (C_1 to C_{10}) make sure that at least one phrase is selected for each of the leadership attributes. The constraint C_{11} makes sure that multiple phrases sharing the same headword are not chosen at a time. Also, single word candidate phrases are chosen only if they are adjectives or nouns with lexical category *noun.attribute*. This is imposed by the constraint C_{12}. It is important to note that all the constraints except C_0 are soft constraints, i.e. there may be feasible solutions which do not satisfy some of these constraints.

Data: N: No. of candidate phrases
Result: K: No. of phrases to select as part of summary
if $N \leq 10$ then
    $K \leftarrow \lfloor N * 0.5 \rfloor$;
else if $N \leq 20$ then
    $K \leftarrow \lfloor getNoOfPhrasesToSelect(10) + (N - 10) * 0.4 \rfloor$;
else if $N \leq 30$ then
    $K \leftarrow \lfloor getNoOfPhrasesToSelect(20) + (N - 20) * 0.3 \rfloor$;
else if $N \leq 50$ then
    $K \leftarrow \lfloor getNoOfPhrasesToSelect(30) + (N - 30) * 0.2 \rfloor$;
else
    $K \leftarrow \lfloor getNoOfPhrasesToSelect(50) + (N - 50) * 0.1 \rfloor$;
end
if $K < 4$ *and* $N \geq 4$ then
    $K \leftarrow 4$
else if $K < 4$ then
    $K \leftarrow N$
else if $K > 20$ then
    $K \leftarrow 20$
end

Algorithm 1: *getNoOfPhrasesToSelect* (For determining number of phrases to select to include in summary)

Table 8. Integer Linear Program (ILP) formulation

Parameters:

- N: No. of phrases
- K: No. of phrases to be chosen for inclusion in the final summary
- $Freq$: Array of size N, $Freq_i$ = no. of distinct peers mentioning the ith phrase
- Adj: Array of size N, $Adj_i = 1$ if the ith phrase contains any adjective
- $Verb$: Array of size N, $Verb_i = 1$ if the ith phrase contains any verb
- $NumWords$: Array of size N, $NumWords_i = 1$ no. of words in the ith phrase
- $NounCat$: Array of size N, $NounCat_i = 1$ if lexical category (WordNet) of headword of the ith phrase is $noun.attribute$
- $InvalidSingleNoun$: Array of size N, $InvalidSingleNoun_i = 1$ if the ith phrase is single word phrase which is neither an adjective nor a noun having lexical category (WordNet) $noun.attribute$
- $Leadership, Team, Innovation, Communication, Knowledge, Delivery,$ $Ownership, Customer, Strategy, Personal$: Indicator arrays of size N each, representing whether any phrase corresponds to a particular performance attribute, e.g. $Customer_i = 1$ indicates that i^{th} phrase is of type $Customer$
- S: Matrix of dimensions $N \times N$, where $S_{ij} = 1$ if headwords of ith and j^{th} phrase are same

---

Variables:

- X: Array of N **binary** variables, where $X_i = 1$ only when i^{th} phrase is chosen to be the part of final summary
- $S_1, S_2, \cdots S_{12}$: **Integer** slack variables

Objective:
Maximize $\sum_{i=1}^{N} ((NounCat_i + Adj_i + Verb_i + 1) \cdot Freq_i \cdot NumWords_i \cdot X_i)$
$\qquad - 10000 \cdot \sum_{j=1}^{12} S_j$

Constraints:

C_0: $\sum_{i=1}^{N} X_i = K$ (Exactly K phrases should be chosen)

C_1: $\sum_{i=1}^{N} (Leadership_i \cdot X_i) + S_1 \geq 1$
C_2: $\sum_{i=1}^{N} (Team_i \cdot X_i) + S_2 \geq 1$
C_3: $\sum_{i=1}^{N} (Knowledge_i \cdot X_i) + S_3 \geq 1$
C_4: $\sum_{i=1}^{N} (Delivery_i \cdot X_i) + S_4 \geq 1$
C_5: $\sum_{i=1}^{N} (Ownership_i \cdot X_i) + S_5 \geq 1$
C_6: $\sum_{i=1}^{N} (Innovation_i \cdot X_i) + S_6 \geq 1$
C_7: $\sum_{i=1}^{N} (Communication_i \cdot X_i) + S_7 \geq 1$
C_8: $\sum_{i=1}^{N} (Customer_i \cdot X_i) + S_8 \geq 1$
C_9: $\sum_{i=1}^{N} (Strategy_i \cdot X_i) + S_9 \geq 1$
C_{10}: $\sum_{i=1}^{N} (Personal_i \cdot X_i) + S_10 \geq 1$
 (At least one phrase should be chosen to represent each leadership attribute)

C_{11}: $\sum_{i=1}^{N} \sum_{j=1, s.t.i \neq j}^{N} (S_{ij} \cdot (X_i + X_j - 1)) + S_{11} <= 0$
 (No duplicate phrases should be chosen)

C_{12}: $\sum_{i=1}^{N} (InvalidSingleNoun_i \cdot X_i) - S_{12} <= 0$
 (Single word noun phrases are not preferred if they are not $noun.attribute$)

But each constraint which is not satisfied, results in a penalty through the use of slack variables. These constraints are described in detail in Table 8.

The objective function maximizes the total *importance* score of the selected candidate phrases. At the same time, it also minimizes the sum of all slack variables so that the minimum number of constraints are broken.

6.1 Evaluation of Auto-Generated Summaries

We considered a dataset of 100 employees, where for each employee multiple peer comments were recorded. Also, for each employee, a manual summary was generated by an HR personnel. The summaries generated by our ILP-based approach were compared with the corresponding manual summaries using the ROUGE [11] unigram score. For comparing performance of our ILP-based summarization algorithm, we explored a few summarization algorithms provided by the Sumy package[4]. A common parameter which is required by all these algorithms is number of sentences keep in the final summary. ILP-based summarization requires a similar parameter K, which is automatically decided based on number of total candidate phrases. Assuming a sentence is equivalent to roughly 3 phrases, for Sumy algorithms, we set number of sentences parameter to the ceiling of $K/3$. Table 9 shows average and standard deviation of ROUGE unigram f1 scores for each algorithm, over the 100 summaries. The performance of ILP-based summarization is comparable with the other algorithms, as the two sample t-test does not show statistically significant difference. Also, human evaluators preferred phrase-based summary generated by our approach to the other sentence-based summaries.

Table 9. Comparative performance of various summarization algorithms

Algorithm	ROUGE unigram F1	
	Average	Std. Deviation
LSA	0.254	0.146
TextRank	0.254	0.146
LexRank	0.258	0.148
ILP-based summary	0.243	0.15

7 Conclusions and Further Work

In this paper, we presented an analysis of the text generated in Performance Appraisal (PA) process in a large multi-national IT company. We performed sentence classification to identify strengths, weaknesses and suggestions for improvements found in the supervisor assessments and then used clustering to discover

---

[4] https://github.com/miso-belica/sumy.

broad categories among them. As this is non-topical classification, we found that SVM with ADWS kernel [16] produced the best results. We also used multi-class multi-label classification techniques to match supervisor assessments to predefined broad perspectives on performance. Logistic Regression classifier was observed to produce the best results for this topical classification. Finally, we proposed an ILP-based summarization technique to produce a summary of peer feedback comments for a given employee and compared it with manual summaries.

The PA process also generates much structured data, such as supervisor ratings. It is an interesting problem to compare and combine the insights from discovered from structured data and unstructured text. Also, we are planning to automatically discover any additional performance attributes to the list of 15 attributes currently used by HR.

References

1. Apte, M., Pawar, S., Patil, S., Baskaran, S., Shrivastava, A., Palshikar, G.K.: Short text matching in performance management. In: Proceedings of the 21st International Conference on Management of Data (COMAD 2016), pp. 13–23 (2016)
2. Carvalho, V.R., Cohen, W.W.: Improving "email speech acts" analysis via n-gram selection. In Proceedings of the HLT-NAACL 2006 Workshop on Analyzing Conversations in Text and Speech, ACTS'09, pp. 35–41 (2006)
3. Cohen, W.W., Carvalho, V.R., Mitchell, T.M.: Learning to classify email into "speech acts". In: Proceedings of the Empirical Methods in Natural Language Processing (EMNLP-2004), pp. 309–316 (2004)
4. Cotterill, R.: Question classification for email. In: Proceedings of the Ninth International Conference on Computational Semantics (IWCS 2011) (2011)
5. Deshpande, S., Palshikar, G.K., Athiappan, G.: An unsupervised approach to sentence classification. In: Proceedings of the International Conference on Management of Data (COMAD 2010), pp. 88–99 (2010)
6. Godbole, S., Sarawagi, S.: Discriminative methods for multi-labeled classification. In: Dai, H., Srikant, R., Zhang, C. (eds.) PAKDD 2004. LNCS (LNAI), vol. 3056, pp. 22–30. Springer, Heidelberg (2004). https://doi.org/10.1007/978-3-540-24775-3_5
7. Hachey, B., Grover, C.: Sequence modelling for sentence classification in a legal summarisation system. In: Proceedings of the 2005 ACM Symposium on Applied Computing (2005)
8. Karypis, George: Cluto-a clustering toolkit. Technical report, DTIC Document (2002)
9. Khoo, A., Marom, Y., Albrecht, D.: Experiments with sentence classification. In: Proceedings of the 2006 Australasian Language Technology Workshop (ALTW2006), pp. 18–25 (2006)
10. Levy, P.E., Williams, J.R.: The social context of performance appraisal: a review and framework for the future. J. Manag. **30**(6), 881–905 (2004)
11. Lin, C.-Y.: Rouge: a package for automatic evaluation of summaries. In: Text summarization branches out: Proceedings of the ACL-04 workshop, vol. 8. Barcelona, Spain (2004)

12. McKnight, L., Srinivasan, P.: Categorization of sentence types in medical abstracts. In: Proceedings of the American Medical Informatics Association Annual Symposium, pp. 440–444 (2003)
13. Murphy, K.R., Cleveland, J.: Understanding Performance Appraisal: Social, Organizational and Goal-Based Perspective. Sage Publishers, Thousand Oaks (1995)
14. Osinski, S., Stefanowski, J., Weiss, D.: Lingo: search results clustering algorithm based on singular value decomposition. In: Intelligent Information Processing and Web Mining, Proceedings of the International IIS: IIPWM'04 Conference held in Zakopane, Poland, pp. 359–368, 17–20 May 2004
15. Palshikar, G.K., Deshpande, S., Bhat, S., Quest: discovering insights from survey responses. In: Proceedings of the 8th Australasian Data Mining Conference (AusDM09), pp. 83–92 (2009)
16. Pawar, S., Ramrakhiyani, N., Palshikar, G.K., Hingmire, S.: Deciphering review comments: identifying suggestions, appreciations and complaints. In: Biemann, C., Handschuh, S., Freitas, A., Meziane, F., Métais, E. (eds.) NLDB 2015. LNCS, vol. 9103, pp. 204–211. Springer, Cham (2015). https://doi.org/10.1007/978-3-319-19581-0_18
17. Qadir, A., Riloff, E.: Classifying sentences as speech acts in message board posts. In: Proceedings of the Empirical Methods in Natural Language Processing (EMNLP-2011) (2011)
18. Ramrakhiyani, N., Pawar, S., Palshikar, G.K.: A system for classification of propositions of the Indian supreme court judgements. In: Proceedings of the 5th 2013 Forum on Information Retrieval Evaluation (FIRE 2013), pp. 1–4 (2013)
19. Ramrakhiyani, N., Pawar, S., Palshikar, G.K., Apte, M.: Aspects from appraisals!! a label propagation with prior induction approach. In: Métais, E., Meziane, F., Saraee, M., Sugumaran, V., Vadera, S. (eds.) NLDB 2016. LNCS, vol. 9612, pp. 301–309. Springer, Cham (2016). https://doi.org/10.1007/978-3-319-41754-7_28
20. Schraeder, M., Becton, J., Portis, R.: A critical examination of performance appraisals. J. Quality Particip. **30**(1), 20–25 (Spring 2007)
21. Stolcke, A., Ries, K., Coccaro, N., Shriberg, E., Bates, R., Jurafsky, D., Taylor, P., Martin, R., Van Ess-Dykema, C., Meteer, M.: Dialogue act modeling for automatic tagging and recognition of conversational speech. Comput. Linguist. **26**(3), 339–373 (2000)
22. Viswesvaran, C.: Assessment of individual job performance: a review of the past century and a look ahead. In: Anderson, N., Ones, D.S., Sinangil, H.K., Viswesvaran, C. (eds.) Handbook of Industrial, Work and Organizational Psychology. Sage Publishers, Thousand Oaks (2001)
23. Yamamoto, Y., Takagi, T.: A sentence classification system for multi biomedical literature summarization. In: Proceedings of the 21st International Conference on Data Engineering Workshops, pp. 1163–1168 (2005)

Automatic Detection of Uncertain Statements in the Financial Domain

Christoph Kilian Theil[(✉)], Sanja Štajner, Heiner Stuckenschmidt,
and Simone Paolo Ponzetto

Data and Web Science Group, University of Mannheim, Mannheim, Germany
{christoph,sanja,heiner,simone}@informatik.uni-mannheim.de

Abstract. The automatic detection of uncertain statements can benefit
NLP tasks such as deception detection and information extraction. Furthermore, it can enable new analyses in social sciences such as business
where the quantification of uncertainty or risk plays a significant role.
Thus, for the first time, we approached the automatic detection of uncertain statements as a binary sentence classification task on the transcripts
of spoken language in the financial domain. We created a new dataset
and – besides using bag-of-words, part-of-speech tags, and dictionaries –
developed rule-based features tailored to our task. Finally, we analyzed
systematically, which features perform best in the financial domain as
opposed to the previously researched encyclopedic domain.

Keywords: Automatic uncertainty detection
Binary sentence classification · Financial domain

1 Introduction

In linguistics, the use of uncertain statements is described by the phenomenon of
"hedging" which is defined as "any linguistic means used to indicate either (a) a
lack of complete commitment to the truth value of an accompanying proposition,
or (b) a desire not to express that commitment categorically" [1, p. 1]. As can be
seen, this definition is centered on a speaker or writer. For the scope of this paper
– an application of uncertainty detection in social sciences, and more specifically,
the financial domain – we adjust this definition slightly. As we bear in mind to
predict market reactions in future work, we establish a definition of uncertainty
which also keeps in mind the recipient's side of the communication process.

1.1 Uncertainty as Opposed to Linguistic Hedging

In addition to the sentences encompassed by the aforementioned linguistic definition, we also classify sentences as uncertain:

© Springer Nature Switzerland AG 2018
A. Gelbukh (Ed.): CICLing 2017, LNCS 10762, pp. 642–654, 2018.
https://doi.org/10.1007/978-3-319-77116-8_48

- If their truth value cannot be determined (e.g. statements about the future)
- If they refer to uncertain factors (e.g. statements about market volatility)
- If they show uninformedness (e.g. statements conveying lack of knowledge)

How uncertainty could be further broken down into more granular categories will be introduced in Sect. 4.2.

1.2 Opportunities of Automatic Uncertainty Detection

Automatic detection of uncertain statements can benefit NLP tasks such as deception detection [2,3], information extraction [4,5], and summarization [6]. Furthermore, automatic uncertainty detection can enable new analyses in social sciences where the quantification of uncertainty or risk plays a significant role. Disciplines like business and economics would profit from an automatically extractable measure of uncertainty which does not depend on manual analysis.

As of now, automatic uncertainty detection has been limited to detecting hedges (as opposed to our broader concept of uncertainty) in biomedical scientific texts and Wikipedia articles. The results of the CoNNL-2010 shared task [7, pp. 6–8] indicate that this task is easier to solve for the former than for the latter.

Loughran & McDonald specifically proposed to investigate "whether or not managers using high levels of uncertain or weak modal [...] words during conference calls experience worse subsequent stock or operating performance" [8, p. 43]. Within this paper, we address the first part of this suggestion by providing a classifier of uncertainty suited to analyze earnings calls.[1]

1.3 Contributions

For the first time in the financial domain, we performed the classification of uncertain statements. For this purpose, we gathered and annotated a new financial domain dataset. Adapting the definition of "linguistic hedging", we developed a new concept of uncertainty which fits the domain-specific needs. In contrast to previous work, our concept of uncertainty encompasses how the use of uncertain statements can have an impact on other social agents and thus enables predictions of market reactions.

To achieve our goal of comparing the automatic uncertainty detection in the financial to the encyclopedic domain, we pose four research questions (RQ 1–4):

- **RQ 1**: How do all of our feature sets separately perform on both our new financial and the existing Wikipedia datasets?
- **RQ 2**: How do lexical and syntactic features perform on both the financial and the encyclopedic domain?
- **RQ 3**: How do knowledge-poor and knowledge-rich features perform on both the financial and the encyclopedic domain?
- **RQ 4**: How do our new, domain-specific rules contribute to the classification for the financial domain? Are they applicable to the encyclopedic domain?

---

[1] Earnings calls are publicly accessible teleconferences or webcasts in which executives of a public company present the financial results of the last quarter.

2   Related Work

The NLP task of detecting uncertain statements has already been addressed in the biomedical and the encyclopedic domain – for example in the CoNLL-2010 shared task [7]. However, the topic has not been explored in the social sciences, let alone finance. Hence, within this section, we give a short overview about the existing approaches of uncertainty detection in NLP and then cover the closest related applications in the financial domain.

2.1   Approaches of Uncertainty Detection in NLP

The first to perform uncertainty detection in the biomedical domain where Light *et al.*, which have shown that a substring matching approach (with 14 manually selected hedge cues) slightly outperforms an SVM classifier with bag-of-word (BoW) vectors in terms of accuracy (95% vs. 92%) [9, p. 22]. Following up, Medlock & Briscoe presented a weakly supervised machine learning approach which outperformed Light *et al.*'s best classifier (76% vs. 60% accuracy) [5, p. 998]. Yielding better results than both of these, Szarvas presented a maximum entropy classifier achieving an F_1 score of 0.85 [4, p. 287]. He showed that a classifier using only unigrams instead of bi- and trigrams performs significantly worse ($F_1 = 0.80$) [4, p. 286]. Furthermore, he proved that even only a slight out-of-domain application (bioinformatics articles instead of biomedical papers) yields a high drop in performance ($F_1 = 0.75$) [4, p. 286].

The first classifier of Wikipedia sentences was presented by Ganter & Strube and made use of both corpus statistics and syntactic patterns [10, p. 173]. Subsequently, uncertainty detection both within the biomedical and the encyclopedic domain has been resumed in the CoNLL-2010 shared task, where an extensive array of features was used: dictionaries, orthographic token information, lemmas/stems, part-of-speech (POS) tags, syntactic chunk information, dependency parsing, and the position of the token within the document were addressed with sequence labeling (SL), token classification and BoW approaches [7, p. 9]. While the best performing system for the biomedical domain made use of an SL approach, the best classifier for the encyclopedic domain used a BoW approach with a dictionary [7, p. 8].

These insights motivated our proposed feature sets. We evaluated the features that performed best for the Wikipedia dataset (BoW vectors in combination with a dictionary-based approach) on our new financial domain dataset. Furthermore, we enriched the BoW vectors with POS tags and explored the possibilities of applying hand-written syntactic rules. A systematic overview of all features is provided in the beginning of Sect. 4.

2.2   NLP in the Financial Domain

While already a few surveys provide a literature overview of NLP in finance [11–13], the most recent one was presented by Loughran and McDonald [8].

Most applications of NLP in finance focus on formal disclosures such as 10-Ks or 10-Qs[2] as opposed to earnings calls, e.g. Li [14,15] or Loughran and McDonald [16–18]. Larcker and Zakolyukina [2] summarize the limitations of the former textual forms such as a relative uniformity of the content over time and little spontaneity [2, p. 499]. Hence, we are further motivated to investigate earnings calls instead of formal disclosures.

Loughran and McDonald [17] extracted their own dictionary of uncertainty triggers from a sample of ∼50,000 10-Ks [17]. As using a dictionary has shown to yield good results for classifying the uncertainty of Wikipedia sentences [7, p. 9], we use this domain-specific uncertainty dictionary for our experiments.

3   New Uncertainty Dataset

An earnings call consists of one or more executives (e.g. the CEO or CFO) presenting the company's financial results of the ending quarter to the public via a teleconference and/or a webcast. In a second section, the call is opened for a question-and-answer-session (Q&A) with investors and banking analysts. In addition to the mentioned protagonists, an operator takes care of technical requirements such as opening and ending the call or moderating the Q&A.

Since the first part of the call closely follows the accompanying press release, it is highly formalized and provides little opportunity for the executives to speak freely. Hence, our analyses solely focus on the second part of the call, the Q&A session. Moreover, as we are interested in obtaining data that might help to gain insight about the company's financial (un)certainty itself, we only include sentences uttered by the executives (the answers), instead of the analyst's questions or the operator's technical remarks.

As we analyze free speech instead of written, formalized text, we expect our problem to be more challenging to solve than e.g. classifying biomedical or encyclopedic sentences. Consider, for example, the following statement:

Example 1. "And increasingly look as you are sort of describe us [sic] as well, we look to focus where we can really make a difference [...]"

In spite containing the hedge "sort of", this sentence was annotated as *certain*, according to our methodology. In this case, the adverb of degree "sort of" is used in a colloquial sense as inherent in any free speech. The first part "as you [...] sort of describe us" is highly unspeculative and easily verifiable/falsifiable by a potential listener. In contrast, consider the following example:

Example 2. "Now, what we don't know is what's going to happen at the end of the third quarter."

---

[2] 10-Ks/10-Qs are standardized annual/quarterly reports providing an overview of a company's financial results, which are required by the U. S. Securities and Exchange Commission.

While this sentence does not contain any hedges such as adverbials of degree or of possibility, it indicates a lack of knowledge of the speaker, which is why we annotated it as *uncertain*. These two examples might give an idea why the task of detecting uncertain statements in spoken language within the financial domain is of particular complexity.

As basis for the dataset, we used the Standard & Poor's 500 Index (S&P 500)[3] as one of the most important equity indices. Since the webpage Seeking Alpha supplies a large database of publicly available earnings call transcripts,[4] we obtained all data from there. For our dataset, we took a total of 7,725 transcripts from 217 different S&P 500 companies belonging to a wide array of industries such as Financials, Industrials or Information Technology.

Out of the dataset, we randomly sampled 1,800 sentences and annotated them to either be *certain* or *uncertain*. Habitual utterances such as greetings, expressions of thanks, farewells etc. were excluded from the sampling process, as they would dilute the results of our task. Out of these 1,800, 100 sentences were randomly selected and independently annotated by a second annotator of financial background. As the inter-annotator agreement measured as Cohen's kappa (κ) [19] was 0.81, which – depending on the source – can be considered as "almost perfect" [20, p. 165] or "excellent" [21, p. 218], the rest of the annotation was carried out only by the first annotator, which is of linguistic background. Afterwards, we split the set in two: 800 sentences (683 *certain*, 117 *uncertain*) were taken to develop the syntactic rules (see Sect. 4.2), while the remaining 1,000 (829 *certain*, 171 *uncertain*) were used for the classification experiments.

4   Methodology

We addressed the problem of automatic uncertainty detection as a binary sentence classification task. As BoW vectors, POS tags, and a comprehensive list of uncertainty cues have been used before [7], we explored the possibility of adding novel features. Due to its domain-specificity, we used Loughran and McDonald's uncertainty dictionary [17]. In addition, we applied a set of hand-written rules specifically designed for our task. All features can be classified along lexical vs. syntactic and knowledge-poor vs. knowledge-rich dimensions, which yields a feature set matrix as depicted in Fig. 1.

We lemmatized the BoW vectors with NLTK's WordNet implementation [22] and normalized them via tf-idf weighting. Additionally, we extracted POS tags with NLTK 3.2.1's standard POS tagger, which is based on Honnibal's implementation of the Averaged Perceptron tagger [23,24].[5] The following Sects. 4.1 and 4.2 further elaborate on the features unique to our approach.

---

[3] http://us.spindices.com/indices/equity/sp-500
[4] http://seekingalpha.com/earnings/earnings-call-transcripts
[5] This tagger reached an accuracy of 96.80% when applied to an evaluation set of 130,000 words taken from The Wall Street Journal [25].

4.1 Lists of Speculation Triggers

Within the experiments, we used the following lists of speculation triggers:

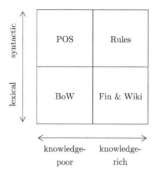

Fig. 1. Feature set matrix.

- **Fin**: Loughran & McDonald's list of 297 unigrams indicating uncertainty in the financial domain (e.g. "fluctuation", "recalculation") based on a sample of ~50,000 10-Ks.[6] After lemmatization, the list totaled 192 items.
- **Wiki**: 1,984 uncertainty triggers of arbitrary length (e.g. "a matter in dispute", "some prehistoric cultures") were extracted from the CoNLL-2010 shared task's Wikipedia training set.[7] After lemmatization, the list totaled 1,868 unique items.

4.2 Rules

We developed a set of 95 hand-written rules according to which a sentence can be classified as *uncertain* based on 800 randomly selected sentences. A rule is always characterized by syntactic criteria (POS tags, phrase chunks), which can additionally be refined by lexical features (lemmas, word lists). The word lists define more granular word classes such as *adverbs of degree* (e.g. "kind of", "quite"), *adverbs of probability* (e.g. "potentially", "probably"), *fuzzy quantifiers* (e.g. "about half of", "close to 100"), and *verbs of expectation* (e.g. "anticipate", "expect"). All rules can be assigned to seven different categories which are defined as presented in Table 1. According to our methodology, *Example 2* presented in Sect. 3 is captured by the rules category "Uninformedness".

In addition, we applied the rules to 30 random samples of 1,000 Wikipedia test sentences to check their applicability for a general domain as opposed to the domain-specific context of our new dataset. To guarantee the greatest possible comparability, for each of the samples, the class-distribution of 829 *certain* to 171

---

[6] http://www3.nd.edu/~mcdonald/Word_Lists.html
[7] http://rgai.inf.u-szeged.hu/conll2010st/download.html

uncertain sentences of our dataset was maintained. The results of this experiment are shown in Table 2.

As expected, the rules match substantially fewer sentences (9.70 on average) in the Wikipedia test set than in the financial domain dataset (54 matches).

Table 1. Categorization of the rules.

Category	Count	Example
Expectation	29	"I expect our maintenance capital [...] to probably be"
Assumption	25	"I think it's pretty mature"
Probability	12	"perhaps by the end of this year"
Uninformedness	10	"we really don't know what ultimately it's going to sell"
Subjunction	9	"it might be a few hundred thousand dollars"
Volatility	6	"the volatility of where we are"
Unspecificity	4	"somewhere in the 40% range"

Table 2. Descriptive statistics for the number of times the rules match 30 random samples of the Wikipedia test set.

n	Min	Max	Mean	Median	Mode	SD	SK
30	3.00	15.00	9.70	9.50	8.00	2.76	0.04

4.3 Experiments

For each sentence in the dataset, we defined a vector containing each feature's occurrences. Afterwards, we applied seven machine learning algorithms in WEKA experimenter [26] using a 10-fold cross-validation setup with 10 repetitions: Logistic Regression [27], Naïve Bayes [28], Support Vector Machines (SVM) [29], k-Nearest Neighbors [30], JRip [31], C4.5 [32], and Random Forest [33]. We evaluated the performance for all eleven feature sets used in the subsequent experiments and compared the weighted average F_1 scores. Since SVM achieved the best results in all cases, we used this algorithm for the subsequent experiments.

Addressing our research questions (see Sect. 1.3), we carried out four sets of experiments as shown in Table 3. We applied the SVM algorithm to both our dataset and the Wikipedia test set with different feature set combinations across the matrix presented in Fig. 1. Thus, we evaluated the performance of our domain-specific classifier on the general domain. To ensure comparability of the data, we used the 30 random samples of the Wikipedia test set as shown in Table 2 and calculated the means of the respective performance measures.

As Farkas *et al.* have summarized [7], pure BoW vectors have proven to be a strong feature set in the encyclopedic domain, which is why we used it, too, and additionally contrasted it to POS-enriched BoW vectors ("POSBoW"). Apart from all individual features (RQ 1, see Sect. 5.1), we were interested in how the dimensions lexical vs. syntactic (RQ 2, see Sect. 5.2) and knowledge-poor vs. knowledge-rich (RQ 3, see Sect. 5.3) would compare. Lastly, we investigated how the rules benefit the overall performance of the classification task (RQ 4, see Sect. 5.4).

5   Results and Discussion

In this section, we present the results for each set of experiments. We conducted corrected paired t-tests with $\alpha = 0.05$ to check for significant differences in classification performance. The performance was evaluated in terms of precision (P), recall (R), and F_1 score (F).

Table 3. Sets of experiments.

Set	Description	Feature sets
1	Separate features (RQ 1)	BoW, POSBoW, Fin(+Wiki), Wiki, Rules
2	Lexical vs. syntactic (RQ 2)	BoW+Fin+Wiki, POSBoW+Rules
3	Knowledge-poor vs. -rich (RQ 3)	POSBoW, Fin+Wiki+Rules
4	Contribution of the rules (RQ 4)	Rules, POSBoW+Fin+Wiki(+Rules)

5.1   Separate Features (RQ 1)

With P = 0.77, the rules significantly outperform the other individual features of the *uncertain* class (see Table 4). As expected, this comes at the cost of a relatively low recall of 0.13. BoW reaches a recall significantly higher (0.37) than all features apart from POSBoW (0.35). The latter receives the highest F_1 score (0.41) which is insignificantly higher than the former's (0.40), yet significantly higher than the value of all other feature sets.

For the encyclopedic domain (see Table 5), the Wiki dictionary outperforms all other features. This is not surprising, as it was designed specifically for this domain. The domain-specificity can also explain why the rules prove to be the weakest feature set by far. As previously shown in Table 2, they rarely match any of the Wiki sentences which is reflected in a generally poor performance.

5.2   Lexical vs. Syntactic Features (RQ 2)

For the new dataset (see Table 4), syntactic features prove to perform only slightly better than lexical ones with no significant improvement across all performance measures. For the Wikipedia test set (see Table 5), in contrast, the lexical features perform noticeably better, especially in terms of recall (0.39 vs. 0.31) and F_1 score (0.47 vs. 0.41) of the *uncertain* class. This can probably be attributed to the high performance of the Wiki dictionary when treated separately.

As Wikipedia attempts to provide an unbiased source of encyclopedic knowledge, the sentence structure is highly formalized. Hence, rule-based and other features leaning more towards the syntactic side are likely to have little applicability. Instead, lexical choices seem to reflect degrees of uncertainty better in this case. Since the opposite case holds for our dataset (i.e. the sentence structure is relatively free and spontaneous), the results are in line with our expectations.

Table 4. Results of the classification task on our financial domain dataset (the best results are presented in bold).

Features	Uncertain			Certain			Accuracy
	P	R	F	P	R	F	
RQ 1: Separate Features							
BoW	0.46	**0.37**	0.40	0.87	0.91	0.89	81.54%
POSBoW	0.53	0.35	**0.41**	**0.88**	0.93	0.90	83.18%
Fin	0.53	0.14	0.21	0.85	0.97	**0.91**	82.90%
Wiki	0.48	0.17	0.23	0.84	0.97	**0.91**	82.68%
Fin+Wiki	0.53	0.26	0.34	0.87	0.95	**0.91**	83.09%
Rules	**0.77**	0.13	0.21	0.84	**1.00**	**0.91**	**84.58%**
RQ 2: Lexical vs. Syntactic							
BoW+Fin+Wiki	0.53	**0.39**	**0.44**	**0.88**	0.92	0.90	83.34%
POSBoW+Rules	**0.56**	0.37	0.43	0.87	**0.94**	**0.91**	**84.01%**
RQ 3: Knowledge-Poor vs. -Rich							
POSBoW	0.53	**0.35**	**0.41**	**0.88**	0.93	0.90	83.18%
Fin+Wiki+Rules	**0.58**	0.24	0.32	0.86	**0.96**	**0.91**	**83.76%**
RQ 4: Contribution of the Rules							
Rules	**0.77**	0.13	0.21	0.84	**1.00**	0.91	84.58%
POSBoW+Fin+Wiki	0.57	**0.40**	0.46	**0.88**	0.93	0.91	84.25%
POSBoW+Fin+Wiki+Rules	0.59	**0.40**	**0.47**	**0.88**	0.94	**0.92**	**84.71%**
Majority Class (certain)	0.00	0.00	0.00	0.83	1.00	0.90	82.90%

Table 5. Results of the classification task on 30 random samples of the Wikipedia test set (the results are averages and the best results are presented in bold).

Features	Uncertain			Certain			Accuracy
	P	R	F	P	R	F	
RQ 1: Separate Features							
BoW	0.59	0.34	0.42	0.87	0.95	0.91	84.59%
POSBoW	0.63	0.31	0.41	0.87	0.96	0.91	85.00%
Fin	0.41	0.05	0.09	0.83	0.99	0.90	82.97%
Wiki	**0.66**	0.40	**0.49**	**0.89**	0.96	**0.92**	**86.16%**
Fin+Wiki	**0.66**	**0.41**	**0.49**	**0.89**	0.95	**0.92**	86.10%
Rules	0.13	0.01	0.02	0.83	**1.00**	0.91	82.89%
RQ 2: Lexical vs. Syntactic							
BoW+Fin+Wiki	**0.63**	**0.39**	**0.47**	**0.88**	0.95	**0.92**	**85.46%**
POSBoW+Rules	**0.63**	0.31	0.41	0.87	**0.96**	0.91	85.01%
RQ 3: Knowledge-Poor vs. -Rich							
POSBoW	0.63	0.31	0.41	0.87	**0.96**	0.91	85.00%
Fin+Wiki+Rules	**0.65**	**0.41**	**0.49**	**0.89**	0.95	**0.92**	**86.04%**
RQ 4: Contribution of the Rules							
Rules	0.13	0.01	0.02	0.83	**1.00**	0.91	82.89%
POSBoW+Fin+Wiki	**0.66**	0.37	**0.47**	**0.88**	0.96	**0.92**	**85.83%**
POSBoW+Fin+Wiki+Rules	**0.66**	**0.38**	**0.47**	**0.88**	0.96	**0.92**	**85.83%**
Majority Class (certain)	0.00	0.00	0.00	0.83	1.00	0.90	82.90%

5.3 Knowledge-Poor vs. Knowledge-Rich Features (RQ 3)

On both datasets, the knowledge-rich approaches perform slightly better than the knowledge-poor ones. However, the difference is again more noticeable for the encyclopedic domain. For the financial domain dataset (see Table 4), the relatively high precision of the knowledge-rich (0.58) compared to the knowledge-poor features (0.53) comes at the cost of a significantly lower recall (0.24 vs. 0.35) and an insignificantly lower F_1 score (0.32 vs. 0.41). For the Wikipedia test set (see Table 5), a slightly higher precision of the knowledge-rich features (0.65 vs. 0.63) is accompanied by a distinctively higher recall (0.41 vs. 0.31) and F_1 score (0.49 vs. 0.41).

We argue, again, that the comparably good performance in case of the Wikipedia test set can be attributed to the Wiki dictionary. It is not only specific to domain and written (instead of spoken) language but also considerably larger than our tailored set of rules (1,984 vs. 95 features), which results in a relatively high recall.

5.4   Contribution of the Rules (RQ 4)

The rules' relatively low recall of 0.13 gets outperformed by a combination of all knowledge-poor features (POSBoW+Fin+Wiki) yielding a recall of 0.40. Combining both yields the strongest feature set in terms of F_1 score of the *uncertain* class (0.47). However, this improvement in performance is rather small, with the slight increase of F_1 score and accuracy being only insignificant.

As proven in Sects. 4.2 and 5.1, the rules are not applicable to the Wikipedia test set. This is also why – when being added to POSBoW+Fin+Wiki – they do not yield a noticeable performance change.

6   Conclusions

In this paper, we addressed the automatic detection of uncertain statements as a binary sentence classification task on the transcripts of spoken language in the financial domain. We presented a newly annotated dataset and introduced rule-based features specific to our task. Furthermore, we have proven that the SVM algorithm with a combination of BoW, POS, a general-domain as well as a domain-specific dictionary, and our handcrafted rules performs best.

We have shown that a rule-based approach is not applicable to the general encyclopedic domain. What is more, the domain-specific rules neither increase the classification performance of our in-domain dataset noticeably. Hence, we argue that the efforts of future research should focus on developing an in-domain dictionary – possibly enriched with POS tags. This recommendation is in line with the relatively high performance of the POSBoW feature set. Added value compared to Louhgran and McDonald's dictionary of uncertainty [17] could be generated by also incorporating n-grams with $n > 1$. This idea is partly motivated by Szarvas, who found that his biomedical classifier's performance dropped significantly when using only unigrams [4, p. 286].

In the future, classification performance could be additionally improved by optimizing the parameters of the SVM – indeed, the best performing approach applied to the CoNLL-2010's Wikipedia test set did precisely this [7, p. 9]. Moreover, the high dimensionality of feature set combinations such as the ones discussed in Sect. 5.4 indicates that a feature selection could decrease the risk of overfitting.

Given our concept of uncertainty, incorporating real-world knowledge into the classifier might also prove as another fruitful avenue of research. Lastly, the classifier could be applied to a larger-scale set of unseen data thus enabling the prediction of market dynamics such as stock performance.

Acknowledgments. We thank Alexander Diete for his help with the data acquisition and technical advice as well as Clemens Müller for his help with the annotation. This work was supported by the SFB 884 on the Political Economy of Reforms at the University of Mannheim (project C4), funded by the German Research Foundation (DFG).

References

1. Hyland, K.: Hedging in Scientific Research Articles. John Benjamins, Amsterdam/Philadelphia (1998)
2. Larcker, D.F., Zakolyukina, A.: Detecting deceptive disucssions in conference calls. J. Account. Res. **50**, 494–540 (2012)
3. Bachenko, J., Fitzpatrick, E., Schonwetter, M.: Verification and implementation of language-based deception indicators in civil and criminal narratives. In: Proceedings of the 22nd International Conference on Computational Linguistics, Manchester, pp. 25–32 (2008)
4. Szarvas, G.: Hedge classification in biomedical texts with a weakly supervised selection of keywords. In: Proceedings of ACL-08: HLT, Columbus, OH, pp. 281–289 (2008)
5. Medlock, B., Briscoe, T.: Weakly supervised learning for hedge classification in scientific literature. In: Proceedings of the 45th Annual Meeting of the Association of Computational Linguistics, Prague, pp. 992–999 (2007)
6. Riloff, E., Wiebe, J., Wilson, T.: Learning subjective nouns using extraction pattern bootstrapping. In: Proceedings of the Seventh Conference on Natural Language Learning, Edmonton, pp. 25–32 (2003)
7. Farkas, R., Vincze, V., Móra, G., Csirik, J., Szarvas, G.: The CoNLL-2010 shared task: learning to detect hedges and their scope in natural language text. In: Proceedings of the Fourteenth Conference on Computational Natural Language Learning: Shared Task, Uppsala, pp. 1–12 (2010)
8. Loughran, T., McDonald, B.: Textual analysis in accounting and finance: a survey. J. Account. Res. **54**, 1187–1230 (2016)
9. Light, M., Qiu, X.Y., Srinivasan, P.: The language of bioscience: facts, speculations, and statements in between. In: HLT-NAACL 2004 Workshop: BioLINK 2004, Linking Biological Literature, Ontologies and Databases, Boston, MA, pp. 17–24 (2004)
10. Ganter, V., Strube, M.: Finding hedges by chasing weasels: hedge detection using wikipedia tags and shallow linguistic features. In: Proceedings of the ACL-IJCNLP 2009 Conference Short Papers, Singapore, pp. 173–176 (2009)
11. Li, F.: Textual analysis of corporate disclosures: a survey of the literature. J. Account. Lit. **29**, 143–165 (2010)
12. Kearney, C., Liu, S.: Textual sentiment in finance: a survey of methods and models. Int. Rev. Financ. Anal. **33**, 171–185 (2014)
13. Das, S.R.: Text and context: language analytics in finance. Found. Trends Financ. **8**, 144–261 (2014)
14. Li, F.: Annual report readability, current earnings, and earnings persistence. J. Account. Econ. **45**, 221–247 (2008)
15. Li, F.: The information content of forward-looking statements in corporate filings: a naïve bayesian machine learning approach. J. Account. Res. **50**, 494–540 (2012)
16. Loughran, T., McDonald, B., Yun, H.: A wolf in sheeps clothing: the use of ethics-related terms in 10-K reports. J. Bus. Ethics **89**, 39–49 (2009)
17. Loughran, T., McDonald, B.: When is a liability not a liability? Textual analysis, dictionaries, and 10-Ks. J. Financ. **66**, 35–65 (2011)
18. Loughran, T., McDonald, B.: Measuring readability in financial disclosures. J. Financ. **69**, 1643–1671 (2014)
19. Cohen, J.: A coefficient of agreement for nominal scales. Educ. Psychol. Meas. **20**, 41–48 (1960)

20. Landis, J.R., Koch, G.G.: The measurement of observer agreement for categorical data. Biometrics **33**, 159–174 (1977)
21. Fleiss, J.L.: Statistical Methods for Rates and Proportions, 2nd edn. John Wiley, New York (1981)
22. Bird, S., Loper, E., Klein, E.: Natural Language Processing with Python. O'Reilly Media, Sebastopol (2009)
23. Bird, S., Loper, E.: Natural language toolkit: taggers (2017). https://github.com/nltk/nltk/blob/develop/nltk/tag/__init__.py. Accessed 27 Jan 2017
24. Honnibal, M.: Averaged perceptron tagger (2013). https://github.com/nltk/nltk/blob/develop/nltk/tag/perceptron.py. Accessed 27 Jan 2017
25. Honnibal, M.: A good part-of-speech tagger in about 200 lines of python (2013). https://explosion.ai/blog/part-of-speech-pos-tagger-in-python. Accessed 27 Jan 2017
26. Hall, M., Frank, E., Holmes, G., Pfahringer, B., Reutemann, P., Witten, I.H.: The WEKA data mining software: an update. SIGKDD Explor. Newsl. **11**, 10–18 (2009)
27. Le Cessie, S., van Houwelingen, J.: Ridge estimators in logistic regression. Appl. Stat. **41**, 191–201 (1992)
28. John, G.H., Langley, P.: Estimating continuous distributions in bayesian classifiers. In: Proceedings of the Eleventh Conference on Uncertainty in Artificial Intelligence, pp. 338–345 (1995)
29. Platt, J.C.: Fast training of support vector machines using sequential minimal optimization. In: Advances in Kernel Methods-Support Vector Learning (1998)
30. Aha, D., Kibler, D.: Instance-based learning algorithms. Mach. Learn. **6**, 37–66 (1991)
31. Cohen, W.W.: Fast effective rule induction. In: Proceedings of the Twelfth International Conference on Machine Learning, pp. 115–123 (1995)
32. Quinlan, R.: C4.5: Programs for Machine Learning. Morgan Kaufmann Publishers, San Mateo (1993)
33. Breiman, L.: Random forests. Mach. Learn. **41**, 5–32 (2001)

Automatic Question Generation From Passages

Karen Mazidi[✉]

Department of Computer Science, University of Texas at Dallas,
Dallas, TX 75080, USA
karen.mazidi@utdallas.edu

Abstract. Prior work in automatic question generation typically creates questions from *sentences* in a text. In contrast, the work presented here creates questions from a text *passage* in a holistic approach to natural language understanding and generation. Several NLP techniques including topic modeling are combined in an ensemble approach to identify important concepts, which then are used to create questions. Evaluation of the generated questions revealed that they are of high linguistic quality and are also important, conceptual questions, compared to questions generated by sentence-level question generation systems.

1 Introduction

The past decade has witnessed increased interest in automatic question generation (QG), largely motivated by the need to supply questions for intelligent tutoring systems [16]. Examination of prior work in QG can be seen in varying levels of specificity of the source text: words and their meaning [3,6], words and/or phrases based on their importance in the text [2,4], combinations of key words in an ontology [1,11,15], and sentences [7,9,12]. Most recent work in QG generates questions from sentences by syntactic manipulation of the source text from a declarative statement to one or more interrogatives [5]. Unfortunately, this progressive expansion of the QG scope from smaller to larger portions of text has largely stalled. The next logical approach would be to generate questions from textual units larger than one sentence. That is precisely the aim of the work presented here. At the passage level, Natural Language Understanding (NLU) analysis determines what the text is trying to communicate, so that questions can be generated about these key ideas. This work advances the state of the art of automatic question generation to the point that the overwhelming majority of output questions are meaningful, quality questions.

Research Question: Can generating questions from a passage as a whole lead to higher quality questions compared to approaches that employ syntactic manipulation of sentences, and can these techniques move the state of the art closer to the quality of human-authored questions?

© Springer Nature Switzerland AG 2018
A. Gelbukh (Ed.): CICLing 2017, LNCS 10762, pp. 655–665, 2018.
https://doi.org/10.1007/978-3-319-77116-8_49

2   NLU Modeling of Text Comprehension

In prior work [13] we demonstrated that NLU analysis of individual sentences in text could greatly increase the percentage of acceptable questions compared to previous state-of-the-art systems. This was done by identifying sentence structures that conveyed predictable meaning and then matching those sentences to templates designed to hone in on the key semantic point of the sentence.

This observation of the role of various structures as indicators of meaning inspired the present QG system; however, the search for structure-scaffolded meaning has been extended beyond the sentence into the passage as a whole. A good reader knows how to glean greater understanding from text by using headings and frequently used terminology to identify what is important. The QG system presented here is modeled after the NLU approach of good readers. Specifically, the QG system extracts important information from the NLU analysis of the passage and uses this extracted material for question generation. The sentence-level approach of prior work in QG only requires the student to identify the missing portion of the sentence in order to answer the question, either from memory or by re-reading the text. In contrast, these passage-level questions require a student to synthesize information from multiple sentences, thus invoking a higher level comprehension process [8].

3   Approach

Four methods of NLU analysis were applied to text passages: topic modeling, noun phrase extraction, terminology extraction, heading analysis. Each is described below, and Fig. 1 gives an overview.

Development and evaluation texts were extracted from open source college-level textbooks available in electronic form. The texts were from various domains: anatomy, biology, economics, history, and psychology. A text *passage* is considered to be one subsection of one textbook chapter but could also be a Wikipedia or other online article. Each passage is further divided into *sections* by the headings in the passage. Sections average around 15 sentences in length. Each of the four techniques is applied to each section so that questions are generated section by section.

3.1   Topic Modeling

The R topic modeling package was used[1] to determine the top topics for each passage section, as well as the words for each topic. Topic modeling is run on the sections of the passage, so that each section is considered a document in the topic modeling paradigm.

Topic modeling is a generative bag-of-words model that learns topics and topic words from frequency measures in the corpus documents. Topic modeling

---

[1] https://cran.r-project.org/web/packages/topicmodels/vignettes/topicmodels.pdf.

assumes that some probabilistic generative process created the documents. There are actually two levels of probabilistic generation: (1) each topic is a distribution of words $w = (w_1, ... w_N)$ from the corpus vocabulary V, and (2) each document is a distribution of topics. So we must find the term distribution β for each topic, as well as the the proportions θ of the topics for document w. A Dirichlet model gives us a mathematical way to assign prior probabilities to all possible models that could have generated the observed texts. In other words, we see the result of the distribution in the documents and the Dirichlet analysis lets us discover the distributions that resulted in the observed results. Mathematically this is expressed as follows:

$\beta \sim Dirichlet(\delta)$

$\theta \sim Dirichlet(\alpha)$

For a given topic z_i, chosen from

$z_i \sim Multinomial(\theta)$

and a word w_i is chosen from a multinomial distribution conditioned on the topic

$z_i : p(w_i | z_i, \beta)$

An iterative two-step Expectation-Maximization process is used to improve the estimates until a certain threshold is reached. In the maximization phase, the sum over the log-likelihoods of the documents is maximized with respect to parameters α and β. For the expectation step, for each document, optimal values are found for parameters in an LDA (Latent Dirichlet Allocation) model. Also, a Gibbs sampling technique is used for estimating the unseen processes creating the documents. For this application, the burnin parameter was set to 4000, the number of iterations to 2000, and the number of starts to 5, letting the system pick the best of the 5.

The results of topic modeling are very sensitive to the number of topics, k. This parameter is set in a configuration file, which allows the user to select k based on their observations of the data and topic modeling results. If the number of topics, k, is set to X, then the system will determine an appropriate value of k as follows. In an initial run of the R topic modeling script, $k = 2 \times |sections|$ since each section will have at most 2 topics used for question generation. Then, if there are n unused topics in the top 2 topics for all sections, k will be reduced by n and topic modeling will be rerun with this new k.

The configuration file also sets the number of terms per topic. The top six stemmed words of the top two topics are stored as the topic words for that section.

Topic Modeling Questions. Phrases are gathered from the most important topic modeling words of a section to generate summary questions such as: *Explain what you learned about* `topic phrase` *in this passage.*

The potential topic phrases are gathered by starting with the most frequently occurring topic terms. If combinations of these terms are found in the section, they are used for question generation.

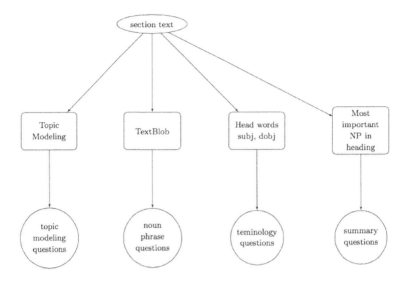

Fig. 1. Sources of passage-level generated questions

An optional "hint" feature can be enabled that looks through the section sentences, extracting sentence constituents that contain the phrase. An example follows:

```
Explain what you learned about brain waves in this passage.
Your discussion may include the following phrases:
- higher amplitude brain waves than alpha waves;
- a rapid burst of higher frequency brain waves that may be important for
learning and memory.
```

3.2 Noun Phrase Extraction

Noun phrases are extracted using the Python package TextBlob [2] which is built on top of the better-known NLTK package.

Noun Phrase Questions. For noun phrases that have the same head word, a compare question is generated. Example: *Differentiate between alpha waves and theta waves.*

For leftover noun phrases that were not matched with another noun phrase, a description question is generated which asks the student to relate the noun phrase to the section topic. Example: *Describe the relation between brain waves and stages of sleep.*

---

[2] https://textblob.readthedocs.io/en/dev/

3.3 Terminology Extraction

Each sentence in the section is searched to find head words of subject or direct object constituents. This list of important words is filtered in two ways. First, words that are capitalized are excluded since these probably represent named entities that will be identified through other means. Second, the list is checked against a list of 100k common English words. Words that remain are likely to be important words. These words are then categorized into one of 5 types: people, living, physical, event, and abstract by working through the WordNet[14] hierarchy until the appropriate top-level category is found.

Terminology Questions. If a term is a person according to the WordNet analysis described above, a describe question is generated. Example: *Describe the role of neuroscientists as discussed in this section.* The other categories are not used at this time as their successful application is likely to be domain dependent and the goal was to keep this version of the question generater domain independent.

From all terms that are not people, pairs of similar terms are gathered where similarity is a Levenshtein distance. The Levenshtein distance measures how similar two items are by the number of insertions, deletions and substitutions required to change one item into another, normalized by item length. For example, the terms mesoderm and ectoderm meet the threshold of similarity (0.4) and so one of three compare questions will be generated. Three question forms are used on a rotating basis to provide variety. Examples:

- Compare and contrast term1 and term2.
- Describe the difference(s) between term1 and term2.
- Is there a relationship between term1 and term2? Explain.

Any item in the list of possible terms that did not find a close term is used to form a definition question: *Provide a definition for* `term` *and discuss its relation to* `section_topic`. For example: *Provide a definition for epithelium, and discuss its relation to epithelial tissue.*

3.4 Section Topic

The section topic is the noun phrase in the section heading that occurs most frequently in the section text. The first section heading is used for subsequent sections if the system cannot find a noun phrase in the heading.

Summary Question. A summary question is asked about the section topic, for example: *Summarize what you learned about epithelial tissues in this section.*

4 Sample Questions

In order to examine the types of questions generated by the system, this section provides a sample text section and questions generated from it.

4.1 Sample Text

The following sample text is one section of a passage from an Anatomy textbook chapter on the heart.

Membranes

The membrane that directly surrounds the heart and defines the pericardial cavity is called the pericardium or pericardial sac. It also surrounds the roots of the major vessels, or the areas of closest proximity to the heart. The pericardium, which literally translates as around the heart, consists of two distinct sublayers: the sturdy outer fibrous pericardium and the inner serous pericardium. The fibrous pericardium is made of tough, dense connective tissue that protects the heart and maintains its position in the thorax. The more delicate serous pericardium consists of two layers: the parietal pericardium, which is fused to the fibrous pericardium, and an inner visceral pericardium, or epicardium, which is fused to the heart and is part of the heart wall. The pericardial cavity, filled with lubricating serous fluid, lies between the epicardium and the pericardium. In most organs within the body, visceral serous membranes such as the epicardium are microscopic. However, in the case of the heart, it is not a microscopic layer but rather a macroscopic layer, consisting of a simple squamous epithelium called a mesothelium, reinforced with loose, irregular, or areolar connective tissue that attaches to the pericardium. This mesothelium secretes the lubricating serous fluid that fills the pericardial cavity and reduces friction as the heart contracts.

4.2 Generated Questions

The most frequent topic modeling words in this passage were: heart, pericardium, pericardial, and cavity. Checking all permutations of these top words discovered that the phrase *pericardial cavity* was present in this passage. This resulted in the question: *Explain what you learned about the pericardial cavity in this passage.*

The noun phrase extraction component identified the following important noun phrases: *pericardial sac, connective tissue, fibrous pericardium, simple squamous epithelium, heart wall, pericardial cavity, serous pericardium.* From these noun phrases, sample generated questions include: *Differentiate between fibrous pericardium and serous pericardium* and *Describe the relation between the fibrous pericardium and membranes.*

From the terminology extraction component, the system identified numerous terms, some of which were used to generate the following terminology questions. Terms which were similar to other terms were used for questions such as: *Compare and contrast mesothelium and epithelium* and *Describe any difference(s) between epicardium and pericardium.*

Finally, the passage heading was used to generate the question: *Summarize what you learned about membranes in this section.* Not all passages result in all question types. This short passage resulted in at least one question in each of the four categories.

5   Evaluation

The questions produced by the system were evaluated in three ways: question breadth and depth, question linguistic quality, and question importance.

5.1   Evaluation 1: Question Breadth and Depth

Although there are numerous ways to classify questions, a straightforward and intuitively understandable scheme is to divide questions into factual comprehension questions and conceptual comprehension questions. Questions generated from sentences are by their very nature likely to be factual comprehension questions because the answer is typically one phrase from one sentence. In contrast, the passage-level questions can be considered to be conceptual comprehension questions because they require the student to synthesize material from multiple sentences. The system presented here is the first automatic question generation system to successfully generate questions spanning multiple input sentences and to quantify this question breadth.

Questions generated by the system were evaluated programmatically for question breadth as follows. Each generated question either asks about one key phrase, or the comparison of two key phrases. In order to answer the question from the section text, a student will need to re-read the relevant sentences and construct an answer synthesizing information from these sentences. This can be considered *question breadth*. Programmatically, a count of relevant sentences was calculated for each generated question by searching for the key phrase(s) in the sentences. Table 1 shows the average counts of the number of relevant sentences, as well as the percentage of sentences in the section that this count represents, for three representative evaluation texts.

Table 1. Question breadth

Topic	Sentences	Questions	Avg. Count	Avg. Percent
Epithelial tissue	146	87	4	28%
Monetary policy	87	41	6	47%
Stages of sleep	71	21	4	25%
Average			4.7	33.4

A student would need to re-read an average of nearly 5 sentences in the section to adequately construct an answer. This is about 33% of the sentences.

5.2   Evaluation 2: Question Linguistic Quality

There is no standard way to evaluate automatically generated questions. Recent work in QG and other NLP applications favors evaluation by crowdsourcing

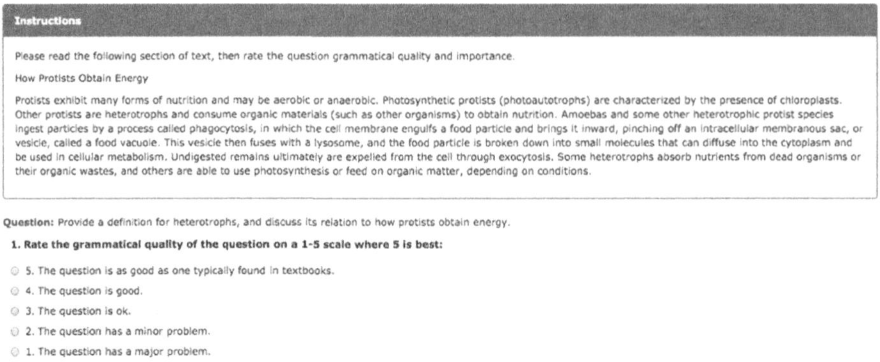
Instructions

Please read the following section of text, then rate the question grammatical quality and importance.

How Protists Obtain Energy

Protists exhibit many forms of nutrition and may be aerobic or anaerobic. Photosynthetic protists (photoautotrophs) are characterized by the presence of chloroplasts. Other protists are heterotrophs and consume organic materials (such as other organisms) to obtain nutrition. Amoebas and some other heterotrophic protist species ingest particles by a process called phagocytosis, in which the cell membrane engulfs a food particle and brings it inward, pinching off an intracellular membranous sac, or vesicle, called a food vacuole. This vesicle then fuses with a lysosome, and the food particle is broken down into small molecules that can diffuse into the cytoplasm and be used in cellular metabolism. Undigested remains ultimately are expelled from the cell through exocytosis. Some heterotrophs absorb nutrients from dead organisms or their organic wastes, and others are able to use photosynthesis or feed on organic matter, depending on conditions.

Question: Provide a definition for heterotrophs, and discuss its relation to how protists obtain energy.

1. Rate the grammatical quality of the question on a 1-5 scale where 5 is best:

5. The question is as good as one typically found in textbooks.
4. The question is good.
3. The question is ok.
2. The question has a minor problem.
1. The question has a major problem.

Fig. 2. Sample Amazon Mechanical Turk HIT for question linguistic quality.

which has proven to be both cost and time efficient and to achieve results comparable to human evaluators [10,17]. In this evaluation, all questions generated from the system were evaluated using Amazon's Mechanical Turk Service. A sample HIT (Human Intelligence Task) is shown in Fig. 2. Workers were selected with at least 90% approval rating on their prior work and who were located in the US and proficient in US English. To monitor quality, work was submitted in small batches, manually inspected, and run through software to detect workers whose ratings did not correspond well with fellow workers. Each worker was presented with the section text and one question at a time. Each question was rated on a 1–5 scale by 4 workers. The four scores were averaged and a mean of at least 3.5 was considered acceptable. For comparison with the new Passage-Level QG system, questions were also generated from two other QG systems: our prior sentence-level QG system [12] and Heilman and Smith's system [9] which is the most frequently cited prior work in QG. For this evaluation, three chapter passages were randomly chosen from anatomy, biology and economics books. Questions were randomly chosen from the passage QG system, and randomly chosen from among the highest rated questions in the other two systems.

Table 2. Average scores for linguistic quality

Text	Passage QG	Sentence QG	Heilman & Smith
Anatomy	4.2	3.7	2.8
Biology	4.0	3.2	3.1
Economics	4.1	3.8	3.6
Average	4.1	3.5	3.2

The results in Table 2 show that our questions improve 16% over our prior sentence-level QG system, and 29% over the Heilman and Smith system. Taking the average scores over 4 workers for each question and considering an average of 3.5 or above as an acceptable question, 91% of the Passage QG questions were acceptable, compared to 57% of the Sentence QG questions, and 48% of the Heilman and Smith questions. The results were statistically significant, $p < 0.01$.

5.3 Evaluation 3: Question Importance

Next we asked workers to rate the questions on *importance* relative to the text. A 1–5 scale was again used, with 5 being the best:

- 5. The question is important.
- 4. The question is somewhat important.
- 3. The question is possibly important.
- 2. The question is trivial.
- 1. The question does not make sense.

The same input files were used for this evaluation. Table 3 shows that the workers scored the passage QG questions as being more important and less trivial compared to the other systems. The passage-QG questions were rated as 8% more important than the sentence QG questions and 14% more important than those of the Heilman and Smith system. Using the same threshold of 3.5 or above, 94% of the passage QG questions were important, compared to 74% of the sentence QG questions and 79% of the Heilman and Smith system. This importance evaluation is perhaps more subjective than the linguistic quality evaluation. The concept of importance does not differentiate between lower-level factual questions and higher-level conceptual questions.

Table 3. Average scores for question importance

Text	| Passage QG |	Sentence QG |	Heilman & Smith
Anatomy	4.5	4.0	3.4
Biology	4.0	3.7	3.7
Economics	4.4	4.2	4.2
Average	4.3	4.0	3.8

5.4 Comparison to Human-Authored Questions

There were a limited number of end-of-chapter questions available for one text on epithelial tissues. A similar evaluation as evaluation 3 above was performed, but the word *importance* was replaced with the word *meaningfulness*. In future work, a more robust evaluation will be done by finding or creating a large set of quality human-authored questions for a text. Comparison of the results in Tables 3 and 4 indicates problems with subjective measures such as importance and meaningfulness. Variations are also to be expected as the input source text changes. It is interesting that the results in Table 4 show that the human-authored end-of-chapter questions were not rated highly for meaningfulness. This would be expected to change based on the input source. Many textbooks have thought-provoking end-of-chapter questions but unfortunately many are not of high quality. Although the results in this limited evaluation show that the passage QG questions were competitive with human-authored questions, the sample is too small to be statistically significant.

Table 4. Average scores per source

Passage QG	Sentence QG	Heilman & Smith	Human
3.9	3.6	2.6	3.4

6   Discussion

The research question driving this exploration is that generating questions from a passage as a whole could lead to higher quality questions compared to approaches that employ syntactic manipulation of sentences, and that the passage-generated questions might even approach the quality of human-authored questions. The first part of the question can be answered affirmatively, but results comparing to human-authored questions will likely vary with the quality of the human-authored questions. The QG system presented here has demonstrated that infusing the question generation process with NLU analysis leads to the majority of output questions being high quality questions linguistically and semantically. Further, the system is the first to successfully break through the sentence barrier, reliably generating questions from textual units larger than one sentence. In prior QG systems, the answer to a question is found within one sentence. In contrast, the answer for these passage-generated questions must be synthesized from an average of 33% of the passage sentences. Evaluations demonstrated that question quality rated higher than prior state-of-the-art question generation systems when considering both linguistic quality and question importance.

References

1. Afzal, N., Mitkov, R.: Automatic generation of multiple choice questions using dependency-based semantic relations. Soft Comput. **18**, 1269–1281 (2014)
2. Agarwal, M., Mannem, P.: Automatic gap-fill question generation from text books. In: Proceedings of the 6th Workshop on Innovative Use of NLP for Building Educational Applications. Association for Computational Linguistics (2011)
3. Aist, G.: Towards automatic glossarization: automatically constructing and administering vocabulary assistance factoids and multiple-choice assessment. Int. J. Artif. Intell. Educ. **12**, 212–231 (2001)

4. Becker, L., Basu, S., Vanderwende, L.: Mind the gap: learning to choose gaps for question generation. In: Proceedings of the 2012 Conference of the North American Chapter of the Association for Computational Linguistics: Human Language Technologies. Association for Computational Linguistics (2012)
5. Boyer, K., Piwek, P.: In: Proceedings of QG2010: The Third Workshop on Question Generation (2010)
6. Brown, J., Frishkoff, G., Eskenazi, M.: Automatic question generation for vocabulary assessment. In: Proceedings of the Conference on Human Language Technology and Empirical Methods in Natural Language Processing. Association for Computational Linguistics (2005)
7. Gates, D.: Generating look-back strategy questions from expository texts. In: Workshop on the Question Generation Shared Task and Evaluation Challenge (2008)
8. Grabe, W., Stoller, F.: Teaching and Researching. Reading, Routledge (2013)
9. Heilman, M.: Automatic factual question generation from text. Ph.D. Thesis, Carnegie Mellon University (2011)
10. Heilman, M., Smith, N.: Rating computer-generated questions with Mechanical Turk. In: Proceedings of the NAACL HLT 2010 Workshop on Creating Speech and Language Data with Amazon's Mechanical Turk, ACL (2010)
11. Jouault, C., Seta, K.: Content-dependent question generation for history learning in semantic open learning space. In: Trausan-Matu, S., Boyer, K.E., Crosby, M., Panourgia, K. (eds.) ITS 2014. LNCS, vol. 8474, pp. 300–305. Springer, Cham (2014). https://doi.org/10.1007/978-3-319-07221-0_37
12. Mazidi, K., Tarau, P.: Automatic question generation: from NLU to NLG. In: Micarelli, A., Stamper, J., Panourgia, K. (eds.) ITS 2016. LNCS, vol. 9684, pp. 23–33. Springer, Cham (2016). https://doi.org/10.1007/978-3-319-39583-8_3
13. Mazidi, K., Tarau, P.: Infusing NLU into automatic question generation. In: International Conference on Natural Language Generation (2016)
14. Miller, G.: Wordnet: a lexical database for English. Comm. ACM **38**(11), 39–41 (1995)
15. Olney, A., Graesser, A., Person, N.: Question generation from concept maps. Dialogue Discourse **3**(2), 101–124 (2012)
16. Rus, V., Cai, Z., Graesser, A.C.: Experiments on generating questions about facts. In: Gelbukh, A. (ed.) CICLing 2007. LNCS, vol. 4394, pp. 444–455. Springer, Heidelberg (2007). https://doi.org/10.1007/978-3-540-70939-8_39
17. Snow, R., O'Connor, B., Jurafsky, D., Ng, A.: Cheap and fast–but is it good?: evaluating non-expert annotations for natural language tasks. In: Proceedings of the Conference on Empirical Methods in Natural Language Processing, ACL (2008)
18. Woolf, B.: Building Intelligent Interactive Tutors: Student-Centered Strategies for Revolutionizing e-Learning. Morgan Kaufman, San Francisco, CA (2010)

Author Index

Adel, Heike   I-3
Adrasik, Frank   I-17
Aker, Ahmet   II-495
Alexeyevsky, Daniil   I-488
Aleyxendri, Andrea   II-506
Al-Khatib, Amr   II-105
Alnemer, Loai M.   II-115
Andruszkiewicz, Piotr   I-405
Ariss, Omar El   II-115
Asgari, Ehsaneddin   I-3
Atkinson, Katie   I-418

Bahloul, Raja Bensalem   I-170
Bandyopadhyay, Sivaji   I-317
Banjade, Rajendra   I-17
Baroni, Marco   I-209
Batita, Mohamed Ali   I-342
Ben Mesmia, Fatma   I-475
Berment, Vincent   I-81
Besançon, Romaric   I-329
Bhattacharyya, Pushpak   I-225, II-3
Biemann, Chris   I-377
Bigot, Benjamin   I-503
Binnig, Carsten   I-527
Blache, Philippe   I-170
Bobed, Carlos   II-506
Boitet, Christian   I-81
Boleda, Gemma   I-209
Bolívar Jiménez, Margarita   II-71
Bollegala, Danushka   I-418
Bölücü, Necva   I-110
Bontcheva, Kalina   II-495
Borgne, Hervé Le   I-329
Borin, Lars   I-550
Bosco, Cristina   II-46
Bouabidi, Kaouther   I-475
Bouziri, Ahlem   II-563
Bui, Marc   II-324

Calvo, Hiram   II-303
Cambria, Erik   I-503, I-564, II-90, II-166, II-549
Can, Burcu   I-87, I-99, I-110
Cardiel, Oscar   II-506

Charhad, Mbarek   II-575
Chen, Jiajun   II-366
Chhaya, Niyati   II-20
Chourasia, Saheb   II-628
Ciobanu, Alina Maria   I-576, I-591
Claveau, Vincent   I-30
Clematide, Simon   II-141
Coenen, Frans   I-418
Cotterill, Rachel   II-391

Daher, Hani   I-329
Dai, Xin-Yu   II-366
Daquo, Anne-Laure   I-329
Dasgupta, Tirthankar   I-463
Dash, Sandeep Kumar   I-354
Dashtipour, Kia   II-129
Deksne, Daiga   II-445
Dewdney, Nigel   II-391
Dey, Lipika   I-463
Dinarelli, Marco   I-44, I-249
Dinu, Liviu P.   I-576, I-591
Du, Jiachen   II-35
Dufour, Richard   II-404
Dupont, Yoann   I-44, I-249

Ekbal, Asif   I-317
El-Beltagy, Samhaa R.   II-105
Elmongui, Hicham G.   II-353

F. Samatova, Nagiza   II-195
Faili, Heshaam   II-469
Feldman, Anna   I-291
Ferret, Olivier   I-329
Fischer, Andreas   II-324
Friburger, Nathalie   I-475
Fujita, Sumio   I-391
Fürnkranz, Johannes   I-527

Gambäck, Björn   I-276, II-58
García-Flores, Jorge   II-303
Garrido, Angel Luis   II-506
Gaussier, Eric   II-563
Gautam, Dipesh   I-17
Ge, Weiyi   II-366

Gelbukh, Alexander   I-354, II-129
Geraci, Filippo   II-585
Giang Do, Hoang   II-549
Giménez, Maite   II-313
Graesser, Arthur C.   I-17
Gridach, Mourad   I-264
Gui, Lin   II-35
Guilbaud, Jean-Philippe   I-81
Gurevych, Iryna   I-527

HaCohen-Kerner, Yaakov   II-241
Haddad, Hatem   I-264
Haddar, Kais   I-170, I-475
Hamidi, Mansour   II-324
Hamzah, Diyana   I-564
Hashimoto, Masayuki   II-180, II-482
Hazan, Rafał   I-405
Hazarika, Devamanyu   II-166
Hemmati, Nima   II-469
Heracleous, Panikos   II-180, II-482
Hernández Farías, Delia Irazú   II-46
Hernandez, Nicolas   I-305
Hernández-Castañeda, Ángel   II-303
Hiranandani, Gaurush   II-605
Ho, Danyuan   I-564
Hosseinia, Marjan   II-255
Huang, Shujiang   II-366
Hussain, Amir   II-129, II-166

Inui, Kentaro   II-379
Ishikawa, Akio   II-180, II-482

Jagaluru, Darshan Siddesh   I-363
Jaidka, Kokil   II-20
Jhamtani, Harsh   II-605
Jiménez Pascual, Adrián   I-391
Jones, Paul   II-195

Kadri, Nesrine   I-170
Kalivoda, Ágnes   I-123
Kanneganti, Silpa   I-158
Kapočiūtė-Dzikienė, Jurgita   I-81
Kawashima, Hiroyuki   II-180
Kedia, Sanket   II-20
Kersting, Kristian   I-527
Khusainov, Aidar   I-515
Kijak, Ewa   I-30
Klenner, Manfred   II-141
Kohail, Sarah   I-377

Komiya, Kanako   I-195
Kompatsiaris, Ioannis   I-450
Krišlauks, Rihards   II-445
Kurfalı, Murathan   I-87
Kurohashi, Sadao   II-535

Labatut, Vincent   II-404
Lafourcade, Mathieu   II-617
Laki, László János   II-435
Lalitha Devi, Sobha   I-233
Lark, Joseph   II-211
Latiri, Chiraz   II-563
Lautier, Christian   I-249
Lee, Kuei-Ching   I-134
Lee, Lung-Hao   I-134
Lhioui, Chahira   II-575
Li, Lei   II-522
Li, Rumeng   I-145
Liebeskind, Chaya   II-241
Liebeskind, Shmuel   II-241
Linarès, Georges   II-404
Liparas, Dimitris   I-450
Litvak, Marina   II-522
Liwicki, Marcus   II-324

Ma, Yukun   I-503
Maharjan, Nabin   I-17
Mahesh, Kavi   I-363
Mahmoud, Adnen   II-338
Malaka, Rainer   I-354
Maleki, Jalal   II-469
Mandya, Angrosh   I-418
Maneriker, Pranav   II-605
Mansour, Riham   II-353
Mansouri, Sadek   II-575
Markov, Ilia   II-289
Masuichi, Hiroshi   II-156
Matsuda, Koji   II-379
Matsumoto, Kazunori   II-482
Maurel, Denis   I-475
Mavropoulos, Thanassis   I-450
Mazidi, Karen   II-655
Melgar, Andrés   I-185
Meyer, Christian M.   I-527
Miks, Toms   II-445
Mondal, Dibyendu   II-3
Morin, Emmanuel   I-305, II-211
Mujadia, Vandan   I-158
Mukherjee, Arjun   II-224, II-255

Nagata, Masaaki   I-145
Nallagatla, Manikanta   II-20
Nand, Parma   I-435
Naskar, Abir   I-463
Naskar, Sudip Kumar   I-317

Okazaki, Naoaki   II-379
Okumura, Manabu   I-195
Oncevay-Marcos, Arturo   I-185
Onishi, Takeshi   II-156
Ozen, Serkan   I-99

Padó, Sebastian   I-209
Pakray, Partha   I-354
Palshikar, Girish Keshav   II-628
Paolo Ponzetto, Simone   II-642
Papegnies, Etienne   II-404
Papini, Tiziano   II-585
Paredes, Roberto   II-313
Patti, Viviana   II-46
Pawar, Sachin   II-628
Peña Saldarriaga, Sebastián   II-211
Peng, Haiyun   II-90
Peng, Jing   I-291
Perera, Rivindu   I-435
Pinnis, Mārcis   II-445
Poria, Soujanya   II-166
Porzel, Robert   I-354
Pradhan, Manali   I-291
Pu, Pearl   II-71

Quilez, Ruben   II-506

Ræder, Johan G. Cyrus M.   II-58
Ramadier, Lionel   II-617
Ramrakhiyani, Nitin   II-628
Ross, Joe Cheri   I-225
Rosso, Paolo   II-46, II-313
Roth, Stefan   I-527
Rus, Vasile   I-17

Saikh, Tanik   I-317
Sakaguchi, Tomohiro   II-535
Salim, Soufian   I-305
Sanchan, Nattapong   II-495
Sango, Mizuki   II-379
Sasaki, Minoru   I-195
Sastry, Chandramouli Shama   I-363

Satake, Koji   II-156
Saxena, Anju   I-550
Sayadi, Karim   II-324
Schuller, Björn W.   II-275
Schütze, Hinrich   I-3
Sgarro, Andrea   I-576
Shang, Di   II-366
Sharma, Dipti M.   I-158
Sharma, Raksha   II-3
Sheremetyeva, Svetlana   I-67
Shindo, Hiroyuki   I-145
Shinnou, Hiroyuki   I-195
Sidorov, Grigori   II-289
Sikdar, Utpal Kumar   I-276
Simpson, Edwin   I-527
Singh, Thoudam Doren   II-457
Sintsova, Valentina   II-71
Smeddinck, Jan   I-354
Sobrevilla-Cabezudo, Marco Antonio   I-185
Solorio, Thamar   II-457
Štajner, Sanja   II-642
Stamatatos, Efstathios   II-289
Stuckenschmidt, Heiner   II-642
Subramanyam, R. B. V.   II-166
Sudoh, Katsuhito   I-145
Sugaya, Fumiaki   II-180, II-482
Suzuki, Shota   I-195
Symeonidis, Spyridon   I-450

Tamaazousti, Youssef   I-329
Tellier, Isabelle   I-44, I-249
Theil, Christoph Kilian   II-642
Tran, Ha-Nguyen   II-549
Tsatsaronis, George   II-420
Tseng, Yuen-Hsien   I-134
Tuggener, Don   II-141

Üstün, Ahmet   I-87

Vanetik, Natalia   II-522
Varlamis, Iraklis   II-420
Ventirozos, Filippos Karolos   II-420
Vijay Sundar Ram, R.   I-233
Virk, Shafqat Mumtaz   I-550
Vrochidis, Stefanos   I-450

Wadbude, Rahul   II-20
Wang, Xuan   II-35

Wang, Xun   I-145
Wang, Yiou   II-156
Wright, Bianca   I-291

Xing, Frank Z.   I-564
Xu, Mingyang   II-195
Xu, Ruifeng   II-35

Yang, Fan   II-224
Yang, Ruixin   II-195
Yang, Zijian Győző   II-435
Yasuda, Keiji   II-180, II-482

Zhang, Yifan   II-224
Zrigui, Ahmed   II-338
Zrigui, Mounir   I-342, II-338, II-575

Printed in the United States
by Bookmasters